Intelligence in a Small Materials World

*Selected Papers from IPMM-2003
The Fourth International Conference on
Intelligent Processing and Manufacturing of Materials*

Edited by

John A. Meech
*Mining Engineering
University of British Columbia
Vancouver, BC, Canada*

Yoshiyuki Kawazoe
*Institute for Materials Research
Tohoku University
Sendai, Japan*

Vijay Kumar
*Institute for Materials Research
Tohoku University
Sendai, Japan*

John F. Maguire
*Materials Directorate
US Air Force Research Laboratories
Wright-Patterson Air Force Base, Ohio, USA*

DEStech Publications, Inc.

Intelligence in a Small Materials World

DEStech Publications, Inc.
1148 Elizabeth Avenue #2
Lancaster, Pennsylvania 17601 U.S.A.

Copyright © 2005 by DEStech Publications, Inc.
All rights reserved

No part of this publication may be reproduced, stored in a
retrieval system, or transmitted, in any form or by any means,
electronic, mechanical, photocopying, recording, or otherwise,
without the prior written permission of the publisher.

Printed in the United States of America
10 9 8 7 6 5 4 3 2 1

Main entry under title:
  Intelligence in a Small Materials World: Selected Papers from IPMM-2003—The Fourth International Conference on Intelligent Processing and Manufacturing of Materials

A DEStech Publications book
Bibliography: p.
Includes index p. 965

ISBN No. 1-932078-19-3

HOW TO ORDER THIS BOOK

BY PHONE: 866-401-4337 or 717-290-1660, 9AM–5PM Eastern Time

BY FAX: 717-509-6100

BY MAIL: Order Department
DEStech Publications, Inc.
1148 Elizabeth Avenue #2
Lancaster, PA 17601, U.S.A.

BY CREDIT CARD: American Express, VISA, MasterCard

BY WWW SITE: http://www.destechpub.com

Table of Contents

Preface xiii

Chapter 1: Philosophy

Welcome to IPPM'03—Nanotechnology: Do Good Things
Really Come in Small Packages?........................ 3
YOSHIYUKI KAWAZOE and JOHN A. MEECH

On the Mechanics of Morals......................... 12
JOSHUA GREENE

Emergence and Complex Systems: Towards New Practices for
Industrial Automation............................ 28
HEIKKI HYÖTYNIEMI

PART 1: COMPUTATIONAL INTELLIGENCE AND SOFT COMPUTING

Chapter 2: Fuzzy Systems

On Kernel-Based Fuzzy Cluster Loading 65
MIKA SATO-ILIC

Fuzzy Evolutionary Approach to Determination of Dietary
Composition for Diabetics......................... 77
T. VAN LE

Intelligent Information Technology: From Autonomic to
Autonomous Systems............................ 86
LORNA STROBEL STEWART

Design of a Fuzzy Controller for a Microwave Monomode
Cavity for Soot Trap Filter Regeneration............... 100
M. PAPPALARDO, A. PELLEGRINO, M. D'AMORE, P. RUSSO
and F. VILLECCO

Optimal Operating Parameters in Fuzzy-Assisted Microwave Removal of Carbon Particulate from Ceramic Filters 110
V. PALMA, A. PELLEGRINO, S. ROBUSTELLI and F. VILLECCO

Fuzzy Probability for Modelling Particle Preparation Processes . . 119
ANNA WALASZEK-BABISZEWSKA

Prediction of Charpy Impact Properties Using Fuzzy Modelling on Ship Structural Steels 130
M. CHEN and D. A. LINKENS

Chapter 3: Artificial Neural Networks

Artificial Intelligence in Large-Scale Simulations of Materials . . . 141
JOHN F. MAGUIRE

Application of an Artificial Neural Network to the Control of an Oxygen Converter Process 149
J. FALKUS, P. PIETRZKIEWICZ, W. PIETRZYK and J. KUSIAK

Improving the Prediction of the Roll Separating Force in a Hot Steel Finishing Mill 160
Y. FRAYMAN, B. F. ROLFE, P. D. HODGSON and G. I. WEBB

A Solution to the N-Body Problem for Particles of Non-Spherical Geometry Using Artificial Intelligence 175
MARK BENEDICT and JOHN F. MAGUIRE

Distributed Backpropagation Neural Network System for Pattern Recognition Using JavaSpaces 183
MOHAMED GALAL, HESHAM ELDEEB and SALWA NASSAR

Eccentricity and Hardness Control in Cold Rolling Mills with a Dynamically Constructed Neural Controller 199
Y. FRAYMAN and B. F. ROLFE

Chapter 4: Process Control and Simulation for Optimization

Level Control Model by Numerical Fluid Dynamics Method 217
DAI SUZUKI and KEISUKE FUJISAKI

H-infinity Control of Micromolecular State Transition of a Sodium Ion Channel Gating System 230
HIROHUMI HIRAYAMA

Defeasible Deontic Traffic Light Control Based on a Paraconsistent Logic Program EVALPSN 247
KAZUMI NAKAMATSU, TOSHIAKI SENO, JAIR MINORO ABE and ATSUYUKI SUZUKI

Development of an Electrorheological Bypass Damper for Railway Vehicles—Estimation of Damper Capacity from the Prototype ... 261
S. CHONAN, M. TANAKA, T. NARUSE and T. HAYASE

A Concept for Simulation-Based Optimization of Sheet Metal Forming Processes ... 277
M. GRAUER, G. STUFF, T. BARTH, P. NEUSER, O. REICHERT and M. GERDES

Multiple Layer Control Method ... 286
SHIGEKI SUGIYAMA

Chapter 5: Intelligence in Environmental Applications

Morphological and Morphometrical Characteristics of Ornamental Stone Airborne Dusts: Capture and Filtration. ... 301
GIUSEPPE BONIFAZI, VINCENZO GIANCONTIERI, SILVIA SERRANTI and FABIO VOLPE

Advanced Sampling and Characterization Techniques of Nanoparticle Products Resulting from Thermal Decomposition Processes ... 316
GIUSEPPE BONIFAZI and STEFGANO DI STASIO

Development of a Magnetically Levitated Hoisting System for Use in Underground Mines. ... 326
RYAN ULANSKY and JOHN A. MEECH

Bio-Nanotechnology and Phytomining: The Living Synthesis of Gold Nanoparticles by Plants ... 336
CHRIS ANDERSON, BOB STEWART, CAREL WREESMANN, GEOFF SMITH and JOHN MEECH

Developing a Plug to Seal Mine Openings for a Thousand Years: The Millennium Plug Project ... 343
BRENNAN LANG, RIMAS PAKALNIS and JOHN A. MEECH

Chapter 6: Intelligence in Energy Systems

Multiple Sensor Surface Vibrations Analysis for Monitoring Tumbling Mill Performance ... 359
S. J. SPENCER, J. J. CAMPBELL, V. SHARP, K. J. DAVEY, P. L. PHILLIPS, D. G. BARKER and R. J. HOLMES

Fracture Toughness and Surface Energies of Covalent Materials: Theoretical Estimates and Application to Comminution ... 376
DESMOND TROMANS and JOHN A. MEECH

Usable Heat from Mine Waters: Coproduction of Energy and Minerals from "Mother Earth" ... 401
MORY M. GHOMSHEI and JOHN A. MEECH

Al Clusters as a Tool for Hydrogen Storage 412
A. GOLDBERG, M. MORI and A. BICK

**Theoretical Study of Clathrate Hydrates with
Multiple Occupation** 423
V. R. BELOSLUDOV, T. M. INERBAEV, R. V. BELOSLUDOV, MARCEL SLUITER,
Y. KAWAZOE and J. KUDOH

Chapter 7: Intelligent Instrumentation and Metrology

**Image Mining of Evanescent Microwave Data for Nondestructive
Material Inspection** 439
SOICHI OKA and STEVEN LeCLAIR

Magnetohydrodynamic Stability in Pulse EMC 448
KEISUKE FUJISAKI

**Ductile Machining Phenomena of Nominally Brittle Materials at
the Nanoscale** 459
JOHN PATTEN, H. CHERUKURI, GEORGE PHARR, RON SCATTERGOOD,
WEI GAO and JIWANG YAN

**Fractal Dimension: A New Machining Decision-Making
Parameter** 470
A. M. M. SHARIF ULLAH, MD. REZAUR RAHMAN,
VORATAS KACHITVICHYANUKUL and KHALIFA HAMAD HARIB

**EB-Reinforcement by Formation of Atomic Scale Defects
in Inorganic Glasses** 487
YOSHITAKE NISHI, NAOKI YAMAGUCHI, ATSUSHI KADOWAKI,
KAZUYA OGURI and AKIRA TONEGAWA

**Texture Observations of Ferromagnetic Shape Memory of
Nanostructured Fe-Pd Alloy by Laser and
Electronic Microscope** 495
T. OKAZAKI, T. KUBOTA, H. NAKAJIMA, Y. FURUYA, S. KAJIWARA,
T. KIKUCHI and M. WUTTING

**New Advanced Diffusion Simulators for Boron
Ultra-Shallow Junction** 502
YUZURU SATO, MASAYASU MIYATA, MASAMITSU UEHARA,
HIROSHI NAKADATE, GYEONG S. HWANG, EUGENE HEIFETS,
TAHIR CAGIN and WILLIAM A. GODDARD, III

**Nano-Sensors for Gas Determination Based on Heterostructure
SnO_2-Si** 508
V. V. IL'CHENKO, A. I. KRAVCHENKO, V. P. CHEHUN, A. M. GASKOV
and V. T. GRINCHENKO

Electrochemical Detection of Urea Using Self-Assembled Monolayers on a Porous Silicon Substrate 511
DONG-HWA YUN, JOON-HYUNG JIN, NAM-KI MIN and SUK-IN HONG

Chapter 8: Intelligence in Materials Science

Atomic Scale Simulations for Compositional Optimization 523
ROBIN W. GRIMES, MOHSIN PIRZADA, PATRICK K. SCHELLING, SIMON R. PHILLPOT, KURT E. SICKAFUS and JOHN MAGUIRE

Effects of Different Initial Conditions of Liquid Metals on Solidification Microstructures 537
JI-YONG LI, RANG-SU LIU, KE-JUN DONG, CAI-XING ZHENG and FENG-XIANG LIU

Combined Approach of Statistical Moment and Cluster Variation Methods for Calculation of Alloy Phase Diagrams 545
K. MASUDA-JINDO, VU VAN HUNG and R. KIKUCHI

Fabrication and Mechanical Properties of TiNi/Al Smart Composites 559
GYU CHANG LEE, JUN HEE LEE and YOUNG CHUL PARK

Mechanochemical Doping to Prepare a Visible-Light Active Titania Photocatalyst 570
QIWU ZHANG, JUN WANG, SHU YIN, TSUGIO SATO and FUMIO SAITO

Influence of Mean Interatomic Distance on Magnetostriction of Fe-Pd Alloy Films 576
HIROMASA YABE and YOSHITAKE NISHI

Chapter 9: Intelligent Materials Processing

Sub-Nanoscale Structure-Controlled Alloys Produced by Stabilization of Supercooled Liquid 585
AKIHISA INOUE

A Novel Method—Hot Quasi-Isostatic Pressing to Fabricate a Metallic Cellular Material Containing Polymer using an SPS System. 599
ZHENLUN SONG, SATOSHI KISHIMOTO and NORIO SHINYA

Consideration of AC-Component Force of Electromagnetic Stirring in the MHD Calculation 606
SHOUJI SATOU, KEISUKE FUJISAKI and TATSUYA FURUKAWA

**Fabrication of Ceramic Hip Joint by Bingham Semi-Solid/Fluid
Isostatic Pressing Method** 614
FUJIO TSUMORI, NAOYA YASUDA and SUSUMU SHIMA

**Control of Surface Layer Formed in Fe-Si-0.4mass%Mn Alloys
by Annealing under Low Partial Pressure of Oxygen** 622
SHIGERU SUZUKI, HAIJME HASEGAWA, SHOZO MIZOGUCHI
and YOSHIO WASEDA

**Development of a Method to Fabricate Metallic Closed Cellular
Materials Containing Organics** 637
SATOSHI KISHIMOTO, ZHENLUN SONG and NORIO SHINYA

Bulk AxiSymmetric Forging of Magnesium Alloys 646
MARGAM CHANDRASEKARAN, CHOY CHEE MUN and
JOHN YONG MING SHYAN

Chapter 10: Manufacturing Systems

**A Knowledge Management System for Manufacturing
Design** 661
Y. NAGASAKA and S. KISANUKI

**Promising Information Technology in the Near Future and
Current Obstacles: UWB/PLC and Cyber-Security
in the Broadband Age** 679
YOSHIYASU TAKEFUJI

3D Interactive Input System 685
HESHAM ELDEEB, HALA ELSADEK and ESMAT ABDALLAH

PART 2: NANOTECHNOLOGY

Chapter 11: Fullerene Materials Science

**Novel Self-Assembled Material: Fullerenes Self-Intermixed
with Phthalocyanines** 701
M. DE WILD, S. BERNER, H. SUZUKI, A. BARATOFF, H.-J. GUENTHERODT
and T. A. JUNG

**Electronic Structure of Self-Localized Excitons in a One
Dimensional C_{60} Crystal** 704
V. R. BELOSLUDOV, T. M. INERBAEV, R. V. BELOSLUDOV, Y. KAWAZOE
and J. KUDOH

A Simple Route to New 1D Nanostructures 721
A. HUCZKO, H. LANGE, J. PASZEK, M. BYSTRZEJEWSKI, S. CUDZILO,
S. GACHET, M. MONTHIOUX, Y. Q. ZHU, H. W. KROTO and D. R. M. WALTON

Structures and Intermolecular Interactions of Cocrystallites
Consisting of Metal Octaethylporphyrins with Fullerene C_{70} 730
TOMOHIKO ISHII, RYO KANEHAMA, NAOKO AIZAWA,
MASAHIRO YAMASHITA, KEN-ICHI SUGIURA, HITOSHI MIYASAKA,
TAKESHI KODAMA, KOUICHI KIKUCHI and ISAO IKEMOTO

Ab Initio Molecular Dynamics Simulation of Foreign Atom
Insertion into C_{60} 742
KAORU OHNO, KEIICHIRO SHIGA, TSUGUO MORISATO, SOH ISHII,
MARCEL F. SLUITER, YOSHIYUKI KAWAZOE and TSUTOMU OHTSUKI

Formation of Radioactive Fullerenes by Using Nuclear Recoil ... 749
TSUTOMU OHTSUKI, KAORU OHNO, KEIICHIRO SHIGA, TSUGUO MORISATO,
SOH ISHII, MARCEL F. SLUITER, HIDEYUKI YUKI and YOSHIYUKI KAWAZOE

Chapter 12: Nano-Machines and Biological Nanotechnology

Engineering with the Engines of Creation 761
CARLO D. MONTEMAGNO and JACOB J. SCHMIDT

Calculation of the Characteristics of a Molecular Single-Electron
Transistor with Discrete Energy Spectrum 768
V. V. SHOROKHOV and E. S. SOLDATOV

Electronic Transport through Benzene Molecule and DNA
Base Pairs 781
A. A. FARAJIAN, R. V. BELOSLUDOV, H. MIZUSEKI and Y. KAWAZOE

Realization of "Molecular Enamel Wire" Concept for
Molecular Electronics 786
RODION V. BELOSLUDOV, HIROYUKI SATO, AMIR A. FARAJIAN,
HIROSHI MIZUSEKI, KYOKO ICHINOSEKI and YOSHIYUKI KAWAZOE

Mechanical Properties of a Thermoelectric SMA Manipulator ... 793
KUMIKO YAKUWA, YUN LUO and TOSHIYUKI TAKAGI

Single-Electron Tunneling in Planar Molecular Nanosystems. ... 801
E. S. SOLDATOV, S. P. GUBIN, V. V. KHANIN, G. B. KHOMUTOV, V. V. KISLOV,
I. A. MAXIMOV, L. MONTELIUS, L. SAMUELSON, A. N. SERGEYEV-CHERENKOV,
M. V. SMETANIN, O. V. SNIGIREV, D. B. SUYATIN

Integration of Smart Materials in Si to Create Opto- and
Micro-Electronic Hybrid Devices 807
SEBANIA LIBERTINO, MANUELA FICHERA and A. LA MANTIA

High Performance Thermo-Elastic Metallic Materials by Rapid
Solidification and their Applications in "Smart"
Medical Systems 820
Y. FURUYA, Y. SHINYA, M. YOKOYAMA, T. YAMAHIRA, S. TAMOTO,
T. OKAZAKI and Y. TANAHASHI

Systemic Evaluation for Replication Processing of DNA Double Strand as a Zippering Micro-Machine 827
HIROHUMI HIRAYAMA

High Resolution Detection and Coating Method for the Functionalisation of Nanometre Separated Gold Electrodes with DNA 845
CHRISTOPH WÄLTI, RENÉ WIRTZ, W. ANDRÉ GERMISHUIZEN, MICHAEL PEPPER, ANTON P. J. MIDDELBERG and A. GILES DAVIES

Chapter 13: Nanotubes

Magnetic Properties of Doped Silicon Nanotube 857
ABHISHEK K. SINGH, TINA M. BRIERE, VIJAY KUMAR and YOSHIYUKI KAWAZOE

Computer Simulation of Single Wall Carbon Nanotube Crystals 861
VIJAY KUMAR, MARCEL H. F. SLUITER and YOSHIYUKI KAWAZOE

A Study of the Pressure Dependence of the Raman Spectrum of C_{60} and Single-Walled Nanotubes in Methanol-Water Mixtures 869
MOSTAFA EL-ASHRY, MAHER AMER and JOHN F. MAGUIRE

Chapter 14: Nano-Scale Surface Phenomena

Fundamental Studies of Large Organic Molecules on Metallic Substrates by High Resolution STM 877
FEDERICO ROSEI

Dielectrophoretic Manipulation of Surface-Bound DNA 888
W. ANDRÉ GERMISHUIZEN, CHRISTOPH WÄLTI, PAUL TOSCH, ADAM E. COHEN, RENÉ WIRTZ, MICHAEL PEPPER, ANTON P. J. MIDDELBERG and A. GILES DAVIES

Comparative Study on Interactions of Polypeptides of Bacteriorhodopsin in the Langmuir-Blodgett Film Based on Hydrogenated Amorphous Silicon Thin Film 896
YUTAKA TSUJIUCHI, JUNYA SUTO, KENJI GOTO, SHIGEKI SHIBATA, MANABU IHARA, HIROSHI MASUMOTO and TAKASHI GOTO

Nano-Granular Co-Zr-O Magnetic Films Studied by Lorentz Microscopy and Electron Holography 904
DAISUKE SHINDO, ZHENG LIU, GAO YOUHUI, SHIGEHIRO OHNUMA and HIROYASU FUJIMORI

Electrochemomechanical Deformation of Polypyrrole Film in Complex Buffer Media 910
SHYAM S. PANDEY, WATARU TAKASIMA and KEIICHI KANETO

**Nano-Metallization on Surface Nano-Sized Granules of
Polytetrafluoroethylene Matrix** 922
M. S. KOROBOV, G. YU. YURKOV and S. P. GUBIN

Chapter 15: Nano-Clusters

Cluster Expansion Method for Chemisorption Geometries 931
MARCEL H. F. SLUITER and YOSHIYUKI KAWAZOE

All-Electron *GW* Calculations for Small Clusters 940
SOH ISHII, KAORU OHNO and YOSHIYUKI KAWAZOE

**Theoretical Study of Unimolecular Rectifying Function of a
Donor-Spacer-Acceptor Structure Molecule** 949
H. MIZUSEKI, Y. KIKUCHI, K. NIIMURA, C. MAJUMDER, R. V. BELOSLUDOV,
A. A. FARAJIAN and Y. KAWAZOE

**Fully-Coordinated Silica Nanoclusters: Building Blocks for
Novel Materials** 954
S. T. BROMLEY, M. A. ZWIJNENBURG, E. FLIKKEMA and TH. MASCHMEYER

**Reductive Selective Deposition of Ni-Zn Nanoparticles onto
TiO_2 Fine Particles in the Liquid Phase** 958
HIDEYUKI TAKAHASHI, YOJI SUNAGAWA, SARANTUYA MYAGMARJAV,
KATSUTOSHI YAMAMOTO, NOBUAKI SATO and ATSUSHI MURAMATSU

Author Index   965

About the Editors   969

Preface

Intelligence in a Small Materials World has been produced to share ideas being generated by participants in IPMM—Intelligent Processing and Manufacturing of Materials—with other researchers involved in developing software and hardware systems for materials production who might benefit from some of the so-called "intelligent" technologies. IPMM was founded in 1997 and 4 international conferences have been held in Gold Coast-Australia, Honolulu-Hawaii, Vancouver-Canada, and Sendai-Japan. The 5th conference is scheduled for July 19-23, 2005 in Monterey-California.

The members of IPMM define "intelligence" in its broadest sense to include all types of Artificial Intelligence methods as well as the conventional numerical and phenomenological approaches to modeling and simulation of processes and materials. **Intelligence in a Small Materials World** contains 88 papers carefully chosen from the 156 papers presented at the Fourth International IPMM Conference, which took place from May 18th to May 23rd, 2003 at the Excel Hotel Tokyu in Sendai, Japan and the Hotel Taikanso in Matsushima, Japan. The theme of the 4th international conference was Nanotechnology for the 21st Century: do good things really come in small packages?

The scope of topics range from software methods such as fuzzy logic, neural networks, genetic algorithms, finite element and finite difference modeling, thermodynamic modeling, first-principles modeling, and data-correlation techniques such as Linear Regression and Principal Component Analysis, to hardware approaches such as industrial automation, robotics, optical and hyperspectral sensors, as well as conventional and new processes to solidify, form or machine metals and materials.

The papers were selected by a committee consisting of the executive of IPMM in consultation with a number of referees, contributors, and experts in the field. A number of the papers have been edited for clarity and continuity. The responsibility for the information contained in this volume remains totally that of the Editorial Committee of IPMM.

Chapter 1 deals with Philosophy. The first paper introduces the reader to IPMM and its organization. The paper discusses the field of Nanotechnology—its promised benefits and possible dangers. The second paper presents some moral dilemmas that might be faced as we engineer new strategies to overcome problems or to save the world from certain epidemics or pandemic. The third paper presents an argument for modeling using fundamental science rather than empiricism.

Chapter 2 deals specifically with applications that rely on Fuzzy Logic while Chapter 3 describes a number of successful systems based on Artificial Neural Networks.

Process Control and Optimization software applications are described in Chapter 4.

Environmental systems are presented in Chapter 5. The subject areas as expected are broadly based with papers on particulate systems, biotechnology for phyto-reclamation of waste dumps, magnetic levitation hoisting, and sealing abandoned mines with bulk materials

Chapter 6 examines intelligent applications in Energy Systems including sensor-based monitoring of grinding mills, fracture toughness of covalent materials, geothermal energy from mine effluents, hydrogen storage, as well as a theoretical analysis of methyl hydrates.

Chapter 7 contains a number of important developments in the field of metrology and intelligent instrumentation. Topics covered include hyperspectral non-destructive materials inspection, magneto-hydraulic stability in continuous casting, ductile machining phenomena, using fractal dimension to control machining operations. The Chapter also contains some important examples of nano-scale monitoring in the areas of shape-memory alloys, boron-diffusion, gas-detection systems, as well as nano-scale detection of urea.

Materials Science research is contained in Chapter 8. Subject areas include materials property estimation using simulation, solidification modeling, calculating alloy phase diagrams, development of smart composites, titania photocatalysts, and magnitostriction of Fe-Pd alloys.

Aspects of Intelligent processing of Materials are describes in Chapter 9 including: sub-nanoscale alloy structure control, quasi-isostatic pressing of cellular material, electromagnetic stirring in continuous casting, fabrication of a ceramic hip-joint, fabricating closed cellular materials, and forging of magnesium alloys.

Chapter 10 deals with Intelligent Manufacturing with papers on managing knowledge for design, ultra-wide-band and cyber security, and a description of a novel 3-D Input System.

Although a number of papers on nanotechnology are sprinkled through Part 1 of the book, Part 2 of the book deals with specific nanotechnological breakthroughs. Chapter 11 presents work on fullerene materials science while in Chapter 12, one can find papers on nano-machines and biological nano-scale systems. Chapter 13 deals specifically with nano-tubes with one paper describing novel silicon-based nanotube structures. Chapter 14 is reserved for nano-scale surface phenomena, an exciting area on measuring properties of monolayers. Finally Chater 15 presents several unique studies on nanoclusters.

Acknowledgement
The editors are very grateful to the following colleagues for helpful assistance and suggestions regarding selection of the papers and for assisting in the review of various versions of the manuscript: Mory Ghomshei, Junko Isoda, Yukiko Miyahara, Hiroshi Mizuseki, Kitsuko Motosugi, Suichi Oka, Marcel Sluiter, Yoshiyasu Takefuji, Jing-Zhi Yu, and Lotfi Zadeh. We wish to thank DEStech Publications and in particular, Anthony Deraco and Steve Spangler for their help, patience, and encouragement.

We are grateful to colleagues in the Department of Mining Engineering at the University of British Columbia for their support. In particular, we wish to thank Amy Wang for formatting the manuscript.

John A. Meech,
Editor-in-Chief,
Vancouver, British Columbia

Yoshiyuki Kawazoe,
Sendai, Japan

Vijay Kumar,
Sendai, Japan

John F. Maguire,
AFRL/ML, Ohio, USA

September 30, 2004

Chapter 1: Philosophy

Welcome to IPMM'03—Nanotechnology: Do Good Things Really Come in Small Packages?

Y. KAWAZOE
The Centre for Computational Materials Science, Institute for Materials Research, Tohoku University, Sendai, Japan

J. A. MEECH
The Centre for Environmental Research in Minerals, Metals, and Materials, The University of British Columbia, Vancouver, B.C., Canada

ABSTRACT

IPMM'03 is the fourth in a series of biannual conferences dealing with the field of Intelligent Processing and Manufacturing of Materials. The conference focuses on the area of Nanotechnology—materials and processes that exploit unique properties and applications that operate at the nano-scale. This sub-field offers many exciting opportunities and promises many innovative solutions to many of mankind's problems. Medical breakthroughs are expected and the overlaps into biotechnology, information science, and industrial processes are extensive.

And yet, as with any new science, the unknowns and uncertainties are fraught with potentially unintended, negative side-effects that may generate unforeseen problems. As we enter the Machine Age and begin creating nano-scale devices to manipulate molecules and atoms, there are dangers that must be recognized. Reactions with passive species present in the system can lead to new diseases or other unanticipated negative effects. Similar to concerns expressed on genetically engineered insect-resistant crops or ones with increased productivity, so too can the creation of new organisms that act as machines that may be able to clone themselves, lead to the introduction of dangerous and misunderstood entities into the world. Just as a virus is able to genetically change its structure as it transfers between species, so too can molecular machines with self-reproductive properties release new molecular forms that exceed our ability to control them.

This paper introduces the reader to the field of Intelligent Processing and Manufacturing of Materials with particular reference to nano-scale systems. The potential benefits and pitfalls are discussed to provide some balance to the hype from both sides of the issue of safety.

BACKGROUND ON IPMM

Intelligent Processing and Manufacturing of Materials (IPMM) is a formal organization of about 500 members in many diverse disciplines from over 50 countries around the world. These include, but are not limited to: materials science; metals properties and processing;

intelligent systems; artificial intelligence; computational intelligence; thermodynamic modeling; First-Principles modeling; materials property prediction; biotechnology; nanotechnology; information technology; bioinformatics; metrology; and manufacturing science. The connection that binds our members together is a keen interest in sharing new techniques and collaborative opportunities in intelligent methodologies (both software and hardware). IPMM defines the term "intelligence" in its broadest sense—from a software viewpoint, it can be a way in which a computer is programmed to mimic the human thought-process (fuzzy expert systems, artificial neural networks, SWARM intelligence, etc.)—from a hardware viewpoint, it can be a way that a machine or device is designed or built to adapt to its surroundings or to environmental changes (SMART materials, adaptive control systems, robotics, nano-scale machines, etc.).

Our organization has held four conferences. The first meeting took place in 1997 in Gold Coast, Australia with a focus on Intelligent Systems and Materials. In 1999, we traversed half-way across the Pacific Ocean to Honolulu, Hawaii where the theme was Intelligent Materials for the 21st Century. In 2001, the odyssey was completed with the third conference in Richmond, British Columbia (Vancouver) that dealt with the Cross-Disciplinary Aspects of Intelligent Materials Research. We crossed the Pacific once again in 2003 to Sendai, Japan where our group was hosted by the Centre for Computational Materials Science in the Institute for Materials Research at Tohoku University. The theme for IPMM'03 focused on Nanotechnology.

BACKGROUND ON NANOTECHNOLOGY

In a speech given in 1959 at the Annual General Meeting of the American Physics Society [1], Nobel Laureate Richard Feynman provided the foundation from which the field of nanotechnology has risen. Entitled "There's Plenty of Room at the Bottom—An Invitation to Enter a New Field of Physics", Feynman gave the public its first real insight into the intrigue, challenges, opportunities, and curiosities that exist in studying sub-micron size (or nano-scale) materials and processes. To fully appreciate of what Feynman spoke, one must realize that miniaturization of engines and machines had already commenced, yet he referred to these innovations as "…most primitive, halting steps…" in a science that today has evolved into actual molecular machines.

The field moved along at a relatively slow pace until the appearance of a number of popular books and texts in the late 1980s and early 1990s [2,3,4] following which a virtual explosion of research activity has taken place. Originally perceived as a field for physicists, Nanotechnology began to receive increased notice following the discovery of C60, the "third form of carbon" in 1985 [5], which opened up the field of "designer" chemicals and materials to the chemists and material scientists of the world (see Figure 1).

This crystal is named after the famous architect, R. Buckminster Fuller, who invented the geodesic dome. C60 was discovered by Richard Smalley and Robert Curl of Rice University, USA together with Sir Harold Kroto of the University of Sussex, U.K., while they were conducting experiments to help understand formation of long-chain carbon-nitrogen molecules in interstellar space. If ever there was an example of the importance to stay abreast of developments in other fields of science, this is clearly it! The discovery of Buckyballs and the impact of this finding on so many "unrelated" fields are absolutely astonishing. Who could have imagined that while performing research on the chemistry of

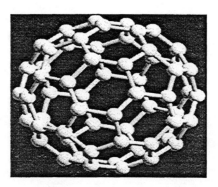

Figure 1. C60 structure of carbon – buckminsterfullerene (buckyballs) [5].

interstellar space, these scientists would stumble onto a brand-new crystal form of carbon that actually exists here on earth in common soot?

As more and more laboratories have developed methods to rapidly synthesize fullerenes the price has dropped appreciably. Today, ultra-pure C60 can be purchased for $90–100 per gram while C70 is around $400 per gram. Less pure varieties can be obtained for far less.

One of the first extensions of C60 was the discovery that carbon can be manipulated into a whole series of related structures in which the number of atoms range considerably away from that of a "Buckyball". Some of the structural varieties depicted in Figure 2 include a rugby ball as opposed to a soccer ball, but there are other complex structures such as toroids and dumb-bells.

Nanotechnology Research Institutes

Nanotechnology is a research-intensive field that relies on public and private funding. As a result, many companies have begun making substantial investments in nano-research. Large corporations such as IBM, Lucent Technologies, Hewlett-Packard, Samsung, and Siemens have put significant research resources into nanotechnology. Numerous startup companies have obtained venture capital funding to engage in a variety of nanotech initiatives. But, a significant amount of the research in nanotechnology takes place within universities and national laboratories.

Most groups specialize in a specific aspect of nanotechnology. For example, research in **nanoelectronics and photonics** can be found at the Albany Institute of Nanotechnology, at Cornell University, and at Columbia University. Universities specializing in **nanopatterning and assembly** include Northwestern University and the Massachusetts Institute of Technology (MIT). **Biological and environmental-based studies** exist at the University of Pennsylvania, Rice University and the University of Michigan.

In 2003, more than 100 U.S. universities had departments or research institutes specializing in nanotechnology. National laboratories, such as the Center for Integrated Nanotechnology at Sandia National Laboratories in New Mexico and the Center for Nanoscale Materials at Argonne National Laboratory, just outside Chicago, are undertaking major research efforts. International institutes such as the Max-Planck Institute in Germany, the Centre National de la Rescherche Scientifique in France, and the National Institute of

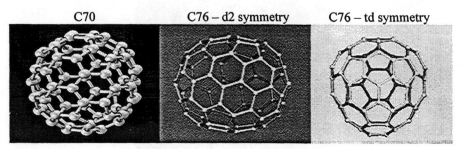

Figure 2. Other crystal structures of carbon spawned by the discovery of C60.

Advanced Industrial and Technology of Japan are all engaged in extensive nanotechnology research programs.

New **multi-disciplinary research centers** are being set up to better meet the challenges posed by nanotechnology. Major centers of nanoscience and nanotechnology around the world have been founded to combine scientific knowledge from a variety of disciplines. For example, in the U.S., there is the National Nanotechnology Initiative while Germany has established six Virtual Nanotechnology Competency Centres funded by the Ministry of Education and Research.

Partnerships between private and public entities have been set up to advance nanotechnology with a focus on areas often beyond existing areas of competence. The California Nanosystems Institute was created in 2000 as a joint enterprise between UCLA and the UC Santa Barbara to partner with companies such as IBM, Hewlett Packard and small biotech firms.

Traditional technology uses a **top-down approach** when constructing materials. Most objects are created through a process of shaping or molding pieces of materials until they precisely form the desired configuration. Results of such processes may be small (e.g., integrated circuits with structures measured in microns) or very large (ocean liners or jumbo jets). Nanotechnology has followed two pathways. The first is the traditional **top-down approach** to directly manipulate matter atom by atom and/or molecule by molecule. This approach depends heavily on high tech instruments such as the atomic force microscopy. Alternatively, the **bottom-up approach** recognizes that the building blocks of life (enzymes and other components of each living cell), already act as machines at the nanoscale. Applying these elements in new ways is the main focus.

Carbon nanotubes

The next important development was the carbon nanotube [6, 7] in which an infinitely long, hollow chain of carbon atoms can be formed that provide a channel to transport atoms or other tiny agents (see Figure 3). These hollow fibres provide exceptional strength char-acteristics (100 times that of steel) and so, the field of composite materials received a major boost in applying these fibrous materials. Development of carbon nanotubes, has led to major breakthroughs in fuel cell batteries, ultra-wide screen television and micro-sensor technologies.

Endohedral fullerenes

Having a molecule in the shape of a cage invites the question: can something be put inside? And so was born the sub-science of Endohedral Fullerenes. As these structures were studied, it became apparent that certain metallic ions and atoms could

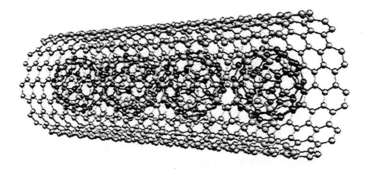

Figure 3. Large-scale Carbon nanotube with C_{60} crystal inside.

be inserted inside a fullerene structure without any direct bonding to the atoms that make up the structure. Intriguing and unusual properties are being discovered with these unique chlathrate-like "compounds". Extension of the technology into such areas as methyl hydrates shows promise to develop new energy sources [8].

Si-based Fullerenes

The breakthrough in fullerene chemistry has led others to investigate structural manipulation of other materials. Silicon-based materials have now been found to exhibitsimilar families of structures as do the carbon-based ones with the research beginning with endohedral applications [9] and extending into silicon nano-tubes [10].

Designer Chemistry and First Principles Modeling (Quantum Mechanics)

As scientists have realized the potential to manipulate materials at the molecular, atomic, and even electronic level, a new field of chemistry has evolved known as Chemical Design [11]. Using Quantum Mechanics, the physicists have become involved in developing modeling techniques based on first-principles [12]. We can now create virtual materials that possess properties required for new applications. These include surface chemistry of thin films, the properties of nano-clusters and development of nano-machines.

Biotechnology and Bioinformatics

BioSPICE [13] provides life science researchers with a set of open source software modules to create computer models of different cellular processes allowing investigators to research numerous questions not amenable to direct experimentation. BioSPICE can simulate cell-division, circadian rhythms, bacterial sporulation, and gene transcription networks. It may prove particularly useful in simulating the effects of unknown pathogens so we can rapidly respond to new biological threats.

The DARPA BioComputation program aims to develop a computational framework (Bio-SPICE), analytical techniques, and bioinformatics tools able to predict cellular processes and control their spacio-temporal behaviour. The system can explore how external pathogens or agents invade and disrupt normal cell functions to help design new drug interventions. Algorithms are being validated using in-vitro and in-vivo experiments. A second goal is aimed at how biomolecules can be used in computing and engineering applications in which scalable nucleotide (DNA) manipulations can solve highly complex, NP-complete problems (designing ultra-high density information storage systems and developing programmable DNA/RNA nano-structures).

POTENTIAL PITFALLS

There is great fear in some circles that nanotechnology may release new materials and/or agents into the environment that will create major health and environmental problems. Michael Creighton's novel, *Prey*, has done much to propagate such concern. Research already suggests that aquatic biota is affected by these tiny carbon balls. A level of 800 ppb was found sufficient to kill 50% of water fleas in a 3-week trial while young largemouth bass presented cellular injury to brain tissue at a rate of 17-fold at a concentration of 500 ppb. This damage, called "lipid peroxidation", impairs normal cell membrane functioning and is linked to illnesses such as Alzheimer's disease in humans [14]. In 2003, researchers found damaged lung tissue in mice that inhaled nanotubes [15].

But one of the fundamental questions raised by nanotechnologies relates to the safety of humans and the potential cultural impact as "nhancement" of human nature blurs the distinction between a human and a machine [16]. Human life expectancy is likely to increase significantly as nanotechnology and biotechnology combine to prevent current diseases and replace failed biological parts with synthetic machines. Organizations such as the World Transhumanist Association, the Institute on Biotechnology and the Human Future, and the Extropy Institute have been formed to examine the benefits and problems that the Machine Age may pose for Mankind.

Dangers in New Ideas

The ability to manipulate material structure at the level of the molecular bond opens the door to new methods to deliver drugs; new ways to build computers; new types of more productive genetically-engineered foods resistant to pests; new wear-resistant surface coatings of diamond; etc. These discoveries must rank as the most-convincing demonstrations that fundamental science with its intricate links to diverse areas provides the true source to advance technology in unpredictable ways. On its own, an idea has no intrinsic value, but value cannot exist without ideas. Value materializes (as if by magic) from applying linked ideas one after another or in parallel to generate new wealth (or profit) through improved living, life-style, or by extending life.

Ideas are a metaphor for the electrons that zoom along inside wires connected together to form electric circuits. On its own, each electron has little to offer at the macro-scale, but the flow of electrons and the harnessing of this stream into controlled systems have given our species the ability to develop machines and a myriad of household products that we take for granted each and every day. Do we care where the electrons come from? Whether they begin their journey at some hydroelectric dam in Northern Canada to which they return over and over again? Do we pay attention to the fact that until Faraday discovered electricity, its utility remained in the hands of the gods–lightning and thunder? Did Faraday imagine that one day his discovery would evolve into the many important products and applications that we share today?

Of course not! Faraday carried out his research for the pure joy of curiosity and discovery. Had the society of the day been able to control his work, he might have been denied the opportunity to apply his intelligence and knowledge to the subject. The practical nature of his work only became widely recognized as more and more scientists and engineers applied his discovery in combination with other ideas to develop products and methods. Does that mean we would never have discovered electricity if Faraday had been impeded in his work? Of course not, again! Development and application of electricity would simply have begun somewhere else and been done by someone else at another time and place with an evolutionary path connecting ideas to new products and methods in a sequence different than we know occurred today.

The role of science can be thought of as the "progression of applying logical ideas towards an unknown goal to achieve an unknown purpose that is ultimately aimed at benefiting society". At each step along this journey, a specific sub-goal is in mind and milestones are attained by individual scientists who share in and apply the collective ideas of their colleagues. These sub-goals combine at different stages to evolve new methods or products that are "of interest". Such interest can range from the simple curiosity of a single scientist to the generation of incredible wealth by an industry.

In some cases these steps may be driven by particular negative characteristics of the human spirit–greed, envy, fear, anger, control of others, etc. Sometimes the research may open up a Pandora's Box of great difficulties resulting in significant trouble for the individual and perhaps, for others in society. The discovery of radiation by Marie and Pierre Curie, for example, led to their relatively early demise from anemia brought on by significant and continuous exposure to their discovery. Many others had to suffer a life shortened by radiation exposure before we found out how to harness this "interesting" energy for good applications. Even today we are still learning to use nuclear power in a beneficial way–so we can store dangerous by-products in a safe fashion and attempt to prevent them from being used to create weapons of mass destruction.

Should the work performed by the Curies have been stopped? It is doubtful that rational people would agree with this, even in hindsight. But what is evident from this example is the need for caution, the need to pay attention to some of the likely negative side-effects of one's work and to put into place measures to attempt to control such effects or at least, minimize their impact.

Such is the case with the field of nanotechnology wherein Science is moving into a realm that has considerable uncertainty with respect to unknown dangers–dangers that could impact in ways that are irretrievable–dangers that may require a total change in how our society operates and develops. Often such work causes possible harm to the environment even when the benefits of the discovery are obvious. Table 1 presents some guidelines that might be applied in making a decision to implement a new idea. They involve a trade-off of economic or social benefits against harm to a local or global environment. Obviously the methods used to measure and/or monitor the benefits, costs, and level of harm must be well-understood and accepted widely.

The moral dilemma to study and develop techniques and materials tries to separate those outputs that produce a "profit" from those that do not. Profit can be quantified in a monetary sense or in other ways, but over-riding this economic analysis is the need to determine the impact that the new product or process may have on the environment or on populations that inhabit that environment. Two examples of situations in which benefits far outweigh costs and yet maximum harm is "apparently" inflicted on an environment are: 1. production of energy using fossil fuels; 2. eradication of malaria-carrying mosquitoes using DDT.

In the former case, the world is just beginning to accept the reality, uncovered through modeling and measurement, that our environment is significantly threatened by the accumulation of Greenhouse Gases in the atmosphere. As we grapple with the "obvious" decision to dispense with fossil fuels as an energy source, the need for a balanced transition is necessary so that the harm does not become a permanent outcome as the cost of change approaches the benefits at the lowest levels of our economic systems. The Kyoto Accord [17]

Table 1. Hard and Easy Decisions that involve trade-offs between Economic/Societal Benefits and Possible Harm to an environment or population (adapted from [12]).

	Minimum or No Harm	Maximum Harm
Benefits Far Outweigh Costs	Morally Required	Morally Permitted in cases of Extreme Emergency
Costs Far Outweigh Benefits	Morally Permitted in cases of Extreme Emergency	Morally Prohibited

is an attempt to establish a carbon-reduction trading system to provide a logical and orderly decline in fossil fuel use around the world without causing unbalanced economic suffering for developing countries. As harm to our environment is recognized as a new cost with which the world's economy must deal, the moral imperative is obvious that transformation to alternate sources of energy (renewables) must be accelerated even if the cost/benefit analysis is negative.

The second case is equally complex [18]. The banning of DDT was one of the first "victories" claimed by the environmental movement in the 1970s. Prior to the world-wide edict, DDT was winning the battle against malaria in many parts of the Third World. Today, a reemergence of this deadly disease that kills more people annually than AIDS/HIV is occurring. In the past decade, malaria cases have escalated at an alarming rate, especially in Africa. An estimated 300 to 500 million cases each year cause 1.5 to 2.5 million deaths—more than 90% of which are children. Widespread spraying with DDT to control mosquitoes was well on the way to eliminating the disease by containment. Today the use of insecticide nets is recognized as effective, but providing such protection to the majority of the population is far too costly. The banning of DDT was driven by the claim that it weakened the eggshells of the Peregrine Falcon and other birds. In hindsight it appears that the harm to one environment (which has been challenged today by numerous scientific studies) was miniscule compared to the terrible harm that this banning has had on eradicating malaria. Once a decision like this is made, it appears our society finds it very difficult, if not impossible, to revisit the evidence and perhaps, admit that the choice was wrong.

CONCLUSION

The evolution of the field of Nanotechnology provides us with lessons on the development of new fields and ideas:
1. Science involves application of ideas, one at a time, to provide understanding or new knowledge about a subject. Ideas are the life-blood of science in all fields.
2. Nanotechnology is a broadly-based science involving manipulation of atoms, electrons, protons, and neutrons in a variety of ways to generate new understanding of how materials can be developed to solve many problems in medicine, engineering, agriculture, biology, chemistry, surface science, space exploration, ocean and marine science, geography and geology. In some cases the research will develop new machines or ways to deliver new products. In others, the research will simply serve to provide new understanding about how matter interacts at the very smallest of levels.

3. There is certain danger in nanotechnology and its applications. Differences in the properties of materials at this small size are just now being understood. Caution is an element in any exploration, but so too, is the ability to take risk and try something new.
4. In taking risk, it is important that the individual scientist shows respect for some of the possible negative outcomes of an experiment and attempts to design fail-safe backup systems to prevent "accidental" release of harmful substances into the environment that could pose greater danger than the future benefits to be derived from the research itself.
5. An individual scientist or researcher should not be held responsible or accountable for the potential application of the work of others for negative purposes (whether intentional or otherwise). Cause and effect chains can be direct or indirect indicators of the proliferation of an idea or set of ideas, but the instigator of any one of these ideas shouldn't necessarily feel guilty nor be blamed for the inappropriate application of these ideas by others.

REFERENCES

[1] R. Feynman, 1959. There's plenty of room at the bottom—an invitation to enter a new field of science. Engineering and Science, California Institute of Technology, February, 1960. Authorized web version: http://www.zyvex.com/nanotech/feynman.html
[2] K. Eric Drexler, 1987. Engines of Creation: the Coming Era of Nanotechnology, Anchor Books, New York, pp. 320.
[3] K. Eric Drexler, 1992. Nanosystems: molecular machinery, manufacturing, and computation Wiley Interscience, New York, pp. 576.
[4] E. Regis, 1996. Nano! The Emerging Science of Nanotechnology. Little Brown and Company, Back Bay Books, pp. 336.
[5] H.W. Kroto, J.R. Heath, S.C. O'Brien, R.F. Curl and R.E. Smalley, 1985. C60: Buckminsterfullerene, Nature, 318(6042), 162-163.
[6] S. Iijima, 1991. Helical micro-tubules of graphitic carbon, Nature, 354, 56–58.
[7] S. Iijima, P.M. Ajayan, and T. Ichihashi, 1992. Growth model for carbon nano-tubes, Phys. Rev. Lett., 69, 3100–3103.
[8] V.R. Belosludov, T.M. Inerbaev, R.V. Belosludov, M. Sluiter, Y. Kawazoe and J. Kudoh, 2003. Theoretical Study of Clathrate Hydrates with Multiple Occupation, Proceedings of IPMM-03, Sendai, Japan, pp.9.
[9] F. Hagelberg and C. Xiao, 2003. Computational Study of Endohedral $IrSi_9^+$ Isomers, Structural Chemistry, 14(5), 487–497.
[10] A.K. Singh, T.M. Briere, V. Kumar, and Y. Kawazoe, 2003. Magnetic Properties of Doped Silicon Nanotube, Proceedings of IPMM'03, Sendai, Japan. pp.4.
[11] A. Newman, 1994. Designer Chemistry, Environmental Sci. and Tech., 28(11) 463A.
[12] K. Ohno, K. Esfarjani, and Y. Kawazoe, 1999. Computational Materials Science—From Ab Initio to Monte Carlo Methods, Springer-Verlag, New York, pp. 325.
[13] S.P. Kumar, J.C. Feidler, 2003. BioSPICE: A Computational Infrastructure for Integrative Biology, OMICS: A Journal of Integrative Biology. 7(3), 225–225.
[14] NewScientist.com news service, 2004. Buckyballs cause brain damage in fish, March 29.
[15] C.-W. Lam, J.T. James, R. McCluskey and R.L. Hunter, 2004. Pulmonary Toxicity of Single-Wall Carbon Nanotubes in Mice 7 and 90 Days After Intratracheal Instillation, Toxicological Sciences, 77, 126-134.
[16] F. Fukuyama, 2002. Our Posthuman Future: consequences of the biotechnology revolution, Picador, New York, pp. 272.
[17] http://www.climatechange.gc.ca/cop/cop6_hague/english/overview_e.html
[18] T.C. Nchinda, 1998. Malaria: A Reemerging Disease in Africa, Emerging Infectious Diseases, Special Issue, 4(3), 398–403.

On the Mechanics of Morals

J. GREENE
Department of Psychology, Center for the Study of Brain, Mind, and Behavior, Princeton University, Princeton, NJ 08544, USA

ABSTRACT

Traditional Western philosophy and theology regard human beings as immaterial souls conjoined with material bodies. Modern science has rendered this "dualist" conception of human nature increasingly untenable, replacing it with a materialist conception according to which human beings are extremely complicated machines. One of the many implications of this conclusion is that human beings and everyday machines lie not on opposite sides of an uncrossable metaphysical chasm, but along a single continuum of functional complexity, a Great Continuum of Material Being. We are now beginning to understand the processes that produce human moral judgments in physical terms, and these results illustrate the material nature of humanity in a way that is accessible to the general public. Just as people are confronting the fact that humans are complicated machines, we will increasingly be faced with machines that are more and more human. These may include machines that are sufficiently sophisticated to be objects of moral concern, or even morally concerned themselves. In addition, we may soon be faced with natural creatures who have been created or modified in ways that are morally significant. Whether or not such technology is on the immediate horizon, it is worth considering the implications of such technology as preparation for the future and as a means of bringing into focus three questions of current interest:

> *What makes an entity a fit object of moral concern?*
> *What does it mean to be a moral agent?*
> *What does it mean for a material being that obeys the laws of physics to have free will?*

REVERSE ENGINEERING

Most of you are engineers, and as such your work begins with two things: a functional *goal* and a set of *tools*. Your goals derive from human desires, say, to explore the surface of the moon or to replace failing human hearts. As for your tools, you have at your disposal a set of physical materials and devices as well as some intellectual ones: concepts, principles, formulae, and so on. Thus, you start small and end big. You begin with relatively simple things that you understand and attempt to fashion them into larger, more complex things that you do not understand (yet).

I am a *reverse engineer* [1]. I begin with the most complex device in the known universe, the human brain/mind, and try to understand what it does and how it does it: What are its functions? What are the smaller devices out of which it is built? What are its organizing principles and what concepts do we need to understand them?

Fortunately, I, and my colleagues, don't have to start from scratch. The first question—What is the brain's function?—was answered in a general way by Charles Darwin. While we have a fair amount of leeway in choosing which functions we will perform with our brains, the basic *design* of our brains, if I may call it that, is a product of natural selection. Our brains are the way they are because brains like ours proved very useful in helping our ancestors spread their genes.

This answer, of course, is only a start. Darwin's Theory tells us what general biological function the human brain performs, but it doesn't tell us how it is performed. That is, it doesn't tell us anything about the functions of the brain's many *parts*. Some of these parts, generally the smaller ones, we understand reasonably well thanks to the mature sciences of physics, chemistry, molecular biology, and traditional neuroscience. But this understanding is also just a start. It's one thing to understand what makes a neuron fire, quite another to understand what makes one human being fire his weapon at another.

Thus, we are in the process of bridging two radically different sets of concepts: the concepts of everyday psychology (beliefs, desires, feelings, thoughts) and the concepts of the physical sciences (molecules, cells, circuits, etc.). The name for this bridge-building project is "cognitive neuroscience," the merging of cognitive psychology and traditional neuroscience. Within cognitive neuroscience there exists a brand new sub-discipline, "social cognitive neuroscience", which, as the name suggests, integrates the methods and subject matter of neuroscience, cognitive psychology, and the social sciences. Social cognitive neuroscience holds the promise of explaining *in the mechanistic language of the physical sciences* why one human being fires his weapon at another.

As I've said, this ambition exists today in the form of a promise, but one of the things I hope to convey to you is that this promise is not empty. We are indeed on our way to understanding the "nuts and bolts" of our humanity and, just as important, our inhumanity. Today I will tell you about some of my own work on the cognitive neuroscience of moral judgment. This work is aimed at piecing together one small part of the larger puzzle that is human nature, but I think it's an illustrative case.

It's illustrative because moral judgment is central to our humanity. According to traditional "dualist" conceptions of human nature, moral judgment is a function of the soul, not the body. According to Christian theology, for example, it is the quality of a given soul's moral judgments that determines its final destination. In other words, moral judgment is widely viewed as an essential component of the soul's "job description". Thus, if your soul is not in the moral judgment business, it's not clear that it has any business at all.

And that is why the neuroscientific study of moral judgment is so very interesting, and, to many people, is threatening. If we can demonstrate that moral judgment can be accomplished by a *machine*, a device that operates according to the same mechanical principles as one's car, then many people will have to revise their understanding of what it means to be a person.

Thus, the terms of the classic debates over human and animal nature are changing. These debates won't be resolved any time soon, but the burdens of proof and the weightings of the evidence are shifting. And as old questions eke closer to resolution, new ones gain prominence: If Free Will is not the action of the soul upon the matter of the body, what is it? Is it just an illusion? [2] If so, how do we decide when to hold people responsible for their actions? What about the moral status of non-humans? Is it okay to kill and eat other species simply because they're not human? [3] Can we imprison chimpanzees in zoos simply because we enjoy watching them? What about our obligations to the artificial life forms that lie on the horizon? Professional philosophers have been seriously debating these issues for decades, but few others have taken more than a passing interest in such questions. As advances in the sciences of human nature work their way into the public consciousness, these questions will cease to be merely academic. Ordinary people whose worldviews will have been shaped by the new sciences of the mind will fail to be satisfied by the moral common sense of the past.

MORAL JUDGMENT UNDER THE MICROSCOPE

Consider the following dilemma (the trolley dilemma):

> A runaway trolley is headed for five people who will be killed if it proceeds on its present course. The only way to save them is to hit a switch that will turn the trolley onto an alternate set of tracks where it will kill one person instead of five. Ought you to turn the trolley in order to save five people at the expense of one? Most people (educated Americans, at any rate) say yes.

Now, consider a similar problem (the footbridge dilemma):

> As before, a trolley threatens to kill five people. You are standing next to a large stranger on a footbridge spanning the tracks, between the oncoming trolley and the five people. This time, the only way to save the five people is to push this stranger off the bridge and onto the tracks below. He will die if you do this, but his body will stop the trolley from reaching the others. Ought you to save the five other persons by pushing this stranger to his death? Most people say no. [4]

Taken together, these two dilemmas create a puzzle for moral philosophers: What makes it morally acceptable to sacrifice one life to save five in the trolley dilemma, but not in the footbridge dilemma? Many answers have been proposed. For example, one might suggest, in a Kantian vein, that the difference between these two cases lies in the fact that in the footbridge dilemma, one literally *uses* a fellow human being as a means to some independent end, whereas in the trolley dilemma, the unfortunate person just happens to be in the way. This solution, however, runs into trouble with a variant of the trolley dilemma in which the track beneath the one person loops around to connect with the track beneath the five people [4]. Here we will suppose that without a body on the alternate track, the trolley would, if turned that way, make its way to the other track and kill the five people. In this variant, as in the footbridge dilemma, you would use someone's body to stop the trolley from killing the five. Most agree, nevertheless, that it is still appropriate to turn the trolley in this case, in spite of the fact that that here, too, we have a case of "using".

This is just one proposed solution and one counterexample to this problem, but together they illustrate the sort of dialectical difficulties that all proposed solutions have encountered. If this problem has a solution, its solution is not obvious. That is, there is no set of consistent, readily accessible moral principles that captures people's intuitions concerning what behavior is or is not appropriate in these and similar cases. This leaves psychologists with a puzzle of their own: How is it that nearly everyone manages to conclude that it is acceptable to sacrifice one life for five in the trolley dilemma but not in the footbridge dilemma, in spite of the fact that a satisfying justification for distinguishing between these two cases is remarkably difficult to find?

My collaborators and I investigated this question using functional magnetic resonance imaging (fMRI), testing a theory about why people respond differently to the trolley and footbridge dilemmas [5]. Our theory, which was suggested to us by some earlier neuroscientific work (e.g. that of Damasio and colleagues [6]), was that the "up close and personal" nature of the harm in the footbridge case (pushing someone to his death with one's bare hands) tends to evoke a strong emotional response against performing that action whereas the relatively impersonal harm described in the trolley cases (hitting a switch that causes a train to run people over) tends not to engage people's emotions as strongly. Thus, we predicted that brain areas associated with emotion would be more active during contemplation of dilemmas such as the footbridge dilemma as compared to during contemplation of dilemmas such as the trolley dilemma. That was, in fact, what we found. The contemplation of "personal" moral dilemmas, ones that involve an "up close and personal" violation as in the footbridge dilemma, produced increased activity in three brain areas associated with emotional response. At the same time, the contemplation of "impersonal" moral dilemmas, ones like the original trolley case in which there is no "up close and personal" violation, produced increased activity in brain areas associated with working memory, suggesting a more "cognitive" and perhaps more "rational" style of response to these cases.

Our hypothesis concerning emotional engagement made a second prediction, this time concerning people's behavior. According to our theory, people tend to say that the action in the footbridge case is "inappropriate" because of a negative emotional response to the thought of committing a personal moral violation. But what about the handful of people who judge this action "appropriate"? If such individuals have normal emotional responses, then it would follow that they would have to overcome those responses in order to judge the action "appropriate", in order to say "yes" when their emotions say "no". Thus, we would predict that in response to personal moral dilemmas like the footbridge case, the subjects who answer that the action is "appropriate" will on average, take longer to give their answers than the ones who answer that the action is "inappropriate". In contrast, our hypothesis gives us no reason to predict a difference in reaction times between judgments of "appropriate" and "inappropriate" in response to impersonal moral dilemmas. As predicted, judgments of "appropriate" took longer than judgments of "inappropriate" in response to personal moral dilemmas, but this pattern was not observed for impersonal moral judgments. These results suggest, once again, that people have negative emotional responses to personal violations, responses which must be overcome when one judges a personal moral violation to be appropriate.

In sum, it appears that two moral dilemmas that are very similar in structure—Is it okay to sacrifice one person's life to save five by performing some simple physical intervention that will result in death by trolley?—can be very different in terms of how we tend to think about them. More specifically, it seems that the crucial difference is a matter of emotional engagement. What's striking however, is how little people are aware of the difference in emotional engagement and its effects. If you ask someone why she said that it's okay to hit the switch, but not okay to push the man off the footbridge, she will most likely try to give you a rational argument like the Kantian one described above. Almost no one says, "I don't really know why I judged them differently. Pushing the man just feels more wrong to me than hitting the switch".

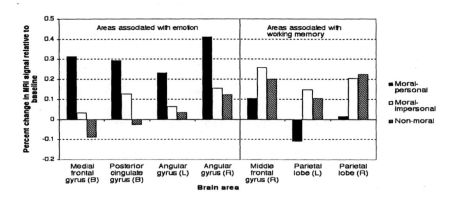

Figure 1. Neural activity in brain areas as a function of type of moral dilemma.

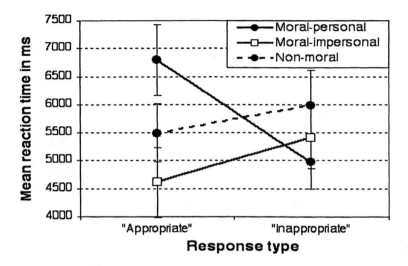

Figure 2. Average reaction times of judgments of "appropriate" vs. "inappropriate" for three types of dilemmas.

This pattern in our moral thinking (and thinking about our thinking) is puzzling, but I think it makes sense when viewed from an evolutionary perspective. Consider chimpanzees as a stand-in for our pre-linguistic primate ancestors. Chimps live intensely social lives, guided by what one could reasonably describe as moral instincts [7]. Chimps are quite capable of harming one another, and sometimes they do, but they don't harm one another indiscriminately. Nor do they harm one another whenever they can get away with it. And not only do chimps refrain from harming one another when it might otherwise benefit them to do so, they will sometimes exhibit a kind of righteous indignation when the social rules have been violated. Consider the following story reported by de Waal [8] and summarized by Haidt [9]:

> [The] zoo-keepers created the rule that no chimp would be fed until all members of the colony had moved from the outdoor island to the indoor sleeping enclosure. The chimps themselves worked to ensure that stragglers obeyed the rule, in part by showing hostility to latecomers. One evening two adolescent females stayed out for hours, delaying the feeding of the whole colony. The zoo-keepers finally isolated the two errant adolescents, fearing for their safety if they faced the angry colony. The next day, when they were released into the company of the colony, the other members vented their hostility on them. In other words, the members of the colony were aware that specific individuals were the cause of the delayed feeding, they remembered this information overnight, and they inflicted punishment the next day.

This story illustrates an important point, namely that moral instincts need not be grounded in an ability to reason or use language. (That is, if one assumes that the chimpanzees in the story above are responding emotionally rather than applying moral principles.) It also tells us something about our own moral instincts, namely that they are not so much instincts to reason as they are instincts to feel and react [9]. But what engages these social-emotional instincts? Our primate ancestors did not live in a world in which individuals could be killed or saved by means of actions mediated by complex mechanical devices such as trolleys on train tracks. Among our primate ancestors, all moral violations were, presumably, "up close and personal". With this in mind, it should not be surprising if our moral instincts respond most intensely to the sorts of moral violations that filled our evolutionary past. At the same time, however, it's clear that our moral thought should not be limited to such instincts. We have a general reasoning capacity that we can use to think about anything, including moral matters.

With this in mind, I propose that our capacity for moral judgment isn't so much one capacity, but two capacities that often work together and sometimes against each other [10]. On the one hand we have the basic, social-emotional instincts that we inherited from our primate ancestors, instincts that respond most strongly to "personal" violations of the kind exhibited in the footbridge dilemma. On the other hand, we have a general reasoning capacity that can be used to think about moral issues in the same way that we would think about any number of other topics. I propose that because the trolley dilemma, with its mechanically mediated action at a distance, sets off fewer of our instinctive moral alarm bells, we tend to think about it in a cooler, more rational way, opting for a cost-benefit analysis that favors saving five at the expense of one. In contrast, the footbridge dilemma, with its "up close and personal" violation, sets off those emotional alarm bells rather strongly and typically elicits a swift judgment of "inappropriate".

Building on this theory, my collaborators and I are currently doing some follow-up work aimed at dissociating the functions of the various brain areas identified in Figure 1. Our strategy is to focus on the personal moral dilemmas (the ones like the footbridge case), comparing the difficult ones to the easy ones. Consider the following case:

> It's wartime, and you and your fellow villagers are in a basement hiding from the enemy soldiers outside. The soldiers have orders to kill all remaining civilians, including women and children. Your baby starts to cry. If you do nothing, the soldiers will hear the baby, find you and the others in the basement, and kill you all. However, if you cover the baby's mouth, you and the others will be saved, although your baby will smother to death. Is it okay to smother your baby in order to save yourself and the others? Most people find this question very uncomfortable and take a long time to answer it. Some people say yes, and some say no.

Now consider a different case:

> You are a fifteen-year-old girl who has become pregnant. By wearing loose clothing and gaining weight you have kept your pregnancy a secret. When the baby is ready to be born, you hide in the bathroom and deliver it by yourself. You are sure that you

do not want to have a baby to take care of. Part of you wants to throw the baby in the trash and pretend that nothing happened. Is it okay for you to throw your baby in the trash in order to move on with your life? Most people (again, based only on the responses of educated Americans) say that this is not okay, and they say so very quickly.

What makes the first case difficult and the second case easy? My theory is that in the difficult case of the crying baby in the basement, the two different kinds of moral thinking mentioned above, personal and impersonal, are pitted against one another. Our basic social-emotional instincts balk at the horrible thought of killing one's own baby, but our rational side recognizes that there is nothing to be gained by not killing one's baby under such dire circumstances and much to lose. An internal moral conflict ensues, accounting for the long reacting times and feelings of discomfort exhibited by our subjects. In the case of the teenage mother, one has the same negative emotional reaction to killing the baby, but there is relatively little in the way of rational considerations to oppose that reaction. The emotional response dominates from the start, and people quickly say that the action is wrong.

My colleagues and I had people respond to cases such as these two while in the brain scanner, and what we have found so far makes sense. The data from difficult dilemmas like the crying baby dilemma suggest a conflict between "emotional" and "cognitive" processes. More generally, from studies like this one, we are getting a better sense of the role that different parts of the brain play in moral judgment. Some parts of the brain appear to be involved most crucially in the initial coding of morally relevant information. Others appear to be involved in the emotional-cognitive battle right up to the end. Some brain areas seem to work closely together. Others appear to function more independently. These results are preliminary and will no doubt be refined with future research, but the broader significance of these results is, I think, clear. We're on our way to understanding the physical mechanisms that produce moral judgments.

SO WHAT?

Yes, so what? Long before the advent of brain scanning we knew the brain had to accomplish moral judgment somehow. So what in the end, does it matter if the brain does it *this way* instead of *that way*?

First, if it turns out that the way the brain *actually* makes moral judgments is different from the way we generally *assume* it makes them, that could affect the attitudes we take toward the judgments we make. Consider, for example, the fact that you don't blindly believe everything you read. In our efforts to sort fact from fiction, we pay attention to the nature of our sources. A scandal reported in the *New York Times* carries much more weight than one reported in a trashy tabloid. In a similar way, knowing more about the sources of our moral judgments may lead us to question (or affirm) their validity.

Take, for example, our moral position with respect to the world's most needy people. Most of us spend money on things for ourselves that we don't really need: dinners in restaurants, electronic gadgets, etc. At the same time, there are people in the world whose lives can be saved by relatively small donations. Nevertheless, we choose to spend our disposable income on ourselves. The philosopher Peter

Singer dramatizes the issue with a striking example [11]. Suppose you're walking along when you spot a baby drowning in a pond. You could wade in and save the baby, but in doing so you would ruin your Italian leather shoes. Is it acceptable to let the baby drown for the sake of your shoes? Certainly not! But what's the difference between refusing to save the drowning baby for the sake of your shoes and refusing to save a sick or starving baby somewhere else in the world for the sake of your dinners in restaurants and electronic gadgets? Common sense morality tells you that it's monstrous to let the baby drown, but perfectly acceptable to keep your disposable income to spend on luxuries for yourself, in spite of the tremendous good it could do elsewhere. What is the basis for this common sense judgment? Surely, you might think, there must be *some good argument* for why it's okay to treat these two cases differently.

But what if there isn't? [12] What if our moral instincts feel like they're insights into Moral Truth, when really they're (largely) biological adaptations, full of quirks and inconsistencies? What if our moral instincts are driven not by good arguments but by emotional responses inherited from our primate ancestors? [9] What if we're insensitive to the suffering of people far away because our ancestors lived in a world in which interaction with faraway people was impossible?

It's been said that consumers of public policy, like consumers of sausages, are better off not knowing how those products are made. One might be tempted to say the same of consumers of moral judgments. I trust, however, that knowledge of where our moral judgments come from and how they are produced won't lead us to give up on morality entirely, but I do think that such knowledge, however unsettling, will dramatically change the way think about moral issues.

Thus, our moral opinions may change in response to knowledge of the details of moral psychology. But other changes may arise not in response to the details themselves, but simply in response to the fact that the details, whatever they are, are now *visible and vivid.*

Tell a professional philosopher that modern neuroscience is just now revealing that the human mind is nothing more than a complicated machine, and he'll laugh at you. This is old news. Modern neuroscience hasn't suddenly given us a good argument for materialist philosophy (the view that all things, including the human mind, are made of ordinary matter). To the extent that we have such an argument now, we had one before. What science is now giving us are *powerful illustrations* in the service of *old arguments.* It's one thing to argue for a materialist view of the mind on the basis of esoteric philosophical principles and scientific hunches, quite another to see (with the help of a brain scanner) that the material human brain can perform feats that were supposed to be performed only by the immaterial human soul. In the past, one has had to receive an extensive scientific and/or philosophical education to feel the force of materialist philosophy. Starting now, one need only open the newspaper, in which colorful pictures of deli-sliced brains drive the point home: "You are the hunk of matter between your ears—nothing more, nothing less!"

I do not mean to disparage the new sciences by claiming that they provide us with vivid illustrations in the service of old arguments. In this case, the illustration, the ability to convey to the general public the significance of what has been learned, is all too important. Here we see a sharp contrast with fields that benefit humanity primarily through the development of new technology. Engineers can hand the public a faster, more powerful computer or a new antiviral drug, and the public can benefit from these things without understanding the science behind them. In contrast, the work of philosophers and reverse engineers such as myself isn't aimed at producing new technology. Rather, our work is aimed at producing a new understanding of ourselves as human beings. But (in my opinion) this work is all for nothing if our hard-won knowledge remains in the heads of a few specialists. Unlike engineers, we can't use what we've learned to make a boatload of pills that we may then ship them off to the people who need them. Our chief export is ideas. And if these ideas are to take hold, if they are to affect the way we live, they must be packaged in such a way that the consumers can consume them. And that is why those pretty brain pictures matter. They force people to confront the fact that the human mind is the human brain, an earthbound hunk of fatty tissue, made of the same basic stuff and operating according to the same basic principles as everything else in the world, from rocks to rats to chimpanzees.

THE GREAT CONTINUUM OF MATERIAL BEING

We are animals that evolved in imperceptible steps from the inanimate ingredients of primordial soup. There was no magic moment when the lifeless came to life or when the soulless became ensouled. Humans, animals, and the lifeless matter out of which they arose are all part of a Great Continuum of Material Being with no sharp lines of separation, only more or less distant family relations.

I accept this, as do many of you. Some of you don't, but I predict that your descendants will, thanks to the new sciences of human nature. Your children's children will grow up in a world in which things like moral judgment are understood down to the last neuron and in which children begin to study such things in elementary school. As a result they will accept the fact that humans are part of the Great Continuum of Material Being as comfortably as you and I accept the roundness of the Earth. What are the implications of this conclusion and our appreciation of it? For one thing, it means many of our common sense moral concepts will be stretched, revised, and in some cases discarded. Let's examine a few of them.

A basic question about morality is: Who's in, and who's out? Which beings count as moral beings? According to a traditional dualist view, the answer may be rather simple. The beings that matter morally are those with immaterial souls. Humans have souls, so they're in. Chickens don't, so they're out. Human infants are in, and human fetuses probably are, too. Computers don't have souls, so they're out. And so on.

In light of our materialist conclusions, however, these answers will no longer do. If we don't believe in immaterial souls, we are going to have to make our moral distinctions based on the observable properties of the entities under consideration. Questions like "Does it have a soul?" will be replaced by a variety of more complex questions: "Does it understand the consequences of its actions for others?" "Can it be deterred by threat of punishment?" "Does it feel remorse?" "Does it feel pain?" "Does it suffer from the pain it feels?" And these questions will lead to more questions: "What do we mean by 'understand'?" "What do we mean by 'pain'?" And so on.

In making sense of this question "Who's in, and who's out?" it's worth drawing a familiar distinction between moral *agents* and moral *patients*. A moral agent is, roughly, a creature that can make moral decisions and can reasonably be held morally responsible for the decisions it makes. A moral patient, in contrast, is one who has legitimate moral interests, a being that ought morally to be treated one way, rather than another. To illustrate the distinction, consider the case of animals, and primates in particular. It seems reasonable to say that they cannot be moral agents because they lack the requisite understanding and self-control, but that they can be moral patients because they are capable of suffering and enjoyment [13]. This may seem obvious enough, but notable thinkers have disagreed. Immanuel Kant argues that morality's foundation lies solely in the capacity to reason. From this it follows that animals, who are not rational by Kant's standards, are not moral creatures. Thus, he concludes that it's not wrong in itself to inflict pain on animals. He hastens to add, however, that we ought to avoid such behavior for fear that it may encourage the mistreatment of more rational creatures [14]. (Talk about putting the cart before the horse!)

The moral agent/patient distinction is useful for speculative engineers to keep in mind since artificial moral patients are likely to arrive well before artificial moral agents. What does it take to be a moral patient? It's not clear what sort of standard one must use to answer such a question, but a reasonable answer was suggested above. If a creature can feel pain or pleasure, if it has, so to speak, negative and/or positive qualities associated with its experience, and then it deserves moral consideration. The question then becomes, what, in engineering terms, does it take to build such a creature? How far are we, and how will we know when we get there?

There are two ways to think about experiential qualities such as pain, in functional terms and in more intuitive, experiential terms. In functional terms it seems that pain is something of an alarm signal to the organism that experiences it. The creature detects an object or event that is harmful to itself and responds to it, typically by withdrawing in some way. Understood this way, pain-experiencing creatures may be fairly easy to create. Single-celled organisms respond to noxious stimuli, and perhaps we're not all that far from building artificial creatures that are as complicated as single-celled organisms. And it's possible that single-celled organisms are more complex than necessary for pain. Consider a simple artificial control system such as a thermostat [15]. A thermostat has but one belief (regarding the current temperature in the room) and but one desire (regarding the temperature it has been programmed to achieve). When the temperature drops below its desired level it "knows" that it's too cold and "wants" to make it warmer. Does your thermostat not *yearn* for a room at twenty-five degrees? Well, probably it doesn't. But on what grounds do we reach that conclusion? Here we're making use of the other way of thinking about pain, thinking in intuitive, experiential terms. It does seem a bit crazy to think that the thermostat on your wall or a creature without a brain (or something like it) could be in pain. The challenge, then, is to explain in functional terms what's missing from such simple systems that prevent them from having genuine positive and negative experiences. The answer is not at all obvious, and yet, if we deny that such experiences are activities of an immaterial soul, there must be an answer. How will we know when we've crossed that line?

There is some evidence to suggest that we're not close to crossing that line. For one thing, it seems that pain (the perception of a noxious stimulus) and pain (the emotional state, the suffering that noxious stimuli cause) can be distinguished and appear to be subserved by different brain systems in humans, the former largely in primary sensory cortex and the latter primarily in more recently developed portions of the frontal lobes [16]. Thus, it seems quite likely that the sort of pain that really matters morally, the unpleasant experience of pain, requires something far more complex than simple systems such as that possessed by thermostats and bacteria. How much more complex? We don't know, and I believe that it's important to find out. Unless we take the science of conscious experience seriously (as increasingly many scientists do) we could, in the not-too-distant future, find ourselves unwittingly employing robots that suffer day in and day out as they toil on our behalf.

What about moral agents? Since almost no one regards present day non-humans as moral agents, questions concerning who is or isn't a moral agent tend to center around humans with characteristics that may disqualify them from moral agenthood, characteristics such as mental illness, youth, a history of abuse, and so on. For the most part, these questions are not of particular interest to engineers, and, as noted above, wholly artificial moral agents, along the lines of Commander Data from *Star Trek: The Next Generation*, are not going to arrive any time soon. However, wholly artificial moral agents are not the only ones that we could engineer. Genetic engineering holds the promise (or threat) of creating humans and other animals with unusual features that may be morally relevant.

When people speculate about the moral status of human clones, they are most often concerned with the clones as moral patients: Is it unfair to them that they will not be (genetically) unique? Will they have a second-class status? Will they experience life with all the fullness and richness that we do, or will they be somehow "empty", like the living dead zombies from horror movies? I don't think these questions are terribly difficult to answer. They are primarily the concerns of people who have yet to accept that human life and human consciousness have a purely material basis. In time, I propose, we will get used to clones just as we are used to identical twins. (A clone is just an identical twin born substantially later than its twin sibling.) We won't be bothered by the fact that clones are not genetically unique any more than we're bothered by identical twins for that reason. Nor will anyone, aside from bigots akin to racists of today, doubt their humanity. In my opinion, the more interesting questions about clones are questions about their status as moral agents, and ultimately about what it means to be a moral agent in general.

Many of you will recall the classic film *The Boys From Brazil*. In the movie, Nazi expatriates living in South America create a group of baby Hitler clones using DNA collected from the Fuehrer. Moreover, they attempt to raise these little Hitler clones in environments that, in certain crucial respects, approximate the environment of Hitler's formative years. Thus, through a combination of complete genetic control and partial environmental control, they attempt to bring their beloved Hitler, or something like him, back to life.

The genetic component of this plan is entirely plausible, but the environmental component is more than far-fetched. Nevertheless, it's not at all inconceivable that such a plan could succeed if the goal of bringing a recognizable individual back to life were substituted for something a bit less ambitious. Genes seem to matter a great deal, much more than many people think, or would like to think. In a recent paper in *Science*, Caspi and colleagues report on a single gene that dramatically reduces the likelihood of abusive behavior among those who were abused as children [17]. And, as Steven Pinker notes, identical twins separated at birth are freakishly similar in spite of their different environments [18]. They have similar attitudes regarding things like the death penalty, religion, and modern music. They resemble each other in terms of consequential behavior such as gambling, divorcing, crime, getting into accidents, and watching television. They share idiosyncracies such as giggling incessantly, giving long answers to simple questions, and dipping buttered toast into coffee. I don't mean to suggest that identical twins, like clones, are not distinct individuals. The point is simply that genes can go a long way in predisposing people to various kinds of important and complex behaviors.

Suppose, then, the goal of a project like the one from the movie was not to recreate a specific evil individual who will do some specific evil things, but simply to create an individual who will do some evil things. With your favorite villain's genes and a well-chosen environment, you might have a pretty good shot at success. Suppose the project did succeed, and suppose the little Hitler, rather than leading a nation to war and genocide, ended up getting into a fight in a bar and killing a man in a narcissistic rage. And let's suppose further that we actually managed to make one hundred little Hitlers and that ninety-eight of them ended up committing some kind of serious crime. Ought we hold these cloned individuals responsible for their actions? After all, they were deliberately bred to be evil. They had no control over their genes. They had no control over the general shape of their environments. And it seems that controlling those two things was enough to make a criminal in ninety-eight out of one hundred trials. Should we treat these individuals the way we might treat dangerous animals (lock them up, but don't blame them, don't hold them personally responsible) or like genuine criminals (lock them up, but *do* blame them, hold them responsible for their actions)? My guess is that many of you would take some pity on these hapless miscreants and want to treat them less like full-blown criminals and more like victims of a devilish plot.

But here's the rub. What makes the Hitler clones less responsible than the original Hitler, or any other criminal for that matter? Hitler had no choice about his genes, nor did he have any choice about the general shape of his environment. (Hitler's crimes may have been more severe, but the question here is about how much people are responsible for what they do, regardless of the magnitude.) And it's not just Hitler, or even just people who do bad things to whom this worry applies. The courses of my life and your life are just as shaped by external forces as those of our imaginary little Hitlers. The only difference is that in the case of the little Hitlers it's all part of someone else's evil plan whereas in our cases (we assume) there's no grand plan behind these influences. The point remains that, plan or no plan, the external influences are there in our lives as well, and no less influential.

Lurking within this little thought experiment is the age-old problem of Free Will. One might say that the real question here is simply (simply!) one of determining who in this discussion (Big Hitler, little Hitlers, ordinary criminals, you, and me) has free will. If you have free will, you're a responsible moral agent. If not, not. So far so good, but now the problem is that it's not so clear what we mean by "free will" once we've accepted that we humans are part of the Great Continuum of Material Being. There is absolutely no evidence that anything special or magical takes place in the brains of decision-making humans. Human brains seem to operate according to the same mechanistic laws as everything else. (Yes there may be indeterministic events at the quantum level occurring in human brains (and everywhere else), but that sort of indeterminism won't save us from determinism. You can hardly call actions your own if they are the results of microscopic coin-flips, the outcomes of which are no more under your control than the genes with which you were born.)

This observation, in various forms, got the philosophical problem of free will going. In response to this problem, philosophers in recent years have developed a number of "compatibilist" positions, i.e. ones according to which free will and determinism are compatible. The most elegant one, in my opinion, comes from Harry Frankfurt, who argues that free will is not a matter of having one's actions caused in some physically special way, but rather a matter of bearing a certain kind of psychological relationship to one's actions, namely one of "identification" [19]. To act freely is to have your action be one that you want to perform and that you *want to want* to perform. For example, a drug addict who hates her addiction is not free. She is a slave to her addiction because, even though she wants to take the drug, she doesn't *want to want* to take it. In contrast, a willing addict, though equally compelled to seek his drug, is acting freely. He identifies with his uncontrollable cravings, endorses them fully, and thus is the author of his actions, a free man.

I think there is something right about this account of free will, but I don't think it can solve our problem. Think of the little Hitlers. They may identify fully with their actions. They may have been bred not only to be hateful and violent but also to identify with their hatred and violence whole-heartedly, to want to want to be hateful and violent. And yet we are still reluctant to say that they are fully free and fully responsible for their actions. The demon of determinism returns, and not just for genetically engineered, would-be moral agents but also for *all of us*. Once again, our actions are no less subject to the influences of genetics and environment than those of the boys from Brazil.

Perhaps, one of the great consequences of accepting that we are part of the Great Continuum of Material Being is that we will have to revise our understanding of free will and moral responsibility. In an excellent recent book, Daniel Wegner argues that Free Will as we understand it, is an illusion [2]. He argues, however, that we need something like this illusion to make social life and the responsibility it requires possible. I'm inclined to disagree. I think that there are conceptions of freedom and responsibility that can work quite well without illusions, ones that derive from the utilitarian philosophies of Jeremy Bentham, John Stuart Mill, and Oliver Wendell Holmes. If our goal in holding people responsible and punishing them when necessary is not to give people their "just desserts", but simply to prevent future crimes, we can have our social order without buying into any fictions about the nature of human moral agency. But that's a topic for another occasion.

SUMMARY AND CONCLUSION

The new sciences of human nature are illustrating a point that some philosophers and scientists have been making for a long time: Human beings are physical beings, fantastically complicated machines, nothing more and nothing less. Even such essentially human processes as moral judgment are physical processes that can be described in detailed mechanistic terms. Many people don't like this picture of human nature, and for a long time such individuals have had ample room to doubt it, but now the mechanical details are starting to come in. As our scientific knowledge of the moral mind accumulates and trickles down into common sense, the conclusion that we are part the Great Continuum of Material Being will be irresistible to all but the most stubborn thinkers.

The details surrounding the nature of moral cognition may be important in themselves. They may show us that our moral judgments are not what we think they are, that they are far more quirky and impulsive and far less rational than we had thought. Combine this knowledge of moral psychology with an understanding of its evolutionary basis, and we have a recipe for some serious revisions to our moral outlook. In particular, I hope that a better understanding of how our moral minds work will lead us to take more seriously the need to help those less fortunate than ourselves, especially when they are out of sight and therefore easily rendered out of mind.

While the details of moral cognition are important, the great significance of studying this and other topics may lie in the dissemination of a more accurate worldview, one that does justice to our purely material nature. Without immaterial souls to form the bases of our moral distinctions, we will have to make them on other grounds. In deciding who's in and who's out of the moral domain, we would be wise to distinguish between moral patients and moral agents. Of interest to engineers, is the fact that artificial moral patients are likely to arrive long before artificial moral agents. To know when they've arrived we will have to invest in the science of conscious experience and pay attention to its results. Completely artificial moral agents are not going to be here for a while, but we may soon encounter genetically engineered human agents that raise deep questions about the nature of moral agency and responsibility. So deep, in fact, that they call into question many firmly held beliefs about the free will and moral responsibility of ordinary humans.

I began this discussion by distinguishing myself from you. Most of you, I said, are engineers, while I am a reverse engineer. This is not an entirely accurate description. For one thing, many of you are reverse engineers as well. In designing your products you learn from nature's engineering secrets, whether they lie in the way insects walk over rough terrain or the way neurons process information in parallel. And I, like you, am a bit of an engineer as well. As a reverse engineer I am trying to understand how the parts that compose the human mind work, but each human mind is also a part of something larger, namely a human society. As I suggested above, I hope that a better understanding of how the individual human mind works will help us engineer a better society. I hesitate, however, to use the term "engineer" in this context given the Twentieth Century's history of disastrous exercises in social engineering. Nevertheless, I think that this is more of a problem with the word than the underlying idea. Every law, every bit of policy we make is an attempt to engineer our society. The difference between good laws and policies and dangerous totalitarian ones is not so much a matter of willingness to engineer as it is a matter of favoring the light touch over the heavy hand. Like all engineers, social engineers learn from their mistakes, although the costs of their mistakes are too often very high. The intelligent processing of materials, human and otherwise, is clearly a work in progress.

REFERENCES

[1] D. C. Dennett, 1995. Darwin's dangerous idea: evolution and the meanings of life, Simon & Schuster.
[2] D. M. Wegner, 2002. The illusion of conscious will, MIT Press.
[3] P. Singer, 1993. Practical ethics, Cambridge University Press.
[4] J. J. Thomson and W. Parent, 1986. Rights, restitution, and risk: essays, in moral theory, Harvard University Press.
[5] Greene, J.D. et al., 2001. An fMRI investigation of emotional engagement in moral judgment, Science 293 (5537), 2105–2108.
[6] A. R. Damasio, 1994. Descartes' error: emotion, reason, and the human brain, G.P. Putnam.
[7] F. de Waal, 1996. Good-natured: The origins of right and wrong in humans and other animals, Harvard University Press.
[8] F. de Waal, 1982. Chimpanzee politics, Harper & Row.
[9] J. Haidt, 2001. The emotional dog and its rational tail: A social intuitionist approach to moral judgment. Psychological Review 108, 814–834.
[10] J. Greene, and J. Haidt, 2002. How (and where) does moral judgment work? Trends Cogn Sci 6 (12), 517–523.
[11] P. Singer, 1972. Famine, affluence and morality. Philosophy and Public Affairs 1, 229–243.
[12] P. K. Unger, 1996. Living high and letting die: our illusion of innocence, Oxford University Press.
[13] M. D. Hauser, 2000. Wild minds: what animals really think, Henry Holt.
[14] I. Kant, 1963. Lectures on Ethics, Harper & Row.
[15] D. C. Dennett, 1987. The intentional stance, MIT Press.
[16] P. Rainville et al., 1997. Pain affect encoded in human anterior cingulate but not somatosensory cortex, Science 277 (5328), 968–971.
[17] A. Caspi et al, 2002. Role of genotype in the cycle of violence in maltreated children. Science 297 (5582), 851–854.
[18] S. Pinker, 2002. The blank slate: the modern denial of human nature, Viking.
[19] H. Frankfurt, 1971. Freedom of the Will and the Concept of a Person, Journal of Philosophy 67, 5–20.

Emergence and Complex Systems: Towards New Practices for Industrial Automation

H. HYÖTYNIEMI
Helsinki University of Technology, Control Engineering Laboratory, P.O. Box 5400, FIN-02015 HUT, Finland

ABSTRACT

There exist various complex systems theories. Common to these ideas is that they construct fancy abstractions about complexity—so fancy that it is difficult to see the underlying domain fields any more. However, there will not exist applications of the theories if the abstractions cannot be concretized. This paper tries to show that something practical can really be reached: New approaches and conceptual tools can be developed, the application field here being industrial automation systems.

INTRODUCTION

The systems being analyzed in engineering fields are becoming more and more complicated, and more and more complicated approaches are being proposed to attack these problems. It seems that a practicing engineer can have considerable difficulty mastering all of the new trends in his/her own field. This is one of the main reasons that so much hope has been put on the *theory of complex systems*: This particular theory promises to unite the diversity by defining a new level of understanding so that different domain fields and different theoretical methodologies can be mastered in the same unified framework.

One of the new approaches that promise wonderful solutions to these problems is the theory of *complex systems*. The idea of complex systems theory is intuitively appealing. We see various examples of complex systems all around us, and we understand that there is something fundamentally similar beneath the surface. However, the systems are so multifaceted that it is difficult to see this similarity. In complex systems theory it is assumed that there exists some underlying very simple process that when iterated massively, something qualitatively different comes out. So the observed complexity is an *emergent phenomenon*, and it is enough just to reveal the underlying simple single function to completely understand the fundamentals concerning the system and its behaviour. Well, this is at least what the advocates of the New Science tell us.

But, intuitively, can a theory possibly exist that would apply to all systems, big and small?

In this paper, I discuss why such a panacea probably will never be reached. However, the case is not black-and-white—useful results can be found even if the goals are not so ambitious. Concentrating on certain specific cases makes it possible to reach new results—results that implement the *meaning* of the domain field. Words like *semantics* and *understanding* are discussed throughout the text—it turns out that results from *artificial intelligence* (AI) play a central role in these studies about complex systems (CS).

The contents of this paper are as follows:
- First, today's mainstream approach to complexity research is presented with a view to demonstrate that some *other* approach is needed to avoid deadlock.
- Second, the role of *systemic* considerations is emphasized: Trying to understand the future, knowing the past is essential; there is considerable dynamics and inertia in science.
- Third, the role of *mechanized understanding* is concentrated on to show how such understanding can be implemented in the field of industrial automation.
- Next, concrete studies on *simple* complex data and then, *more complex* complex data are carried out to show how new approaches and tools can be found.
- Then, an approach is presented to combine *qualitative* with *quantitative*—after all, there usually exists some known structure, and new tools just cannot be based merely on data.
- Finally, I discuss the somewhat *philosophical consequences* of the new approaches in which I claim that, even though general theory has been forgotten, one still can address the *whole world*.

This paper summarizes a presentation given during the IPMM'03 Conference in Sendai, Japan. Comparing this work to the original version of the paper [18], one can see that in less than a year, some important developments have taken place ...

APPROACHES TO COMPLEXITY

There are different views of what complex systems are, and what are the right approaches to attack such systems (for example, see [2]). Here, only one prototype of the current approaches is presented, and the emerging problems are identified.

Wolfram's World

One year has passed since the publication of the Stephen Wolfram's book "A New Kind of Science" [24]. There has been very much discussion about this book—not only on its scientific merits—and it is still too early to estimate what will be the lasting value of this contribution.

Wolfram proposes that all complex systems, natural and man-made alike, can be represented as a combination of simple, interacting pieces of program code, or *cellular automata* (or CA for short). Wolfram gives a wealth of examples where complex-looking natural forms can simply be reproduced using simple CA's (as an example, the patterns on a mollusc shell are presented in the book, see Fig. 1). Indeed, Wolfram goes further: he proposes that all complex systems fundamentally *are* cellular automata.

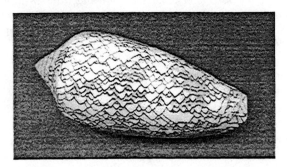

Figure 1. Prototypical example of how CA can explain Nature.

Wolfram's book shows that within the simplest of CA's there lives the power of *universal computation*. For such systems the *Gödel's Incompleteness Theorem* [5] applies. This famous theorem states that all systems that are so powerful are either *incomplete* or *inconsistent*, meaning that no analysis can reveal every detail of their behaviour.

Wolfram explains that practically all natural systems have the same *computational complexity* (in other sources, this is known as Kolmogorov–Chaitin complexity). A complex system cannot be represented in a more compressed form—all artificial representations of the system behaviour are *more complex* than the original meaning that the best system model is the system itself, and the best behavioural information one can obtain is to study a system directly by running a *simulation*.

Wolfram concludes that a New Science is needed, a simulations-based study of nature, and that the old science must be abandoned altogether. No wonder the scientific community has reacted aggressively. What is more, Wolfram's conclusion is inevitable, no matter if one selects the cellular automaton paradigm or something else, since by definition any formalism that can present *anything* has the power of the Turing machine. Then, inevitably, Gödel's result applies. Is there any way to avoid collisions between paradigms?

Modeling View

So, either one needs a completely New Science, or one must forget the ideal of having a general theory of all complex systems. Why be prejudiced, why not go courageously for the new?

Looking at the long and successful history of "Old Science", one is still perhaps better off to stick with traditional mathematics[1]. New Science would mean the End of Science as we know it, and as long as the new approaches do not offer us practical tools that outperform the traditional ones, at least in special cases, it is just reasonable to wait and see. In concrete terms, the assumption here is that *models still are useful*.

Wolfram starts from the assumption that a general framework must exist for all complex systems—and this generality assumption inevitably results in a set of too powerful modelling tools. The key question is to somehow *limit* the power of the modelling tool so that a reasonable compromise between expressional power and analysability/applicability of the model can be reached. The goal is to determine a modelling environment where the relevant phenomena can be represented, but *only those*, so that the universality problems do not become acute.

In this paper, I am not trying to answer the metaphysical questions—whether or not the adopted structures really convey the essence of the underlying system or not—rather, I am only constructing models. It must be admitted that models are abstractions, and they are *always false*—but they can be useful, helping to gain intuition, and helping to construct new tools for practical use (However, even though the starting point here is humble, by the end of the paper I will return to these eternal questions...but from a different viewpoint).

How does the theory (and practice) that we are pursuing differ from the age-old modelling approaches? Here I assume that the main point in these new approaches is the idea of *emergence*. When a large number of simple operations or structures are combined, something qualitatively new may pop up: *The whole is more that the sum of the parts*. This is one of the slogans of *General System Theory* [4]. But when trying to capture the holistic idea of emergence using reductionistic tools, it is evident that a wider perspective is needed.

SYSTEMS VIEW

It is clear that complex systems theory is far from mature. It is a science being made right now. Where is this field going—how can we possibly predict the future of this field? In the following, I discuss a few viewpoints in order to reveal the challenges faced when making prophesies.

About Scientific Theories

General System Theory is a framework for finding general underlying laws and analogies between different systems. The problem here is the same as in the case of general theory of complex systems: Trying to be too universal, one loses the connection to low-level practice, and the conclusions are too loose to be of any practical use. In this case, the ideas of system theory are employed rather freely, only obeying the basic ideas of *system dynamics*. In Fig. 2, different kinds of

---

[1] Remember the *Gaia Hypothesis*: The incomprehensible appropriateness of the conditions and processes on our planet (geological, climatological, etc.) has motivated divine explanations—*it must be the Goddess of Earth that simply wants to support life here*. Similarly, one can define the *Athene Hypothesis*: The marvellous improvements in science point to the conclusion that the Goddess of Science just wants Nature to be mathematically tractable!

"systems" are depicted, each of them affecting the future of complex systems theory and applications, as will be discussed below.

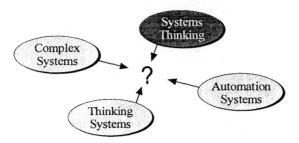

Figure 2. What does the future look like? Some viewpoints.

For example, a *system of humans* is a system (and there is a truly complex system!). From the viewpoint of complex systems theory and applications, there are many consequences.

It is often assumed that science steadily proceeds as new results are found. Scientists are assumed to be unbiased, making objective conclusions on observations. However, research is a human endeavour: How things evolve depends on individual human actors and their preferences, and the route towards progress is not unique. There is considerable reluctance among researchers to accept something new—often a revolution must take place. If enough evidence accumulates such that the new ideas offer a simpler (or more complete) explanation of the observations, only after that, does a transition between paradigms takes place, as presented by Thomas Kuhn [21].

However, the above view of scientific revolution is extremely static assuming that one simply steps from one equilibrium point to another. Complex systems theory contributes here by offering a more dynamic view of a paradigm shift: it has been recognized that interesting developments take place on the *borderline between order and chaos*. So it is not necessary that only the absolutely best ideas survive; in our current quickly-developing scientific world the "hot" areas are in a continuing state of turmoil.

Understanding the dynamic nature of these scientific paradigms helps us understand one of the basic criticisms of Wolfram's theory. The emphasis of complex systems theory has evolved from *chaos theory* in the 1980s to *complexity theory* in the 1990s. It has become recognized that within almost all nonlinear systems there lurks chaos, and so, previous theory is too trivial. What becomes interesting then is how there can emerge some form of order *out* of the chaos. Wolfram spent ten years in his chambers preparing his book, missing contacts with colleagues—and what he came out with was already outdated: It only showed how chaos and complexity can be generated, something we already knew, but it gave no tools for analyzing the emergent patterns.

When trying to look beneath the turmoil in complex systems theory, the key questions to be asked are: How can we find the borderline between *order* (mathematics) and *chaos* (non-mathematics) in modeling? How can we combine the two to reach *analyzable emergence* or *holistic mathematics*? Too wild a theory that proposes a completely non-mathematical approach drops us into chaos, whereas too much mathematical rigor easily turns into *rigor mortis*!

Control Engineering Practice

So, theory is human and so too is *practice* which brings us to industrial applications: Process operators either approve of new tools and theories, or they do not. A good example is the current situation: When the modern state-space modeling and control methods were introduced in the late 1960s and early 1970s, they never became widely used. They would have been a powerful framework to attack multi-input/multi-output control problems; however today we find it is still the simple PID controller that is almost exclusively used in practice. The necessary shift between paradigms never took place on the factory floor.

In a PID controller, there is one tuning knob for Proportional action, one for Integrative, and one for Derivative—all of these actions being intuitive to a practicing operator. It can be claimed that this simplicity is the key to the success of PID: a controller must be *intuitively understandable* to be widely accepted. This intuitive way of managing complex processes should not be underrated: Operators have plenty of expertise that cannot be explicated—although this knowledge is subjective, it is still valuable and should be supported.

Complex Systems vs. Artificial Intelligence

One way to apply a system theoretical approach is to recognize similarities between the two paradigms, CS and AI research—artificial intelligence just happens to be some decades older (for example, see [22]). It seems that the same kind of open-minded researchers and audience constitute the *infosphere* in both cases—the promises of huge development sound so familiar. However, the big promises made in the AI field have not yet been fulfilled, if they ever will. Are there some lessons to be learned?

In both fields, researchers and financers alike are adventurers. This means that interest is easily aroused, but it also decays fast. This fluctuation between euphoria and depression is reflected in financing, making both fields extremely turbulent.

In CS and in AI, the same kinds of objectives are pursued: In a complex system, emergence is searched for, whereas in an artificial cognitive system, the corresponding objectives are intelligence or consciousness. Intelligence and consciousness are emergent phenomena which cannot be captured using reductionistic tools. In CS and in AI the goals are escaping: the feel of natural emergence or intelligence vanishes when one explicitly knows how the trick is done. Indeed, AI is a special case of complex systems in general.

In the beginning of the AI era, the goal was defined in terms of the "Turing test": *Something is intelligent if it successfully mimics intelligence*. This has resulted in the "shallow view" of AI where machines are explicitly programmed to do some intelligent-looking tricks. It is this starting point that has plagued AI research ever since. In the same way, the Wolframian approach to CS may lead to a "shallow view" of complex systems: *Something is relevant in CS if it only looks interesting*. Looking simply at the surface and not at the underlying functions is intuitively not the right approach to complexity (examine the mollusc shell in Fig. 1).

But perhaps one should not be too sceptical about mollusc studies. For example, take one example from Finland:

> Charles Darwin published his Origin of Species in 1859, and his ideas soon spread throughout the scientific community. However, in 1860, when the theory arrived in Finland, Fredrik Wilhelm Mäklin, later to become the Professor of Zoology at the University of Helsinki, condemned the theory, and concluded that "No scientist, perhaps excluding some **sea shell** collectors [referring to local enthusiasts], could ever believe it"!

Artificial intelligence is not just an example of complex systems; AI also offers conceptual tools to attack the CS challenges. One of these conceptual tools is the notion of *semantics*.

ABOUT MACHINE UNDERSTANDING

To try to utilize the explosion in computing capacity to attack the explosion in available data, we need automatic modelling, or automatic abstraction and compression of data. To realize such an abstraction mechanism, a kind of *formalized understanding* is necessary.

Boundary Conditions
It has been claimed that "When more computing capacity is available, the computer automatically becomes intelligent". However, this is not true: If data is just trashed mindlessly, only trash comes out. The issue is how to make a computer do something reasonable, how to make a computer *understand* what it is doing?

It turns out that conceptual tools, developed in the field of *philosophy of mind*, offer the ideas and concepts needed for discussing the problems that emerge.

There are some fundamental limitations for what a computer can do. Indeed, these limitations are common to all data processing mechanisms, and such questions have been analyzed by empirist philosophers. It was David Hume who first recognized that *causal relations* cannot be deduced from data alone. Only correlation structures can be directly observed and trying to go further from this by assuming some dependency structures between observations, is always risky. This is especially true in a complex automation system, where feedback loops may totally mix up the virtual dependencies between signals. All "emergent models" can only be descriptive. The nasty truth is that the human perception machinery seems to be based on constructing causal structures. In this sense, it is clever to be less ambitious and admit that machine-generated models probably should have simpler structure than mental models. Having no causality integrated into the created model, the final analysis and application, for example, controller construction, must be carried out by the human expert.

Another problem with data-based approaches is that reasonable models will be found only if the data is representative and rich, so that system phenomena are excited and visible in the data. This is not the case, for example, if there are some closed feedback loops to control the system—and when the final goal here is to model large-scale systems, how can this be assured? Not all controls at a plant can be suspended. The nice thing is that, informally speaking, those models that we are struggling to produce are of a new type and because such models have not existed before, they cannot have been utilized, and so no feedback structures can exist in these systems!

Having defined the ultimate limits, we can now redirect our development efforts: to implement some level of understanding, we must capture and process the *meaning* or *semantics* of the data. The problem of semantics is an age-old issue in the "deep" sense of AI where the problem of synthetic understanding has been studied for a long time. In this context, implementing some level of understanding in the machine is *not* regarded as impossible. In concrete terms, there are two concepts that make it possible to discuss the dimensions of meaning in a practical way: *Naturalistic* and *contextual* semantics. In this way the meaning of a signal is determined in terms of its connections to the outside world, or to other signals. Within the narrow world determined by the observation data, this simple starting point makes it possible, as the constructs are nested deeply enough, to reach interesting results as will be shown below.

What is evident, once again, is that pursuing a general theory of all complex systems is futile: The semantics is domain area specific, so that implementations of smart systems should probably differ from one application to another. In what follows, to reach some concrete results, we will concentrate strictly on *industrial automation systems* and the models being employed in that field.

Semantics in an Automation System

Is it possible to create a theory and modelling methodology for all industrial processes, then? It seems that even this is a too ambitious goal—see Fig. 3: When modelling a system involving gases, for example, the appropriate modelling approach changes when the scale is changed. Moving up the hierarchy, the time scales get longer, whereas the numbers of contributing entities become less. The complexity that is faced on the lower level (temporarily) vanishes when new emergent concepts are employed – but this shift between levels also involves adopting new ideas and new ways of thinking about the phenomena. Each successive level seems to change from the need for a stochastic approach to a deterministic model.

Current modeling techniques used in industrial automation environments concentrate on the component level—the highest deterministic level shown in Fig. 3. There is a need for a higher-level approach since when there are dozens of simple models such as ideal mixers, the essence of the overall system behavior is blurred. Looking at the structure of the hierarchy in Fig. 3, one can assume that this higher level should perhaps be *stochastic*. The question that arises is *how to reach the next level of abstraction*. How to represent the essence of the structured system in a statistically applicable form, and how to utilize such statistical representation? These problems will now be elaborated on exclusively in the rest of this paper.

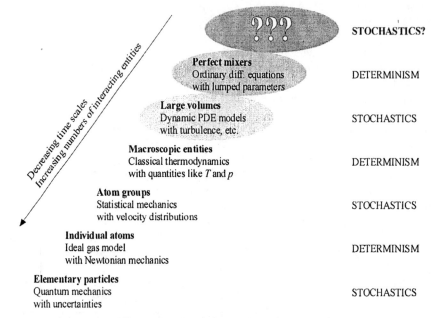

Figure 3. Different levels of abstraction when studying **gases**.

The relevant structures of the higher level are not visible to the human eye; the assumption is that only after heavy preprocessing carried out by a computer does some pattern become visible. For a pattern to emerge, the problem of integrating semantics with data manipulations becomes acute. When looking at the component models to be abstracted, two crucial points are apparent:

- Semantic atoms can be acquired by *simulation*. The contextual semantics of a subsystem is obtained by observing how the environment affects its behaviour, and how it affects its environment—these effects being determined by input/output signal pairs.
- Interpretations of the behaviors can be expressed in *mathematics*. Expert understanding can typically be implemented in the form of explicit mathematical formulae, and the *goodness* of the system behavior, for example, can be expressed in a compact numeric form.

New Abstractions

We are now ready to present the key idea, the framework for implementing emergence. Look at Fig. 4, representing the coupling between lower and higher abstraction levels. Rather than looking at the actual signals, symbols u and y, which denote inputs and outputs, a model is constructed between the system *quantifiers* and *qualities*. Quantifiers are variables that affect a system behaviour, such as system parameters, set points, and raw material properties, etc. Qualities are those quantities that one is especially interested in such as process accuracy, robustness, and speed. If there are n quantifiers and m qualities, the corresponding vectors are constructed as:

$$\theta = \begin{pmatrix} \theta_1 \\ \vdots \\ \theta_n \end{pmatrix} \quad \text{and} \quad q = \begin{pmatrix} q_1 \\ \vdots \\ q_m \end{pmatrix}. \tag{1}$$

The data space in the quantifier/quality space can be spanned by Monte Carlo simulations, repeating the simulations with the quantifier vector being changed more or less randomly, and recording the results. Actually, rather than speaking of simulation, one should speak of *information pursuit*. Data can be acquired not only through simulation, but it can also be real process data; rather than being a model, the inner block in Fig. 4 can be the actual process, so that *hardware-in-a-loop* structures can be supported. Indeed, it is possible that the information comes from some completely separate source such as *Semantic Web* [17].

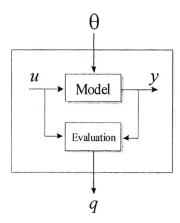

Figure 4. The next level for looking at a dynamic system.

When studying the consequences of a higher-level system view, some advantages are evident:

- **Generality.** All kinds of systems can be studied in the same framework, no matter what the physical systems are like, or what kind of methodologies (neural networks, fuzzy systems, standard mathematics) are applied, as long as they can deliver input/output data.
- **Homogeneity.** The structure of the representation (vector of real-valued data) remains the same, no matter what the complexity of the underlying system happens to be; hierarchical structures also have the same representation regardless of the hierarchy level.
- **Compactness.** Counterintuitively, if appropriate statistical tools are applied, the data-based representation can be more compact than a first-principles model because only relevant phenomena that are really visible in the data become manifested.
- **Simplicity.** Individual signal realizations can be forgotten; the originally dynamic systems can be studied using static methods, ignoring the time axis, meaning that considerable simplification in modelling techniques is available.

Of course, there are also challenges: First, the data is very noisy because of the stochastic nature of the individual signal realizations. Second, the data is typically of high-dimensionality. This means that the traditional control engineering tools become obsolete, and new ones are needed. It turns out that *multivariate statistics* offers a powerful framework to attack the new problems [11]. Some concrete examples are presented below.

"SIMPLE COMPLEX" DATA

To show that the lengthy philosophical discussions above actually give guidelines for reaching useful results, some very concrete examples will now be presented.

Unimodal Data

Assume that the observed data is *unimodal*, so that it spans a single multivariate Gaussian distribution in the data space. Because of the Gaussianity assumption, the maximum likelihood dependency between the variables can be assumed to be linear, that is, one can write $q = F^T\theta$, where F is a compatible matrix (assume the variables have been mean-centered). Further, assume that we have a set of k quantifier/quality vector pairs available, $\theta(1)$ to $\theta(k)$ and $q(1)$ to $q(k)$, respectively. These observations can compactly be written in matrix form as

$$\Theta = \begin{pmatrix} \theta^T(1) \\ \vdots \\ \theta^T(k) \end{pmatrix} \quad \text{and} \quad Q = \begin{pmatrix} q^T(1) \\ \vdots \\ q^T(k) \end{pmatrix}, \tag{2}$$

and they can be used to "calibrate" the model. The linearity assumption means that a mapping between these matrices can be written as

$$Q = \Theta \cdot F. \tag{3}$$

Regression analysis attempts to find a good mapping matrix F. Given the training data matrices Θ and Q, the dependencies between them can be captured by F, so that later, when only the input variables θ are known, the output estimates can be calculated from $\hat{q} = F^T\theta$. Even though the objective can be expressed in such a simple form, the problem is far from trivial: It turns out that the simple approach known as *multilinear regression*, defined by the *pseudoinverse* formula

$$F_{MLR} = \Theta^\dagger \cdot Q = \left(\Theta^T \Theta\right)^{-1} \Theta^T Q, \tag{4}$$

is typically not robust; that is, noisy measurements can result in large errors in estimates, especially if the number of samples k is low as compared to n. As explained in [11], better results *can* be reached if the input data is first projected onto a lower-dimensional subspace, assuming that this subspace has been selected in a clever way (see Fig. 5).

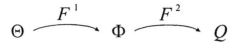

Figure 5. The dependency structure refined.

How to determine the subspace basis ϕ so that noise is rejected but "information" gets through in the mapping? It turns out that a good approach is to select the basis such that the maximum of the cross-correlation between the quantifiers and qualities is captured. This idea results in the *PLS regression*; it turns out that one way to implement PLS is to define the basis vectors ϕ_i as the *eigenvectors* of the following eigenproblem (see [11]):

$$\frac{1}{k^2} \cdot \Theta^T Q Q^T \Theta \cdot \phi_i = \lambda_i \cdot \phi_i. \tag{5}$$

The directions revealed by the eigenvectors that correspond to the largest eigenvalues, represent the maximum correlation directions in the data space. Collecting the most significant of these into ϕ should give good regression results. Since the basis provided by PLS is *orthonormal*, the matrix Φ of *latent variables* can be calculated from the original quantifier variables using the mapping:

$$F^1 = \phi, \tag{6}$$

so that

$$\Phi = \Theta \phi. \tag{7}$$

Further, the mapping from the latent variables to qualities (after the noise has assumedly been ripped off) can reliably be based on the normal pseudoinverse

$$F^2 = \left(\Phi^T \Phi\right)^{-1} \Phi^T Q, \tag{8}$$

so that the combined mapping from Θ to Q in the PLS sense is defined by

$$F_{PLS} = \phi \left(\phi^T \Theta^T \Theta \phi\right)^{-1} \phi^T \Theta^T Q. \tag{9}$$

When the modelling of the data is structured in this way, emphasizing the cross-correlations between the "quantifier input" Θ and "quality output" Q appropriately, it turns out that the emerging linear subspace within the single unimodal data cluster captures those directions in the parameter space where changes in the quality variables are maximal. Being based on latent variables, the model is robust against high dimensionality. Extra variables can be included among the quantifiers; if they have no contribution in explaining the qualities, they will simply not be weighted in the model. This means the model can be utilized in parameter optimization, for example, as shown below.

Tuning of Parameters

Assume there is only one quality measure to be optimized, so that $m = 1$. Because the rank of a matrix product cannot exceed the minimum rank of its component matrices, and because the rank of Q (consisting of a vector of scalars) must be 1, there can be only one non-zero eigenvalue in (5). This means that analysis of the correlation matrix structure has become especially simple: The sole remaining eigenvector is the only reasonable candidate for the subspace basis vectors. The dimension of the latent variable also must be 1, basis ϕ consisting of a single vector, so that the mapping vector between Θ and Q can be written as

$$f = \frac{\phi \phi^T \Theta^T Q}{\phi^T \Theta^T \Theta \phi}. \tag{10}$$

Using this model, the relationship between the quantifiers and corresponding quality estimates becomes

$$\hat{q} = f^T \cdot \theta = \frac{Q^T \Theta \phi \phi^T}{\phi^T \Theta^T \Theta \phi} \cdot \theta. \tag{11}$$

If one wants to optimize the behaviour, one must first calculate the estimated gradient,

$$\frac{d\hat{q}}{d\theta} = \frac{d}{d\theta}(f^T \theta) = f. \tag{12}$$

One can write the steepest descent algorithm as

$$\theta \leftarrow \theta - \mu \cdot \frac{d\hat{q}}{d\theta} = \theta - \mu \cdot f, \tag{13}$$

where μ is a small scalar ("step size"). This algorithm updates parameters so that one gradually goes towards the optimum. However, because the signal values within the system being studied are stochastic, one cannot be sure that the direction truly is correct. In this way, the algorithm should be regarded as a stochastic gradient algorithm and it is clever to only take rather short adaptation steps at a time repeating the information pursuit process around the new parameter values. Various gradient descent steps are needed to reach the (local) minimum. Note however, that because of the sophisticated regression approach, the number of samples around the nominal values need not be very high (as compared with the basic multilinear regression which typically must hold $k = n$).

No matter where the parameters are actually located in the underlying structures, the scheme presented above can be applied for parameter optimization, assuming that the mapping between the qualifiers and the quantities is differentiable (see Fig. 6). Two different examples are presented in [12]: The first one illustrates a modeling problem where an approximate model is optimized to better match the observed behavior; it is like having the model and the actual process integrated with each other so the process delivers accurate up-to-date information with the model being updated in real time. The other example optimizes the PID controller parameters within the process itself according to the observed closed loop performance. In this sense, the proposed approach is a generalization of the *iterative feedback tuning* (*IFT*) method originally suggested in [7]. The main difference here is the use of multivariate regression methods rather than traditional control engineering approaches. The proposed approach is more general, and in the case of excessive parameters, it is more robust, being based on the relevance-motivated latent variables. If some parameter seemingly does not affect the quality, its value is unchanged.

The presented approach offers new perspectives on adaptive control. Typically, when adaptive control is applied to a linear system, the resulting structures are highly nonlinear:

$$y(k) = \theta^T(k) \cdot x(k). \tag{14}$$

When analyzing such *bilinear* systems, the theoretical problems are overwhelming. Now, however, when the system is studied at two distinct levels, this kind of problem is avoided because there are two separate linear models: On the lower level, there is the original linear system

$$y(k) = \theta^T \cdot x(k), \tag{15}$$

and, on the higher level, there is the mapping between the parameters and qualities:

$$q(t) = \phi^T \cdot \theta(t). \tag{16}$$

However, stability of the closed-loop system cannot be guaranteed if the proposed method is used as an on-line parameter adaptation strategy. That is why, it would be better to implement some kind of *gain scheduling* strategy, where the spectrum of appropriate parameter values are tabulated beforehand.

If the above optimization is carried out using an accurate process simulator, plant-wide optimization of quantifiers (parameters, reference levels, etc.) can be done. It is a typical problem when a new industrial plant is being set up in that finding a good combination of controller parameters over the whole plant can take a long time. If the operators have to carry out the process tuning, this more or less random search process may even last for years although new approaches using consistent search directions can considerably simplify the initial tuning of the process performance.

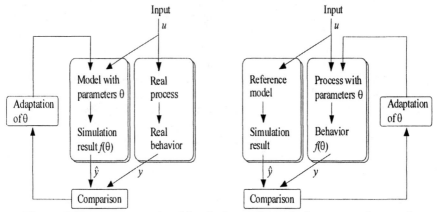
Figure 6. Parameters can be either in the model or in the process (see text).

Factory-Floor View

In practice, the controller tuning cannot be carried out once and for all. During run-time operation, the properties of the raw materials can vary, for example, and the operators may want to tune the process performance accordingly. All cases cannot be taken into account beforehand—a good automation system should perhaps only offer the operators efficient tools for manipulating the fine-tuning of the process behaviour. As was presented before, today this tuning is typically carried out by adjusting the parameters in the PID controllers in individual loops. However, changing a single parameter in a single subsystem is clearly a non-optimal policy. What do the highly technical concepts like integration times, etc., have to do with the actual product quality?

Using the technique presented in the previous section, one can determine the dependencies between quality measures and a larger set of parameters. If these parameters are the tuning knob readings in the individual controllers, a new controller scheme can readily be created: Determine the "slopes" of the quality measures as functions of the underlying controller parameters, and supply the higher-level controller with means of sliding along these slopes. For example, depending of the process being controlled, the quality measures could be something like

- Accuracy,
- Robustness, and
- Speed.

Such an "ARS" controller can be seen as an extension of PID, and it just might be that this kind of control strategy would be intuitive enough for the operators to become accepted also in practice.

It needs to be recognized that the quality measures are typically contradictory (like "robustness" and "speed"). What is more, no closed loop stability is guaranteed any more than when using traditional PIDs. In this sense, the role of expert knowledge does not change; it is still the operator who assesses the appropriateness of the plant behaviour and who makes the tuning decisions.

If the process changes over time, the internal models of ARS can be kept up to date by slightly varying the underlying controller parameters every now and then. Minor variations in the quality measures can be statistically analyzed, and the internal model can be adapted accordingly. Problems will be encountered in the vicinity of instability regions, where all quality measures typically explode and the linearity assumptions do no more hold.

INTEGRATING STRUCTURE

To have real possibilities of becoming accepted, a new methodology should embrace old approaches. For example, in process models dynamics is essential, after all—how can such thing be represented in the proposed static framework? And how to express something more realistic than simple linear relationships?

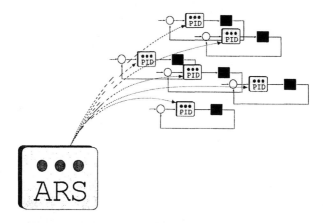

Figure 7. ARS Controller—the "higher-level PID".

Elaborating on Data

In what follows, the simplicity in the data structure will be utilized; to emphasize homogeneity, in what follows, rather than having separate vectors for quantifiers and qualities, there is just one data vector for each observation sample:

$$\chi = \begin{pmatrix} q \\ \theta \end{pmatrix}. \tag{17}$$

How can anything practical be presented in this framework? As an example, let us study how dynamic systems can be expressed, like the first-order AR model

$$y(k) = a \cdot y(k-1). \tag{18}$$

This can be written in vector form as

$$0 = -y(k) + ay(k-1) = \begin{pmatrix} -1 \\ a \end{pmatrix}^T \cdot \begin{pmatrix} y(k) \\ y(k-1) \end{pmatrix} = \Upsilon^T \cdot \chi(k). \quad (19)$$

Assume that the data vector is now defined as above, so that it consists of the present and previous y values. Vector Υ contains the model information. All combinations of $y(k)$ and $y(k-1)$ that result in 0 in the above expression fulfill the original dynamics of the given system. For higher-dimensional dynamics, or if there are, for example, exogenous inputs, the dimensions of the vectors are increased. Indeed, it is the number of free variables that is actually the crucial point. Each new variable increases the degrees of freedom in the system by one, spanning a new dimension in data space, whereas each constraint (the above dynamic equation) binds the variables together, so that the degrees of freedom decrease. The key observation is that only a *subspace* Ψ in the high-dimensional data space is employed (in the above model (19), vector Υ determines the *null space*, or the "opposite" of the one-dimensional subspace). *All linear dependency structures, static and dynamic alike, can readily be implemented in this framework.* If needed, regression to unknown variables can readily be implemented also in such an "associative" framework (see [14]).

However, there are complications: Noise is added in the measurements, and typically the data can not be exactly matched with the model. A new framework is needed to manipulate subspaces and uncertainty in a flexible way. It turns out that multivariate statistical methods are exactly what is needed: Indeed, *principal component analysis (PCA)* [3] gives an efficient method for detecting the dependency structures in the data, and for identification of the parameters of such model. The idea in PCA is to capture most of the "information" within the data as revealed the *covariation* among variables.

The PCA basis vectors φ_i determining the most relevant subspace axes in the $\eta = n+m$ dimensional space of the combined data are given as the most significant eigenvectors of the data covariance matrix (here, $\bar{\chi}$ denotes the mean):

$$\frac{1}{k}\sum_{\kappa=1}^{k}(\chi(\kappa) - \bar{\chi})(\chi(\kappa) - \bar{\chi})^T \cdot \varphi_i = \lambda_i \cdot \varphi_i. \quad (20)$$

Now, assume that N of the most important principal components are collected in the matrix Ψ, and the corresponding eigenvalues in Λ, meaning that

$$\Psi = (\varphi_1 \mid \cdots \mid \varphi_N) \quad \text{and} \quad \Lambda = \begin{pmatrix} \lambda_1 & & 0 \\ & \ddots & \\ 0 & & \lambda_N \end{pmatrix}. \quad (21)$$

The basis is orthonormal, so that, in principle, the vector z of latent variables can be calculated as in the PLS case above

$$z = \Psi^T \cdot (\chi - \bar{\chi}). \quad (22)$$

However, in this expression there is no cost for the size of z; because the data is assumed to be Gaussian, a more appropriate formula can be determined in the *maximum likelihood* sense. No matter how the latent variables are determined, once they are known, the reconstruction becomes

$$\hat{\chi} = \bar{\chi} + \Psi \cdot z, \qquad (23)$$

and the reconstruction error is

$$e = \hat{\chi} - \chi = \bar{\chi} - \chi + \Psi \cdot z. \qquad (24)$$

Let us study the data from the probabilistic point of view, assuming that the Gaussianity hypothesis applies. Using the Bayes rule, the density function can be decomposed as

$$f(e, z | \chi) = f(z | \chi) \cdot f(e | z, \chi)$$

$$= \frac{1}{\sqrt{(2\pi)^N \cdot |\Lambda|}} \cdot \exp(-\tfrac{1}{2} z^T \Lambda^{-1} z) \cdot \frac{1}{\sqrt{(2\pi)^n \cdot |\Sigma|}} \cdot \exp(-\tfrac{1}{2} e^T \Sigma^{-1} e), \qquad (25)$$

where Σ stands for the error covariance, and (according to theory) Λ is the diagonal covariance of the latent variables. In practice, because of the uninvertibility problems, some simplified version of Σ has to be utilized (for example, the diagonal only). Forgetting the constant parameters, the log-likelihood criterion becomes

$$J(z) = z^T \Lambda^{-1} z + e^T \Sigma^{-1} e. \qquad (26)$$

This is a criterion that combines Hotelling's T^2 and the (generalized) Q statistics, measuring the fit as seen from *inside* and *outside* of the model, respectively (see [11]). It turns out that this measure follows the χ^2 distribution, with the degrees of freedom being η. The cost criterion can readily be minimized, and it results in

$$z = \left(\Lambda^{-1} + \Psi^T \Sigma^{-1} \Psi \right)^{-1} \Psi^T \Sigma^{-1} (\chi - \bar{\chi}). \qquad (27)$$

When there is some real structural dependency beneath the data, the covariation in the data is much more significant in the directions involved in the appropriate subspace as compared to directions outside it, and such dependencies will be manifested also in the data-based model. The originally structural representation changes into a more or less numeric form; however, there is no hard limit between structural and numeric phenomena. For example, an existing subsystem can be dropped from the system model if its effects are negligible in the data, so that model reduction automatically takes place; on the other hand, there may exist some unstructured correlations among noise ("colored noise"), and this redundancy can then automatically be utilized in the model. Additional variables not explaining other variables are also automatically ignored.

Mixture Models

The above approach cannot represent nonlinearities. The question that arises is what kind of an extension to introduce so that only minimum additional complexity results, but the "natural" nonlinearities would still be captured. In this context, it is assumed that the classes of natural nonlinearity often existing in real data are the following (see [10]):

- **Smooth nonlinearities.** Typically system properties change in a continuous, differentiable fashion when the parameters or variables are modified.
- **Clustered data.** There also exist non-continuous dependencies constituted of different operating modes or changes in the underlying system structure, so that the dependencies between variables change altogether.

When new ways of thinking are introduced, they should not change the traditional thinking too much. In control engineering, the ideas of operating points and related linearized models is the mainstream approach.

It has been shown that for Gaussian unimodal data linear models are the maximum likelihood solution to finding dependencies between variables. This can be inverted: In the stochastic framework, a linear model can be seen as a Gaussian distribution, and a set of locally linear models can be seen as a set of Gaussian sub-distributions. Correspondingly, in the case of clustered data, a set of Gaussian sub-distributions can explain the global multimodal data distribution. This kind of combination of sub-distributions is known as a *Gaussian mixture model*.

From the mathematical point of view, the mixture model is piecewise linear. It can also be seen as a *sparse* linear model, only a subset of candidate constructs (or latent variables) being employed. Also the idea of *independent components* can readily be implemented in the presented framework; however, the efficiency of PCA techniques cannot be utilized, because it is not simply the variation in the data but higher statistical properties that are needed in ICA [8].

As an example, the construct in Fig. 8 has been modeled using a cyclic chain of Gaussian clusters. Separate clusters are constructed for different operating modes: The system dynamics is essentially different depending on which of the legs touches ground. Within the clusters, the dependencies are smooth but highly nonlinear, and they can be modeled by a succession of linearized submodels. When the relevant state variables are stored in the mixture model with the appropriate control signals (momenta in joints), the learned control behavior could be reconstructed in simulation studies.

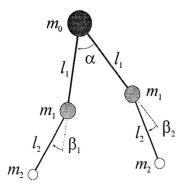

Figure 8. Walking mechanism can be modelled using mixture models.

In principle, the whole world can be approximated by a successively dense grid of Gaussians. However, having an infinite number of Gaussians is not much better than having an infinite number of data samples. The level of individual Gaussians is just one step towards higher abstraction levels.

Remember the importance of selecting the appropriate abstraction level when pursuing understandable models (see Fig. 3; also [23]). There exist different scales also among Gaussians – when selecting a longer time span, the scarce connections become tight, and it is natural to see the set of submodels as constituting a single model. The interactions within one subsystem, on the other hand, become so tight that it is no more relevant to distinguish between the submodel components. At each level the submodels are based on mixture models, so that one effectively has *mixtures of mixtures*. How to functionalize this understanding? It is easy to speak using general abstractions, but in order not to speak in too vague terms here we try to keep discussions on a concrete level. Only then one can see the practical relevance and possibilities there perhaps exist.

AND/OR Graphs

When looking closely at how individual mixture models relate to each other, it is instructive to study an example. For this purpose, study the simple pH process in Fig. 9, consisting of the linear dynamics (ideal mixer with first-order dynamics) and static nonlinearity (titration curve).

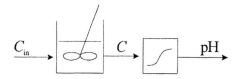

Figure 9. pH process: Linear dynamics + static nonlinearity.

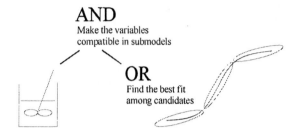

Figure 10. A simple example of an AND/OR hierarchy.

It is the whole pH process that is to be modeled; however, it must be studied in terms of its two submodels. First, there is the ideal mixer model

$$C(k) = aC(k-1) + bC_{in}(k-1), \qquad (28)$$

with three variables $C(k)$, $C(k-1)$, and $C(k-1)$, and two degrees of freedom. Because of the linearity, this degenerated mixture model consists of a single Gaussian represented by its two principal components. Second, the static titration curve gives the relationship between x and y:

$$y = f(x). \qquad (29)$$

The curve can be locally linearized, so that each subpart defines a one-dimensional line segment in the two-dimensional x–y space. If the titration curve is divided in three parts, the corresponding mixture model consists of three distinct Gaussians.

The connection of the individual mixtures is now easy to motivate: The variables in the sub-distributions have to matched so that they do not have contradictory values. Whereas in a mixture model the best submodel is selected ("OR"), the coordination between mixtures is carried out by exposing constraints to variables ("AND"). It seems that this kind of AND/OR structures constitute a useful basic block for implementing models for higher-level abstractions; such blocks can further be combined. One can say that the proposed framework gives new substance to the age-old AI idea of *AND/OR graphs* (see Fig. 10).

To functionalize the AND/OR graphs, the ideas of *hierarchic modelling* turn out to be handy. It has been shown that larger (quadratic) optimization problems can be divided in subproblems; when simpler problems are solved, simultaneously taking care of the coordination between optimization processes appropriately, the global optimum is reached. To study these issues closer, first define some symbols; there are now information structures on three different levels:

1. Individual Gaussian distributions within mixtures—indexed with γ below.
2. Individual mixtures within a block—these distributions are indexed with μ.
3. Individual blocks within the world model—these are indexed with β.

When the hierarchic modelling ideas are applied to our case, it is assumed that the global optimization problem can be decomposed as follows, coordination being carried out in terms of linear constraints:

$$\text{Minimize } \sum_{\mu \in \beta} J^\mu(z^\mu)$$
$$\text{when } \sum_{\mu \in \beta} (G^\mu)^T \hat{\chi}^\mu = g. \qquad (30)$$

Except for the constraints the mixtures are assumed to be independent of each other. The Hamiltonian can be written as

$$J^\beta = \sum_{\mu \in \beta} J^\mu(z^\mu) + \lambda^T \cdot \sum_{\mu \in \beta} (G^\mu)^T \hat{\chi}^\mu - \lambda^T \cdot g. \qquad (31)$$

The dimension of the vector λ containing the Lagrangian multipliers equals the number of constraint equations in (30). The unconstrained criterion (31) can be decomposed, giving a set of unconstrained optimization problems to be solved on the lower level:

$$\bar{J}^\mu(z^\mu) = J^\mu(z^\mu) + \lambda^T (G^\mu)^T \hat{\chi}^\mu. \qquad (32)$$

When the optimization problems have been solved for some (more or less arbitrary) λ, this vector can be refined by noting that

$$\frac{dJ^\beta}{d\lambda} = \sum_{\mu \in \beta} (G^\mu)^T \hat{\chi}^\mu - g. \qquad (33)$$

The theory says that (under rather general conditions) the maximum of λ corresponds to the minimum of the cost criterion, so that one can write the gradient ascent iteration as

$$\lambda \leftarrow \lambda + \delta \cdot \left(\sum_{\mu \in \beta} (G^\mu)^T \hat{\chi}^\mu - g \right). \qquad (34)$$

Here, δ denotes the step length. Whereas the adaptation of λ takes place on the coordination layer, in an individual submodel one now has to minimize the following criterion (here indices are being ignored):

$$\bar{J}(z) = z^T \Lambda^{-1} z + e^T W^T \Sigma^{-1} W e + L^T \cdot \hat{\chi}. \qquad (35)$$

Here, the first term is the same as in (26); in the second term, additional weighting has been included. The additional matrix W contains the "certainties" of given variables. Zero weighting means, for example, that the corresponding variable is completely unknown: It is to be interpreted as an output variable to be associatively determined according to the other (input) variables. Another way to utilize this extra weighting is to take into account the boundary effects (the details are not elaborated on here). The third, linear term in (35) is something new—it takes care of the higher-level coordination between submodels. Vector L is used to select from the data vector those elements that are referred to in the constraint equations coming from all of the above AND blocks. Note that the data is matched against all Gaussians within a mixture, but only the best is transferred to the upper layer; also note that in different mixtures the contents of the data vector χ can differ, whereas within a mixture it remains the same in all Gaussians.

The cost criterion (35) can be minimized in a closed form:

$$z = \left(\Lambda^{-1} + \Psi^T W^T \Sigma^{-1} W \Psi\right)^{-1} \Psi^T \left(W^T \Sigma^{-1} W (\chi - \bar{\chi}) + \frac{1}{2} \cdot L\right). \qquad (36)$$

To use the above framework efficiently, the AND/OR blocks have to be given a probabilistic role so the blocks themselves can be regarded as basic entities as above. It is the "appropriateness" of the block, or its ability to explain the data, that is used to characterize the block. This explaining capability can be measured as the probability of given data sample to belong to that distribution, as optimally fitted against the model. This needs to be studied closer.

Study one Gaussian sub-distribution within a mixture. Because the cost measure J^γ in the Gaussian γ has χ^2 distribution, the probability for a data vector χ to be explained by that distribution can be measured as

$$p^\gamma(\chi^\gamma) = 1 - F_{\chi^2}\left(\frac{J^\gamma(\chi^\gamma)}{2}, \frac{\eta^\gamma}{2}\right) = 1 - \int_0^{J^\gamma(\chi^\gamma)} f_{\chi^2}\left(\frac{v}{2}, \frac{\eta^\gamma}{2}\right) dv . \qquad (37)$$

The cumulative distribution function reveals how probably a data point really belonging to the distribution is nearer to the distribution center than the data sample being studied is; its inverse can be used as a measure for the fit against the model. The degrees of freedom is denoted η. Further, the total probability of a mixture model μ can be calculated in terms of its component Gaussians (remember that $\chi^\mu = \chi^\gamma$):

$$p^\mu(\chi^\mu) = \max_{\gamma \in \mu}\left\{ p^\gamma(\chi^\gamma) \right\}. \qquad (38)$$

If it is assumed that the individual mixtures are uncorrelated, so that the global data also follows χ^2 distribution, the overall "fitness" of the block β can be approximated as

$$p^\beta(\chi^\beta) = 1 - \int_0^{J^\beta(\chi^\beta)} f_{\chi^2}\left(\frac{v}{2}, \frac{\eta^\beta}{2}\right) dv , \qquad (39)$$

where the parameters J^β and η^β can be calculated from the corresponding submodel parameters:

$$J^\beta(\chi^\beta) = \sum_{\mu \in \beta} J^{\arg\max_{\gamma \in \mu}\{p^\gamma(\chi^\gamma)\}}(\chi^\mu), \text{ and } \eta^\beta = \sum_{\mu \in \beta} \eta^\mu . \qquad (40)$$

This means that the combination of best Gaussians are selected, and the corresponding cost function values are summed; because all Gaussians within a mixture share the same data vector, it is assumed that the degrees of freedom within a mixture model are equal in all underlying Gaussians, and $\eta'' = \eta^\gamma$. Functions for computing the χ^2 distribution values are readily available in the `Matlab` environment, for example.

To conclude, the procedure that takes place when data is matched against the model has the following iterative form.

4. **OR:** In each individual mixture model, relay the appropriate cost criterion to all submodels, matching the data against each individual Gaussian, and select the best of them to represent the whole mixture.
5. **AND:** On the coordination level, recalculate the Lagrangian multipliers, reformulating the cost criteria, and unless the accuracy goal has not been reached, go back to 1.

This procedure takes place separately on each level of the AND/OR graph. It is evident that the pattern matching process is time-consuming and complex—indeed, this nonlinear iteration is the process traditionally seen as the *essence* of complex systems. Now, on the other hand, the iteration is more or less irrelevant; it is just a side-effect, the means of carrying out the actual computations in practice. Rather than concentrating on the dynamics of this iteration, one can *forget* about it, because the important thing is the static pattern, the fixed point where the iteration will finally get to. In this way, when looking at the system in a wider perspective, much simpler analysis of complex systems can be reached (see [13]). Mastering the structure makes all the difference.

System Grammar

It seems that sooner or later complexity again pops up when AND/OR blocks are nested. Indeed, this is something we are already expecting—there is no end to higher-level abstractions. Sooner or later the presented stochastic framework also becomes exhausted: Could we perhaps tell something about the yet higher, now assumedly *deterministic* abstraction level? Can we find tools for conceptually mastering the new level of complexity?

It turns out that the mixtures or mixtures, or blocks, above can be interpreted as some kind of *concepts*. Depending on their complexity, they are either subsymbolic, or they may also have some intuitive interpretation in mental terms. The framework for mastering concepts is a *language*. As Ludwig Wittgenstein put it (or what he probably meant):

"Whatever you cannot express in a language, you cannot think about!"

If Wittgenstein had ever seen a computer, he would probably have spoken about *formalisms* in general, not only natural languages. Indeed, mathematics is a language that is specially powerful when discussing complex systems: For example, high dimensionality, parallelity, and fuzziness of concepts can readily be manipulated within a language that is based on mathematics.

Figure 11 presents the basic syntax and semantics of the AI-VO ("Artificial Intelligence with Versatile Ontologies") language. This formalism is a direct formalization of the AND/OR graph ideas above: The syntax makes it possible to express mixtures of mixtures, and the semantics couples the constructs directly with the underlying mathematics. Compilation of the language constructs produces `Matlab` code that can readily be used for matching given data vectors with the defined mixture of mixtures model.

When looking at the structure of the language, one can see various interesting properties. Running AI-VO codes is more like static pattern matching and "associative regression", based on matrix calculation and linear algebra, rather than traditional sequential processing. As compared with traditional object oriented languages, the difference is that rather than having crisp classes and methods, the data structures are *numeric*. When the data structures are adapted to better match the observed data, all the distributions are modified; in this sense, two-way inheritance takes place.

"Programming" the language can be carried out in a standard way, coding the expert knowledge directly into data structures. However, being so closely connected to the actual data, the program structures can be adapted according to the observed data. Based on data distributions on each mixture level, the most appropriate distributions can be accurately updated, facilitating fast adaptation. In standard programming languages, where the symbolic program constructs are too far from the data, such adaptation cannot be carried out.

To avoid lack of conformity, there is a need for finding connections between data-based and structure-oriented approaches. This has always been a challenge, and, indeed, a painstaking problem in AI research, where the qualitative approaches (like expert systems) and quantitative approaches (like neural networks) never really met. There is the same risk in the area of complex systems research. In the previous version of this paper, for example, the traditional ideas of complexity research—letting the high-level structures automatically emerge from low-level processing—was studied [18]. Even if such purely data-based considerations are interesting from the point of view of general complex systems theory, they are not very relevant from the point of view of engineering practices. There will never exist enough data or time so that everything that is relevant would emerge; on the other hand, there always exists some a priori structural understanding, and it is not wise to ignore this expertise in modelling.

```
MIXTURE_OF_MIXTURES
        := AND_OR_STRUCTURE *

AND_OR_STRUCTURE
        := <structure_name>
        "{"
        OR_BLOCK *              %Invidual mixtures
        AND_BLOCK *             %Higher-level coordination
        "}"

OR_BLOCK
        := <mixture_name>
        "(" <Gaussian_distribution_definition> * ")"
        "{"
        ENTRY *                 %Definition of data elements
        "}"

AND_BLOCK
        := <mixture_name>.<entry_name> "="
           <mixture_name>.<entry_name>

ENTRY := <state_variable_name>
      := <entry_name> "=" function( VARIABLE * )

VARIABLE
        := <state_variable_name>
        := "A"*":"<state_variable_name>   %Steps towards future
        := "B"*":"<state_variable_name>   %Steps back in time
```

This BNF formalism gives the AI-VO language syntax using a context-free grammar. The asterisks * mean that any corresponding structures are possible; strings are given between quotes, and terminal symbols are between "<" and ">". For a complete description, see [19].

The domain is divided into a set of AND-OR structures; i.e., the basic construct consists of lower-level mixtures (OR) with common variables among the substructures being identified (AND). Additional AND-OR structures can be in a hierarchy, or the hierarchies can be tangled. The structure of each mixture (OR) is characterized by a set of multivariate Gaussian distributions that connect between the symbolic and numeric representations. In practice, these data structures refer to Matlab level matrices; when the language is compiled, the definitions become operational functions in Matlab making it possible to adapt structure definitions. Running the compiled program becomes a process of matching data against the model, so that the final pattern or result *emerges* from the iteration.

The data manipulated by the functions, or the observed "world", consist of a set of *snapshot* vectors. In addition to traditional continuous state variables, a snapshot can contain class variables that reveal the probability of submodel, given the observations. Typically with dynamic process models, the snapshots constitute a time sequence of the system. State vectors are organized as a "whiteboard" where they can be manipulated in a parallel fashion by the computational mechanism. How state variables are manipulated depends on the rules; each snapshot is iteratively selected as the "hot spot" where the data is being fitted. This process is iterative as rules can refer to neighbouring snapshots (A and B above), and the structure between constructs need not be hierarchic: Cyclicity can also exist in definitions.

A preliminary version of the AI-VO language has been implemented using Python.

Figure 11. Simplified syntax and semantics of the AI-VO language.

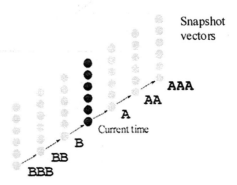

Figure 12. One-dimensional whiteboard representing time axis.

Now that there is a framework to manipulate large amounts of data, the data need to be delivered to the process machinery in a practical form (compare to [15]). It turns out that one can give new contents to the age-old *blackboard technique* employed in AI: Separate entities (agents) independently operate on different data units and modify them in parallel fashion. Now the traditional idea is slightly polished, so that let us call the canvas being operated on a *whiteboard*. When studying processes along the time axis, for example, each time point is represented by a *snapshot vector*, containing the information that determines the system state (see Fig. 12). Each of the snapshot is processed independently by the data matching machinery, but there may exist interactions: The entries in submodels can be defined using neighbouring snapshots (see examples below). Being based on such data representations, dynamic models can easily be implemented and modelled in this framework.

Application Examples

The simple dynamic system model presented in (18) can be coded using the AI-VO language as follows:

```
AR_model {
    dynamics (M) {
        y
        _Prev = B:y
    }
}
```

On the first row, the name of the block ("AR_model") is given. On the second row, the name of the first (and the only) mixture is given ("dynamics"). The list of Gaussians is given in parentheses, so that the names refer to data structures defined on the Matlab level. Because of the simplicity (linearity) of the example model, the mixture consists of only one Gaussian M. The next two rows determine how the entries in the data vector are constructed: The first element in χ is the variable y from the current snapshot, whereas the second element is the y variable from the neighbouring previous snapshot (symbol "B" standing for "before", and

"A" for "after"; note that A and B can be concatenated to refer to farther away states). This variable is locally renamed ("Prev"); while variables that are preceded by "_" are interpreted as internal variables which are not stored in the snapshot vector.

To determine the most relevant features about a Gaussian distribution, the mean vector, the principal component structure, and the error covariance need to be given; that is, the following quantities determine the distribution to a sufficient degree:

$$\bar{\chi}, \Sigma, \text{ and } \Psi, \Lambda. \tag{41}$$

Let us study some practical details concerning the implementation. To represent the above data structures in a compact, Matlab compatible form, one can first notice that the eigenvalues and eigenvectors can be uniquely separated even though only the product $\Psi\Lambda$ is stored—this is due to the orthonormality of Ψ. Second, the whole error covariance is useless; it is better to store only its diagonal vector σ. This means that in Matlab the underlying numeric data structure corresponding to the above AR description reads (note that distribution center is in the origin; there is now only one degree of freedom, so that the dimension of Ψ is 1):

$$M = (\bar{\chi} \mid \sigma \mid \Psi\Lambda) = \begin{pmatrix} 0 & \sigma_1 & \lambda a/\sqrt{a^2+1} \\ 0 & \sigma_2 & \lambda/\sqrt{a^2+1} \end{pmatrix}. \tag{42}$$

The pH process in Figs. 9 and 10 can be described as follows:

```
pH_process {

    Mixer (M) {
        Conc
        _PrevConc = B:Conc
        _Incoming = B:ConcIn
    }
    TitrationCurve (T1 T2 T3) {
        _X = Conc
        _Y = pH
    }

    Mixer.Conc = TitrationCurve._X;
}
```

The pH_process contains two mixtures, Mixer and TitrationCurve. The mixer is linear, so that only one Gaussian is defined; the data is three-dimensional, whereas there are only two degrees of freedom. The titration curve is determined by three separate Gaussians; each of these spans a one-dimensional subspace in the two-dimensional data space. When data is being matched against the model, the mixer concentration output and the titration curve input are kept equal. The snapshot vectors will have the following structure in each time point:

$$s = \begin{pmatrix} \texttt{ConcIn} \\ \texttt{Conc} \\ \texttt{pH} \\ \texttt{pH\_proc} \end{pmatrix}. \tag{43}$$

If the pH model is being simulated, only the first element in the snapshot vectors ("`ConcIn`") is filled in initially (the corresponding diagonal element in W having a relatively high value, reveals that this entry should be trusted); the matching process then finds the other variables that best explain the given data. If the model is to be adaptively identified, the "`pH`" entries are typically fixed; after snapshot convergence, the matrices determining these distributions can be updated. Whereas the continuous-valued state variables are resolved associatively using the underlying Gaussians, the more or less binary class variables emerge from the mixture blocks themselves: The snapshots contain an entry for each AND/OR structure with values between 0 and 1 revealing how relevant (or probable) that substructure happens to be when the snapshot is being judged ("`pH_proc`" above). These values are determined using Equation (39).

If the data structures are to be updated according to observations, it is a challenge to keep the mixture models from degenerating, i.e., preventing some Gaussians from becoming too strong and having some vanish. A robust way to assure stability is to apply a Self-Organizing Map (SOM) [20], i.e., defining a one-dimensional map from the first Gaussian to the last to keep all prototypes involved. In complex cases, when there are alternative paths in the AND/OR graph, adaptation must be based on the relevant of each block in explaining the data sample. Relevance values, or block probabilities (estimated or dictated) determine the adaptation rate within each block.

Only the distributions can be adapted in a consistent way. A challenge would be to implement mechanisms for modifying also the structures—numbers of variables in Gaussians, numbers of Gaussians within mixtures, numbers of mixtures within blocks, and numbers of blocks within the model. However, even though implementing such mechanisms is not easy, it can be claimed that the proposed framework with domain-oriented hierarchical structures offers a good environment for implementing Genetic Algorithms (for example, see [1]).

The idea of mixtures in mixtures can be applied to a wide range of different modelling problems. It can even be claimed that constructing such models is characteristic of our perception machinery (see below). For example, visual views have been experimentally modelled: Line segments are mixture models based on pixel distributions, etc. Now the whiteboard becomes two-dimensional (three-dimensional, if image sequences are studied!), and other differences exist as compared to the above examples such as it is mostly category variables (block probabilities) that are utilized, not continuous ones; the models are also not strictly hierarchic but cyclic in this case.

FROM "NEW SCIENCE" TO "NEW WORLD"

Languages are a way to structure a world. There exists a wide spectrum of different kinds of formalisms; however, almost all are strictly symbolic. At the other extreme, there are approaches where all structure is completely dropped. As a prototypical example, Lotfi Zadeh's idea about "computing with words" [25] is a case in point in which concepts are expressed in numerical form. In [16], a language capable of universal computation is presented without any data structures.

However, even though modelling based on extreme homogeneity sounds like a good idea, the truth is that *apples and bananas do not fit in the same basket*. Different data items have different roles, and ignoring this fact ruins the scalability of those approaches: When modelling larger systems, unstructured data is simply not enough to determine the syntactic categories. The proposed formalism seems to offer a rather practical compromise between conceptual and connectionist ways of representing phenomena. As compared to traditional programming languages, it seems very weak, it is not universal. Not all things can be represented in this formalism—only the *relevant* ones can!

In this formalism, all computation is implemented as associative, statistical pattern matching of data against the model. Applications like reasoning, etc., have to be interpreted as multivariate regression problems. The representations are basically differentiable, so that minor changes in the data to be processed typically result in small changes in the emergent patterns. However, in some cases abrupt transitions to different data regimes takes place, so that continuity vanishes altogether. This kind of behaviour—continuity and uncontinuity residing in the same framework—resembles the human *categorization* process: Minor changes in the observed data just modify the features of the perceived object, but at some point, when the data has been distorted sufficiently, the interpretation of what the object is in the first place (horse or camel, for example) changes (see [9]). In this sense, one can claim that the presented formalism supports the human way of structuring the world.

One can go even further along these lines. Remember that—with the exception of the lowest level hardwired functionalities—the mental representations that humans adults have did not yet exist in his/her brain in the childhood. The mental machinery has created the appropriate constructs based on the observed data. In this sense, the computer seeing measurement data is facing the same problem as the human cognitive machinery is: It should construct appropriate data structures representing the observed phenomena in its environment. Senses are substituted with sensors, so that the sensations can be of non-natural origin, and the world can look very different from what we are familiar with—but it can be assumed that also in this world the same kinds of basic principles apply to the properties of data.

Being based on observation data, the mental machinery can only be based on some kind of optimization, compression of the data, coding the relevant correlation structures among the observations. When constructing smart machines, capable of carrying out (more or less) automated modelling process, the same kind of data compression has to be carried out. If this optimization is carried out using the same structural framework, there must exist similarity between the obtained data structures and the corresponding mental representations, assuming that the computer and the human both have the same input data.

It does not even matter if the assumed data structures do not really exist in the environment, or if they are not the optimal ones for representing it. The human constructs the mental images representing the world through the filters determined by the mental machinery, imposing the predestinated structure on the data, perceiving more or less what he/she has been expecting. The principles of decomposing observations is mirrored to all systems being observed. In this sense, the computer can see the same things as the human does, even if this view would not reflect the real world in an unbiased way!

It was admitted in the beginning of the paper that we are now only constructing models. We were not claiming that the models we find would have some fundamental correspondence with the real world. Of course, this is the only way that an engineer can proceed—speaking of the underlying realm where one never can see is only metaphysics and philosophy. But, as explained above, it is the *subjective*, not the objective world that is the only realm for each individual human, too! Since Immanuel Kant, the possibility of intersubjectivity among people has been discussed; now this discussion can be extended to the human–machine interaction. Applying good modelling tools the machine can perhaps reach understanding in the deepest sense, and real knowledge mining (rather than mere data mining) becomes possible.

To conclude—it can be said that the "Old Science" (mathematics, etc.) will remain, but instead, there will be a totally "New World" of ways to see data. What is the nature of creatures inhabiting such a world—there exists plenty to research. The End of Science as prophesized in [6] is not in sight!

ACKNOWLEDGEMENT

During the conduct of this work, the author received financial support from Nokia Research Center (Helsinki, Finland).

REFERENCES

[1] W. Banzhaf, P. Nordin, R.E. Keller, and F.D. Francone, 1998. Genetic Programming—An Introduction. Morgan Kaufmann Publishers, San Francisco.
[2] A.-L. Barabasi, 2002. Linked: The New Science of Networks. Perseus Books, Cambridge, MA.
[3] A. Basilevsky, 1994. Statistical Factor Analysis and Related Methods. J.Wiley & Sons, NY.
[4] L. von Bertalanffy, 1976 (revised edition). General System Theory: Foundations, Development, Applications. George Braziller, NY.
[5] K. Gödel, 1992. On Formally Undecidable Propositions of Principia Mathematica and Related Systems. Dover, NY.

[6] J. Horgan, 1997. The End of Science: Facing the Limits of Knowledge in the Twilight of the Scientific Age. Helix Books, NY.
[7] H. Hjalmarsson, S. Gunnarsson, and M. Gevers, 1994. A convergent iterative restricted complexity control design scheme. Proceedings of the 33rd IEEE Conference on Decision and Control, Orlando, Florida, December 14–16, 1994, 1735–1740.
[8] A. Hyvärinen, J. Karhunen, and E. Oja, 2001. Independent Component Analysis. J. Wiley & Sons, NY.
[9] H. Hyötyniemi, 1998. On mental images and "computational semantics". Proc. 8th Finnish Artificial Intelligence Conference STeP'98, Jyväskylä, Finland, Sept., 1998, 199–208. (http://www.control.hut.fi/hyotyniemi/ publications)
[10] H. Hyötyniemi, 1999. From intelligent models to smart ones. Proceedings of IPMM'99—The Second International Conference on Intelligent Processing and Manufacturing of Materials, Honolulu, Hawaii, July 10–15, 1999, 1, 179–184. (http://www.control.hut.fi/ hyotyniemi/publications)
[11] H. Hyötyniemi, 2001. Multivariate regression—From static models to subspace identification. Proceedings of IPMM'01—The Third International Conference on Intelligent Processing and Manufacturing of Materials, Vancouver, Canada, July 29–August 3, 2001 (CD-ROM). (http://www.control.hut.fi/hyotyniemi/publications)
[12] H. Hyötyniemi, 2002. Towards new languages for systems modelling, and On emergent models and optimization of parameters. Proceedings SIMS 2002—The 43rd Conference on Simulation and Modelling, Oulu, Finland, September 26–27, 2002, 77–82 and 45–50. (http://www.control.hut.fi/ hyotyniemi/ publications)
[13] H. Hyötyniemi, 2002. Studies on emergence and cognition—Parts 1 and 2: Low-level functions, and High-level functionalities. Proceedings of the 10th Finnish Artificial Intelligence Conference STeP'02, Oulu, Finland, December 2002, 286–299 and 300–312. (http://www.control.hut.fi/ hyotyniemi/publications)
[14] H. Hyötyniemi, 2002. Life-like control. Proceedings of the 10th Finnish Artificial Intelligence Conference STeP'02, Oulu, Finland, 15–17 December 2002, 124–139. (http://www.control.hut.fi/hyotyniemi/ publications)
[15] H. Hyötyniemi, 2002. Towards Perception Hierarchies. Proceedings of the 10th Finnish AI Conference STeP'02, Oulu, Finland, 15–17 December 2002, 111–123. (http://www.control.hut.fi/hyotyniemi/publications)
[16] H. Hyötyniemi, 2002. On the universality and undecidability in dynamic systems. Helsinki University of Technology, Control Engineering Laboratory, Report 133. (http://www.control.hut.fi/hyotyniemi/ publications)
[17] H. Hyötyniemi, 2002. Reality and Truth in the Semantic Web. In Eero Hyvönen (ed.): Semantic Web Kick-Off in Finland—Vision, Technologies, Research, and Applications. HIIT Publications 2002–01, Helsinki, Chapter 9, 199–211. (http://www.control.hut.fi/ hyotyniemi/publications)
[18] H. Hyötyniemi, 2003. Emergence and Complex Systems—Towards a New

Science of Industrial Automation? Preprints of IPMM'03—The Fourth International Conference on Intelligent Processing and Manufacturing of Materials, Sendai, Japan, May 18–23, 2003 (CD-ROM format). (http://www.control.hut.fi/hyotyniemi/publications)

[19] H. Hyötyniemi, 2003. Modeling Mixtures of Mixtures. Helsinki University of Technology, Control Engineering Laboratory. (http://www.control.hut.fi/hyotyniemi/publications)

[20] T. Kohonen, 2001. Self-Organizing Maps (3rd edition). Springer-Verlag, Berlin.

[21] T. Kuhn, 1962. The Structure of Scientific Revolutions. University of Chicago Press.

[22] S. Russell and P. Norvig, 1995. Artificial Intelligence – A Modern Approach. Prentice Hall International, Englewood Cliffs, New Jersey.

[23] H.A. Simon, 1996 (third edition). Sciences of the Artificial. MIT Press, Cambridge, MA.

[24] S. Wolfram, 2002. A New Kind of Science. Wolfram Media, Champaign, Illinois.

[25] L. Zadeh, 1999. From Computing with Numbers to Computing with Words—From Manipulation of Measurements to Manipulation of Perceptions. IEEE Transactions on Circuits and Systems, 45, 105–119.

PART 1: COMPUTATIONAL INTELLIGENCE AND SOFT COMPUTING

Chapter 2: Fuzzy Systems

On Kernel-Based Fuzzy Cluster Loading

M. SATO-ILIC
Institute of Policy and Planning Sciences, University of Tsukuba, Tsukuba, Ibaraki, Japan

ABSTRACT

We have proposed an estimation method for fuzzy cluster loading using the kernel method [3]. The kernel based fuzzy cluster loading will change depending on the result of fuzzy clustering. This paper investigates how the kernel-based fuzzy cluster loading is changed by the result of fuzzy clustering and finds the basis for the relationship between the kernel-based fuzzy cluster loading and the fuzzy clustering.

INTRODUCTION

Conventional clustering means classifying a given observation into exclusive subsets (clusters). So, we can discriminate clearly if an object belongs to a cluster or not. However, such a partition is not sufficient to represent many real situations. So, a fuzzy clustering method is offered to contract clusters with uncertainty boundaries, this method allows one object to belong to some overlapping clusters with some grades. From this, it is known that fuzzy clustering is an efficient technique for real complex data.

However, replaced by the representativeness of fuzzy clustering to real complex data, the interpretation of such a fuzzy clustering causes us some confusion, because we sometimes think that objects which have a similar degree of belongingness can together form one more cluster. In order to have the interpretation of the obtained fuzzy clusters, we have proposed a fuzzy cluster loading [4] which can show the relationship between the clusters and the variables. The degree of the belongingness of objects to the fuzzy clusters also represents the state of the irregularity of the data structure, so the estimate of the fuzzy cluster loading and the estimate of the regression coefficients of the weighted regression analysis [2] are closely related to each other. It has been shown that the fuzzy cluster loading is obtained by the same method used to estimate the regression coefficient of the weighted regression analysis.

Regression analysis is one widely used and well known data analysis method.

If the data performs irregularly in the spatial variables, then the conventional regression analysis can not extract the data structure. So, a geographically weighted regression analysis was proposed for spatial data which are not stationary according to the geographical area. The main difference between conventional regression analysis and weighted regression analysis is the consideration of the difference among the area by the weight that shows the degree of the relationship of objects to each area.

In this paper, we present a method to obtain the fuzzy cluster loading in a higher dimension space than the data space and show that we can extract the data structure more efficiently. In order to extend the data space to a higher dimension space that is nonlinearly related with the data space, we use the kernel method. This model is for the nonlinear case of the fuzzy cluster loading model. We show how we can extend the model for the nonlinear case. Moreover, the estimate of the kernel based fuzzy cluster loading depends on the result of the fuzzy clustering. So, we show the relationship between the kernel based fuzzy cluster loading and the fuzzy clustering.

Several numerical examples show the validity of the kernel based fuzzy cluster loading and better performance when compared with the estimate of data space. Also, the results show the advantage of the use of the fuzzy cluster structure for the kernel based fuzzy cluster loading and several properties of this method based on the result of the fuzzy clustering.

KERNEL METHOD

The kernel method was originally developed in the context of support vector machines [3], the efficient advantage of which has been is widely recognized in many areas. The essence of the kernel method is arbitrary mapping from lower dimension space to higher dimension space. Note that the mapping is an arbitrary mapping, so we do not need to find the mapping, this is called the kernel trick.

Suppose an arbitrary mapping Φ:

$$\Phi: R^p \to F,$$

where F is a higher dimension space than R^p.

We assume

$$k(\mathbf{x}, \mathbf{y}) = \Phi(\mathbf{x})^t \Phi(\mathbf{y}),$$

where k is the kernel function defined in R^p and $x, y \in R^p$.

The typical examples of the kernel function are as follows:

$$k(\mathbf{x},\mathbf{y}) = \exp(-\frac{\|\mathbf{x}-\mathbf{y}\|}{2\sigma^2}).\qquad 1.$$

$$k(\mathbf{x},\mathbf{y}) = (\mathbf{x}\cdot\mathbf{y})^d.\qquad 2.$$

$$k(\mathbf{x},\mathbf{y}) = \tanh(\alpha(\mathbf{x}\cdot\mathbf{y})+\beta).\qquad 3.$$

Expression 1 shows gaussian kernel, expression 2 is the polynomial kernel of degree d, and expression 3 is sigmoid kernel. By the introduction of this kernel function, we can analyze the data in F without finding the mapping Φ explicitly.

WEIGHTED REGRESSION ANALYSIS

The geographically weighted regression was proposed by C. Brunsdon et al. in 1998 [2], and the model is represented as follows:

$$\mathbf{y} = V_h X \boldsymbol{\beta}_h + \mathbf{e}_h, \qquad 4.$$

where

$$V_h = \begin{pmatrix} v_{h1} & 0 & \cdots & 0 \\ 0 & v_{h2} & \cdots & \vdots \\ \vdots & \vdots & \vdots & \vdots \\ 0 & \cdots & \cdots & v_{hn} \end{pmatrix},\ X = \begin{pmatrix} 1 & x_{11} & \cdots & x_{p1} \\ 1 & x_{12} & \cdots & x_{p2} \\ \vdots & \vdots & \vdots & \vdots \\ 1 & x_{1n} & \cdots & x_{pn} \end{pmatrix},\ \mathbf{y}=\begin{pmatrix} y_1 \\ y_2 \\ \vdots \\ y_n \end{pmatrix},\ \boldsymbol{\beta}_h = \begin{pmatrix} \beta_{0h} \\ \beta_{1h} \\ \vdots \\ \beta_{ph} \end{pmatrix},\ \mathbf{e}_h = \begin{pmatrix} e_{1h} \\ e_{2h} \\ \vdots \\ e_{nh} \end{pmatrix},$$

in which $\mathbf{y}$ is a vector of dependent variables and $\boldsymbol{\beta}_h$ a vector of regression coefficients at the h-th area. V_h shows a matrix whose diagonal elements v_{hi} are the weights of the i-th object to the h-th area and are estimated explicitly. e_h is an error vector. Roughly speaking, the main difference between a conventional regression model and the weighed model is consideration of differences among the geographical areas. The estimate of $\boldsymbol{\beta}_h$ is obtained as

$$\boldsymbol{\beta}_h = (X'V_h^2 X)^{-1} X'V_h \mathbf{y}.\qquad 5.$$

FUZZY CLUSTER LOADING

In order to obtain an interpretation of fuzzy clustering result, we have proposed the following model [4]:

$$u_{ik} = \sum_{a=1}^{p} x_{ia} z_{ak} + \varepsilon_{ik},\quad i=1,\cdots,n,\quad k=1,\cdots,K,\qquad 6.$$

where, ε_{ik} is an error.

An observed 2-way data set which is composed of n objects, p variables is denoted as

$$X = (x_{ia}), \quad i = 1, \cdots, n, \quad a = 1, \cdots, p.$$

z_{ak} shows the fuzzy degree which represents the amount of loading of cluster k to variable a and we call this fuzzy cluster loading. This parameter will show how each cluster can be explained by each variable.

u_{ik} shows the obtained fuzzy clustering result as the degree of belongingness of an object i to a cluster k. The essence of fuzzy clustering is to consider not only the belonging status to the assumed clusters, but also to consider how much the objects belong to the clusters. So, there is a merit to representing the complex data situations which real data usually have. The state of fuzzy clustering is represented by a partition matrix whose elements show the grade of belongingness of the objects to the clusters, u_{ik}, $i=1,\ldots,n$, $k=1,\ldots,K$, where n is number of objects and K is number of clusters. In general, u_{ik} satisfies the following conditions:

$$u_{ik} \in (0,1), \quad \sum_{k=1}^{K} u_{ik} = 1.$$

The fuzzy clustering is to obtain the adaptable partition u_{ik} from the data X.

Then the purpose of Eq. 6 is to estimate z_{ak}, which minimize the following normalized sum of squared errors β^2.

$$\beta^2 = \frac{\sum_{i=1}^{n}\sum_{k=1}^{K}(u_{ik} - \sum_{a=1}^{p} x_{ia} z_{ak})^2}{\sum_{i=1}^{n}\sum_{k=1}^{K}(u_{ik} - \bar{u})^2},$$

where,

$$\bar{u} = \frac{1}{nK}\sum_{i=1}^{n}\sum_{k=1}^{K} u_{ik}.$$

Eq. 6 is then rewritten as

$$1 = U_k X \mathbf{z}_k + \mathbf{e}_k, \qquad 7.$$

using

$$U_k = \begin{pmatrix} u_{1k}^{-1} & 0 & \cdots & 0 \\ 0 & u_{2k}^{-1} & \cdots & \vdots \\ \vdots & \vdots & \vdots & \vdots \\ 0 & \cdots & \cdots & u_{nk}^{-1} \end{pmatrix}, \quad X = \begin{pmatrix} x_{11} & x_{12} & \cdots & x_{1p} \\ x_{21} & x_{22} & \cdots & x_{2p} \\ \vdots & \vdots & \vdots & \vdots \\ x_{n1} & x_{n2} & \cdots & x_{np} \end{pmatrix}, \quad \mathbf{1} = \begin{pmatrix} 1 \\ 1 \\ \vdots \\ 1 \end{pmatrix}, \quad \mathbf{z}_k = \begin{pmatrix} z_{1k} \\ z_{2k} \\ \vdots \\ z_{pk} \end{pmatrix}, \quad \mathbf{e}_k = \begin{pmatrix} e_{1k} \\ e_{2k} \\ \vdots \\ e_{nk} \end{pmatrix}.$$

From Eq. 4, Eq. 5, and Eq. 7, we obtain the estimate of z_k as

$$\mathbf{z}_k = (X'U_k^2 X)^{-1} X'U_k \mathbf{1}, \quad k = 1, \cdots, K. \qquad 8.$$

KERNEL FUZZY CLUSTER LOADING

From Eq. 8, we can obtain the following:

$$\mathbf{z}_k = (X'U_k^2 X)^{-1} X'U_k \mathbf{1} = ((U_k X)'(U_k X))^{-1}(U_k X)'\mathbf{1} \equiv (C_k' C_k)^{-1} C_k' \mathbf{1}. \qquad 9.$$

where

$$C_k = (c_{ia(k)}), \quad c_{ia(k)} \equiv u_{ik}^{-1} x_{ia}, \quad i = 1, \cdots, n, \quad a = 1, \cdots, p. \qquad 10.$$

Using $\mathbf{c}_{a(k)}^t = (c_{1a(k)}, \ldots, c_{na(k)})$, we can represent Eq. 9 as follows:

$$\mathbf{z}_k = (\mathbf{c}_{a(k)}^t \mathbf{c}_{b(k)})^{-1} (\mathbf{c}_{a(k)}^t \mathbf{1}), \quad a, b = 1, \cdots, p, \qquad 11.$$

where

$$C_k' C_k = (\mathbf{c}_{a(k)}^t \mathbf{c}_{b(k)}), \quad C_k' \mathbf{1} = (\mathbf{c}_{a(k)}^t \mathbf{1}), \quad a, b = 1, \cdots, p.$$

Then we consider the following mapping Φ:

$$\Phi: R^p \to F, \quad \mathbf{c}_{a(k)} \in R^p. \qquad 12.$$

From Eq. 11 and Eq. 12, the fuzzy cluster loading in F is as follows:

$$\mathbf{z}_k' = (\Phi(\mathbf{c}_{a(k)})' \Phi(\mathbf{c}_{b(k)}))^{-1} (\Phi(\mathbf{c}_{a(k)})' \Phi(\mathbf{1})), \quad a, b = 1, \cdots, p, \qquad 13.$$

where z_k' shows the fuzzy cluster loading in F.

Using the kernel representation $k(x,y) = \Phi(x)^t \Phi(y)$ mentioned above, Eq. 13 is rewritten as follows:

$$\mathbf{z}_k' = (k(\mathbf{c}_{a(k)}, \mathbf{c}_{b(k)}))^{-1} (k(\mathbf{c}_{a(k)}, \mathbf{1})), \quad a, b = 1, \cdots, p. \qquad 14.$$

From this, using the kernel method, we can estimate the fuzzy cluster loading in F.

COMPARISION OF FUZZY CLUSTER LOADING AND KERNEL FUZZY CLUSTER LOADING

From the definition of the kernel fuzzy cluster loading shown in Eq. 14, we can see that kernel fuzzy cluster loading depends on the result of fuzzy clustering, that is u_{ik}. Also, the essential difference between kernel fuzzy cluster loading and fuzzy cluster loading is the difference between the product $k(\mathbf{c}_{a(k)}, \mathbf{c}_{b(k)}) \equiv \Phi(\mathbf{c}_{a(k)})^t \Phi(\mathbf{c}_{b(k)})$ in F and the product $\mathbf{c}_{a(k)}{}^t \mathbf{c}_{b(k)}$ in R^p. Since $\mathbf{c}_{a(k)}{}^t = (c_{1a(k)}, \ldots, c_{na(k)})$, $c_{ia(k)} \equiv u_{ik}^{-1} x_{ia}$, these products are closely related to the result of fuzzy clustering, that is the degree of belongingness of the objects to the clusters u_{ik}. So, we must investigate the property of these products to establish the change in the fuzzy clustering result. From Eq. 10:

$$\mathbf{c}_{a(k)}^t \mathbf{c}_{b(k)} = (u_{1k}^{-1})^2 x_{1a} x_{1b} + \cdots + (u_{nk}^{-1})^2 x_{na} x_{nb} = \sum_{i=1}^{n} (u_{ik}^{-1})^2 x_{ia} x_{ib}. \qquad 15.$$

If k is the polynomial kernel with degree 2 (see Eq. 2 with $d=2$), then

$$k(\mathbf{c}_{a(k)}, \mathbf{c}_{b(k)}) = \Phi(\mathbf{c}_{a(k)})' \Phi(\mathbf{c}_{b(k)}) = (\mathbf{c}_{a(k)}' \mathbf{c}_{b(k)})^2 = (\sum_{i=1}^{n} (u_{ik}^{-1})^2 x_{ia} x_{ib})^2. \qquad 16.$$

For fixed k, a, and b, when $n=2$, we can simplify, respectively Eq. 15 and Eq. 16 as follows:

$$\mathbf{c}_{a(k)}^t \mathbf{c}_{b(k)} = w_1 x + w_2 y, \qquad 17.$$

$$k(\mathbf{c}_{a(k)}, \mathbf{c}_{b(k)}) = \Phi(\mathbf{c}_{a(k)})' \Phi(\mathbf{c}_{b(k)}) = (w_1 x + w_2 y)^2. \qquad 18.$$

where,

$$w_i \equiv (u_{ik}^{-1})^2, \quad x \equiv x_{1a} x_{1b}, \quad y \equiv x_{2a} x_{2b}, \quad i = 1,2.$$

Table 1 shows the changing situation of both Eq. 17 and Eq. 18 according to the change in the fuzzy clustering result, that is the values of w_1 and w_2.

Table 1. Comparison of Eq. 17 and Eq. 18 according to the Change in Fuzzy Clustering Result (w_1 and w_2).

w_1	w_2	Equation 17 (Product in R^p)	Equation 18 (Product in F)
1.0	0.0	x	x^2
0.0	1.0	y	y^2
0.5	0.5	0.5(x+y)	$0.25(x+y)^2$
0.3	0.7	0.3x+0.7y	$(0.3x+0.7y)^2$

Figure 1 shows the value of Eq. 17 in R^p with respect to x and y in the case of $w_1=w_2=0.5$. Figure 2 shows the value of Eq. 18 in F with respect to x and y in the case of $w_1=w_2=0.5$. In this case, the clustering structure is homogeneous in comparison to other cases in Table 1.

Figure 3 shows the intersecting lines between the plane or surface shown in Figure 1 and Figure 2 and the plane $x+y = 1$, respectively. The solid line shows the intersection between the plane in Figure 1 and the plane $x+y = 1$ while the dotted line shows the intersection between the surface in Figure 2 and the plane $x+y = 1$. Figure 4 shows the value of Eq. 17 in R^p with respect to x and y in the case of $w_1=0.3$, $w_2=0.7$. Figure 5 is the value of Eq. 18 in F with respect to x and y in the case of $w_1=0.3$, $w_2=0.7$. In this case, the clustering structure is heterogeneous compared to the other cases in Table 1.

Figure 6 shows the intersecting line and curve between the plane or the surface shown in Figure 4 and Figure 5 and the plane $x+y=1$, respectively. The solid line shows the intersection between the plane in Figure 4 and the plane $x+y=1$ while the dotted line shows the intersection between the surface in Figure 5 and the plane $x+y=1$. In comparing Figure 3 and Figure 6, we can see the more complex difference between the two products in Figure 6, that is, if the clustering result can capture the heterogeneous structure of the given data, then the difference between the two products are affected more sensitively compared to that case in which the cluster structure of the data is homogeneous. In other words, if the data has a structure that can be classified into a small number of clusters clearly, the difference tends to be monotonic with respect to the value of the observation. However, if the data is distributed among a large number of clusters, the difference is more complex with respect to the value of the observation.

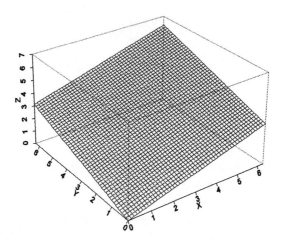

Figure 1. Equation 17 in R^p when $w_1=w_2=0.5$.

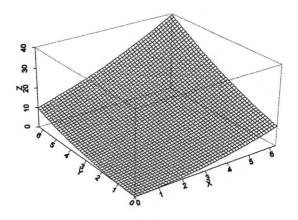

Figure 2. Equation 18 in F when $w_1=w_2=0.5$.

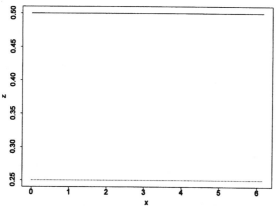

Figure 3. Intersecting Lines.

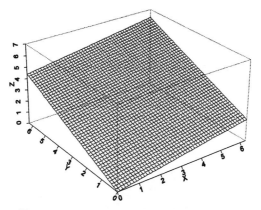

Figure 4. Equation 17 in R^p when w_1=0.3, w_2=0.7.

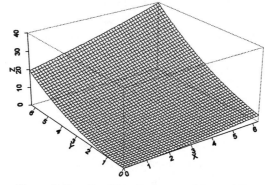

Figure 5. Equation 18 in F when w_1=0.3, w_2=0.7.

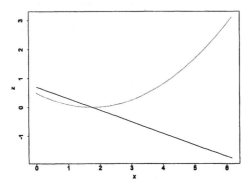

Figure 6. Intersecting Line and Curve (see text).

NUMERICAL EXAMPLE

We will now show an example that uses the kernel fuzzy cluster loading shown in Equation 14 when the kernel function is the polynomial kernel shown in expression 2. The data is made up of the measurements of rainfall from 328 locations around Japan over a 12-month period [5]. The degree of belongingness of a location to each cluster, u_{ik}, is obtained by using the fuzzy k-means method with $m=2.0$. The data was classified into 5 clusters, Sapporo/Sendai, Tokyo, Osaka, Fukuoka, and Okinawa areas.

Figure 7 shows the result of fuzzy cluster loadings in Eq. 14 when k is a polynomial kernel with $d=1$. Note when $d=1$ in Eq. 2, then Eq. 14 is reduced to Eq. 11. So, this is the same as finding the solution of fuzzy cluster loading in R^p. In Figure 7, the abscissa shows each month (variable) while the ordinate is the value of fuzzy cluster loadings. Each line shows each cluster. From this diagram, we can see that the Sapporo/Sendai area has a situation opposite to the other areas. Especially, in the month of February, Sapporo/Sendai area does not have as much rainfall, but they receive snow due to lower temperatures.

Figure 8 shows the result of Eq. 14 using a polynomial kernel with $d=2$. In this case, the estimated fuzzy cluster loading is a solution which is obtained in F (mapped higher dimension space). From this diagram, we can see that the same feature of February, that is, Sapporo/Sendai has a remarkable difference compared to the other four areas. Also, we can see clearer properties in Figure 8, compared with those shown in Figure 7. For example, in May, Sapporo/Sendai also has a clearly different feature from the other four areas in Figure 7, but in Figure 8, we see the difference is small and in fact, Sapporo/Sendai, Tokyo, and Osaka are similar to each other with the next smaller value being Fukuoka while the smallest is Okinawa. Since those five area are located from north to south according to the order, Sapporo/Sendai, Tokyo, Osaka, Fukuoka, Okinawa, the values for fuzzy cluster loading in May are arranged in order from north to south. So, the result seems to be reasonable.

Moreover, in November, we can see the similarity between Sapporo/Sendai and Okinawa in Figure 7, this is difficult to explain because the location of these two areas are completely different, that is, to the north and to the south. In Figure 8, we cannot find any remarkable similarity of those two areas. From the comparison of these two results, it seems clearer to use the result in Figure 8 to explain the data.

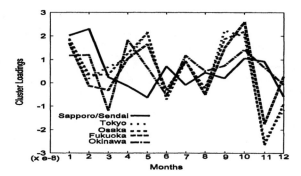

Figure 7. Fuzzy Cluster Loadings for Rainfall Data using Polynomial Kernel ($d=1$).

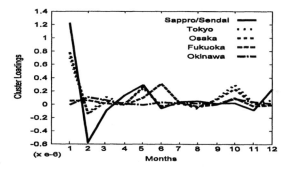

Figure 8. Fuzzy Cluster Loadings for Rainfall Data using Polynomial Kernel ($d=2$).

CONCLUSION

In this paper, we propose kernel-based fuzzy cluster loading and investigate the feature of the kernel based fuzzy cluster loading that changes the results of fuzzy clustering. Fuzzy clustering can capture data structures which show the homogeneous/heterogeneous structure of the given data. So, this feature can show the property of the kernel-based fuzzy cluster loading that provides improvement over conventional fuzzy cluster loading. The main difference between kernel based fuzzy cluster loading and fuzzy cluster loading is the product space that can be measured regarding the similarity among variables. That is, in the case of kernel based fuzzy cluster loading, the product is defined in a higher dimensional space. In the case of fuzzy cluster loading, the product is in p dimension space where p is the number of variables. So, comparison of these loadings is reduced to a comparison of the two products which are in different dimensional space.

REFERENCES

[1] J.C. Bezdek, J. Keller, R. Krisnapuram, N.R. Pal, 1999. Fuzzy Models and Algorithms for Pattern Recognition and Image Processing, Kluwer Academic Publishers.

[2] C. Brunsdon, S. Fotheringham and M. Charlton, 1998. Geographically weighted regression-modelling spatial non-stationarity, Journal of the Royal Statistical Society, 47, Part 3, 431–443.

[3] N. Cristianini and J. Shawe-Taylor, 2000. An Introduction to Support Vector Machines and Other Kernel-Based Learning Methods, Cambridge University Press.

[4] M. Sato-Ilic, Fuzzy Cluster Loadings for Weighted Regression Analysis, 2001. Japanese Classification Society, 71–78.

[5] Annual Report of Automated Meteorological Data Acquisition System (AMeDAS), 1999. Meteorological Business Support Center.

Fuzzy Evolutionary Approach to Determination of Dietary Composition for Diabetics

T. VAN LE
School of Computing, University of Canberra, University Drive, Bruce, ACT 2617, Australia

ABSTRACT

Dietary therapy is an essential component of the treatment of diabetes. Diabetics must adopt a strict diet with low intake of sugar, salt, and fat. It is necessary, however, to ensure that the diet provides adequate energy to the patient. The task of selecting carbohydrate foods and determining the composition of a diabetic diet is very complex and time consuming. This paper presents a computer system that assists dieters in establishing their own diets with great variety and under their full control. The system allows the user to select their favourite foods and indicate their required intake of energy. Based on the user's specification of strictness on minimizing the amounts of fat, sugar, cholesterol and salt, the system works out the food composition in order to provide the required intake of energy with minimum total amounts of fat, sugar, cholesterol, and salt. The method employed by the system is based on a combination of fuzzy modelling of the problem and an evolutionary approach to constrained optimisation problems. Several experiments were carried out and the results show the system is highly effective and efficient.

INTRODUCTION

Dietary therapy is an essential component of the treatment of diabetes. Diabetics must adopt a strict diet with low intake of sugar, salt, and fat. It is necessary, however, to ensure that the diet provides adequate energy to the patient. The task of selecting carbohydrate foods and determining the composition of a diabetic diet is very complex and time consuming. There have been several reports ([2], [6]) indicating the great difficulty of patients to follow dietary instructions due to the restrictedness and inflexibility of their prescribed diets. In order to assist diabetics in adhering to diet and still enjoy their meals, it is necessary to provide dieters with some means to select the food and determining the food composition by themselves.

This paper presents a computer system that is designed to assist dieters in establishing their own diets with great variety and under their full control. The system allows the user to select their favorite foods and indicate their required intake of energy. Based on the user's specification of strictness on minimizing the amounts of fat, sugar, and salt, the system works out the food composition in order to provide the required intake of energy. The system also advises the user on the total amounts of fat, sugar, cholesterol, and salt in the established diet. The method employed by the system is based on fuzzy evolutionary approach to constrained optimization problems that the author has presented in previous publications ([3], [4]). The tables of foods' nutritional composition used in this paper are obtained from the documents published by the Australian Government's Department of Community Services and Health ([1]).

A FUZZY MODEL OF THE DIETARY COMPOSITION PROBLEM

Consider a table of N food items selected by a dieter for the purpose of establishing a suitable diet. Let M be the expected weight of the meal and assume that the required energy intake is between A and B kilo Joules. The task at hand is to determine the food components x_i, where $1 \leq i \leq N$ and $\sum_{i=1}^{N} x_i = M$, so that the diet provides the required energy intake with minimum total amounts of fat, sugar, cholesterol, and salt.

More precisely, let e_i, f_i, g_i, c_i, s_i be respectively the composition of energy, fat, sugar, cholesterol, and salt in 100g of the ith food item in the table. Also let $\alpha, \beta, \gamma, \delta$ be respectively the weight factor of fat, sugar, cholesterol, and salt restriction. That is, the higher value α is, say, the more strictly the amount of fat needs to be reduced. Thus, if $\alpha = 0$, then fat need not be restricted at all. The problem of determining the food components x_i to provide the required energy intake with minimum total amounts of fat, sugar, cholesterol, and salt can be expressed as follows:

$$\text{Minimize} \quad \frac{\alpha \sum_{i=1}^{N} x_i f_i}{M \sum_{i=1}^{N} f_i} + \frac{\beta \sum_{i=1}^{N} x_i g_i}{M \sum_{i=1}^{N} g_i} + \frac{\gamma \sum_{i=1}^{N} x_i c_i}{M \sum_{i=1}^{N} c_i} + \frac{\delta \sum_{i=1}^{N} x_i s_i}{M \sum_{i=1}^{N} s_i} \quad (1)$$

$$\text{Subject to} \quad A \leq \sum_{i=1}^{N} x_i e_i \leq B \quad (2)$$

$$\sum_{i=1}^{N} x_i = M \quad (3)$$

Fuzzy Evolutionary Approach to Determination of Dietary Composition for Diabetics

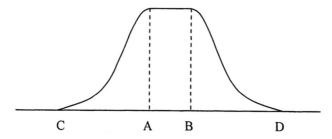

Figure 1. Fuzzy constraint of required energy intake.

$$x_i \geq 0, \quad i = 1, \ldots, N. \tag{4}$$

In reality, condition 2 above is normally tolerable. That is, if the provided energy is slightly less than A or slightly greater than B, it is still acceptable (with some degree) to the dieter. Only when it is less than some lower limit C, or greater than some upper limit D will be considered as totally unacceptable. Thus, condition 2 can be replaced with a fuzzy constraint represented by a fuzzy membership function defined below and depicted in Figure 1.

$$\mu_{(A,B,C,D)}(X) = \begin{cases} 1 & \text{if } A \leq X \leq B \\ \dfrac{e^{-p\left(\frac{X-A}{C-A}\right)^2} - e^{-p}}{1 - e^{-p}} & \text{if } C \leq X \leq A \\ \dfrac{e^{-p\left(\frac{X-B}{D-B}\right)^2} - e^{-p}}{1 - e^{-p}} & \text{if } B \leq X \leq D \\ 0 & \text{if } X \leq C \text{ or } D \leq X \end{cases} \tag{5}$$

The problem of determining the food components x_i to provide the required energy intake with minimum total amounts of fat, sugar, cholesterol, and salt is re-modelled as follows:

Maximize

$$\exp\left(-\left(\frac{\alpha \sum_{i=1}^{N} x_i f_i}{M \sum_{i=1}^{N} f_i} + \frac{\beta \sum_{i=1}^{N} x_i g_i}{M \sum_{i=1}^{N} g_i} + \frac{\gamma \sum_{i=1}^{N} x_i c_i}{M \sum_{i=1}^{N} c_i} + \frac{\delta \sum_{i=1}^{N} x_i s_i}{M \sum_{i=1}^{N} s_i}\right)\right) \mu_{(A,B,C,D)}\left(\sum_{i=1}^{N} x_i e_i\right) \tag{6}$$

Subject to
$$\sum_{i=1}^{N} x_i = M \qquad (7)$$

$$x_i \geq 0, \quad i = 1,\ldots,N. \qquad (8)$$

The value of expression 6 is called the *fitness* of the food table which is used to determine the most suitable dietary table for the user. Here, the acceptability degree of the provided energy is used as a weight factor of the proposed table.

EVOLUTIONARY APPROACH TO DETERMINATION OF DIETARY COMPOSITION

The model described in previous section represents a fuzzy constraint optimization problem. Therefore, the technique of fuzzy evolutionary programming [3, 4] can be employed to find an optimal solution for the problem. This method is presented in the following algorithm.

Algorithm 1

Generate a population of K chromosomes $x^k = \{x_i^k, i = 1,\ldots,N\}$, $k = 1,\ldots,K$, with $x_i^k \geq 0$ and $\sum_{i=1}^{N} x_i^k = M$. For each x^k compute its fitness value f^k using formula 6 in the previous section.

Repeat

Make each chromosome x^k reproduce an offspring x^{K+k} that inherits more or less the genes of

its parent depending on the parent's fitness, by letting

$s_i^k = x_i^k$ if sign $-$ is chosen, and $1 - x_i^k$ if sign $+$ is chosen;

$$\delta x_i^k = \begin{cases} \dfrac{s_i^k}{4}\sqrt{\ln\left(\dfrac{1}{r^{1-\rho(f^k)}}\right)} & \text{if } r^{1-\rho(f^k)} > 10^{-6} \\ s_i^k & \text{otherwise} \end{cases}$$

$x_i^{K+k} = x_i^k \pm \delta x_i^k$, (the sign $+$ or $-$ and value r are chosen at random)

and compute the offspring's fitness f^{K+k} using Equation 6 above.

For each $k = 1,\ldots,2K$, select a random set U of indices from 1 to $2K$, and record the number

w^k of $h \in U$ such that $f^h \leq f^k$.

Select K fittest chromosomes (with highest scores w^k) among the existing $2K$ chromosomes to form the next generation.

Until the population is stabilized or the allowed time is exhausted.

In Algorithm 1, the symbol ρ denotes a retracting function that reduces the real line $(0, \infty)$ to the open interval $(0, 1)$, while the expression δx_i^k is designed to yield a (random) small change if the chromosome is highly fit or a larger change otherwise. Algorithm 1 was used to produce a number of optimized dietary tables as described in the next section.

EXPERIMENTAL RESULTS

The system presented here is based on the fuzzy model and evolutionary algorithm (Algorithm 1) described above. The user is allowed to select a list of foods from a database of food compositions and to specify the expected meal weight M, the restricting factors (α, β, γ, δ) of fat, sugar, cholesterol, and salt respectively, as well as the fuzzy parameters (A, B, C, D, P) representing the dieter's required energy intake. The system performs its optimising process to establish a dietary table showing the weights of each food item, the minimum amounts of fat, sugar, cholesterol, and salt contained in the diet, the energy provided by the diet, and also the fitness of the diet (see Figure 2).

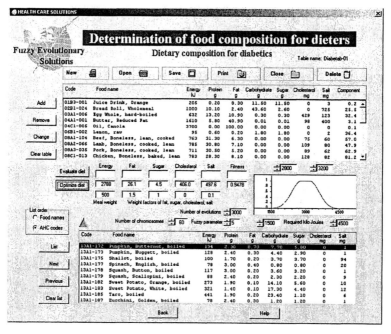

Figure 2. Optimisation of a diet table for diabetics.

Experiment 1: In this experiment, we chose a list of foods consisting mainly of meat and vegetables. The meal weight was set to $M = 500$g; the restricting factors were set to $\alpha = 1.5$, $\beta = 1$, $\gamma = 0$, and $\delta = 0.1$. The required energy intake parameters were set to $A = 1500$, $B = 2800$, $C = 3200$, $D = 4500$ (using the data from [2, 5]). The optimised diet is shown in Figure 2 and Table 1.

The last column of Table 1 shows the food components taken to achieve the required energy intake with minimum amounts of fat, sugar, cholesterol and salt. Figure 2 shows the energy provided by this optimized diet is 2788 kJ and the total amounts of fat, sugar, cholesterol, and salt are 26.1g, 4.5g, 406mg, and 497.6mg respectively. The fitness of this diet has been calculated by the system as 0.9478, which is relatively high. Observe that the proposed components of some food items are less than 1g, which are not worth being included in the meal. The system allows the user to change the component of any food item and will reevaluate the provided energy, the total amounts of fat, sugar, cholesterol, and salt of the changed diet and its fitness. This process can be done repeatedly until the user is completely satisfied with the prescribed diet. The final optimized dietary table can then be saved and/or printed out for use. The system also allows any existing table to be reopened for change or update.

Experiment 2: In this second experiment, we chose a list of foods consisting mainly of fish and vegetables. All other parameters remained the same as in Experiment 1. The optimized diet is shown in Table 2. The energy provided by Table 2 is 2796 kJ and the total amounts of fat, sugar, cholesterol, and salt are 36g, 4g, 130.5mg, and 369.4mg respectively which are relatively lower than those contained in Table 1. The fitness of this diet is 0.9429.

Table 2 also shows some food items with low components such as Apricot (0.0) and Plum (0.8), which can be removed from the table and the system reactivated to reevaluate the table before it is saved for later use.

CONCLUSTION AND CONTINUING WORK

Fuzzy evolutionary programming is a satisfactory solution to the problem of determining the dietary composition for diabetics. The system we built to implement this technique has been shown to be highly effective. The system is currently being used as a test at some health consulting centers in Canberra. It is considered by health-care workers and users to be preferable to pre-prescribed diets due to its high flexibility in meeting individual needs and requirements. We are working on extending the system to cater for more dietary parameters in order to assist a wider range of diabetics.

Table 1. A diet table for diabetics consisting of meat and vegetable.

Food name	Energy	Protein	Fat	Carbo	Sugar	Chol.	Salt	Comp.
Juice Drink, Orange	205	0.2	0.3	11.5	11.5	0	3	0.2
Bread Roll, Wholemeal	1000	10.1	2.4	43.6	2.6	0.01	725	25.5
Egg Whole, hard-boiled	632	13.2	10.9	0.3	0.3	429	123	32.4
Butter, Reduced Fat	1610	5.8	40.9	0.01	0.01	98	400	3.1
Oil, Canola	3700	0.0	100.0	0.0	0.0	0	0.01	0.1
Lemon, raw	95	0.6	0.2	1.8	1.8	0	2	36.4
Beef, Boneless, lean	763	31.3	6.3	0	0	75	60	57.0
Lamb, Boneless, lean	785	30.8	7.1	0	0	109	80	47.9
Pork, Boneless, lean	711	30.5	5.2	0	0	89	62	62.9
Chicken, Boneless, lean	783	28.3	8.1	0	0	128	82	81.2
Cheese, Cheddar	1690	25.4	33.8	0.1	0.1	101	655	9.0
Bean Sprouts, raw	84	3.1	0.1	1.6	1.0	0	1	12.7
Celery, raw	51	0.6	0.1	2.2	1.2	0	88	5.3
Lettuce, raw	27	0.9	0.1	0.4	0.4	0	23	2.7
Mushroom, raw	98	3.6	0.3	1.5	0.3	0	7	1.4
Pea, Green, boiled	203	4.8	0.4	6.4	2.7	0	1	10.6
Radish, Red, raw	53	0.8	0.2	1.9	1.9	0	20	26.4
Chili, Red, raw	115	1.4	0.4	4.2	4.2	0	3	1.4
Potato, Pale Skin	272	2.6	0.2	13.0	0.4	0	7	12.1
Asparagus	79	2.8	0.1	1.6	1.6	0	2	3.5
Beetroot, peeled	173	1.9	0.1	8.4	8.4	0	51	4.0
Broccoli, boiled	101	4.7	0.3	0.4	0.4	0	20	8.9
Cabbage, White	66	1.2	0.1	2.5	2.5	0	13	11.1
Capsicum, Green	75	1.7	0.1	2.6	2.6	0	2	9.2
Carrot, Baby	119	0.8	0.1	6.1	6.1	0	48	4.3
Cauliflower, boiled	80	2.2	0.2	2.0	2.0	0	14	13.0
Garlic, boiled	429	6.8	3.1	11.3	1.7	0	8	12.4
Onion, Brown, boiled	113	1.8	0.1	4.8	4.8	0	10	4.7
Pumpkin, Butternut	194	2.3	0.7	7.7	5.8	0	1	0.5

Table 2. A diet table for diabetics consisting of fish and vegetables.

Food name	Energy	Protein	Fat	Carbo	Sugar	Chol.	Salt	Comp.
Rice, White, boiled	523	2.3	0.2	28.0	0.1	0	5	99.6
Egg, scrambled	682	10.4	13.5	0.4	0.4	314	141	6.5
Butter, Regular	3040	0.8	81.4	1.0	1.0	200	720	4.1
Oil, Olive	3700	0	100.0	0	0	0	0.01	11.2
Tuna, canned in brine	458	22.1	2.2	0	0	43	390	24.5
Salmon, Red, in-brine	722	19.4	10.6	0	0	63	615	6.9
Flathead, steamed	487	26.1	1.2	0	0	78	91	55.4
Gemfish, steamed	942	21.9	15.4	0	0	61	78	71.8
Lemon, raw	95	0.6	0.2	1.8	1.8	0	2	2.7
Apricot, raw	156	0.8	0.1	7.4	6.8	0	2	0.0
Plum, raw	146	0.6	0.1	7.1	6.5	0	2	0.8
Avocado, raw	879	1.9	22.6	0.4	0.4	0	2	27.1
Bean Sprouts, raw	84	3.1	0.1	1.6	1.0	0	1	20.1
Cucumber, Common	45	0.4	0.1	2.1	2.1	0	21	11.0
Mushroom, raw	98	3.6	0.3	1.5	0.3	0	7	17.7
Radish, Oriental	72	0.7	0.3	2.9	2.9	0	28	4.8
Tomato, Egg, raw	61	1.0	0.1	2.3	2.3	0	8	7.0
Potato, Red Skin	261	2.4	0.1	12.6	0.7	0	2	4.0
Asparagus, boiled	79	2.8	0.1	1.6	1.6	0	2	8.2
Bean, Kidney, boiled	480	12.8	0.4	14.2	1.2	0	3	19.5
Capsicum, Green	75	1.7	0.1	2.6	2.6	0	2	3.0
Carrot, Baby, boiled	119	0.8	0.1	6.1	6.1	0	48	1.8
Celery, boiled	56	0.7	0.1	2.4	1.3	0	84	2.1
Chilli, Red, boiled	127	1.6	0.4	4.7	4.7	0	3	1.6
Garlic, boiled	429	6.8	3.1	11.3	1.7	0	8	27.7
Ginger, boiled	121	0.9	0.4	5.3	1.9	0	10	5.1
Onion, Brown, boiled	113	1.8	0.1	4.8	4.8	0	10	3.3
Shallot, boiled	100	1.7	0.2	3.7	3.7	0	94	30.1
Spinach, English	78	3.0	0.4	0.8	0.8	0	20	11.8
Sweet corn, Young	77	1.6	0.2	2.5	1.6	0	350	10.8

REFERENCES

[1] R. English, J. Lewis, 1992. Nutritional Values of Australian Foods. Austral. Gov. Pub. Serv., Canberra.
[2] H. Keen and B.J. Thomas, 1978. Diabetes Mellitus. Nutrition in the Clinical Management of Disease. J.W.T. Dickerson and H.A. Lee (Eds.), Edward Arnold, London, 167.
[3] T.V. Le, 1996. A fuzzy evolutionary approach to constrained optimization problems. Proc. Third Int. Conf. on Evolutionary Computation, Nagoya, 274–278.
[4] T.V. Le, 2001. A fuzzy evolutionary approach to investment decision. Mathematics and Simulation With Biological, Economical and Musicoacoustical Applications, C.E. D'Attellis & N.E. Mastorakis (Eds.), WSES, 268–271.
[5] M. Scott and A.I. Pryke, 1973. So I'm a diabetic. Angus and Robertson.
[6] A.S. Truswell, B.J. Thomas and A. M. Brown, 1975. Survey of dietary policy and management in British diabetic clinics. British Med. J. 4, 7.

Intelligent Information Technology: From Autonomic to Autonomous Systems

L. S. STEWART
Decyde Ware Inc., 157 Adelaide Street W., Suite 610, Toronto, Ontario,
M5H 4E7 Canada

ABSTRACT

People have evolved internal information-processing structures that allow them to live autonomously in a world that is inexact, and full of the unexpected. They survive on their reflexes and wits. While their involuntary systems keep them ticking, they use their motor-sensory systems to do the physical tasks—get from one place to another, avoid obstacles, build bridges, drive cars, operate machines. And, they use their brains to do the cognitive tasks—form opinions, change opinions, do sums, make plans, hypothesize, interpret, decide, learn, remember, improvise and invent.

Human inventions—machines, IT systems, businesses and organizations—lack the autonomy of their inventors. They cannot survive in a complex, unpredictable world without the help of humans. They depend on people to build them, service them, repair them, replace them, and think for them.

The ultimate goal of intelligent information technology (IIT) is to make human inventions more autonomous—to mimic the neurological processes that keep biological systems going. It is an ambitious goal. It requires IT systems that will—similar to involuntary nervous systems—keep mechanical and social systems ticking. Autonomic computing [1] is a first tentative start.

It requires IT systems that will—like motor-sensory systems—let mechanical and social systems operate themselves. Fuzzy technology has made significant progress in the area of process and industrial control.

It also requires IIT systems that will—like the biological brain—help mechanical and social systems think for themselves. No existing technology successfully addresses this. The missing piece in the IIT puzzle is an algorithm that captures the ability of people to hypothesize. That is, extrapolate or infer conclusions from one or two experiences—and do it with sparse data.

When lacking experience or information, people extrapolate. Even with only one experience, and/or one piece of evidence—whether it is a hunch, impression or fact—people can begin to form opinions, draw conclusions, and make judgement calls. This use of judgement makes it possible for people to deal with uncertainty as it surfaces—on the ground—in the real world. It is an important and pervasive survival skill.

The paper refers to an algorithm that incorporates the hypothesizing process, and makes a case for embedding it in existing business and corporate IT systems—in those rudimentary IT "nervous" systems that currently deliver information to decision-makers who, in turn, make the judgement calls on behalf of the organization.

It is argued,—in the context of the principles of self-organization—that by using this hybrid IIT to make professional judgment more consistent and mathematically rigorous across the enterprise, businesses and organizations will become more autonomous and less vulnerable to poor judgement on the part of their decision-makers

Finally, an instantiation of the algorithm is provided and along the way, the evolution of technology is described. It is a saga of invention.

INTRODUCTION

Human Invention

Humans invent. That is their genius. They invent to overcome their limitations. The list is long. The inventions vary.

From the very beginning people invented tools to carry out tasks beyond their physical capabilities. The evolution of these mechanical inventions has been long and steady. Now a legion of machines do things that are well beyond the normal capabilities of humans—fly, move at high speeds, travel to the moon, live under water, lift and carry great loads, do menial, repetitive jobs without complaining, kill from long distances.

Social organization is another early human invention. People have always formed tribes, clans, and villages to overcome the limitations of individuals to fend for themselves. Now the globe is divided into collections of people—cities, nations, alliances, and corporations—co-operating, competing, trading, warring.

Humans also invented language. From the very start, people have used sounds and signs to communicate—to exchange information, ideas, values, warnings, and threats. Now, thanks to their inventive genius, there are IT infrastructures that enable people to communicate across oceans, continents, and space.

Underpinning all invention, either implicitly or explicitly, is the most significant human invention of all—mathematics.

TOWARD INTELLIGENT INFORMATION TECHNOLOGY

It is mathematics that has given information-processing technology (IT) a language. It is mathematics that is driving the evolution toward intelligent information technology (IIT).

Originally, the ambition for computers was modest—to improve the reliability of counting and adding. Now, in their present form, computers are used pervasively to collect, store, and process huge amounts of numerical data at speeds and accuracies far beyond human capabilities. Existing computing technology displays an incipient intelligence that mimics certain ways that people reason. Each has a mathematical basis.

Existing Binary Computing Technology

People can draw precise yes/no conclusions.

"The sun is shining."

"The door is locked."

"There are seven pebbles."

Binary computing technology captures this process.

The language of binary computing technology is Boolean—yes/no, on/off, 0/1, true/false, black/white. As a result, binary technology can only process data that is precise, and/or quantifiable. And, there must be a rule for every contingency. If there is no rule to cover the situation at-hand, the technology cannot reason. This makes such an approach code-intensive.

Existing Fuzzy Computing Technology

People can draw conclusions that are imprecise.

"The car is travelling very fast."

"The box feels kind of heavy."
"The room seems to be cooling."

Fuzzy computing technology captures this.

Because fuzzy technology uses fuzzy mathematics and the very linguistic terms that people use to summarize numerical data—"fast", "heavy", "cool"—it can, like people, compute with imprecise measurement-based information [2]. This makes a fuzzy approach less code-intensive. And the method is beginning to learn to compute with concepts where the perception-based information cannot be quantified—"honest", "reliable", "risky". It uses fuzzy logic to infer conclusions—the same experienced-based rules people use when they reason intuitively.

But fuzzy technology needs a complete set of overlapping rules otherwise it cannot reason.

NEW FUZZY COMPUTING TECHNOLOGY

Even though existing IT systems display a rudimentary intelligence, they are far from doing what people must do regularly in order to survive. That is, use judgement.

Making Judgement Calls

People hypothesize. They use their wits to extrapolate from experience. People do not need to wait until they have experienced every contingency—or all the evidence is in—to form an opinion, or take action. Even if the current situation does not match experience, or they have only one or two experiences to go on, people form immediate impressions by hypothesizing. Then they modify those impressions as they gain more experience, or as more evidence comes in. This is an important and pervasive survival skill—central to making a judgement call. This gives humans a unique autonomy. They can act and react quickly, and they can learn—but there are risks.

The Risks

First, much of the information used in making judgement calls is "soft". That is, it is imprecise, vague, ambiguous or probabilistic. It is subject to miscalculation, misinterpretation, and misunderstanding—and to conflicting interpretations.

Secondly, assessing information is inherently arbitrary. It depends on the degree to which the decision-maker "believes" something is "true". This contributes to the risk of a misjudgment.

There are also subtle, but significant risks that come from both belief-based and plausibility-based reasoning [3]. In belief-based decision-making, actions are taken based on the preponderance of evidence supporting an assertion. In the real world, strong actions can lead to strong reactions, and an incorrect assumption that a condition is true can lead to disaster. It may turn out that the cure is worse than the condition.

Plausibility-based decision-making, in contrast, is based on "reasonableness". The less evidence against an assertion the more plausible, or reasonable, it seems. An incorrect assumption that a condition is false can also lead to disaster. Small causes can have large effects, or lead to unexpected consequences.

Finally, some decision-makers make better judgement calls than others, or they are less affected by mood swings or the time of day.

Because businesses and corporations are completely dependent on the judgement of their decision-makers, any of these cognitive limitations [4] can put them at risk.

A NEW ALGORITHM

An algorithm has been invented [5] that captures the cognitive processes that people use when they make decisions. A new fuzzy implication operator drives the inferencing process [6]. The engine can extrapolate consistent and mathematically rigorous conclusions from one or two rules, and it can do it with minimum information.

The Daams Implementation Operator

The Daams implication operator provides a mathematical tool to duplicate the intuitive reasoning that people use when they make judgement calls. It generalizes the fuzzy entailment rule of inference [7] to include situations where the current circumstance and experience do not match exactly and the decision-maker only has one or two heuristics to go on.

The Mathematics

An explicit assumption of continuity is used to generate a fuzzy implication operator that yields an envelope of possibility for the conclusion. A single fuzzy rule A B entails an infinite set of possible hypotheses A' B' whose degree of consistency with the original rule is a function of the "distance" between A and A' and the "distance" between B and B'. This distance may be measured geometrically or by set union/intersection. As the distance between A and A' increases, the possibility distribution B* spreads further outside B—somewhat like a bell curve—corresponding to common sense reasoning about a continuous process.

The manner in which this spreading occurs is controlled by parameters encoding assumptions about (a) the maximum possible rate of change of B' with respect to A'; (b) the degree of conservatism or speculativeness desired for the reasoning process, and (c) the degree to which the process is continuous or chaotic.

In general terms, the mathematics [8] used in this invention bridges the gap between non-matching rules and rule inputs by creating envelopes of possibility for an output—the output having different shapes and rates of spreading where the rate of spreading is a function of the distance between user input and rule output. The desired shape of the envelope of possibility is a system parameter determined at set up by an expert. The similarity between the user input and the rule input may be measured by existing measures, or by a novel measure.

The rate of spread is a function of the dissimilarity between the user input and the expert determined rule input. It may also depend on the location of the input in the input space, or other parameters of the input and the rule input. In multidimensional inputs, a weight function makes it possible for one input dimension to "compensate" for another.

The invention also provides a way to eliminate the requirement of a complete set of overlapping rules. It is possible to calculate the degrees of similarity between disjoint fuzzy sets using a distance function in order to interpolate or extrapolate from sparse rules. Fuzzy limits can be set on the vaguely known possible rate of change. It is possible to reconcile contradictory inputs, and choose the appropriate pattern from which to interpolate or extrapolate.

This invention also provides a solution to the problem where the concepts of belief and plausibility are only applied to assertions and not to propositions. Using the kernel of the new fuzzy implication operator, one can arrive at a degree of plausibility for an entailed proposition and an envelope of possible conclusions for a given input.

Using set intersection or other distance measures, the strength of the chain of evidence and reasoning linking the data to the conclusion can be calculated, and an envelope of belief obtained. The difference between the envelopes of belief and possibility measures all the vagueness, uncertainty gaps, contradiction, and probabilistic nature of the rules and the input data as well as the mismatch between the inputs and the rules inputs. The degree to which an assertion is proven and the degree to which it is merely possible can be quantified.

Embedding the algorithm in existing IT creates a new IIT that can be used to make decision-making more consistent and mathematically rigorous across the enterprise. This gives businesses and corporations increased autonomy by making them less vulnerable to the risks inherent in the judgement calls of people.

MAKING BUSINESSES AND CORPORATIONS MORE AUTONOMOUS

At the macro-level businesses and corporations exhibit the characteristics of autonomous, self-organizing systems [9]:

1. **Companies create themselves.** They begin as a gleam in the eye of an entrepreneur, and grow up from there.

2. **Companies regulate themselves.** The use structures such as boards of directors, and executive committees to formulate policies and plans.

3. **Companies operate themselves.** They have structures such as plants and branch offices to implement policy and carry out procedures and operations. HR departments look after the employees.

4. **Companies maintain and repair themselves.** Structures like maintenance departments replace equipment and light bulbs. Managers conduct performance reviews. Hiring and firing committees replace employees.

5. **Companies reproduce themselves.** They franchise themselves or open new branches in the corporate image.

6. **Companies learn.** They build new and innovative structures in response to changing circumstances—new or modified policies, procedures, updated IT systems, robotized operations, new international corporate headquarters.

The result of this self-organization is a set of complex social inventions—apparently autonomous, but at the same time totally dependent on the people that serve them. It is people who negotiate the creating, regulating, operating, maintaining, repairing, reproducing, and adapting of corporate structures [10]. This means that people are to social organization as leaves are to a tree, or neurons to a nervous system—networks of individual, self-similar units that sustain the structure. It is the collective behaviour of these networks that defines the dynamics of the system.

Leaves, neurons, and people carry out similar processes in the systems they inhabit. Leaves photosynthesize. Neurons process and transmit messages. People make decisions. In each case, the individuals act together to the benefit of the system—biological or social, but there is a difference. Leaves faithfully repeat the same process as they photosynthesize. Neurons repeat the same chemical processes over and over as they transmit messages. But, even though the process is the same for all decision-making, people are not as conscientious about following the steps. This puts companies at risk.

The new algorithm captures decision-making in a structured way. By incorporating it into existing IT structures, IIT systems can be created that make decision-making processes more consistent and reliable. Used by decision-makers at specific decision points across the enterprise, this IIT technology can make businesses less vulnerable to the inconsistencies of individual decision-makers, and, therefore, more autonomous.

Even though the process of decision-making is similar for all decisions, its specific application is determined by context—in the same way that some neurons populate the motor-sensory system, and some the brain. In the instance that follows, a health worker uses the technology to draw conclusions and make decisions about an industrial injury case.

AN INSTANTIATION

Making Disability Case Management More Reliable

In this instance, the application of the algorithm is in the field of disability case management. The decision-maker is a caseworker. The process begins with the inputs. First the caseworker chooses a primary factor for evaluation. This generates a set of secondary factors.

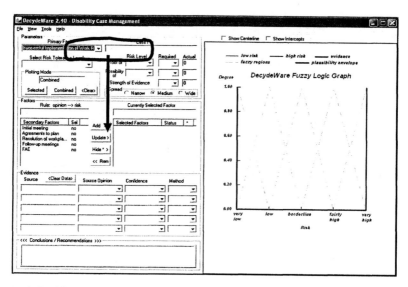

Figure 1. In this case the case manager chooses one of four primary factors, "Successful implementation of return to work plan". The "Secondary Factors" list is automatically populated with the five relevant secondary factors.

Next the user sets the risk tolerance level—the degree of proof required, the limit on the allowable degree of outside risk, the strength of evidence required, and, the rate of spread of the plausibility envelope. Then the logic is established. The user completes the inputs by collecting and entering the evidence. Given the inputs, the engine generates the outputs. It calculates the risks based on the evidence, and displays them graphically. Finally, the technology draws conclusions and makes recommendations on a course of action.

Inputs

1. Picking the factors to be used:
The decision-maker must first decide, "What are the factors to be addressed in the evaluation?"

2. Setting the risk tolerance:
Corporate decisions can range from very conservative to very speculative. Risk tolerance parameters are either set intuitively or according to company policy.

94 FUZZY SYSTEMS

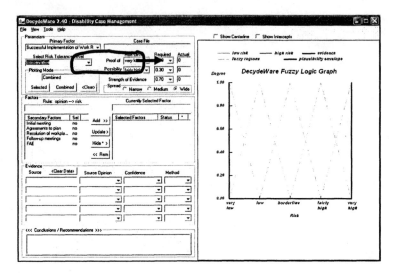

Figure 2. In this case, the caseworker sets the risk tolerance according to company policy, which is conservative. The settings for required proof, allowable outside risk, spread of envelope of possibility, and strength of evidence are set automatically.

3. Choosing the logic:

The corporate decision-maker chooses the informal "if...then" logic to be used to arrive at a conclusion. This logic can be based on experience, knowledge, expertise, or company policy. If experience or expertise is limited, there may only be one or two rules-of-thumb with which to reason.

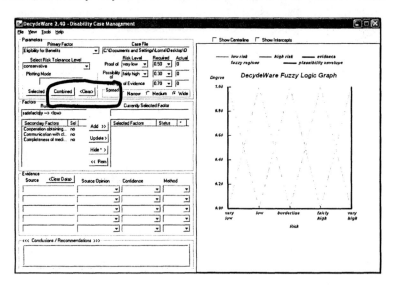

Figure 3. In this case there is one rule. Here the caseworker goes with the default rule, "If, in the opinion of the **Source**, the **Factor** is satisfactory, then the risk is low. The logic can be changed if necessary.

Intelligent Information Technology: From Autonomic to Autonomous Systems

4. **Collecting and entering the evidence:**
   Next, the decision-maker collects and enters evidence on each of the factors. The information could be hard, i.e., precise, clear, unambiguous; or soft, i.e., imprecise, fuzzy, ambiguous. Encoding is done in numbers, i.e., quantitative, statistical; or in words, i.e., verbal or written opinions, hunches, points-of-view, intuition.

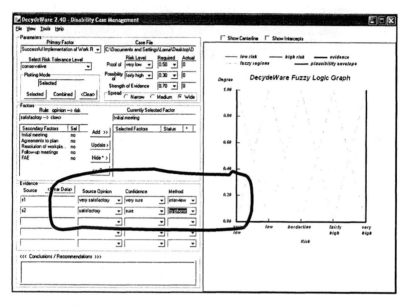

Figure 4. Here the caseworker has collected 2 pieces of evidence. In this case, the information is soft and encoded in words. Asking the same question of every respondent is a tenet of good enquiry.

Outputs

5. **Summing up the evidence:**
   Normally corporate decision-makers do much of the summing up in their heads. They mentally combine the mixture of hard and soft information that they have collected, i.e. the opinions, hunches, points-of-view, rumor, guesstimates, numbers and statistics, then compare it to experience, and form an opinion about the current situation.

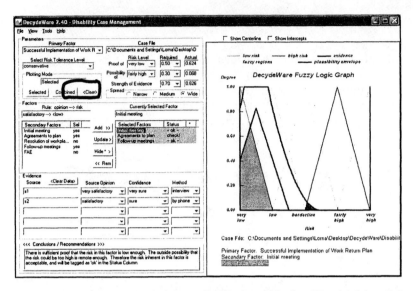

Figure 5. The outputs from the engine are displayed graphically. The risk inherent in the selected secondary factor "Initial meeting" is shown. The blue triangle is the aggregated evidence. The solid green overlap is the amount of proof for "very low" risk. The solid red is the degree to which "fairly high" risk is possible. The conclusion is that the risk is acceptable. The risk status is green, or "OK".

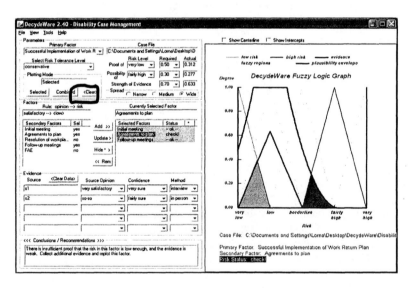

Figure 6. This window shows the risk inherent in the selected secondary factor "Agreement to plan". Again, the blue triangle is the aggregated evidence. The solid green overlap is the amount of proof of "very low" risk. The solid red is the degree to which "fairly high" risk is possible. The conclusion is that the risk inherent in this secondary factor is unacceptable. The risk status is red, or "Check".

Intelligent Information Technology: From Autonomic to Autonomous Systems 97

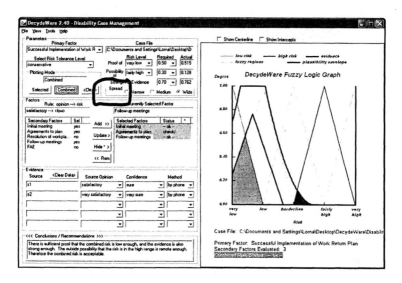

Figure 7. This window shows the combined risk, i.e. the risk inherent in the primary factor, "Successful implementation of return to work plan". This time the blue triangle is the combined risk from two secondary factors. The solid green overlap is the amount of proof for "very low" risk. The solid red is the degree to which "fairly high" risk is possible. The conclusion is that the combined risk inherent in this primary factor is acceptable. The risk status is green, or "OK".

6. **Drawing the conclusions:**
   Decision-makers use professional judgement to draw conclusions. The reasoning is generally based on experience. Sometimes the current situation matches experience exactly, and the conclusions are easy. Sometimes the two do not match exactly, and some hypothesizing—or speculation—is necessary. The less experience and/or the weaker the evidence the more speculation is required.

SUMMARY OF INSTANTIATION

The instantiation described here is an application of the algorithm to a specific operation in an organization. In this case, it was embedded in existing IT to standardize disability case management around exemplary practice. This new IIT has made a specific category of decisions more consistent, reliable and transparent. Now, the organization is less vulnerable to poor judgement in its disability case management department.

But, every minute of every day thousands of conclusions are drawn, and decisions made, across a business or corporation. A few are made by technology. Most are made by people. Some are made in ordinary circumstances—as part of daily routines. Others are made in extraordinary circumstances—where there are new challenges, or the organization is threatened. As noted earlier, whether routine or non-routine, each decision is subject to the same process. This is the key to helping businesses and corporations become more autonomous in general, i.e. less dependent on the arbitrary judgement calls of their decision-makers.

The goal requires generalizing, to decision points across the enterprise, the IIT that made the specific operation of disability case management more consistent and mathematically rigorous.

CONCLUSION

Generalizing the Instantiation

If businesses and corporations are—like their inventors—to survive on their reflexes and wits, internal information-processing structures must to be constructed that allow them to live relatively autonomously in a world that is inexact and full of the unexpected. This is a tall order.

Yes, companies will need the equivalent of involuntary and motor-sensory nervous systems—systems that keep the internal works of businesses and corporations ticking—and the machines of production operating by themselves.
People have started combining mathematics and silicon to build such structures. Grid computing, neural nets and fuzzy systems are used together and separately to discover patterns, mine databases, let IT systems run themselves, control industrial processes, and operate machines. But these structures can only react. They have a limited, pre-programmed repertoire of behaviours—behaviours supplied by experts. They need a complete set of rules. They cannot improvise.

If businesses and corporations are really to become more autonomous, they will need IIT networks that help them think, not just react. IIT that corporate decision-makers can work with to make more consistent and reliable judgement calls, no matter where or what the circumstances. IIT that can be adapted to any decision-making situation—on-the-ground, and in real-time—and the IIT should not be code-intensive.

New IIT Structures

Each person in a business or corporation is a decision point—sometimes deciding alone—sometimes deciding collaboratively. Whether alone, or in groups, each person contributes to a complex decision-making network. Currently, organizations manage this complexity by self-organizing into decision-making hierarchies. At the top are the people who plan, make policy, and design procedures. Next are the people who supervise the implementation of plans, policies and procedures. Finally, there are the people who actually carry out the plans, policies and procedures. The fact is that the decision-making process is the same at the top as it is at the bottom—only the nature of the decisions vary. This is the key to building IIT networks that will help make businesses and corporations more autonomous.

The algorithm and the embedded mathematics described in this paper open the door to such networks. They make it possible to construct a computing unit that captures the ability of people to hypothesize, to extrapolate or infer conclusions from one or two experiences or before all the evidence is in—and it does it with consistency and mathematical rigor without being code-intensive. In principle, thousands of these computing units or cells can be combined—like leaves or neurons—to make IIT networks that are congruent with the decision-making hierarchies of the business or corporation. And, borrowing further from the biological analogy, these computing cells can be plugged in anywhere, any time, and as many times as necessary—as trees grow leaves, or neurons grow synapses—to keep the system ticking, thinking, and learning. Used in practice, such IIT networks would make decision-making across the enterprise more consistent, rigorous, and less dependent on the arbitrary judgement of its decision-makers—and by doing so, move toward the goal of autonomous businesses and corporations.

REFERENCES

[1] Paul Horn, 2001. Autonomic Computing, www.ibm.com/research/autonomic.
[2] Lotfi Zadeh, 2000. Toward the Concept of Generalized Definability, Lecture presented at the Rolf Nevanlinna Colloquium, University of Helsinki, Helsinki, Finland.
[3] J. Daams and L. Strobel Stewart, 1997. Fuzzy Logic and the Uncertain Art of Audit Enquiry, Proc. Seventh IFSA World Congress, Prague.
[4] Refer to "bounded rationality" as defined in W.R. Scott, 1981. Organizations—Rational, Natural and Open Systems, Prentice-Hall New Jersey.
[5] N.S. Sutherland, 1992. Irrationality—Why We Don't Think Straight, Rutgers University Press, N.J.
[6] Letters of Patent issued March, 2003.
[7] J. Daams, 1999. Envelope of Plausibility Defined by a New Fuzzy Implication Operator, Proc. NAFIPS '99, New York
[8] C-T Lin, C.S.G. Lee, 1996. Neural Fuzzy Systems—A Neuro-Fuzzy Synergism to Intelligent Systems, Prentice Hall New Jersey.
[9] J. Daams and L. Strobel Stewart, 1999. Applying a New implication Operator to Professional Judgement in Risk Assessment, Proc. NAFIPS '99, New York
[10] E. Jantsch, 1980. The Self-Organizing Universe, Pergamon Press, New York
[11] Refer to the definition of process structures in Ilya Prigogine, 1980. From Being to Becoming—Time and Complexity in the Physical Sciences, W. H. Freeman and Company, San Francisco.

Design of a Fuzzy Controller for a Microwave Monomode Cavity for Soot Trap Filter Regeneration

M. PAPPALARDO and A. PELLEGRINO
Department of Mechanical Engineering, University of Salerno, Fisciano, Italy

M. D'AMORE, P. RUSSO and F. VILLECCO
Department of Chemical and Food Engineering, University of Salerno, Fisciano, Italy

ABSTRACT

As an innovative and efficient tool for thermal treatment, microwaves are attracting increased interest for several promising industrial applications. Among these, removal of a hydrocarbon from an inert matrix is one of these intriguing issues.

A microwave monomode cavity with a cylindrically-symmetric electromagnetic field has been designed and built to regenerate soot trap ceramic filters. Advantages arise from the possibility of selective heating of the irradiated materials and a reduced time to regenerate. However, perfect control of filter temperature is essential as temperature gradients can induce mechanical stresses and/or fractures that prove fatal to a long filter life.

The complexity of a system in which selective energy dissipation, combustion, mass and heat transfer phenomena interfere with one another to produce a highly-nonlinear, hard-to-model behaviour hinders application of conventional control such as PID or even adaptive control. Having selected a superficial filter temperature (measured variable) and microwave power (manipulated variable), i.e. the magnetron duty cycle, as control parameters, a system with variable set-point has been designed and realized based on the fundamentals of fuzzy logic.

INTRODUCTION

Filtration is one of the most widely used unit operations in chemical engineering adopted into different processes that range from separation of water from slurries to air cleaning. In the automotive industry, filtration is the most common way that the diesel engine exhaust is purified of carbon particulates generated during combustion.

Carbon particulates are one of the major drawbacks of the diesel engine. In spite of continuous improvements over the past few decades, these engines generate considerable soot, even in the new Common Rail technology. A filtration system is thus necessary to attain the future standards that will be required for vehicle emissions. A porous ceramic monolith is generally accepted as the most suitable material to use as a soot trap filter. However, progressive blocking affects the operation until the soot occupies all the channels suitable for gas passage, rendering the filter unusable. Periodic cleaning of the filter is thus necessary. Self-regeneration is not possible because of the high ignition temperature of the diesel soot, typically in the range of 773–923 K as compared to the exhause gas temperature, generally below 673 K. Therefore, an additional methodology is needed to purify the filter and remove the trapped soot.

Filter regeneration involves incineration which is typically done off-line taking the blocked filter out of production, putting it into an oven, and heating it to a required temperature. However, high temperatures and non-uniform heating of the filter can lead to breakage or melting [1].

Microwaves recently have come into attention as an innovative tool for selective, efficient heating. In fact, microwave heating greatly differs from conventional heating which is effectively a surface phenomenon, since the heat reaches the inner part of the material passing through the exposed surface. On the contrary, microwave heating is a volume phenomenon, since microwaves penetrate the material to a depth depending on the dielectric permittivity, and so the filter heats up from the inside-out. Different material permittivities lead to the possibility of selective heating. Water and soot are both strong absorbers of microwaves whereas most ceramics are virtually transparent to this form of radiation.

Microwave irradiation has been proposed as an effective means for particulate trap regeneration [2–7], since this method can overcome many of the difficulties encountered with other methods. In particular, instantaneous penetration of microwaves into the filter and their selective absorption makes the soot burn while the ceramic filter remains undisturbed.

The soot-filter-microwave system is fairly complex. On one hand, microwaves are dissipated by the soot, whereas the ceramic is almost transparent to them. The workload however influences the electromagnetic field itself which in turn changes the heat released by the energy dissipation. On the other hand, the heat of reaction due to soot combustion combined with heat transfer to determine the temperature profile inside the filter, and thus, determines the efficiency of regeneration. All this leads to a polyhedric situation where a coupled system of heat and mass transfer equations must face the uncertainty of parameters such as thermal conductivity, reaction rate constant, heat and mass transfer coefficients, dielectric permittivities of the materials and their change with the reaction. Any control system must thus overcome difficulties related to these interlinked parameter effects.

In this work, fuzzy temperature control has been designed and tested to assist a single mode microwave applicator to regenerate ceramic foam used as a filter for soot emissions from a gas-oil burner exhaust.

THE FUZZY CONTROLLER

Since heat losses are relatively high, temperature can be controlled through the amont of heat generated. As outlined above, heat is generated by microwave dissipation until the ignition temperature is acheved. The combustion reaction begins and heat is further released by burning of the soot. The idea for the controller is based on the fact that microwaves are generated by the magnetron at maximum power with its duty cycle controlled according to fuzzy logic [8–10]. The controlled parameter is the filter temperature. Two input variable are defined:

- A is the difference between the ith temperature desired temperature, which in turn, is a function of time:

$$A = T_i - T_i^* \tag{1}$$

- B is the difference between A evaluated at the $(i-1)^{th}$ time step and A at the ith time step

$$B = [(T_{i-1} - T_{i-1}^*) - (T_i - T_i^*)] \tag{2}$$

The output variable, C, is the time fraction for which the magnetron is switched on to generate the microwaves in a duty cycle of given length of time. As with a typical fuzzy control system, the rules adopted are logical rather than mathematical. The following blocks describe the states of the input variables and the way the system should consequently be varied according to the rules [11].

Table 1. List of the fuzzy rules adopted in the system.

Rules	Block antecedent	Block consequent
I	If A is negative large and B is positive large	then C has to be positive large
II	If A is negative medium and B positive medium	then C has to be positive medium
III	If A is negative small and B positive medium	then C has to be positive small
IV	If A is zero and B is zero	then C has to be zero
V	If A is positive small and B is negative medium	then C has to be negative small
VI	If A is positive medium and B is negative medium	then C has to be negative medium
VII	If A is positive large and B is negative large	then C has to be negative large

Seven labels have been used to define the conditions of the system for each variable as follows:

NL = negative large
NM = negative medium
NS = negative small
ZE = zero

PL = positive large
PM = positive medium
PS = positive small

For the input variables, triangular membership functions (MF) were used since our experience with this simple shape has been satisfactory. The membership function of the output variable is a singleton, i.e., a series of unit pulses as the center of gravity is easier to calculate when defuzzifying. A and B vary over the temperature interval -30 < T < +30 °C, while C ranges between 0 and 100%. Seven membership functions are assigned to A and C with five used with B. Overlap of these functions are essential since the system has a Fuzzy behaviour, not a Boolean one. Figures 1 through 3 show the fuzzy membership functions used for variables A, B, and C.

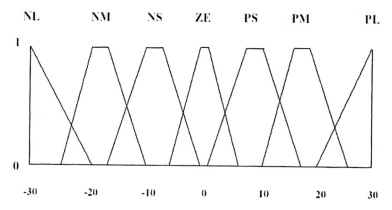

Figure 1. Membership functions of variable A.

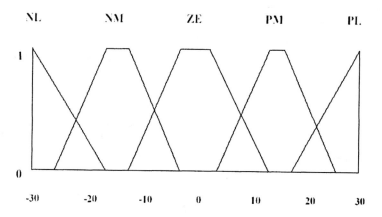

Figure 2. Membership functions of variable B.

104   FUZZY SYSTEMS

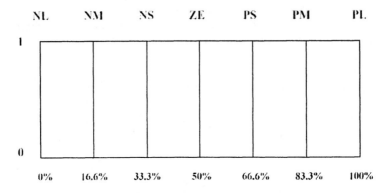

Figure 3. Membership functions of variable C.

In this system, where 7 MF have been adopted for A and 5 MF for B, we could have 35 input combinations and 245 different rules, using 7 output functions but in fact, the number of required rules is lower. In our case, a FAM (Fuzzy Associative Memory) matrix has been used. In the matrix, the first line and the first column report the input variables, whereas the output variable is derived according to the rule:

$$A \text{ and } B \rightarrow C$$

The matrix that combines these functions into the rules in Table 1 are shown in Table 2:

Table 2. FAM (Fuzzy Associative Memory) matrix.

A \ B	NL	NM	NS	ZE	PS	PM	PL
NL	PS	NS	NM	NL	NL	NL	NL
NM	PM	PS	NS	NM	NM	NL	NL
ZE	PL	PM	PS	ZE	NS	NM	NL
PM	PL	PL	PM	PM	PS	NS	NM
PL	PL	PL	PL	PL	PM	PS	NS

EXPERIMENTAL DETAILS

The experimental apparatus is briefly described together with the characteristics of the ceramic foam filter [12].

Apparatus

Figure 4 shows the monomode applicator used for the ceramic foam filter regeneration. The applicator consists of a stainless steel resonant cavity equipped with a 900W magnetron (2.45 GHz) as a power source and a thermocouple properly shielded. CO and CO2 concentrations at the reactor outlet are measured by NDIR continuous analysers (Hartmann & Braun Uras 10E). A paramagnetic analyser (Hartmann & Braun Magnos 6G) performed continuous monitoring of O_2 concentrations at the reactor outlet. Finally, signals from the analysers were acquired and processed by a personal computer.

The system and controller are independent. The temperature is continuously checked and kept at the desired set point by means of the fuzzy controller. The ceramic foam filter is placed inside the cavity which is connected to the outside world in two different ways. The first one is the feed gas inlet which is in correspondence with the microwave source. The second one is the exhaust gas outlet. The foam disk is placed in the microwave cavity so as to have its axis coincident with the cavity axis. Being, as above outlined, the electromagnetic field in the monomode cavity independent of axial coordinate—the position along the axis can be arbitrarily chosen.

The fuzzy controller is interfaced with the microwaves monomode cavity by a 16 input/2 output 200 kS/s 16-bit multifunction I/O DAQ National Instruments card inserted into a laptop computer.

Materials

The filter was a 76 mm diameter, 15 mm thick alumina foam disk (Vesuvius Hi-Tech Ceramics) with 92 % porosity and 65 pores per inch (Figure 5). Soot particulate was generated and deposited directly on the filter from the exhaust of a gas-oil burner.

Procedure

The filter was placed inside the cavity at room temperature. Then the cavity was sealed and microwave irradiation began. A given programmed temperature rise is imposed on the system through the fuzzy controller. Runs were performed at various oxygen concentrations. Gas concentrations were measured at the outlet and used to monitor the regeneration process.

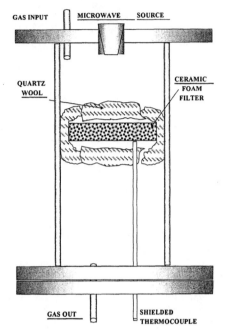

Figure 4. The single mode microwave cavity.

Figure 5. An image of the ceramic foam disks.

RESULTS

Figure 6 shows two different heating runs performed in nitrogen atmosphere on a ceramic filter with a soot deposit. The first curve (circles) is a programmed 10 C°/min temperature raise. The experimental points lie on the set point curve. The same agreement between the programmed temperature curve and the experimental results has been obtained in the second test run (triangles), where a temperature plateau was required and achieved after an initial 30 C°/min programmed temperature rise up to 550 °C. The plateau holds until the magnetron is purposely turned off.

Figure 7 shows the results obtained during regeneration of a catalytic filter in the microwave cavity controlled by fuzzy logic. In the diagram the concentrations of O_2, CO_2 and CO are reported in the exhaust as a function of time. It is interesting to note the symmetry of the CO_2 and O_2 curves, Moreover, the end of regeneration is clearly revealed by the gas concentration values. The programmed temperature curve is also shown.

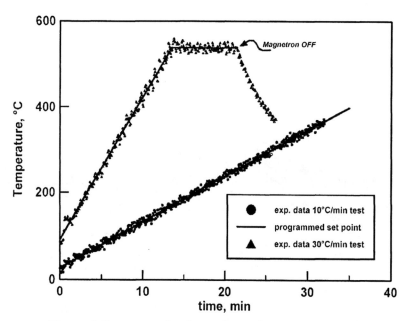

Figure 6. Programmed and experimental temperature curves.

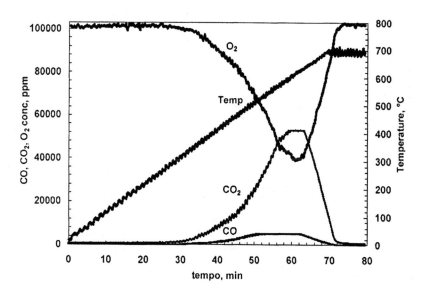

Figure 7. Filter temperature and O_2, CO_2 and CO levels in the exhaust as a function of time.

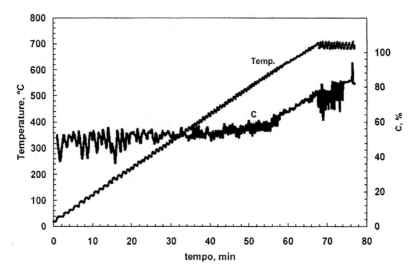

Figure 8. Filter temperature and active fraction of the magnetron duty cycle as a function of time.

In Figure 8, the fraction of the duty cycle for which the magnetron is on, is reported as a function of time and compared with the temperature curve. Note that when the temperature reached 700 °C and the soot has been almost completely burned out, the magnetron was kept on for more than 80% of the time to guarantee the required temperature level was met.

CONCLUSIONS

In spite of the complexity of the system monomode microwave cavity—soot dirty filter, the fuzzy controller gave excellent answers to all requests. The temperature set-up curve can be any type and can be changed as necessary, as this information is directly input to the control program. The filter temperature is perfectly controlled, as long as the heat losses can be counterbalanced by the microwave dissipation. Should fast combustion take place due to soot ignition at high temperature, it is easy to add a second fuzzy control level based on the reacting gas concentration. The controller is actually a virtual one making it versatile and inexpensive. Filter regeneration takes place under the desired operating conditions preserving the integrity of the ceramic filter.

REFERENCES

[1] E.T. Thostenson, T.W. Chou, 1999. Composites: Part A 30:1055–1071.
[2] A. Metaxas, R. Meredith, 1993. Industrial Microwave Heating, IEE Power Eng. Ser.4, 130.
[3] J.P.A. Neeft, 1995. PhD Thesis, TU Delft, Delft, The Netherlands.
[4] F.B. Walton, P.J. Hayward, D.J. Wren, 1990. SAE Paper 900327.
[5] C.P. Garner, J.C. Dent, 1989. SAE Paper 890174.

[6] Z. Chunrun, M. Jiayi, C. Jiahua, L. Lunhui, L. Junmin, L. Chengbin. SAE P941774.
[7] Z. Ning, Z. Guangiong, L.Yong, L. Junmin, G. Xiyan, L. Iunhul, C. Jiahua. SAE P952058.
[8] J. Ma, M. Fang, P. Li, B. Zhu, X. Lu, N.T. Lau, 1997. Appl. Catal. A: General 159: 211.
[9] M. Gautam, S. Popuri, B. Rankin, M. Seehra, 1999. SAE Paper 1999-01-3565.
[10] H. Zimmermann, 1992. Fuzzy Set Theory and its Applications, Kluwer Academic Pub.
[11] L.X. Wang, 1994. Adaptative Fuzzy Systems and Control—Design and Stability Analysis, PTR Prentice Hall.
[12] M. Sugeno, 1985. Industrial Application of Fuzzy Control, North Holland.
[13] L. A. Zadeh, 1992. The calculus of fuzzy if-then rules, AI Expert, March.
[14] V. Palma, M. d'Amore, P. Russo, A. D'Arco, P. Ciambelli, 2002. Comb. Sci. Tech. 174(11&12): 295–308.

Optimal Operating Parameters in Fuzzy-Assisted Microwave Removal of Carbon Particulate from Ceramic Filters

V. PALMA

Department of Chemical and Food Engineering, University of Salerno, Fisciano, Italy

A. PELLEGRINO

Department of Mechanical Engineering, University of Salerno, Fisciano, Italy

S. ROBUSTELLI and F. VILLECCO

Department of Chemical and Food Engineering, University of Salerno, Fisciano, Italy

ABSTRACT

Ceramic filters are of basic importance in the abatement of pollutants in the exhaust of combustion processes in both industrial plants and mobile sources. Increasing filter life is one of the major issues in the ceramic industry. A method to clean-up a filter without affecting the ceramic support is the subject of a number of research projects. Microwaves treatment of filters has been tested recently as a possible way to remove carbon particulate from the ceramic filter by selective heating based on the tremendous difference between ceramic and soot dielectric characteristics.

A microwave monomode cavity, with a specially designed fuzzy temperature control, has been used to regenerate catalytic and non-catalytic ceramic filters previously used to trap the soot in the exhaust of a diesel powered engine. The process efficiency has been measured on-line using gas (CO, CO_2, O_2) concentration analyses and mass balances. Risk of structural failure by thermal stresses was minimised by attaining a compromise between regeneration speed, i.e., soot reaction rate and control of the ceramic temperature.

INTRODUCTION

The interest of the worldwide market in the diesel engine is witnessed by increasing diffusion into the marketplace due to the characteristics of economics and safety. However, these vehicles still suffer from a major drawback related to their exhaust which is claimed to be potentially carcinogenic because of the presence of soot particulates [1]. Both engine modifications and after-treatment filtration of the exhaust are ways to reduce particulate matter [2–6]. Engine modifications with development of the Common Rail technology characterised by higher fuel injection pressure and better control of the injected fuel volume, has greatly contributed to the conformity to legislative demands. However, development of a filtration system will be necessary to meet the 2007 standards for particulate matter reduction.

A number of devices, including honeycomb ceramic monolith, ceramic foam, fibrous mesh, candle filter have been extensively used for this purpose. However, as they get progressively blocked, a rise in the total backpressure penalizes the engine's performance. Hence, periodic cleaning of the filter is necessary. It must be noted that self-regeneration is not possible because of the high ignition temperature of diesel soot, typically in the range of 773–923 K, compared to the exhaust temperature generally below 673 K. Therefore, an additional method is needed to purify the filter of the trapped soot.

One possible method is thermal incineration by fuel burners or electrical heaters to bring the collected soot up to the ignition temperature. However, some intrinsic difficulties are still to be overcome. For example, the fuel burner needs a set of ancillary equipment that is fairly complex and expensive, such as fuel injector, air supplier, ignitor, exhaust by-pass and control system. On the other hand, the electric heater has energy utilisation and regeneration efficiency lower than the filter. Moreover, the power supply (in the order of kilowatts) is hard to meet using a vehicle battery. As a matter of fact, control and location of energy input to a ceramic filter are of basic importance for successful filter regeneration. External heat sources such as fuel burners or electrical resistance heaters initiate soot combustion at the inlet face of the filter. In this case, the high temperatures required for soot oxidation and the uneven or uncontrolled combustion can cause either incomplete regeneration and filter breakage or melting [3].

Among various proposed applications [7] microwave heating has recently come into attention as an innovative tool to accelerate chemical reactions [8,9]. Microwave heating takes advantage of the characteristics of some materials that can quickly dissipate the electromagnetic energy. Microwave heating differs greatly from conventional heating. This difference can be seen as a surface phenomenon since the heat reaches the inner part of the material by passing through the exposed surface. On the contrary, microwave heating is a volume phenomenon, since microwaves penetrate the material at depth depending on the dielectric permittivity, thus heating the material from the inside-out. Furthermore, the different permittivity of materials leads to the possibility of selective heating.

Filter regeneration by microwave heating should be an effective operation. Soot is a strong absorber of microwaves whereas most ceramics are virtually transparent to them. Therefore, it is possible to provide a more uniform energy input to the soot trapped on the filter using microwaves to heat it up to the ignition temperature. This method allows improved control of soot combustion and lower thermal stresses inside the filter [10-14]. Preliminary tests performed in a monomode microwave cavity showed the feasibility of this method for the regeneration of non-catalytic [15] and catalytic filters [16] and the advantages of using microwaves.

The special characteristic of microwave ignition is the almost instantaneous response of soot to electromagnetic energy. Because of the large heat exchange surface, soot dispersion on the filter resolves into a large heat loss, the filter temperature can thus be controlled by manipulating microwave generation. The entire system is fairly complicated: heat is generated by microwave dissipation and by soot combustion, whereas heat transfer is due to both convection and conduction. Moreover, dissipation is strictly related to the filter soot load that in turn, influences the electromagnetic field. All of the above strongly justifies the choice of a fuzzy controller which can overcome the difficulties of mathematically and physically modelling such a system.

Experimental Details
Materials
The filter was a 76 mm diameter, 15 mm thick alumina foam disk (Vesuvius Hi-Tech Ceramics) with 92 % porosity and 65 pores per inch. A commercial amorphous Carbon Black CB330 (DEGUSSA) was used as soot substitute in preliminary tests. Soot particulate was generated and deposited directly on the filter at the exhaust of a gas-oil burner.

Apparatus
The burner is a gas-oil burner for home heating equipped with a nozzle giving a gas-oil mass flow rate of 1.9 kg/h. [15] Commercially available gas-oil (H/C molar ratio of 1.75, S content of 0.05 wt%).) was employed. Air flow rate to the burner was adjusted in the range 28.5–45.6 m^3/h (STP) to obtain an air/fuel mass feed ratio (α) ranging from 23 to 37. Soot deposition over ceramic foam was carried out at the burner exhaust. For this purpose, the soot filter was placed into a stainless steel trap holder (20 cm long, 10 cm ID) located 2 m from the burner outlet. The overall burner exhaust was fed to the filter. An external heater allowed the trap holder temperature to be controlled. The trap inlet and outlet gas temperatures and the pressure drop through the trap were monitored by a K-type thermocouple and differential pressure gauge transducers respectively. The soot concentration (C_{BS}) was measured upstream and downstream of the trap by means of a previously calibrated opacimeter (Tecnotest).

In Figure 1, the scheme of the apparatus for filter regeneration is shown. It is described in details elsewhere [16]. It consists of mass flow controllers (Brooks) operating on each gas; a single mode microwave applicator; a NDIR continuous analyser (Hartmann & Braun Uras 10E) for measurements of CO and CO_2 concentrations at the reactor outlet, and a paramagnetic analyser (Hartmann & Braun Magnos 6G) for continuous monitoring of O_2 concentration. The carbon mass balance was verified within a 5% range for all tests. Finally, the signals from the analysers were acquired and processed by a personal computer.

The applicator basically consists of a stainless steel resonant cavity, equipped with a 900W magnetron (2.45 GHz) as a power source, a cooling fan, a specially designed fuzzy controller and a thermocouple properly shielded (Figure 1). The system and the controller are independent. The system temperature was continuously checked and kept under the desired set point value. The ceramic foam filter was allocated inside the cavity. The cavity was connected to the outside in two different ways. The first one is the feed gas inlet or the microwave source; the second one, facing the inlet, is the exhaust gas outlet. The foam disk was placed in the microwave cavity so as to have its axis coincident with the cavity axis. As outlined, the electromagnetic field in a single mode cavity is independent of axial coordinate and so, position along the axis can be arbitrary.

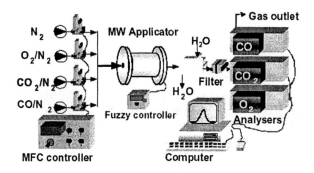

Figure 1. The application.

Fuzzy Control

The microwaves generator was able to regenerate the filter by dielectrical heating. Since heating must be performed gradually to reduce thermal stresses on the material, temperature control was crucial. Temperature measurements were acquired and elaborated on-line by a 16 input/2 output 200 kS/s, 16-bit Multifunction I/O DAQ Card National Instruments inserted into a laptop PC.

The control algorithm was based on the following steps:

- Temperature measurement and data acquisition
- Definition of the control action
- Output generation

Temperature Measurement and Data Acquisition

A K thermocouple was used for temperature measurements. To reduce noise, values used to define the control action were averaged on the basis of 30 data points.

Definition of the Control Action: The fuzzy controller

The measured variable was the filter temperature, whereas the manipulated variable was the magnetron duty cycle at maximum power. Fuzzy logic was used to control the filter temperature. The input logic rules used the last two temperature averages:

A—difference between the ith temperature and the set-up temperature which is a function of time

$$A = T_i - T_i^* \qquad (1)$$

B—difference between A evaluated at time $(i-1)^{th}$ and A at time i^{th}

$$B = [(T_{i-1} - T_{i-1}^*) - (T_i - T_i^*)] \qquad (2)$$

The following blocks thus describe the states of the input variables and the way the system should consequently be varied according to the rules described in detail elsewhere.

For the input variables, triangular membership functions were used. C was defined as the time fraction that the magnetron is on its duty cycle and seven functions were assigned to A and C, with five used with B; A and B vary over the interval $-30 < T < +30$ °C, while C ranges from 0–100%.

In our case 7*5 = 35 logic rules [17] were defined, i.e., all possible permutations of A and B. These rules are reported in a FAM (Fuzzy Associative Memory) matrix (see Table 1). In this matrix, the first row and first column refer to the inputs A and B, while output C is derived according to the rule:

$$A \text{ and } B \rightarrow C$$

Table 1. FAM Map used to Manipulate C according to A and B.

A\B	NL	NM	NS	ZE	PS	PM	PL
NL	PS	NS	NM	NL	NL	NL	NL
NM	PM	PS	NS	NM	NM	NL	NL
ZE	PL	PM	PS	ZE	NS	NM	NL
PM	PL	PL	PM	PM	PS	NS	NM
PL	PL	PL	PL	PL	PM	PS	NS

NL = negative large
NM = negative medium
NS = negative small
ZE = zero
PL = positive large
PM = positive medium
PS = positive small

Output Generation

The fuzzy control determines the fraction of duty cycle that the magnetron was on. The program is fairly flexible since for a given cycle duration, the status of the magnetron can be varied as often as required by a voltage change (0–4 volt). This helps to simulate a cycle at constant power as the sum of a series of on-off steps which increases the stability of the system.

Experimental Procedure

Preliminary tests were performed in the microwave cavity without any control of temperature using the maximum microwave power. Tests were performed to investigate the influence of the electromagnetic field on the thermocouple used for temperature measurement. A beaker containing 50 ml of water, positioned in the cavity, was heated by microwave while the gas temperature was monitored by a K-type thermocouple. The same test was carried out with acetone. After a transient period due to the achievement of saturation conditions, the measured temperature reached a constant value corresponding to the boiling temperatures of water and acetone, respectively 373 and 330 K. The result outlines the feasibility of the thermocouple use in a microwave cavity.

Eventually, temperature programmed combustion tests of the soot deposited on the filter were performed in the single mode cavity using the previously described temperature controller. The inlet gas flow rate was maintained at 2000 Ncm^3/min. The inlet gas stream contained 0–5 vol% O_2 in nitrogen. In each test, the temperature was raised from room temperature up to 973 K using a heating rate of 2 to 30 K/min. The operating pressure was 101 kPa. CO and CO_2 concentrations at the reactor outlet were measured by NDIR continuous analysers (Hartmann & Braun Uras 10E). Carbon mass balance was verified within a 5% tolerance range for all tests. A paramagnetic analyser (Hartmann & Braun Magnos 6G) performed continuous monitoring of O_2 concentrations at the reactor outlet. Finally, signals from the analysers were processed by a personal computer.

RESULTS

Temperature programmed regeneration tests by microwave heating were performed in the presence of oxygen and with a heating rate of 10 K/min on both catalytic and non-catalytic filters in the monomode microwave cavity using the fuzzy control system described elsewhere.

In Figure 2 the outlet concentration of CO, CO_2 and O_2 and the temperature evolution as a function of time are reported for the non-catalytic case. Figure 3 shows the fraction of time that the magnetron is on, as a result of the action of the fuzzy controller together with the relevant programmed temperature curve.

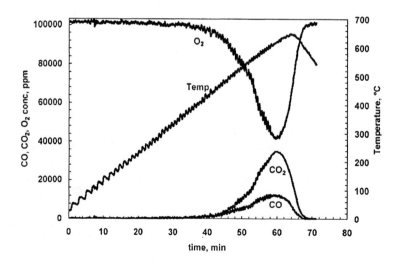

Figure 2. Temperature programmed test of soot loaded ceramic foam: non catalytic case.

At temperatures above 650 K, CO and CO_2 concentrations in the outlet gas can be seen to increase whereas O_2 decreases confirming the occurrence of soot oxidation. When the temperature reaches the maximum value of 850 K, CO and CO_2 concentrations were also at a maximum which corresponds to the minimum in the oxygen outlet concentration. After this point, the conversion rate of soot oxidation decreases indicating that the oxidation rate and the ultimate temperature reached are functions of the mass of soot present on the filter. It was found that when the soot mass was too low, the filter could not attain a temperature of 850K and in some cases, the ultimate temperature was below that at which soot oxidation occurs. Nevertheless, regardless of the soot load, the temperature followed the programmed curve as required. A comparison of the effectiveness of regeneration in the case of a soot loaded catalytic filter was also performed under the same operating conditions although a higher amount of soot was used.

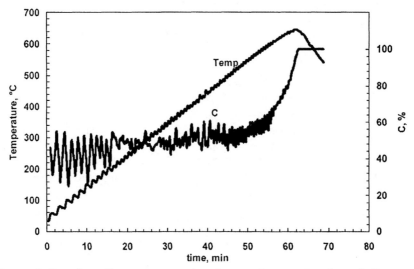

Figure 3. Fraction of magnetron active duty cycle as a function of time: non-catalytic case.

The results in Figure 4 indicate that a catalytic filter promotes soot combustion at a reduced ignition temperature of about 120 K with respect to a non-catalytic filter. The catalyst enhances the rate of soot oxidation over the entire range of temperature investigated. Moreover, the catalyst strongly increases the production carbon dioxide with respect to carbon monoxide, allowing a CO_2/CO ratio eight time greater than that of the noncatalytic filter. These findings are in complete agreement with previous results concerning the effect of the presence of the catalyst on the soot combustion rate obtained by experiments carried out in a conventional heating system [18–21].

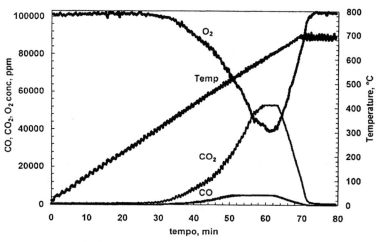

Figure 4. Temperature programmed test of soot loaded ceramic foam: catalytic case.

Figure 5 shows for the catalytic case, the fraction of time that the magnetron was on and the programmed temperature curve over time. Note that the filter temperature remained at the desired level of 700 C even after the soot was completely burned out. In fact, the catalyst itself dissipates the microwaves generating the heat necessary to keep the filter temperature at the desired value. This can be also seen by comparing the curve representing the active fraction of the magnetron duty cycle to the same curve for the non-catalytic case in Figure 4. The catalyst continuously heats the filter smoothing the effects of ignition/extinction due to the magnetron switching on and off.

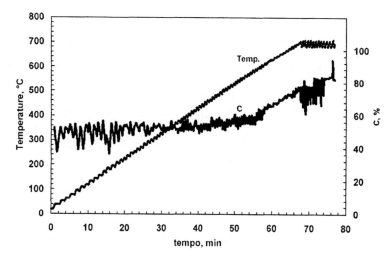

Figure 5. Fraction of magnetron active duty cycle as a function of time: catalytic case.

CONCLUSIONS

Soot oxidation with microwaves is shown to be very fast. Filter regeneration, even under a cold reacting gas, is completed faster than that attained by conventional methods. A catalyst helps filter regeneration since it allows the soot combustion to begin at a lower temperature, thus reducing the thermal stresses to which the filter is subjected. The apparatus in the present arrangement is perfectly controlled by the designed fuzzy system acting on the microwaves generator duty cycle.

REFERENCES

[1] J.L. Mauderly, 1992. Human exposure and their health effects, in Environmental toxicants, Ed. M. Lippmann, van Nostrand Reinhold, New York, 119.
[2] S.T. Gulati, 1998. in Structured Catalysts and Reactors, A. Cybulski, J.A. Moulijn (Editors), Marcel Dekker Inc., New York, 1998, 510.
[3] J.P.A. Neeft, 1995. PhD Thesis, TU Delft, Delft, The Netherlands.
[4] M.V. Twigg, A.J.J. Wilkins, 1998. in Structured Catalysts and Reactors, A. Cybulski, J.A. Moulijn (Editors), Marcel Dekker Inc., New York, 1998, 91.
[5] Y. Teraoka, K. Nakano, W. Shangguan, S. Kagawa, 1997. Catalysis Today 27: 107.
[6] P. Stobbe, J.W. Høj, 2000. European Patent No. W O0014639A1.
[7] D.M.P. Mingos, D.R. Baghurst, 1991. Chem Soc. Rev. 20, 1.
[8] A.K. Bose, M.S. Manhas, B.K. Danik, E.W. Robb, 1994. Res. Chem. Intermed., 20, 1.
[9] L. Seyfried, F. Garin, G. Marin, J.M.. Thiebaut, G. Roussy, 1994. J. Catal. 148, 281.
[10] F.B. Walton, P.J. Hayward, D.J. Wren. SAE P 900327.
[11] C.P. Garner, J.C. Dent. SAE P 890174.
[12] Z. Chunrun, M. Jiayi, C. Jiahua, L. Lunhui, L. Junmin, L. Chengbin. SAE P 941774.
[13] Z. Ning, Z. Guangiong, L. Yong, L. Junmin, G. Xiyan, L. Iunhul, C. Jiahua. SAE P 952058.
[14] M. Gautam, S. Popuri, B. Rankin, M. Seehra. SAE P 1999-01-3565.
[15] V. Palma, M. d'Amore, P. Russo, P. Ciambelli, 2002. Comb. Sci. Tech. 174(11), 295–308.
[16] P. Ciambelli, V. Palma, P. Russo, M. D'Amore, 2002. TOCAT-4, 4th Tokyo Conference on Advanced Catalytic Science Technology, Tokyo, Japan, July 14-19.
[17] L. A. Zadeh, 1992. The calculus of fuzzy if-then rules. AI Expert, March.
[18] P. Ciambelli, V. Palma, P. Russo, S. Vaccaro, B. Vaglieco, 2001. Top. Catal., 16/17, 279.
[19] P. Ciambelli, V. Palma, P. Russo, S. Vaccaro, 2002. Catal. Today 73, 363.
[20] V. Palma, M. d'Amore, P. Russo, A. D'Arco, P. Ciambelli, 2002. in press, Comb. Sci. Tech.
[21] P. Ciambelli, P. Corbo, M.R. Scialò, S. Vaccaro, 1990. Thermochimica Acta, 162, 83.

Fuzzy Probability for Modelling Particle Preparation Processes

A. WALASZEK-BABISZEWSKA
University of Zielona Gora, Institute of Control and Computation Engineering, Zielona Góra, Poland

ABSTRACT

In many industrial processes the material to be prepared consists of a population of different types of particles. This is true in chemical and biochemical research, mineral preparation processes, numerous food processes, etc. In a traditional approach, the entire characteristics of a material can be symbolized by a certain multidimensional random variable $X \in R^r$ containing r components $X_1, X_2,...,X_r$, that each represent a certain selected property of a single particle in the population to be tested and a certain probability density function $f_r(x_1, x_2,...,x_r)$ which is non-negative and normalised such that:

$$\iiint_X ... \int f_r(x_1, x_2,...,x_r) dx_1...dx_r = 1 \qquad (1)$$

Material properties can feature different spatial and temporal variability during preparation. Considering the different conditions used in practice to prepare particles, a number of variations in the description are possible:

- *descriptions characterized by time-independent distributions of random variables applied to describe materials, separation products, mixture of materials, etc.;*
- *descriptions covering the dynamics of a preparation process.*

In this paper two views using fuzzy probability will be adopted to present these situations:

    a) *probability of a fuzzy even—as a crisp number, according to Zadeh,*
    b) *probability of an even—as a fuzzy set, according to Zadeh and Yager.*

Some example calculations and experimental results are included.

INTRODUCTION AND BASIC CONCEPTS

In many industrial processes, including physics and chemistry, it is considered that the population of elements are described by individual characteristics and a variety of these characteristics apply in different technological situations. For the observer, the aim of the analysis is the effect of the behavior of particles on a macroscopic scale. The properties of particles, essential in such processes, whose particles fall within the technological operations, represent a certain multidimensional random variable $(X_1, X_2,...,X_r) \in X = R^r$. What is also considered is the existence of a probability distribution, i.e., a certain non-negative and normalised density function $f_r(x_1, x_2,...,x_r)$ fulfilling Eq. 1.

Both in physics [2] and in industrial practice, the approach of a probability distribution of a certain parameter of an element can be replaced by the ensemble of particles with different values of their parameters in the same population. To put it another way, the probability $f(x)dx$ that the scalar variable X assumes the value from the interval $(x, x+dx)$, is equivalent to the quotient n/N, where n defines the power of subset of the particles, whose feature X determines the value over the interval $(x, x+dx)$, and N is the population size.

In addition it is assumed that a certain finite number J_i of disjoint variation intervals exists

$$[x_{i,jmin}, x_{i,jmax}), \; j=1,2,..., J_i; \; i=1,2,...,r \tag{2}$$

for each variable X_i, such that the probability of a simultaneous event

$$P(X_1 \in [x_{1,j_1 \min}, x_{1,j_1 \max}), X_2 \in [x_{2,j_2 \min}, x_{2,j_2 \max}),..., X_r \in [x_{r,j_r \min}, x_{r,j_r \max})) =$$
$$= P_{X_1, X_2,...X_r}(x_{1,j_1}, x_{2,j_2},..., x_{r,j_r}) \tag{3}$$

is constant and equal to the quotient

$$P_{X_1, X_2,...X_r}(x_{1,j_1}, x_{2,j_2},..., x_{r,j_r}) = \frac{n_{1,j_1}, n_{2,j_2},..., n_{r,j_r}}{N} \tag{4}$$

where

$n_{1,j_1}, n_{2,j_2},..., n_{r,j_r}$ is the number of particles whose features $X_1,...,X_r$ take their values from proper intervals,
N is the total number of the particles of population.

Defined in this case, the empirical distribution should fulfill the following dependence:

$$\sum_{j_1=1}^{J_1} \sum_{j_2=1}^{J_2} ... \sum_{j_r=1}^{J_r} \frac{n_{1,j_1}, n_{2,j_2},..., n_{r,j_r}}{N} = 1 \tag{5}$$

In the space of the features of particles $X=R^r$, we can define a fuzzy event A as a couple set

$$A = \{(\mu_A(x), x)\}, \forall x \in X \quad (6)$$

where
  $x = [x_1, x_2, ..., x_r]^T$ - an element in space X,
  $\mu_A(x)$ = membership function of the fuzzy set A that assigns the level of affiliation of each element x to the fuzzy set A.

According to [1] if the number of elements x in the space X is finite, then the set A might be described by the finite sum:

$$A = \sum_{m=1}^{M} (\mu_A(x_m)/x_m) \quad (7)$$

If the number of elements x in the space of particle features X is infinite, then we can write the set A as follows:

$$A = \sum_{x \in X} (\mu_A(x)/x) \quad (8)$$

In this notation, the sum sign is considered as a set character, not an arithmetic one.

The probability $P(A)$ of the fuzzy event A, that is the fuzzy subset of the probabilistic space mentioned above, is referred to as the expression [1,5,6]:

$$P(A) = \sum_{x \in X} p(x)\mu_A(x) \quad (9)$$

where
  $x = [x_1, x_2, ..., x_r]^T$ - an element in the space X.

In order to define this fuzzy probability, let us suppose that A is a fuzzy event defined in r-dimensional space X representing the features of particles. Let p be the probability function which assigns to each Borel set in area X the number $p \in [0,1]$. Then the fuzzy probability of a fuzzy event A, $P(A)$, is defined, according to [4] and [1], as:

$$P(A) = \sum_{\alpha \in [0,1]} \alpha / p(A_\alpha) \quad (10)$$

where $p(A_\alpha)$ is the know probability in the classical sense of the non-fuzzy event A_α - the α-cut of set A:

$$A_\alpha = \{x \in X; \mu_A(x) \geq \alpha\} \quad (11)$$

Taking into account the fact that the probability of event A is also defined as the quotient of cardinality of sets, we assume to note the definitions of fuzzy probability as:

$$P(A) = \frac{|A|}{|X|} = \frac{\sum_i \mu_A(x_i)}{|X|} \qquad (12)$$

and

$$P(A) = \frac{\sum_{\alpha \in [0,1]} \alpha / |A_\alpha|}{|X|} = \sum_{\alpha \in [0,1]} \alpha / \frac{|A_\alpha|}{|X|} \qquad (13)$$

where

A is a fuzzy event defined in the space X,
$|A|, |A_\alpha|, |X|$ are cardinalities of particular sets.

FUZZY PROBABILITY OF A 2-D RANDOM BARIABLE AS INPUT MATERIALS CHARACTERISTICS

Traditional Approach

To show practical aspects and to simplify the notation, we consider two physical features of particles: volume V and density D. Moreover, we analyse certain quality features of particles such as the contents of a certain component β.

In a traditional approach, the material characteristics which is used by process engineers, is the discrete empirical probability distribution of a two-dimensional random variable (V,D) as a probability of simultaneous events:

$$p_{ij}(v, d) = P\{V \in [v_{i,min}, v_{i,max}), D \in [d_{j,min}, d_{j,max})\}, \qquad (14)$$
$$i=1,2,...,I; \quad j=1,2,...,J.$$

where I is the number of particle size classes and J is the number of particle density fractions, and

$$p_{ij}(v, d) = \frac{N_{ij}}{N}, \quad i=1,2,..., I; \quad j=1,2,...,J; \qquad (15)$$

N_{ij} - number of particles in the parent population, the volume of which belongs to i-th interval (v_i) and the density of which belongs to j-th interval (d_j),

N - total particles number in the population.

In engineering practice, a different measure of the probability in Eq. 14 is more often applied - the quotient of the respective mass:

$$\pi_{ij}(v,d) = \frac{M_{ij}}{M}, \quad i=1,2,...,I; \quad j=1,2,...,J; \tag{16}$$

where
M_{ij} - mass of particles in the parent population, the volume of which belongs to i^{th} interval (v_i) and the density of which belongs to j^{th} interval (d_j),
M - total mass of the population.

There is a strict relationship between two measures of probability [3]:

$$\pi_{ij} = a_{ij} p_{ij}; \quad i=1,2,...,I; \quad j=1,2,...,J; \tag{17}$$

where:

$$a_{ij} = \frac{v_i d_j}{v_m d_m} \tag{18}$$

$$M_{ij} = v_i d_j N_{ij} \tag{19}$$

$$M = v_m d_m N \tag{20}$$

The variables v_m and d_m are mean values of volume and density of particles in the population, and

$$N = \sum_{i=1}^{I} \sum_{j=1}^{J} N_{ij} \tag{21}$$

$$M = \sum_{i=1}^{I} \sum_{j=1}^{J} M_{ij} \tag{22}$$

Table 1 presents the values of probability $p(v_i, d_j) = p_{ij}$ for certain material. Each of the ranges of volume and density of particles is divided into 4 intervals. The smallest value of the indexes i,j concerns the smallest value of density and volume, i.e.,

$$v_1 < v_2 < v_3 < v_4; \quad d_1 < d_2 < d_3 < d_4.$$

Table 1 also gives calculated values of marginal probabilities that define the events:

$$p_{i.}(v) = P\{ V \in [v_{i,min}, v_{i,max}) \}, \quad i=1,2,...,I \tag{23}$$

$$p_{.j}(d) = P\{ D \in [d_{j,min}, d_{j,max}) \}, \quad j=1,2,...,J. \tag{24}$$

Table 1. The probability distribution $p_{ij}(v,d)$ of particle features.

Density fraction number j	Probability distribution $p_{ij}(v, d)$				Marginal probability $p_{.j}(d)$
	Size class number				
	i=1	i=2	i=3	i=4	
1	0.001902	0.007356	0.030429	0.628992	0.668679
2	0.000330	0.001092	0.004410	0.133848	0.139680
3	0.000243	0.001068	0.004949	0.053352	0.059612
4	0.000525	0.002484	0.009212	0.119808	0.132029
Marginal probability $p_{i.}(v)$	0.003000	0.01200	0.049000	0.936000	$\sum_i p_{i.} = $ $\sum_j p_{.j} = 1$

Table 2. Membership coefficients $\mu_A(v_i, d_j)$ of fuzzy event A: "small and light particles."

j/i	1	2	3
1	1	0.7	0.2
2	0.7	0.5	0
3	0.2	0	0

Table 3. Membership coefficients $\mu_B(d_j)$ of fuzzy event B: "heaviest particles."

j	2	3	4
$\mu_B(d_j)$	0	0.5	1

Table 4. Membership coefficients $\mu_C(v_i, d_j)$ of fuzzy event C: "middle-sized and heaviest particles."

j/i	1	2	3	4
3	0	0.5	0.5	0
4	0	1	1	0

Probability of a Fuzzy Event and Probability as a Fuzzy Set

On the basis of the probability distribution in Table 1, we can define some fuzzy events that might have the form of certain linguistic variables such as:

        A: "small and light particles"
        B: "heaviest particles"
        C: "middle-sized and heaviest particles"

The membership functions for each fuzzy event are presented in Tables 2, 3 and 4. Using Eq. 9, we can calculate the probability of each particular fuzzy event A, B, and C:

$$P(A) = \sum_{i=1}^{3}\sum_{j=1}^{3} p(v_i, d_j)\mu_A(v_i, d_j) = 0.013963$$

$$P(B) = \sum_{j=3}^{4} p_{.j}(d_j)\mu_B(d_j) = 0.161835$$

$$P(C) = \sum_{i=2,3}\sum_{j=3,4} p(v_i, d_j)\mu_C(v_i, d_j) = 0.0147045$$

So, the fuzzy conditional probability of event D: "middle-sized particles between the heaviest ones" is:

$$P(D) = P(C)/P(B) = 0.09086$$

Each of the calculated values $P(A)$, $P(B)$, $P(C)$, $P(D)$ is a real number. Nevertheless, with respect to Eq. 10 which gives the probability of the fuzzy set and using the probability distributions in Table 1 and membership coefficients in Tables 2, 3, and 4 respectively, we get for each fuzzy event A, B, and D, the appropriate fuzzy set that describes their probability:

$$P(A) = \sum_{\alpha \in \{1, 0.7, 0.5, 0.2\}} \alpha / p(A_\alpha) =$$

$$= 1/p_{11} + 0.7/(p_{11}+p_{12}+p_{21}) + 0.5/(p_{11}+p_{12}+p_{21}+p_{22}) + 0.2/(p_{11}+p_{12}+p_{21}+p_{22}+p_{13}+p_{31}) =$$
$$= 1/0.0019 + 0.7/0.009588 + 0.5/0.01068 + 0.2/0.041352$$

$$P(B) = 1/p_{.4} + 0.7/[p_{.3}+p_{.4}] = 1/0.132029 + 0.5/0.191641$$

$$P(D) = 1/\frac{p_{24}+p_{34}}{p_{.4}} + 0.5/\frac{p_{23}+p_{24}+p_{33}+p_{34}}{p_{.3}+p_{.4}} = 1/0.0886 + 0.5/0.0924$$

Quality Parameters as Fuzzy Mean Values

We define a particular quality variable β to represent the exact content of a selected component in each particle. Furthermore suppose there is the possibility to measure the mean value $\beta_{ij} = \beta(v_i, d_j)$ in each elementary fraction of particles, then the mean value of a tested substance in a whole material or in, e.g., "small and light particles" can be calculated as follows:

$$\beta = \frac{\sum_{i=1}^{I_A}\sum_{j=1}^{J_A} \beta_{ij} p_{ij} \mu_A(v_i,d_j)}{\sum_{i=1}^{I_A}\sum_{j=1}^{J_A} p_{ij} \mu_A(v_i,d_j)} \qquad (25)$$

where

I_A, J_A – numbers of size particle classes and density fractions belonging to the event A: "small and light particles".

FUZZY PROBABILITY CHARACTERISTICS IN PREPARATION PROCESSES OF PARTICLES

Mixing Materials: Fuzzy Characteristics of Heterogeneity

Engineers often consider heterogeneity of materials that result from mixing materials with different characteristics and proportions. To do a theoretical analysis of this problem, it is assumed that certain quantities of two different types of particle populations are mixed. The two populations differ in their characteristics $\{p_i^{(1)}\}, \{p_i^{(2)}\}, i = 1,...I$, which describes the probability of selecting a particle with features
$X \in [x_{i,\,min},\, x_{i,\,max})$, $i=1,2,...,I$

Characteristic $\{p_i^{(s)}\}, i = 1,...,I$ of the mixture is expressed as [3]:

$$p_i^{(s)} = p_i^{(1)}Q + p_i^{(2)}(1-Q) \qquad (26)$$

where

Q – proportion of materials mixed, i.e., the probability of the occurrence a particle of material (1) in a mixture; $0 \leq Q \leq 1$.

The measure of probabilities $p_i^{(1)}, p_i^{(2)}, p_i^{(s)}, Q$ is a quotient of particle numbers in their respective populations.

Let Q, the proportion of mixing, be a fuzzy number "low" with membership function

$$\mu_{low}(q) = \begin{cases} 1, & 0 \leq q \leq 0.2 \\ 1 - \dfrac{q-0.2}{0.1}, & 0.2 < q \leq 0.3 \\ 0, & 0.3 < q \leq 1 \end{cases} \qquad (27)$$

and $1-Q$ is a fuzzy number "high" with a membership function:

$$\mu_{high}(q) = \mu_{low}(1-q)$$

$$\mu_{high}(q) = \begin{cases} 1, & 1 \geq q > 0.8 \\ 1 - \dfrac{0.8-q}{0.1}, & 0.8 \geq q > 0.7 \\ 0, & 0.7 \geq q \geq 0 \end{cases} \qquad (28)$$

Then the characteristics from Eq. 26, $\{p_i^{(s)}\}, i=1,2,...,I$ is a fuzzy probability distribution in the mixture. This is a linear combination of fuzzy numbers Q and 1-Q and the real numbers $p_i^{(1)}, p_i^{(2)}, i=1,2,...,I$.

Static Characteristics in Separation Processes of Particles

In separation processes of particles, the static characteristic is a certain distribution of conditional probabilities defining for a particle having the features (v_i, d_j):

R_{ij} – probability that particle is situated in the chosen product,
$1-R_{ij}$ – probability of the opposite event.

One of these products might be defined as a concentrate (or valuable product), and one as a tailing (or waste product).

Prognostic probability distribution of particles having features (v_i, d_j) in a concentrate of the mth separation process might be defined as follows [3]:

$$p_{ij}^{(m)} = \dfrac{R_{ij}^{(m)} p_{ij}^{(m-1)}}{\sum_{i=1}^{I}\sum_{j=1}^{J} R_{ij}^{(m)} p_{ij}^{(m-1)}}, m = 1,2,...,M \qquad (29)$$

where

$$\sum_{i=1}^{I}\sum_{j=1}^{J} p_{ij}^{(m)} = 1, \; m=1,2,...,M \qquad (30)$$

$R_{ij}^{(m)}$ – process static characteristics (conditional distribution curve for a given facility),
$p_{ij}^{(m-1)}$ – probability distribution of particles properties (v_i, d_j) in the material from the preceding stage, which is the material feed for m-th process.

The probability distribution of properties (v_i, d_j) of the particles in the waste material of the mth process:

$$q_{ij}^{(m)} = \frac{\left(1 - R_{ij}^{(m)}\right) p_{ij}^{(m-1)}}{\sum_{i=1}^{I} \sum_{j=1}^{r} \left(1 - R_{ij}^{(m)}\right) p_{ij}^{(m-1)}}, \; m = 1, 2, \ldots, M \tag{31}$$

where

$$\sum_{i=1}^{I} \sum_{j=1}^{J} q_{ij}^{(m)} = 1, \; m = 1, 2, \ldots, M \tag{32}$$

Let us define the fuzzy event G that determines "undesirable particles in the concentrate of the mth process" with membership function $\mu_G(v_i, d_j)$. According to Eq. 9, we can express the probability of this event as:

$$P(G) = \sum_{i=1}^{I_G} \sum_{j=1}^{J_G} p_{ij}^{(m)}(v_i, d_j) \mu_G(v_i, d_j) \tag{33}$$

where

I_G, J_G – numbers of size particle classes and density fractions belonging to the event G: "undesirable particles in the concentrate of m-process".

Process Dynamics in Laboratory Separators

The process dynamics in a laboratory separator has been widely recognised and discussed in many professional publications. By making some assumptions, we can write a system of two differential equations that describe changes in the occurrence of particles in a certain part of a separator:

$$\frac{d}{dt} p_{\alpha-}(t) + a(t) p_{\alpha-}(t) = F_{\alpha-}(t) \tag{34a}$$

$$\frac{d}{dt} p_{\alpha+}(t) + a(t) p_{\alpha+}(t) = F_{\alpha+}(t) \tag{34b}$$

where the function $p(t)$, $t \in T = R$ is a fuzzy-valued function meaning that $\forall \alpha \in (0,1]$, the α-level function is as follows:

$$p_\alpha(t) = [p_{\alpha-}(t), p_{\alpha+}(t)] \tag{35}$$

is an interval-valued function on T and

$$p(t) = \bigcup_{\alpha \in (0,1]} \alpha [p_{\alpha-}(t), p_{\alpha+}(t)] \tag{36}$$

where

$$-\infty < p_{\alpha-}(t) = \inf p_\alpha(t), \quad p_{\alpha+}(t) = \sup p_\alpha(t) < +\infty.$$

If $\forall \alpha \in (0,1]$ and derivatives: $\frac{d}{dt}p_\alpha(t), \frac{d}{dt}p_{\alpha-}(t), \frac{d}{dt}p_{\alpha+}(t)$ exist then [7], the fuzzy-valued function $p(t)$ is differentiable and

$$\frac{d}{dt}p(t) = \bigcup_{\alpha \in (0,1]} \alpha(\frac{d}{dt}p_\alpha(t)) \qquad (37)$$

or

$$\frac{d}{dt}p(t) = \bigcup_{\alpha \in (0,1]} \alpha[\frac{d}{dt}p_{\alpha-}(t), \frac{d}{dt}p_{\alpha+}(t)]. \qquad (38)$$

Coefficient $a(t)$ and a fuzzy-valued function $F(t)$ depend on conditions of separation.

CONCLUDING REMARKS

In the paper some chosen possibilities of fuzzy modelling of particles preparation processes have been considered. On the basis of traditionally known probability distributions, fuzzy events and the fuzzy probability of particle features in certain technological situations have been determined. Both rules and examples are included. Research in this area is continuing.

REFERENCES

[1] J. Kacprzyk, 1986. Fuzzy sets in system analysis, (Polish edition). Warsaw, PWN.
[2] N.G. van Kampen, 1990. Stochastic processes in physics and chemistry, (Polish edition). Warsaw, PWN.
[3] A. Walaszek-Babiszewska, 1996. Stochastic models and optimisation of sampling operation in preparation plants, In: New Trends in Coal Preparation Technologies and Equipment. Proc. of 12th International Coal Preparation Congress, 639–643. Amsterdam, Gordon and Breach Publishers.
[4] R.R. Yager, 1979. A note on probabilities of fuzzy events, Inf. Sci., 18, 113–129.
[5] L.A. Zadeh, 1968. Probability measures of fuzzy events, J. Math. Anal. Appl., 23, 421–427.
[6] L.A. Zadeh, 1999. From computing with numbers to computing with words— from manipulation of measurements to manipulation of perceptions, IEEE Trans. on Circuits and Systems – I: Fundamental Theory and Applications, 45(1), 105–119.
[7] Zhang Yue, Wang Guangyuan, 1998. Time domain methods for solutions of N-order fuzzy differential equations, Fuzzy Sets and Systems, 94, 77–92.

Prediction of Charpy Impact Properties Using Fuzzy Modelling on Ship Structural Steels

M. CHEN and D. A. LINKENS
Institute for Microstructural and Mechanical Processing Engineering, The University of Sheffield, Sheffield, UK

ABSTRACT

The present research aims to link steel composition and microstructure to a prediction of Charpy impact properties of ship structural steel, including Charpy impact energy and transition temperature (ITT). In this paper, adaptive fuzzy modelling techniques were applied to develop general models to predict Charpy impact properties for Grade A and AH36 ship steels. Using the proposed fuzzy modelling approach, a fuzzy rule-based model was generated and optimised automatically from numerical data without prior knowledge. The fuzzy prediction models used chemical composition, grain size, tensile strength, and Charpy impact energy obtained from industrial data from Lloyds Register. The investigation of the influence factors on Charpy impact properties is presented in the paper. Numerical analysis shows that using the obtained fuzzy models we are able to predict Charpy toughness for given steel compositions and microstructures, and also reveal useful qualitative information linking composition-process conditions to Charpy impact properties.

INTRODUCTION

There is continuing concern regarding the susceptibility of carbon-manganese steel ship structures to brittle fracture. At low temperatures, such material undergoes a ductile to brittle transition. This can be determined by the Charpy test. In general the fine grain size microstructures generate both high yield strength and very good notched impact properties, but a matter of concern is that nominally identical samples taken from thermomechanically processed products sometimes exhibit high variability in their notched Charpy impact responses. The variability is most plausibly related to variations in the steels' microstructure. The aim of the present work is to establish the link between compositions, microstructure and Charpy toughness for high strength structural steels by multi-disciplinary expertise and modelling techniques.

Since steel companies highly value achieving the required levels of mechanical properties of rolled steel products, much effort has been expended to predict tensile properties from steel chemistry, microstructure and processing variables [1]-[3]. Some work on Charpy test modelling in a specific temperature region for a particular steel has been reported [4]-[6]. However, little analogous work has been done on establishing generic composition-processing-impact toughness models so far. The work described here aims at linking steel composition and microstructure to a prediction of Charpy impact energy for Grade A and AH36 ship steels. In this paper, after introducing the general scheme of adaptive fuzzy models, the effects of chemical composition and microstructure on impact toughness are discussed. The paper highlights the interactions between some interested elements such as C, Mn and S etc., using fuzzy model response surfaces.

A GENERAL SCHEME OF ADAPTUVE FUZZY MODELLING

A fuzzy model is a system description with fuzzy quantities, which are expressed in terms of fuzzy numbers or fuzzy sets associated with linguistic labels. Adaptive fuzzy systems can be viewed as fuzzy logic systems whose rules are automatically generated through fuzzy neural network training. The major advantage of neural-fuzzy modelling is its ability to integrate the logical processing of information with attractive mathematical properties of general function approximators capable of representing complex non-linear mappings. Also, the if-then rule mechanism is easy to manipulate, understand and, to a certain extent, is domain independent. In material engineering, it is beneficial to develop accurate, transparent and computationally efficient structure-property models for materials development.

Consider a collection of N data points $\{P_1, P_2,..., P_N\}$ in a $m+1$ dimensional space that combines both input and output dimensions. Without loss of generality, we consider a MISO fuzzy logic system as a general presentation of fuzzy systems. A generic fuzzy model is presented as a collection of fuzzy rules in the following form

$$R_i: \text{ If } x_1 \text{ is } A_{i2} \text{ and } x_2 \text{ is } A_{i2} \ldots \text{ and } x_m \text{ is } A_{im} \quad \text{then } y = z_i(x)$$

where $x = (x_1, x_2, \ldots, x_m) \in U$ and $y \in V$ are linguistic variables, A_{ij} are fuzzy sets of the universes of discourse $U_i \in R$, and $z_i(x)$ is a function of input variables. Typically, z takes the following three forms: singleton, fuzzy set or linear function. Fuzzy logic systems with centre average defuzzifier, product-inference rule and singleton fuzzifier are of the following form:

$$y = \sum_{i=1}^{n} z_i [\prod_{j=1}^{m} \mu_{ij}(x_j)] \bigg/ \sum_{i=1}^{n} \prod_{j=1}^{m} \mu_{ij}(x_j)$$

where $\mu_{ij}(x)$ denotes the membership function of x_j belonging to the ith rule.

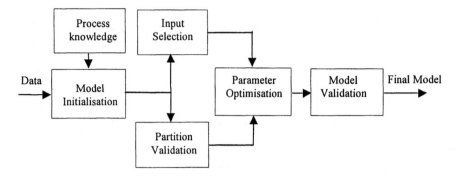

Figure 1. General scheme of adaptive fuzzy modelling.

According to the fuzzy modelling paradigm proposed by Chen and Linkens [7, 8], a fuzzy modelling problem is equivalent to satisfying the following requirements: 1) generating an initial fuzzy rule-base from data; 2) selecting the important input variables; 3) determining the optimal number of fuzzy rules; and 4) optimising the parameters both in the antecedent part and consequent part of the rules. The general scheme of the neural fuzzy modelling is depicted in Figure 1.

Clearly, data-driven fuzzy modelling consists of four stages. In the first stage, data pre-processing including data cleaning, data transformation and normalisation, and model initialisation should be done. In this stage, data processing techniques and prior knowledge about the modelled process are needed for model initialisation, such as determining the type of the fuzzy model, and choosing the type of membership functions. The second stage involves the structure identification. There are two challenging problems in this stage: (1) input selection, that is to select the important inputs that affect the system output significantly among all possible input variables; (2) fuzzy partition validation, which is to determine the optimal number of rules for the fuzzy model. The task of parameter optimisation is carried out in stage 3. An effective learning strategy should be used to find the optimal parameters for the model. Stage 4 concerns the task of model validation. The acquired fuzzy model should be validated under certain performance indices, such as accuracy, generality, complexity, interpretability, etc. If the model performance is not good enough, further modification including structure and parameter optimisation would be required. Once the model performance achieves the pre-defined criteria, the final model is produced.

CHARPY TOUGHNESS MODELLING

The proposed fuzzy modelling approach has being used to construct composition-microstructure-property models for Charpy impact energy prediction of Grade A and AH36 ship steels. In the process of modelling, 60% data were used for model training and 40% data were used for testing.

CHARPY MEDELLING FOR SHIP STEEL GRADE A

A total of 38 LR certified grade A plates were tested at test temperature of –20, 0 and 20 C using longitudinal specimens. The thickness of the individual plates was in the range of 12 to 16mm. Most of the plates were delivered in the as-rolled condition and the plates were all satisfactorily fine-grained with a typical ferrite/pearlite microstructure. The database available for the analysis contains chemical compositions, ASTM grain size, tensile strength, measurements of Charpy impact energy and corresponding test temperature. A total of 342 data for Grade A, provided by Lloyds Register, were used to develop the prediction models.

As the processing information is not available for this ship steel data set, microstructure and tensile strength can be used for impact property prediction. Firstly, a composition-microstructure-impact property fuzzy prediction model with the inputs of C, Mn, S, P, Al, Grain Size, Test Temperature was developed for Grade A ship steels. The prediction result with Root Mean Square Error RMSE=22.19 is shown in Figure 2. It is seen that Charpy impact energy can be predicted well based on provided chemical compositions, microstructure. As tensile strength strongly links to compositions, processing conditions and microstructure of materials, it is beneficial to introduce UTS (Ultimate Tensile Strength) in the toughness prediction model due to the lack of processing information. Figure 3 shows the prediction result with UTS as an added input variable in the fuzzy prediction model. Clearly, the prediction accuracy was improved compared to the model without UTS. In the case of UTS data not being available, a tensile strength prediction model can be developed first, and then used to further predict impact energy. Figure 4 shows the tensile strength prediction result using C, Si, Mn, Al, and Grain Size as input variables. Using the predicted UTS to predict impact energy, we obtained a very similar prediction result with RMSE=21.11.

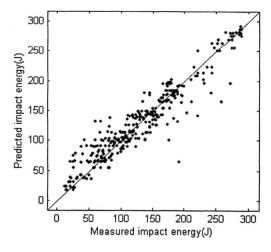

Figure 2. Impact energy prediction without UTS. RMSE=22.19.

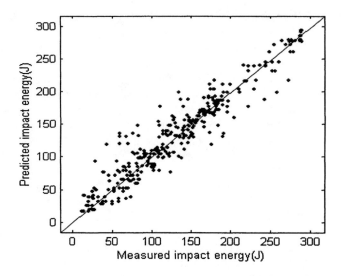

Figure 3. Impact energy prediction with UTS. RMSE=20.75.

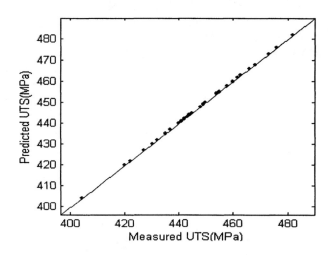

Figure 4. UTS prediction RMSE=1.31.

In order to investigate the influence of chemical compositions and microstructure on Charpy properties, model response surface analysis with respect to the link between two specific factors and Charpy energy has been conducted on the fuzzy models at a test temperature of 0 °C for Grade A steel. Through the model response surfaces, we could reveal not only the influence of individual variables but also the interactions between some specific factors such as C/Mn and Mn/S. The corresponding response surfaces are displayed in Figure5. It can be

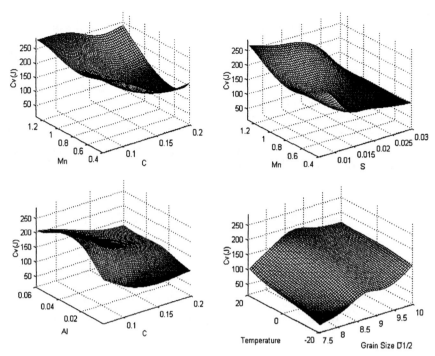

Figure 5. Response surface of the fuzzy model for Grade A ship steels.

seen that low carbon and high manganese usually produce better impact properties. Low sulphur residual and high Mn/S ratio generally improve impact toughness. The increase of aluminium generally increases Charpy impact energy. It is also clear that fine grain size generates good Charpy v-notch impact property, which is deemed to be consistent with existing metallurgical knowledge.

CHARPY MEDELLING FOR SHIP STEEL GRADE AH36

Charpy impact tests for Grade AH36 steel were undertaken on specimen sets from 22 different plates originating from worldwide steel makers. The test temperature ranged from 0 to −80 °C. A total of 214 data for Grade AH36, containing information on chemical compositions, tensile strength and impact properties (including impact energy and 27J transition temperature), were used to develop the fuzzy prediction model. Using the proposed adaptive fuzzy modelling method, a 6-rule fuzzy model with C, Si, Mn, S, Ni, Al, UTS and Test temperature as input variables was developed for the prediction of impact energy and ITT_{27J}. The model prediction results are displayed in Figure 6 and Figure 7 respectively. It can be seen that both Charpy impact energy and ITT_{27J} can be predicted well using a simple fuzzy model with a small number of rules.

To reveal the relationship between individual variables and Charpy toughness, model response surfaces were also obtained. Figure 8 shows the effect of compositions and tensile strength on Charpy impact energy in the brittle-ductile transition region (–40°C). It can be seen that lower sulphur residual improves Charpy properties. It is also seen that addition of Ni content is generally beneficial to Charpy toughness, and the additions of Cr, Mo and Al did not create significant changes in impact energy. In Grade AH36 ship steels, higher tensile strength generally corresponds to higher impact energy, which might be produced by finer grain size. It should be pointed out that different combination of elements have different effects in the transition region and upper shelf energies, hence making the problem very complex. Further investigation of influence factors on Charpy toughness in different regions is needed.

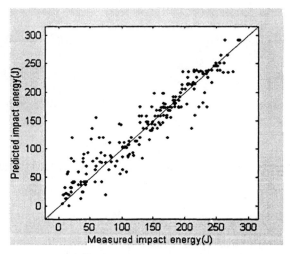

Figure 6. Impact energy prediction for Grade AH36 ship steel RMSE=25.06.

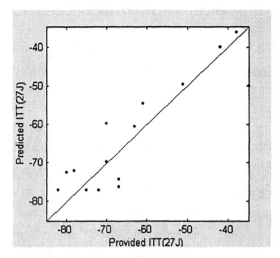

Figure 7. ITT(27J) prediction for Grade AH36 ship steel RMSE=8.51.

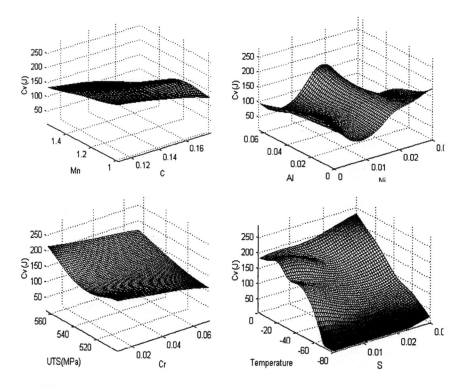

Figure 8. Response surfaces of the fuzzy prediction model for Grade AH36.

CONCLUSIONS

Numerical analysis shows that using the developed neural-fuzzy models we can predict Charpy toughness by given steel compositions, microstructure and/or tensile strength, and also reveal useful qualitative information linking composition-microstructure to Charpy impact properties. Fuzzy modelling results also show that the impact energy of all test plates in Grade A steels are well above 27J at 20 °C, while 27J transition temperatures (ITT_{27J}) for all plates in Grade AH36 steels are well below −20 °C, which are well above the required LR toughness criteria. It is clear that the proposed fuzzy modelling method is effective for Charpy toughness prediction and assessment. The model predicted trends are consistent with the known metallurgical knowledge. Further investigation of interactive influence factors and incorporation of expert knowledge and data-driven modelling techniques will be beneficial in the prediction of Charpy properties.

ACKNOWLEDGEMENT

The authors are grateful to the Engineering and Physical Sciences Research Council for their financial support and to the collaborating company, Lloyds Register, for their support and the supply of data and tested Charpy samples.

REFERENCES

[1] P.D. Hodgson, 1996. Microstructure modelling for property prediction and control, Journal of Materials Processing Technology, 60, 27–33.
[2] C. Chen, Y. Cao, S.R. LeClair, 1998. Materials strucutre-property prediction using a self-architecting neural network, Journal of Alloys and Compounds, 279(1), 30–38.
[3] B.R.Bakshi and R. Chatterjee, 1998. Unification of neural and statistical methods as applied to materials structure-property mapping, Journal-of-Alloys-and-Compounds. 279(1), 39–46.
[4] R. Moskovic and P.E.J. Flewitt, 1997. An overview of the principles of modeling Charpy impact energy data using statistical analyses, Metallurgical and Materials Transactions A; 28(12) 2609–2623.
[5] M.T. Todinov, M. Novovic, P. Bowen, J.F. Knott, 2000. Modelling the impact energy in the ductile/brittle transition region of C-Mn multi-run welds, Materials-Science-&-Engineering-A-(Structural-Materials:-Properties,-Microstructure-and-Processing). AA287(1), 116–124.
[6] M. Tahar, R. Piques, P. Forget, 1996. Modelling of the Charpy V-notch test at low temperature for structural steels, ECF 11. Mechanisms and Mechanics of Damage and Failure. Proc. 11th Biennial European Conf. on Fracture - ECF 11. Eng. Mater. Advisory Services, Warley, UK; 3, 1945–1950.
[7] M.Y. Chen and D.A. Linkens, 1998. A Fast Fuzzy Modelling Approach Using Clustering Neural Networks, IEEE World Congress on Intelligent Computation, Proc. Fuzzy-IEEE'98, 1406–1411.
[8] M.Y. Chen and D.A. Linkens, 2001. A Systematic Neurofuzzy Modelling Framework with Application to Material Property Prediction, IEEE SMC Trans., Part B, 31(5), 781–790.

Chapter 3: Artificial Neural Networks

Artificial Intelligence in Large-Scale Simulations of Materials

J. F. MAGUIRE
Materials and Manufacturing Directorate, Air Force Research Laboratory, Dayton, Ohio, USA

ABSTRACT

In this talk I review the development and application of artificial intelligence as it pertains to the processing and manufacturing of materials. The emphasis will be on how new developments in machine intelligence have enabled realistic very large scale computer simulation of materials and processes. These new methods have the near term potential to provide real breakthroughs across the spectrum from basic science to manufacturing technology. The emphasis is on polymers and polymeric composites that are used in aerospace applications but the paradigm is quite general and could be applied to many materials and processing operations. While the AI approach has proven successful at the level of both material process development and in fundamental statistical mechanical model development, fundamental limitations are encountered in the area of discovery. It is not yet obvious how 'inspired serendipity' might be programmed into a machine.

INTRODUCTION

In a recent monograph [1] Robert Cahn (Robert Cahn, 2001. The Coming of Materials Science, Pergammon, 360) has traced the emergence of materials science from its origins in solid-state physics, condensed matter physics, physical chemistry, physical metallurgy and polymer science. These core subjects provide the firm intellectual foundation on which the discipline is established. As its origins suggest, materials science is an inherently multidisciplinary and interdisciplinary activity of considerable scope and reach.

The core academic activity is the search for new knowledge and a better fundamental understanding of how the macroscopic properties of matter are related to microscopic structure. At the basic science level there is, in common with chemistry and physics, the drive to understand the nature of matter over a range of distance and energy scales. However, a key differentiating, perhaps even defining, element of materials science is that the subject concentrates on processing and producing new forms of matter that are enabling or have enhanced performance in engineering applications. This applied dimension anchors the materials science and engineering approach firmly in the industrial and applied science arena. In this sense it could be argued that MS&E is characterized, not so much by what is done, for in operational detail that is largely the chemistry, physics, polymer science, etc., of the classical core disciplines, but rather how the work is driven and how it is executed. It is driven, i.e. funded, by the need to produce a materiel-based solution to an operational need. It is executed by interdisciplinary teams, often through the collaboration of university, government and industrial laboratories. Materials science and engineering provides what might be termed a "systems" approach that integrates teams in particular specialties towards an engineering goal. [2] Good (Mary L. Good, "Materials through Chemistry" American Chemical Society, pp 217, 1998) has pointed out the degree to which the materials research enterprise now embraces a large cross section of research activity ranging from synthetic organic chemistry on the one hand through statistical mechanics and thermodynamics to chemical engineering and process control.

For example, in aerospace applications there is a ubiquitous requirement to produce stronger lighter materials from which to fabricate airframes and spacecraft. Engine development is far advanced and expected gains in the thrust/weight ratio from increasing the thrust through better aerodynamic designs are of order 10 %. However, new polymers with fiber reinforcements have the potential to halve the weight, hence the high importance placed on materials development in the aerospace field. A largely "plastic" airplane is shown in Figure 1.

Another example from the author's laboratory is the development and processing of new lightweight polymeric materials for application in large (>100 m diameter) space borne mirrors, a prototype of which is shown in Figure 2. Such mirrors will allow us to see to almost the start of time from observatories in outer space; like the Hubble observatory only very much larger. Glass mirrors of this size are much too heavy to launch into orbit but one might think of a large inflatable polymeric mirror as shown in Figure 2 (the men in the figure are included to give it a scale) that is folded during launch and deployed on station when needed. If the polymer is coated with reflective materials and has just the right mechanical response to form the required parabolic profile when deployed then these concepts might be used to produce exceedingly large space-based optics or even very cheap large telescopes for the amateur astronomer.

Telescopes based on these concepts will provide an unprecedented look back in time but they will require the development of fundamentally new sorts

Figure 1. The advanced tactical fighter is made from very light weight high strength reinforced polymeric materials and has coatings that make it more difficult to detect by radar and protect it from attach by lasers.

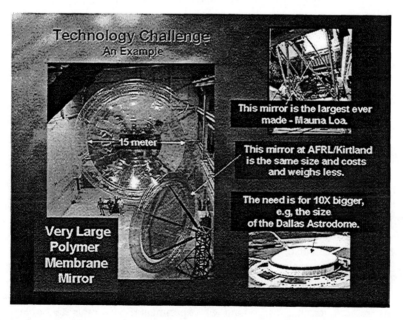

Figure 2. A prototype of a large (10m) space optic under development at AFRL. Such structures will require active control of the optical surface over mesoscopic distance scales.

of adaptive soft matter for active control of the reflective surface. These advanced applications require novel structures of matter that have performance characteristics, which are far in excess of current capabilities. Similar examples could be selected from a number of fields. In the broadest sense MS&E represents the work of human minds expressed by human hands to add value and to create wealth from the natural resources of the earth. The subject is concerned with how to turn the very "dirt" of the earth, the metal ores, the crude oil, into the automobiles, aircraft, aircraft engines, computers, materials and medicines of our advanced civilization. This cycle requires the accumulated interdisciplinary knowledge of generations of scientists and engineers and represents the real or bedrock "knowledge environment" on which our economy is based and against which progress should be measured.

In order to produce these materials in the required quantities and quality and cost it is necessary

(a) Develop materials characterization techniques that probe the relevant structural property of matter over an appropriate domain of energy and momentum.
(b) Explore theoretical and computer modeling approaches that can help rationalize observed behavior and response and, more importantly, predict materials response in situations where direct measurement would be difficult or impossible.
(c) Integrate the experimental and theoretical knowledge gathered (a) and (b) above and provide a seamless transition of this knowledge into the engineering and manufacturing environment using advanced techniques such as computational methods in artificial intelligence.

COMPUTER SIMULATION AND MODELING

The idea that mathematics could be used to describe the properties of matter originated with the fourteenth child of the Earl of Cork, Robert Boyle. In the 300 years since Boyle the subject has matured considerable and we now have well developed statistical mechanical theories and computer simulation methods can be used to calculate the properties of matter with quite high accuracy in some cases. In materials science these models fall into two broad classes, both of which are addressed in the present work.

One class of calculation uses very simple intermolecular potential functions and uses the computer as a kind of experimental apparatus to solve the equations of motion of the coupled many-body problem and return an equation of state for the model material. This class of calculation can provide deep physical insight into the fundamental properties of matter. A second type of simulation again solves the equation of motion directly but does so for an intermolecular potential function that is "engineered" to give the correct experimental properties. This class of calculation is open to criticism at a fundamental level but it has the saving grace that it often provides a reasonable representation, in fact the best available representation of materials for which the intermolecular function are not known. Again this can be very useful in engineering applications where the response of materials is needed under conditions in which experiments are very difficult or very costly.

ARTIFICIAL INTELLIGENCE IN MATERIALS PROCESSING

The final papers in this compilation relate to developing and applying machine intelligence, often called artificial intelligence, to the process control of materials. We are concerned in these papers with developing a "materials and processing robot". This robot does not have a head or legs of the sort depicted in movies and science fiction novels but it does have the capability to fulfill certain functions. First it has an ability to sense critical variables in the environment, in this case the physical and chemical state of a polymerization reaction. Second, it must have an ability to reason, i.e. it must have some theoretical basis on which to consider and weigh incoming sensor information and make a decision. Third, it must have the ability to take action based on this decision. If these functions are enabled, then the computer can function as a very intelligent and efficient materials processing engineer.

CONCLUSIONS

With regard to sensor development, the availability of high power gas lasers, notably the argon ion, made possible the initial studies in light-scattering and spectroscopy, but the real advance came with solid state detectors, precision microdrives and fiber optics. This work is continuing in this laboratory and others to develop nanometer-scale Raman and Rayleigh-Brillouin probes that further reduce the size of the scattering volume so that individual functional groups or moieties can be investigated with spatial resolution of the order of 1.0 nm. As well as providing the "ultimate" in surface spectroscopy, it will be recognized that such technology will also allow, for example, very rapid mapping of genomes using massively parallel fiber optic nanoprobes. If accurate Rayleigh-Brillouin studies can be conducted using confocal interfacial scattering at this level of spatial resolution, it may well be possible to measure experimentally the *local* free energy as a function of position through an interface. Such experimental information is absolutely vital if we are to understand materials transport and reaction in interfaces and thin films, including the cell wall.

On the modeling front our understanding of the fundamental nature of intermolecular and inter-particulate forces has increased considerably. It can now be said with some justification that the essential physics of simple dense systems with rapid ($\sim 10^{-12}$ s) relaxations are reasonably well understood. This is not the case for complex molecular fluids or polymers. Here relaxation times are relatively slow ($>10^{-3}$s) and the phenomena highly cooperative and long range. Prior to development of the PAANDA and N5 approaches which rely on artificial intelligence, such systems were outside the realm of exact machine calculations. With development of these new machine simulation techniques, such systems are now amenable to more rigorous treatment and many interesting results will undoubtedly unfold. Coupling of AI with massively parallel machines brings the solution of a number of seminal problems within reach if not yet quite within grasp. This should allow, for example, the first direct simulation of nucleation and crystal growth from equilibrium along the melting curve. Similarly, our early work has shown the potential importance of external fields on surfaces and how the presence of a surface affects the conformation and adsorption of individual molecules. Again, new methods will allow the study of larger ensembles where the fully coupled nature of the phenomena, possible near phase transitions, will undoubtedly enable serious investigation in applications ranging from catalysis to the nature of protein interaction on cell walls.

In the area of machine intelligence for process control, progress has been fitful but the work has been great fun. You get a really weird feeling watching a machine make its own decisions and beat a human being in the process, especially when the human is actually a team of graduate engineers! There can be little doubt that the paradigm developed in our laboratory will prove useful in the more widespread development of intelligent computer systems. The expansion of such systems to diverse processes or even complete factories would seem to be relatively straightforward and a number of large companies, notably General Electric and Texas Instruments have sizable groups currently working in the area.

On the other hand, the future of machine intelligence in the area of materials development is an open question. The problem is fundamental and cuts to the core of the very nature of machine intelligence. In all current approaches, the computational problem is reduced to a generalized optimization procedure possibly with combinatorial searching using a genetic algorithm or related technique. One arranges matters so the composition and process variables are correlated with desired properties in such a way as to minimize deviations from a quantity of merit that measures how well the system has "learned". While this has proven successful in optimizing the performance of given classes of known materials, the methods are clearly incapable of discovery. It is the aspect of discovery, of observing the unexpected result, not infrequently through serendipity, that is the hallmark of true scientific advance. It is in careful experimental observation of a phenomenon that does not fit the theory where progress is made. Those studies that serve to provide confirmatory evidence of existing theory or fill in minor details are, of course, valuable and necessary but they are of a derivative nature. It is not obvious how computers will be programmed for discovery!

It is as well to be cautious here. When AI approaches are fully implemented in quantum and statistical mechanics, possibly coupled with genetic algorithms, it may be possible with massively parallel machines to come close to the situation where Dirac's famous comment is no longer the case:

> "a large part of physics and the whole of chemistry are thus completely known, and the difficulty is only that the exact application of these laws leads to equations much too complicated to be soluble". We will have to wait and see.

The interdisciplinary nature of the modern research effort has been a recurring theme throughout much of our work at AFRL. While this is certainly true in MS&E, it is also the case in many subjects that support major application areas such as engineering and medicine. As evidence of the fruitfulness of the interdisciplinary approach we can cite the spectacularly successful work that has resulted at the interface between chemistry and biology called "molecular biology". Work in this field has given rise to the new discipline of molecular science that has resulted in tremendous advances in medicine, genetics, and agriculture. At the interface between chemistry and physics, there has also been a fusion of new ideas in an area variously called materials chemistry, soft matter, and sometimes surface and interfacial science. Here the focus is to understand the forces between atoms and molecules in dense medium and to use this knowledge to design new and useful forms of matter.

As Philip Ball writes [3] (quoted by Rita Colwell, Director of the National Science Foundation in an address to the Materials Research Society (http://www.nsf.gov/ od/lpa/forum/colwell/rc81202.htm)

> "we can make synthetic skin, blood, and bone. We can make an information superhighway from glass. We can make materials that repair themselves, that swell and flex like muscles, that repel any ink or paint, that capture the energy of the sun."

There is hardly a better example of how the old barriers have crumbled. The power and momentum of materials research lies in no small measure in its porous boundary between physics, chemistry, and biology. Thus in the real world of today, we routinely turn chemists into physicists and physicists into materials scientists; no longer recognizing divisions that are not nature-made but rather man-made. The science writer Ivan Amato in his book "Stuff" [4] argues that seminal points in the ascent of mankind are correlated closely with breakthroughs in understanding materials. His opinion is that

> "we are coming to the point where they are gaining the ultimate level of control over the material world. Contemporary materials science is likely to have as profound an effect on posterity as did that original act of materials engineering in eastern Africa's rift valley, where the sound of stone against stone first snapped into the Paleolithic air."

While this enthusiasm is admirable and there has indeed been considerable progress it is essential to temper the enthusiasm with the reality that we are only starting to glimpse the complexity of matter at the mesoscopic level. In 1900 [5] James Clarke Maxwell (J.C. Maxwell, quoted by J.W. Gibbs. Collected Works, vol. 12, Longmans N.Y., 1928; J.S. Rowlinson, Chem. Soc. Rev., 12, no.3, 1983) noted, in relation to classical thermo-dynamics, that what was needed was "to create a science with secure foundations, clear definitions and distinct boundaries."

The dilemma is that when the foundations are absolutely secure, the definitions crystal clear, and the boundaries rigidly distinct, the field may have matured to the point of sterility. On the one hand, our challenge will be to preserve, protect, and encourage development of our core disciplines in Maxwellian fashion. On the other hand, many of the most interesting applied problems arise in interdisciplinary areas. These need to be addressed effectively, if societal needs are to be met, and the charge against the public purse justified. A major challenge for the Universities will be to design courses that have both breadth and depth to meet the challenge or perhaps we will find it easier to give the job to intelligent machines!

REFERENCES

[1] Robert Cahn, 2001. The Coming of Materials Science, Pergammon, 360.
[2] M. L. Good, 1998. Materials Through Chemistry, American Chemical Society, 217.
[3] R. Colwell, 1998. An Alchemy of Disciplines, address to the Materials Research Society, Materials Research Society Annual Meeting.
[4] Ivan Amato, 1977. Stuff: the Materials the World is Made of, Banic Books, Harper Collins.
[5] J. C. Maxwell, 1983. J. W. Gibbs, Collected Works, 12, Longmans, NY, J. S. Rowlinson, Chem. Soc. Rev., 12(3).

Application of an Artificial Neural Network to the Control of an Oxygen Converter Process

J. FALKUS, P. PIETRZKIEWICZ, W. PIETRZYK and J. KUSIAK

Akademia Górniczo-Hutnicza, Mickiewicza 30, 30-059 Kraków, Poland

ABSTRACT

The oxygen converter process is the most important one used in steel production. Therefore, any improvement in steelmaking technology has significant economic value. Development of control systems for oxygen converter processes brought about the creation of a general system, the main components of which are static and dynamic models. Implementation of a static model into the control system allows the undertaking of adequate decisions about the composition of the metallic charge of each heat, i.e., determination of the necessary amount of hot metal and scrap. The right decisions regarding the proper converter charge result in regular converter operation and have significant influence on process economy.

The majority of static models presently applied are based on statistical calculations using a multiple regression equation or calculation of the material and thermal balance of the process. Intermediate processes containing both types of models are also known to be in use.

The main objective of this work is to develop a static model of a new type, using artificial neural networks (ANN). The ANN-based static model consists of two parts. First, the final temperature of steel is predicted. Secondly, the volume of the oxygen blow is predicted. Training and testing data come from real oxygen converter industrial processes. A series of tests were executed, aimed at defining the optimum network structure that would enable prediction of the parameters of the converter charge. Different configurations of the input variables were tested. The achieved results indicate the great potential of models based on artificial neural networks. It is notable that the prediction accuracy of this type of model significantly exceeds the precision of conventional models presently in use. The main drawbacks of statistical or balance-type models are their considerable low *precision of predictions in cases which are not typical for considered conditions. Neural networks solve this problem much better. Neural network based models are simple to use. They can be implemented into the existing structure of on-line control systems.*

INTRODUCTION

The oxygen converter process is the most important one used in steel production. Approximately the 80% of the world total steel production comes from the oxygen converter (LD) process. Therefore, any improvement in steelmaking technology has significant economic value. The crucial part of the LD process is the adequate and efficient control system. Taking into account that the steel refining time is from 15 up to 20 minutes, the inappropriate control algorithm decreases the effectiveness of the steelmaking process. The main goal of many researches is to elaborate the effective control systems of the LD process.

The control system of the LD process has 2 main parts. First one is the off-line static control system, which has to ensure the accurate composition and amount of the converter charge materials. The second part of the control system is the dynamic one. Its goal is to control the oxygen blow during the steelmaking process. The dynamic control systems are based on measurements of the carbon content and the steel bath temperature as well as measurements of the composition of the exhaust gases.

The main goal of the work was to elaborate the static model of the converter control system based on the artificial neural network (ANN) approach. The ANN model can predict the temperature of the steel bath and the required volume of the oxygen necessary for the prime blow of the LD process. The ANN prediction of these variables is based on the knowledge of such input variables as: mass and temperature of liquid hot metal, mass and composition of scrap, mass of ore, etc.

OXYGEN CONVERTER PROCESS

The principal goal of the oxygen converter process is to blow through the hot metal using the technically pure oxygen. The oxygen is blown at the ultrasonic speed, thanks which it penetrates deeply the metal bath. The main component of the metal is carbon, which during the oxidation gives the gas product creating the metal-slag emulsion. The oxidation products of the other additives as Si, Mn or P go directly to slag.

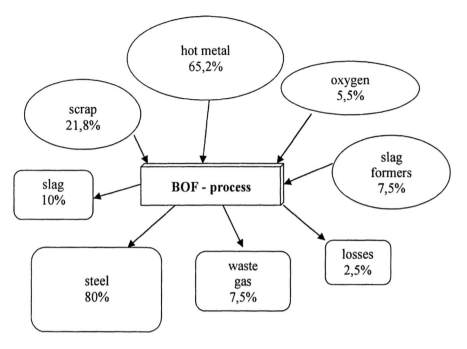

Figure 1. Flowchart of mass flow in an oxygen converter process.

The converter process is difficult to control. It is almost impossible to correct the energy of the process in an on-line system. The energy source is the physical heat of the liquid hot metal and the hidden chemical heat of the input additives. Therefore, the adequate composition of the input materials for every steelmaking cycle is the crucial problem for the economy of the whole process. The total of physical heat of hot metal and chemical components exceeds the required final temperature. Therefore, it is necessary to correct the composition of metallic input materials by adding some scrap. Additionally, some slag formers are added: lime, dolomite, etc. Sometimes, the iron ore is added as a cooling material. If the amount of the hot metal is not sufficient, some coke is added [5].

The final product of the converter process is the metallic bath of the required chemical composition and temperature. The slag and exhaust gases are the side-product of the converter process. The flow chart of the mass flow in the oxygen converter process is presented in Figure 1.

STATIC MODEL OF THE OXYGEN CONVERTER PROCESS

The static model of the oxygen converter process based on the artificial neural network (ANN) requires the collection of the process data sets. The process input parameters necessary for the correct model of the converter process are very well known, however, it is very difficult to measure all of them and even, some of them can not be measured. The values of these parameters can be evaluated indirectly.

For example, it is impossible to evaluate the heat losses. These values are obtained from the correlated parameters: the temperature of the previous heat, the time breaks between the steelmaking cycles. The similar is the problem of the strong dependencies of the oxidation of slag and the oxidation of converter dusts.

The database consists of two campaigns of 330 Mg converter and one campaign of 130 Mg converter. The converter campaign is the number of all heats during the time between to successive overhauls of the converter during which a ceramic converter linings is being replaced.

ANN CONVERTER MODEL

Artificial neural networks have become an effective tool in modelling metallurgical processes. The first author's work focused on applying neural networks in modelling metal forming processes and modelling mechanical and microstructural properties of metals [3,4,6–8]. The results obtained inspired further research in the field of steelmaking processes. The results of the application of ANN in the ladle furnace, blast furnace and steel refining in the RH-degassing process were published in [1,2,9–13].

Two separate models were built in this work. The first one predicts the final temperature of the metal bath after the primary oxygen blow. The second predicts the required volume of the primary oxygen blow. The ANN models were built with the Statistica Neural Networks software package.

Data were collected in 2 different converters: 330 Mg (2 campaigns—1446 and 1263 records, respectively) and 130 Mg (849 records). The initial vector of the input variables was analysed by the automatic design module of the Statistica Neural Networks software package. Some variables were eliminated and the results are shown in Table 2. Different topologies of the neural networks were tested. The vector of input variables had different sizes from 11 to 19 parameters, while the hidden layer had 1 to 16 neurons. The RMS error of the metal bath prediction for the first converter ranged from 26.7°C to 32°C, while the range of the RMS error for the volume of the oxygen blow was between 620 m^3 and 701 m^3. The RMS error of the test of the second converter for the temperature and oxygen volume predictions were much smaller: 19°C and 180 m^3, respectively. The prediction results for temperature and oxygen volume in the primary blow for optimal topology of the neural network are presented in Table 3.

Table 1. Vector of selected variables of the static model of the converter process.

number of the heat	x_1
metal mass, t	x_2
metal temperature, °C	x_3
C contents in metal, %	x_4
Mn contents in metal, %	x_5
Si contents in metal, %	x_6
P contents in metal, %	x_7
S contents in metal, %	x_8
final C content in steel, %	x_9
mass of heavy scrap, kg	x_{10}
mass of light scrap, kg	x_{11}
mass of rolled scrap, kg	x_{12}
mass of W15 scrap, kg	x_{13}
mass of pig iron scrap, kg	x_{14}
mass of other scrap, kg	x_{15}
mass of lime, kg	x_{16}
mass of dolomite, kg	x_{17}
dolomite stone, kg	x_{18}
ore, kg	x_{19}
coke, kg	x_{20}
mass of solid fuel, t	x_{21}
temperature of a previous steelmaking cycle, °C	x_{22}
time break between the heats, min	x_{23}
volume of the blown oxygen, m^3	x_{24}
temperature after the prime blow, °C	x_{25}

Table 2. Vector of input variables for two considered artificial neural networks.

Variable	Temperature of the metal bath			Oxygen volume in the prime blow		
	Campaign I	Campaign II	Campaign III	Campaign I	Campaign II	Campaign III
heat number	X			X	X	
metal mass, t	X	X	X	X	X	X
metal temperature, °C	X	X	X	X	X	X
C content in metal, %					X	
Mn content in metal, %	X		X	X	X	X
Si content in metal, %	X	X	X	X	X	X
P content in metal, %	X	X	X		X	X
S content in metal, %		X	X			
final C content in steel, %			X			X
mass of heavy scrap, kg	X	X	X*	X	X	X*
mass of light scrap, kg	X	X		X	X	
mass of rolled scrap, kg	X	X		X	X	
mass of W15 scrap, kg	X	X		X	X	
mass of pig iron scrap, kg	X			X	X	
mass of other scrap, kg				X		
lime, kg		X	X	X	X	X
Dolomite, kg			X		X	
Dolomite stone, kg						
mass of solid fuel, t			X			X
ore, kg	X	X		X	X	
coke, kg	X	X		X	X	
temperature of previous steelmaking cycle, °C			X			X
break between heats, min			X			X
volume of oxygen, m³	X	X	X			
temperature after primary blow, °C				X	X	X

*total mass of heavy and light scrap

Table 3. Topologies of artificial neural networks and corresponding errors of temperature and oxygen volume predictions.

Campaign	ANN	Structure	Temperature, °C RMS of the test	Structure	Oxygen volume, m³ RMS of the test
Campaign I	MLP	14 : 8 : 1	26.9	15 : 8 : 1	686
Campaign II	MLP	13 : 9 : 1	29.9	17 : 9 : 1	661
Campaign III	MLP	14 : 6 : 1	18.8	14 : 4 : 1	196

INDUSTRIAL DATA-SELECTION STRATEGY

It was observed that the data collected in the industrial conditions were very noisy, and it was necessary to eliminate these records. An example of the distribution of the metal charged to the first converter is presented in Figure 2.

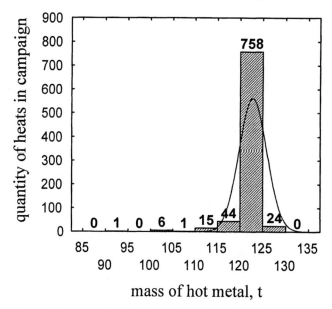

Figure 2. Distribution of the metal charge of convertor I.

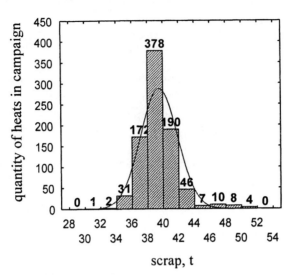

Figure 3. Distribution of mass of scrap in the charge of converter I.

Table 4. Selection of the data base according to the distribution of the input variables.

Variable	Number of eliminated records
metal mass	8
scrap mass	12
time break between cycles	3

Table 5. RMS error for different testing data sets: before and after selection (filtering according to the technological reasons)

Database	Temperature RMS error	Oxygen volume RMS error
Initial data set	18,8°C	196,3 m^3
After filtration of the metal mass	17,6°C	195,5 m^3
After filtration of the scrap mass	18,5°C	188 m^3
After filtration of the time break between cycles	17,9°C	188,2 m^3
After filtration of 3 variables: metal mass, scrap mass and time break between cycles	18,3°C	189 m^3

The data presented in Table 5 are mean values of the RMS errors for 6 random experiements. The filtered data improved the prediction of the networks. However, the degree of improvement is not very significant. The RMS error was calculated by the following relation:

$$RMS = \sqrt{\frac{\sum_{i=1}^{N}(y_{rz} - y_m)^2}{N}} \qquad (1)$$

where: y_{rz} – measured value of the output variable,
y_m – value calculated from the ANN model,
N – number of records in the data base.

STATISTICAL APPROACH

The elaborated ANN model was next verified using the statistical approach. The following linear regression model was considered:

$$y_m = b_0 + b_1 x_1 + ... + b_N x_N \qquad (2)$$

In the first step, all data (not filtered data set) were used. The second step was the filtering procedure. Each record was examined and the value of so-called residual error was evaluated. The variation of this parameter was observed and records, which give the residual errors of temperature and oxygen volume exceeding 1°C and 20 m³, respectively, were eliminated. 53 records for the temperature model and 16 records for the oxygen model were eliminated as the result of this analysis. The filtered data set was validated next in the neural network approach. Similar topologies as previously were used (14:6:1 for temperature model and 14:4:1 for oxygen model). The obtained results are presented in Table 6.

Table 6. The average values of RMS error for all and filtered data sets (the filtration on the base of residual error).

Data base	Temperature	Oxygen volume
	RMS error	RMS error
before filtration	18,8 °C	196,3 m³
after filtration	16,2 °C	182,5 m³

CONCLUSIONS

The analysis of obtained results allows formulating the following conclusions:
- The static model of the oxygen converter process based on the ANN approach does not require more than one hidden layer.
- The input parameters such as: metal mass, metal temperature, silicon content in metal, heavy scrap, light scrap, scrap W15, pig iron scrap, ore, coke, oxygen in prime blow and the temperature after the prime blow give the most interesting information used in the ANN training,
- The RMS error value of the model based on the ANN approach is slightly lower than this obtained by the regression model.
- Simple technological analysis of the input data is not sufficient to eliminate the false records in the data base.
- Filtration of the input data set according to the statistical analysis of the residual error gives much better results, but does not assure the required accuracy of the model records.

ACKNOWLEDGEMENTS

The work was financed by the Polish Research Committee, project No. 7T08B 06220.

REFERENCES

[1] J. Chen, 2001, A predictive system for blast furnace by integrating a neural network with qualitative analysis. Eng. Appl. of Artificial Intelligence, 14, 77–85.
[2] J. Falkus, P. Pietrzkiewicz, 2001. Neural networks in statical controlling of oxygen converter process, Proc. Conf. High. Technologies in Advanced Metal Science and Engineering', St. Petersburg, Russia.
[3] J. Kusiak, D. Svietlichnyj, M. Pietrzyk, 2001. Application of Artificial Neural Networks in the On-Line Control of Hot-Flat Rolling Processes. Int. J. Engineering Simulation, 1, 3, 17–23.
[4] Kuziak R., Zalecki W., Kusiak J., 2001, Zastosowanie sieci neuronowych do oceny wpływu pierwiastków domieszkowych na własności mechaniczne wyrobów stalowych. Informatyka w Technologii Materiałów, 1, 44–51.
[5] M. Kruciński, J. Falkus, 1988. Materiały do ćwiczeń audytoryjnych i projektowych z metalurgii stali. Skrypt 1140 AGH, Kraków.
[6] M. Pietrzyk, J. Kusiak, S.M. Roberts, 2001. A history dependent constitutive model for the hot forming of Waspaloy. Proc. Conf. ECCM-2001, Kraków, CD-ROM.
[7] S.M. Roberts, J. Kusiak, Y.L Liu, Forcellese, A., Withers, P.J., 1998, Prediction of damage evolution in forged aluminium metal matrix

composites using a neural network approach, J. Materials Processing and Technology, 80-81, 507–512.
[8] S.M. Roberts, J. Kusiak, 2001. The application of artificial intelligence to the constitutive and microstructural behaviour of nickel-base superalloys. Proc. Conf. ESAFORM2001, Liege, 441–444.
[9] J. Falkus, H. Gutte, 1999, Verwendung neuronaler Netze für die statische Steuerung des Sauerstoffblasprozesses Barbaratagung. Moderne Materialtechnik, Hüttentechnik, Gießereitechnik, Mat. Konf. Glastechnik und Keramik, Duisburg, 52–58.
[10] 10. S.Y.Yuan, K.S Chang, S.M. Byun, 1996. Dynamic Prediction Using Neural Network for Automation of BOF Process in Steel Industry, I&SM, 37–42.
[11] 11. J. Falkus, W. Wajda, P. Pietrzkiewicz, 1999. Zastosowanie sztucznych sieci neuronowych do statycznego sterowania tlenowym procesem konwertorowym, Mat. Seminarium NeuroMet'99, AGH Kraków, 71–76 (in Polish).
[12] 12. H. Whitaker, R. Mulder, M. Hartwig, M. Poschman, 2001. Neural networks for process condition prediction in basic oxygen steelmaking, Final Report, European Commission, EUR 19467.
[13] J. Falkus, P. Pietrzkiewicz, P. Drożdż 2001. "Ocena możliwości zastosowania sztucznych sieci neuronowych do sterowania pracą urządzenia RH". XXI Konf. Naukowo-Techniczna Huty Katowice "Automatyzacja i informatyzacja w hutnictwie—doświadczenia, problemy i kierunki rozwoju, Rogoźnik, 24-26.10.2001, s. 1–8.

Improving the Prediction of the Roll Separating Force in a Hot Steel Finishing Mill

Y. FRAYMAN and B. F. ROLFE

School of Information Technology and School of Engineering and Technology, Deakin University, Geelong, VIC 3217, Australia

P. D. HODGSON

School of Engineering and Technology, Deakin University, Geelong, VIC 3217, Australia

G. I. WEBB

School of Computer Science and Software Engineering, Monash University, Clayton, VIC 3168, Australia

ABSTRACT

This paper focuses on the development of a hybrid phenomenological/inductive model to improve the current physical setup force model on a five stand industrial hot strip finishing mill. We approached the problem from two directions. In the first approach, the starting point was the output of the current setup force model. A feedforward multilayer perceptron (MLP) model was then used to estimate the true roll separating force using some other available variables as additional inputs to the model.

It was found that it is possible to significantly improve the estimation of a roll separating force from 5.3% error on average with the current setup model to 2.5% error on average with the hybrid model. The corresponding improvements for the first coils are from 7.5% with the current model to 3.8% with the hybrid model. This was achieved by inclusion, in addition to each stand's force from the current model, the contributions from setup forces from the other stands, as well as the contributions from a limited set of additional variables such as: a) aim width; b) setup thickness; c) setup temperature; and d) measured force from the previous coil.

In the second approach, we investigated the correlation between the large errors in the current model and input parameters of the model. The data set was split into two subsets, one representing the "normal" level of error between the current model and the measured force value, while the other set contained the coils with a "large" level of error. Additional set of data with changes in each coil's inputs from the previous coil's inputs was created to investigate the dependency on the previous coil.

The data sets were then analyzed using a C4.5 decision tree. The main findings were that the level of the speed vernier variable is highly correlated with the large errors in the current setup model. Specifically, a high positive speed vernier value *often correlated to a large error. Secondly, it has been found that large changes to the model flow stress values between coils are correlated frequently with larger errors in the current setup force model.*

INTRODUCTION

Maintaining product consistency and quality in the manufacturing process has become a widespread concern as a result of increasing competition in the world markets. Increasing demands on the quality of rolling mill products have led to great efforts to improve the control and automation systems of the rolling process [1–3]. Hot steel rolling is one of the most important steel manufacturing processes. Hot rolling is the first metal shaping process after the slab has been cast, in flat products such as plate, strip and sheet. The final shaping stage of hot rolling steel strip is normally performed on a tandem mill known as a finishing mill consisting typically of two to six stands. Here the final thickness, flatness and profile of the work-piece are determined. It is important to have a sound understanding of the behavior of the roll gaps in the finishing mill for design, scheduling and control purposes [2]. In particular, accurate predictions of the roll separating force are necessary to meet the current and the future quality standards of final product dimensions and flatness.

This paper focuses on the development of a hybrid phenomenological/inductive model to improve the current physical setup force model on a five stand industrial hot strip finishing mill. The motivation for the application of inductive learning-based methodologies lies in the fact that they do not require the expert development of phenomenological models [4–7]. This technology could provide a powerful tool for accurate prediction of the roll separating force, thereby ensuring that the products manufactured conform to target specifications and thus contribute to enhanced business benefits.

The mill settings are determined from the physical models based on expert metallurgical and mechanical knowledge. The set-up of these models is crucial since it determines, to a large extent, the thickness of the final product. In practice, it is sometimes observed that the roll-gap settings produced by set-up models are not as accurate as those required by increasing consumer product-quality demands. Although small errors can usually be compensated for by the mill controllers, larger errors lead to quality degradation and potentially out-of-specification product. This is particularly prominent in the case of first coils in a rolling campaign whenever there has been a change in width or thickness or steel grade [3].

The mill set-up errors arise since the set-up model only uses factors whose exact physical relationships are understood. Unfortunately, the rolling process involves many additional factors that affect the elastic/plastic material deformation in the roll gap, particularly due to the stochastic nature of the rolling process. In this sense, the physical model is far from perfect.

EXPERIMENTAL RESULTS

We are interested in improving a current setup force model on five stand hot strip industrial finishing mill using the input, intermediate and output values of various setup models and the measured values of the force at each stand. We approached the problem from two directions. Firstly, we analyzed which variables may add more information to the estimation of the true force for each stand. Secondly, we investigated which variables were correlated to large errors in the current setup force model.

Improving the Estimation of the Stand Forces

A feedforward multilayer perceptron (MLP) model [8] was used to investigate possible improvements to the setup force model for each stand. The starting point was the output of the current setup force model for the available 10,000 records of production coils. An MLP model was then used to estimate the true force at each stand using some other available variables as additional inputs to the model.

The MLP model enables us to measure whether inclusion of certain variables, or groups of variables, in the current setup force model can:

a) provide additional information about the rolling stands that has not been included in the current model;
b) achieve a better estimation of the true roll separating force.

The sensitivity of each of the MLP's input variables gives an indication of how much information each variable can add to the current model. The input parameters to the MLP model were chosen based on a sensitivity analysis where only the most important variables were retained. The final selection of the input variables can be found in the Table 1.

The results obtained for both the current setup and the hybrid models are in Table 2.

As can be seen from the Table 2, it is possible to significantly improve the estimation of a roll separating force from 5.3% error on average with the current setup model to 2.5% error on average with the hybrid model by adding some additional variables to the MLP model. The corresponding improvements for the first coils are from 7.5% with the current model to 3.8% with the hybrid model. This was achieved by inclusion, in addition to each stand's force from the current model, the contributions from setup forces from the other stands, as well as the contributions from a limited set of additional variables in Table 1.

Table 1. The final input parameters to the MLP model.

Variable Name
Aim width (current and the previous coil)
F4 roll balance reference (current and the previous coil)
F5 roll balance reference (current and the previous coil)
Setup Force (F1–F5) (current and the previous coil)
Delayed Measured Force (F1–F5) (previous coil only)
Setup Forward Slip (F1–F5) (current and the previous coil)
Setup Speed (Entry and F1–F5) (current and the previous coil)
Setup Temp (Entry and F1–F5) (current and the previous coil)
Setup Thickness (Tbar and F1–F5) (current and the previous coil)
Arc of Contact (F1–F5) (current and the previous coil)
Strain Rate (F1–F5) (current and the previous coil)
Carbon (current and the previous coil)
Molybdenum (current and the previous coil)
Niobium (current and the previous coil)
Phosphorus (current and the previous coil)
Sulphur (current and the previous coil)
Speed Vernier (current and the previous coil)
Thickness Vernier (current and the previous coil)

Table 2. The results of the estimation of the roll separating force by the current setup and the hybrid model for test data set (3000 coils). First Bar results are for first bars only for the test data set (752 coils). Here, F1-F5 are the roll separating forces for the respective stands, and Relative Error = Abs(Predicted Force - Measured Force)/Measured Force.

		Relative Error (Current Setup Model)					Relative Error (Hybrid Model)				
		F1	F2	F3	F4	F5	F1	F2	F3	F4	F5
Overall	Avg	0.036	0.049	0.061	0.044	0.073	0.018	0.020	0.025	0.024	0.043
	Stdev	0.034	0.045	0.045	0.040	0.063	0.018	0.021	0.025	0.026	0.043
First Bars	Avg	0.054	0.073	0.078	0.066	0.104	0.027	0.032	0.038	0.037	0.064
	Stdev	0.043	0.057	0.058	0.051	0.083	0.025	0.028	0.034	0.034	0.053

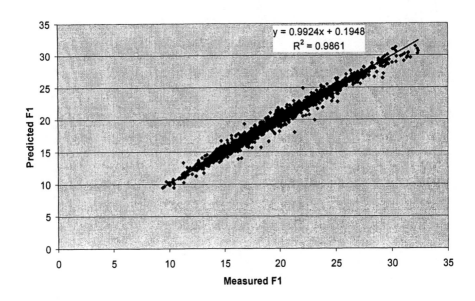

Figure 1. Predicted rolling force F1 [t] versus measured rolling force F1 [t] with a hybrid model (overall).

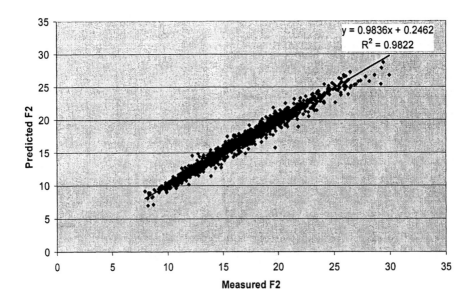

Figure 2. Predicted rolling force F2 [t] versus measured rolling force F2 [t] with a hybrid model (overall).

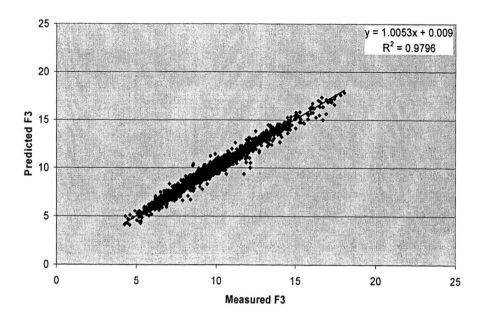

Figure 3. Predicted rolling force F3 [t] versus measured rolling force F3 [t] with a hybrid model (overall).

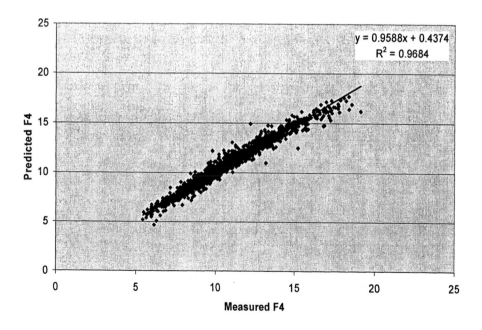

Figure 4. Predicted rolling force F4 [t] versus measured rolling force F4 [t] with a hybrid model (overall).

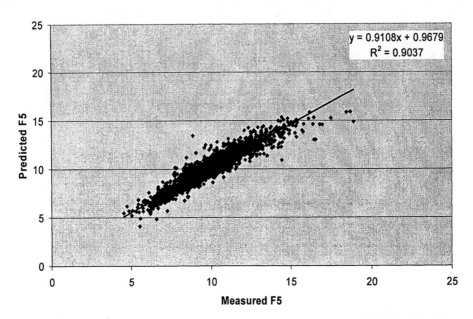

Figure 5. Predicted rolling force F5 [t] versus measured rolling force F5 [t] with a hybrid model (overall).

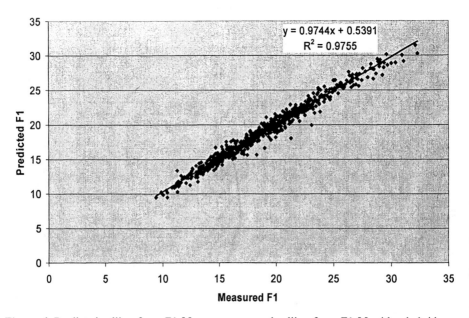

Figure 6. Predicted rolling force F1 [t] versus measured rolling force F1 [t] with a hybrid model (first bars).

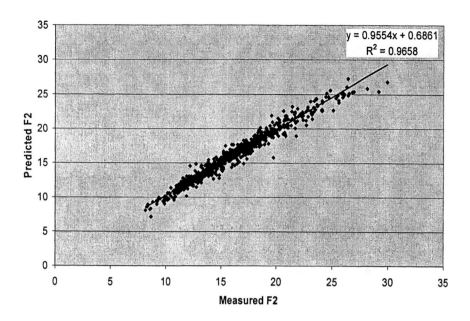

Figure 7. Predicted rolling force F2 [t] versus measured rolling force F2 [t] with a hybrid model (first bars).

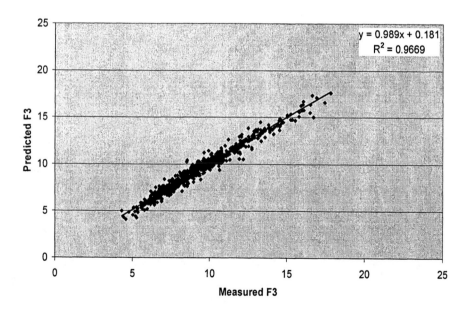

Figure 8. Predicted rolling force F3 [t] versus measured rolling force F3 [t] with a hybrid model (first bars).

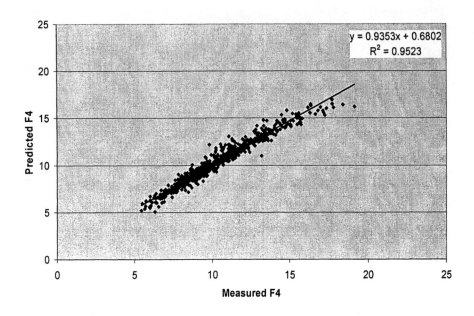

Figure 9. Predicted rolling force F4 [t] versus measured rolling force F4 [t] with a hybrid model (first bars).

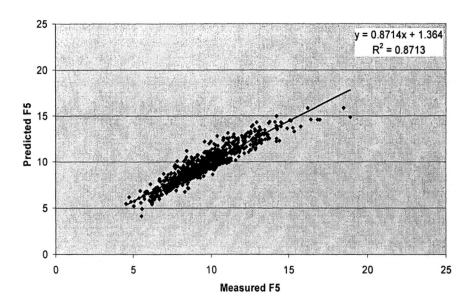

Figure 10. Predicted rolling force F5 [t] versus measured rolling force F5 [t] with a hybrid model (first bars).

Table 3. The ranking of variables in order of their correlation with F1–F5 overall.

Variable Name	Relative importance
Aim Width	0.208
Setup Thickness	0.178
Arc of Contact	0.149
Setup Strip Temp	0.129
Strain Rate	0.120
Setup Force	0.114
Delayed Measured Force	0.113
Setup Roll Speed	0.071
Carbon	0.061
Setup Predicted Forward Slip	0.059
Niobium	0.045
Molybdenum	0.039
F5 Roll Balance Ref	0.027
Phosphorus	0.020
F4 Roll Balance Ref	0.015
Sulphur	0.011
Thickness Vernier	0.011
Speed Vernier	0.010

Table 4. The input parameters of the current setup force model for correlation analysis.

Variable Name
F4 Roll Balance Ref
F5 Roll Balance Ref
Model Flow Stress F1–F5
Base Temp Entry F1–F5
Lower thread speed limit
Upper thread speed limit
Hardness code
Aim Width
Aim Thickness
Aim Exit Temp
Aim Coil Temp
Chemical composition
Work Roll Diameter F1-F5
Setup Draft
Tension ref looper 1–4
Angle ref looper 1–4
Thickness Vernier
Speed Vernier
Gauge Meter Error F1–F5

The scatter plots of the estimated forces versus measured forces for each stand with the hybrid model are in Figures 1–10.

The ranking of the input variables to the hybrid model in order of their importance are in the Table 3.

Correlating Current Setup Force Model Errors

The second approach to improving the current setup force model was to investigate the correlation between the large errors of the model and a restricted set of input parameters (or initial conditions) of the current setup force model. The intermediate variables, such as Arc of Contact, were not considered. The setup force, which is an output of the current model, was also chosen as a (input) variable of interest. The list of variables used in this analysis can be found in Table 4.

The data set of production coils was split into two subsets (for each stand), one representing the "normal" level of error between the current model and the measured force value, while the other set contained the coils with a "large" level of error. A second set of data was created to see if there was any dependency on the previous coil. A change in each coil's inputs from the previous coil's inputs was recorded (the inputs of the very first coil in the data set were set to zero).

The data sets of the input parameters and the output variable for each stand were then combined and analyzed using a C4.5 decision tree. The decision tree is useful in finding a model that can classify data into two or more discrete classes. The decision tree can also be converted into rule sets for each class; these rules are often easier to interpret than the tree diagram.

The data distribution of the "normal" sets and the corresponding "large error" sets is highly skewed as each "normal" set contained on average 95% of the data. To alleviate this problem, the decision trees were grown using an over-sampled "large error" set and then tested on the unmodified sample distribution. This provides a bias for the model to the "large error" data to extract any significant, but small in frequency, relationships. A cross validation was used to prevent the model from over-fitting. Next, a rule set was extracted for each condition.

The main findings from the analysis of the errors between the output of the current model and the measured force for each stand are that the level of the Speed Vernier variable is highly correlated with the large errors in the current setup force model. Specifically, it was discovered that a high positive Speed Vernier value often correlated to a large error. Secondly, it has been found that large changes to the Model Flow Stress values between coils are correlated frequently with larger errors in the current setup force model.

The minor findings are that the large errors in the first four stands are not influenced by a coil's chemical composition. The last stand was slightly influenced by the Carbon content and the coil to coil change in the Chromium levels. There were also minor consistent influences across most stands by the Lower Thread Speed Limit and the Aim Exit Thickness variables. The rules are shown below. Only the two most accurate rules for Stands 4 and 5 are shown for space reasons. This can lead to anomalies (or interesting results) because often the most accurate rules are not the most frequently used rules. A rule (IF-THEN format) is made up of ANDed expressions. A second or third coil is a proceeding coil following a first coil (without any changes in the input conditions). The rules for the current condition variables use the following labels:

- Low for values that are one standard deviation or more below the average value for that variable;
- Medium for values that are within one standard deviation from the average value of the variable;
- High for values that are above one standard deviation from the average value of the variable.

The first order difference variables are treated differently due to them being mainly zero average, so the low, medium and high labels do not apply in the same manner. The first order difference variable ranges from large negative changes to large positive changes. Slight positive or negative changes are those that are much smaller than one standard deviation.

Stand 4

Combined Current Condition and First Order Difference data sets

Rule set error = 18.9%
18 rules (6 for normal condition, 12 for large errors)

Normal Condition:
Rule 1:
- IF Set-up Force F3 is at low or medium levels
- AND Δ F4 Roll Balance Ref is slightly negative or is positive
- AND Δ Model Flow Stress F2 is contained within less than one standard deviation around zero
- AND Δ Set-up Force 1 is slightly negative or is positive
- AND Coil is not a First Bar
- THEN Not a Large Error

Rule 2:
- IF Set-up Force F3 is at a high level
- AND Set-up Force F4 is at a high level
- AND Δ Model Flow Stress F2 is contained within less than one standard deviation around zero
- AND Coil is not a First Bar
- THEN Not a Large Error

Large Error Rules:
Rule 1:
- IF Set-up Force F3 is at low or medium levels
- AND Set-up Force F4 is at a high level

- AND Gauge Meter Error is at a high level
- AND Coil is not a First Bar
- THEN Error is Large

Rule 2:
- IF Aim Exit Temperature is at a medium temperature and above
- AND Carbon is at a medium to high carbon steel level
- AND Δ Set-up Force 5 is very negative
- THEN Error is Large

Primary variable	Secondary variables	Tertiary variables
Coil Counter (reset at each First Bar)	Δ Model Flow Stress F2	Δ Model Flow Stress F2
	Aim Exit Temperature	Speed Vernier

Stand 5

Combined Current Condition and First Order Difference data sets

Rule set error = 15.1%
16 rules (3 for normal condition, 13 for large errors)

Rule 1:
- IF F5 Roll Balance Ref is at a low or medium levels
- AND Lower Thread Speed Limit is at a low or medium levels
- AND Set-up Force F5 is at a medium level and above
- AND Δ Set-up Force F1 is slightly negative or is positive
- AND Coil is not a First Bar
- THEN Not a Large Error

Rule 2:
- IF Lower Thread Speed Limit is at a low or medium levels
- AND Set-up Force F1 is at a medium level or above
- AND Set-up Force F5 is at a medium level or above
- AND Δ Set-up Force F1 is slightly negative or is positive
- AND Coil is not a First Bar
- THEN Not a Large Error

Large Error Rules:

Rule 1:
- IF Aim Exit Thickness is at a medium level and above
- AND Carbon is at a high carbon steel level
- AND Δ Set-up Force F5 is slightly negative or positive
- THEN Error is Large

Rule 2:
- IF Aim Exit Thickness is at a medium level or above
- AND Carbon is at a high carbon steel level
- AND Δ Model Flow Stress F4 is very negative
- AND Δ Set-up Force F1 is very positive
- THEN Error is Large

Primary variable	Secondary variables	Tertiary variables
Coil Counter (reset at each First Bar)	Set-up Force F5	Lower Thread Speed Limit
	Base Temp Entry F1	Δ Chromium

CONCLUSION

We were interested in improving a current setup force model on five stand hot strip industrial finishing mill. We approached the problem from two directions. Firstly, we analyzed which variables may add more information to the estimation of the true force for each stand. Secondly, we investigated which variables were correlated to large errors in the current setup force model.

The most promising direction found for improvement of the current setup model is to include in addition to each stand's force from the current setup model a weighted contributions of setup forces from the other stands; as well as contributions from a limited set of additional variables. It was also found that the improvements made to force estimation on First Bars were similar to the overall improvements made to the current setup model with the exception of the Stand 5. Stand 5 required additional minor variables to improve its force estimation on First Bars.

The second direction of this investigation has revealed that the Speed Vernier variable is highly correlated to the large errors in the current setup model. Moreover, large changes to the Model Flow Stress values between coils are correlated with large errors in the current setup model.

REFERENCES

[1] Y. Bissessur, E. B. Martin, A. J. Morris and P. Kitson, 2000. Fault detection in hot steel rolling using neural networks and multivariable statistics, IEE Proc.-Control Theory Applications, 147(6), 633–640.

[2] S. Duysters, J. A. J. Govers and A. J. J Van Der Weiden, 1994. Process interactions in a hot strip mill: Possibilities for multivariable control, IEEE, 1557.

[3] D. C. Martin and P. D. Hodgson, 1998. Friction modeling in the hot rolling of steel strip, Proceedings of the Biennial Conference of the Institute of Material Engineering, Materials 98, 253–258.

[4] O. Wiklund, 1995. Rolling force models for temper rolling using finite elements and artificial neural networks, Proc. Steel Strip'96.
[5] Y. J. Hwu and J. G. Lenard, 1996. Application of neural networks in the prediction of roll force in hot rolling, Proceedings of the 37th Mechanical Working and Steel Processing Conference, 549–554.
[6] J. Leven, N.G. Johnsson and O. Wiklund, 1995. An artificial neural network for rolling applications, Steel Times, April, 137.
[7] R. Pichler and M. Pffaffermayr, 1996. Neural networks for on-line optimisation of the rolling process, Iron and Steel Review, August, 45–56.
[8] S. Haykin, 1999. Neural networks: A comprehensive foundation. Prentice Hall, 2nd ed.

A Solution to the N-Body Problem for Particles of Non-Spherical Geometry Using Artificial Intelligence

M. BENEDICT and J. F. MAGUIRE
Air Force Research Laboratory, Materials and Manufacturing Directorate, Wright-Patterson AFB, Ohio, USA

ABSTRACT

In this paper we describe a new approach, based on artificial intelligence methods to solve the N-Body problem for particles of non-spherical geometry. This is illustrated by solving the dynamics of the N-body problem for an ensemble of hard smooth impenetrable triangles in two dimensions, but the method may readily be extended to large numbers of objects of essentially arbitrarily complex geometry or interaction potential. The approach, based on a particulate artificial neural network dynamic algorithm (PANNDA) is more than two orders of magnitude faster than existing methods when applied to large systems and is only marginally slower (~10%) than the theoretical lower limiting case of hard spheres. Applications to granular media, sand piles, slurries, and avalanches are briefly discussed.

INTRODUCTION

In this paper our objectives are twofold. First, to give a brief overview of how methods in machine intelligence have been used in materials processing. Secondly, to illustrate how similar approaches may be used to overcome serious problems in computational statistical mechanics and quantum molecular dynamics. In this regard it is interesting to note that the breakthrough in computational methodology came in the applied area before being introduced into so-called fundamental research.

OVERVIEW

Artificial intelligence (AI) techniques in materials science fall into three categories. First, intelligent process control is now an established technology and is being increasingly applied in real world applications [1]. It is now generally accepted that significant gains can be made in the area of materials processing through the paradigm of integrating appropriate materials sensors [2] with quite sophisticated materials transformation models (process models) that capture the coupled effects of chemical reactivity and transport phenomena [3]. In the most sophisticated implementations [4, 5], information from both real sensors and virtual sensors (models) is interpreted by a decision support hierarchy that manages the process control machinery [6]. This represents a fundamental shift in industrial materials process control philosophy in that the focus is now to measure, interpret, and control the structure of the material *in situ* during processing rather than the traditional approach of controlling the external parameters (pressure, temperature, heating rate, etc.) of the processing equipment.

In order that a machine can process chemical knowledge, it is essential that the machine obtain information on the physico-chemical state of the system. This information is supplied by some suitable number and types of sensors. Just as the human senses supply information on the world around us to enable us to make decisions, so too suitably chosen physical and chemical sensors can be engineered to interface with specialized materials and process control computers that possess "senses" uniquely suited to processing a particular class of material. These sorts of sensors, their number and type are clearly a matter determined by the specifics of the particular process operation.

The second area in which AI is increasing in use is in the discovery of useful new materials. Materials discovery is a major challenge with high risk but with the potential for commensurately high return. On the one hand this area can and has, been attacked by applying AI to data mining, to rapid mapping of phase diagrams, and to the design of new materials. A number of contributions in these proceedings and previous work [7, 8] address this important area, and so only a few general comments are appropriate here. Clearly it is important to realize that discovery is an essentially artistic or creative endeavor of the human intellect and there arises deep philosophical questions as to what degree a machine might achieve this end. However, as a practical matter it may well prove possible to extract patterns and trends and in a "robotic" fashion to discover useful new compositions of matter.

The final area to which AI may be applied is at the leading edge of advanced computational research. It is to this area that the remainder of this paper is devoted. Here, the objective is to use the computer as a sort of experimental apparatus to probe nature in a way that cannot be done by other methods. This kind of fundamental scientific research can be enabled by AI approaches and though the work is fundamental, the pathway to fairly rapid application is clear. For example, a range of advanced applications including high power radars, cloaking (stealth and chameleon) technologies, ultra-light weight airframe structures, photonic band gap materials, and very large adaptive space-based optics requires development of a range of new materials with various measures of performance that are far in excess of current capabilities. Recent discoveries in the area of soft, interfacial, and granular materials [9-11] hold much promise in that it may be possible to design new nanomaterials with engineered properties and adaptively-controlled structure. It will be appreciated, for example, that these materials may not be at the thermodynamic limit so that they are technically small systems for which macroscopic theory either does not apply or is in need of revision [12]. It is critical to future developments in this field that the well-known limitations of distance and time scale [13, 14] in conventional "molecular dynamics" calculations be overcome if the full promise of nanomaterials is to be realized. It is in this area that AI approaches will prove useful.

THE N-Body PROBLEM AND MOLECULAR DYNAMICS

Poincare [15] won a prize of 2500 crowns from King Oscar II of Sweden and Norway for his mathematical proof that there is no analytic solution to the dynamics of the three (and higher) body problem. Recent advances in chaos theory [16] and the physics of many body systems, the cosmos, liquids, and moving heaps of sand underline the necessity to employ numerical methods. It is for this reason that the understanding of disordered systems, notably liquids, made little real progress until the advent of digital computers. Numerical methods based on Molecular Dynamics and Monte Carlo simulation have achieved a degree of maturity and are now widely employed. Ciccotti, Frenkel, and McDonald [17] have assembled a useful compilation of the early papers and a number of books [13-14] discuss the technique in detail.

Notwithstanding the success of these methods, there remain two well-known problems. First it is only possible to simulate rather small systems and secondly, it is possible to do so only for rather short times. For simple monoatomic fluids far from phase transition, this is not a serious limitation, but for many other systems the limitation is serious. It is possible to get a rough order of magnitude of the problem from a simple example. In a polycrystalline metal with grain size of 1 micron, each grain contains a number of atoms, N, of order 10^{10}. A realistic simulation will require perhaps 10^3 grains, n, so that an atomistic approach will need to track the position and momentum of a total nN or about 10^{13} atoms. Because interatomic potentials are very repulsive at short range, it is necessary in such calculation to use a time step of the order of 10^{-15} s in order to conserve total energy. Now calculation of the force requires that the distance between each atom be evaluated and since there are $\approx (nN)^2/2$ pairs (or nN log(nN), if various tricks are used) one is led to the realization that about 10^{14} forces must be computed every femtosecond. Bearing in mind that complex macroscopic systems such as plastics and polycrystalline metals relax over millisecond timescale, this mammoth computation would need to be repeated 10^{12} times in order to capture the physics of the deformation or relaxation process! Even by using the current generation of teraflop computers, this would take over ten thousand years assuming that we had the memory for the task, i.e., this is a very hard problem!

Attacking this class of problem from an atomistic viewpoint is perverse in that the problem is not only computationally intractable, but more importantly, an atomistic approach misses the essential physics of the problem. While we are not particularly interested in how individual atoms vibrate and diffuse, we do want to understand the cooperative response and flow properties of the polyhedral grains. What is needed is a method to capture the geometry and topological constraints of each particle and to follow the dynamics of the space packing of polyhedra as a function of time.

PARTICULATE ARTIFICIAL NEURAL NET DYNAMICS ALGORTLUMS (PANNDA)

While such a scenario is not trivial, the overall strategy has shown much promise. The simplest model that might capture the essential physics would be the hard smooth polygon. In order to simulate such a system it is necessary to predict where and when the next contact will be made for a system of translating and rotating polygons. For hard-spheres this is straightforward [18] but for granular particles with asperities the situation is considerably more complex. The difficulty has been discussed for the case of hard line segments [19, 20]. The equations of motion for the line segments that form the edges of the grain are complicated transcendental functions of the angular frequencies leading to a set of equations that require a time consuming numerical root-finding procedure of the kind discussed in reference [20]. Moreover if we consider a polygonal smooth hard solid as the simplest model of a grain, then edge-edge and vertex-face collisions are possible. So for tetrahedral grains in three dimensions, a combinatorial factor of 104 enters (72 edge-edge and 32 vertex-face possibilities), while for squares, the number is 384 since only edge-edge and vertex-face collisions are allowed in three dimensions. Even for hard triangles in two dimensions (vertex-edge), a factor of 9 is required.

It is important to recognize that when solving the equations of molecular dynamics, we solve (or rather the computer solves) the same equation many millions of times for input parameters that vary only slightly. For hard systems, the potential is pair-wise and no three-body terms exist. It is sensible therefore to arrange matters such that we consider only two particles and calculate the point and time of contact for all accessible relative positions, orientations, and angular and linear momenta. This can be done for a large but finite training set that spans the Hilbert space available to the two-particle system. For a two-particle system, a net in 24 dimensions is required - a large number, but one that is by no means intractable. When such a net is trained "off-line", it can then be used "on-line" in the inner loop of the dynamics calculation to return the required parameters in a time that is essentially independent of the complexity of the two objects.

The artificial neural net functions as a very rapid and accurate curve fitting and interpolation tool in such applications. While this in itself has no particular value in a scientific sense, fitting a neural net to the curve of heat capacity versus temperature provides a good numerical parameterization but one that will not give any hint of the underlying quantum physics in the sense of Einstein's theory (one parameter!) of heat capacity. For this reason the net has little value in physics *per se*, but the approach does provide a useful numerical procedure to enable new discoveries in machine calculations.

RESULTS

Figure 1 shows a plot of relative execution speed as a function of system size for a system of hard smooth triangles. For small systems, the PANNDA approach is slightly less efficient than the standard method but as the system size increases, the role of the AI method increases steadily so that once 5000 particles are reached, the method is faster by about a factor of four. Moreover the increase in speed of execution is approximately linear so that for very large systems of, say five million grains, over two orders of magnitude in computational speed might be expected. Figure 2 shows the equation of state of the hard-triangle fluid for a reduced packing fraction of 0.25.

The hard-triangle system will pack to fill space and a number of phase transitions both translational and rotational may occur as the packing fraction is increased. In these simulations the packing fraction is defined as the total area occupied by the triangles divided by the total area of the system. This simulation used periodic boundary conditions and a single configuration is illustrated in Figure 3. As is customary, a system of reduced units is used where the unit of length is the diameter of the outscribed circle, and Boltzmann's constant is taken as unity. It is clearly possible to use such models to explore the rich classical physics of such systems numerically. A detailed discussion of this aspect will be published elsewhere [21].

A striking aspect of Figure 3 is that the triangles do not appear randomly distributed in space even though the area available per triangle is very much greater than the area of a triangle, i.e., the free volume is large. The origin of this tendency to clustering at low and intermediate density is still an open question.

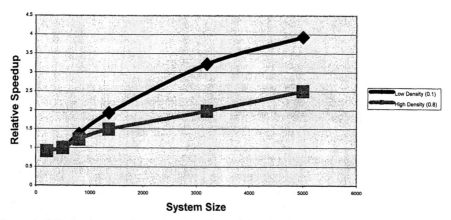

Figure 1. Relative increase in computational speed for simulation of hard triangles in two dimensions.

Figure 4 shows the angle-averaged radial distribution function at a very low packing fraction (0.01). Here each triangle has one hundred times its own area in which to move. If these particles were discs of unit diameter, this figure would depict the unit step function starting at unity. The spike in g(r) is due to the tendency of particles to associate even though they interact purely repulsively. They are in a sense bound by mutual repulsion.

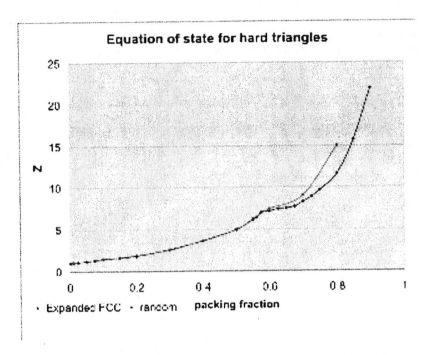

Figure 2. Equation of state of the hard triangle fluid obtained from PANNDA molecular dynamics simulation up to a reduced packing fraction of 0.9.

CONCLUSIONS

This paper extends our results reported earlier [22]. There is intense interest in granular systems as evidenced by the roughly two thousand recent citations listed in reference [23], but there is still no consensus as to the classical physics. Granular media are formally thermodynamically "small" systems that may not be in equilibrium and that exhibit dissipative interactions which has led Kadanoff [24] to speculate that attempts to describe granular materials using continuum approaches analogous to Navier-Stokes formalism for liquids [25] may not be well-founded. The ability to construct exact computer models in which interparticle interactions, shape, and polydispersity are all well-defined and from which the thermo-mechanical and transport coefficients can be calculated exactly is likely to be of value in a wide range of computational experiments.

In conclusion, the main advantage of AI-based algorithms is that the speed of simulation is *independent of the complexity* of the potential once the "net" has been trained. The program runs as fast with many-body forces, as it does with only a pair-force without any loss in accuracy. The algorithm may be used with any types of equations of motion, i.e., for Stokesian, Brownian, or granular dynamics of colloids. Finally we should also mention that there many applications such as air traffic control or intelligent automotive guidance systems where an ability to predict where and when two rapidly moving objects of complex geometry may collide can be of great benefit. An ability to do this calculation more rapidly may help in the general development of intelligent control systems.

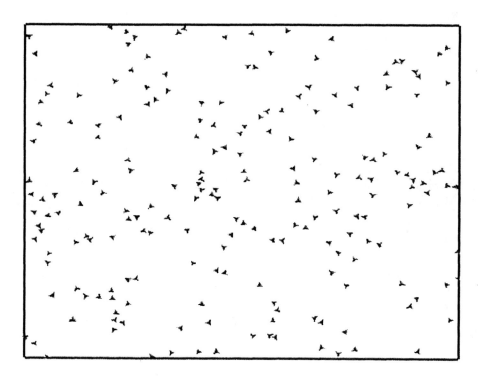

Figure 3. A configuration of hard triangles at a low packing fraction (0.2).

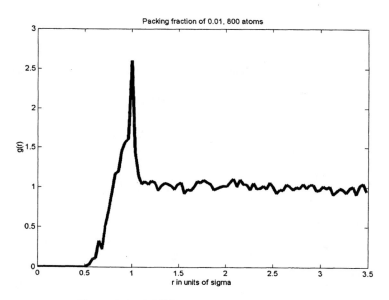

Figure 4. Radial distribution function at low density.

ACKNOWLEDGEMENT

This work was partially supported by the Air Force Office of Scientific Research under the "Electronic Prototyping" initiative.

REFERENCES

J.F. Maguire and P.L. Talley, 1995. J. Advanced Materials, 27–39.
J.F. Maguire, M.A. Miller, S. Venkatesan, 1998. Engineering Applications of AI, 11, 605–618.
S. Guessasma, G. Montavon, C. Coddet, 2001. Paper S 8.2, MRS Symposium, Boston.
P.K. Schenk, D.L. Kaiser, et al., 2001. Paper S1.2, MRS Symposium, Boston.
J.F. Maguire, J.D. Busbee, D. Liptak, D. Lubbers, S.R. LeClair and R. Biggers, 2000. Process Control for Pulsed Laser Deposition Using Raman Spectroscopy, U.S. Patent 6038525.
Y-H Pao, Y.H. Zhao, S.R. LeClair, 2001. Neural-Network Based Exploration in the Design and Discovery of Materials, IFAC Symp. on A.I. in Real Time Control, Budapest, Hungary.
Y.H. Pao, 1989. Adaptive Pattern Recognition and Neural Networks, Addison-Wesley, Reading, MA.
D. Henderson, 1992. Fundamentals of Inhomogeneous Fluids, Marcel Dekker, New York.
M. Adam et. al., 1988. Nature, 393.
1999, Physical Chemistry in the Mesoscopic Regime, Faraday Discussion 112.
J. S. Rowlinson and B. Widom, 1982. Molecular Theory of Capillarity, University Press, Oxford.
D. Frenkel and B. Smit, 1996. Understanding Molecular Simulations, Academic Press, New York.
M.P. Allen and D.J. Tildesley, 1987. Computer Simulation of Liquids, Clarendon Press, Oxford. www.uncwil.edu/people
P. Cvitanović, 1989. Universitality in Chaos, Adam Hilger, Bristol, 2nd edition.
G. Ciccotti, D. Frenkel, I.R. McDonald, 1989. Simulation of Liquids and Solids, Elsevier, Amster.
B.J. Alder and T.E. Wainright, 1959, J. Chem. Phys., 31, 459.
D. Frenkel and J.F. Maguire, 1981. Phys. Rev. Lett., 47(15), 1025–1028.
D. Frenkel and J.F. Maguire, 1983. Molecular Physics, 49(3), 503–541.
J.F. Maguire and M. Benedict, to be published.
J.F. Maguire and L.V. Woodcock, to be published.
J.F. Maguire, M. Benedict, L.V. Woodcock, and S.R. LeClair, 2001, MRS Proceedings, 700. www.ical1.uni.stuttgart.de/~lui/REFS/references.html
L.P. Kadanoff, 1999. Rev. Mod. Phys. 71, 435.
C. Bizon, M.D. Shattuck, J.B. Swift, and H.L. Swinney, 2001. Physical Review E, 1

Distributed Backpropagation Neural Network System for Pattern Recognition Using JavaSpaces

M. GALAL
National Center for Examination and Educational Evaluation (NCEEE), Egypt

H. ELDEEB and S. NASSAR
Electronic Research Institute (ERI), Egypt

ABSTRACT

In this paper, we describe a system that distributes the computation process of artificial neural networks of type MLP. The system uses a standard supervised MLP-learning algorithm in conjunction with the well-known Backpropagation (Backpro) algorithm for pattern recognition. The Backpropagation algorithm has been used in many applications especially in pattern recognition field, but one of its weaknesses is the time consuming computation. Our system uses a parallel Backpropagation version that runs on distributed system using JavaSpaces technology.

JavaSpaces technology provides a good platform for heterogeneous computing that uses single processor, multiprocessor, LAN, or internet networks. JavaSpaces use lookup service of the Jini technology that makes a network more dynamic by allowing the plug and play of devices and without the need for configuring each device. JavaSpaces uses the RMI activation daemon to manage the states of all services. System implementation is written in the Java language. Java is a well-designed object-oriented language, its most important feature being portability. Java compilers generate code in an architecture-independent format so the same copy can be executed on many different machines.

For pattern recognition, we use The MNIST database of handwritten digits. It has a training set of 60,000 examples, and a test set of 10,000 examples. The digits have been normalized by size and each one is centered into a fixed-size image. We use some statistical methods to map the pattern space into a reduced feature space, and extract the most important features. We have reduced the pattern attributes from 28x28 pixels and 256 different shades of gray for each pixel to a meager set of 9 features that intrinsically capture relationships along the rows and down columns. In this paper, we describe the design and implementation of our system and present evaluation performance results.

INTRODUCTION

Over the last few years, machine-learning techniques, particularly when applied to neural networks, have played an increasingly important role in the design of pattern recognition systems. In fact, it could be argued that the availability of learning techniques has been a crucial factor in the recent success of pattern recognition application such as handwriting recognition [1].

When it is determined that an object from a population P belongs to a known subpopulation S, we say that pattern recognition is done. The recognition of an individual object as a unique singleton class is called identification. Classification is the process of grouping objects together into classes (subpopulations) according to their perceived likenesses or similarities. The subject area of pattern recognition includes both classification and recognition and belongs to the broader field of machine intelligence that is, the study of how to make machines learn and reason to make decisions, as do humans [2].

Java, as a programming language, has seen phenomenal growth in the past few years. The initial appeal of Java was related to its connection to the Web. However, its benefits as a good programming language are also being recognized. Its clearly designed object oriented, platform independent, robustness, Internet based programming, and have some built-in mechanism allows for parallel programming with the bounds of shared memory. Also late research papers conclude that Java could well become a dominant language in science and engineering. It also offers some additional features that have lead several groups to begin research on parallel and distributed programming with Java.

The Java programming language is an object-oriented language that is used both for Web based Internet and general-purpose programs. It is a software platform for network-based computing and it is architecturally neural. Java was designed to support applications that run on a network. This could be the Internet or corporate intranets.

JavaSpaces services are tools for building distributed protocols. They are designed to work with applications that can model themselves as flows of objects through one or more servers [3]. JavaSpaces technology is a departure from conventional distributed tools which rely on passing messages between processes or invoking methods on remote objects. JavaSpaces technology provides a fundamentally different programming model that views an application as a collection of processes cooperating via the flow of objects into and out of one or more spaces. The power of Java language merging with JavaSpaces services is a good environment to design parallel and distributed applications.

Artificial neural networks have been applied in many fields, including engineering, physics, psychology and mathematics. The multilayer perceptron, trained by the Backpropagation algorithm, is currently the most widely used neural network. The capabilities of these multilayer networks are for function approximation and pattern recognition.

In this implementation, we are using the power of Java language combining with JavaSpaces services to design a distributed backpropagation system for recognizing patterns.

The paper is organized as follows: First we present an overview of JavaSpaces. Next pattern recognition and neural network techniques and several different types of backpropagation parallelism are discussed. E then describe our proposed system environment, how it was. implemented and what design stages were used. The performance of the new system is given and evaluated by conclusions.

OVERVIEW OF JavaSpaces

Distributed applicationds in Java can be written at many programming levels:

- Low level programming using networking tools in core Java.
  Java applications have the capability to communicate with each other through the sockets protocol, which is a widely used communication protocol. By using sockets, applications can implement distributed functionality by sending requests and data back and forth. This approach is very low level.
- Basic level programming using Remote Method Invocation (RMI).
  An application level communication protocol in an agreed upon format for communicating data between two processes. The purpose of RMI is to provide an abstraction that enables an application designer to invoke methods on remote objects instead of having to communicate at a lower level.
- High level programming using Jini technology.
  Jini extends RMI discovery in a code centric manner to dynamic network environments. It provides many advantages such as:
    o Entries are leased not bound.
    o Jini lookup service is discovered, not in well known place.
    o Search by interface not by name.
- Top level programming using JavaSpaces technology.
  Javaspaces are object repositories where any platform architecture can share objects, and Javaspaces are implemented as a Jini service [4].

Distributed Persistence

Implementations of JavaSpaces technology provide a mechanism for storing a group of related objects and retrieving them based on a value-matching lookup for specified fields. This allows a JavaSpaces service to be used to store and retrieve objects on a remote system.

Distributed Algorithms as Flow of Objects

Many distributed algorithms can be model as a flow of objects between participants. This is different from the traditional way of approaching distributed computing, which is to create method-invocation-style protocols between participants. In this architecture's "flow of objects" approach, protocols are based on the movement of objects into and out of implementations of JavaSpaces technology.

Application Model and Terms

JavaSpaces service holds *entries*. An entry is a typed group of objects, expressed in a class for the Java platform. An entry can be written into a JavaSpaces service, which creates a copy of that entry in the space that can be used in future lookup operations. We can look up entries in a JavaSpaces service using templates that are entry objects with some or all of its fields set to specified values that must be matched exactly. Remaining fields are left as wildcards- these fields are not used in the lookup.

There are two lookup operations: *read* and *take*. A read request to a space returns either an entry that matches the template on which the read is done, or an indication that no match was found. A take request operates like a read, but if a match is found, the matching entry is removed from the space. (see Figure 1).

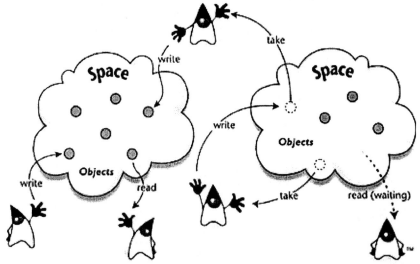

Figure 1. Processes use spaces and simple operations to coordinate [4].

JavaSpaces services can provide distributed object persistence with objects in the Java programming language. Because code written in the Java programming language is downloadable, entries can store objects whose behavior will be transmitted from the writer to the readers, just as in RMI using Java technology.

An entry in a space may, when fetched, cause some active behavior in the reading client. This is the benefit of storing, not just data, in an accessible repository for distributed cooperative computing.

JavaSpaces service is not a relational or object database. JavaSpaces services are designed to help solve problems in distributed computing, not to be used primarily as a data repository. JavaSpaces services functionality is some where between that of a file system and a database, but it is neither [3].

JavaSpaces Requirements and Dependencies

The requirements for JavaSpaces technology are Java Remote Method Invocation, Java Object Serialization, Jini Entry, Jini Entry Utilities, Jini Distributed Event, Jini Distributed Leasing, and Jini Transaction [3].

In this research, we use top level programming with JavaSpaces technology to solve two related problems: distributed persistence and the design of distributed algorithms. JavaSpaces services use Remote Method Invocation (RMI) and the serialization feature of the programming language to accomplish these goals.

PATTERN RECOGNITION AND NEURAL NETWORK TECHNIQUES:

Pattern Recognition

Statistical pattern recognition is the oldest method used in the field. It used measurements of attributes to classify patterns by making overall statistical decisions [5]. The statistical pattern recognition method can be categorized as a decision-theoretic category; the other categories in pattern recognition are structural and associative mapping. Associative mapping can be a neural or fuzzy mapping [2].

With feedforward neural networks, we can use one of the following three methods for pattern recognition: multiple layered perceptrons, functional link nets, and radial basis function networks. We use a multiple layered perceptron approach in this research as the pattern recognition tool to achieve better accuracy. It provides a nonparametric technique to conduct a nonlinear estimation of the data. It has the ability to learn complex, highly dimensional, non-linear mappings from large collections of examples making it an obvious candidate for pattern recognition.

Feature Extraction

Feature extraction is the name given to a family of procedures for measuring the relevant shape information contained in a pattern so that the task of classifying the pattern is made easy. In other words, Feature extraction is the process of mapping the original attributes into fewer features which include the main information of the data structure. Some papers try to learn network by directly fed with real image to recognize without a large, and complex preprocessing stage requiring detailed engineering to feature extraction [6,7]. A large variety of feature extraction methods based on statistical models or on artificial neural networks [8,9].

The need for mapping the pattern space into a reduced feature space is to:
- Retain as much of the original information as possible by selecting a set of features that contains essentially all of the information of patterns necessary for recognition and classification, but in a more efficient form.
- Remove as much as possible of the redundant and irrelevant information that can cause extraneous noise and degrade the performance of recognition by eliminate the redundancy to reduce both the complexity and the error rate. It is known that more independent features provide greater discrimination power, but more features that are correlated can actually increase the error rate.
- Render the measurement data to variables that are more suitable for decision-making by transforming the pattern space into more suitable features that have more meaning to humans [2].

In this work, the pattern attributes are 28x28 pixels and 256 different shades of gray for each pixel, this means matrices with size more than 200,000. Based on that, we need to design implementing features (data engineering). For successful recognition, we use MLP that takes the feature space as input and the recognition decision as its output.

MLP Artificial Neural Networks

The MLP network consists of several layers. Each layer has its own weight matrix **W**, its own bias vector **b**, a net input vector **p** and an output vector **a**. We use superscripts to identify the layer number. The equation that describes the outputs of network with four hidden layers is:

$$a^4 = f^4(\,W^4 f^3(\,W^3 f^2(\,W^2 f^1(\,W^1 p + b^1\,) + b^2\,) + b^3\,) + b^4\,) \qquad (1)$$

Where **f** is the transfer function (sigmoid in this work), and is defined as following

$$f(p) = 1 / (\,1 + \exp(-p)\,) \qquad (2)$$

For a given set of training data {(p,t)} the network is able to learn the mapping p → t, and input p and target t usually are vectors. This training set is normally representative of a much larger class of possible input/target pairs. It is important that the network successfully generalize what it has learned to the total population.

There are several approaches to automatic machine learning for MLP, but one of the most successful approaches is Backpropagation. It is by far the most widely used neural-network learning algorithm, and probably the most widely used learning algorithm of any form [10].

Parallel Backpropagation Algorithm

Several different degrees of Backpropagation parallelism exists: node, pipelining, and training set parallelism [11]:

- Node parallelism: This type of parallelism can take many forms:
  o Nodes of entire network are partitioned among different processors, and each processor keeps a copy of all weights of the network [11].
  o Weight matrices are partitioned into sub-matrices and multiplied by the input [12].
  o Each node with its receptive and projective weights and bias is calculated by a different processor [13].
- Pipelining parallelism: Processors pipeline the training patterns. While the output layer processors calculate output and error values, the hidden layer processors concurrently process the next training pattern. Pipelining requires delayed update of the weight matrices between layers.
- Training set parallelism: This method is well suited for parallelization. It replicates the entire network on each processor and presents different patterns in parallel. Each processor accumulates weight change values for the given training patterns. At the end of all patterns, weights are updated in a global operation [11,12,13].

Based on the degree of parallelism, two implementations are developed in this work. First, a node parallelism is implemented using on-line training or steepest-descent training. It is a very fine-grained method and needs a lot of communication. Secondly, training set parallelism is implemented using a batching method that does not impose any constraints on the network topology. All results in this paper are based on the second implementation.

The MNIST Database of Handwritten Digits

The database used to train and evaluate the performance of the proposed system was constructed from a mixture of two NIST datasets SD-1 and SD-3. SD-1 contains 58,527 digit images written by 500 different writers. It was collected among high-school students. 30,000 patterns from SD-1 were held in the training set while 5,000 patterns were retained in the test set. SD-3 was collected among Census Bureau employees. 30,000 patterns from SD-3 were also placed in the training set with 5,000 patterns put into the test set. The resulting database was called the Modified NIST, or MNIST dataset. It contains 60,000 training examples, and 10,000 test examples. The images were size-normalized to 20x20, centered in a 28x28 image by computing the center of mass of the pixels, and translating the image so as to position this point at the center of the 28x28 field. The database is available at http://www.research.att.com/yann/ocr/mnist.

SYSTEM ENVIRONMENT

The proposed system can be used on a single processor or a machine with multiprocessors. It can also be used in a distributed computer environment without any modification. The research experiments were implemented on different machines connected to a 10/100 Mbps Ethernet local area network. The computers used were three 1.2-GHz Pentium IV PCs, and six 1-GHz Pentium IV PCs. They ran the Windows 2000 operating system, and used Java 2 SDK standard edition version 1.2.1, and Jini version 1.2 including JavaSpaces technology.

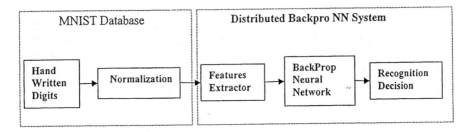

Figure 2. The proposed system block diagram with MNIST database as input data.

SYSTEM IMPLEMENTATION

In this section, we present system design and implementation. The system input data is a MNIST database of handwritten digits. An MLP neural network trained by parallel backpropagation batch mode is our tool for pattern recognition in a distributed environment. The Java language combined with Jini and JavaSpaces technology is the programming language used in this research. A block diagram of the system with its input data is shown in Figure 2.

Preprocessing Stage

We decided to extract features from images instead of feeding all image points as system input. The images were size-normalized, centered, and collected in database files by MNIST staff. At this point, the size of a digit is typically 28 by 28 pixels. We use the idea proposed in [2] to convert multiple gray levels to binary values. This converts all gray levels below a threshold value of 127 into 0 and every thing from 127 to 255 into 1; So a single bit (0 or 1) represents each pixel. Now, we must find a way to design and implement a few features to eliminate the redundancy and irrelevant information found in 784 image pixels, provide greater discrimination, and summarize the original image attributes. We use the relationships along rows and columns by counting the number of times a change occurs from black to white or white to black (a change of values of pixels from 0's to 1's or from 1's to 0's) as shown in Figure 3.

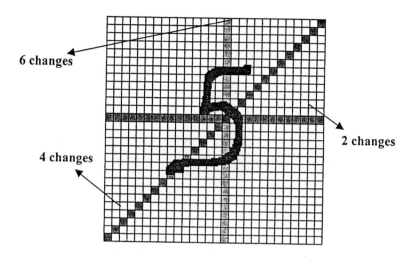

Figure 3. Map of digit 5 written by a hand in 28x28 matrix of pixels with some pre-features selection.

We numbered the rows from top to bottom as 0,1,...,27 and columns from left to right as 0,1,...,27 and use only rows 1, 3, 5, 7, 9, 11, 13, 14, 15, 17, 19, 21, 23, and 25, columns 2, 4, 6, 8, 10, 12, 14, 15, 16, 18, 20, 22, 24, and 26, and the two diagonals. These counts are the prefeatures of which there are 14 for rows, 14 for columns, and 2 for diagonals, for a total of 30 counts. They do not depend upon the thickness of the lines.

This method reduces the pattern attributes from 786 pixels to a meager set of 30 features that intrinsically capture the relationships along the rows and columns. We use a random sample of 1,000 patterns from the 60,000 patterns training set to establish a statistical method that takes one pattern out of every 60. The ten digits from digit 0 to digit 9 are presented in a sample of the following frequencies as shown in Table 1. This extracts a sample in an unbiased manner from the population.

For each digit, we select a vector of more important features for that digit by two criteria; first, the feature has a high frequency value among the sample (a very small example from the frequencies table is presented in Table 2), and secondly, the feature is highly correlated with other features selected by the first criteria. We use the bivariate Pearson correlation in SPSS statistical package release 10.0.1. Table 3 contains a subset of the complete correlations matrix for the digit 0 in the sample. The Correlations Table displays the Pearson correlation coefficient, significance values (Sig.), and number of cases (N). The Pearson correlation coefficient is a measure of linear association between two variables. In Table 3, the correlation between row 12 (H12) and Column 15 (v15) is 0.334 which is a positively correlated relationship. If the significance value is very small (less than 0.05) then the correlation is significant and the two variables are linearly related.

Table 1. Frequencies of ten digits 0-9 in the sample.

Digit	0	1	2	3	4	5	6	7	8	9
Frequency	99	103	98	93	95	99	103	105	112	93
Percent	9.9	10.3	9.8	9.3	9.5	9.9	10.3	10.5	11.2	9.3

Table 2. Frequencies of count values of top-right diagonal feature with digit 0 in the random sample.

No of changes	2	4	6
Frequency	2	88	9
Percent	2.0	88.9	9.1

Table 3. correlation of three features for digit 0 in the random sample.

Variables	Statistics	D2	H12	V15
D2	Pearson Correlation	1.000	.189	.287**
	Sig. (2-tailed)	.	.062	.004
	N	99	99	99
H12	Pearson Correlation	.189	1.000	.344**
	Sig. (2-tailed)	.062	.	.000
	N	99	99	99
V15	Pearson Correlation	.287**	.344**	1.000
	Sig. (2-tailed)	.004	.000	.
	N	99	99	99

. Correlation is significant at the 0.01 level (2-tailed).

The results from the last processes are merged to produce 16 features. Three more steps needed to obtain the final input data for the system are as follows:

- Order the features from most important to least important in distinguishing digits. Feature must discriminate clearly between two or more classes of digits. After ordering the 16 features, we assign weights to them that reflect their importance. The following formula is offered in [2].

$$w_n = 1 / \ln(n+1) \qquad (3)$$

- The correlations between features with a high frequency value in each digit were computed using the bivariate Pearson correlation method to see which are dependent on others based on the sample. The two dependent features with significant correlation mean that we should keep the one with the highest important weighting and drop the other. By performing this process on all 16 features, we get a final set of 9

essentially independent features without redundancy but with enough separating power.
- For the training set, we converted all examples from raw binary data to features. Many patterns are redundant in the data. We take only one pattern from repeated patterns, and store all patterns in descending order according to their frequency. The final data set is reduced from 60,000 to 2414 examples.

The basic component of the proposed system is a feature extractor application which automatically drives the features set from the database of handwritten digits. It extracts the 9 features required to separate the ten patterns. The features extractor outputs are the network inputs to learn it how to recognize digits.

Synchronization in the Proposed System

Synchronization plays a crucial role in designing a distributed application. In a distributed environment, asynchronous processes run independently at their own pace and there is no central controller that manages their activities and interactions. We will present some synchronization techniques that were implemented in the proposed system:

- Semaphore: A semaphore is a synchronization construct that was first used to solve concurrency problems in operating systems back in the 1960s [4]. We implement this construct by using a shared variable entry created by the master application and controlled by two operations: down and up. All workers (processes) can read the semaphore entry and check its value simultaneously. If the semaphore is down, then the workers wait for new tasks and the system is in training mode.
- Atomic Update: Without an atomic means to access and alter shared variables, having multiple processes share access to data can lead to race conditions and corrupt values. In classic distributed applications, designers follow certain steps for exclusive access and atomic update:
  - First, ensure the lock is in the unlocked state and then switch it to the locked state.
  - Next modify the variable by reading its value and writing a new value.
  - Finally release the lock for the variable by switching the lock to the unlocked state so that others may have access

  With JavaSpaces technology we get this atomicity as a key feature of the space. Only one process can hold the entry locally and update it at any given time, but many processes can read the entry while it is in the space.
- Fairly Sharing a Resource: Deadlock and starvation are two potential problems that happen in a distributed application. A deadlock occurs when each process in a collection is waiting for an action or response that only another process in the same collection can generate [4]. Starvation occurs when a process cannot access an available resource because other

processes have some competitive advantage. The proposed system design prevents workers from entering in either a deadlock or starvation state.
- Barrier Synchronization: A barrier is a particular point in a distributed computation that every worker (process) in the system must reach before they can process further. In the proposed system, we force each worker to wait until all other processes have reached the barrier. All entry tasks and weight matrices needed for the next epoch will be in the space after the master collects outputs from the space.

In the proposed system design, the master object runs on one machine and a worker object runs on each of the other machines. The system consists of eight classes that are extractor, master, worker, network, weight matrix, bias vector, dataset, and gradient. All of these classes work together to form the distributed backpropagation system. Its algorithm is as follows:

Master.class

Only one object of the master class runs on a machine and one object of the worker class or more runs on one or more machines to form the distributed backpropagation system.
- Master initializes all system parameters, creates a space, and writes the network object that contains the architecture of the network into the space as a reference for all workers.
- Master creates a weight matrix object and bias vector object for every network layer with random values and puts all of them into the space.
- Master iterates the following steps up to convergence.
    o It creates worker tasks by splitting the training set across the pre-defined number of datasets objects that will put into the space.
    o It collects gradient objects as entries from the space. Gradient objects are the worker outputs.
    o It calculates outputs average to produce an accurate estimate of the gradient.
    o It takes out all weight matrices and bias vectors from the space, then updates the weight matrices and bias vectors with the new values.
    o Master puts weight matrices and bias vectors into the space to be read by all workers in the next epoch. Finally it checks for stop training, or it continues with the next iteration.

Worker.class

Worker connects to the space, reads network class, and waits to start the training. The worker iterates the following steps up to ending the training by the master.

- Worker waits to take a dataset, and reads current weight matrices and bias vectors from the space.
- It computes the forward phase, and computes the gradient for every parameter in the network.
- It creates a gradient object for every layer as the worker output, and it puts them into the space.

PERFORMANCE EVALUATION FOR THE NEW SYSTEM

First: Simple Case Study

Here we describe a simple experiment made to demonstrate the performance of increasing the number of processors. We consider a simple pattern recognition problem presented in [2] with the label "The Digit12 Feature Vectors". This comes from hand-printed digits for machine recognition of ZIP codes. Each digit is represented by one example. The problem consists of a set of 10 examples, 12 inputs, and 2 outputs. The proposed system uses an MLP network 12-12-2 architecture and a backpropagation algorithm in batch-mode learning (steepest descent) [10] to get the results shown in Figure 4.

Second: Practical Case Study

In this experiment, we trained the system using the data set prepared on the preprocessing stage that containing 2414 examples with 9 inputs, and 10 outputs. We used an MLP network with 9-9-10 architecture and a backpropagation algorithm in batch-mode learning. We got an accuracy higher than 80% on the testing data with a MSE (mean square error) of 1.0e-2. We measured the performance in MCPUS (Million Connections Updated per Second) to indicate performance during backpropagation training, as shown in Figure 5.

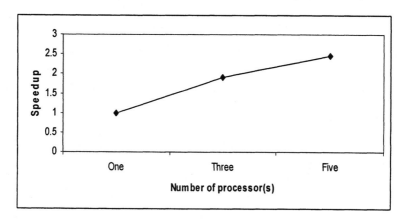

Figure 4. Diagram of the simple experiment performance with different processors number.

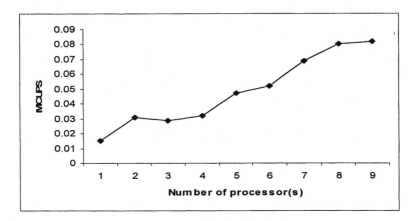

Figure 5. CUPS for the practical experiment performance with different processors number.

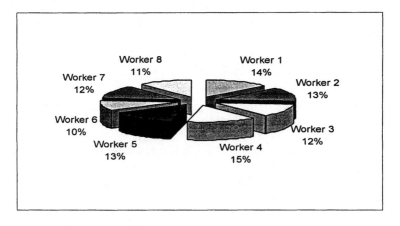

Figure 6. The partitioning of tasks by 8 workers.

In the case of using one machine, the master and the worker shared the same processor. For more than one processor, the master program ran on its own machine while the worker programs ran on the other machines. For example, with three processors, we have two workers, and for five processors we have four workers and so on.

Figure 6. shows that workers synchronize the execution of the tasks in the space and the system prevents them from entering into a deadlocked or starvation states. The chance of executing more tasks depends only on the speed of the processor.

CONCLUSION

We developed an efficient distributed backpropagation neural network system for pattern recognition using JavaSpaces. JavaSpaces is robust and a good tool to design distributed algorithms. The new system uses the capability of separation of master, workers, and space in JavaSpaces to provide intrinsic forms of dynamic scalability and fault tolerance. New workers could be instantiated and the master itself could be replaced by another to collect results from space in case of the master dying.

We applied our system on a practical case study of real handwritten digits of MNIST database. The performance evaluation showed that the proposed system obtained an accuracy higher than 80% on testing data with MSE (mean square error) of 1.0e-2.

The preprocessing on the MNIST database showed how to extract a set of 9 features only from the input data. The statistical methods were used to extract features to separate the ten hand-written digits. The features extractor outputs are the network inputs to learn how to recognize digits.

For the first implementation of the system, node parallelism was used in the design of distributed backpropagation. To speed-up the system, we changed to training set parallelism (batch mode), in which the gradients are accumulated over the entire training set and parameters are updated after the exact gradient has been computed. The problem in using batch mode with a data set that has many duplicated patterns is the performance of redundant computations. We solved this problem and produced a very active training data set. The new version is compressed by more than 95% of the original data set and it includes the features of all patterns.

REFERENCES

[1] Y. LeCun, L. Bottou, Y. Bengio, and P. Haffner, 1998. "Gradient-based learning applied to document recognition". Proc. of the IEEE November 1998.
[2] C. G. Looney, 1997. "Pattern recognition using neural networks". Oxford University Press, New York.
[3] Sun Microsystems, 1999. "JavaSpaces TM Specification". White paper, Sun Microsystems, Available on web: http://www.sun.com/jini/specs/js.pdf .
[4] E. Freeman, S. Hupfer, K. Arnold, 1999. "JavaSpaces: Principles, Patterns, and Practice". Addison-Wesley.
[5] R. A. Fisher, 1934. "The use of multiple measurements in taxonomic problems". Ann. Eugenics, vol.7, part II, 179–188.
[6] Y. LeCun, B. Boser, J. S. Denker,D. Henderson, R. E. Howard, W. Hubbard, L. D. Jackel,1990. "Handwritten Digit Recognition with a Back-Propagation Network". In David Touretzky, editor, Advances in neural Information Processing Systems, vol. 2,404 -396, Denver 1989, Morgan Kaufman.
[7] Y. W. Teh, 2000. "Learning to Parse Images". Master thesis, Graduate Department of Computer Science, Toronto University.
[8] A. S. Ahmad, A. Zulianto, E. Sanjaya, 1999. "Design and Implementation of Parallel Batch-mode Neural Network on Parallel Virtual Machine". Proceedings, Industrial Electronic Seminar 1999 (IES'99), Graha Institut Teknologi Sepuluh Nopember, Surabaya.

[9] B. Lerner, H. Guterman, M. Aladjem, I. Dinstein, 1996. "Feature Extraction by Neural Network Nonlinear Mapping for Pattern Classification". The 13th International Conference on Pattern Recognition, ICPR13, Vienna, (4) 320–324.

[10] M. T. Hagan, H. B. Demuth, M Beale, 1996. "Neural Network Design". PWS publishing company.

[11] A. Singer, 1990. "Implementation of Artificial Neural Networks on the Connection Machine". Parallel Computing, (14) 305–315.

[12] J. Torresen, S. Mori, H. Nakashima, S. Tomita, O. Landsverk, 1994.Parallel back propagation training algorithm for MIMD computer with 2D-torus network. Proceedings of International Conference on Neural Information Processing, Seoul, Korea, (1) 140–145.

[13] F. M. Thiesing, U. Middelberg, O. Vornberger, 1994. Parallel Back-Propagation for the Prediction of Time Series. 1st European PVM Users' Group Meeting, Rome.

Eccentricity and Hardness Control in Cold Rolling Mills with a Dynamically Constructed Neural Controller

Y. FRAYMAN and B. F. ROLFE

School of Information Technology and School of Engineering and Technology, Deakin University, Geelong, VIC 3217, Australia

ABSTRACT

The main objective of a steel strip rolling process is to produce high quality steel at a desired thickness. Thickness reduction is the result of the speed difference between the incoming and the outgoing steel strip and the application of the large normal forces via the backup and the work rolls. Gauge control of a cold rolled steel strip is achieved using the gaugemeter principle that works adequately for the input gauge changes and the strip hardness changes. However, the compensation of some factors is problematic, for example, eccentricity of the backup rolls. This cyclic eccentricity effect causes a gauge deviation, but more importantly, a signal is passed to the gap position control so to increase the eccentricity deviation. Consequently, the required high product tolerances are severely limited by the presence of the roll eccentricity effects.

In this paper a direct model reference adaptive control (MRAC) scheme with dynamically constructed neural controller was used. The aim here is to find the simplest controller structure capable of achieving an optimal performance. The stability of the adaptive neural control scheme (i.e. the requirement of persistency of excitation and bounded learning rates) is addressed by using as the inputs to the reference model the plant's state variables. In such a case, excitation is due to actual plant signals (states) affected by plant disturbances and noise. In addition, a reference model in the form of a filter with a desired transfer function using Modulus Optimum design was used to ensure variance in the desired dynamic characteristics of the system. The gradually decreasing learning rate employed by the neural controller in this paper is aimed at eliminating controller instability resulting from over-aggressive control. The moving target problem (i.e. the difficulty of global neural networks to perform several separate computational tasks in closed-loop control) is addressed by the localized architecture of the controller. *The above control scheme and learning algorithm offers a method for automatic discovery of an efficient controller.*

The resulting neural controller produces an excellent disturbance rejection in both cases of eccentricity and hardness disturbances, reducing the gauge deviation due to eccentricity disturbance from 33.36% to 4.57% on average, and the gauge deviation due to hardness disturbance from 12.59% to 2.08%.

PROBLEM DESCRIPTION

The main objective of a steel strip rolling process is to produce high quality steel at the desired thickness. Such strip is typically 600–1800 mm wide with the thickness ranging from 0.3 to 3 mm. The mills run at a very high speed with the strip exit speeds of the order of 25 m/sec. The mill is either a single stand or consists of multiple stands where each stand is made up of the two backup rolls and the two work rolls. The backup rolls relative to the work roll diameters have a ratio of about 3:1.

The roll force measurements are taken by a load cell, and the mill actuator is an electro-hydraulic servo-valve operated ram. This technology, that has now superseded electrically driven screw-down actuators in most modern mills, can give a fast, high precision roll positioning. The steel strip is progressively reduced as it goes through each stand or each pass of the same stand for the single stand mills. Thickness reduction is the result of the speed difference between the incoming and the outgoing steel strip and the application of the large normal forces via the backup and the work rolls. The main variables to be controlled are the inter-stand strip tension and the gauge (thickness) with secondary issues being the strip shape (stress distribution) and finish. The available mill measurements are the tension, the roll force via the load cells and the gauge measurements via the radiation (x-ray) based technology. Gauge control of a cold rolled steel strip is achieved using the gaugemeter principle, also known as the Automatic Gauge Control (AGC). This principle is based on the phenomenon of the mill stretch. An increase or a decrease in the gauge causes an increase or a decrease in the roll force. Thus, the roll force measurements can be used to correct the output gauge errors.

In the idealized situation, the mill stand can be represented as a very large and stiff spring [1]. The outgoing gauge h_{out} is given by:

$$h_{out} = s + E(f) \tag{1}$$

Here s is the initial roll gap setting, and $E(f)$ is a spring extension, f is a roll force. Linearization about a nominal $h^0{}_{out}$, s^0, f^0 produce:

$$\Delta h = \Delta s + \frac{\Delta E}{\Delta f} \times \Delta f = \Delta s + \frac{1}{M_m} \Delta f \tag{2}$$

Here M_m is a mill modulus. If the gap-setting is under a perfect control, where $\Delta s = 0$, then

$$\Delta h = \frac{1}{M_m} \Delta f \qquad (3)$$

It can be seen that an increase or a decrease in the gauge Δh causes an increase or a decrease in the roll force Δf. Thus, the roll force measurements can be used to correct for the output gauge errors. The gaugemeter principle uses the correlation of the small changes to calculate a gauge setting change. Hence, if $\Delta h \neq 0$, then to obtain $\Delta h = 0$ requires:

$$\Delta h = 0 = \Delta s + \frac{1}{M_m} \Delta f \qquad (4)$$

Consequently,

$$\Delta s = -\frac{1}{M_m} \Delta f \qquad (5)$$

Here Δf is a measured roll force change. Thus:
1. If $\Delta h > 0$ then $\Delta f > 0$ and $\Delta s < 0$ is required to close the roll gap.
2. If $\Delta h < 0$ then $\Delta f < 0$ and $\Delta s > 0$ is required to open the roll gap.

The implementation of the gaugemeter principle known as the mill spring compensation takes the form of a tuned compensation:

$$\Delta s = -\frac{1}{M_c} \Delta f \qquad (6)$$

Here M_c is tuned to equal the mill modulus M_m. Thus, the control system regulates the strip exit gauge by responding to the entry gauge changes and the hardness variations in the strip. The gaugemeter system uses a roll force measurement to compensate for the roll gap changes due to the mill stretch. The mill stretch characteristics are used with the roll force measurement to provide a feedback signal to the roll positioning actuator.

The gaugemeter principle works adequately for the input gauge changes and the strip hardness changes. However, the compensation for some factors is problematic, for example:
  a) Eccentricity of the backup rolls.
  b) Gap in the oil-bearing film varying with the rolling speed.
  c) Roll hardness, expansion, wear and an inaccurate roll profile.

In this paper, our concern is with the eccentricity effects of the backup roll. For various reasons such as the thermal expansion, the roll wear, and the non-perfect grinding, the backup and the work rolls are not perfectly circular. However, due to the larger diameter of the backup rolls their non-circularity has a more significant effect. This causes the variation of the actual roll gap and the rolling load. The period of the eccentricity signal is equal to the period of the revolution of the backup roll.

This cyclic eccentricity effect causes a gauge deviation, but more importantly, a signal is passed to the gap position control so to increase the eccentricity deviation. Given the perfect geometry and the rolling conditions the roll force and the gauge signals would be constant in a perfect steady state. However, when the eccentricity is causing the output gauge to decrease, this is accompanied by an increase in the roll force caused by the non-circular rolls. The gaugemeter principle takes the increase in the roll force and reduces the roll gap further. Thus, not only does a natural eccentricity deviation result in the gauge variation but this is also exaggerated by the action of the gaugemeter loop. Consequently, the gaugemeter principle is unable to distinguish between the various types of the disturbances that cause the variation of the roll force.

A modern steel mill is a highly automated industrial process. The required high product tolerances are severely limited by the presence of the roll eccentricity effects. Consequently, a viable scheme for the eccentricity compensation is of substantial interest to the industry and considerable industrial research efforts have been devoted to finding the best possible solution[1–6].

Neural networks are a promising recent technology that is becoming more popular for control applications. However, most of the existing research in neural control has been concentrating on indirect control schemes, where the neural network is used to identify the process, and a controller is subsequently synthesized from this model. Such an approach is very prominent in the applications of neural networks to steel manufacturing, where the main research efforts are concentrated on using neural network models to predict the process parameters such as the rolling force (for example, [7,8]).

The existing work on eccentricity control with neural networks is aimed at designing an eccentricity filter based on the periodic nature of the eccentricity signal [9,10]. It is based on the reasoning that if this signal can be predicted satisfactory, it then can be subtracted from the roll force measurements to reconstruct the true roll force signal. However, the behaviour of this filter needs to be investigated within the whole system, as the eccentricity filter output influences the filter input signal via the feedback control loop [9]. It was noted in a later work that experiments on a real steel mill found this method to be non-robust [11].

THE CONTROL SCHEME

A general multi-input multi-output (MIMO) nonlinear dynamical process can be represented by the following state-space representation:

$$\vec{x}(t+1) = F[\vec{x}(t), \vec{u}(t), \vec{d}(t)] \qquad (7)$$

$$\vec{y}(t) = G[\vec{x}(t), \vec{\eta}(t)] \qquad (8)$$

where $\vec{x} = [x_1, x_2, ..., x_n]^T$ is a vector of process inputs, $\vec{u} = [u_1, u_2, ..., u_m]^T$ is a vector of the control signals (the manipulated variables), $\vec{d} = [d_1, d_2, ..., d_p]^T$ is a vector of disturbance inputs, $\vec{y} = [y_1, y_2, ..., y_m]^T$ is a vector of process outputs, $\vec{\eta} = [\eta_1, \eta_2, ..., \eta_m]$ is a vector of measurement noise, and t is the sampling time. Here n, m, and p are the dimensions of corresponding vectors.

The indirect control schemes mentioned in the previous section are attempting to control such a process following the traditional route where better process models lead to better control. However, there is no guarantee that even a very good process model will lead to a good control. The inherent conflict between the identification and the control is a well-recognized one. The objective of the control for a typical set-point regulation is to minimize (make zero) the error between the actual process outputs and the desired outputs (set-points). However, when the error is zero, or a constant, one has no influence on the outputs and thus cannot identify the process parameters [12]. This refers to the lack of the persistency of excitation.

The typical control law attempts to minimize the approximation error by driving the parameter estimations towards the manifold where the error becomes zero, thus allowing for the bursting phenomena. In the worst case, an appropriate arbitrarily small disturbance can induce a parameter drift on this manifold and thereby cause a persistent bursting. The parameters drift can be interpreted as a non-robustness of an ill-posed optimization problem. The error bursts in this framework are the immediate consequence of a Lipschitz continuity of the parameter approximation (finite adaptation gains) [13]. Consequently, burst suppression in a general case requires controllers with the infinite adaptation gains or the injection of the excitation.

A fundamental obstacle in overcoming the persistency of the excitation problem is that a designer has limited or no control over external inputs and, consequently, the level of excitation. This means a high level of excitation (frequency-rich and large-amplitude signals) are required to obtain accurate parameters of the controller. However, a low-level excitation required by a typical control objective, for example, the regulation, the disturbance rejection, and the tracking of the low-frequency reference signals [13].

It would be more practical to exploit the noise/disturbances already existing in the process to provide such persistent excitation. One may look on the inference canceling in the adaptive signal processing for the analogy. To filter the noise from the signal an optimal Kalman or Wiener filter are not well suited as they introduce some inevitable phase distortion. A better solution is to introduce an additional reference input x_{nr} containing the noise which is correlated with the original corrupting noise x_n. The reference noise x_{nr} is then filtered the to produce an estimate of the actual noise x^*_n. Then the noise is subtracted from the primary input $s + x_n$ which acts as the desired response to produce the estimate of the signal s^* [14].

Successful control, therefore, requires the whole environment that acts systematically towards the goal of accumulating the knowledge and using it. If one tries to optimize both objectives (the control and the identification), the identification solution alone would indeed play the role of the persistent excitation [12]. Consequently, the control should imply both the learning (identification) and the tracking (control in its traditional form). In this paper, an attempt is made to design such environment that includes not only the learning in the controller but also the overall control scheme that permit such learning to occur.

In view of the above discussion, the squared difference between the desired output set-point, $\vec{y}^{sp}$, and the process output, $\vec{y}$, cannot be used as the objective function to be minimized for the learning in a neural control:

$$\varepsilon_k(t) = \tfrac{1}{2}\left(y_k^{sp}(t) - y_k(t)\right)^2 \tag{9}$$

where $k = 1, 2, \ldots, m$. With this objective function, neural controller learning may proceed to a physically unrealizable situation, since the set-points are obviously not persistently exciting. In order to overcome the lack of the persistency of excitation, one may obtain the desired output response, $\vec{y}_d$, from the output of a reference model with the state variables, $\vec{x}_n$, being inputs to the model. In this case the excitation is due to the actual process signals (states) affected by the disturbance signals, $\vec{d}$, and the noise, $\vec{\eta}$.

The objective function in such a case becomes the squared difference between the measured outputs of the process, $\vec{y}$ and the reference model (desired process responses), $\vec{y}_d$.

$$\varepsilon_k = \tfrac{1}{2}\left(y_k - y_{dk}\right)^2 \tag{10}$$

The learning algorithm is designed to obtain the correct control signals (manipulated variables), $u_k (k = 1, 2, \ldots, m)$, corresponding to the desired process outputs, $\vec{y}_{dk}$, by minimizing the learning error,

In such a way, we try to exploit a natural level of excitation that exists in the controlled process by applying the same noisy inputs (states) signals to both the controller and the reference model. In addition, a reference model in the form of a filter with a desired transfer function may be used to ensure a variance in the desired dynamic characteristics of the process. The frequency response of the closed-loop may be adapted in line with the changes in the frequency response of the filter. Such reference models may be used when the required performance of the time-varying process is to be achieved by the change in the overall process transfer function. This is an optimization-based design method using a Modulus Optimum, also called a loop-shaping method [15].

One may use a linear stable reference model, for example, a Butterworth filter, since the general well-behaved nonlinear models are not yet available. The coefficients of a Butterworth filter are selected to correspond to the Modulus Optimum criteria for a desired performance in terms of the standard control objectives such as the overshoot, the settling time, and the steady-state error. The output of such model would be a desired signal with the acceptable level of the noise in it. While the use of such a general reference model would not permit us to achieve an ideal control, this should guarantee the adequate control performance for a wide range of the processes in which it is to be used. The controller is designed to transfer the original process transfer function to a desired one between the noisy inputs (states) and the desired outputs of the reference model. In such a case a reference model represents our desired process (the controller plus the original process).

From another perspective, most of the reported work in the neural control is concerned only with the problem of the process modeling or the identification with an open-loop mode of the operation using the static Multilayer Perceptron (MLP) with a back-propagation learning algorithm. For the static processes such an open-loop strategy may be a viable solution. However, for the dynamical processes such a solution is not practical since it is not robust with respect to the disturbances and the parameter uncertainties [16].

Therefore, in a real-world process control, the ability of the controller to adapt to the process changes is vital. While it is possible to tune a standard proportional-integral-derivative (PID) controller to the current state of the process, its inability to deal effectively with the non-stationarity of the process is a serious problem. In practical control applications it is necessary to use the external recurrent (feedback) networks (equivalent to the closed-loop control) to deal with the temporal changes in the process parameters in contrast to the feed-forward networks (equivalent to the open-loop control). The external recurrent neural networks are also called the Jordan networks or the taped delay line networks [17].

In this paper, a feedback neural network controller (NN) is employed to deal with the temporal changes in the process parameters. An outer recurrent feedback loop [17] provides the possibility to include the temporal information, that is, the network produces a dynamic input-output mapping, in contrast to the static feed-forward neural networks.

Following the above discussion, we utilize in this work a direct model reference control (MRAC) scheme with the same noisy inputs (states) signals being

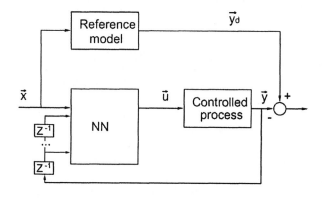

Figure 1. A block diagram of the overall control system.

supplied to both the controller and the reference model in a form of a Butterworth filter as presented in Figure 1.

THE CASCADE CORRELATION NEURAL NETWORK

Most of the existing efforts in the neural control are concentrated on a parameter tuning, selecting the structure on a trial-and-error basis. At the same time, insufficient efforts have been made concerning a structural tuning, that is, to find the simplest controller structure capable of achieving an optimal performance. The need to find a neural network architecture that can fit the training data well has to be guarded against too much flexibility that can result in the over-fitting of the training data, consequently a smaller neural network is required. However, there are no simple ways to determine in advance the minimal size of the hidden layer necessary to achieve the desired performance. This type of the structure selection, which defines the control hyper-surface, may require the in-depth knowledge about the underlying nonlinear process that is rarely available [18]. It is not uncommon to test many architectures to find the appropriate one by trial-and-error, although there are some algorithms for constructing an MLP during learning [19].

In addition, large neural networks are able to generalize as well as the smaller ones for the problems where the data is uniformly distributed. However, for the cases where the data samples are concentrated within a small area of the mapping space, the large networks may fail to generalize at all, especially for the analog (continuous) data [20]. In the closed-loop control the data cannot be selected freely, as the actual states and the outputs of the process are constrained by the process dynamics [21]. In addition, the desired process outputs are constrained by the specification of the control problem. Under these conditions, the training data often remain in the small regions of the domain for the extended periods. This data fixation can have the negative effects in the situations where the parameter adjustments can affect the input/output map globally. For example, if a parameter that has a global effect on the mapping is repeatedly adjusted to correct the mapping in a particular input region, this may cause the map in other regions to deteriorate. Thus, it can effectively erase the effects of the previous learning [21].

A manifestation of this problem is the *moving target problem* [22]. The *moving target problem* (also called the *herd effect*) appears if several separate computational tasks have to be performed by a singular neural network (for example controlling two or more process outputs simultaneously). In case of a global sigmoidal neural network (such as a MLP with a back-propagation learning algorithm), each hidden node of the network decides which problem to handle. If one task generates a larger or a more coherent error signal than the other, there would be a tendency for all hidden nodes to concentrate on the first task while ignoring the second.

Once the first problem is solved, the hidden nodes would concentrate on the second problem, however, if all hidden nodes concentrate on the second problem, the first problem reappears. Eventually, the herd of nodes would split up to deal with both sub-problems at once, but only after a long period of indecision.

One way to deal with the moving target problem is to allow only few hidden nodes to change at once, keeping the rest constant. This effectively aims on making the network a local one.

The above discussion also corresponds to problem of using a single multi-output sigmoidal neural network to control all outputs [23]. The main difficulty here was in the adjustments of the parameters of such network as the errors in the outputs were observed at the different time instances. To alleviate the problem the authors used as many networks as there were outputs. The cause of the above problem was the *moving target problem* (i.e.,, a global network had difficulty following several different outputs at the same time).

Consequently, the above characteristic of the control problems requires the spatially localized architectures and the learning rules [21]. In such architectures, the learning in one part of the input domain has marginal effects on the existing knowledge acquired in other parts of the mapping. Several networks do have such spatially localized characteristic including a cascade-correlation neural network (CCNN) [22], whereas the sigmoidal networks are the global networks. Other examples of the spatially localized networks are the radial basis function networks (RBF), and the fuzzy neural networks. In this paper, we use a dynamically constructed CCNN in view of the above discussion.

A small and gradually decreasing learning rate is also needed in order to prevent the controller from the instability problem resulted from the over-aggressive control. For the incremental learning, the learning rate is strictly bounded by the stability requirements of the stochastic approximation theory [24]. The gradual increase in the size of a growing (constructive) network, such as CCNN, makes the learning even more stochastic. Accordingly the learning rate must be small and diminishing in order to maintain the stability of the learning algorithm. This also corresponds to the observation that the gradient methods with the small step sizes did not cause the instability [23]. In addition, a controller in our case relies only on the measurements without any additional knowledge of the process dynamics. Consequently, the CCNN starts with a relatively large learning rate to enhance the learning speed. Whenever the error starts increasing, the learning rate is reduced. Such gradually decreasing learning rate prevents the controller instability problem resulting from the over-aggressive control.

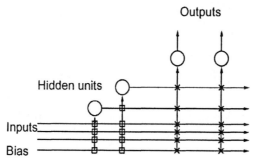

Figure 2. The structure of the cascade-correlation neural network (CCNN).

Figure 2 shows the architecture of the CCNN, used as a controller in current work, whose construction and learning algorithms can be summarized as follows:

1. Starts with a minimal network consisting only of an input layer and an output layer. Both layers are fully connected with adjustable weights. There is also a bias unit, set permanently to +1. Linear output units are used.
2. Train all connections to the output layer using the quick-propagation learning algorithm [22] until the overall error of the network no longer decreases.
3. If the network performance satisfies a prescribed accuracy target, the algorithm stops. In such case, as there is no hidden layer, the problem at hand is linear. We note that CCNN can thus be used to test if the problem at hand is *really* a nonlinear one. There is no benefit in applying neural controller to a linear or linearizable plant, as this will result in degradation of performance in terms of computation time and controller performance: the solution should not be more complex than the problem at hand [25].
4. If the network performance is not satisfactory (and therefore the problem is *really* a nonlinear one), generate candidate nodes. Every candidate node receives trainable connections from all inputs nodes and from all pre-existing hidden nodes. There are no connections between the candidate nodes and the output nodes.
5. Maximize the correlation between the activation of the candidate nodes and the residual error of the network by training all connections leading to a candidate node. The training stops when the correlation no longer improves.
6. Choose the candidate node with the maximum correlation and add it to the network. To change the candidate node into a hidden node, connect it to all output units. Return to step 2.

The algorithm is repeated until the overall error of the network falls below a pre-specified threshold.

SIMULATION MODEL

A thorough study of the electro-hydraulic-mechanical system produced the following transfer function block diagram in Figure 3. The parameter values were selected based on the experimental and the physical property data [3].

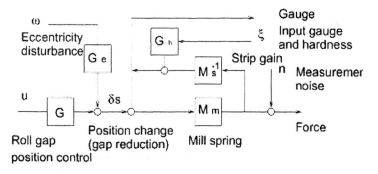

Figure 3. Single-stand model for the cold rolling mill.

The frequencies and the gain values of the model were scaled appropriately to match the experimental data.

In Figure 3, G represents a hydraulic roll-gap model for the closed-loop position control:

$$G = 4056.32/(s^2 + 63.58s + 4123.93) \qquad (11)$$

where $s = \partial x/\partial t$.

G_e is an eccentricity disturbance model:

$$G_e = 852.224/((s^2 + 0.0314s + 985.96) \times (s^2 + 0.0294s + 864.36)) \qquad (12)$$

G_h is an input thickness and hardness disturbance model:

$$G_h = 0.333/s + 0.333 \qquad (13)$$

Experimental testing verified the use of an eccentricity model comprising of two lightly damped oscillators driven by zero-mean white noise with the covariance $E\{\omega(t)\omega(\tau)\} = x_1^2$, where $x_1 = 0.00012$.

The high cost of a physical data for the hardness variation in the cold rolled strip mills lead to the use of a first order lag driven by zero-mean white noise with the covariance $E\{\xi(t)\xi(\tau)\} = x_2^2 \delta(t-\tau)$, where $x_2 = 0.00007$ and δ is the Kronecker delta function.

The disturbances and the noise were applied concurrently to represent a real situation when the disturbances and the noise in the rolling mill are present at the same time. A combination of these models results in a complete model representation of the combined mill and the disturbance system.

To generate the roll force and the gauge variations a simple linear small-change model of the mill is used [3]. Here, the gauge $h(t)$ satisfies the following:

$$\Delta h(t) = \frac{M_m M_s^{-1}}{1 + M_m M_s^{-1}} \Delta s(t) + \frac{1}{1 + M_m M_s^{-1}} \Delta H(t) \quad (14)$$

where $M_m = 1.039 \times 10^9$ N/m and $M_s = 9.81 \times 10^8$ N/m are the mill and the strip moduli, respectively, and $s(t)$ is the roll gap setting.

The measured roll force $z(t)$ is:

$$\Delta z(t) = M_m (\Delta h(t) - \Delta s(t)) + \eta(t) \quad (15)$$

Here $\eta(t)$ represents the measurement noise with the covariance $E\{\eta(t)\eta(\tau)\} = x_3^2 \delta(t-\tau)$, and $x_3 = 1000$.

The control of a rolling mill requires the output gauge to be regulated in the presence of the eccentricity and the input hardness disturbances using a measured roll force. This is a control problem where the measured variable is not controlled directly, but the measured variable is used to achieve the control of another system variable - the strip gauge, resulting in the inferential control.

SIMULATION RESULTS AND DISCUSSION

As a reference model a Butterworth characteristic equation [26] for a 5th order system was used:

$$F(s) = s^5 + 3.24\omega_n s^4 + 5.24\omega_n^2 s^3 + 5.24\omega_n^3 s^2 + 3.24\omega_n^4 s + \omega_n^5 \quad (16)$$

Here ω_n is a natural frequency of the system.

The input-output data generated using this reference model was used to train the CCNN with on-line (pattern) learning. A 1000 input-output data pairs were generated using a fifth order Runge-Kutta integrator. The data was split in two sets, that is, 500 samples used for training, another 500 for the validation of the neural network generalization ability, and the whole data set was used for the final testing of a trained neural network to evaluate the performance of the controller for both the known and the unknown data. While the performance of the neural controller is traditionally demonstrated on the testing (unseen) data only this prevents the comparison of any degradation of the performance when the controller is used on unseen data in comparison to its performance on the known (training) data. The controller ability to generalize is at it best where there is a minimal difference between its performance on known data and its performance on unseen data. While this is well known, it is rarely evaluated. Simulation results are presented in Figure 4 and Figure 5 and summarized in Table 1. Figure 4 shows only the last 500 samples in order to increase its readability.

Table 1. Root Mean Squared Error (RMSE) for the eccentricity and the hardness for the uncontrolled and the CCNN controlled cases.

	RMSE	
	Uncontrolled	CCNN
Eccentricity	0.3336	0.0457
Hardness	0.1259	0.0208

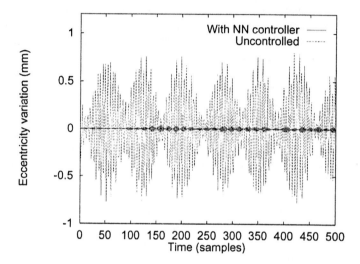

Figure 4. Time response for the eccentricity disturbance rejection for the CCNN.

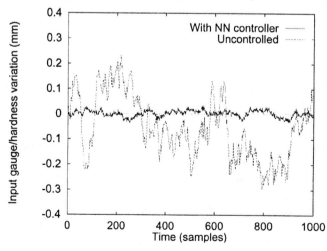

Figure 5. Time response for the input hardness disturbance rejection for the CCNN.

As it can be seen, a neural controller produces an excellent disturbance rejection in both cases of eccentricity and hardness disturbances, reducing the gauge deviation due to eccentricity disturbance from 33.36% to 4.57% on average, and the gauge deviation due to hardness disturbance from 12.59% to 2.08%.

CONCLUSION

In this paper a direct model reference adaptive control (MRAC) scheme with dynamically constructed neural controller was used for control of eccentricity and hardness disturbances. The aim here was to find the simplest controller structure capable of achieving an optimal performance. The stability of the adaptive neural control scheme (i.e. the requirement of persistency of excitation and bounded learning rates) is addressed by using as the inputs to the reference model the plant's state variables. In such a case, excitation is due to actual plant signals (states) affected by plant disturbances and noise. In addition, a reference model in the form of a filter with a desired transfer function using Modulus Optimum design was used to ensure variance in the desired dynamic characteristics of the system. The gradually decreasing learning rate employed by the neural controller in this paper is aimed at eliminating controller instability resulting from over-aggressive control. The moving target problem (i.e. the difficulty of global neural networks to perform several separate computational tasks in closed-loop control) is addressed by the localized architecture of the controller. The control scheme and learning algorithm offers a method for automatic discovery of an efficient controller. The resulting neural controller produces excellent disturbance rejection in both cases of eccentricity and hardness disturbances, reducing the gauge deviation due to eccentricity disturbance from 33.36% to 4.57% on average, and the gauge deviation due to hardness disturbance from 12.59% to 2.08%.

REFERENCES

[1] S. G. Choi, M. A. Johnson, and M. J. Grimble, 1994. Polynomial LQG control of back-up-roll eccentricity gauge variations in cold rolling mills', Automatica, 30(6), 975–992.
[2] S. S. Garimella and K. Srinivasan, 1996. Application of repetitive control to eccentricity compensation in rolling, Transactions of the ASME, Journal of Dynamic Systems, Measurement and Control, 118, 657–664.
[3] M. J. Grimble, 1995. Polynomial solution of the standard H_∞ control problem for strip mill gauge control, IEE Proceedings, part D-Control Theory and Applications, 142(5), 515–525.
[4] I. Postlethwaite and J. Geddes, 1994. Gauge control in tandem cold rolling mills: a multivariable case study using H-infinity optimization, Proceedings 3rd IEEE Conference on Control Applications, 3, 1551–1556.
[5] C. C. de Wit, L. Praly, 2000. Adaptive eccentricity compensation, IEEE Transactions on Control Systems Technology, 8(5), 757–766.
[6] A. Kugi, W. Haas, K. Schlacher, K. Aistleitner, H. M. Frank, and G. W. Rigler, 2000. Active compensation of roll eccentricity in rolling mills, IEEE Transactions on Industry Applications, 36(2), 625–632.

[7] Y. J. Hwu and J. G. Lenard, 1996. Application of neural networks in the prediction of roll force in hot rolling, Proceedings of the 37th Mechanical Working and Steel Processing Conference, 549–554.

[8] R. Pichler and M. Pffaffermayr, 1996. Neural networks for on-line optimisation of the rolling process, Iron and Steel Review, August, 45–56.

[9] T. Fechner, D. Neumerkel, and I. Keller, 1994. Adaptive neural network filter for steel rolling, Proceedings 1994 IEEE International Conference on Neural Networks, 6, 3915–3920.

[10] K. Aistleitner, L. G. Mattersdorfer, W. Haas, A. Kugi, 1996. Neural network for identification of roll eccentricity in rolling mills, Journal of Materials Processing Technology, 60(1–4), 387–392.

[11] D. Neumerkel, R. Shorten, A. Hambrecht, 1996. Robust learning algorithms for nonlinear filtering, Proceedings 1996 IEEE International Conference on Acoustics, Speech and Signal Processing, 6, 3565–3568.

[12] A. Guez, I. Rusnak, and I. Bar-Kana, 1992. Multiple objective optimization approach to adaptive and learning control, International Journal of Control, 56(2), 469–482.

[13] K. S. Tsakalis, 1997. Bursting scenario and performance limitations of adaptive algorithms in the absence of excitation, Kybernetika, 33(1), 17–40.

[14] A. Zaknich and Y. Attikiouzel, 1995. Application of artificial neural networks to nonlinear signal processing, In M. Palaniswami, Y. Attikiouzel, R. J. Marks, D. Fogel and T. Fukuda (Eds.) Computational intelligence, a dynamic system perspective, IEEE Press, New York.

[15] T. Hägglund and K. Åström, 1996. Automatic tuning of PID controllers, In W. S. Levine (Ed.), The Control Handbook, CRC Press, 817–826.

[16] A. U. Levin and K. S. Narendra, 1996. Control of nonlinear dynamical systems using neural networks – Pt. 2: observability, identification, and control, IEEE Transactions on Neural Networks, 7(1), 30–42.

[17] M. I. Jordan, 1986. Attractor dynamics and parallelism in a connectionist sequential machine, Proceedings Eighth Annual Conference of the Cognitive Science Society, 531–546.

[18] S. Tan and Y. Yu, 1996. Adaptive fuzzy modeling of nonlinear dynamical systems', Automatica, 32(4), 1996, 637–643.

[19] T.Y. Kwok and D.Y. Yeung, 1997. Constructive algorithms for structure learning in feedforward neural networks for regression problems, IEEE Transactions on Neural Networks, 8(3), 630–645.

[20] C. Ji and D. Psaltis, 1997. Network synthesis through data-driven growth and decay, Neural Networks, 10(6), 1133–1141.

[21] J. A. Farrell and W. Baker, 1993. Learning control systems, In P. Antsaklis and K. Passino (Eds.), An introduction to intelligent and autonomous control, Kluwer Academic, 237–262.

[22] S. E. Fahlman and C. Lebiere, 1990. The cascade-correlation learning architecture, In D. S. Touretzky (Ed.), Advances in Neural Information Processing Systems 2, Morgan Kaufmann, San Mateo, CA, 524–532.

[23] K. S. Narendra and S. Mukhopadhyay, 1994. Adaptive control of nonlinear multivariable systems using neural networks, Neural Networks, 7(5), 737–752.

[24] T. Hrycej, 1997. Neurocontrol: towards an industrial control methodology, John Wiley and Sons.
[25] P. Mars, J. R. Chen, and R. Nambiar, 1996. Learning algorithms: theory and applications in signal processing, control and communications, CRC Press, Boca Raton, FL.
[26] K. Åström and B. Wittenmark, 1990. Computer controlled systems, theory and design, 2nd ed., Prentice-Hall.

Chapter 4: Process Control and Simulation for Optimization

Level Control Model by Numerical Fluid Dynamics Method

D. SUZUKI and K. FUJISAKI

Environmental & Process Technology Center, Nippon Steel Corp., 20-1, Shintomi, Futtsu-City, Chiba, 293-8511, Japan

ABSTRACT

This paper presents a coupled model and a numerical fluid dynamics analysis of level control in a continuous casting mold. Three-dimensional free surface flow exists in the mold while the free surface level is detected by a sensor and controlled by the supply of liquid steel. Large Eddy current Simulation (LES) and the Volume of Fluid method (VOF) are used for the plant model to capture free surface motion over time as a distributed parameter system. The combination of these methods enables us to obtain exact results and conduct evaluations based on the non-linear and distributed characteristics of the process. The calculated results such as free surface shape and fluid velocity vectors are useful to understand the physical phenomena in the process.

INTRODUCTION

Development and improvement of process control is very important for obtaining high productivity, high quality and low costs. It requires precise understanding of the process characteristics. Many industrial processes have non-linear and distributed characteristics such as heat and flow problems. These process characteristics must be faithfully reflected in any modeling and simulation work since they are fundamental to design and evaluate the process control system. Today's computers are significantly advanced in their computing power and storage capacity enabling one to conduct large-scale physical simulation by numerical analysis. In addition, user requirements of product qualities have recently been especially intense demanding advanced modeling and simulation technologies to design the process control system properly. We have applied a dynamic, three-dimensional distributed plant model to a process control simulator in order to obtain exact results and perform an evaluation [1]. As an example, level-flow control in a continuous slab caster has been studied, which is a key technology for product quality and operating safety in continuous casting. Since continuous casting is a type of fluid process, it has a free surface and its position is maintained at constant level by feedback control to obtain good quality steel.

Level control has been studied conventionally using a centralized model although the practical process is a non-linear and distributed parameter system. In other words, the centralized model assumes that the free surface is static and that only the balance between pouring velocity and casting speed decides the free surface position. However, the practical free surface should be dynamic and distributed because of turbulent flow in the mold. In order to account for the effects of turbulence, we propose a level control model using numerical fluid dynamics of which the Large Eddy simulation approach (LES) is used to characterize turbulence and the Volume of Fluid method (VOF) is used to track the free surface. A combination of these methods can capture the three-dimensional free surface motion over time as a distributed parameter model. Level control simulation by numerical fluid dynamics is conducted as follows: As the free surface fluctuates from fluid volume changes in the mold, the free surface level is detected by the sensor at an interval of 0.1 seconds and is fed-back to the controller. The controller adjusts the fluid supply into the mold so variations in detected level are reduced. These sequences are repeated in turn to analyze a level-controlled free surface motion. Figure.1 shows the block diagram of the proposed distributed level control simulator by numerical fluid dynamics method.

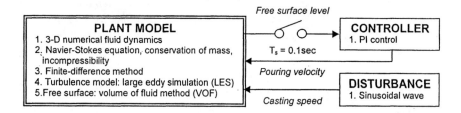

Figure 1. Block diagram of the distributed level control simulator by numerical fluid dynamics method.

In this paper, we discuss features of the distributed level control model compared to a centralized one. The effect of the casting speed is investigated and the calculated results are introduced such as the free surface shape and the control response. Increase in casting speed makes the flow more turbulent and the free surface more variable. As a result, level control becomes less stable at higher speed casting. The effect of the turbulent flow is impossible to take into account in the centralized model. It is shown that the distributed model has advantages of obtaining exact results and evaluation in comparison to the centralized model since the distributed model simulates the practical process from the viewpoint of turbulent fluid dynamics.

LEVEL CONTROL IN CONTINUOUS CASTING

The slab mold is used in a steel-making plant to solidify molten steel into solid in large sizes. The molten steel is injected into the slab mold from a tundish, which is a pool of the molten steel located upper of the mold, through a bifurcated submerged entry nozzle. There is a free surface in the slab mold and its level is full-time monitored by a sensor and controlled by liquid steel supply.

The level control to keep the free surface position fixed is much important for surface qualities because the initial solidification begins at the free surface line. In addition, the free surface contacts casting powder and its large fluctuations will cause powder entrapment into the molten steel. To prevent these product defects such as cracks and powder inclusions, variations of the free surface position are maintained less than 5mm empirically. An eddy current type sensor is placed above the mold to detect the free surface level and a sliding valve at the submerged entry nozzle is controlled to adjust the liquid steel supply. Figure.2 shows a schematic diagram of level control system in a continuous slab caster.

The free surface position usually fluctuates from 'bulging'. It is because the solidified shell cannot withstand hydrostatic pressure of the molten steel and would bulge if unconstrained. Moreover the nozzle inside is clogged with alumina and flux coefficient of the nozzle is unsteady. These phenomena, which are difficult to forecast, will cause the level fluctuations. Many studies have been conducted on robustness of the controller to time-variant characteristics and disturbances [2]. However, they have been assuming that the controlled object is a centralized parameter system.

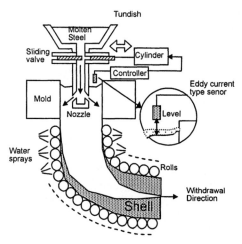

Figure 2. Schematic diagram of level control system in a continuous slab caster.

MODEL DESCRIPTION

NUMERICAL METHOD

The plant model is described as a non-linear and distributed system by numerical fluid dynamics. The governing equations used in this analysis are the conservation of mass and momentum as follows:

$$0 = \nabla \cdot u \qquad (1)$$

$$\frac{\partial u}{\partial t} + u \cdot \nabla u = -\frac{1}{\rho}\nabla P + v_e \nabla^2 u + g + F \qquad (2)$$

where u, t, P, ρ, v, g and F are fluid velocity, time, pressure, density, kinetic viscosity, gravity, wall friction, respectively.

It is assumed that the liquid steel is incompressible and has constant density. The suffix e denotes the effective value, which is a sum of the laminar term and the turbulence term. The large eddy simulation model (LES) is used as the turbulence model to capture the large scale of eddies and their motion with time. In this analysis, the viscosity of small eddies that are not resolved in computational grids is modeled by the sub-grid scale model (SGS):

$$v_e = v + v_{SGS} \qquad (3)$$

$$v_{SGS} = (C_s \Delta V^{1/3})^2 \sqrt{2 \cdot e_{ij}^2} \qquad (4)$$

$$e_{ij} = \frac{1}{2}\left(\frac{\partial u_j}{\partial x_i} + \frac{\partial u_i}{\partial x_j}\right) \qquad (5)$$

where C_s, ΔV and e_{ij} are Smagorinsky constant, volume of the cell and strain rate tensor, respectively. Wall friction F is proportional to roughness and the second power of velocity:

$$F = -\frac{\lambda}{2}\frac{|u|}{\delta}u \qquad (6)$$

where λ and δ are wall roughness and distance from the wall in the cell.

The roughness λ depends on material. The nozzle is assumed to have a smooth rigid of which roughness is calculated from a confine pipe flow model:

$$\lambda = \frac{1}{4} \times 0.3164 \mathrm{Re}^{-1/4} \qquad (7)$$

$$\mathrm{Re} = \frac{u_c d}{v} \qquad (8)$$

where Re, u_c and d are Reynolds number, central velocity and diameter of the pipe flow, respectively.

On the other hand, the solidified shell is assumed to have a stationary rough rigid of which roughness is deduced from experimental results [3]. The volume of fluid method (VOF) is used to track fluid interface:

$$\frac{\partial f}{\partial t} + \nabla(u \cdot f) = 0 \qquad (9)$$

where f is a fractional volume of the cell occupied by the fluid.

These discrete equations are solved using finite difference method (FDM) and the successive over-relaxation method (SOR) as a pressure iteration solver. The combination of these methods is able to simulate a three-dimensional free surface flow with time. Table.1 shows standard input conditions for model calculation.

Table. 1. Standard input conditions for model calculation.

Input Variable	Unit	Value
Density of the liquid steel	[kg/m^3]	7000
Viscosity of the liquid steel	[kg/m/s]	0.00444
Surface tension of the liquid steel	[N/m]	1.645
Gravity constant	[m/s^2]	9.81
Smagorinsky constant	[-]	0.10

BOUNDARY CONDITIONS

The free surface position is balanced between the inflow and the outflow that are pouring velocity at the nozzle inlet and casting speed at the caster bottom, respectively. These velocities are uniform in their distribution. The controller controls the pouring velocity, while the fluid volume in the mold changes from variations in the casting speed. The free surface level is detected as an average of the level distribution in the sensor scope. Two sensors are located cetrally in thickness of the mold and 1/4 width away from the narrow sides. In this study, only the sensor in the right side is used for level control. The detected level is defined as a distance from the initial free surface. The computational domain is perfectly straight in reference to the gravitational field. The nozzle used here is a bifurcated one with a well at the bottom, of which the flow channel is orthogonal and the outside is a column. Submergence depth of the nozzle is defined as a disitance between the initial free surface and the top of the port. Figure.3 shows a schematic diagram of the boundary conditions of the slab caster. Table.2 shows dimensions of the slab caster.

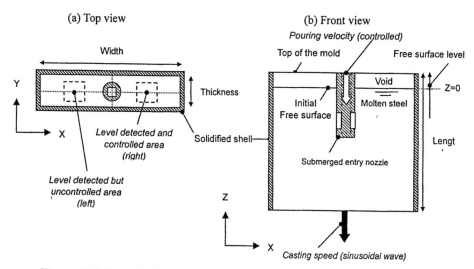

Figure 3. Schematic diagram of the boundary conditions of the slab caster.

Table 2. Dimensions of the slab caster (calculated).

Input	Unit	Value
Width of the caster	[m]	1.500
Thickness of the caster	[m]	0.240
Length of the caster	[m]	5.100
Side of the sensor scope (rectangle)	[m]	0.180

CASTING AND CONTROL CONDITIONS

The level control is conducted after the flow in the mold is sufficiently developed where the inflow balances with the outflow. To imitate the free surface fluctuations from 'bulging', the casting speed V_c varies with time as a sinusoidal wave:

$$V_c(t) = V_{co} + B \cdot \sin(2\pi f \cdot t) \qquad (10)$$

$$f = V_{co}/L \qquad (11)$$

where B, f and L are amplitude and frequency of variations in the casting speed and interval of the support rolls, respectively.

Note that the frequency becomes higher at higher speed casting. The amplitude B is calculated so that amplitude of the level variations becomes 5 mm without control in a centralized model. The controller K(s) is given in Laplace transformation:

$$K(s) = K_p (1 + \frac{1}{sT_i}) \qquad (12)$$

where s, K_p and T_i are Laplace operator, proportional coefficient and integral time, respectively.

Dynamics of the sensor and the actuator are not considered. The nozzle flux coefficient is time-invariant though it is time varying in practice. These omitted factors are main issues of the controller design. However, we focus on only the characteristics of the distributed model. Table.3 shows casting and control conditions for model runs.

Table 3. Casting and control conditions for model runs.

Input	Unit	Case 1	Case 2	Case 3	Case 4
Initial pouring velocity	[m/s]	1.60	1.60	2.24	2.24
Initial casting speed	[m/s]	0.025	0.025	0.035	0.035
Amplitude	[m/s]	0.0030	0.0030	0.0042	0.0042
Frequency	[Hz]	0.10	0.10	0.14	0.14
Proportional coefficient	[-]	0.5	2.0	0.5	2.0
Integral time	[s]	∞	1.0	∞	1.0
Nozzle flux coefficient	[1/s]	60	60	60	60
Reference level	[mm]	-5	-5	-5	-5

NUMERICAL RESULTS AND DISCUSSION

Here are 4 cases in a 2 x 2 matrix of the casting speed and the controller gain. The effect of the turbulent flow is investigated and discussed. These cases are also simulated using a centralized model to compare with the distributed model results. Case.1 and 2 are of V_{co}=1.5m/min the slower, while case.3 and 4 are of V_{co}=2.1m/min the faster. The odd numbers are of the low gain controller while the even numbers are of the high gain controller.

CENTRALIZED MODEL

The centralized model results are presented here. The control responses of the centralized model are smooth and basic because only the mass balance is calculated from the inflow and the outflow. Figure.4 and Figure.5 show a block diagram of the centralized level control simulator and calculated time responses of the centralized model, respectively.

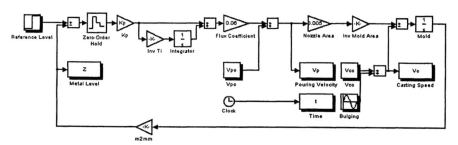

Figure 4. Block diagram of the centralized level control simulator.

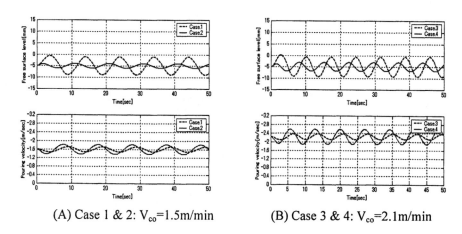

(A) Case 1 & 2: V_{co}=1.5m/min   (B) Case 3 & 4: V_{co}=2.1m/min

Figure 5. Calculated time responses of the centralized level control model.

DISTRIBUTED MODEL

The distributed model results are presented here. The control responses of the distributed model are similar to the centralized ones except their 'ripples' and 'asymmetry'. The ripples are evidence that the free surface fluctuates by turbulent flow. As the flow becomes more turbulent, the ripples are more obvious in the detected levels. Concurrently, the ripples in the control responses are apparent of the higher gain controller. The detected level by the right sensor is maintained at the reference level however the left side is not controlled to the same extent. This is since the level sensor detects local variations in metal level. As a result, the controller controls only the local area of the free surface but not the whole. These results are realistic and more exact than using a centralized one. Figure 6 shows the calculated time responses of a distributed model.

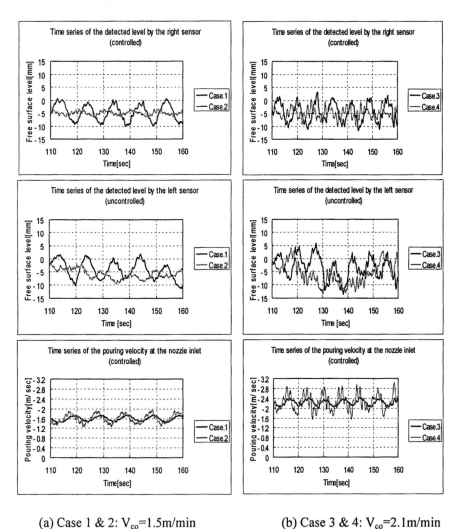

(a) Case 1 & 2: V_{co}=1.5m/min   (b) Case 3 & 4: V_{co}=2.1m/min

Figure 6. Calculated time responses of the distributed level control model.

The distributed model is able to capture the whole free surface motion with time. To evaluate the free surface characteristics, the free surface shape and the vorticity are introduced. The vorticity used here is defined as follows:

$$\omega_z = \left| \frac{\partial u_i}{\partial x_j} - \frac{\partial u_j}{\partial x_i} \right| \tag{13}$$

The vorticity represents intensity of the vortex that will entrap the power into the molten steel. It is supposed that the vortices are strong where the free surface velocity is mostly large and has large variations. As the strong flow passes by the nozzle, the wake flow generates small but strong vortices behind the nozzle. These vortices will catch the powder and it will be mixed into the jet flow of the nozzle and transported deep inside the molten steel. If the level fluctuations are kept small, the vortices will not arise near the nozzle. The level control effectively reduces the level fluctuations and the vortices in the lower speed casting. However, such vortex control becomes more difficult in the higher speed casting because the flow in the mold is more turbulent and the level control cannot kill the turbulence itself. As a result, the possibilities of the powder entrapment must increase in the higher speed casting. Figure.7 shows contours of average and standard deviation of the free surface level. Figure.8 shows contours of the maximum vorticity at the free surface.

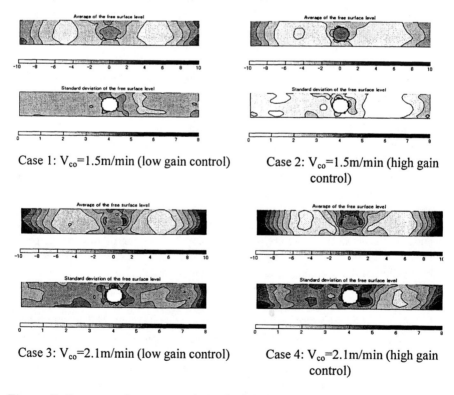

Case 1: V_{co}=1.5m/min (low gain control)   Case 2: V_{co}=1.5m/min (high gain control)

Case 3: V_{co}=2.1m/min (low gain control)   Case 4: V_{co}=2.1m/min (high gain control)

Figure 7. Contours of average and standard deviations of the free surface level (unit: [mm]).

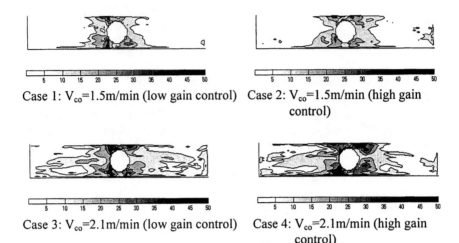

Case 1: V_{co}=1.5m/min (low gain control)   Case 2: V_{co}=1.5m/min (high gain control)

Case 3: V_{co}=2.1m/min (low gain control)   Case 4: V_{co}=2.1m/min (high gain control)

Figure 8. Contours of the maximum vorticity at the free surface (unit: [1/s]).

The free surface has slopes at the narrow sides of the mold where the jet flow arrives directly to the surface. The level variations at the narrow sides are larger and the slopes are steeper as the casting speed increases. These results coincide with the practical phenomena qualitatively. As mentioned above, the metal level is well controlled in the right side, while less controlled and fluctuant in the left. In the lower speed casting, such asymmetry is not obvious and the high gain control well controls the whole free surface. The level fluctuations around the nozzle are reduced and the vorticity is small. It explains that the vortices are generated by large variations in the metal level, particularly around the nozzle. Thus the powder entrapment will not occur if the free surface level is well controlled. On the other hand, the high gain control rather promotes the asymmetry of the free surface in the higher speed casting. The free surface around the nozzle is more fluctuant and biased. The vorticity around the nozzle is also stronger. These results show that the level control will be more difficult in higher speed slab casting.

To help understand the process characteristics discussed above, the fluid velocity vectors in the mold are presented here. The cutting planes are the central wide face in the mold and the X-Y plane at the free surface. Figure 9 shows distribution of the averaged fluid velocity vectors.

Here can be seen the results of Case 1 and 3 while the others are almost the same. Typical flow patterns can be observed: Jet flow from the nozzle impinges to the narrow sides and bifurcates into upstream and downstream. The upstream forms a circulation in the longitudinal direction and a strong flow toward the nozzle. Such flow patterns can be obtained by the Reynolds averaged numerical simulation (RANS), however the transient behavior is much more important. It is because the worst possible situation will decide the product qualities. The proposed model succeeds in capturing transitions of the turbulent flow thanks to LES. For example, transient vortices around the nozzle can be observed at the free surface. Figure.10 shows transitions of the instant fluid velocity vectors at the free surface position in Case.1 at the interval of 1.0 second.

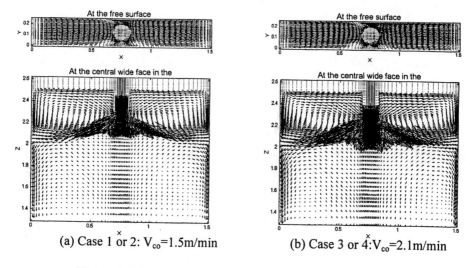

Figure 9. Distribution of the averaged fluid velocity vectors.

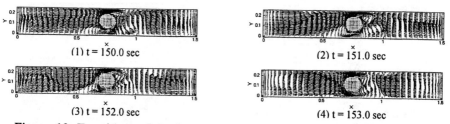

Figure 10. Transitions of the instant fluid velocity vectors at the free surface in Case 1.

Thus the proposed model can estimate statistics value of the turbulent flow. The turbulence of the jet flow region is the strongest in the mold. The jet flow perturbs the whole flow in the mold and the flow fluctuations are larger in the faster casting speed. As mentioned above, the flow around the nozzle is also turbulent since the nozzle blocks the flow and generates the vortices. Figure.11 shows contours of standard deviation of the fluid velocity.

The flow fluctuations cannot be reduced by the level control, however the level control is expected to calm down the free surface. Since the level control can control only the mass but not the turbulence, excessive control will rather disturb the free surface and the whole flow in the mold. Therefore careful control of the liquid steel supply is very important for avoiding the excessive flow perturbations by the jet flow. The appropriate level control will keep the whole free surface less turbulent. These calculated results and physical prospects improve understanding of the process characteristic and indicate how the process control should be. The calculations are qualitatively reliable and some of these phenomena must occur in the practical process.

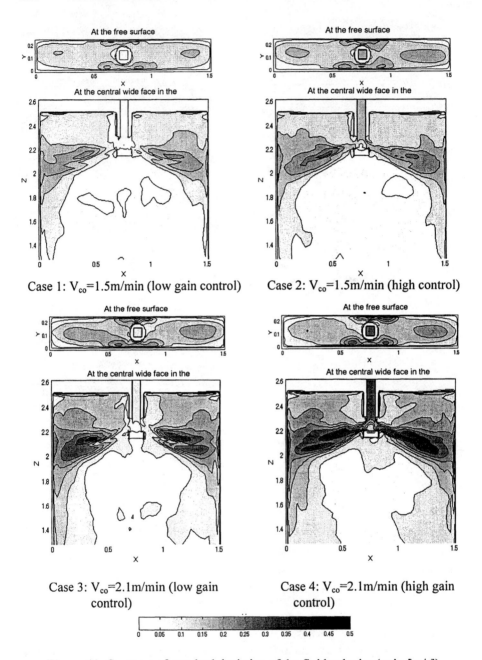

Figure 11. Contours of standard deviation of the fluid velocity (unit: [m/s]).

CONCLUSIONS AND FINAL REMARKS

We have developed an enhanced process simulator of the level control system in continuous casting process. The proposed model can calculate a transient three-dimensional free surface flow with a level control system. The level control can be evaluated not only by the control responses but also the free surface behavior. Consequently, we can obtain the significant results and the physical prospects as follows:

- Level control in high-speed slab casting is influenced very much by turbulent flow because the free surface fluctuates due to turbulence.
- The free surface is asymmetric under level control. It is well controlled on the controlled side, while less controlled and fluctuating on the opposite side. This is most obvious at high slab casting speeds.
- Excessive level control promotes asymmetry of the free surface behavior in high-speed slab casting.
- Powder entrapment mainly occurs near the nozzle because of strong vortices. Level control can reduce these vortices if the free surface flow is less turbulent.

These achievements are unique to the proposed distributed level control model. It proves that the combination of numerical fluid dynamics and control system is very useful and important for evaluation of the process control. Such enhanced simulation helps one understand the process characteristics and the physical phenomena to be controlled.

Some challenge items remaining to be tackled are:

- The dynamics of the sensor and the actuator should be considered for more exact evaluation.
- The effect of electromagnetic force on flow should be considered since electromagnetic techniques are usually installed in slab casting.
- Nozzle clogging should be considered. Clogged nozzles change the jet flow. If one port is closed, the flow is asymmetric and more turbulent.
- The effect of sensor alignment should be studied because the free surface is a distributed parameter system.

These problems are realistic and important for the practical process control. The proposed distributed model can flexibly solve such problems, however the conventional centralized one cannot. We are studying the level control design in details by using the distributed model. These results will be presented and discussed later on in the near future.

REFERENCES

1. K. Fujisaki et al, 1998. Dynamics phase control of electromagnetic stirring. Japan-USA flexible automation symposium, pp.385–388.
2. The Iron and Steel Institute of Japan, 2002. Iron and Steel Handbook 4th Ed. (in Japanese), 7.
3. K.Fujisaki et al, 2000. MHD calculation for free surface velocity. J. Applied Physics, 89(11), 6734–6736.

H-infinity Control of Micromolecular State Transition of a Sodium Ion Channel Gating System

H. HIRAYAMA
Department of Public Health Asahikawa Medical College, Higashi 2-1, Midorigaoka, Asahikawa City, 078, Japan

ABSTRACT

H-infinity control has been applied to evaluate the noise filtering function of a sodium ion selective channel on an excitable biological membrane. Open-close structural transition of the sodium channel is expressed by 20 differential equations, however, for simplicity, we reduced these to 8 equations. By applying the principles of H-infinity control, temporal changes in the amounts of channel species per unit membrane area of the closed, inactivated, and open states were computed. The time trends of these species are sensitive to the electrical voltage gap across the membrane. This present work is presented to provide an evaluation of the noise filtering function of the sodium ion selective channel.

INTRODUCTION

Sodium ion current plays an essential role in the physiological function of excitable cellular membranes such as neurons [1]. Sodium ion channel gatings have been analyzed by many investigators [2,3,4,5]. The channel is composed of four identical subunits each of which is further broken down into four gates. The channel opens and closes in one motion without any intermediate state. This property derives from the concerted movement of four subunits [1]. When the channel has been strongly depolarized, it becomes insensitive to stimuli. This characteristic is called the inactivated state [2,3]. A small blocking biomolecule in the intracellular space has been identified as the origin of this inactivation [3]. The channel opening can only be achieved when all 4 gates in one subunit have taken on the activating position [4,5]. Thus a sodium ion channel has multi-activated, -inactivated, and -closed states. The integrated behavior of one sodium channel is thus, hard to discover if the single patch clamp technique [1] is used. From the standpoint of noise filtering, the sodium ion selectivity originates from the noise selective function of the channel. Thus it should be important to evaluate the noise filtering function of the sodium ion selective channel and so, the investigation described in this paper is intended to introduce H-infinity control principles to evaluate the noise filtering function of a sodium ion selective channel.

BASIC MOLECULAR ANATOMY

One channel molecule is composed of four subunits named I, II, III, IV [1]. Each subunit is further broken down into 6 perforating helical proteins, S1, S2, S3, S4, S5 and S6. The voltage sensitive nature of channel opening and closing is determined by movement of the S4 helixes in the subunit [4,5]. The S4 helix operates as a gate [4,5] for the channel. In the intracellular space, there is a small biomolecule [3] which can invade the channel pore. This ball-like molecule prevents free movement of the gates. Even if only one gate has been blocked, the channel cannot open nor can it transit to a closed state until the ball molecule is pushed out of the channel pore. As a result, channel opening or closing becomes insensitive to excitatory signals for some period. Such an insensitive state is referred to as inactivated.

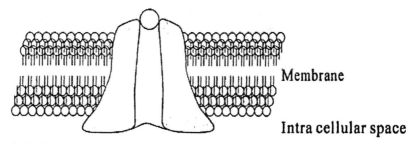

Figure 1. A schematic illustration of the sagital section of a sodium ion selective channel.

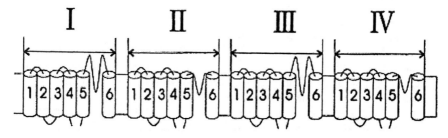

Figure 2. Transverse expansion of subunits, I, II, III, IV of the channel.

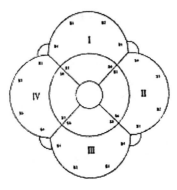

Figure 3. Top view of the channel showing relative positions of six helical polypeptides.

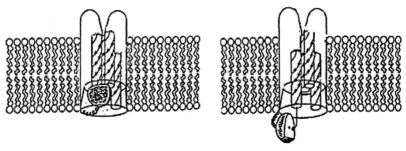

Figure 4. Inactivation of the channel by a blocking ball-like molecule (left) and the activated state (right) reestablished when the blocking ball molecule has been ejected out from the channel pore.

TWENTY POSSIBLE CHANNEL CONFIGURATIONS

According to a statistical analysis [6], there are more than 4 closed states and 3 inactivate ones. There are 4 gates in each subunit. Thus there are 3 possible conFigureurations when one of the 4 gates takes on an activating position in the subunit. We represent these C3 states as C3a, C3b, C3c, and C3d. In each of these C3 state, one of three closed positioned gates can take the activating position in the subunit. Thus, one C3 conFigureuration can have 3 possible transitions to achieve a C2 state in which 2 gates have taken on an activated position. Hence there are 6 possible conFigureurations for C2 consisting of C2a, C2b, C2c, C2d, C2e, and C2f. Each C2 state has 2 gates that can take a closed position in the subunit. Thus each C2 conFigureuration has 2 possible state transitions to achieve a C1 state in which only 1 gate is in a closed position in the subunit, C1a, C1b, C1c, and C1d. Since 3 gates can have open positions in C1, there is only one gate that has taken the closed position in the subunit. To express transient changes in the amount of each of these possible channel conFigureurations, we define the following variables for the amounts of channel state per unit membrane area of the excitable membrane: C4—all 4 gates have taken closed positions; C3—one of the 4 gates has taken an open position; C2—two of the 4 gates have taken an open position; C1—three of the 4 gates have taken an opened position. For the special case in which all 4 gates are open, we use [4] the designation C0. The channel molecule of course, is hydrated in this situation, thus the channel is still closed. O—the channel molecule is dehydrated and open; I1—Three gates have taken an open position while one remains in the rest position; I0—All 4 gates are open, but a blocking particle has invaded from the inside; I—same situation as I0, but the channel is dehydrated.

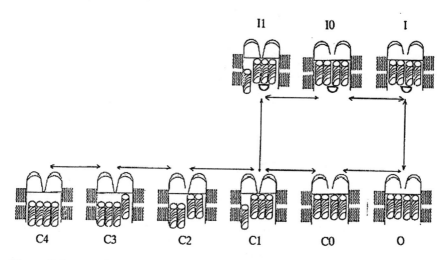

Figure 5. Suspected statistical conFigureuration transitions of a multi-state sodium channel system.

The instantaneous changes in the amounts of channels states per unit membrane area are given by the following 20 linear differential equations:

$$dC4/dt = - (k_1 + k_2 + k_3 + k_4)C4 + k_{-1}C3a + k_{-2}C3b + k_{-3}C3c + k_{-4}C3d + b_1U1 \quad (1)$$

$$dC3a/dt = k_1C4 - (k_5 + k_6 + k_7 + k_{-1})C3a + k_{-5}C2a + k_{-6}C2b + k_{-7}C2d + b_2U2 \quad (2)$$

$$dC3b/dt = k_2C4 - (k_8 + k_9 + k_{10} + k_{-2})C3b + k_{-8}C2a + k_{-9}C2c + k_{-10}C2e + b_3U3 \quad (3)$$

$$dC3c/dt = k_3C4 - (k_{11} + k_{12} + k_{13} + k_{-3}) C3c + k_{-11}C2b + k_{-12}C2c + k_{-13}C2f + b_4U4 \quad (4)$$

$$dC3d/dt = k_4C4 - (k_{14} + k_{15} + k_{16} + k_{-4}) C3d + k_{-14}C2d + k_{-15}C2e + k_{-16}C2f + b_5U5 \quad (5)$$

$$dC2a/dt = k_5C3a + k_8C3b - (k_{17} + k_{18} + k_{-5} + k_{-8})C2a + k_{-17}C1a + k_{-18}C1b + b_6U6 \quad (6)$$

$$dC2b/dt = k_6C3a + k_{11}C3c - (k_{19} + k_{20} + k_{-6} + k_{-11})C2b + k_{-19}C1a + k_{-20}C1c + b_7U7 \quad (7)$$

$$dC2c/dt = k_9C3b + k_{12}C3c - (k_{21} + k_{22} + k_{-9} + k_{-12})C2c + k_{-21}C1a + k_{-22}C1d + {}_8U8 \quad (8)$$

$$dC2d/dt = k_7C3a + k_{14}C3d - (k_{23} + k_{24} + k_{-7} + k_{-14})C2d + k_{-23}C1b + k_{-24}C1c + b_9U9 \quad (9)$$

$$dC2e/dt = k_{10}C3b + k_{15}C3d - (k_{25} + k_{26} + k_{-10} + k_{-15})C2e + k_{-25}C1b + k_{-26}C1d + b_{10}U10 \quad (10)$$

$$dC2f/dt = k_{13}C3c + k_{16}C3d - (k_{27} + k_{28} + k_{-13} + k_{-16})C2f + k_{-27}C1c + k_{-28}C1d + b_{11}U11 \quad (11)$$

$$dC1a/dt = k_{17}C2a + k_{19}C2b + k_{21}C2c + m_1C1a + k_{-29}C0 + k_{-33a}I1 + b_{12}U12 \quad (12)$$

$$dC1b/dt = k_{18}C2a + k_{23}C2d + k_{25}C2e + m_2C1b + k_{-30}C0 + k_{-33b}I1 + b_{13}U13 \quad (13)$$

$$dC1c/dt = k_{20}C2b + k_{24}C2d + k_{27}C2f + m_3C1c + k_{-31}C0 + k_{-33c}I1 + b_{14}U14 \quad (14)$$

$$dC1d/dt = k_{22}C2c + k_{26}C2e + k_{28}C2f + m_4C1d + k_{-32}C0 + k_{-33d}I1 + b_{15}U15 \quad (15)$$

$$dI1/dt = k_{33a}C1a + k_{33b}C1b + k_{33c}C1c + k_{33d}C1d + k_{-36}I0 + m_5I1 + b_{16}U16 \quad (16)$$

$$dC0/dt = k_{29}C1a + k_{30}C1b + k_{31}C1c + k_{32}C1d + k_{-35}O + m_6I1 + b_{17}U17 \quad (17)$$

$$dI0/dt = k_{36}I_1 + k_{-37}I - (k_{-36} + k_{37})I0 + b_{18}U18 \quad (18)$$

$$dI/dt = k_{37}I_0 + k_{-38}O - (k_{-37} + k_{38})I + b_{19}U19 \quad (19)$$

$$dO/dt = k_{38}I + k_{35}C0 - (k_{-38} + k_{-35})O + b_{20}U20 \quad (20)$$

where

$$m_1 = -(k_{29} + k_{33a} + k_{-17} + k_{-19} + k_{-21}), \quad m2 = -(k_{30} + k_{33b} + k_{-18} + k_{-23} + k_{-25})$$

$$m3 = -(k_{31} + k_{33c} + k_{-20} + k_{-24} + k_{-27}), \quad m4 = -(k_{32} + k_{33d} + k_{-22} + k_{-26} + k_{-28})$$

$$m5 = -(k_{36} + k_{-33a} + k_{-33b} + k_{-33c} + k_{-33d}), \quad m6 = -(k_{35} + k_{29} + k_{30} + k_{31} + k_{32})$$

CONTROL INPUTS AND WEIGHTING COEFFICIENTS

U_n is a control input acting on configuration transitions of the channel gating structures and b_n is the weighting of U_n so as to measure the relative potency of minimizing these control inputs. The sources of these control inputs must derive from energy produced by hydrolysis of ATP. However, It is still premature to set these inputs by such biochemical species, because the concerted molecular structural changes among the gates and subunits are too rapid to assume that these processes are active. They may be due to any thermodynamic energy source.

REDUCED 8-STATE MODELING

The above 20-state model was derived on the basis of a statistical analysis. However, it is significantly difficult to apply H-infinity control, since the total number of differential equations becomes 38 even though the conservation law reduces one equation from those above. In addition, the correspondence between each closed state and an inactivate one is unclear. For example, the inactivated state I1 corresponds to C1. There is, however no corresponding inactivated state for C4, C3, and C2. Furthermore, there is no correlation between the C0 state and the I0 state. Since the blocking ball-like molecule can take any arbitrary position on the cross sectional plane of the channel pore, all corresponding inactivated states can exist for C4, C3, C2 and C0. Moreover the biophysical significance and corresponding entity are unclear particularly for I0 and C0 states since these are speculated states that represent the most likely ones to match the statistical probabilistic illustration of channel state transitions. In another words, the I0 and C0 states or the transitional schema were chosen on a stochastic basis. Hence, we reduce this complicated schema to a reasonable and realistic one as first proposed by Hodgkin [1] and modified later [2].

Figure 6 elucidates the spatial positioning of the S4 helixes in terms of channel opening, closed, and inactive states. We have drawn only the S4 helixes; the other are not shown for simplicity. State O is open where all the activating gates S4I, S4II, and S4III are associated with channel opening. The open state is achieved only when all three activating gates take on their open positions (denoted by level a') simultaneously while the inactivating particle stays outside of the channel hole. State C represents closed ones. A closed state can occur when any of the 3 activating gates takes its resting position (denoted by b') and the inactivating particle stays out of the hole. State I is an open, but inactivated one. The only difference between a closed state and an open one is the participation of the inactivating particle which has invaded the channel hole from the intracellular space. The open state has obstacle-free vertical movement of the three activating gates. Movements of these gates along the vertical direction result in channel closing and opening.

Movement of the inactivated particle on the horizontal plane in the cross section of the channel hole prevents free vertical movement of the activating gates. Blocking this inactivated particle results in channel inactivation. State transitions in the upper horizontal direction show up as being among the inactivated states and they describe vertical movement of one of the three activating gates. To the right most part of Figure 6, all 3 activation gates have open state alignments and their vertical levels in the channel hole are denoted by solid line "a". In all of the inactivated states I1 to I4, an inactivated particle plugs the channel pore. The state transitions in the lower horizontal direction indicate those in the closed and open states when the inactivated particle is ejected from the channel pore. In the left side of the Figure, all three gates have a closed alignment and their vertical positions in the channel duct are denoted by dotted line b'.

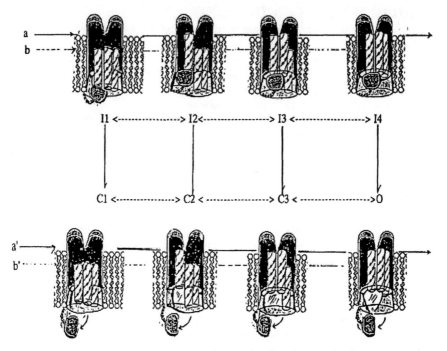

Figure 6. Schema of configureuration changes in the three activation gates and an inactivating particle.

CHANNEL STATE TRANSITION AND CONTROL INPUTS

The state transitions are schematically illustrated in Figure 7., where Xn indicates the state variables. Control inputs were set for the activating process from In to Cn and state transitions from C2 to C3. The control for activation is necessary because once the blocking ball-like molecule has plugged the channel pore, the channel cannot operate until the ball molecule is ejected from the pore. Thus, some active energy must be supplied to activate the channel. This effect is important to understand in order to capture the true voltage sensitivity of a sodium channel. Transitions among the inactivated states, such as I1 to I2, are achieved by horizontal movement of the ball-like molecule on a given cross sectional plane in the channel pore. Only a small horizontal displacement of the ball is necessary to block the free vertical movement of a gate. The minimum displacement is the radius of the gate. Such small movement may occur by thermodynamic mechanisms such as Brownian movement of the blocking ball. Thus, transitions among the inactivated states are set free of the control. The state transitions among closed states are achieved by vertical movement of the gates. It is still unclear whether each gate molecule moves independently or if certain interactions occur between the gates since all the gates have electrical charge on their surfaces. Thus, once a gate has shifted vertically to an upper open position, the electrical potential field around the 3 residual gates is changed. When the second and third gates have trans-located to activated positions, such movement must be accelerated by the sum of the attractive and repulsive forces between the charges on the surfaces of the interacting gates. Thus, control may be needed for only one transition step. We set this transition control on C2 to C3 but there may be other possibilities such as C1 to C2 or C3 to O.

The temporal changes in the amount of channel state per unit membrane area are:

$$\partial[O]/\partial t = k_1[C3] + k_8[I4] - (k_{-1} + k_{-8})[O] + b_1 U1 \tag{21}$$

$$\partial[C3]/\partial t = k_{-1}[O] + k_2[C2] + k_9[I3] - (k_1 + k_{-2} + k_{-9})[C3] + b_2 U2 + b_3 U5 \tag{22}$$

$$\partial[C2]/\partial t = k_{-2}[C3] + k_3[C1] + k_{10}[I2] - (k_2 + k_{-3} + k_{-1})[C2] + b_5 U5 + b_4 U3 \tag{23}$$

$$\partial[C1]/\partial t = k_{-3}[C2] + k_4[I1] - (k_3 + k_{-4})[C1] + b_6 U4 \tag{24}$$

$$\partial[I4]/\partial t = k_{-8}[O] + k_7[I3] - (k_8 + k_{-7})[I4] + b_7 U \tag{25}$$

$$\partial[I3]/\partial t = k_{-9}[C3] + k_{-7}[I4] + k_6[I2] - (k_9 + k_7 + k_{-6})[I3] + b_8 U2 \tag{26}$$

$$\partial[I2]/\partial t = k_{10}[C2] + k_{-6}[I3] + k_5[I1] - (k_{-10} + k_6 + k_{-5})[I2] + b_9 U3 \tag{27}$$

$$\partial[I1]/\partial t = k_{-5}[I2] + k_{-4}[C1] - (k_5 + k_4)[I1] + b_{10} U4 \tag{28}$$

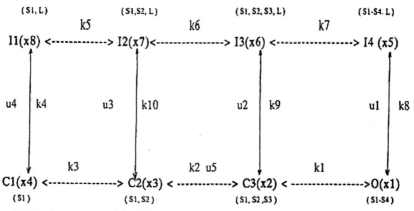

Figure 7. Kinetic representation of the channel by state variables and rate constants and the linear kinetic schema to elucidate the action sites of control inputs. U1, U2, U3, U4, and U5 are control inputs. The k_n terms are rate constants. The arrows indicate opening processes. The terms S1 to S4 indicate the activated gates and those followed by L indicate inactivated gates.

$$C1 \underset{\lambda}{\overset{4\kappa}{\longrightarrow}} C2 \underset{2\beta}{\overset{3\alpha}{\longrightarrow}} C3 \underset{3\beta}{\overset{2\alpha}{\longrightarrow}} C4 \underset{4\gamma}{\overset{\delta}{\longrightarrow}} O.$$

Figure 8. Thermodynamic rate constants among the state transitions.

RATE CONSTANTS AMONG THE CLOSED STATE AND CHANNEL OPENING

On the basis of the observed activations results [4], the following schema would be most adequate at explaining control of the activating process for a sodium channel using four equally-charged gates. Unlike the Hodgkin and Huxley model, movement of these separate gates is not completely independent. Particularly, since the conFigureuration in which all four gates have the same orientation either open or closed are energetically unfavorable.

According to reported data [4], the rate constants are given by:

$\alpha = kT/h \exp(W_\alpha + zdeV/kT)$,

$\beta = kT/h \exp(W_\beta - z(1 - d)eV/kT)$

$\delta = kT/h \exp(W_\delta + d\Delta + zdeV/kT)$

$\gamma = kT/h \exp(W_\gamma + (1 - d)\Delta - z(1 - d)eV/kT)$

$\kappa = kT/h \exp(W_\kappa + (1 - d)\Delta + zdeV/kT)$

$\lambda = kT/h \exp(W_\lambda + d\Delta - z(1 - d)eV/kT)$

where

W is the activation energy in units kT;
k is the Boltzmann constant 1.380658 10^{-23} J/K;
T is absolute temperature in K;
z is the charge on a single gate (we set z = 2.45 [4]);
d is the fraction of total voltage drop across the membrane that the gate traverses before reaching the activation energy peak (we set d = 0.6 [4]);
e is the elementary charge unit 1.60217733 x 10^{-19} C;
V is the potential gap across the membrane;
h is Planck's constant 6.6260755 x 10^{-34} Js;
Δ is the increase in barrier height.

In this work however, we introduce a simpler rate constant to accelerate computation of temporal changes in channel species using H-infinity control, although a precise thermodynamic description is important. We use the following basic schema [2]:

$$C1 \xrightarrow[\beta_m]{3\alpha_m} C2 \xrightarrow[2\beta_m]{2\alpha_m} C3 \xrightarrow[3\beta_m]{\alpha_m} O$$

Figure 9. Basic state transition schema.

The integer on each rate constant indicates the number of gates that have to move to take an activating position within the channel pore. We suppose that each gate moves in an independent manner and there is no mutual interaction between the gates.

For a closed state to an open state transition (α_m), and for an open state to a closed state transition (β_m), the elementary rate constants per unit time per unit membrane area are given by [2,4,5]:

$$\alpha_m = 0.1 \ (V_m + 35)/[1 - \exp(-(V_m + 35)/10)] \tag{29a}$$

$$\beta_m = 4 \exp(-(V_m + 60)/18) \tag{29b}$$

where V_m is the membrane potential in mV. We set $V_m = -50$ mV as the standard potential for the excitable membrane so the channel opening probability reaches its highest value around -50 mV [5].

RATE CONSTANT FOR THE CHANNEL ACTIVATION

The rate constants for membrane voltage dependent transition between an inactivated state to an open one (α_h) and for open to inactivated (β_h) are reported [2, 4] as

$$\alpha_h = 0.07 \exp(-(V_m + 60)/20) \tag{30a}$$

$$\beta_h = 1/[1 + \exp(-(V_m + 30)/10)] \tag{30b}$$

Biochemical Noise and their Biophysical Interpretation

This paper is interested in evaluating the filtering function of the channel against noise. The Na ion selective channel is usually under the influence of bio-molecules whose radii are similar to that of a sodium ion. Such bio-molecular mimetic access to the entrance zone of a Na ion channel takes place by nonspecific mechanisms such as diffusion or microfluidics. These mechanisms do not involve recognition ability for the Na channel specific molecular structure such as Van der Waals force. Hence the bio-molecular mimetic can be understood as noise that competes for the channel with Na ions. The significance of the physiological opening of a Na ion channel gating under noise are:

1. to elucidate undisturbed outputs that are channel states activated purely from Na ion current and
2. to minimize the influence of biochemical noise on the outputs.

In the mathematical expression, the performance of the channel can be interpreted as the extent to which the magnitude of the closed loop transfer function from the disturbance to the regulated output is minimized. In general, we assume that all channel species suffer from noise w. One such biochemical agent is TTX (TeTrodo) toxin because it acts selectively on the Na channel externally. Other biochemical species such as tris-hydroxy-methyl-amino-methane [2] and tetra-methyl-ammonium are reported to suppress the ionic current and can be regarded as noise for the channel opening. Moreover there must be some antiarrythmic agents in cardiovascular medicine that act as a Na-channel blocker. Thus we set the vector state equation to be as follows:

$$\partial x(t)/\partial t = A\, x + B1\, w + B2\, U \tag{31}$$

The elements of matrix **A** are [a_{ij}]. **U** is a matrix for the control input, while **w** is a matrix for the noise. **B1** and **B2** are weighting matrices to characterize the relative amounts of disturbing noise and control inputs acting on the different channel species. **B1** is a (7, 7) matrix while **B2** is a (7, 7) diagonal matrix.

CONTROLLED OUTPUT AND INPUT FOR OBSERVERS

We suppose that the noise acts on an entire state variable. The vector form of the equation for controlled output **z** is a (38,38) matrix:

$$z = C1\, x + D12\, u = [\, q_1 x_1,\, q_2 x_2,\, q_3 x_3,\, \ldots,\, q_7 x_7,\, q_8 u_1,\, q_9 u_2,\, q_{14} u_7\,]^T \tag{32}$$

where $C1 = [q_{ij}]$ is a (14, 7) matrix and $D12 = [q_{kL}]$ is a (14,7) matrix.

z describes the optimally controlled channel states of all possible conformations. u_n is the electrical motive force and are functions of membrane potential because an S4 molecule carries a high charge. u_n must be under the control of an electronic potential which is evoked by local excitatory change in the excitable biological membrane. The vector form of the equation for an input **y** to an observer is given by

$$y = C2\, x + D21\, w \tag{33}$$

where **C2** is a [7, 7] diagonal matrix and **D21** is the [7, 7] unit diagonal matrix

H-INFINITY NORM AND ITS BIOLOGICAL SIGNIFICANCE

Under these mathematical preparations, we interpret that the Na channel must operate even under the worst disturbance noise or must be ready for the worst situation of signal transmission. In another words, the Na ion selective filtering channel works to achieve the best channel state (maximum output) under the influence of the worst-case noise. This might be, in general, the basic survival principle for all biological systems. Each system works to maintain its essential function simultaneously while preparing to overcome the worst-case. A typical example is the self-defence mechanism, such as blood clotting for bleeding or the immune system attack on exogenous antigens. The Na channel system must be organized to save its essential function, namely filtering against the worst-case noise so as to maintain noiseless high-speed signal transmission of neural systems. On the basis of these philosophical considerations, we propose to apply the H-infinity control principle to evaluate the noise-filtering function of the Na ion selective gating channel systems. The present problem of minimization control of noise in the Na channel gating process is formalized by the ordinal H-infinity control as [7]:

"Given a finite value γ, synthesize an internally stabilizing proper controller $K(s)$ so that the closed-loop transfer matrix from noise w to output z, Tzw satisfies the H-infinity norm $\|T_{zw}\|_\infty < \gamma$."

The norm of a transfer function is defined by

$$\| T_{zw}(j\omega) \|_\infty = \sup_{0 \leq \omega \leq \infty} \sigma\{ T_{zw}(j\omega) \} \quad (34)$$

σ is the largest singular value of the transfer function Tzw(jω). The H∞ can also be described by the ratio of output z(t) against the input w(t) signals, namely the induced norm such that

$$\| T_{zw}(j\omega) \|_\infty = \sup_{0 \leq \omega \leq \infty} [\{\int \infty_z T(t)z(t)dt\}/\{\int \infty w T(t)w(t)dt\}] = \sup_{0 \leq \omega \leq \infty} [\|z\|2/\|w\|2] \quad (35)$$

ASSUMPTIONS FOR H-INFINITY CONTROL SYSTEM AND THEIR BIOLOGICAL SIGNIFICANCE

The assumptions are:

[A1]. (**A**, **B2**) can be stabilized and (**C₂**, **A**) is detectable
[A2]. (**A**, **B1**) can be stabilized and (**C₁**, **A**) is detectable.
[A3]. **C1^T D12** =0 and **B1D21^T** = 0
[A4]. **D12** has full column rank with **D12^{T}D12** = **I** and **D21** has full row rank with **D21 D21^T** =**I**.

T denotes transposition of the matrix.

We compute temporal changes in the channel states, the corresponding observers, the worst-case disturbance (γ-2)**B1^{T}X**∞x, and the optimal control input U = F∞x where F∞ = -**B2^{T}X**∞ and X∞ is the solution of the related Riccati equations. The control input minimizes the H-infinity norm of the system transfer function from noise to output.

RESULTS

Figure 10 shows temporal changes in channel state species (the first row), their observers (the second row), the optimal control inputs (the third row), and the worst-case disturbances (the fourth row). The membrane potential was set to a resting potential V of -40 mV. As closed states decrease, the inactivated ones increase. This behavior seems to be counteractive. The magnitude of change in observers (denoted by Es) was significantly more potent than those changes in the state variables. Rapid change in the optimal control is manifested in the period during which the state variables change most rapidly (t = 0.5 to 0.9). The worst-case disturbances on the closed states continue to increase until near the end of the reaction, then they decrease rapidly. Changes in the inactivated states continue to increase.

Figure 11 shows temporal changes in channel state species for the case where the membrane potential was raised to +25 mV. The amounts of channel species increased markedly compared to those of the resting states. There were long resting periods during which there was no evident change in the amount of channel species. Rapid change in all species appeared from t = 0.7. These changes were not only rapid, but also appeared to be synchronized among the state variables, their observers and the optimal controls for the closed states. Only the time trend of the worst-case disturbance showed a mirror image pattern against those of other species. For inactivated states, on the other hand, the time trends among the state, observer, and worst-case disturbance were similar while those of the optimal control was in contrast.

Figure 12 shows temporal changes in the worst-case disturbances for the open state when V= -50 mV (left) and when V = +25 mV (right). When the membrane was resting, the worst-case disturbance shows a clear peak at t= 0.75. When the membrane was excited, the amount of the worst disturbance increased markedly and lasted for a long time.

DISCUSSION

This paper has provided a modeling approach for Sodium ion selective channels on an excitable membrane. The channel opening and closing are achieved by four identical subunits. Each individual subunit is also composed of six cylindrical polypeptides that perforate through the membrane. Among them, S4 cylindrical polypeptide has a high electrical charge on its surface enabling it to be highly sensitive to changes in the membrane potential (see Figure 13). There are four channel gates that dominate the channel opening and closing. In addition a ball-like molecule that invades from the intracellular space can block free vertical movement of the gates. The ball-like molecule is also sensitive to changes in electrical circumstance in the intracellular space. Thus, movement of the ball-like molecule is influenced by the membrane potential. Once it has plugged the channel pore, it reduces the channel to an inactive state in which free vertical movement of the gates are inhibited. The difference between the closed and inactivated states is determined by the position of the ball-like molecule.

There have been many modeling investigations [2,3,4,5] to explain voltage-dependent Sodium channel behaviour. The first was by Hodgkin and Huxley [2]. There have been extended models proposed which are based on the statistical analysis of wood. Even though such highly correlative statistical models may be available, it is premature to take such statistically determined model as the most realistic because there is a discrepancy for the actual structure of the channel species. For example, even though there are four kinds of closed states C0, C1, C2, C3 and C4, all accountable inactivated states such as I0, I1, I3 and I4 are reported from the statistical stochastic analysis, but only I0, I1 and I2 exist. Since there are four gates in the channel pore, it is natural to assume four corresponding inactivated states. Fundamentally, the spatial positions of the gate molecules are the same between the closed and open states. The only difference was the invasion of a ball-like molecule and its horizontal displacements within the cross sectional plane of the channel pore. Taking into consideration the unrealistic nature of these statistically-determined models, we have preferred to start with the original Hodgkin and Huxley model [2]. The shortcoming of this approach is that movement of the gates are assumed to be independent of each other. The gate molecules have significant charge on their surfaces and so, movement of such a gate will generate an electrical influence on the movement of other gates. Thus there must be interaction of vertical gating movement among the gate molecules. It is, however, premature to set the control inputs as a function of these electrical charge dependent movements of the channel gating molecules.

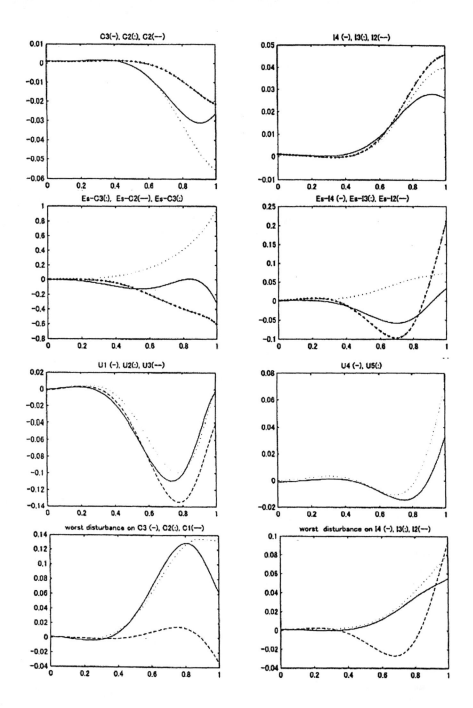

Figure 10. Temporal changes in the amount of channel states, the optimal control, and the worst-case disturbances for V = -50 mV. The time scale is normalized to unity. Initial condition = 0.001.

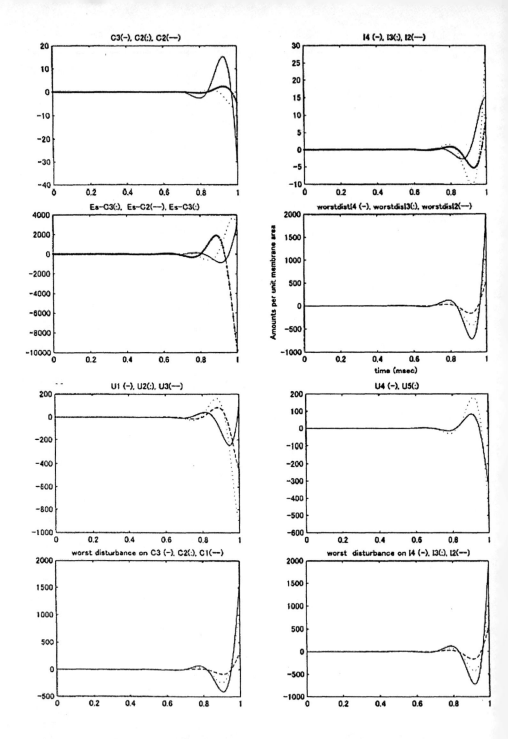

Figure 11. Temporal changes in the amount of channel states, the optimal control, and the worst-case disturbances for V = +25 mV. The time scale is normalized to unity. Initial condition = 0.001.

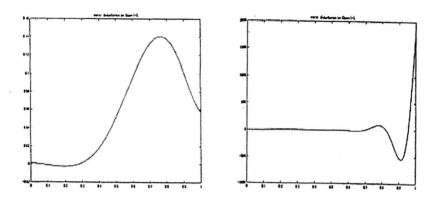

Figure 12. The worst-case disturbances on the open states for V = -50 mV (left) and +25 mV (right).

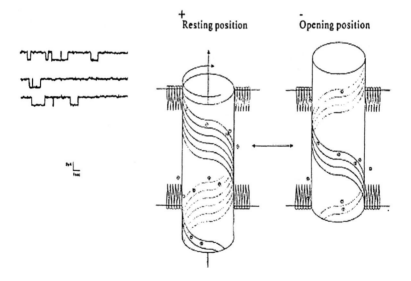

Figure 13. Recorded Sodium ion current by the patch clamp method

Figure 14. Modeling for vertical movement of the cylindrical polypeptide perforating the channel pore.

The computed results showed that temporal changes in the amounts of channel species depend on the membrane voltage. When the voltage is in a resting state, the rate of changes in channel states, observers, control inputs and the worst-case disturbances are small compared to those of the excited state when we set V=+25mV. These differences show that channel states and circumstances are significantly difference when the cell is excited. We still do not know the exact time trends of each of the channel states, control inputs, or the worst-case disturbance because we do not have any biological experimental technique to isolate the temporal change of a given channel state and the worst-case disturbance due to technical limitations in instrumentation.

This paper permits the evaluation of the noise minimization filter function of the Sodium ion selective channel when more extensive biological experiments have been achieved.

CONCLUSION

The H-infinity control principle was applied to a Sodium channel gating system with detailed modeling. The ability of the present approach will be examined in the future by comparing to experimental research.

REFERENCES

[1] L. Stryer, 1991, Biochemistry (3rd Edition), Freeman and Company, N. Y.
[2] A.L. Hodgkin and A.F. Huxley, 1952. Quantitative description of membrane current and its application to conductance and excitation in nerve. J. Physiology, 117, 500–544.
[3] C.M. Armstrong, 1981. Sodium channels and gating currents. Physiological. Rev. 61, pp. 644–683.
[4] J. Patlak, 1991: Activation kinetics of sodium channels. Physiological Rev. vol 71, pp 1047–1080.
[5] R.D. Keynes and H. Meves, 1993. H. Properties of the voltage sensor for the openinf and closing of the sodium channels in the squid giant axon. Proc. Royal Soc. London. B253, pp. 61–68.
[6] C.A. Vandenberg and F. Bezanilla. 1991. A sodium channel-gating model based on single channel macroscopic ionic and gating currents in the squid giant axon. Biophysical J., 60, pp. 1511–1533.
[7] K.M. Xhou and J.C. Doyle, 1998. Essentials of Robust Control. Prentice Hall.

Defeasible Deontic Traffic Light Control Based on a Paraconsistent Logic Program EVALPSN

K. NAKAMATSU

Himeji Institute of Technology, Shinzaike 1-1-12, Himeji 670-0092, Japan

T. SENO

Shizuoka University, Johoku 3-5-1, Hamamatsu 432-8011, Japan

J. M. ABE

Dept. Inf. ICET, Paulista University, R Dr. Bacelar 1212 04026-002 Sao Paulo-SP-Brazil

A. SUZUKI

Shizuoka University, Johoku 3-5-1, Hamamatsu 432-8011, Japan

ABSTRACT

Traffic jams caused by inappropriate signal control are big problems that must be resolved. Recently, optimization of traffic signal control using a genetic algorithm has been proposed and implemented. However, the method takes considerable computation time. In this paper, we introduce a control system for intelligent real-time traffic signals based on a paraconsistent logic program called EVALPSN (Extended Vector Annotated Logic Program with Strong Negation), that can deal with contradiction and defeasible deontic reasoning.

After reviewing EVALPSN formally, we show how the traffic signal control is described in EVALPSN with taking a simple intersection example in Japan. Simulation results for comparing EVALPSN traffic signal control to fixed-time traffic signal control are also provided.

Keywords: *traffic signal control, paraconsistent logic program, intelligent control, defeasible deontic reasoning.*

INTRODUCTION

Previously, we have proposed a paraconsistent logic program called EVALPSN (Extended Vector Annotated Logic Program) that can deal with contradiction and defeasible deontic reasoning [1,2]. Roughly speaking, reasoning in which one thing is chosen among many conflicting things under certain given conditions is called **defeasible reasoning**. There are many cases where agents must decide their actions based on norms such as laws, policies, regulations etc., so that deontic notions such as obligation, permission, forbiddance etc. are required in the reasoning process which is called **deontic reasoning**. Combination of these two kinds of reasoning is called defeasible, deontic reasoning and these approaches have been studied by Nute et al. [3]. Some applications based on EVALPSN such as robot action control, automatic safety verification for railway interlocking and air traffic control have already been introduced [4,5]. We introduce another interesting application of EVALPSN.

Traffic jams caused by inappropriate traffic signal control are serious problem that we must resolve. In this paper, we introduce an intelligent real-time traffic signal control system based on EVALPSN as one proposal to solve these problems. Suppose that you are waiting for a front traffic signal to change from red to green at an intersection. You demand the change in your mind. This demand can be regarded as permission for the change. On the other hand, if you are going through the intersection with a green signal, you must demand that the signal stay green. This demand can be regarded as forbiddance of a change. So, we see there is conflict between permission and forbiddance. The basic idea of traffic signal control is that conflict must be managed by defeasible, deontic reasoning of EVALPSN. In this work we show how to formalize traffic signal control using this method.

The paper is organized as follows:
First after reviewing EVALPSN formally, we use an simple intersection example in Japan and
formalize traffic signal control for the intersection using defeasible, deontic formulae;
Next, we translate these defeasible, deontic formulae into an EVALPSN;
Finally, we use a cellular automaton traffic model to show simulation results of the traffic
signal control.

EVALPSN

Generally, a truth-value, called an *annotation*, is explicitly attached to each literal in an annotated logic program. For example, let p be a literal, μ an annotation, then $p:\mu$ is called an *annotated literal*. The set of annotations constitutes a complete lattice. An annotation in VALPSN [2, 4] is a 2-dimensional vector called a *vector annotation* such that each component is a non-negative integer and the complete lattice T_v of vector annotations is defined as:

$$T_v = \{(x,y) \mid 0 \leq x \leq n, 0 \leq y \leq n, x, y \text{ and } n \text{ are integers}\} \quad (1)$$

The ordering of the lattice T_v is denoted by a symbol $\preceq$ and defined as: let $\vec{v}_1 = (x_1, y_1)$ and $\vec{v}_2 = (x_2, y_2)$, then,

$$\vec{v}_1 \preceq \vec{v}_2 \text{ iff } x_1 \leq x_2 \text{ and } y_1 \leq y_2 \quad (2)$$

For a vector annotated literal, $p:(i,j)$, the first component i of the vector annotation denotes the amount of positive information to support the literal p, while the second subscript j denotes the set of negative information. We assume the integer n is 2 throughout this paper. For example, a vector-annotated literal p:(2, 1) can be informally interpreted such that the literal p is known to be true of strength 2 and false of strength 1. On the other hand, an annotation in EVALPSN called an *extended vector annotation* has the form of $[(i, j), \mu]$ such that the first component (i, j) is a 2-dimensional vector as well as a vector annotation in VALPSN while the second element:

$$\mu \in T_d = \{\perp, \alpha, \beta, \gamma, *_1, *_2, *_3, T\} \qquad (3)$$

is an index that represents deontic notion or contradiction. The complete lattice T of the extended vector annotation is defined as $T_v \times T_d$. The ordering of the lattice T_d is denoted by a symbol $\preceq_d$ and described by the Hasse's diagrams given in Figure 1. The intuitive meanings of the members of T_d are: $\perp$ (unknown), α (fact), β (obligation), γ (non-obligation), $*_1$ (both fact and obligation), $*_2$ (both obligation and non-obligation), $*_3$ (both fact and non-obligation) and T(inconsistent). The Hasse's diagram (cube) shows that the lattice T_d is a tri-lattice in which the direction $\overrightarrow{\alpha\beta}$ represents *deontic truth*, the direction $\overrightarrow{\perp *_2}$ represents the amount of *deontic knowledge* and the direction $\overrightarrow{\perp\alpha}$ represents factuality. Therefore, for example, the

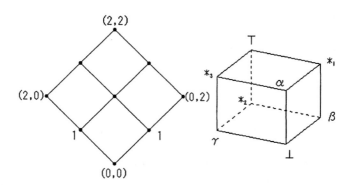

Figure 1. Lattice T_v and Lattice T_d.

annotation β can be intuitively interpreted to be deontically truer than the annotation γ and the annotations $\perp$ and $*_2$ are deontically neutral, i.e., neither obligation nor not-obligation. The ordering over the lattice T is denoted by a symbol $\preceq$ and defined as: let $[(i_1, j_1), \mu_1]$ and $[(i_2, j_2), \mu_2]$ be extended vector annotations:

$$[(i_1,j_1),\mu_1] \preceq [(i_2,j_2),\mu_2] \text{ iff } (i_1,j_1) \preceq_v (i_2,j_2) \text{ and } \mu_1 \preceq_d \mu_2 \quad (4)$$

There are two kinds of epistemic negations, $\neg_1$ and $\neg_2$ in EVALPSN that are defined as mappings over T_v and T_d respectively.

Definition 1: (Epistemic Negations of EVALPSN, $\neg_1$ and $\neg_2$)

$$\neg_1([(i,j),\mu]) = [(j,i),\mu], \quad \forall \mu \in T_d$$
$$\neg_2([(i,j),\bot]) = [(i,j),\bot], \quad \neg_2([(i,j),\alpha]) = [(i,j),\alpha],$$
$$\neg_2([(i,j),\beta]) = [(i,j),\gamma], \quad \neg_2([(i,j),\gamma]) = [(i,j),\beta],$$
$$\neg_2([(i,j),*_1]) = [(i,j),*_3], \quad \neg_2([(i,j),*_2]) = [(i,j),*_2],$$
$$\neg_2([(i,j),*_3]) = [(i,j),*_1], \quad \neg_2([(i,j),\top]) = [(i,j),\top] \quad (5)$$

The epistemic negations ($\neg_1, \neg_2$) followed by extended vector annotated literals can be eliminated by the above syntactic operation. The strong negation ($\square$) in EVALPSN can be defined by the epistemic negations $\neg_1$ or $\neg_2$ as follows and interpreted as classical negation.

Definition 2: (Strong Negation) Let F be an arbitrary formula:

$$\square F =_{def} F \rightarrow ((F \rightarrow F) \wedge \neg(F \rightarrow F)) \quad (6)$$

where, $\neg$ is $\neg_1$ or $\neg_2$.

Deontic notions and facts can be expressed in extended vector annotation as follows:

- a "fact" is expressed by an extended vector annotation - $[(m,0),\alpha]$;

- an "obligation" is expressed by an extended vector annotation, $[(m,0),\beta]$;

- a "forbiddance" is expressed by an extended vector annotation, $[(0,m),\beta]$;

- a "permission" is expressed by an extended vector annotation, $[(0,m),\gamma]$;

where m = 1 or 2. For example, an extended vector annotated literal $p:[(2,0),\alpha]$ can be intuitively interpreted as "the literal p is a fact of strength 2", and an extended vector annotated literal $q:[(0,1),\beta]$ can be also expressed as "the literal q is forbidden of strength 1".

Definition 3: (a well-extended vector annotated literal) Let p be a literal such that $p:[(i,0),\mu]$ or $p:[(0,j),\mu]$ are called *well-extended vector annotated literals*, where $i,j \in \{1,2\}$, and $\mu \in \{\alpha, \beta, \gamma\}$.

Definition 4: (EVALPSN): If $L_0, \ldots, L_n$ are well extended vector annotated literals, then

$$L_1 \wedge \ldots \wedge L_i \wedge \Box\,{}^\sim L_{i+1} \wedge \ldots \wedge \Box\,{}^\sim L_n \to L_0$$

is called an *extended vector annotated logic program clause with strong negation* (EVALPSN clause). An *Extended Vector Annotated Logic Program with Strong Negation* is a finite set of EVALPSN clauses.

Generally, EVALPSN has stable model semantics [9] and the computation of stable models takes long time. However, if an EVALPSN is stratified, it has well-founded models [8] and strong negation in the EVALPSN can be treated as the Negation of Failure in PROLOG. The EVALPSN that we use is a stratified program and so, we do not have to account for stable models. EVALPSN can be implemented as a usual logic program. We now will show a simple example of EVALPSN programming.

Example 1: Suppose we have an EVALPSN as follows:

$$P \;\Box\Box \quad p:[(1,0),\alpha],$$
$$\Box\,{}^\sim p:[(0,3),\alpha] \to q:[(0,3),\beta]$$
$$\Box\,{}^\sim q:[(0,2),\beta] \to r:[(0,2),\gamma] \quad \}.$$

Generally, in annotated logic programs, an annotated literal $p:\mu$ is evaluated to be true if the literal p is mapped by an interpretation to an annotation λ such that $\mu \preceq \lambda$, and a strongly negated literal is evaluated as usual. Since $p:[(0,3),\alpha]$ does not exist, we have $q:[(0,3),\beta]$. However, as there exists $q:[(0,3),\beta]$, $\Box\,{}^\sim q:[(0,2),\beta]$ does not hold, so $r:[(0,2),\gamma]$ cannot be derived in the EVALPSN P.

TRAFFIC SIGNAL CONTROL IN EVALPSN

Preliminary

Consider an intersection in which two roads are crossing as shown in Figure 2. We suppose an intersection in Japan meaning that "cars must keep to the left". The intersection has four traffic signals $T_{(1,2,3,4)}$ that have four types of lights—green, yellow, red, and right-turn arrow. Each lane connected to the intersection has a sensor to detect the amount of traffic. Each sensor is described as $S_i (1 \leq i \leq 8)$ in Figure 2. For example, the sensor S_6 detects the right turn traffic level confronting the traffic signal T_1. Basically, the signal control is performed based on the traffic sensor values. The chain of signaling is:

$$\rightarrow \text{red} \rightarrow \text{green} \rightarrow \text{yellow} \rightarrow \text{right-turn arrow} \rightarrow \text{all red} \rightarrow$$

In this paper, for simplicity, we assume that the lengths of yellow and "all red" signaling times are constant, therefore, the signaling times of yellow and "all red" are simply included in the length of the green and right-turn arrow, respectively. Therefore, we have the following signaling chain:

$$T_{1,2} \rightarrow \text{red} \rightarrow \text{red} \rightarrow \text{green} \rightarrow \text{arrow} \rightarrow \text{red}$$
$$T_{3,4} \rightarrow \text{green} \rightarrow \text{arrow} \rightarrow \text{red} \rightarrow \text{red} \rightarrow \text{green}$$

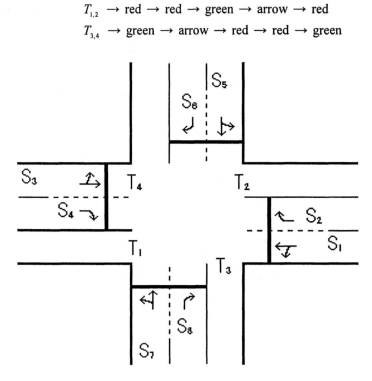

Figure 2. Intersection.

Basic Ideas of Traffic Signal Control

We now describe the basic ideas of traffic signal control. Only changes from green to arrow and from arrow to red are usually controlled. The change from red to green of the front traffic signal follows the change of right-turn arrow to red of the neighboring signal. The signaling is also controlled at each time unit $t \in \{0,1,2,\ldots,n\}$. The traffic level in each lane is regarded as permission for or forbiddance of a signal change such as green to right-turn arrow. For example, if there are many cars waiting for a signal change with green to right-turn arrow, this can be regarded as permission for a signaling change from green to right-turn arrow. On the other hand, if there are many cars going through the intersection under a green light, this can be regarded as a forbiddance of a signal change from green to right-turn arrow. So, there is a possible conflict between permission and the forbiddance. By EVALP defeasible, deontic reasoning if permission is derived, the signal change is performed. On the other hand, if forbiddance is derived, the signal change is not performed. We also assume that minimum and maximum signaling times are established for each signal sequence. The length of each signal time must be controlled between its minimum and maximum signaling times.

Defeasible and Definite Rules for Traffic Signal Control

In order to clarify defeasible, deontic traffic signal control based on EVALPSN, we must express the traffic signal control using defeasible, deontic formulas and translate them into EVALPSN.

Generally in defeasible logic, there are two kinds of rules, definite rules ($\rightarrow$) and defeasible rules ($\Rightarrow$). There may be superiority relations between conflicting defeasible rules. For example, definite rules can be treated as strong rules in terms of facts. Defeasible reasoning is performed based on superiority relations between the defeasible rules. The details of defeasible reasoning are in [3].

We consider four states of the traffic signal sequence:

case 1: ($T_{1,2}$ are red and $T_{3,4}$ are green),
case 2: ($T_{1,2}$ are red and $T_{3,4}$ are right-turn arrow),
case 3: ($T_{1,2}$ are green and $T_{3,4}$ are red),
case 4: ($T_{1,2}$ are right-turn arrow and $T_{3,4}$ are red).

Due to space restriction, we will only discuss case 1 in order to introduce the traffic signal control by EVALPSN. We consider the following conditions and create defeasible or definite rules.

1. If one of the sensors $S_{1,3}$ detects a traffic level greater than the criterion at time t, permission to change the front or back traffic signals $T_{1,2}$ from red to green and permission to change the neighboring traffic signals $T_{3,4}$ from green to right-turn arrow are derived. However, since we focus on only the change from green to right-turn arrow, we have the following two defeasible rules :

$$S_1^{rs}(t) \wedge T_{1,2}(r,t) \wedge T_{3,4}(g,t) \Rightarrow \neg \Box \mapsto T_{3,4}(a,t), \qquad (7)$$

$$S_3^{rs}(t) \wedge T_{1,2}(r,t) \wedge T_{3,4}(g,t) \Rightarrow \neg \Box \mapsto T_{3,4}(a,t), \qquad (8)$$

where, $S_1^{rs}(t)$ expresses that the traffic amount detected by the sensor S_1 is over the criterion at the time t and $S_3^{rs}(t)$ also ; $T_{1,2}(r,t)$ expresses that the traffic signals $T_{1,2}(r,t)$ is signaling red at the time t and $T_{3,4}(g,t)$ also ; $\circ$ is a modal symbol to express obligation, therefore, $\neg \Box \mapsto$ expresses permission, and $\Box \mapsto$ expresses forbiddance.

2. If one of the sensors $S_{2,4}$ detects a traffic level above the criterion at time t, permission to change the front or back traffic signals $T_{1,2}$ from red to green and permission to change the neighboring traffic signals $T_{3,4}$ from green to right-turn arrow are derived. Thus, we have the following two defeasible rules :

$$S_{21}^{rs}(t) \wedge T_{1,2}(r,t) \wedge T_{3,4}(g,t) \Rightarrow \neg \Box \mapsto T_{3,4}(a,t), \qquad (9)$$

$$S_4^{rs}(t) \wedge T_{1,2}(r,t) \wedge T_{3,4}(g,t) \Rightarrow \neg \Box \mapsto T_{3,4}(a,t), \qquad (10)$$

3. If one of the sensors $S_{5,7}$ detects a traffic level above the criterion at time t, forbiddance to change the neighboring traffic signals $T_{1,2}$ from red to green and forbiddance to change the front or back traffic signals $T_{3,4}$ from green to right-turn arrow are derived. Thus, we have the following two defeasible rules :

$$S_5^{rs}(t) \wedge T_{1,2}(r,t) \wedge T_{3,4}(g,t) \Rightarrow \Box \mapsto T_{3,4}(a,t), \qquad (11)$$

$$S_7^{rs}(t) \wedge T_{1,2}(r,t) \wedge T_{3,4}(g,t) \Rightarrow \Box \mapsto T_{3,4}(a,t). \qquad (12)$$

4. If one of the sensors $S_{6,8}$ detects a traffic level above the criterion, and one of the sensors $S_{7,5}$ also detects traffic above the criterion thus disturbing a right turn at time t, permission for changing the neighboring traffic signals $T_{1,2}$

from red to green and permission for changing the front or back traffic signals $T_{3,4}$ from green to right-turn arrow are derived. Thus, we have the following defeasible rules:

$$S_6^{rg}(t) \wedge S_7^{rg} \Box(f) \wedge T_{1,2}(r,t) \wedge T_{3,4}(g,t) \Rightarrow \neg \Box \vdash T_{3,4}(a,t), \tag{13}$$
$$S_8^{rg}(t) \wedge S_5^{rg} \Box(f) \wedge T_{1,2}(r,t) \wedge T_{3,4}(g,t) \Rightarrow \neg \Box \vdash T_{3,4}(a,t), \tag{14}$$

where, $S_7^{rg} \Box(f)$ expresses that the traffic amount detected by the sensor S_7 is over the criterion at the time t and $S_5^{rg} \Box(f)$ also.

Moreover, we need the following definite rules to control the traffic signals.

Minimum Signaling Time Rule

Each signaling is guaranteed its minimum signaling time ($MIN_i(\mu,t)$, $\mu \in \{r, g, a\}, i \in \{1,2,3,4\}$. If a signaling time is less than its minimum signaling time at time t, then a signaling change is definitely forbidden. Thus, we have the following definite rules:

$$MIN_{1,2}(g,t) \wedge T_{1,2}(g,t) \rightarrow \Box \vdash T_{1,2}(a,t), \tag{15}$$
$$MIN_{1,2}(a,t) \wedge T_{1,2}(a,t) \rightarrow \Box \vdash T_{1,2}(r,t), \tag{16}$$
$$MIN_{3,4}(g,t) \wedge T_{3,4}(g,t) \rightarrow \Box \vdash T_{3,4}(a,t), \tag{17}$$
$$MIN_{34}(a,t) \wedge T_{3,4}(a,t) \rightarrow \Box \vdash T_{3,4}(r,t). \tag{18}$$

Maximum Signaling Time Rule

Each signaling is also restricted by its maximum signaling time ($MAX_i(\mu,t)$, $\mu \in \{r, g, a\}, i \in \{1,2,3,4\}$. If a signaling time is greater than its maximum signaling time at time t, then a signaling change is definitely permitted. Thus, we have the following definite rules :

$$MAX_{1,2}(g,t) \wedge T_{1,2}(g,t) \rightarrow \neg \Box \vdash T_{1,2}(a,t), \tag{19}$$
$$MAX_{1,2}(a,t) \wedge T_{1,2}(a,t) \rightarrow \neg \Box \vdash T_{1,2}(r,t), \tag{20}$$
$$MAX_{3,4}(g,t) \wedge T_{3,4}(g,t) \rightarrow \neg \Box \vdash T_{3,4}(a,t), \tag{21}$$
$$MAX_{3,4}(a,t) \wedge T_{3,4}(a,t) \rightarrow \neg \Box \vdash T_{3,4}(r,t), \tag{22}$$

Among the above defeasible and definite rules, there is a conflict between the permission and forbiddance rules. If permission for the signaling change is defeasibly derived at time t, the signaling change must be performed as an obligation at time $t+1$. On the other hand, If forbiddance of the signaling change is defeasibly derived, the signaling change does not have to be performed at time $t+1$. Therefore, we have the following permission and forbiddance derivation as definite rules:

Permission Derivation

$$T_{1,2}(g,t) \wedge \neg\Box\rightarrowtail T_{1,2}(a,t) \rightarrow \Box T_{1,2}(a,t+1), \tag{23}$$

$$T_{1,2}(a,t) \wedge \neg\Box\rightarrowtail T_{1,2}(r,t) \rightarrow \Box T_{1,2}(r,t+1), \tag{24}$$

$$T_{3,4}(g,t) \wedge \neg\Box\rightarrowtail T_{3,4}(a,t) \rightarrow \Box T_{34}(a,t+1), \tag{25}$$

$$T_{3,4}(a,t) \wedge \neg\Box\rightarrowtail T_{3,4}(r,t) \rightarrow \Box T_{3,4}(r,t+1), \tag{26}$$

Forbiddance Derivation

$$T_{1,2}(g,t) \wedge \Box\rightarrowtail T_{1,2}(a,t) \rightarrow \Box T_{1,2}(g,t+1), \tag{27}$$

$$T_{1,2}(a,t) \wedge \Box\rightarrowtail T_{1,2}(r,t) \rightarrow \Box T_{1,2}(a,t+1), \tag{28}$$

$$T_{3,4}(g,t) \wedge \Box\rightarrowtail T_{3,4}(a,t) \rightarrow \Box T_{34}(g,t+1), \tag{29}$$

$$T_{3,4}(a,t) \wedge \Box\rightarrowtail T_{3,4}(r,t) \rightarrow \Box T_{3,4}(a,t+1), \tag{30}$$

Moreover, we need more definite rules for synchronization between neighboring traffic signals and the front or back traffic signals. For example, if the front traffic signal T_1 must signal red, the back traffic signal T_2 also must signal red, and the front traffic signal T_1 must signal red while the neighboring traffic signal T_3 must signal green or right-arrow.

Front, Back and Neighboring Traffic Signal Synchronization

$$\Box T_{1,2}(r,t) \rightarrow \Box T_{2,1}(r,t), \quad \Box T_{3,4}(r,t) \rightarrow \Box T_{4,3}(r,t), \tag{31}$$

$$\Box T_{1,2}(g,t) \rightarrow \Box T_{2,1}(g,t), \quad \Box T_{3,4}(g,t) \rightarrow \Box T_{4,3}(g,t), \tag{32}$$

$$\Box T_{1,2}(a,t) \rightarrow \Box T_{2,1}(a,t), \quad \Box T_{3,4}(a,t) \rightarrow \Box T_{4,3}(a,t), \tag{33}$$

$$\Box T_{1,2}(g,t) \rightarrow \Box T_{3,4}(r,t), \quad \Box T_{3,4}(a,t) \rightarrow \Box T_{3,4}(r,t), \tag{34}$$

$$\Box T_{3,4}(g,t) \rightarrow \Box T_{1,2}(r,t), \quad \Box T_{3,4}(a,t) \rightarrow \Box T_{1,2}(r,t). \tag{35}$$

EVALPSN for Traffic Signal Control

We now will translate the defeasible and definite rules into EVALPSN. In order to do this, we suppose that superiority relationships ($\Box$) exist between the defeasible rules. If there is no priority between the two roads, we assume that maintaining the present signaling state is superior to other signaling states. Therefore, in case 1, forbiddance of a signaling change of traffic signals $T_{3,4}$ from green to right-arrow is superior to the permission for such a change. That is to say, the defeasible rules 11 and 12 are superior to the defeasible rules 7, 8, 9, 10, 13 and 14. Thus, we have the following superiority relations

$(7) < (11)$, $(8) < (11)$, $(9) < (11)$, $(10) < (11)$, $(13) < (11)$, $(14) < (11)$,

$(7) < (12)$, $(8) < (12)$, $(9) < (12)$, $(10) < (12)$, $(13) < (12)$, $(14) < (12)$,

Moreover, we assume that definite rules, **Minimum Signaling Time Rule** and **Maximum Signaling Time Rule**, are superior to all other rules.

Here we introduce a method to translate defeasible deontic formulas into EVALPSN using an example. Suppose there are two defeasible rules in conflict: $R1: a \Rightarrow \neg \Box \succ q$ and $R2: b \Rightarrow \Box \succ q$, and a superiority relation $R1 < R2$. If both a and b hold, then by the superiority relation $\neg \Box \succ q$ is derived. If a holds and b does not hold then $\neg \Box \succ q$ derives since the conflicting rule $R2$ does not fire. If a does not hold and b holds then $\Box \succ q$ derives since the conflicting rule $R1$ does not fire. Details of the translation method from defeasible and definite rules into EVALPSN are described in [1].

The defeasible rules, 7 to 14 are translated into:

$$S_1^{rb}(t):[(2,0),\alpha] \wedge T_{1,2}(r,t):[(2,0),\alpha] \wedge T_{3,4}(b,t):[(2,0),\alpha] \wedge \Box`MIN_{3,4}(b,t):[(2,0),\alpha] \wedge$$
$$\Box`S_5^{rb}(t):[(2,0),\alpha] \wedge \Box`S_7^{rb}(t):[(2,0),\alpha] \rightarrow T_{3,4}(a,t):[(2,0),\alpha], \quad (36)$$

$$S_3^{rb}(t):[(2,0),\alpha] \wedge T_{1,2}(r,t):[(2,0),\alpha] \wedge T_{3,4}(b,t):[(2,0),\alpha] \wedge \Box`MIN_{3,4}(b,t):[(2,0),\alpha] \wedge$$
$$\Box`S_5^{rb}(t):[(2,0),\alpha] \wedge \Box`S_7^{rb}(t):[(2,0),\alpha] \rightarrow T_{3,4}(a,t):[(2,0),\alpha], \quad (37)$$

$$S_2^{rb}(t):[(2,0),\alpha] \wedge T_{1,2}(r,t):[(2,0),\alpha] \wedge T_{3,4}(b,t):[(2,0),\alpha] \wedge \Box`MIN_{3,4}(b,t):[(2,0),\alpha] \wedge$$
$$\Box`S_5^{rb}(t):[(2,0),\alpha] \wedge \Box`S_7^{rb}(t):[(2,0),\alpha] \rightarrow T_{3,4}(a,t):[(2,0),\alpha], (38)$$

$$S_4^{rb}(t):[(2,0),\alpha] \wedge T_{1,2}(r,t):[(2,0),\alpha] \wedge T_{3,4}(b,t):[(2,0),\alpha] \wedge \Box`MIN_{3,4}(b,t):[(2,0),\alpha] \wedge$$
$$\Box`S_5^{rb}(t):[(2,0),\alpha] \wedge \Box`S_7^{rb}(t):[(2,0),\alpha] \rightarrow T_{3,4}(a,t):[(2,0),\alpha], \quad (39)$$

$$S_6^{rb}(t):[(2,0),\alpha] \wedge T_{1,2}(r,t):[(2,0),\alpha] \wedge T_{3,4}(b,t):[(2,0),\alpha] \wedge S_7^{rb}\Box(f):[(2,0),\alpha] \wedge$$
$$\Box`MIN_{3,4}(b,t):[(2,0),\alpha] \wedge S_5^{rb}(t):[(2,0),\alpha] \wedge \Box`S_7^{rb}(t):[(2,0),\alpha] \rightarrow T_{3,4}(a,t):[(2,0),\alpha]$$
$$(40)$$

$S_8^{rb}(t):[(2,0),\alpha] \wedge T_{1,2}(r,t):[(2,0),\alpha] \wedge T_{3,4}(b,t):[(2,0),\alpha] \wedge S_5^{rb}(t):[(2,0),\alpha] \wedge$
$\Box ' MIN_{3,4}(b,t):[(2,0),\alpha] \wedge S_5^{rb}(t):[(2,0),\alpha] \wedge \Box ' S_7^{rb}(t):[(2,0),\alpha] \to T_{3,4}(a,t):[(2,0),\alpha]$
(41)

$S_5^{rb}(t):[(2,0),\alpha] \wedge T_{1,2}(r,t):[(2,0),\alpha] \wedge T_{3,4}(b,t):[(2,0),\alpha] \wedge$
$\quad MAX_{3,4}(b,t):[(2,0),\alpha] \to T_{3,4}(a,t):[(0,1),\beta]$,   (42)

$S_7^{rb}(t):[(2,0),\alpha] \wedge T_{1,2}(r,t):[(2,0),\alpha] \wedge T_{3,4}(b,t):[(2,0),\alpha] \wedge$
$\quad MAX_{3,4}(b,t):[(2,0),\alpha] \to T_{3,4}(a,t):[(0,1),\beta]$.(43)

The definite rules 17, 21, 25 and 29 are also translated into :

$MIN_{3,4}(g,t):[(2,0),\alpha] \wedge T_{3,4}(g,t):[(2,0),\alpha] \to \wedge T_{3,4}(a,t):[(0,2),\beta]$,(44)

$MAX_{3,4}(g,t):[(2,0),\alpha] \wedge T_{3,4}(g,t):[(2,0),\alpha] \to \wedge T_{3,4}(a,t):[(0,2),\gamma]$,(45)

$T_{3,4}(g,t):[(2,0),\alpha] \wedge T_{3,4}(a,t):[(0,1),\gamma] \to T_{3,4}(a,t+1):[(2,0),\beta]$, (46)

$T_{3,4}(g,t):[(2,0),\alpha] \wedge T_{3,4}(a,t):[(0,1),\beta] \to T_{3,4}(g,t+1):[(2,0),\beta]$, (47)

We omit the translation of the other definite rules in {(15),...,(35)} due to space restriction.

EXAMPLE AND SIMULATION

Example 2

Suppose that the traffic signals $T_{1,2}$ are red and the traffic signals $T_{3,4}$ are green, and the minimum signaling time of green has already passed.

- If the sensors $S_{1,3,5}$ detect more traffic than the criteria and the sensors $S_{2,4,6,7,8}$ do not detect any cars at time t, then EVALPSN clause 42 fires and the forbiddance statement $T_{3,4}(g,t):[(0,1),\beta]$ is implemented. Furthermore, EVALPSN clause 47 also fires and the obligatory result $T_{3,4}(g,t+1):[(2,0),\beta]$ is obtained.

- If the sensors $S_{1,3}$ detect more traffic than the criteria and the sensors $S_{2,4,5,6,7,8}$ do not detect any cars at time t, then EVALPSN clause 43 fires and the permission statement $T_{3,4}(a,t):[(0,1),\gamma]$ is implemented Furthermore, EVALPSN clause 46 also fires and the obligatory result $T_{3,4}(a,t):[(0,1),\beta]$ is obtained.

SIMULATION

We used a cellular automaton model for the flow of traffic and compared EVALPSN traffic signal control with fixed-time traffic signal control in terms of the numbers of cars stopped and moved under the following conditions:

Condition 1

We supposed that cars flow into the intersection under the following probabilities from all 4 directions:

- right-turn 5%, left-turn 5% and straight 20%;
- fixed-time traffic signal control: green 30, yellow 3, right-arrow 4, and red 40 unit times;
- the length of green signaling is between 3 and 14 time units while the length of right-arrow signaling is between 1 to 4 time units.

Condition 2

We suppose that cars are flowing into the intersection under the following probabilities:

- from the South, right-turn 5%, left-turn 15% and straight 10% ;
- from the North, right-turn 15%, left-turn 5% and straight 10% ;
- from the West, right-turn, left-turn and straight 5% each ;
- from the East, right-turn and left-turn 5% each, and straight 15% ;
- all other conditions are the same as in **Condition 1**.

We measured the sum of cars stopped and moved for 1000 time units, and repeated the simulation 10 times under the same conditions. The average number of cars stopped and moved are shown in Table.1. These results show that: the number of cars moved under EVALPSN control is larger than that when a fixed control system is used while the number of cars stopped under EVALPSN control is smaller than that under fixed time control. Taking the simulation into account, it is concluded that the EVALPSN control is more efficient than a fixed time control.

CONCLUSION AND FUTURE WORK

We have proposed an EVALPSN-based real-time traffic signal control that depends on sensor inputs. The practical implementation of real-time traffic signal control being planned is under the assumption that EVALPSN can be easily implemented on a microchip, although we have not addressed this here.

As future work, we are considering multi-agent intelligent traffic signal control based on EVALPSN. The basic idea is that traffic signals at each intersection can be regarded as an intelligent agent and each agent demands traffic control with its neighboring agents.

Table 1. Simulation Results.

	fixed-time control		EVALPSN control	
	cars stopped	cars moved	cars stopped	cars moved
Condition 1	17,690	19,641	16,285	23,151
Condition 2	16,764	18,664	12,738	20,121

REFERENCES

[1] A.V. Gelder, K.A. Ross, and J.S. Schlipf, 1991. The Well-Founded Semantics for General Logic Programs, Journal of the Association for Computing Machinery, 38, 620–650.

[2] M. Gelfond and V. Lifschitz, 1989. The Stable Model Semantics for Logic Programming, Proc. 5th International Conference and Symposium on Logic Programming, 1070–1080.

[3] K. Nakamatsu, J.M. Abe, and A. Suzuki, 1999. Defeasible Reasoning Between Conflicting Agents Based on VALPSN, Proc. AAAI Workshop Agents' Conflicts, AAAI Press (1999) 20–27

[4] K. Nakamatsu, J.M. Abe, and A. Suzuki, 2001. Annotated Semantics for Defeasible Deontic Reasoning, Proc. 2nd International Conference on Rough Sets and Current Trends in Computing, LNAI 2005, Springer-Verlag, 470–478.

[5] K. Nakamatsu, 2001. On the Relation Between Vector Annotated Logic Programs and Defeasible Theories, Logic and Logical Philosophy, UMK Press, Poland, 8, 181–205.

[6] K. Nakamatsu, J.M. Abe, and A. Suzuki, 2001. A Defeasible Deontic Reasoning System based on Annotated Logic Programming, Computing Anticipatory Systems, CASYS2000, AIP Conference Proceedings, AIP Press, 573, 60–478.

[7] K. Nakamatsu, J.M. Abe, and A. Suzuki, 2001. Applications of EVALP-based Reasoning, Logic, Artificial Intelligence and Robotics, Frontiers in Artificial Intelligence and Applicatiions, IOS Press, 71, 174–185.

[8] K. Nakamatsu, H. Suito, J.M. Abe, and A. Suzuki, 2002. Paraconsistent Logic Program-based Safety Verification for Air Traffic Control, 2002 IEEE Conf. on Systems, Man and Cybernetics, CD-ROM.

[9] D. Nute, 1992. Basic Defeasible Logic, Intentional Logics for Programming. Oxford Sci. Pub., 126–154.

[10] D. Nute, 1997. "Apparent Obligation - Defeasible Deontic Logic", Kluwer Academic Publishers, 287–316.

Development of an Electrorheological Bypass Damper for Railway Vehicles—Estimation of Damper Capacity from the Prototype

S. CHONAN, M. TANAKA and T. NARUSE
Department of Mechatronics and Precision Engineering, Tohoku University, Sendai, Miyagi 980-8579, Japan

T. HAYASE
Vehicle Structure Technology Division, Railway Technical Research Institute, Kokubunji, Tokyo 185-8540, Japan

ABSTRACT

This paper studies the development of an active buffer for railway vehicles using an electrorheological (ER) fluid as a functional material. The coupler force is controlled by adjusting the applied electric field to the ER fluid. The fabricated prototype buffer consists of a hydraulic cylinder, an ER bypass slit valve and a PID feedback controller. The damping force develops in the ER fluid through the bypass slit and at the piston-cylinder gap. In the theoretical analysis, it is assumed that the ER fluid flowing through the bypass slit follows the Bingham plastic mode flow while in the piston-cylinder gap, Newtonian mixed-mode flow prevails. The force opposed by the buffer is generated as a function of the shaft velocity and the applied electric field to the ER fluid. The theoretical and experimental results are compared. Furthermore, an active control test using a PID feedback controller was carried out on the problem where the shaft reaction force is retained at a constant prescribed level while the shaft is translating at a constant velocity. Both results show that the coupler force of the railway vehicle can be controlled effectively by using the proposed electrorheological bypass damper.

INTRODUCTION

Electrorheological (ER) and magnetorheological (MR) fluids are suspensions that exhibit rapid, reversible and significant changes in their material properties when subjected to external electric or magnetic fields. The most significant change is associated with their rheological properties such as viscosity and shear modulus change as a function of field. This unique property makes ER fluids very suitable for active control of smart systems and structures. Application areas of these materials include dynamic dampers and absorbers [1–5], vehicle seat suspensions

[6], clutches [7–8], aircraft landing gears [9], robot actuators [10] and wall motion actuators [11]. Clearly many endeavors as represented by these reports have been done to apply the ER and MR fluids in a variety of research and engineering fields. However, to the authors' knowledge, no attempt has been made to apply the functional materials to railway vehicle engineering.

This paper is a study on the development of a new ER hydraulic draft gear whose function is to control the coupler force. The design purpose of this device ("relief buffer") is to reduce the coupler force acting between a failed and a relief train set which is an important primary design factor for the structure of Sinkansen/Bullet Train lightweight electric car. The relief buffer assuming an idealized displacement -force/velocity-force characteristic is to be introduced into the coupling device of a failed train set. To this end, a prototype buffer consisting of a hydraulic cylinder, an ER bypass valve and a PID feedback controller have been fabricated and its primary function as an active buffer were investigated both theoretically and experimentally. The results obtained show that the basic performance of the prototype is satisfactory to control the coupler force.

PROTOTYPE BUFFER

A prototype buffer was fabricated and is shown in Figure 1. It consists of a hydraulic cylinder, a bypass duct and an ER bypass valve. The three are all filled with an ER fluid (NIPPON SHOKUBAI TX-ER8). The bypass valve is essentially a partition with three parallel slits through which the ER fluid flows. The ER fluid is activated by the electric field created by applying a voltage across the electrodes of the slits. The piston shaft is connected to a linear drive actuator (MATSUTAME KM30A-A06-010). The buffer develops the rate dependent damping force due to pressure drops through the bypass slit and at the piston-cylinder gap as the velocity is applied to the buffer shaft. The force opposed by the buffer is measured by a load cell (SSK LT18-15) installed between the shaft and the linear actuator. At the same time, the data is sent to a personal computer (NEC LW450J/1) via an amplifier (MINEBEA DAS-406B) and a DA/AD card (NATIONAL INSTRUMENTS DAQCARD-6024E). The hydraulic cylinder and piston measures are as follows: the cylinder has an inner diameter of 48 mm; the piston head diameter and thickness, L_m, are 47.12 mm and 30 mm respectively; the piston-cylinder gap, d_m, is 0.44 mm; the shaft diameter is 14 mm; the piston head area, A_p, is 506π mm^2; and the stroke is 100 mm. The viscosity of the ER fluid, μ, measures 115 mPa.

The ER bypass valve is presented in Figure 2. It is a stainless steel partition with three parallel slits insulated by acrylic plastic. The slit gap, d_f, is 1.0 mm, and the width, b_f, is 25 mm with the depth, L_f, being 30mm.

Development of an Electrorheological Bypass Damper for Railway Vehicles   263

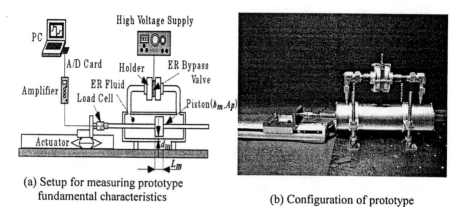

(a) Setup for measuring prototype fundamental characteristics

(b) Configuration of prototype

Figure 1. Prototype buffer.

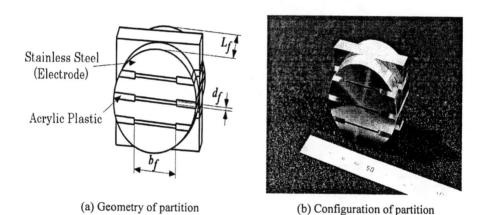

(a) Geometry of partition

(b) Configuration of partition

Figure 2. Partition of ER bypass valve.

THEORETICAL ANALYSIS

In the following analysis, the piston-cylinder gap as well as the bypass valve slit assumes parallel plate geometry [4]. Considering that the ER fluid is subjected to an electric field only at the bypass slit, the flow of the ER fluid through the slit is modeled as Bingham plastic flow-mode while in the piston-cylinder gap Newtonian mixed-mode flow prevails.

The quasi-steady governing equations of Bingham plastic flow for the approximate parallel plate analysis are as follows:

$$\frac{d\tau}{dy} = \frac{dp}{dx} \tag{1}$$

$$\tau = \tau_y + \mu \frac{du}{dy} \tag{2}$$

where τ is the shear stress, p is the pressure developed by the piston head motion, u is the velocity, and τ_y is the dynamic yield stress. The governing equations for Newtonian flow are realized by setting $\tau_y = 0$ in the above equations.

BINGHAM FLOW-MODE FLOWTHROUGH BYPASS SLIT

For the flow through the bypass slit, the pressure gradient is assumed to vary linearly along the length of the electrodes. Thus, equation (1) is modified as

$$\frac{d\tau}{dy} = \frac{\Delta P}{L_f} \tag{3}$$

where $\Delta P = -F/A_p$. Here, F is the applied force to the piston shaft, and A_p is the piston head area. Substituting equation (2) into equation (3) and solving give

$$u(y) = \frac{\Delta P}{2\mu L_f} y^2 + Ay + B \tag{4}$$

The flow-mode flow in the bypass slit is illustrated in Figure 3. The velocity profile in each region is determined with the use of the boundary conditions. Regions 1 and 3 are post-yield regions, while region 2 is the pre-yield region. In the diagram, δ is the plug thickness, which is given by

$$\delta = -\frac{2L_f \tau_y}{\Delta P} \tag{5}$$

The volume flux in each region is calculated as

$$Q_1 = b_f \int_0^{y_{pi}} u_1(y)dy = -\frac{b_f d_f^3 \Delta P}{24\mu L_f}(1-\bar{\delta})^3 \tag{6}$$

$$Q_2 = b_f \int_{y_{pi}}^{y_{po}} u_2(y)dy = -\frac{b_f d_f^3 \Delta P}{8\mu L_f}(1-\bar{\delta})^2 \bar{\delta} \tag{7}$$

$$Q_3 = b_f \int_{y_{po}}^{d_f} u_3(y) dy = -\frac{b_f d_f^3 \Delta P}{24 \mu L_f}(1-\bar{\delta})^3 \qquad (8)$$

where $\bar{\delta}(= \delta/d_f)$ is the nondimensional plug thickness. From Eqs. 6-8, the total volume flux passing one bypass slit is obtained as

$$Q_f^B = Q_1 + Q_2 + Q_3 = -\frac{b_f d_f^3 \Delta P}{24 \mu L_f}(1-\bar{\delta})^2(2+\bar{\delta}) \qquad (9)$$

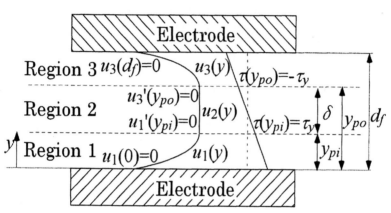

Figure 3. Velocity and stress profiles for Bingham flow-mode flow in bypass slit.

NEWTONIAN MIXED-MODE FLOW IN PISTON-CYLINDER GAP

The governing equations in this case are Eq. 2 where $\tau_y = 0$, and

$$\frac{d\tau}{dy} = \frac{\Delta P}{L_m} \qquad (10)$$

Proceeding in the same way as in the above calculation, one has

$$u(y) = \frac{\Delta P}{2\mu L_m} y^2 + Cy + D \qquad (11)$$

After determining the velocity profile from Eq. 11 and the boundary conditions $u(0) = -v_0$ and $u(d_m) = 0$, the volume flux through the piston-cylinder annulus is calculated from

$$Q_m^N = b_m \int_0^{d_m} u(y)dy = -\frac{b_m d_m^3 \Delta P}{12\mu L_m} - \frac{b_m v_0 d_m}{2} \qquad (12)$$

Here, b_m is the circumference of the annulus formed between the cylinder and the piston.

APPLICATION TO PROTOTYPE BUFFER

For the buffer under consideration, there are two types of flow of the ER fluid depending on the applied electric field and the strength of the axial force on the buffer shaft. The first is the case where the ER fluid flows through both the bypass valve and the piston-cylinder annulus. The other is the case in which a solid plug occupies the whole space of the bypass slit, hence the bypass valve is stuck. For this second case, the ER fluid flows only through the piston-cylinder gap.

The first case means that $\delta < d_f$. Thus, from Eq. 5:

$$\frac{2L_f \tau_y A_P}{d_f} < F \qquad (13)$$

Equating the volume flux through the three bypass slits and piston-cylinder annulus to the volume flux displaced by the piston head, one obtains

$$Q_p = 3Q_f^B + Q_m^N \qquad (14)$$

Here, Q_p is the volume flux displaced by the piston head and is given by $Q_p = v_0 A_p$. Combining Eqs. 5, 9, 12, and 14 leads to the shaft velocity v_0 represented as a function of the yield stress τ_y and the axial shaft force F, i.e.,

$$v_0 = \frac{1}{A_p + \frac{b_m d_m}{2}} [\frac{b_f L_f^2 \tau_y^3 A_P^2}{\mu} \frac{1}{F^2} - \frac{3b_f d_f^2 \tau_y}{4\mu} + (\frac{b_f d_f^3}{4\mu L_f A_P} + \frac{b_m d_m^3}{12\mu L_m A_P})F] \qquad (15)$$

The second case implies that $\delta \geq d_f$, and so, from Eq. 5 one has

$$F \leq \frac{2L_f \tau_y A_P}{d_f} \qquad (16)$$

The volume flux through the piston-cylinder gap equals the volume flux displaced by the piston head, so that

$$Q_p = Q_m^N \qquad (17)$$

Combining Eqs. 12 and 17 leads to the shaft velocity represented by the axial shaft force, i.e.,

$$v_0 = \cfrac{1}{A_P + \cfrac{b_m d_m}{2}} \cfrac{b_m d_m^3}{12\mu L_m A_P} F \tag{18}$$

The dynamic yield stress was measured and found that it is approximated by a quadratic function of the applied electric field:

$$\tau_y = 2.58 \times 10^2 E^2 + 4.22 \times 10^2 E \tag{19}$$

The theoretical axial force F was calculated from Eqs. 15 and 18 and is plotted in Figure 4 as a function of shaft velocity v_0 and applied electric field E.

EXPERIMENTAL RESULTS

Force Versus Velocity Diagram

Given the shaft velocity, v_0, the opposing buffer force, F, was measured experimentally. The applied electric field to the ER bypass valve was increased from 0 to 1.0kV/mm. Furthermore; the piston velocity was increased from 5 to 50mm/s. Three measurements were made at each individual setting and the average buffer force was calculated. Figure 5 shows a comparison of the theoretical force-velocity diagram with the experimental results.

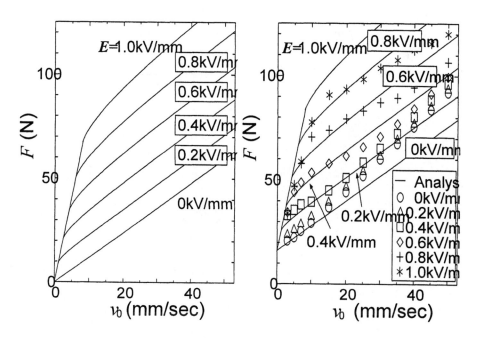

Figure 4. Theoretical force variation as a function of piston velocity and applied electric field.

Figure 5. Comparison of theoretical and experimental force-velocity diagram.

The theoretical buffer force was corrected for friction between the buffer shaft and the cylinder caps which was measured as 16.1N. It can be seen that the prototype buffer shows the expected ER effect depending on the electric field intensity. Some discrepancies are observed between the theoretical and the experimental results. This occurs because of degeneration of the ER fluid as it picks up moisture and because of the repeated applied electric field. Still, the general tendency of the theoretical results corresponds will with the experimental data. The results show that the primary capacity as well as the performance of the buffer can be estimated theoretically.

System Mobility

The mobility of the buffer was also examined. Figure 6 shows the variation of the buffer reaction force when the piston started to move at v_0=30mm/sec with no initial applied electric field. At point A, a field of intensity 0.8kV/mm was applied to the bypass slit and then returned to 0kV/mm at point B. The direction of piston translation was reversed and application of the electric field was repeated at point A' and B'. To find the mobility of the system, the time constant τ (= settling time of 63% of the target) at point B and B' was measured for several values of piston velocity and applied electric field. The results are presented in Figure 7. It can be seen that the time constant is around 100–180 msec, meaning the mobility of the buffer is satisfactory for timely control of buffer force.

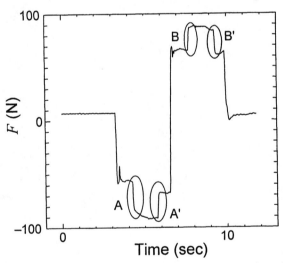

Figure 6. Variation of piston reaction force with increasing/decreasing electric field. v_0 =30mm/sec.

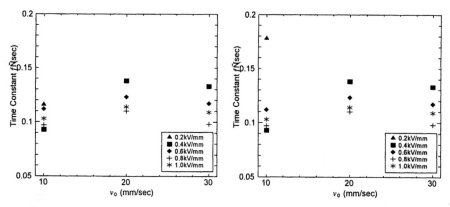

Figure 7. Time constant of shaft reaction force at (left) point B and (right) point B'.

Force Control by Feedback

Control characteristics of the buffer were examined for the problem in which the buffer reaction force was maintained at a constant prescribed level while the shaft was translated at a constant velocity. Both proportional and proportional-integral-differential feedback control were tested. The prototype buffer with feedback controller is illustrated in Figure 8.

P Control

Control Scheme

Given the shaft velocity v_0, the input electric field E necessary to achieve a target reaction force F is determined from Figure 5. For example, for a velocity input of 30mm/sec, the electric field necessary to achieve a reaction force of 70N is found to be 0.6kV/mm. For control purposes, the conversion equation for the shaft force to the electric field was formulated beforehand for each individual input velocity. Control was started after fixing the input velocity v_0 and the target force F_r. The shaft was translated by the linear actuator at velocity v_0 with the electric field initially set to 0kV/mm. After a while, measurement of the buffer reaction force F_y was started by the force sensor and the signal sent to a personal computer every 2msec. Meantime, by using the conversion equation, the force magnitudes F_r and F_y were converted to the strength of the electric field. After calculating the error E_e (= $E_r - E_y$), it is multiplied by the proportional gain K_p and fed back to the electric field applied to the ER bypass valve. Thus, the applied field strength to the valve is given by

$$E_u(t) = E_y(t) + K_p E_e(t) \qquad (20)$$

The corresponding discrete-time equation is

$$E_u(i) = E_y(i-1) + K_p E_e(i) \qquad (21)$$

The closed-loop system with the proportional feedback controller is presented in Figure 9.

The experiment was conducted by increasing the shaft velocity to $v_0 = 30, 35$ and 40mm/sec. The target shaft reaction force and the corresponding force-field conversion equations are listed in Table 1 for each individual shaft velocity. To eliminate arcing at high voltages, the applied electric field was limited to 1 kV/mm.

Results

The experimental results without and with the P control are presented in Figure 10. It is seen from the top figure that the commanded force magnitude cannot be achieved without the feedback control. Even for the cases of the controlled, 2% of deviation is observed on the achieved force compared with the commanded. Bearing this in mind, the feedback control using PID scheme will further be examined in the following.

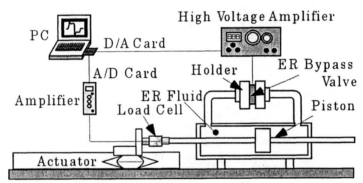

Figure 8. Prototype buffer with a feedback controller.

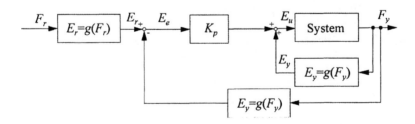

Figure 9. Closed-loop system controller with P control law.

Table 1. Conversion equations of force to electric field.

Piston Velocity v_0 (mm/sec)	Target Force F_r (N)	Corresponding Electric Field
30	80	$E = (F-60.9)/36.3$
35	90	$E = (F-69.0)/35.0$
40	95	$E = (F-78.2)/30.5$

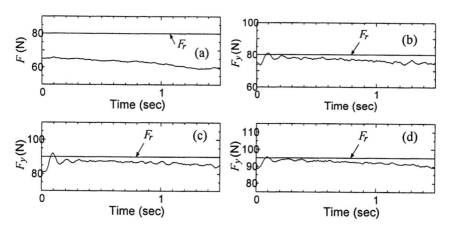

Figure 10. Time response of piston reaction force under P control.
(a) without control, v_0=30mm/sec, F_r=80N,
(b) v_0=30mm/sec, F_r=80N, K_p=2.2,
(c) v_0=35mm/sec, F_r=90N, K_p=2.2, and
(d) v_0=40mm/sec, F_r=95N, K_p=2.4

PID CONTROL

Control Scheme

The input field to the bypass valve using the PID controller was calculated as follows:

$$E_u(t) = E_y(t) + K_p E_e(t) + K_i \int_0^t E_e(\tau)d\tau + K_d \frac{dE_e(t)}{dt} \quad (22)$$

Here, K_p, K_i and K_d are the proportional, integral and differential feedback gains, respectively. The discrete-time form of Eq. 22 is

$$E_u(i) = E_y(i-1) + K_p E_e(i) + K_i \sum_{k=0}^{i} E_e(k) \cdot T + \frac{K_d}{T}[E_e(i) - E_e(i-1)] \quad (23)$$

The system controller with the PID control raw is illustrated in Figure 11.

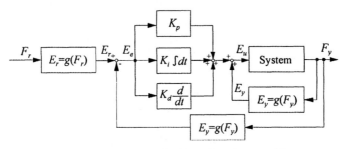

Figure 11. System controller with PID control law.

Evaluation of System Response

Three parameters as shown in Figure 12 are introduced to evaluate the system response. They are T_s, F_{max}, and E_{err}. T_s is the settling time for response, F_y, to fall into the range of ±3% of the target, F_r. F_{max} is the maximum of F_y, and E_{err} is the absolute error $e(=|F_r - F_y|)$ integrated over t = 0-1.5 sec, which is shown with the shaded area in the figure. F_{max} and E_{err} are further nondimensionalized in the form

$$c_f = \frac{F_{max} - F_r}{F_r} \times 100 \quad (24)$$

$$c_e = \frac{E_{err}}{F_r \times 1.5} \times 100 \quad (25)$$

Note that c_f and c_e are indices that show respectively, overshoot and convergence of system response.

Figure 13 is also an example of the system response when the shaft was driven without feedback control. Differing from the response of Figure 10(a), drops of force at A and B are observed in this response. This is due to air bubbles in the ER fluid. The appearance of these drops is optional. C_e is also affected by the drop. In this experiment, the ER fluid was not been pressurized in advance since the differential pressure was low for the operating conditions tested. In the course of design, the balance of the response factors (c_e, c_f and T_s) is a matter of importance. Two cases will now be studied. One is the settings at which feedback gains are chosen to make c_e smaller. The second is the case where the gains are selected so that both c_f and T_s decrease simultaneously.

Results

Optimum feedback gains were determined for the two cases. Time response of the buffer force and the c_e, c_f and T_s measured are presented in Figure 14-16 and Table 2. It can be seen that the buffer force converges to the target rapidly. c_e is less than 3%. Furthermore, c_f is around 2-10%. Except for the case of Figure 14(a), T_s is smaller than 0.20sec. For Figure 14(b), the response amplitude becomes excessive at 0.3sec and 1.0sec. These are disturbances by air bubbles in the fluid. In any event, the effect is small and the problems disappear in a short time.

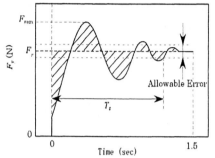

Figure 12. Parameters for evaluating system response.

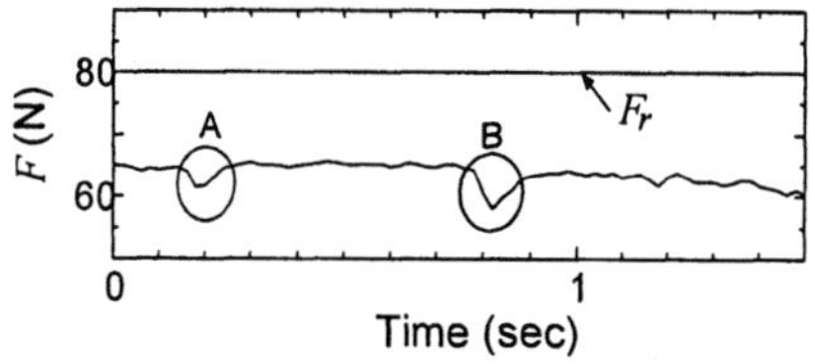

Figure 13. Example of system response without feedback control. v_0=30mm/sec, F_r=80N.

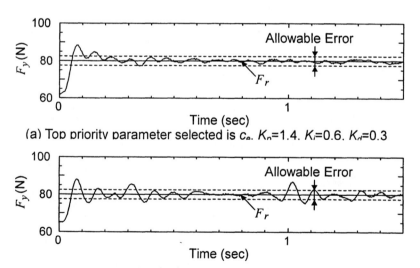

(a) Top priority parameter selected is c_e. K_p=1.4. K_i=0.6. K_d=0.3

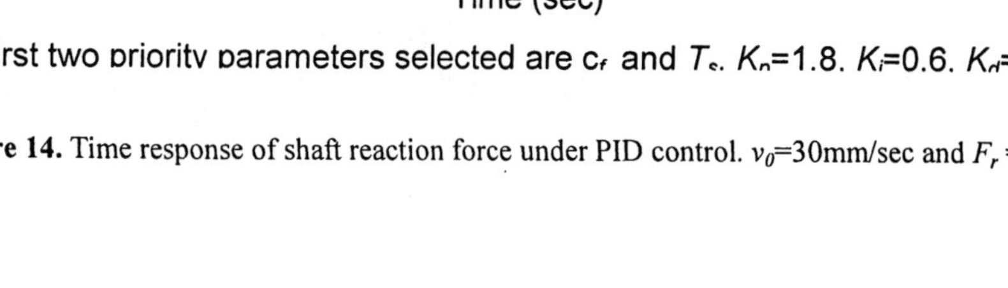

(b) First two priority parameters selected are c_f and T_s. K_p=1.8. K_i=0.6. K_d=0.3

Figure 14. Time response of shaft reaction force under PID control. v_0=30mm/sec and F_r=80N.

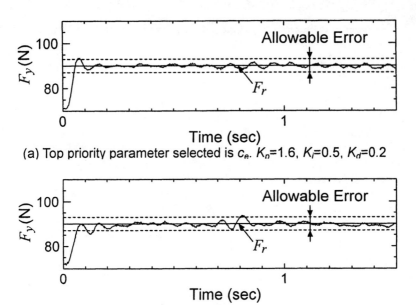

Figure 15. Time response of shaft reaction force under PID control. $v_0=35$mm/sec and $F_r=90$N.

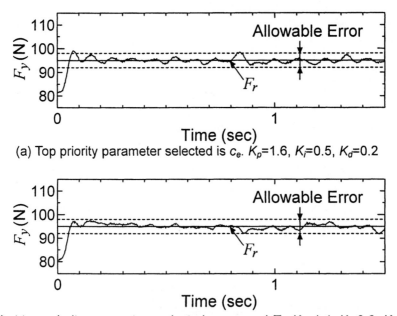

Figure 16. Time response of shaft reaction force under PID control. $v_0=40$mm/sec and $F_r=95$N.

Table 2. C_e, c_f and T_s for time responses of Figure 14–16.

Response	K_p	K_i	K_d	c_e (%)	c_f (%)	T_s (sec)
Figure 14(a)	1.4	0.6	0.3	1.8	10.5	0.36
Figure 14(b)	1.8	0.6	0.3	2.4	10.1	0.18
Figure 15(a)	1.6	0.5	0.2	1.3	3.8	0.08
Figure 15(b)	1.6	0.3	0.2	1.5	0.0	0.14
Figure 16(a)	1.6	0.5	0.2	1.2	4.4	0.09
Figure 16(b)	1.4	0.3	0.2	1.3	2.4	0.05

The results show that the mobility of the valve under feedback control is satisfactory. It is noted, however, that the optimum feedback gains vary depending on shaft velocity. Thus, the gains will be change over time when the buffer force must be regulated under the temporally-varying shaft velocity.

CONCLUSIONS

An active electrorheological buffer consisting of a hydraulic cylinder and a bypass slit valve has been proposed and its primary functions as an ER hydraulic draft gear for railway vehicles have been examined both theoretically and experimentally. The results obtained are summarized as follows:

1. Theoretical results were obtained assuming flow of the ER fluid through the bypass slit follows the Bingham plastic flow while in the piston-cylinder gap, Newtonian mixed-mode flow prevails. Variation of the theoretical buffer force with increasingly applied electric field compares well with the experimental results verifying that the primary capacity and performance of the present buffer can be modeled using properly-applied theory.
2. The time constant of the buffer force following a change in applied electric field is between 100–180msec. Thus, the buffer mobility is suitable for active control of vehicle coupler force.
3. A PID controller was applied to the problem where the buffer force was maintained at a constant prescribed strength while the buffer shaft translated at a constant velocity. The results obtained show that the buffer force settles onto its target very rapidly without significant overshoot when the feedback gains are selected appropriately.

REFERENCES

[1] N. Takasue, G. Zhang, J. Furusho and M. Sakaguchi, 1999. Precise position control of robot arms using a homogeneous ER fluid, IEEE Control Systems, April, 55–61.

[2] S. Hidaka, Y.K. Ahn and S. Morishita, 1999. Adaptive vibration control by a variable-damping dynamic absorber using ER fluid, J. Vibr. Acous, Trans.ASME 121, 373–378.

[3] M. Yalcintas and H. Dai, 1999. Magnetorheological and electrorheological materials in adaptive structures and their performance comparison, Smart Mater. Struct. 8, 560–573.

[4] J. Lindler and N.M. Wereley, 1999. Analysis and testing of electrorheological bypass damper, J. Intel. Mater. Syst. Struct. 10, 363–376.

[5] D. Sakamoto, N. Oshima and T. Fukuda, 2001. Tuned sloshing damper using electro-rheological fluid, Smart Mater. Struct. 10, 963–969.

[6] S.B. Choi, M.H. Nam and B.K. Lee, 2000. Vibration control of a MR seat damper for commercial vehicles, J. Intel. Mater. Syst. Struct. 11, 936–943.

[7] Z. B. Dlodlo and D.J. Brookfield, 1999. Compensator-based position control of an electrorheological actuator, Mechatronics. 9, 895–917.

[8] M. Nilsson and N.G. Ohlson, 2000. An electrorheological fluid in squeeze mode, J.Intel. Mater. Syst. Struct. 11, 545–554.

[9] C.D. Berg and P.E. Wellstead, 1998. The application of a smart landing gear oleo incorporating electrorheological fluid, J. Intel. Mater. Syst. Struct. 9, 592–600.

[10] S.B. Choi, D.W. Park and M.S. Cho, 2001. Position control of a parallel link manipulator using electro-rheological valve actuators, Mechatronics. 11, 157–181.

[11] J.H. Han, J. Tani, J. Qiu, Y. Kohama and Y. Shindo, 1999. A new wall motion actuator using magnetic fluid and elastic membrane for laminar flow control, J. Intel. Mater. Syst. Struct. 10, 149–154.

A Concept for Simulation-Based Optimization of Sheet Metal Forming Processes

M. GRAUER, G. STUFF, T. BARTH, P. NEUSER, O. REICHERT and M. GERDES

University of Siegen, Institute for Information Systems, Hölderlinstr. 3, D-57068 Siegen, Germany

ABSTRACT

Reducing the time-to-market in the early stages of the overall product lifecycle raises a growing demand for techniques supporting virtual prototyping in manufacturing. Virtual prototyping mainly aims at the reducing cost and time by applying numerical simulation techniques to the design of parts as well as to their manufacturing process. In order to obtain an optimal design of a product or its production process, techniques from simulation-based optimization must be utilized. Due to the excessive runtime of numerical simulation of every single deep drawing stage in the whole process, an adequate software environment has to distribute the workload of many hundreds or even thousands of simulations during the course of the optimization across resources in a parallel computation environment (see [1]).

In this contribution, an approach to the solution of a typical optimization problem in sheet metal forming is presented. For the simulation of the forming process the FEM-System INDEED [2] and for the solution of the resulting optimization problem, software based on the concepts and implementation of the workbench OpTiX [3] are used. The feasibility of this approach is demonstrated by solving the industrial optimization problem of determining the minimal blank thickness with respect to a lower limit for the wall thickness of the manufactured part.

INTRODUCTION

Specific requirements must be met by IT-solutions deployed by companies involved in the demanding task of designing and manufacturing complex parts. One of these requirements is an integrated data management system which contains all the necessary information regarding the lifecycle of a product from a customer's request, through its construction, production, development, and maintenance over several years. This includes storing the data in a product data management system (PDM) (s. [4]). Due to the growing use of simulation techniques in the construction of parts or assemblies, further requirements need to be fulfilled. Integrated solutions for the so-called "virtual prototyping" are not generally available.

In the following, requirements and concepts for the design, implementation, and use of an appropriate hardware/software system for "virtual prototyping" using simulation-based optimization will be introduced. Further, the feasibility of the suggested approach will be demonstrated by means of an example from sheet metal forming, a widely applied technique in the automotive supplier industry. In the context of this paper, "virtual prototyping" means all techniques used to carry out product development extensively through computer use. This includes the process of model generation, assuming a high level of expert knowledge about the product and its production process. Using techniques from computer aided design (CAD) for geometrical modeling and numerical simulation, there is the possibility of determining "optimal"—or at least improved—products and production processes based on "virtual products" and "virtual processes". The economic objective for virtual prototyping is the increasing global competition resulting in the need for shorter time spans from the planning of the products through to their production ("time to market"). By using virtual prototyping, companies can develop a product in significantly reduced time and cost.

This paper is organized as follows: in the next section a general mathematical formulation of multi-stage metal forming processes is given. Afterwards, an architecture for an integrated simulation and optimization software supporting virtual prototyping is shown. Then, a principle solution of virtual prototyping in sheet metal forming is given. The data and control flow of the prototypical software system is also presented. As a prerequisite for simulation-based optimization, the precision of the applied simulation software is verified by comparing simulation results to measurements of these practical manufactured parts. The paper is closed with conclusions and some aspects for future work.

MATHEMATICAL FORMULATION OF THE OPTIMIZATION PROBLEM IN SHEET METAL FORMING

Certain relevant classes of design and control problems in engineering can be formulated as optimization problems with objective and/or constraint functions given implicitly by numerical simulation [3]. The problem of optimal sheet metal forming in a multi-stage process can also be specified this way (s. also [1]). In contrast to problems from dynamic programming, the number of stages n of a process is not fixed. It may be necessary to introduce new stages during optimization to satisfy constraints or it may be possible to reduce the total number of stages to decrease cost.

The following formulation specifies an optimization problem of a metal forming process in n discrete stages:

$$\text{minimize } \{f(x_1, x_2, \ldots, x_n)\}$$

subject to

$$s_k = t_k(s_{k-1}, x_k) \qquad \forall k = 1, \ldots, n$$
$$s_0 = a, s_n = b$$
$$s_k \in S_k, \qquad \forall k = 1, \ldots, n-1$$
$$x_k \in X_k(s_{k-1}), \qquad \forall k = 1, \ldots, n$$

where n is the total number of stages which is also to be minimized, the state variables s_k indicate the problem in stage k, S_k denotes the set of feasible states in step k. Start and final state is given by s_0 and s_n. The decision variables of the problem in stage k are x_k belonging to the set $X_k(s_{k-1})$ of feasible decision variables in stage k based on state s_{k-1}. The function $t_k(s_{k-1}, x_k)$ transforms the system from state s_{k-1} into state s_k given that decision variables are x_k.

In the case of sheet metal forming by deep drawing the state of the system in stage k is a representation of the product in this state, e.g. given by the current geometry after k forming steps. Initial and final states represent the geometry of the blank and the complete requested product. The decision variables x_k are the geometrical parameters (e.g. diameters, radii) of tools or blank and other process parameters, e.g. blank holder forces etc. (see [5]). The transformation function represents the simulation of the deep drawing process depending on the current geometry and the parameters for the following forming step.

At first, straightforward approach to minimize also the number of stages n in the multi-stage production process is to hold the number of stages fixed and perform a complete optimization run for an n-stage process. Iteratively, the number of stages can be varied.

AN ARCHITECTURE FOR SOFTWARE SYSTEMS SUPORTING VIRTUAL PROTOTYPING

An essential component of virtual prototyping is distributed simulation-based optimization (see [1]). OpTiX, a software environment for the distributed solution of simulation-based optimization problems in engineering has been successfully applied to a variety of problems [3]. In this context, OpTiX can be understood as a series of "services" which can be used as a basis for the integration of distributed optimization and simulation systems. An architecture based on the concepts of "grid computing" and the "commodity grid toolkits" (CoG, CoG Kits [6] is shown in Figure 1.

Essential components for virtual prototyping are the geometry generation via CAD systems (e.g. CATIA), the simulation system (e.g. INDEED for sheet metal forming), and the optimization algorithms subsumed under an integrated system providing interfaces to simulation and model generation (e.g. OpTiX).

The geometrical data necessary for virtual prototyping as well as the simulation results are stored in a PDM system via appropriate interfaces (e.g. for the different CAD or simulation file formats). All the data are the planed basis for further

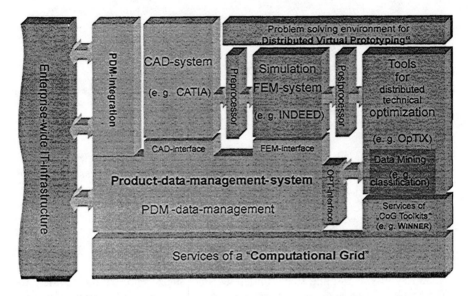

Figure 1. Grid-based system architecture supporting virtual prototyping in sheet metal forming.

analysis, for example to detect similarities between product variations by using classification algorithms from data mining. The services necessary to perform virtual prototyping in a distributed environment are provided by a software layer between the applications and the "grid", i.e. distributed hard- and software of networked workstations. If a flexible, grid-based environment is available, it is possible to integrate and utilize sophisticated methods from optimization and simulation, whether they are sequential or parallel in order to further speed-up the design process (e.g.. Finite Element Tearing and Interconnecting, FETI, [7])

The whole described functionality must be integrated in the enterprise-wide IT-infrastructure in order to ensure consistent data. A PDM system is an appropriate basis for enterprise-wide data integration.

This approach will enable even small and medium enterprises to apply computationally demanding methods from virtual prototyping in their design processes.

AN EXAMPLE OF VIRTUAL PROTOTYPING IN SHEET METAL FORMING

Certain components of the described architecture (see Figure 1) were prototypically implemented in order to demonstrate the feasibility of this approach when solving an industrial example from sheet metal forming in the automotive supplier industry. In this first step, the focus was on the interfaces between the components for optimization in OpTiX and the commercial simulation system

A Concept for Simulation-Based Optimization of Sheet Metal Forming Processes 281

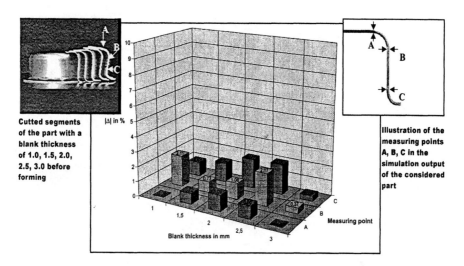

Figure 2. Deviation between the simulation of sheet metal forming by INDEED and the real produced part.

INDEED. As a prerequisite for optimization, the precision of the simulation system INDEED was evaluated by modeling and simulating an exemplary deep drawing process. The results of these simulations were compared to the measurements of the practical manufacturing of these parts. A series with varying blank thicknesses (1, 1.5, 2, 2.5 and 3 mm) was manufactured and also simulated in order to get results for this comparison and to gain an understanding of the problem to be optimized. The results of this study are shown in Figure 2. The error in the comparison between simulation and practice is approximately 2%. Therefore, it is reasonable to apply the simulation software system INDEED for simulation-based optimization in virtual prototyping.

For a simulation, the complete tool set (punch, die, and blank holder) and, as the initial product, the blank, must be generated in the form of geometrical models. Furthermore, process data to initiate and control the simulation system must be prepared as input for the simulation system. For the formulation of the optimization problem, the decision variables can be parameters of the geometry of the tools, the blank, and certain parameters of the forming process. Typical objective functions are measures of the costs of the final product which is to be minimized. Constraints are requirements from construction, for example the minimum wall thickness after forming, or deformation.

The interfaces between simulation and optimization must provide models of tools and blank according to the decision variables. These models are input to the simulation systems. Values of the constraints and objective function are calculated by the simulation and passed to the optimization algorithm using specific functions. The data and control flow of this system is shown in Figure 3.

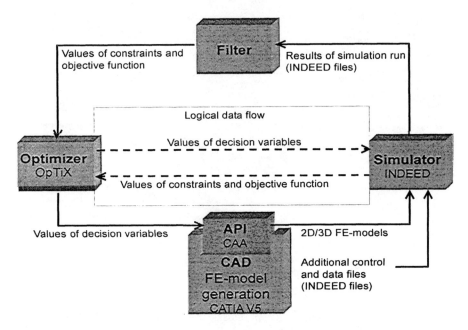

Figure 3. Data and control flow between simulation (INDEED) and optimization (OpTiX) in virtual prototyping for sheet metal forming.

In the case under analysis, the initial blank thickness was selected as the only decision variable. Taking this initial blank thickness as a simple measure for the cost of the final product, it is also the objective function. The optimization of multiple stages of the production process is not necessary because of the given single stage production process. The decision variable x and the objective function f(x) are the blank thickness before forming. The restrictions $g_j(x)$ are the wall thickness of the part after forming over the whole geometry. Wall thickness means the distance interfacial of the part, measured between the corresponding nodes of the surfaces in the model.

This specific problem can be formally stated as follows:

minimize $\{f(x)\}$
subjected to
$g_j(x) \geq 2mm,$
$2mm \leq x \leq 4mm.$

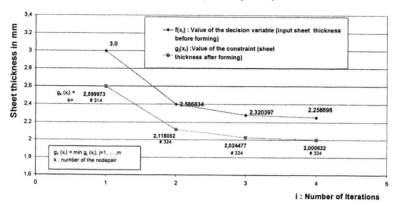

Figure 4. Values of objective and constraint functions during the optimization runs of a single stage deep drawing process (number of the node in the FEM having minimal wall thickness).

The developed procedure to optimize the selected part is a modified Newton-Method. It is proved by general use in OpTiX. This method is not implemented with the demand on best practice. In Figure 4 the history of the objective function values during the optimization runs and the values of the active constraint using the modified Newton method are presented.

In the stated problem, the number of stages is fixed which is a simplification of the situation in industry. Industrial practice demands the optimization of both the product design and the production process simultaneously.

Thus, the problem has to be interpreted in general as a mixed-integer non-linear optimization problem (MINLP). A commonly used strategy for these classes of problems is the Branch&Bound strategy (cp. e.g. [8]). However, referring to both the work in [9] analyzing the huge complexity solving only one of the underlying design problems and the combinatorial character of the given MINLP it is straightforward to see that an intelligent control of the branching and the bounding strategy is necessary in order to prevent excessively long run times for the optimization process.

For that purpose an agent-based control seems to be an adequate approach because of its flexibility and scalability within a distributed computing environment. The application of this technique has been successfully demonstrated in several cases.

However, the problem considered in this paper necessitates integration of agent technology at a lower level. Given the implied MINLP tree structure, each branch is represented by an agent performing an autonomous optimization regarding a certain number of stages. In doing so, different cooperating agents can coordinate the whole optimization process by early intelligent bounding that can be temporal or absolute. Further, according to [10], available computing resources can be effectively managed towards increasing support of promising search directions in order to find an optimal production process and an optimal product design.

CONCLUSIONS AND FUTURE WORK

In this paper, the motivation for the use of simulation-based optimization in virtual prototyping is presented. Major components of an appropriate environment were identified and an architecture was developed, integrating model generation, simulation, optimization, and product data management. Concepts for a platform of such a distributed environment are provided as frameworks for grid computing. As a study in order to demonstrate the feasibility of such an approach, parts of this architecture were implemented and used to solve an exemplary optimization problem from industrial practice in sheet metal forming.

The focus of this contribution was put on the evaluation of the feasibility of the approach rather than giving a complete survey of relevant topics. There are also non-technical aspects such as data security, licensing and billing which must be regarded when designing and developing a commercially applicable system. Another important topic is the development of distributed solution strategies maybe based on agent technology as indicated. After solving successfully the optimization of a single stage deep drawing process the next step forward is to do this for a multiple stage sheet forming process.

ACKNOWLEDGEMENT

The authors wish to thank the Ministry of Science and Research of Northrhine-Westfalia/Germany for its grant to the ViPrO project, the INPRO GmbH, Berlin for the use of the software INDEED, and the FIUKA GmbH & Co. KG, Finnentrop for its support and cooperation.

REFERENCES

[1] M. Grauer, T. Barth, C. Beschorner, J. Bischopink and P. Neuser, 2001. Distributed Scalable Optimization for Intelligent Sheet Metal Forming, Meech, J., Veiga, S., Veiga, M., LeClair, S., Maguire, J. (Eds.) Proc. IPMM'2001, Vancouver, CD-ROM.
[2] K. v. Schoening, M. Mueller and P. Bogon, 1999. Simulation tools in sheet metal forming today's possibilities, tomorrow's demands, Adv. Tech. of Plasticity, III, Proc. 6th ICTP, 1999, Deep drawing simulation, 2113–2130.
[3] T. Barth, M. Grauer, B. Freisleben and F. Thilo, 2000. Distributed Solution of Simulation-Based Optimization Problems on Networks of Workstations, J. Computer Science and Systems, 4(2), 94–105.
[4] J. Stark, 2000. Product Data Management (PDM). http://www.johnstark.com/epwld.html.

[5] J. Chen, X. Shi, and X. Ruan, 2002. Optimization of Sheet Metal Forming Process Using Numerical Simulation and Perturbation Method, D. Yan, S. Oh, H. Huh, Y. Kim (eds.), NUMISHEET 2002 – Proc. 5th Int. Conf. and Workshop on Num. Sim. of 3D Sheet Forming Processes, Jeju Island, 659–663.

[6] I. Foster, C. Kesselman and S. Tuecke, 2001. The Anatomy of the Grid—Enabling Scalable Virtual Organizations, Int. J. of Supercomputer Applications, 15(3), 200–222.

[7] C. Farhat and F. Roux, 1992. An Unconventional Domain Decomposition Method for an Efficient Parallel Solution of Large-Scale Finite Element System, Comp. M. App. Mech. Eng., 113, 367–388.

[8] J.-P. Goux and S. Leyffer2001. Solving Large MINLPs on Computational Grids, Numerical AnalysisReport NA/200, University of Dundee, Department of Mathematics, Dundee, U.K.

[9] M. Gerdes, T. Barth and M. Grauer, 2002. About Performance-Models for the Distributed Solution of Simulation-Based Optimization in Computational Engineering, Proc. of CMMSE'02, I, 132–140, and Journal of Computational Methods for Science and Engineering JCMSE (to be published).

[10] W. Shen, Y. Li, H. Genniwa and C. Wang 2002. Adaptive Negotiation for Agent-Based Grid Computing, Proc. AAMAS2002 Workshop on Agentcities: Challenges in Open Agent Environments, Bologna, Italy, 32–36.

Multiple Layer Control Method

S. SUGIYAMA

Gifu Information and Technology Research Institute, Gifu, Japan

ABSTRACT

In control, there is much information to be processed in order to get a desired output. To obtain the desired output, there are now many kinds of AI methods and related techniques which may be able to handle input information intelligently. However, these methods and their related techniques are not flexible enough to obtain the proper output for a specific input when the output is not the desired one. That is to say, when the present methods have produced an output, it is the final one and it cannot be changed whatever the expectation is about the output. In this situation we need some kind of dynamic knowledge base behavior in the treatment which gives more proper output to the input. What is more, whatever the output situation, that is to say, whether the output is suitable or not, the control processing procedures are the same. So, although we may have achieved the expected output, the full control processing procedures will be taken, which is not necessarily the correct thing to do. So in this paper, the following themes are discussed:

1) The System which has a dynamic behaviour,
2) Multiple Layer Synthesize Method,
3) Multiple Layer Control Method,
4) The System Structure.

INTRODUCTION

We can state the conventional method for control as the following equation:

$$Og = Ig * T \qquad (1)$$

where **Og** : output in general,
T : transfer function,
Ig : input in general.

This method is illustrated in the figure 1.

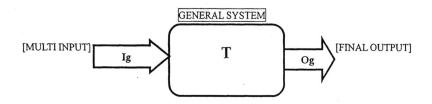

Figure 1. Conventional Method.

Every system, which can be treated in the conventional control methods, can be reduced to the above equation. And it says that once a system has an output, it is the final one and a system cannot do anything for an output to be improved.

We have also Artificial Intelligence methods, and they can be generally reduced into the following form of expression and it can be illustrated as in Figure 2.

$$On = \{ P \rightarrow f(IMn) \} \qquad (2)$$

where
- **On:** Output in **n** processes,
- **P:** Pre-processed data,
- **→:** mapping from **P** to **IM**,
- **IMn:** n Images (any kind) processed data,
- **f():** Image transformed function.

In the above equation, it is obvious that the output **On** is entirely depend upon the input pre-process **P**. The situation for getting the output is the same as the conventional control method that is mentioned above. That is to say, when the present methods have once produced an output, it will be the final one and it cannot be changed whatever the output's expectation is as stated in the conventional methods. So in this paper, we argue for a method in which **O** or **On** can be re-examined to get a desired output, dynamically.

Every system which can be treated by conventional control methods can be reduced to Eq. 1 and Eq. 2. This says that when once a system has an output, it is the final one and the system cannot do anything further to improve the output. And so, procedures taken are always as planned. In order to reduce these problems, we introduce here a new method that can be expressed by the equations that follow. As well, the methods that we use conventionally, have a control behavior that examines every detail of the system regardless of the output situation is (even when a system is stable).

Figure 2. AI methods.

GENERAL MECHANISM OF DYNAMIC BEHAVIOUR

Communication Method in Primitive Elements

In Eq. 1 and Eq. 2, it seems to be very difficult to have proper feedback for an input when an output is not proper. Because a system do not know why it is or is not a proper one and also a system does not know how to communicate with data/database/etc. related for in order to get a proper input data as a feedback.

As we know about the mechanism of the human brain, input information into eye, ear, etc. are broken into pieces at one mediator to another in the brain in order to get very primitive elements. With these very primitive elements, a human brain can recognize/understand things in the outer world. Here, we use this mechanism of the human brain to make a system communicate with data/database/etc. with the intention of obtaining a different, but proper output.

This human mechanism can be simply shown as follows:

An image including seeing, sound, touch, smell, taste, feel, etc. is transformed into very primitive elements by passing one layer of a mediator to other layers of mediators (Input Image (data) and transformed data of **IM1** to **IMn**) in a brain as expressed below:

$$\mathbf{IM1} = \{X1|X11, X12, X13, \cdots, X1n\text{-}1, X1n\}$$
$$\mathbf{IM2} = \{X2|X21, X22, X23, \cdots, X2n\text{-}1, X2n\}$$
$$\mathbf{IM3} = \{X3|X31, X32, X33, \cdots, X3n\text{-}1, X3n\}$$
$$\vdots$$
$$\mathbf{Imm} = \{Xm|Xm1, Xm2, Xm3, \cdots, Xmn\text{-}1, Xmn\}$$
$$\vdots$$
$$\mathbf{IMn} = \{Xn|Xn1, Xn2, Xn3, \cdots, Xnn\text{-}1, Xnn\}$$

This can be rewritten as

$$IM1 \rightarrow IM2 \rightarrow IM3 \rightarrow \cdot\cdot \rightarrow IMm \rightarrow \cdot\cdot \rightarrow IMn$$

$\rightarrow$: Transformation into more primitive elements
And this is expressed by an equation below.

$$Op_n = F_{1 \rightarrow n}(IM_n(X)) \qquad (3)$$

Op : Output in primitive elements

IMn: **n** transformed Image Data in primitive expression in Vector

X : Input Elements in levels of primitive expressions. **m** and **n** are positive integers.

F() : n times Transformations into more primitive elements
$_{1 \rightarrow n}$

<u>By this way, an input data set is transformed into very primitive elements that will be nouns, verbs, (S + V + C, O) form, primitive figures, basic units, etc. These primitive elements are so simple that it makes it possible to communicate with data in a system by using modern AI methods.</u>

This is illustrated in Figure 3.

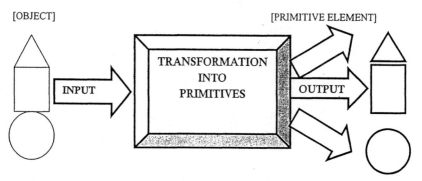

Figure 3. Primitive Element Transformation.

DYNAMIC BEHAVIOR

General System Expression for Reflection

(i) Holon at Present

Holon is a combination of the Greek word holos, meaning "whole", and the suffix "on" meaning particle or part. Keostler observed that in living organisms and in social organizations entirely self-supporting, non-interacting entities did not exist. Every identifiable unit of organization, such as single cell in an animal or a family unit is a society, compresses more basic units while at the same time forcing a part of a larger unit of organization.

The strength of holonic organization, or holarchy, is that it enables the construction of very complex systems that are nonetheless efficient in the use of resources, highly resilient to disturbances, and adaptable to changes in the environment in which they exist. All theses characteristics can be observed in biological and social systems.

(ii) Definition of a HOLON at present

A holon is defined by the Consortium in Europe as shown below:

Holon: An autonomous and co-operative building block of a manufacturing system for transforming, storing and/or validating information and physical objects. Holon consists of an information processing part and often a physical processing part. Holon can be a part of another Holon.

Autonomy: The capability of an entity to create and control the execution of its own plans and/or strategies.

Co-operation: A process whereby a set of entities develops mutually acceptable plans and executes these plans.

Holarchy: A system of Holon that can co-operate to achieve a goal or objectives. The holarchy defines the basic rules for co-operation of the Holons and thereby limits their autonomy.

The understandings of Holon shown above can be restated mathematically in the following definitions.

[Definition 1]: Hierarchy of Multi-levels

Hierarchy of multi-levels (a finite set of whole U, finite sets of parts $\{\sum Pi\}$ and finites sets of subparts $\{\sum SP1j\}$, ... , finites sets of sub-subparts, ... $\{\sum SPnk\})$ of system of organization) is defined mathematically. And the structure of the system's elements' connection at each layer is defined by $\{U \square \sum Pi \square \sum SP1j \square ... \square \sum SPnk\}$. And these U, Pi, $SP1j$, ..., $SPnk$ are correspond to each layer;

Layer-1, Layer-,2 Layer-3, ..., Layer-(n+2) in the system of organization, where i□[0, m0], j□[0, m1], k□[0, m2], n□[0, m3], m0, m1, m2, m3 are positive integers. Hierarchy of multi-levels is made arbitrary by grouping in terms of a function's mechanism, attribute, or physical appearance together.

[Definition 2]: Networking, Regulation, Targeted
The system of organization that has Networking, Regulation, and Targeted factors of being themselves networked with one another in a regulated way for targeted has a characteristic of independent itself as well as dependent to another.

[Definition 3]: SOHO
Self-regulating Open Hierarchic Order (**SOHO**) is defined by a property that each of **U, Pi, SP1j, ..., or SPnk** has its own independent system of organization. And those of **U, Pi, SP1j, ..., SPnk** are independent themselves and also dependent one another as well. And each of those is self-regulating. As defined in above, each set in the system is dependent one another, but the independent occurs when a holarchy is transformed into another form as defined Definition 4.

[Definition 4]: Interchange of Whole and Part
SOHO has a property as below.

Every factor of **U, Pi, SP1j, ..., or SPnk** in a system has a moment to be the whole of it and at another instance the whole may become a part of it. This topological interchange of the whole and a part can occur when U∩Pi∩SP1j∩...∩SPnk =Φ. When the interchange between the whole {U}old and an arbitrarily Ψ⊆{U}old is made, the topological transformation from the old holarchy {U⊇ Σ Pi ⊇ Σ SP1j ⊇... ⊇ Σ SPnk}old to the new holarchy {Unew} ⊇ {Ψ, ΣPi∩Ψ}new occurs.

By using the above Definitions, we can state the Holonic System of Organization as follows. In this case, it needs finites sets of another functions of **R, L, T**; **R** relates to the topology among multi-levels that is defined by Definitions above, **L** relates to the Layers' contents, and **T** relates to the transfer conditions' relationships of among **U, Pi, SP1j, ..., SPnk**.

[Notation 1]
Here we state Holonic System of Organization:

$$SO_\square = \{U \cup \Sigma Pi \cup \Sigma SP1j \cup ... \cup \Sigma SPnk\} \quad (4)$$
$$U = \{U:L,F,T,O \to U(L,F,T,O)\},$$
$$Pi = \{Pi:L,F,T,O \to Pi(L,F,T,O)\},$$
$$SP1j = \{SP1j:L,F,T,O \to SP1j(L,F,T,O)\},$$

$$SPnk = \{Pnk: L, F, T, O \rightarrow Pnk(L, F, T, O)\},$$
$$U_{[R||C]}(Li, Fi, Ti, Oi) = \{Li, Fi, Ti, Oi| \ Li(t) \bigcirc Fi(t) \bigcirc Ti(t) \bigcirc Oi(t) \square U\},$$
$$Pi_{[R||C]}(Li, Fi, Ti, Oi) = \{Li, Fi, Ti, Oi| \ Li(t) \bigcirc Fi(t) \bigcirc Ti(t) \bigcirc Oi(t) \square Pi\},$$
$$SP1j_{[R||C]}(Li, Fi, Ti, Oi) = \{Li, Fi, Ti, Oi| \ Li(t) \bigcirc Fi(t) \bigcirc Ti(t) \bigcirc Oi(t) \square SP1j\},$$
$$\bullet$$
$$SPnk_{[R||C]}(Li, Fi, Ti, Oi) = \{Li, Fi, Ti, Oi| Li(t) \bigcirc Fi(t) \bigcirc Ti(t) \bigcirc Oi(t) \square SPnk\},$$

where

(1) **H**: holonic system which is defined by Definitions above,
(2) **F** specifies finites sets of functions at each part,
(3) **C** specifies finites sets of routes to transfer and this is expressed by using the adjacency matrix of the Graph Theory,
(4) **O** specifies finites sets of processing factors at each part,
(5) **R** specifies finites sets of relations among layers and this is expressed by using the adjacency matrix of the Graph Theory for each Universe, Parts, Part and Subparts, Subparts and ... ,
(6) **L** specifies finites sets of levels of layers of holonic system of organization and **L** has a necessary information for each layer for processing and controlling,
(7) **T** specifies finites sets of transfer methods and time,
(8) $I \in [0, m0], j \in [0, m1], k \in [0, m2], n \in [0, m3]$
   $m0, m1, m2, m3$ are positive integers,
(9) $\bigcirc$ specifies any function in $+, -, \times, /$, etc.

We shall be concerned with mappings whose domains are subsets of some sets already known to us and already named, and they are subjective.

Dynamic Behavior by Reflection

When we think about the human brain, we have the following reflection stages, in general:

    1) Transform data from the outside world into a primitive form;
    2) Find a related area in the knowledge base of the system;
    3) Reflect for a desire objective to be accomplished.

The Reflex Transformation can be summarized as follows with two definitions.

[Definition 5]
    Let **X** denote the set of arbitrary primitive data through the preprocessor from the outside world and assume there are (m x n) real valued functions. **X** is mapped into **R** which is the set of relevant knowledge as defined as below.

$$\Phi ij(x): X \rightarrow R, \ i=1,...,m, \tag{5}$$
$$\Omega j(r): X \rightarrow R, \ j=1,...,n \tag{6}$$

Each element of **X** is producing a primitive data through the preprocessor for each pattern and each element of **R** is producing an output to **X**.

Writing $\Phi j(x)$ to denote the value of the ith primitive data of pattern x, the vector $\Phi j(x)$ is shown as follows,
$$\Phi j(x) = [\Phi 1j(x), \ldots, \Phi mj(x)]^{\square}.$$

Writing $\Omega(r)$ to denote the value of jth knowledge base of entity **r** related with $\Phi j(x)$, vector $\Omega(r)$ is shown as follows:

$$\Omega(r) = [\Omega 1(r), \ldots, \Omega n(r)]^{\square}.$$

The above are feature vectors description of **x** and **r**. There are no restrictions on the forms of the functions Φij; they can be linear or non linear functions of **X**.

[Definition 6]
Let the whole universe {U} consists of $\Psi = \{\Psi 1, \ldots, \Psi n\}$ as shown below.

$$U \supseteq \Psi$$

$\Omega(r)$ is correspond to Ψ each other as shown below:

$$\Omega 1(r) \rightarrow \Psi 1,$$
$$\ldots$$
$$\Omega n(r) \rightarrow \Psi n.$$

The Reflex Transformation is defined as:

$$U \rightarrow \Psi k \supseteq \{U \cap \Psi \cap \Psi k = \phi\}.$$

By this way, the attention to a desired object is able to be targeted and concentrated on and can produce new input for re-processing. This says that any necessary knowledge Ψk is extracted/targeted/concentrated in order to get the desired output.

So by using the above results, it makes it possible to have a dynamic behavior mechanism in situations of a target with reflection which is a method to give an alternative approach to obtain another kind of output to the input as shown in Figure 4.

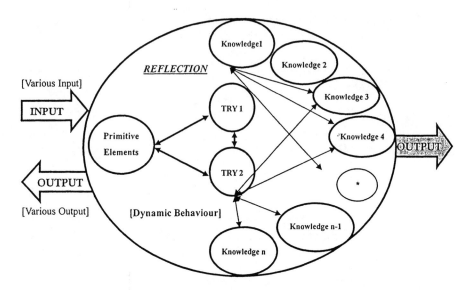

Figure 4. Dynamic Behaviour in Reflection.

SYSTEM STRUCTURE OF ARTIFICIAL INTELLIGENCE DYNAMIC BEHAVIOUR

By using the general system expressions, the notation, and the definitions that were mentioned above, we can have the following theories for artificial intelligence with Dynamic Behavior.

Dynamic Behavior

[Theory 1]
Any image input can be transformed into related knowledge in a system:

$$\Theta_i(w) : W \rightarrow X, (i=1, \ldots, m) \qquad (7)$$
$$\Phi_{jk}(x) : X \rightarrow R, (j=1, \ldots, n), \text{ and } (k=1, \ldots, p),$$
$$\Omega_l(r) : R \rightarrow Y, (l=1, \ldots, q),$$

where m, n, p, and q are integers.

Let **W** denote a set of arbitrary series input data from the outside world and **W** is transformed into the primitive data **X** through the pre-processor. **X** is transformed into the relevant knowledge **R** with **X** and it is assumed that there are (**m x n**) real valued functions. **R** is transformed into the set of related knowledge **Y** with **R**. w, x, and r are feature vectors' descriptions and they have the following features:

1) An element of **W** is producing a primitive data **X** through the pre-processor to each pattern of input.
2) An element of **X** is producing a relevant knowledge **R** and the relevant knowledge **R** is producing a related knowledge **Y**.

Writing $\Phi j(x)$ to denote the value of the **j**th primitive data of pattern x, so the vector $\Phi j(x)$ is shown as follows;
$$\Phi j(x) = [\ \Phi j1(x), \dots, \Phi jk(x)]^T$$
Writing $\Omega l(r)$ to denote the value of the **l**th knowledge base of entity **r** related to $\Phi j(x)$, the vector $\Omega l(r)$ is shown as follows:
$$\Omega l(r) = [\Omega 1(r), \dots, \Omega q(r)]^T$$

There are no restrictions on the form of the functions Θi, Φjk, and Ωl; they can be linear or non-linear.

[Theory 2]
Let the whole universe of knowledge U consists of $\{\Delta, \Lambda, \Sigma, \Psi, \dots\}$. And let $\Omega l(r)$ correspond to an element of U as shown below:
$$\Omega 1(r) \rightarrow \Psi 1$$
$$\Omega n(r) \rightarrow \Psi n$$

By using the above result, it is possible to have the following transformation.

[Reflex Transformation]
$$U \rightarrow \Psi k \square \{U \cap \Psi \cap \Psi k\},$$

where $U \cap \Psi \cap \Psi k = \phi$.

So by using the above theories, it is possible to synthesize any kind of system into Layers with a primitive element, and it is also possible to have a system that has a dynamic behavior mechanism in situations of a targeted output with reflection which is the way to provide an alternative approach to obtain a different kind of output for a particular input. The system is able to deal with Multiple Layers to obtain an expected output, as shown in Figure 5.
From Eq. 4, Eq. 5, Eq. 6, and Eq. 7, we obtain the following equation for acquiring an expected output:

$$SO_H = [\Phi j(x)][\Theta i(w)] * Ig \qquad (8)$$

Eq. 8 says that SO_H can choose any level of the layers which might be suitable to control the system in order to get an expected output. This control situation depends on the output situation and the situation can be changed through dynamic behaviour.

CONCLUSION

Through the examinations and discussions in this paper, the following results have been given:
1) Basic idea of the Primitive Element Method for communication mechanism among processes has been introduced,
2) General mechanism of dynamic behavior of AI and knowledge base has been introduced by showing the two theories on this,
3) Overall system description method in complexity and hugeness by using the idea of Holon and by expanding the idea of Holon mathematically philosophically has been introduced.
4) Multiple Layer Synthesize Method has been introduced,
5) Multiple Layer Control Method has been also introduced.

Also as a simple example for using this method, it is possible to show the dynamic behavior in getting an expected output in a limited knowledge as shown in Figure 6. As future work, it is strongly expected to have more complex example on a computer by using the above results.

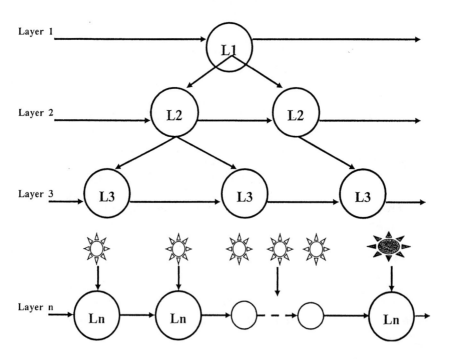

Figure 5. Multiple Layer Synthesize Method

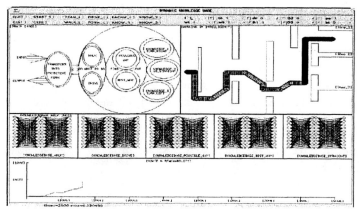

Figure 6. Dynamic Behavior in Obstacle Avoidance.

REFERENCES

[1] Sugiyama, S. 2001. Reflex Transformation of Consciousness, Conference on Consciousness and its Place in Nature, Toward a Science of Consciousness, Sweden.
[2] W. Koehler, 1938. The Place of Values in the World of Fact, Live right.
[3] J. Feibleman, J.W. Friend, 1945. The structure and functionof organization, Philosophical Review, 54.
[4] L.von Bertalanffy, 1950. The theory of open systems in physics and biology, Science, 111.
[5] P. Valckenaers and H.Van Brussel, 2000. Holonic Manufacturing Systems Technical Overview HMS Consorsium Report.
[6] Taro Nawa, 1988. Holonic Management Revolution, Sohko- sha Publishing Co.
[7] R. Saucedo, E.E. Schiring, 1968. Introduction to continuous and digital control systems, Macmillan Publishing Co.
[8] A.G.J. MacFarlane, 1991. Dynamical system models, George G.Harrap & Co.
[9] A. Angyal, 1941. Foundation for a Science of Personality, Harvard University Press.
[10] Shigeki Sugiyama, 1990. Flexible Production and Process Estimation Method, in Proc. Of International Conference on Automation, Robotics and Computer Vision, 104–108.
[11] Shigeki Sugiyama, 1991. Automatic Process Control Method, in Proc. of the IEEE International Symposium on Intelligent Control, pp.334–339.
[12] Shigeki Sugiyama, 1993. Unification of Process Control Method, in Proc. of the IEEE International Symposium on Intelligent Control, 162–167.
[13] Shigeki Sugiyama, 1994. Semantic and Logic Theories in Neuron Networks, in Proc. of IEEE International Conference on System, Man, and Cybernetics, 2448–2452.

[14] Shigeki Sugiyama, 1995. Twofold Type of Back Propagation Neural Network, in Proc. of IEEE International Conference on Neural Networks, 3, 1535–1540.
[15] Shigeki Sugiyama, 1997. A Basic Idea of Consciousness, An International Journal of Computing and Informatics, 21(3), 465–470.
[16] Shigeki Sugiyama, 1998. Concept of Holonic Control (Introduction to Mimic Engineering), Proc. SPIE Sensors and Controls for Intelligent Machining, Agile Manufacturing, & Mechatronics, 3518, 144–153.
[17] Shigeki Sugiyama, 1999. Basic Concept of Holonically Object Oriented System, in Proc. of SPIE Intelligent Systems in Design and Manufacturing, 3833, 44–53.

Chapter 5: Intelligence in Environmental Applications

Morphological and Morphometrical Characteristics of Ornamental Stone Airborne Dusts: Capture and Filtration

G. BONIFAZI, V. GIANCONTIERI, S. SERRANTI and F. VOLPE

Dipartimento di Ingegneria Chimica, dei Materiali, delle Materie Prime e Metallurgia, Universita' di Roma "La Sapienza", Via Eudossiana 18, 00184 Roma, Italy

ABSTRACT

Dust generated during ornamental stone transformation process is considered one of the most dangerous factors affecting the health of workers employed in the stone industry.

Workers are exposed to very-high dust generation during manual work actions, especially during the finishing stage of the production cycle. During such operations tools are handled directly by the workers without using water. This particulate matter (dust) presents different composition and different morphological and morphometrical attributes which strongly influence capture and filtering system behavior. As a consequence, the quantity and the quality of the dust circulating in the work environment directly influences the air being breathed by the workers causing considerable variation in quality.

In this paper, results obtained from a dust collection analysis campaign, carried out in an ornamental stone conversion industry, are analyzed and discussed in order to obtain the best dust reduction adopting an innovative captioning-filtering device. The study was particularly addressed to identify possible correlation between working actions performed, tools utilized, concentration of produced dusts in order to define the best strategy to reduce its production during hand working actions and to improve the air quality of working environment, through the adoption of efficient dust collection devices.

All the tests have been carried out in a controlled environment, directly installed in the factory, equipped with an innovative filtration system and an "on-line" control logic for dust collection. Different types of ornamental stones have been selected for the tests (granite, limestone, sandstone, etc.).

INTRODUCTION

The presence of airborne contaminants in working areas are an occupational problem of increasing interest being dust directly or indirectly responsible of a big number of diseases. More specifically, airborne dusts are well known to be associated with several classical occupational lung diseases, such as the pneumoconiosis, especially when high levels of exposure occur.

In many processes linked to the full utilization of building materials, such as that of ornamental stone transformation industry, there are working phases characterized by a considerable production of dusts, dangerous for workers health. It is very difficult to monitor and quantify such a risk. A correct dust sampling and the following characterization (i.e. composition, morphological and morphometrical attributes) represent a basic step to reach the previous mentioned goal. Such a risk becomes very high when ornamental stone dry working actions are performed. In this case the worst conditions, in terms of produced airborne dust, are realized. The worker is directly involved (Figure 1), not being physically possible to separate him from the source of dust. Concerning the worked materials, the most dangerous stones are those containing quartz, as the inhalation of free crystalline silica particles can cause the silicosis, the well-known fibrotic lung disease that is irreversible, progressive and incurable. Even when quartz is not present the shape of the particulates matter, together with their size class distribution represent an other important group of factors characterising the level of risk. The previous mentioned parameter, in fact, influence the "dust flowability" conditioning the possibility of an efficient capture and filtering.

As the size of dust particles is one of the main critical factors affecting the risk of inhalation exposure [1], a fractional criterion [2] is currently adopted by the American Conference of Governmental Industrial Hygienist (ACGIH), the European Community (CEN) and the International Standard Organisation (ISO). Dust is grouped in three size fractions (*inhalable*, *thoracic* and *respirable*) according to their degree of penetration and deposition inside the respiratory system (Table 1 and Figure 2).

Figure 1. Example of "classical" manual dry operation (stone sawing) performed in an ornamental stone transformation factory. A large amount of airborne dust is produced.

Table 1. Classification of dust particles size for health related purposes.

Inhalable Fraction	Thoracic Fraction	Respirable Fraction
Fraction of dust that can be breathed into the nose or mouth. This class includes particles having a 50% cut-point of 100 µm.	Fraction that can penetrate the head airways and enter the airways of the lung (tracheobronchial region). Dust particles have a 50% cut-point with a median aerodynamic diameter of 10 µm.	Fraction of inhaled airborne particles that can penetrate into the gas-exchange region of the lungs (alveolar region). Dust particles have a 50% cut-point of 4 µm. They include the traditional pneumoconiosis-producing dusts such as silica.

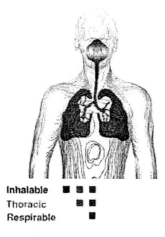

Figure 2. Schematic representation of the degree of penetration of the three different dust fractions in the human airways.

For dust exposure evaluation, the fractions usually considered are the *inhalable* and *respirable*, collected by means of specifically designed devices, for which specific threshold limit values have been fixed (TLV). TLV provide occupational health professionals with a useful tool for assessing health risks and deciding whether a certain exposure situation is acceptable or not, and whether existing controls are adequate. The TLV are a key element in risk management and are often incorporated in legal standards. According to ACGIH, the reference TLV levels adopted for airborne dust in industrial environments are the following:

TLV-TWA for inhalable fraction: 10 mg/m^3,
TLV-TWA for respirable fraction: 3 mg/m^3,
TLV-TWA for free silica in the respirable fraction: 0.5 mg/m^3.

Concerning morphology, with the only exception of asbestos fibers characterization, practically does not exist any specific rule able to quantify the risk associated to this parameter and the consequent actions to apply for its control. This happens for the intrinsic and objective difficulty to realize a systematic and comparable measure of such a factor.

On the bases of what previously reported if it is not possible the control of dust at the source by using, for example, wet methods, the control of dust transmission is usually reached adopting specific "dust sucking" devices or working in a controlled environment, as on aspirating desks or inside aspirating cabins, respectively [3].

In this paper, some of the results obtained inside the *GROWTH Project GRD1-1999-10351 ENVICUT (Human and environmentally friendly cutting and milling of materials)* finance by the European Union are presented. The study was carried out with reference to three main targets, always evaluating the effects of dust morphological and morphometrical attributes, that is:

development of an innovative aspirating cabin for dry hand working actions equipped with a new filtration system, dust caption tests carried out in order to verify the influence of the different dry working actions on the dust concentration, Validation of the new filtration system with reference to the *inhalable* and *respirable* dust fractions and of the free silica (SiO_2) concentration.

DUST ASPIRATING DEVICE

Starting from an available aspirating cabin commonly utilized for manual working actions [4], a new filtration system was designed and developed by SRMP [5] according to the knowledge acquired during the *ENVICUT project* (Figure 3 and Figure 4).

Figure 3. Overview of the front side of the aspirating cabin.

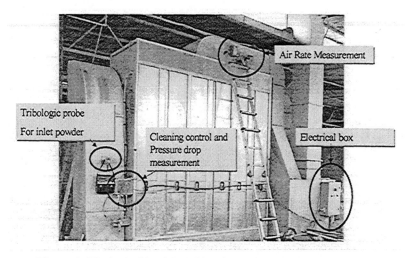

Figure 4. The back side of the cabin with the instrumentation installed.

Table 2. Characteristics of the utilised dust sampler.

Operating Conditions	Sampler for Inhalable Fraction	Sampler for Respirable Fraction
Air Sampling Pump	EGO Base by Zambelli	EGO TT by Zambelli
Sampler Head	Filter holder	Lippman cyclone
Flow Rate	2.7 l/min	2.5 l/min
Filter	Cellulose, 25 mm $\varnothing$, 0.8 µm pore size	Cellulose, 20 mm $\varnothing$, 0.8 µm pore size
Sampling Time	2 min	2 min

A very large filtering surface and a new type of filtering elements were used in order to reach a low-pressure drop and so a high value of air rate and dust collection capacity. To maintain the collection capacity an on-line control system has been installed on the filtration system that controls air rate, pressure drop and pollutant concentration and acts on the cleaning system to obtain the best performance for a long time. The research work led to the following changes on the cabin:

filtering surface: from 73 to 240 m^2,
air rate: from 7000 to 12000 m^3/h,
caption speed: from 2.2 to 6 m/s,
filtration speed: 0.83 m/min,
filter media: from felt to cellulose,
cleaning system: from vibration to air jet (20 l at 8 bar).

Operating this way, a system able to reduce dust levels during dry working activities has been realized. Moreover, it is equipped with an on-line control system, controlled by a PLC, able to monitor its efficiency; it is also coupled with an efficient cleaning system of the filter elements. Such integrated software and hardware architecture allows a strong reduction of power energy loss and maintenance of the filtration units. After 1 year functioning no relevant efficiency decreasing has been registered.

DUST CAPTURE TESTS

The dust has been collected using personal dust samplers for *inhalable* and *respirable* fractions (Table 2) mounted in the breathing zone of the worker. Dust capture tests have been carried out on 14 ornamental stones (Table 3) selected among the most utilized in the market and covering a wide range of features (mineralogical, chemical, physical, etc.). Different hand working actions, the most common usually applied in of the stone finishing stage, have been monitored, that is: edge polishing, sawing, surface polishing (abrasive paper: 50 mesh).

During edge polishing, the dust was collected using tools that operated at two different speed rates, namely low and high. Such an approach was selected because the rate of dust generation increases with the energy associated with the stone dry working process and, as a consequence, a stone working tool will produce more dust when operating at higher speed. On the other hand, it must be understood that decreasing the working speed causes the "production" to decrease. The sawing action has been performed using a low speed tool and an inverse rotation tool, also in this case to achieve the best dust reduction.

Three different operating conditions of the cabin have been investigated: aspirating cabin with the caption system off, aspirating cabin with the old caption system on, that is operating with a fixed cleaning cycle. Filters cleaning was mechanically realized by vibrations, aspirating cabin with the new caption system on. The cleaning cycle was "selected" by the installed control system according to the detected filtration parameters. Filters cleaning was mechanically realized by reverse air.

In order to have comparable results at the end of the sampling campaign, the working period has been fixed at 2 min for each stone and for each working action. A synthesis of dust caption tests is reported in Table 4.

ANALYSES CARRIED OUT ON COLLECTED DUST

The gravimetric analysis has been carried out on dust collected on the filters in order to obtain the dust concentration for both *inhalable* and *respirable* fractions. On the same sample a systematic morphological, digital imaging based, and morphometrical, laser diffraction based, characterization was also applied. Concerning the *respirable* dust fraction, the free silica concentration has been also analyzed by quantitative X-ray diffraction method directly on the filters. For such analysis the filters utilized are those collected during edge polishing with and without the use of the new filtration system.

Table 3. Selected ornamental stones for the hand working tests.

Commercial Name	Quarry Location	Rock Classification	Grain Size and Composition
Travertino (TR)	Italy	Travertine	Fine grained carbonatic
Rosso Verona (RV)	Italy	Limestone	Fine grained carbonatic
Botticino (BO)	Italy	Limestone	Fine grained carbonatic
Bianco Carrara (BC)	Italy	Marble	Fine grained carbonatic
Thassos (TH)	Greece	Marble	Coarse grained carbonatic
Verde Guatemala (VG)	India	Ophicalcite	Fine to medium grained silicatic
Nero Marquina (NM)	Spain	Limestone	Fine grained carbonatic
Grigio Sardo (GS)	Italy	Granite	Coarse grained silicatic
Nero Assoluto (NA)	Zimbabwe	Gabbro	Medium grained silicatic
Labrador (LA)	Norway	Anortosite	Coarse grained silicatic
Rosso Multicolor (RM)	India	Granite	Corse grained silicatic
Basaltina (BA)	Italy	Leucititic Trachyte	Medium grained silicatic
Pietra Dorata (PD)	Italy	Sandstone	Medium grained silicatic
Pietra Serena (PS)	Italy	Sandstone	Medium grained silicatic

Table 4. Synthesis of the dust caption tests carried out inside the aspirating cabin. HS = High Speed   LS = Low Speed   IR = Inverse Rotation.

Cabin with caption system off	Cabin with old caption system on	Cabin with new caption system on
Edge polishing - HS Tool	Edge polishing - HS Tool	Edge polishing - HS Tool
-	-	Edge polishing - LS Tool
Sawing - HS Tool	Sawing - HS Tool	-
-	-	Sawing - LS Tool
-	-	Sawing - IR Tool
Surface polishing - HS Tool	-	Surface polishing - HS Tool

The concentration of dust has been expressed as *normalized-minute-concentration* (*nmc*). In fact, starting from the dust concentration obtained after 1 min sampling, such value can be stretched over the 8 h of an average working day by using the following expression:

$$nmc = 1 \text{ min concentration}/(H_d \times M_h) \qquad (1)$$

being H_d the number of working hours per day, that is 8, and M_h the minutes of each hour, that is 60. When knowing the *nmc* for each working minute, it is possible to compute the global dust concentration during which a worker is exposed during all his working time. In fact, when working in a real environment, the dust concentration to which a worker is exposed when working for *n* minutes with a given action *k* (for example edge polishing or sawing) on the stone *j* is given by:

$$\text{conc}(k,j) = n \times nmc \qquad (2)$$

The *nmc* has been used for the visualization of the test results, and for an easy computation of the allowed working time for a given working action without exceeding the TLV stated by law.

RESULTS AND DISCUSSION

Working Actions Influence on Stone Dust Generation

The dust concentration data have been evaluated to analyse the impact of the different working actions on the generation of dust during stone processing, in order to identify the most dangerous actions for workers health. The results can be summarized in Figure 5, where it has been reported the dust concentration measured without the use of dust filtration system. From Figure 5 it appears that edge-polishing produces a larger quantity of dust compared to sawing and surface polishing actions. Thus, edge polishing can be considered the worst action for workers health. Sawing produces a minor amount of dust whereas surface polishing has intermediate values of dust emissions. These results appear to be independent of stone-type. They are in good agreement to the morphological and morphometrical characteristics of the stones, both at textural level (Figure 6) and with reference to fine particles produced by the different hand made actions (Figure 7).

Dust Inhalable Fraction Reduction

The results of the dust capture tests carried out inside the cabin under different modalities during manual edge polishing are summarized in Figure 8. The results plotted in Figure 8 show how easily detection of the average decrease of generated dust moves from 55% when operating the aspirating cabin with the old settings to 84% of initial by adjusting the aspirating device settings. The decrease can be further improved by operating with a lower speed tool: in this case average reduction is 92%, with several spikes over 95%.

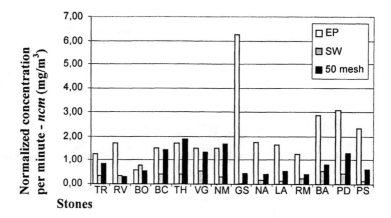

Figure 5. Comparison of dust *nmc* (mg/m^3) detected during manual edge polishing (EP), sawing (SW) and polishing with 50 mesh abrasive paper for the 14 different ornamental stones. Dust caption tests have been carried out using high speed tools and the cabin with the caption system off. See Table 3 for stones legend.

The relative average decrease in dust concentration outlined in Figure 8, agrees well with particle size and shape as detected by SEM analyse for the different stones, according to different capture strategies adopted during the most risky action (edge polishing). An example is reported in Figure 9. The results concerning edge polishing can be summarized as follows: i) the cabin, both in the old and in the new configuration, introduces a strong decrease of dust generation, ii) better results have been obtained by the improvement of the aspirating device, as the cabin equipped with the new system introduces a further decrease in dust generation and iii) with reference to the different speed of the manual tools, it appears that with the use of a lower speed tool, the amount of produced dust is further reduced.

In order to compare the inhalable dust fraction reduction for the three different working actions, in Figure 10 it has been reported the normalized minute concentration (*nmc*) measured when activating the new filtration system of the cabin.

Dust capture tests carried out inside the cabin, operating in different modalities, during manual sawing showed as: i) sawing action produces less dust than edge polishing action, ii) the aspirating system introduces an average decrease in dust concentration but not for all the stones and iii) when working with a lower speed tool, the amount of produced dust is in some cases further reduced.

Concerning surface polishing action, results showed as: i) surface polishing produces a quantity of dust that is intermediate between edge polishing and sawing, ii) when switching on the new filtration system of the cabin, the dust concentration decrease for surface polishing with 50 mesh abrasive paper is greater than that obtained for sawing, but it is of minor entity of the one obtained with edge polishing.

6a 6b 6c

Figure 6. Microphotographs of different rocks presenting a different level of "dustiness" according to the different manual working actions performed. 6a: rosso verona, compacted limestone characterized by the presence of fossils (ammonites). 6b: grigio sardo, granite composed mainly of quartz (SiO_2), microcline ($KalSi_3O_8$), plagioclase ($NaAlSi_3O_8$ to $CaAl_2Si_2O_8$) and biotite $(K(Mg, Fe)_3AlSi_3O_{10} (OH, F)_2$. The rock has a hypidiomorphic-granular texture (most mineral grains are subhedral) and an isotrope structure. 6c: pietra dorata, sandstone composed mainly of calcite ($CaCO_3$), quartz (SiO_2), K-feldspar ($KalSi_3O_8$) and plagioclase ($NaAlSi_3O_8$ to $CaAl_2Si_2O_8$). The rock has a medium grained texture. Thin section digital images in plane polarised light (Reichert Polyvar II Infrapol, image resolution: 1280x1024).

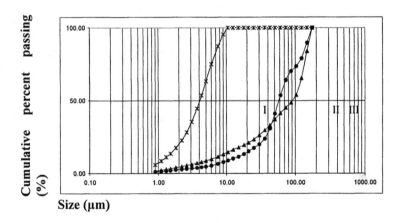

Figure 7. The size class distribution of the 3 different stones (I: rosso verona, II: grigio sardo and III: pietra dorata), presenting different values of *nmc* according to the same work action (edge polishing) clearly show the correlation existing with their textural attributes and mineralogical composition.

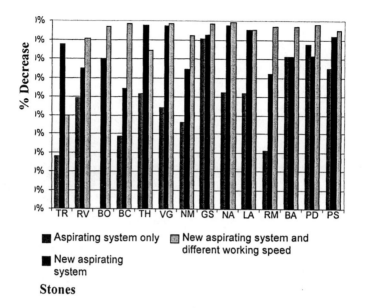

Figure 8. Comparison of the dust concentration decrease (%) by using different working strategies during manual edge polishing of the 14 different ornamental stones. See Table 3 for stones legend.

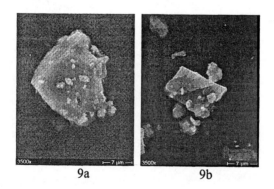

Figure 9. Scanning electron microscopy (SEM) evidencing the micro-morphological and morphometrical attributes of dust particles, such a characteristics strongly affects both the "flying properties" and the risk associated with their inhalation by the workers. 9a: travertino and 9b: pietra dorata, details of calcite ($CaCO_3$) particles (Secondary electron images. Magnification: 3500x - Image resolution: 800x1000).

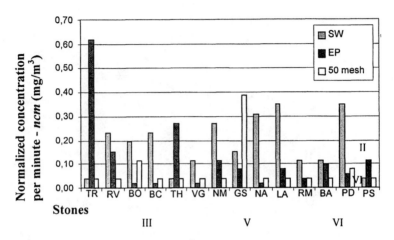

Figure 10. Comparison of dust *nmc* (mg/m^3) detected during manual sawing (SW), edge polishing (EP) and surface polishing with 50 mesh abrasive paper of the 14 different ornamental stones. Dust capture was carried out using low speed tools and the cabin caption system on. See Table 3 for legend.

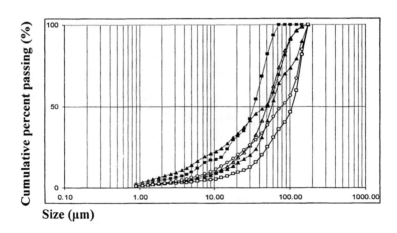

Figure 11. Example of size class distribution related to different airborne dust products for different working actions (EP: edge polishing, SW: sawing and POL50: polishing with 50 mesh abrasive paper) performed on grigio sardo. According to the different aspirating cabin operative conditions (on-off) airborne dust size class distributions change. The risk for workers changes accordingly. (I: SW-off, II: EP-off, III: POL50-off, IV: SW-on, V: EP-on and VI: POL50-on).

The strong reduction of dust directly interesting the worker is accompanied, also thanks to the increase and to a more efficient cabin suction system (optimized air speed flow), by a corresponding increase of dust morphometrical characteristics (Figure 11). These two aspects both strongly contribute to reduce health risk for the workers. Best performance of the new filtration system is obtained for the edge polishing action.

Free Silica Concentration Reduction

From the gravimetric analysis of the collected respirable dust fraction, the same result obtained for the inhalable fraction has been achieved: the edge polishing action produces a bigger quantity of respirable dust compared to sawing and surface polishing actions for all the stones. For this reason, the analysis of free silica concentration has been carried out on dust collected during edge polishing, being identified as the most dangerous action for workers health.

The free silica was not detected (< MDL) in the dust belonging to the following stones: travertino, rosso verona, botticino, bianco carrara, thassos, verde guatemala and nero marquina, according to the mineralogical composition of such stones, as determined by the petrographic and mineralogical study. On the contrary, quartz has been detected in the dust belonging to the following stones in decreasing order of abundance: pietra dorata, grigio sardo, basaltina, rosso multicolor, nero assoluto, pietra serena, labrador. This is in agreement with the mineralogical composition of the stones.

In Figure 12, the comparison among the free silica concentration detected in dust with and without the use of the aspirating device is shown. As it can be seen, when the cabin is off, the TLV stated by law (0.05 mg/m^3) is exceeded for almost all the stones. When switching on the cabin, the free silica dust concentration is strongly decreased. The efficiency of the aspirating device must be considered very high, since the level of free silica contained in the generated dust is, for six of the seven stones, below the sensitivity of the measuring instrument (Figure 12).

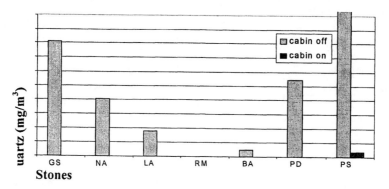

Figure 12. Comparison of the free silica concentration (mg/m^3) in respirable dust fraction generated during edge polishing of 7 different ornamental stones obtained with and without the use of the aspirating cabin and working with high speed tools. See Table 3 for stones legend.

CONCLUSIONS

Several tests have been carried out on 14 different stones by applying the three working actions that are mainly used in the finishing stage of the ornamental stone transformation process, that is: edge polishing, sawing, surface polishing. Comparison tests have been carried out in similar standard operative conditions, with or without operating an aspiration system on. Dust generated in all working conditions on each stone was sampled and analyzed in terms of concentration, morphological and morphometrical particle characteristics. Results have been investigated and compared to identify the best operative conditions and quantify dust reduction. Correlation existing between dust particle size and shape was clearly evidenced.

The development of a new control system for the aspirating cabin, allowed to reach the following benefits:

a substantial airborne dust reduction (up to 80% for certain operations and stones),
a reduced power energy loss,
a reduced air consumption for filter elements cleaning,
a reduced need of maintenance of the filtration elements,
the possibility to apply a continuous monitoring of the operations performed inside, recording parameters such as dust concentration and workers exposure time.

We identified that the working action can be considered more dangerous for workers health. It is also possible to verify the impact of aspirating devices and best practice methods on generated dust concentration. New measurements, carried out for other working tools, can be added in order to increase the knowledge base and develop working conditions simulations.

The new aspirating system of the cabin introduces an improvement of the air quality, reducing the concentration of airborne dust, especially for edge polishing.

Sawing produces less dust than edge polishing, but this is not necessarily associated with an improvement in the air quality because sawing is less sensitive to aspirating devices. When working with surface polishing with 50 mesh abrasive paper, the results can be considered intermediate.

For what concerns free silica content in generated dust, the aspirating device supplies an excellent solution to its limitation, since the concentration of quartz in the dust is almost completely cut down and is therefore below the TLV stated by regulations.

ACKNOWLEDGMENT

The work described was financially supported within the framework of the GROWTH Project GRD1-1999-10351 ENVICUT (Human and environmentally friendly cutting and milling of materials) of the European Union.

REFERECNCES

WHO (1999). Hazard Prevention and Control in the Work Environment: Airborne Dust. WHO/SDE/OEH/99.14, World Health Organization, Geneva, Switzerland.

S.C. Soderholm, 1989. Ann. Occup. Hyg., 33, 301–320.

G. Bonifazi, V. Giancontieri and S. Serranti, 2002. Airborne dust sampling and analysis in an ornamental stone industry, Proceedings of Air Quality III International Conference on Air Quality III. Trace Elements and Particulate Matter: Measurement. Technical Coordinators: Tom Erickson, William Steiz, John Pavlish, Scott Renniger, Leonard Levin. Arlington, VA, USA.

G. Bonifazi, V. Giancontieri, S. Serranti and F. Volpe, 2002. Airborne dust characterization in harsh environments, SWEMP 2002. 7th International Symposium on Environmental Issues and Waste Management in Energy and Mineral Production. Recycling. Recovery of Useful Components From Industrial Waste. pp. 1187–1194. Cagliari, Sardinia, Italy.

SRMP - Centro Sperimentazione Macchine Movimentazione Polveri (2002). 36 Months Report. GROWTH ENVICUT Project GRD1-1999-10351.

Advanced Sampling and Characterization Techniques of Nanoparticle Products Resulting from Thermal Decomposition Processes

G. BONIFAZI

Dipartimento di Ingegneria Chimica, dei Materiali, delle Materie Prime e Metallurgia, Università di Roma "La Sapienza", Via Eudossiana, 18-00184 Roma, Italia

S. DI STASIO

Aerosol and Nanostructures Laboratory, Italian National Research Council, Via Marconi, 8-80125 Napoli, Italy

ABSTRACT

The morphological analysis of flame generated carbon clusters is object of this work. Some methods to characterize this particulate on the basis of the fractal approach are briefly reported. Experiments are performed by sampling on the axis of ethylene-air diffusion flame. Analysis by Scanning Electron Microscope is worked out by application of the Nesting Square technique.

Results show that the fractal dimension is not constant in correspondence of different residence times within the flame, according to previous studies performed by the authors utilizing static light scattering techniques. A comparison is performed between the aggregate morphology observed in flame experiments at temperatures about 1400°C (Diffusion Limited Cluster-Cluster Aggregation) with respect to the shape of clusters generated by a slower agglomeration process at 70°C temperature (Reaction Limited Cluster-Cluster Aggregation).

INTRODUCTION

The physical and chemical properties (viz., morphology, density and surface reactivity) of organic and inorganic fine and ultrafine aerosols of anthropogenic origin emitted to the atmosphere, such as from automobile combustors and furnaces, possesses the ultimate effects of modifying the earth's climate, worsening visibility and endangering human health. Pollution substances in the atmosphere are represented by gaseous compounds (for instance NOx, SOx, CO) and by a class of particulate compounds. These latter pollutants can be either carbon type (soot, black carbon) or of a different nature (sulfates, metal oxides, etc.). In particular, in real situations some metallic compounds adsorb on soot particles. Very often soot is present in atmosphere as clusters of aggregate nano-sized particles termed "primary units".

The focus of this work is to detect and characterize the size and morphology of carbonaceous nanoparticles obtained from combustion of gaseous hydrocarbons. Each of these properties is crucial for characterization of hazards from particulate fumes. For instance, the size and the shape dictate the deposition and penetration properties of nanoparticles inside human lungs. Fractal characterization of the clusters and aggregates is obtained using mass fractal approach.

THEORY AND METHODS

Forrest and Witten [1] reported previous findings about the evidence for certain physical systems [2], such as a fluid near its critical point, that the fluctuating density $\rho(r)$ of these systems is characterized by spatial correlations which extend to arbitrarily long distance. Such correlations are found typically to obey a characteristic power law

$$<\rho(r)\rho(0)> - <\rho(0)>^2 \propto r^{-A} \quad (1)$$

over a large range of r. The exponents A in such power laws often have two striking features. First, in contrast to most power laws one encounters, A is usually not a simple fraction arising from dimensional consideration: it is 'anomalous'. Second, A is typically unaffected by a continuous alteration of the parameters of the system: it is 'universal'. A is found to be equal to $D_F - d$ where d is the Euclidean dimension. The mass scaling can thus be expressed by the known relation

$$M \propto (R_g/r_P)^{D_F} \quad (2)$$

for agglomerates of sub-units with radius r_P. In the past, 'anomalous' exponents, like A, were discovered in a remarkably broad range of physical systems. The best known examples are extended matter at a *second-order phase transition* such as the liquid-gas critical point mentioned above. From a mathematical point of view the observed power laws are manifestations of symmetry known as the renormalization group. Long polymers [3], the spin glass [4] and the connected clusters in a system near its percolation threshold [5] appear to have *critical* correlations in complete analogy to second-order phase transitions.

In the case of soot, mass (M) fractal-like objects are characterized in terms of the empirical law

$$M \propto (2 R_g/a_P)^{D_F^*} \quad (3)$$

which is considered to hold true in a statistical (not necessarily mathematical) sense. In the previous equation R_g and a_P are the aggregate gyration radius and the primary size and the notation D_F^* accounts for the limit of validity of the fractal law with respect to the universal case (power exponent D_F).

Several methods are available to infer the mass density fractal dimension by the processing of 2-D images.

In the method proposed first by Forrest and Witten [1], called by the authors *Point-Counting* method, the scaling relation is probed *internally* for each aggregate. That is, for a given origin chosen to be on a particle in the central portion of the aggregate, concentric squares are drawn, and within each square, the intensity (assumed now to be linearly proportional to the mass), is summed up. The effects of the arbitrary choice of the origin can be minimized by the repetition and averaging of the procedure for several chosen origins [6].

From log-log plots of mass versus size of square, a curve is obtained, the slope of which gives a value of D_F, the fractal dimension. This procedure is termed also *Nesting Squares* method by Tencé et al. [6].

An alternative method to infer D_F was presented by the same authors [6] called the *total mass/size scaling* method. This procedure seeks to verify the scaling relation between mass and size for all the aggregates of a considered ensemble. The intensity is integrated over the whole aggregate and the result is related to a measure of a size characteristic of the aggregate. This size can be the value of $\sqrt{a^2+b^2}$, where a and b are the sides of the close fitting rectangle that encompasses the agglomerate. From a log-log plot of mass versus size for all the aggregates measured, a set of points are obtained that can be fitted to a straight line, with slope D_F. Other characteristic sizes were also used by the same authors such as $\sqrt{ab}$, $(a+b)/2$, $\max(a,b)$ but it was found that this makes only a very slight difference on the value of the slope D_F (about 10% of the standard deviation of the value).

In the comparison of the results from *Nesting Squares* and *total mass/size scaling* method sometimes an underestimate of the true fractal dimension is reported to occur. The magnitude of this systematic error is quantified as about 0.3 for aggregates with a true fractal dimension close to $D_F = 2$. The physical reason for this effect is hypothesized by Mandelbrot [7,8] to be the consequence of the projection of a 3-dimensional aggregate to 2-dimensional images. Nevertheless, it is found that this projection will not affect the value of the fractal dimension D_F calculated by the "nesting squares" method for values of D_F less than 2. For $D_F > 2$, this method cannot give more than an effective fractal dimension equal to 2, whilst the mass / size method would yield the correct value.

The accuracy of the *Nesting Square* and other image processing methods to retrieve D_F, is basically limited by the finite total number of points in the image analyzed. It can be verified that the statistical variations in the measured exponents decrease when there are more points in the image. Another limitation on precision arises from finite size of the considered image.

The second method of analysis is carried out in terms of the *density autocorrelation function*. This method consists in determining the average density of occupied points in the digital image at a distance r from each occupied point, and repeating this work for all values of r. As already stated above, a set of points with density mass dimension, this average density should fall off with distance as r^{D_F-d}, where D_F, d are the mass fractal and the Euclidean dimensions, respectively. Values of D_F for each image are found by plotting the logarithm of the average density against *log r*. Then D_F is obtained from the initial slope, as determined by equally weighted least-square fitting.

In the present work the Nesting Square method will be applied to infer the fractal dimension of single clusters from soot samples probed at different residence times in the flame. Further, the Ensemble characterization of aggregates is performed in term of the mass fractal power law $M \propto R^{Df}$ according to the SEM/TEM image analysis carried out by previous authors [9–12] at different residence times in the flame. The image analysis of previous authors will be used. In particular, the smaller rectangle including each cluster with side lengths W and L is considered. A geometric dimension is computed as $R = \sqrt{LW}$. The mass M of the aggregate is estimated by the evaluation of the surface projection of the aggregate. This has been demonstrated to be correct in the case of sufficiently open aggregates, namely for D_F less than two, which is sometimes stated [13] by classifying these clusters as "optically transparent". In the case of Diffusion-Limited Cluster Aggregation, such as it occurs for carbon particles in a flame, the simulations show that the fractal dimension is expected to be less than 2. The log-log plot of mass vs. R are obtained and a power interpolation yield a slope which is the fractal dimension

EXPERIMENTS

Soot aggregates are produced by a Bunsen burner (*BB*, inner diameter 10 mm) fueled with ethylene (purity higher than 99%) at a flow rate $v_{\text{cold-gas}}$ = 15 *cm/s*, flow rate about 12 cm^3/s, $L_{\text{flame}} \approx 170$ mm). A significant change of the fractal dimension is expected at different Heights-Above-Burner (HABs) on the basis of previous work [14–16]. To provide reproducibility of measurements the air inlet of the Bunsen burner is blocked.

Soot is collected at different HABs by a stainless steel duct (inner dia = 2 mm, outer dia = 4mm) connected with a pipe (inner diameter 8 mm, length 1 m) to an aspirating pump. The sucking rate is about 40 cm3/min. After the pipe particles enter a thermal precipitator device inside which they are forced by the thermophoresis effect to impinge and stick on a glass support (18 × 18 mm). Samples with collected soot are thereafter subjected to a coating process, which produces the deposition of a graphite film (10 nm thickness) on their surface. Thereafter, samples are positioned inside the test chamber of SEM where high-vacuum conditions are reproduced. The Scanning Electron Microscope used is Hitachi S-2500. The acceleration voltage used is typically 25 kV with an emission current 100 μA and the resolution about 3 nm for a probe current of 100 *pA*. The working distance is about 7–8 mm.

SEM images from soot sample of aggregates generated in flame by Diffusion Limited Cluster-Cluster aggregation (DLC-CA), obtained picking soot directly in flame by the quartz thermophoretic probe, are now compared to the SEM images of the Reaction Limited Cluster-Cluster Aggregation (RLC-CA) obtained as following. The samples of ethylene soot are first collected on a cold metallic plate at 60 mm height above the Bunsen burner. The original soot in quantity of about

12 mg has been diluted in 25 cube centimeters of methyl alcohol and treated with ultrasonic wave for about 30 minutes. This procedure is aimed to the breaking and fragmentation of the soot samples in smaller units. Thereafter, each sample has been stored in oven at 70°C till all the alcohol is evaporated. The effect of this second step is of furnishing, by thermal energy transfer, sufficient energy to overcome the repulsive energy barrier to the soot fragments in alcohol solution. Thus, they are put in motion, migrate and collide each other, producing again new agglomerates. The difference with the process of meet-and-stick, which occurs directly in the flame, is that now the kinetics of the process is much slower. In fact, the governing process is in flame Diffusion Limited Colloidal Aggregation (DLCA) which is characterized by time constant of coagulation of order of ms. On the other hand, in liquid solutions very often the major mechanism is the Reaction Limited Colloidal Aggregation (RLCA) which is much slower, with coagulation times of order of minutes or hours.

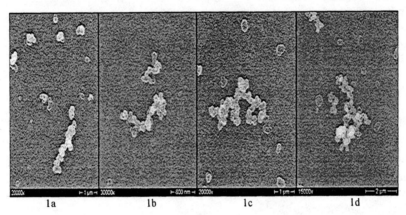

Figure 1. Scanning Electron Microscope (SEM) images of different soot aggregates sampled in an ethylene-air diffusion flame at a different Heights-Above-Burner (HAB). 1a. HAB = 15 mm, 1b. HAB = 20 mm, 1c.HAB = 30 mm and 1d. HAB= 40 mm. The image sequence shows as cluster complexity increases according to residence time increase.

IMAGE ANALYSIS

Figure 1 presents some aggregates collected at different heights above the burner. The larger the residence time, the more branched become the clusters. This is reflected by the fractal dimension, which can be retained as increasing with the HAB. In particular, in Figure 2 and Figure 3 the D_F evaluation of an aggregate ensemble at 20 and 40 mm HAB are reported, respectively.

Figure 2 and Figure 3 show two examples of *ensemble* characterization in terms of fractal dimension at 20 and 40 mm heights above the burner. A comparison with the D_F evaluated by the Nesting Square method for a *single* aggregate sampled at the same H.A.B. = 40 mm is reported in Figure 4. Evidence about the fractal dimension increase with the residence time is found out and it is demonstrated in Figure 5 at HAB = 70 mm.

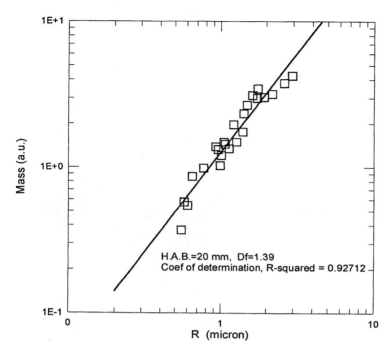

Figure 2. Aggregate ensemble evaluation of D_F =1.39 at HAB=20 mm.

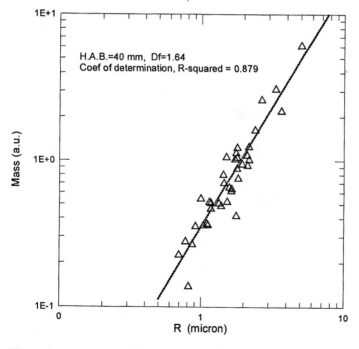

Figure 3. Aggregate ensemble evaluation of D_F =1.39 at HAB=40 mm.

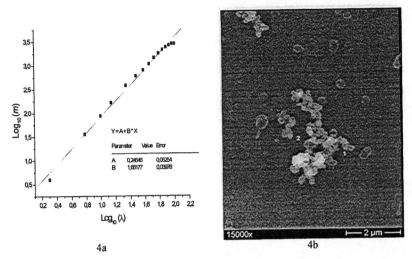

Figure 4. Application of *Nesting Square* method (4a) to single aggregates (cluster 1 in 4b), HAB=40 mm and D_F =1.66.

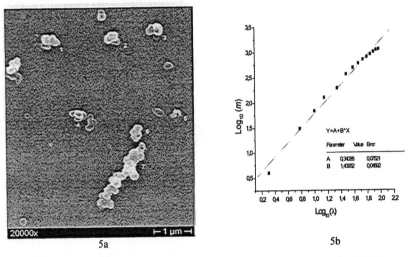

Figure 5. As for Figure 4 at HAB=15 mm, D_F =1.44 for the cluster 7 of the SEM image.

An investigation is then performed on the aggregates formed in flame (DLC-CA) and the clusters formed by breaking the collected soot and inducing a slower agglomeration regime (RLC-CA) as previously described. The agglomerates that are obtained in the latter case (RLC-CA) are characterized by bigger physical dimensions (a few microns), more corrugated surface texture and more compact and filled appearance. This is according the previsions from computer simulations which predict [17,18] for Reaction Limited Cluster-Cluster Aggregation larger mass density fractal dimension ($D_F \approx 2.11$ in 3-D simulations) with respect to Diffusion Limited Cluster-Cluster Aggregation ($D_F \approx 1.38$ to 1.45 in 2-D and $D_F \approx 1.75$–1.80 in 3-D simulations) The physical reasoning which is advocated in the literature to justify these different effects of agglomeration processes, is that the sticking probability is smaller with respect to the DLC-CA case and the primary particles must collide several times before sticking. The net result of such a slower process is that the *screening* from the growing arms of the clusters is less effective.

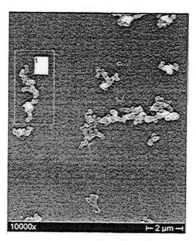

Figure 6. SEM image of ethylene soot collected at 70 mm height-above-burner from central portion of a propane flame from a Bunsen burner with the air inlet blocked (flow-rate of cold ethylene = 12 cm^3/s, vcold-gas=15 cm / s, apparent flame length ≈ 150 mm). The aggregation process is quick (τ_{coag} ≈ 0.2 ms) and governed by Diffusion Limited Colloidal Aggregation (DLCA).

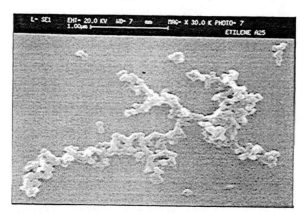

Figure 7. SEM image of ethylene soot collected on a cold plate from flame and treated with ultrasonic shacking bath (half an hour) in methyl alcohol and drying (one hour) in a oven at 70°C. The aggregation process is slow (τ_{coag} ~ 10 min) in comparison to DLCA and governed by Reaction Limited Colloidal Aggregation (RLCA).

Figure 6 and Figure 7 compare the SEM images of the ethylene soot relative to the regimes of quick and slow agglomeration kinetics. In particular, the cluster reported in Figure 6 is relative to the soot collected on the quartz probe by thermophoretic sampling from the inner region of a ethylene-air flame (probed volume on the flame axis) at 70 mm height-above-burner. The flame here considered is obtained burning ethylene at purity better than 99.95% in a Bunsen burner, keeping the air inlet ring of the burner blocked. The flame is characterized by an apparent height L_{flame} of about 170 mm, and a cold-gas velocity at the exit of the burner of 15 cm/s. Inside the flame, as verified for a similar ethylene flame [19], the agglomeration process is quick, with coagulation constants τ_{coag} (time requested to halve the initial number concentration) less than 1 ms depending on size, shape, concentration and density of the solid particles and temperature, pressure and viscosity of the surrounding gas.

The cluster of Figure 7 is obtained via the ultrasonic wave shaking in methyl alcohol and drying in oven, described in the Experiments paragraph. The appearance and the physical outer dimension in the two cases (*Diffusion Limited* Cluster-Cluster vs. *Reaction Limited* Cluster-Cluster aggregation) are markedly different. The soot from flame is chain-like even if different layers of particles are overlapped in the sampling. By contrast, the soot from shaking and drying is a star-like structure with four main legs. Outer dimension of some microns is observed in both cases. For instance a Nesting Square elaboration, not reported here, performed for aggregate number 1 in Figure 6 (DLC-CA) shows that the fractal dimension is 1.30 whereas for the cluster in Figure 7, (RLC-CA) it is 1.90.

CONCLUSION

A study has been performed to characterize ex-situ the soot clusters in a diffusion ethylene-air flame. Sampling at several heights above the burner and analysis at SEM have brought to the following points: i) the fractal dimension is not constant along the flame axis in correspondence of different Heights-Above-Burner (HABs), namely of different residence times; ii) a process of progressive modification of aggregates morphology does occur, starting from the sticking of chain-like aggregates at lower HABs that yield rapidly spherical-like clusters with fractal dimension about 2 at higher HABs. This observation accord the studies reported elsewhere [15,16] about the effect of restructuring of soot clusters observed to occur both inside the flame by light scattering techniques applied to probe directly the flame, and outside the flame by sampling plus TEM analysis at room temperature and, finally, iii) c) a comparison has been made between the morphologies observed in the flame (Diffusion Limited Cluster-Cluster Aggregation) with those in a process of aggregation induced at a considerably lower temperature (Reaction Limited Cluster-Cluster Aggregation). The result is that at parity of outer dimension the aggregates from the RLC-CA have a larger fractal dimension with respect to those from DLC-CA. This result is coherent with the numerical simulations reported in the literature.

REFERENCES

[1] S.R. Forrest and T.A. Witten, 1979. J. Phys. A: Math. Gen., 12, L109-L117.

[2] M.E. Fisher, 1974. Rev. Mod. Phys. 46, 597.
[3] D.S. McKenzie, 1976. Phys. Rep. 27 35.
[4] A.B. Harris, T.C. Lubensky and J-H. Chen, 1976. Phys. Rev. Lett. 36 415.
[5] P.W. Kasteleyn and C.M. Fortuin, 1969. J. Phys. Soc. Japan Suppl. 26 11.
[6] M. Tencé, J.P. Chevalier and R. Jullien, 1989. J. Physique 47 1989–1998.
[7] B.B. Mandelbrot, 1975. Les Objets Fractals (Flammarion: Paris).
[8] B.B. Mandelbrot, 1977. Fractals: Form, Chance and Dimension. Freeman: San Francisco.
[9] R.J. Samson, G.W. Mulholland and J.W. Gentry, 1987. Langmuir 3 272–281.
[10] C.M. Sorensen and G.D. Feke, 1996. Aerosol Sci. Technol. 25 328–337.
[11] A.A. Onischuk, S. di Stasio, V.V. Karasev, V.P. Strunin, A.M. Baklanov, V.N. Panfilov, 2000. J. Phys. Chem. A 104 10426–10434.
[12] C.M. Sorensen, 2001. Aerosol Sci. Technol. 35 648–687.
[13] M.V. Berry and I.C. Percival, 1986. Optica Acta 33 577–591.
[14] S. di Stasio, 2000. Appl. Phys. B 70 635–643.
[15] S. di Stasio, 2001. J. Aerosol Sci. 32 509–524.
[16] A.A. Onischuk, S. di Stasio, V.V. Karasev, A.M. Baklanov, G.A. Makhov, A.L. Vlasenko, A.R. Sadykova, A.V. Shipovalov, V.N. Panfilov, 2003. J. Aerosol Sci. In press.
[17] M.Y. Lin, H.M. Lindsay, D.A. Weitz, R. Klein, R.C. Ball and P. Meakin, 1990. J. Phys. Condens. Matter 2 3093–3113.
[18] D. Asnaghi, M. Carpineti, M. Giglio and M. Sozzi, 1992. Phys. Rev. A 45 1018–1023.
[19] S. di Stasio, A.G. Konstandopoulos and M. Kostoglou, 2002. J. Colloid Interface Sci. 247 33–46.

Development of a Magnetically Levitated Hoisting System for Use in Underground Mines

R. ULANSKY and J. A. MEECH
Centre for Environmental Research in Minerals, Metals and Materials,
The University of British Columbia, Department of Mining Engineering,
Vancouver, BC, V6T 1Z4, Canada

ABSTRACT

Innovations in the mining industry over the past few decades have mainly focused on economies of scale as well as centralized data collection and analysis. Little has taken place in the way of alternative methods to accomplish the various stages of the mining process. If the industry wishes to survive there is a need to make significant revolutionary changes in its operating practices. The integration of batch processes can prove to be one of many types of economic incentives and advantages.

Mining traditionally involves the following separate unit operations:

- *drill and load explosives*
- *blast*
- *muck out the blasted rock*
- *load it into horizontal transportation vehicles and haul it to a central dump point*
- *dump it into an underground storage facility*
- *crush it underground to around 250 cm top size*
- *hoist it to surface using a wire-rope hoisting system and dump it into surface storage facilities*

The evolution of mechanized equipment has led to the integration of the muck-load-haul-dump sequences. Computerization and telerobotics has provided a direction to automate the drill, load, and blast cycles. But the real bottleneck in future evolutions of these component operations is the hoisting facility. This unit is limited by the difficulty in expanding production once the shaft has been sunk together with the fact that only a single skip can operate in the hoist at any one time.

It is time to "think out of the box"! It is time for a radical change! It is time for Magnetic Levitation!

UBC-CERM3 has developed an innovative approach to hoisting that uses a linear motor to pull steel vehicles containing ore all the way from the mining face to the surface storage facility. The shaft is a tube with a series of separate copper wire windings around the exterior of the tube. These consecutive windings are charged in sequence to create a magnetic field that travels along the shaft pulling the skip along with it. Multiple skips can be used in this shaft which is considerable smaller in diameter than a conventional shaft used to deliver the same quantity of ore. The vehicles return underground through an adjacent return tube which follows the delivery tube all the way to the face. Energy is recovered from this tube during the return journey. In this way, underground haulage requirements are significantly reduced since they are essentially integrated into the Magnetic Hoisting system as the vehicles can operate in both vertical and horizontal planes. The capital and operating costs of this system versus a conventional hoist appear to be competitive. This paper will present the results of our quarter-scale prototype system in operation and provide an analysis of the economic and environmental advantages of this approach.

BACKGROUND

The conceptual design of a magnetically-levitated skip for use in an underground mine is presented. A skip is used to move rock from an underground mine to a processing facility located on surface. Conventionally, a skip is raised and lowered by cables (or wire ropes), in a fashion similar to an elevator. In a magnetically-levitated skip, the cable is replaced with a linear induction or synchronous motor to provide propulsion and levitation.

Removal of cables from a magnetically-levitated skip provides many advantages over a conventional system including the ability:

- to negotiate corners,
- to travel vertically in the shaft as well as horizontally to the mining face, and
- to move a very large volume of material using multiple skips in a small shaft.

A magnetically-levitated skip is safer, more environmentally-friendly, and a productive alternative to conventional skipping. It has the added economic benefit of decreasing underground haulage requirements.

The concept presented here describes a skip propelled by a linear induction motor. A conventional skip is propelled by being attached to cables that wind around a drive wheel attached to a counterbalance. Using cables reduces flexibility and limits system productivity. A magnetically-levitated skipping system (MPS) can have multiple containers on a special track inside a tubular-shaped shaft between an underground loading facility and a surface dump point. As no cables are used, the skips are free to negotiate corners, to travel to any depth in the mine, and to go directly to the mining face with advancement of the tubes as the face moves forward through the orebody. With multiple skips in the same shaft, an MPS may be competitive with conventional underground haulage systems.

COMPONENTS

The MPS system consists of skips, track, loading and dump points, and a maintenance/storage facility, each of which is described below:

Skip

The skip comprises a cylindrical tube with a lid to contain the payload. It is made of steel that interacts with a magnetic field around the tube to provide propulsion. The ore is loaded into the skip and the lid is closed to prevent spillage. Guidance of the skip is provided by three sets of wheels that run on three sets of rails located equilaterally around the tube inner surface. The wheels and rails are distributed at angles of 120° around the inner circumference of the skip and tube. These rails are designed so that the lid will close and open automatically as the skip passes through either the dumping or loading stations respectively.

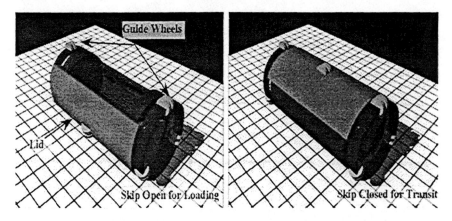

Figure 1. The Skip Design for a magnetically-propelled hoisting system.

Track

The track is cylindrical with three rails located on its interior to guide the skips. The exterior of the tube is wrapped with alternating rings of electrical windings and iron bands which compose the linear motor and generate the magnetic field which lift the skips to surface and control their location and velocity.

Loading Point

The ability of the MPS system to turn corners enables it to run vertically down the shaft and then horizontally along a drift to a location near the face. By approaching the face, a separate material handling system is not required to move material long distances from the face to the shaft. The face is advancing continually, so the loading point is moveable, allowing it to follow the face. The loading system consists of a dump pocket for a scoop-tram to dump its load, a crusher to reduce the muck to top-size suitable for loading, a surge bin, and a loading system for the skip.

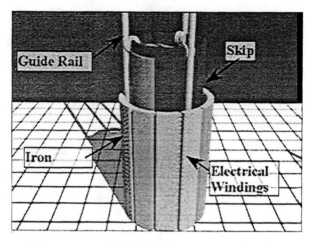

Figure 2. The track system used with the magnetically-propelled hoisting system.

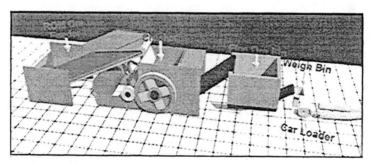

Figure 3. The loading system used with the magnetically-propelled hoisting system.

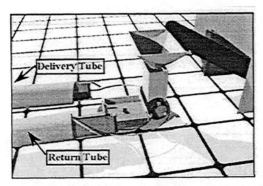

Figure 4. Close-up view of the skip handling system at the loading point.

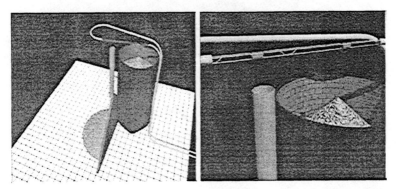

Figure 5. Dumping configuration for a magnetically-propelled hoisting system.

To enable the skip to turnaround inside a narrow drift, a Skip Handler is used. The Skip Handler receives each vehicle from the return tube, rotates it 90° to be loaded with ore, rotates it another 90° to align with the delivery tube, then returns to receive the next vehicle from the return tube. The payload to each skip is weighed to prevent overloading and spillage. The cleanliness of an MPS system will be a prerequisite previously unheard of in conventional mining.

Dump Point

The dump station involves rotating the track to invert the skip above the dump point. To dump a skip, the lid opens automatically as the vehicle rotates allowing the payload to fall into the bin. By controlling when the lid opens, the material can be sent to different locations if required, enabling a mine to sort high grade, low grade, waste, or neighboring mine materials that are all being hoisted within the same system.

Maintenance and Vehicle Storage

A skip can be removed from the system for maintenance and repairs without having to down the entire system. The maintenance facility is located on surface next to the dump point. In operation, a skip destined for maintenance is dumped as shown above, but instead of returning underground, it is diverted by an attached tube to the maintenance facility. For temporary storage, the skips are queued in front of the loading station underground to eliminate the need for continuous storage and recovery of skips.

SUB-SYSTEMS

To travel from a loading point to a dumping point, a skip must negotiate the track, be propelled up the shaft, be braked on descent, and have its speed controlled. Systems to perform these tasks are as follows:

Guidance

Guidance is achieved using rails inside the tube. To load from several locations in a mine at the same time, the skips can be directed to appropriate exits connected to the main return track. Loaded skips are then merged back into the main delivery tube to surface using switching stations much like a railway system.

Propulsion

A linear induction motor located around the circumference of the tube propels the skip along the tube. Only a small portion of the track surrounding the skip is energized at any one time as each skip passes. As the magnetic field in the energized portion of the track moves up the tube, the skips are essentially "pushed" to surface by this magnetic field. The frequency by which adjacent windings are energized will control the speed of travel of the skips up and down the delivery and return tubes.

Braking

Braking is required to control the speed of the skip on its decent back into the mine on the vertical return trip down the shaft. To brake the skip, the same electric and iron core coils are used, but in the reverse sense as a generator. As the skip passes each coil, it induces a current in the coil. The induced current generates a magnetic field resisting the passage of the skip and slowing it down. To control the skip's descent, the electricity generated in these coils can be captured and modified using variable resistors. The current induced in these coils is connected to the main power grid. This energy is run through an inverter converting it from DC to AC current and a variable transformer is used to stabilize and step-up the voltage. In this way this energy is recovered to the main power grid. To provide a fail-safe emergency system, mechanical friction brakes are located along the length of the track. The brake consists of spring-loaded dogs which flip open and prevent skips from falling down the tube should a power failure occur.

Speed Control

Controlling of the speed of a skip is important to avoid collisions and prevent damage should a skip become jammed or go out of control. The speed of each skip while traveling upward or horizontally is limited by the rate that the magnetic wave moves along the shaft. This is controlled directly by the frequency that power is supplied to adjacent coils. For normal operations, the system operates at 60Hz or at the requirements of the local power supply. When moving down the return shaft, sensors monitor the speed and braking power is adjusted accordingly. The control system is designed to maintain a particular spacing between skips. Control of several skips moving together will be achieved using a "convoy" control strategy.

DESIGN PARAMETERS

Table 1 presents some preliminary design ideas of the size and productivity levels of an MLS system.

Preliminary calculations on energy utilization and copper wire requirements suggest that an MLS system could compete with conventional hoisting because of lower overall capital and operating costs. Capital costs are reduced because of reduced shaft-sinking requirements—a raise-boring machine can be used to create a hole size of about 1-2 m in diameter. Reduction in underground haulage equipment is also significant. Depending on the ability to recover energy from the return journey underground, operating costs are also reduced due to lower maintenance and safety requirements and the automation of the entire system. In particular a reduction in underground haulage requirements is likely. Computer control is integral to the overall operation of this complex system. Conservatively speaking, over 14,000 tonnes per day is projected as a possible production rate from a 25 cm (~10 inches) diameter system. This is based on 5 skips passing a particular point every second. The limit to this rate is likely to be the dumping and/or loading cycle times.

Table 1. Proposed Design Parameters.

Skip Diameter	25 cm
Skip Length	50 cm
Skip Volume	0.025 m^3
Skip Capacity (90 % fill factor, ore S.G 3.0, 50 % voids)	35 kg
Skip speed	12 m/s (43 km/h)
Theoretical Capacity (skips back-to-back traveling at 12 m/s)	3000 tph
Practical Maximum Capacity (1/5th maximum capacity)	600 tph

CURRENT RESEARCH ON MAGNETIC LEVITATION

There are several groups performing R&D on magnetic-levitation technology around the world. These applications are focused on creating high-speed trains to move people. Several countries have built full-scale prototypes, with a German group appearing to be in the lead. The German company, Transrapid International, has designed a train that has exceeded 450km/h (125 m/s) on a working test track. It has already carried over 200,000 paying passengers [1].

Vancouver's SkyTrain [2] is the closest application of this technology to the system proposed here. The trains are propelled by a linear induction motor and are guided by wheels on rails. The major difference with our concept is that the windings are located on the car, not the track, and the motor is flat instead of being rolled around a tube.

NASA is also doing considerable work with linear induction motors [3]. They are developing a launch pad to employ a linear induction motor to accelerate the next generation of the Space Shuttle to a lift-off velocity of 600 mph to assist in launching vehicles into space.

In mining, there are several papers on the potential applications of linear induction motors for hoisting, but these are strictly conceptual. In the early 1990s, the United Stated Bureau of Mines investigated a magnetically-levitated material handling system for underground coal mines [4]. It operated in a horizontal mode within a sloped shaft. The project was shelved when the USBM closed in 1994.

CURRENT STATUS OF THE UBC PROJECT

To date much time and effort has been spent working on conceptual details and preliminary design features of the overall system. In 1998, a small working model was built to demonstrate the feasibility of the technique [5]. The model consisted of a single shaft 90 cm in height and 2.5 cm in diameter and was able to demonstrate controlled vertical movement (both up and down) of a 100g payload at speeds equivalent to a production rate of about 5 tpd. In the summer of 2002, the first prototype of an 8 cm model was constructed. Testing and modifications continue to be made to this model.

We are using virtual reality software to design different types of loading and dumping stations, as well as different shape configurations for the shaft. This approach is particularly useful to identify critical issues regarding the geometry and limitations of each design. A 10-cm diameter prototype testbed with a continuous loop in both a vertical and horizontal mode of operation is being constructed to demonstrate productive capabilities and further enhance the "proof-of-concept" of such a system.

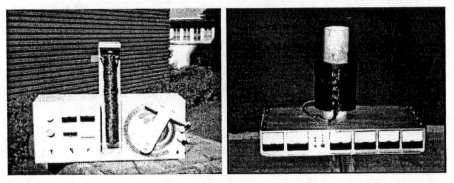

Figure 6. First and Second Prototype Models.

Figure 7. Picture of the Prototype Test Loop: (a) horizontal mode, (b) vertical mode.

CONCLUSIONS

A magnetically propelled hoist can have significant benefits to the mining industry that include:

· Safety
- reduced diesel emissions
- reduced dust as the skip contains all of the dust
- automation keeps personnel out of hazardous locations

· Environment
- zero emission technology
- braking is recaptured as electricity

- Productivity
  - multiple skips in the same shaft will increase the capacity of the system
  - ability to turn corners allows the skip to be loaded at the face
  - system is unaffected by the depth of the mining face
- Economics
  - higher capital costs of the skips are offset by
    - smaller shaft size and better stability, i.e., no shaft reinforcement
    - lower operating and maintenance costs

The significant advantages of this system make it worthy of more detailed investigation. We are currently constructing a 10-cm diameter prototype test track capable of a payload of several kilograms in the new UBC Centre for Environmental Research in Minerals, Metals and Materials (CERM3) within the Mine Automation and Environmental Simulation Laboratory.

REFERENCES

1. http://www.transrapid.de/en/information/technik_txt.html
2. Vancouver's SkyTrain. Rapid Transit Project 2000, SkyTrain Millennium Line, Rapid Transit 2002 (Information Packet).
3. http://liftoff.msfc.nasa.gov/News/2000/News-MagLev.asp
4. J.J. Geraghty, W.E. Wright and J.A. Lombardi, 1995, Magnetic Levitation Transport of Mining Products, United States Bureau of Mines, United States Department of the Interior.
5. Ryan Ulansky, 1998, Magnetically Levitated Skipping, UBC-MMPE Undergraduate Thesis.

Bio-Nanotechnology and Phytomining: The Living Synthesis of Gold Nanoparticles by Plants

C. ANDERSON and B. STEWART
Institute of Natural Resources, Massey University, Palmerston North, New Zealand

C. WREESMANN
Akzo Nobel Chemicals bv, Research-Dept. CFC, P.O. Box 9300, 6800 SB Arnhem, The Netherlands

G. SMITH
Akzo Nobel Chemicals Pte. Ltd., #10-02 SLF Building, 510 Thomson Rd, Singapore 298135

J. A. MEECH
The Centre for Environmental Research in Minerals, Metals, and Materials, University of British Columbia, Mining Engineering, 6350 Stores Rd, Vancouver BC, V6T 1Z4, Canada

ABSTRACT

Phytomining is an intelligent system whereby plants are used to extract valuable metals from soil. These plants are subsequently harvested and the metals recovered from the biomass. Phytomining may be a viable way of recovering gold from waste rock and tailings left after a mining operation. A recent report showed that alfalfa sprouts grown on a gold-rich agar could accumulate gold in shoots. Furthermore, using X-ray spectroscopy and transmission electron microscopy (TEM), gold particles were shown to be stored in the alfalfa biomass as discrete metallic gold nanoparticles.

Using the TEM facilities of Akzo Nobel Chemicals bv in Arnhem, The Netherlands, nano-scale gold particles were also observed in plant material generated through induced-accumulation pot experiments conducted by the authors of this paper in New Zealand (unpublished data). Questions remain regarding the purity of these gold nanoparticles, as elevated concentrations of other metals are also observed in plant tissues (e.g. copper and silver). These metals may alloy with gold during nanoparticle growth.

Field experiments are underway to examine the feasibility of commercial gold phytomining from gold-rich waste soil and ore. Further characterization of the gold stored in the vegetative tissues of harvested plants will allow for better understanding of the form and purity of biologically synthesized nanoparticles, potentially leading to new pathways for the recovery and use of these important metallic structures.

INTRODUCTION

During the 2001 IPMM conference in Vancouver, the idea of phytoextraction as an intelligent environmental system was introduced [1]. Phytoextraction is an application of phytoremediation technology defined as the use of plants and their associated root-bound microbial communities to remove, contain, degrade or render harmless environmental contaminants [2]. A potential use of phytoextraction is for the plant-based mining of precious metals (phytomining). In particular gold has been the subject of several years of phytomining research.

Some plants will naturally accumulate metals. Nickel, for example, is *hyperaccumulated* by 318 known plant species [3,4]. Other metals for which hyperaccumulator species have been identified include Cd, Cu, Co, Mn, Se and Zn. Insoluble metals such as Au, Pb, Pd and Pt are not accumulated by any known plant species. Uptake has to be forced or *induced* by applying a chemical to soil that will increase the soil-solution concentration of the target metal. Chemical irrigation is made once a crop has reached maximum biomass, and must be done under carefully controlled conditions. Induced hyperaccumulation was first reported for lead, using the chemical EDTA.

HYPERACCUMULATION

The term hyperaccumulation was coined by a New Zealand scientist in the 1970's [5], and defined as accumulation to a metal concentration greater than 1,000 mg/kg of biomass (dry weight) in the aerial portions of a plant. This concentration limit was based upon nickel accumulation in hyperaccumulator plants 100 times greater than that expected in non-accumulator plants growing in the same environment. The concentration limit varies for different metals. Cadmium, for example, is a metal with relatively low crustal abundance, hence a cadmium hyperaccumulator has > 100 mg/kg metal, whereas the limit for manganese a much more common metal is 10,000 mg/kg (1%). The definition of hyperaccumulation has been refined since 1977. Today a hyperaccumulator species is one which:

Is tolerant to metal concentrations that would otherwise be toxic to plant life,
Has a metal concentration in aerial tissues 100 times greater than that in normal plants,
Has a greater metal concentration in shoots than roots.

Hyperaccumulation of gold has been induced using thiocyanate and thiosulphate; naturally occurring chemicals that act to increase the soil-solution concentration of this metal [6]. Hyperaccumulation of gold has been defined as accumulation greater than 1 mg/kg, this limit being based upon a normal gold concentration in plants of only 0.01 mg/kg. Gold concentrations are often greater in shoots than roots (unpublished data) although root crops such as carrot have been used in gold uptake studies [7].

STUDIES OF THE FORM AND DISTRIBUTION OF METALS IN PLANT

Compartmentalisation of metal is a key mechanism of hyperaccumulation, but only in recent years due to the advent of analytical instrumentation such as proton-induced X-ray emission (PIXE) and X-ray absorption spectroscopy (XAS) has it been possible to observe and describe the form and distribution of metals within a plant.

For metals to be safely stored in above-ground plant tissues, the contaminant must be sequestered into a non-toxic form. Free metal ions in particular can have toxic effects on a plant. Zinc accumulated by the hyperaccumulator species *Thlaspi caerulescens* is stored within the vacuoles of epidermal and subepidermal cells [8]. Quantitative elemental maps show that nickel in the South African hyperaccumulator species *Berkheya coddii* and *Senecio anomalochrous* is also stored in the epidermal regions of each species leaves and stems [9]. In comparison, the key trace micronutrient elements Zn, K, Ca, Mn and Fe are distributed throughout all tissues. Similarly, Küpper *et al.* [10] reported preferential accumulation of nickel in the intracellular compartment of epidermis cells for three brassicaceous species. The studies mentioned here however, do not describe the crystal structure of nickel or zinc in plants.

Tissues where metals have been found to be sequested such as a vacuole are physiologically inert. That is to say, storage in these regions would allow a plant's critical biological machinery to operate in relative isolation of toxins [11]. Protection against herbation is a possible reason for metal hyperaccumulation. The epidermal tissues of leaf or stem material would the first tissues encountered by grazing animals. A high metal concentration in this herbage could cause an adverse reaction if ingested. Another suggested reason for hyperaccumulation is as a natural sunblock [12]. A concentrated zone of metal in a plants upper cuticle may act as a barrier to ultraviolet radiation, protecting the underlying chlorophyll from solarisation.

Most studies on metal localization in tissues have focused on metals naturally accumulated or hyperaccumulated by certain plant species. In particular nickel has been the focus of many studies, perhaps due to the large number of hyperaccumulator species of this metal. Very little attention has been paid to the storage of metals that are accumulated through induced hyperaccumulation. For example, no study has examined the form and localization of lead accumulated by a plant after EDTA treatment of soil. There has been no study published that looks at the form of gold induced into plant tissues after thiocyanate or thiosulphate treatment.

Girling and Peterson [13] made a number of conclusions on gold storage in plant tissues based upon laboratory experiments. These authors theorized that gold cyanide was stored in an aqueous form in leaf vacuoles, whereas gold chloride was predominantly insoluble and bound to the cell wall. Limited data showing the form of gold in 'natural' field samples collected during biogeochemical surveys can be found in the literature. Dunn [14] published a scanning electron micrograph (SEM) of a black spruce twig (*Picea mariana*) showing nucleation of a gold particle 0.25 microns in diameter. The location of this particle was approximately 100 microns from the outside of the twig, close to the cambial layer between the bark and wood. In a second study Dunn [15] used SEM to identify crystalline native gold attached to a cell wall within the bark of mountain hemlock. Dimensions of the target gold crystal are approximately 1 µm x 0.5 µm, compared to 100 µm long accumulations of manganese phosphate, a compound made up of two elements essential for plant health. In both reports, Dunn used X-ray analytical techniques to confirm the atomic composition of observed particles.

THE DISCOVERY OF GOLD NANOPARTICLES IN PLANT

A group working at the University of Texas at El Paso recently showed that gold accumulated in plant tissues can be present as discrete metallic particles of nanometre scale [16]. In their work, the El Paso team germinated alfalfa seeds on an agar medium enriched with 320 mg/kg gold chloride. After two weeks of growth the alfalfa plants were harvested and analysed using X-ray absorption spectroscopy (XAS) and transmission electron microscopy (TEM). Gold nanoparticles of 2 to 20 nm in diameter were observed preferentially distributed in certain zones of the plant although the tissues for storage were not identified. Larger coalesced particles of 20–40 nm in diameter were also observed. The authors of this study assumed that the differences in particle size indicated growth of nanoparticles over time. Four nm diameter particles had an icosohedral structure, whereas particles of 6–10 nm diameter showed fcc twinned structures.

In a similar study, silver nanoparticles as well as atomic wires or clusters were observed in alfalfa plants grown on silver enriched agar [17]. Spherical nanoparticles of 2–20 nm diameter again had an icosohedral structure, although these nanoparticles often showed internal defects such as twinning, mixed structures and dislocations. Larger coalesced particles indicated nucleation of silver atoms within the plant. The identity of tissues where silver was preferentially stored was again not described in this work.

The nanoparticles described above were made by growing plants on a monometallic artificial medium with a high concentration of gold or silver. No chemical was used to induce uptake in these experiments. Presumably root exudates solubilized a small concentration of metal that was subsequently accumulated as the plant grew. A gold media concentration of 320 mg/kg is much higher than would be expected in soil. Through agar-solution equilibrium reactions a fraction of this gold will be available to the plants (bioavailable). Gold uptake by plants has also been reported from artificial ore made by dispersing gold chloride solution in sand, with subsequent heating of the sand to 100°C to leave a synthetic ore with gold of very high fineness [6]. Artificial ore is usually made to contain 5 mg/kg gold. Gold concentrations in plants as high as 1000 mg/kg dry weight (0.1%) have been obtained in one plant species growing in this material after suitable treatment (unpublished data).

'Real' ores or gold-rich soils are not, however, monometallic. Other precious and non-precious metals will always be present in geochemical systems, and these metals will also be made soluble after treatment to induce the uptake of gold. Table 1 presents selected metal concentrations found in the aerial tissues of *Lupinus* sp. (blue lupin) grown on base-metal mine tailings in New Zealand after suitable treatment to induce uptake.

Table 1. Selected metal concentrations in aerial plant tissues after chemical treatment of soil relative to expected 'normal' metal concentrations.

Metal	Induced plant concentration (mg/kg)	Natural plant concentration (mg/kg)
Pb	58	0.8
Ag	126	0.2
Cu	401	11
Au	6.3	0.1
Fe	154	59
Mg	2,800	nc
Na	4,294	nc
K	23,800	nc
Ca	17,340	nc

Note. No change (nc) in major element concentrations (Mg, Na, K, Ca) was observed in plant tissues after treatment.

Treatment in Table 1 caused an increase in gold uptake by a factor of 63, and an increase in silver uptake by a factor of 620. The plant also accumulated greater amounts of copper and lead as a result of chemical treatment. The quantitative elemental distribution maps described earlier indicate that key micronutrients (e.g. Fe, K and Ca) may be stored in plant tissues independent of contaminating heavy metals [9]. What is not known is how a plant will respond to such high concentrations of a range of other metals, such as Au, Cu and Ag. Will a plant store each of these metals in separate location as discrete nanoparticles? Or will a plant synthesize nano-alloys within certain tissues?

Using the TEM facilities of Akzo Nobel Chemicals bv in Arnhem, The Netherlands, nano-scale gold particles have been observed in plant material generated through induced-accumulation pot experiments conducted by the authors of this paper in New Zealand (unpublished data). Study of these nanoparticles has only recently begun hence, no conclusive data on their form and location can be reported at this time. The plant material being used however, was grown on a poly-metallic gold ore. The potential alloying of Au, Cu and Ag is in particular being examined. Our conclusion at this time is that plants do appear to synthesize gold nanoparticles within living tissues as induced accumulation takes place.

SUMMARY

The Future Application of Gold Phytomining

Supporting the analytical studies described in this paper are field experiments that aim to determine the feasibility of commercial gold phytoextraction from gold-rich waste soil and ore. There are significant commercial opportunities for application of the technology once proven. Large areas of land enriched in gold are potentially available to gold phytoextraction upon successful transition of controlled experiments to the field environment. Economic modeling of gold phytomining has been carried out at Massey University. Consequently, a gold concentration of 100 mg/kg in a harvested dry-weight biomass of 10 t/ha, yielding 1 kg of gold, is the target that has been set for the technology to be economically viable. Research experience shows that obtaining this concentration of gold in a plant is possible, but is a function of the geochemical variables of the substrate. Field tests in Brazil are currently being designed that will integrate laboratory- and computer-based models with suitable plant growth and soil conditions to assess the current feasibility of field-scale phytomining.

Further characterization of the gold stored in the vegetative tissues of harvested plants will allow for better understanding of the form and purity of biologically synthesized nanoparticles potentially leading to new pathways for the use of these important metallic structures. While nanoparticles generated through phytomining may not necessarily have a direct use in nanotechnology, understanding their form and localization is of major importance to the design of effective systems to recover gold and other metals from harvested plant biomass.

ACKNOWLEDGEMENTS

The senior author gratefully acknowledges Akzo Nobel Chemicals Pte. Ltd of Singapore and the New Zealand Foundation for Research, Science and Technology, who are supporting the current development of gold phytomining technology at Massey University.

REFERENCES

[1] C. Anderson, B. Robinson and C. Russell, 2001. Phytoextraction: a plant-based environmental biotechnology, In Proceedings of the Third International Conference on Intelligent Processing and Manufacturing of Materials IPMM 2001, Vancouver, Canada (Ed. J.A..Meech, S.M.Veiga, M.V.Veiga, S.R.LeClair and J.F.Maguire) CD Rom (UBC, Vancouver, Canada).

[2] B. Robinson, C. Russell, M. Hedley and B. Clothier, 2001. Cadmium adsorption by rhizobacteria: implication for New Zealand pastureland, Agriculture, Ecosystems and Environment, 1732: 1–7.

[3] A.J.M. Baker, S.P. McGrath, R.D. Reeves and J.A.C. Smith, 2000. Metal hyperaccumulator plants: A review of the ecology and physiology of a biochemical resource for phytoremediation of metal-polluted soils, In Phytoremediation of Contaminated Soil and Water (Eds. N. Terry, G. Banuelos, J. Vangronsveld) pp 85–107 (Lewis Publishers, Boca Raton, Florida).

[4] R.D. Reeves and A.J.M. Baker, 2000. Metal-accumulating plants, In Phytoremediation of Toxic Metals: Using Plants to Clean Up the Environment (Eds. I. Raskin and B. D. Ensley) pp 193–229 (John Wiley and Sons, New York).

[5] R.R. Brooks, J. Lee, R.D. Reeves and T. Jaffre, 1977. Detection of nickeliferous rocks by analysis of herbarium specimens of indicator plants, Journal of Geochemical Exploration, 7: 49–57.

[6] C.W.N. Anderson, R.R. Brooks, R.B. Stewart and R. Simcock, 1998. Harvesting a crop of gold in plants, Nature, 395: 553–554.

[7] F.A. Msuya, R.R. Brooks and C.W.N. Anderson, 2000. Chemically-induced uptake of gold by root crops: its significance for phytomining, Gold Bulletin, 33(4), 134–137.

[8] M.D. Vázquez, J. Barceló, C. Poschnrieder, J. Mádico, P. Hatton, A.J.M. Baker, and G.H. Cope, 1992. Localization of zinc and cadmium in Thlaspi caerulescens (Brassicaceae), a metallophyte that can hyperaccumulate both metals, Journal of Plant Physiology, 140: 350–355.

[9] J. Mesjasz-Prybyłowicz, W.J. Prybyłowicz, D.B.K. Rama, and C.A. Pineda, 2001. Elemental distribution in Senecio anomalochrous, a Ni hyperaccumulator from South Africa. South African Journal of Science, 97: 593–595.

[10] H. Küpper, E. Lombi, F.J. Zhao, G. Wieshammer, and S.P. McGrath, 2001. Cellular compartmentation of nickel in the hyperaccumulators Alyssum lesbiacum, Alyssum bertolonii and Thlaspi goesingense, Journal of Experimental Botany, 52: 2291–2300.

[11] R.S. Boyd, 1998. Hyperaccumulation as a plant defensive strategy, In Plants that Hyperaccumulate Heavy Metals (Ed. R.R.Brooks) pp. 181-201 (CAB International: Wallingford).

[12] B.H. Robinson, E. Lombi, F.J. Zhao and S.P. McGrath, 2003. Uptake and distribution of Ni and other metals in the hyperaccumulator Berkheya coddii, New Phytologist: in press.

[13] C.A. Girling and P.J. Peterson, 1978. Uptake, transport and localisation of gold in plants, Department of Botany and Biochemistry, Westfield College, University of London.

[14] C.E. Dunn, 1995. Biogeochemical prospecting for metals, In Biological Systems in Mineral Exploration and Processing (Eds. R.R.Brooks, C.E.Dunn, & G.E.M.Hall) pp. 371–426 (Ellis Horwood: Hemel Hempstead).

[15] C.E. Dunn, 1995. Mineral exploration beneath temperate forests: The information supplied by trees, Exploration Mining Journal, 4(3): 197–204.

[16] J.E. Gardea-Torresdey, J.G. Parsons, E. Gomez, J. Peralta-Videa, H.E. Troiani, P. Santiago and M. Jose Yacaman, 2002. Formation and growth of Au nanoparticles inside live alfalfa plants, Nano Letters, 2(4): 397–401.

[17] J.E. Gardea Torresdey, E. Gomez, J.R. Peralta-Videa, J.G. Parsons, H. Troiani and M. Jose-Yacaman, 2003. Alfalfa Sprouts: a natural source for the synthesis of silver nanoparticles, Langumier: in press.

Developing a Plug to Seal Mine Openings for a Thousand Years: The Millennium Plug Project

B. LANG, R. PAKALNIS and J. A. MEECH

Centre for Environmental Research in Minerals, Metals and Materials,
The University of British Columbia, Department of Mining Engineering,
Vancouver, BC, V6T 1Z4 Canada

ABSTRACT

An innovative method to seal mine tunnels has been developed and tested at UBC which has the potential to reduce the cost of mine closure and ensure a long-lasting solution to mine closure issues. Mining companies would prefer a "walk-away" solution because of the high costs associated with long-term monitoring of a site. In most cases, a concrete plug is placed within a tunnel to prevent effluent from being released to the environment and prevent curious individuals from entering into the dangerous conditions in the mine—potential rock falls and lack of adequate air. Unfortunately some effluent conditions are such that the concrete degrades over time—for example, acidic waters will dissolve concrete causing the structure to fail. Another way that failure may occur is by cracking of the concrete when it is subjected to stress over time or to seismic vibrations. Basically, the material is just too rigid and brittle.

CERM3 together with <u>TSS Tunnel and Shaft Sealing</u> *have designed a new approach to sealing that has this problem solved as well as several others. We have dubbed this method THE MILLENNIUM PLUG in recognition of the new century we have entered and to suggest the possible lifetime of the structure.*

INTRODUCTION

Conventional approaches to sealing mine adits and tunnels involves the placement of a concrete plug which fills the opening and is grouted at its interface with the surrounding rock mass. In addition, grout is often pumped into holes drilled into the surrounding rock around the plug to seal up fissures and fractures that may have been created when the tunnel was originally driven into the mine. A relatively short-length concrete plug can be designed to withstand water heads as high as 1000 metres. Typical service pressures range from 10 to 200 metres.

While concrete is a strong and "easy-to-work-with" material with high strength and reasonable cost, in certain environments, it is not a solution of choice. If the mine waters to be retained within the tunnel contain contaminants that react with the concrete, the plug can degrade over time and eventually fail. Such is the case with acidic mine waters that are typically found in sulfide hardrock mines. Salt, chloride and sulfate can also corrode concrete over time eventually leading to failure of the containment system.

The ability of these rigid structures to withstand seismic events is also poor. Extreme vibrational loading can result in cracks into which corrosive waters may flow accelerating the process of erosion. Mining companies are seeking "walk-away" solutions to mine closure – methods in which their liability can be reduced significantly and the uncertainty of plug survival is minimized. Such is the case of using a plug manufactured from bulk materials. Earth dam structures on surface are examples of the potential for bulk materials to stand up to high loads and seasonal changes for long periods of time. In addition, proper selection of the bulk materials can enhance the ability of the plug to be inert to its aquatic environment.

UBC-CERM3 has developed such an approach to seal mine tunnels. Dubbed the Millennium Plug in recognition of the new millennium we have just entered and as a description of the expected lifetime of the structure, we have installed a full-scale unit in the 2200 Level of the derelict Britannia Mine located on Howe Sound about 45 kilometres north of Vancouver (see Figure 1 and Figure 2). The plug consists of graded layers of bentonite, sand, gravel, and rip-rap and the facility has been designed so we can pressure-test the plug under a variety of high-head conditions. Our research has been able to establish design parameters that allow us to say with confidence that the plug will stand-up to chemical attack, seismic forces and high head conditions for at least 1000 years.

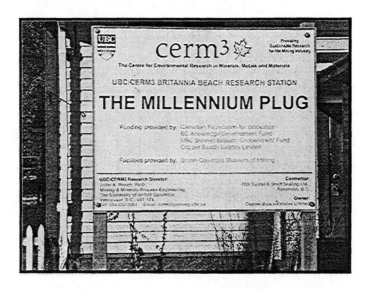

Figure 1. The Millennium Plug Project at Britannia Beach, British Columbia.

(a) (b)
Figure 2. (a) The Britannia Beach townsite, (b) No. 3. Mill Building.

THE PROBLEM AT BRITANNIA BEACH

The mine ceased operation in 1974 and was officially closed in 1976. Since that time, over 10,000 tonnes of copper, zinc, iron, aluminum, arsenic and cobalt have been discharged into Howe Sound contained in an effluent known as Acid Rock Drainage. Prior to Dec. 2001, the ARD flowed from two mine adits located at the 2200 level and the 4100 level. The flowrates and metal levels are shown in Table 1. As can be seen the average flow conditions are considerable with about 600-1000 kg of combined metal values entering Howe Sound each and every day. Figure 3 shows the pollution plume emerging from the mouth of Britannia Creek and flowing up the Howe Sound towards Squamish. This picture was taken in 1994.

Because of ownership disputes and poor management, no one has assumed responsibility for cleaning up the site. The mining industry view Britannia as not only an eyesore that tarnishes the reputation of the industry, but as something that stands as a major liability hindering the establishment of a positive image of mining in Canadian society. Environment-Canada has dubbed this site the "worst ARD pollution site in North America" and virtually all NGOs who pay close attention to mining activities view Britannia as one of the worst legacy mine sites in the world. Despite this common belief, no one has been able to find a "formula" to bring every stakeholder to the table to create a plan to reclaim and restore the site to something that is sustainable and acceptable—until now!

Table 1. Typical Effluent Conditions at Britannia Mines pre-December 31, 2001.

Item	4100 Level	2200 Level
Average Flowrate (m^3/hr)	400	100
PH	3.5–4.4	2.7–3.0
Cu content (mg/L)	12–22	30–100
Zn content (mg/L)	25–30	28–35
Total Tonnes of Cu & Zn per year	~ 300	

BRIEF MINE HISTORY

Britannia Mine is located on Howe Sound about 45 km. North of Vancouver on the Sea-to-Sky Highway. The current site is an abandoned underground mine inside Britannia Mountain. The workings began at the top of the mountain some 4500 ft. (~1370 m.) above sea level and extend to 2000 ft. (~685 m.) below sea level. The mine operated between 1902–1963 by the Britannia Mining and Smelting Co. and between 1963–1974 by the Anaconda Mining Company. During its lifetime over 47 million tonnes of ore were mined producing copper and zinc concentrates. Over 150 km of underground development took place along with the early production of 5 open pits or glory holes at the top of the mountain.

Britannia Mine and the town of Britannia Beach have played a very important role in the development of British Columbia. Some key events at the site are listed below:

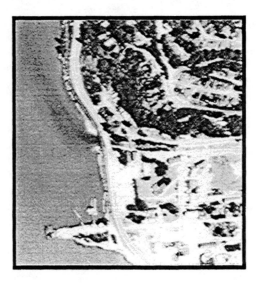

Figure 3. Aerial Photograph of Britannia Beach showing the pollution plume entering Howe Sound.

1859 - Captain Richards, on a survey mission for British Admiralty sails into Howe Sound and anchors off what today is known as Britannia Creek. He names Britannia Mountain after his 100-gun frigate.
1888 - Dr. A. Forbes discovers a copper mineralization outcrop near the summit of Britannia Mountain.
1899 - A mining engineer, George Robinson, begins raising capital. Eventually the site is controlled by the Britannia Mining and Smelting Company, a subsidiary of the Howe Sound Company which eventually operates the site for the next 60 years.
1904 - Under Robinson's direction, the first ore is shipped to the Crofton Smelter on Vancouver Island.
1905 - Britannia Mine achieves full operation.
1912 - James W. Dunbar Moodie is authorized to upgrade the operation and increase mine production.
1915 - On March 21, an avalanche destroys the Jane Camp at the top of the mountain killing 60° men, women and children.
1916 - The No. 2 mill is completed with the new flotation process in place. Production is increased to 2000 tons of ore per day.
1917 - A new, "safer" town site is completed at the 2200 Level (2300 ft. above sea level). The town becomes know as "Mount Sheer".
1918 - The infamous influenza epidemic strikes Britannia and dozens die.
1921 - During a brief period of shut down, Mill No. 2 burns to the ground.
1921 - In October, a massive flood destroys the "safer" community on the banks of Britannia Creek killing 37 people.
1922 - The new town and the No. 3 Mill are constructed. This mill still stands today as the largest heritage building in British Columbia.
1929 - Britannia Mine ranks as the largest copper producer in the British Commonwealth.
1935 - Britannia Mine pioneers the first cementation process in the world recovering over one third of its copper production using this unique process. Water is encouraged to flow into the mine to leach copper into solution for recovery in launders located at Mount Sheer and at the 4100 Level of the mine near to the Britannia Beach townsite. The waters run at high volume, low pH and extremely high copper and zinc levels (> 1.2 g/L).
1942 - Due to low copper prices, the mine is advanced subsidies by the Canadian government to maintain production during the Second World War.
1946 - Britannia mines are unionized and the town suffers through its first strike.
1956 - The rail line is completed from Squamish to North Vancouver.
1958 - The Vancouver-Squamish highway is completed and named the Sea to Sky Highway. Low copper prices lead to tough times and the mine goes into and out of production several times.
1958 - Mount Sheer is emptied, all operations move to Britannia Beach, and the company is reduced to 7 employees.
1959 - Britannia Mining and Smelting Co. Ltd. is voluntarily liquidated and purchased by the Howe Sound Corporation.

1963 - Anaconda Canada Ltd. purchases the Britannia mine and recommences operation.
1973 - A new ore vein is found and a contract is struck with the miners allowing for increased production.
1974 - The first order to collect and treat all ARD is issued to Anaconda Canada in October. The mine closes in November.
1975 - The BC Museum of Mining opens under management of the Britannia Beach Historical Society.
1977 - The property and assets of the mine change ownership to 400091 BC Limited a holding company that mortgages the property to a company called Copper Beach Estates Limited (CBEL).
1981 - The order is revised naming the new owner, Copper Beach Estates Ltd. managed by Tim Drummond.
1988 - The BC Museum of Mining is designated a National Historic Site.
1995 - Environment Canada and the BC government begin joint cooperation to attempt to define liability and to gain an understanding of the ARD problem and the required works and costs for treatment.
1998 - BC Ministry of Environment, Land, and Parks meet with 3 Potentially Responsible Parties (PRPs), Canzinco Ltd., Arrowhead/Ivaco, and Atlantic Richfield, but doesn't add them to the order.
1999 - BC Ministry of Energy and Mines issues a permit to CBEL to conduct mine reclamation including a 2200 level diversion, a 4100 level WTP, reclamation and monitoring.
2000 - CBEL institutes a 180 million dollar lawsuit against the 3 PRPs.
2000 - In Nov., BCMELP hold closed-door talks with the PRPs and negotiate an agreement in which the companies provide the province with $30 million dollars in exchange for a total indemnity. CBEL's lawsuit is withdrawn.
2000 - CBEL is purchased by Alex Tsakumis and Ian Hagemoen
2001 - The 2200 Level plug is installed by UBC. The flow is turned off on December 31, 2001 for the first time since 1974.
2001 - BCMELP becomes the Ministry of Water, Land and Air Protection
2002 - BCMWLAP begin to implement a remediation plan for the site which includes studies on the stability of the 4100 Level plug, the status of contamination on the fan area at Britannia Beach, a flood risk assessment, and the requirements for an ARD water treatment plant.
2003 - Following a detailed design, BCMWLAP announce the construction of a high-density lime sludge treatment plant to be completed by the spring of 2004.

ASPECTS OF THE REMEDIATION PLAN

There are basically three major components of the plan to remediate the Britannia Mine site. These include: plugging the 2200 Level portal, reclaiming the glory holes and pits at the top of the mountain, and installing a water-treatment-plant at the 4100 Level to treat all ARD coming from that level. Figure 4 shows a topographical map of the Britannia Mine region and details about the problem and necessary clean-up elements.

THE MILLENNIUM PLUG

UBC-CERM3 was in the process of building a laboratory to research design guidelines for earth plugs to seal mine tunnels. We came to the conclusion that it made more sense to build a full-scale prototype instead of developing a laboratory that would be fraught with scale-up difficulties. Accordingly we approached Copper Beach Estates to collaborate on building a research facility at the 2200 Level of the Britannia Mine. This installation produced two synergetic outcomes—UBC.

The plug is built with locally available materials that are relatively cheap to obtain and transport to a mine site. Construction is simple and straightforward using conventional mining equipment. The cost of installation is competitive with pouring cement and building a conventional concrete plug. As such, we believe this approach is a universal solution that can be put into place at numerous sites around the world.

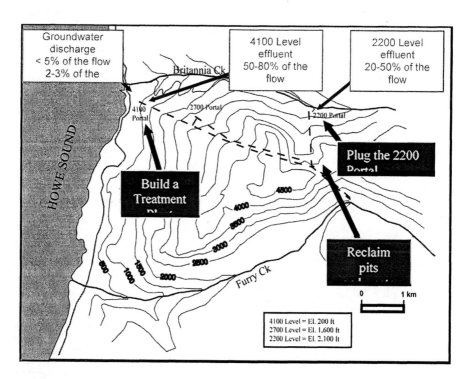

Figure 4. Topology of the Britannia Mine site and the requirements to eliminate/control pollution.

The Millennium Plug has the following advantages:

- it is built with locally-available materials
- it is cheaper to build than a concrete plug
- it does not require grouting of the surrounding rock mass
- it is much more resistant to chemical attack than concrete
- it is much more resistant to seismic events
- it can be designed to last for a 1000 years (a walk-away solution)

In December 2001, the facility was installed. This facility provides UBC-CERM3 with the capabilities to conduct research into these earth plugs and derive design guidelines for other operations around the world. By installing this Research Station in a full-scale site, scale-up issues from the laboratory are eliminated providing significant credibility to the work. In addition by locating the facility in the 2200 Level tunnel, all of the pollution to Britannia Creek is diverted back into the mine to flow with all other effluent down to a single point source discharge at the 4100 Level just behind the BC Museum of Mining.

The following environmental benefits have accrued because of this installation:
- all effluent exits the mine from a single point source
- all effluent has been eliminated from Britannia Creek and the surface waters of Howe Sound
- about 15 to 20 percent of the copper contained in the effluent is now precipitating inside the mine
- the first stage of the remediation plans for Britannia Beach is now completed

The 2200 Level effluent discharged into Jane Creek, a mountain tributary of Britannia Creek. Fig 5 shows the horrible iron precipitate which forms immediately upon the confluence of the two waters.

The Millennium Plug Research Station was built in two stages. First a conventional concrete was installed to divert the ARD effluent back into the mine and to act as a wall against which we could test the Millennium Plug. The concrete plug is considered to be a temporary solution, as it will eventually corrode in the acid environment in the mine. It will likely last only about 50–60 years. Between the two plugs, we left a chamber into which we can pump pressurized water to establish how the Millennium Plug will respond to high heads of water and to seismic events. The plug is instrumented to measure deformation and leakage of water to establish how it responds to load changes due to water build-up or earthquakes.

The concrete plug was designed to hold back up to 300 m of water despite the fact that its service pressure is only 9 m. This allows us to pump water into the chamber and try to blow out the Millennium Plug. The concrete plug is only 4 m in length while the Millennium Plug is ~ 25 m in length giving it superior leakage protection and requiring no grouting to be used into the surrounding rock mass. Figure 6 shows a schematic diagram of the installed system.

Developing a Plug to Seal Mine Openings for a Thousand Years 351

The Concrete Plug was poured on December 17, 2001 in a blinding snowstorm over an 18-hour period (see Figure 7).

A total of 16 cement trucks had to be pulled up to the 2200 Level using a front-end loader and a bulldozer.

The effluent was turned off for the first time in over a quarter of a century at 3:30 pm on Dec. 31, 2001 once the cement had set up.

Figure 5. Jane Creek at the confluence with the 2200 Level effluent discharge (November 2001) prior to installation of the Millennium Plug Research Station.

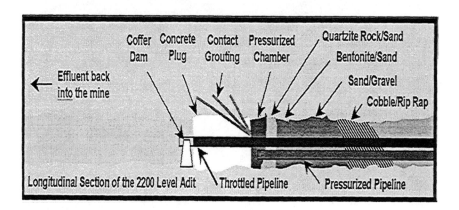

Figure 6. The Millennium Plug Research Station. The ARD effluent is diverted back into the mine workings.

The pressurized chamber allows UBC researchers to study the Millennium Plug under variety of pressure and seismic conditions up to 300 m of water head. The service pressure on the concrete plug is only 9 m, but it was designed for 300 m in order to conduct our studies. Upon completion of the research, the Millennium Plug will be left in place to "back-up" the concrete plug which is expected to fail in 60–70 years through chemical attack by the acidic mine waters (pH = 2.5–3.0).

As of January 1, 2002, all acid rock drainage from the 2200 Level tunnel into Jane and Britannia Creek has been halted. There are at least 4 major benefits to the installation of this system:

1. All effluent now exits the mine from a single point source,
2. No pollution now is entering Britannia Creek, which is showing signs of coming back to life.
3. All pollution flowing directly into the surface waters of Howe Sound has been eliminated.
4. Between 15–20% of the copper that previously flowed out of mine is now precipitating inside the mine workings as the 2200 Level effluent becomes neutralized as it flows down into the mine.

Figure 7. Cement trucks at the 2200 Level portal on December 17, 2001.

Table 2. Before and After Metal Levels in Britannia Creek.

Element	Britannia Creek* before the 2200 Level Plug	Britannia Creek* after the 2200 Level Plug
Copper (mg/L)	0.5	0.041
Zinc (mg/L)	0.5	0.052
Iron (mg/L)	0.024	0.009
pH	6.3	6.9

* Measurements taken at the Townsite Bridge

Although not a recommended practice, Figure 8 provides dramatic evidence that the water in Britannia Creek is now drinkable with respect to metal levels. The very low quantity of copper still in evidence in the creek waters is due to Jane Creek flowing through a glory hole at the top of the mountain. Once this creek is diverted from this ARD, essentially no metal will be in the water.

Figure 8. Some people like to drink the Britannia Creek water and it is now safe to do so, although like all natural water streams in the wild, it is not recommended without prior chemical treatment.

FUTURE PLANS

Over the years, many engineers and miners worked and lived (and died) at Britannia Mines where they raised their families at Mount Sheer and Britannia Beach. The BC Museum of Mining, which has operated at the site since closure, is dedicated to preserving the memory of these people—our ancestors—and to providing the public with knowledge about past mining practices and how they have changed over time. More than 20,000 individual artefacts and photographs are preserved in the Museum and approximately 30,000 visitors come to the site each year as they travel between Whistler and Vancouver on the Sea-to-Sky Highway. Vancouver is currently attempting to gain the right to host the 2010 Winter Olympic Games. Britannia will be the gateway to these Games and has promise to provide a major positive image of the Mining Industry to the world if the site can be changed from the eyesore it is today—a clear liability to all who work in our industry—to a vibrant, sustainable community with multiple activities. The presence of a research facility dedicated to finding environmental solutions in mining can do much to demonstrate the importance of research in addressing society's needs and desires.

About 250 people still reside at Britannia Beach living in a company-controlled town of about 100 homes north of the fan area. There are two major landowners: the Britannia Beach Heritage Society and Britannia Mines and Reclamation Corporation (formerly Copper Beach Estates Ltd), who operation the townsite facilities (power, water, sewage, site maintenance). The Museum is a major public attraction with a walking tour of the buildings on the fan including No. 3 Mill, as well as a train ride through one of the mine tunnels. The land on which the Museum resides is considered contaminated with high copper and zinc in the sediments and ground water, but these flows amount to ~5% of the total emissions from the mine workings. Since installation of the Millennium Plug Research Station, metal levels in Britannia Creek have dropped by over an order of magnitude and the pH has increased to neutral. For the first time in over 25 years and perhaps longer, the waters are drinkable and aquatic life can be seen returning to the lower reaches of the stream.

UBC is planning to build a research centre at the Britannia Beach site to conduct innovative research for the mining industry with respect to environmental problems. Natural Resources Canada is also planning to build a centre at the site dedicated to promoting the industry. The plan is to integrate the museum with the Research Centre with a Dialog and Conference Centre to produce a major tourist attraction that will showcase Canadian Mining and the Environment technologies through exhibits on Mining: Past, Present and Future. The attraction will house an attractive video system in a cavern blasted into the side of the mountain at the top of the Mill Building. Figure 9 provides a perspective view of the planned Innovation Centre while a sectional view is presented in Figure 10. In Figure 11, an overall plan view of the site following the construction of all facilities is shown. The Water Treatment Plant will be tied into the research centre and a constructed wetlands will be used to establish new research into passive ARD treatment processes.

Figure 9. Perspective view of the planned NRCan Britannia Centre for Mining Innovation.

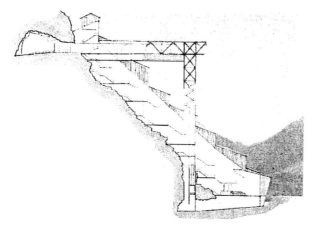

Figure 10. Section view of the planned NRCan Britannia Centre for Mining Innovation.

Figure 11. Overall site layout showing the integration of UBC's Research Centre, the BC Museum of Mining and NRCan's Dialog and Conference Centre.

CONCLUSIONS

This project demonstrates a number of innovative and creative opportunities that exist in the development of research projects.

- Research can be applied to actually mitigate an existing problem directly if one uses sustainable thinking in formulating a creative approach to problem-solving.

- The Millennium Plug Research Station allows CERM3 to design innovative, inexpensive, walk-away solutions for mine closure problems involving Acid-Rock-Drainage.

- Britannia Creek is now free of pollution thanks to the installation of the Millennium Plug Research Station.

- CERM3 will continue to search for other synergetic applications while conducting its research and setting-up its laboratory infrastructure.

- Britannia Beach will evolve into a site of incredible beauty and an asset to the mining industry from its current status as a derelict legacy site.

Chapter 6: Intelligence in Energy Systems

Multiple Sensor Surface Vibrations Analysis for Monitoring Tumbling Mill Performance

S. J. SPENCER, J. J. CAMPBELL, V. SHARP, K. J. DAVEY, P. L. PHILLIPS, D. G. BARKER and R. J. HOLMES

CSIRO Minerals, Private Mail Bag 5, Menai, NSW 2234, Australia

ABSTRACT

Tumbling mills are large-scale grinding devices commonly used in mineral processing. The operation of some of these devices, such as Autogenous/Semi-Autogenous Grinding (AG/SAG) mills is difficult to control and optimise. Direct monitoring of the comminution process is not feasible due to the hostile environment inside the mill. However, surface vibration monitoring of the external shell has been shown to be a valuable tool for soft-sensor monitoring of AG/SAG mill "hidden" process and performance variables. Collision events associated with the AG/SAG mill charge (ore slurry and grinding media) motion and resultant comminution processes strongly contribute to acoustic emissions (transient stress waves generated by deformations in a body) that propagate throughout the mill structure. Accelerometers are mounted such that they detect the component of surface displacement normal to the mill shell, predominantly due to the propagation of surface vibration waves. A key problem is the localisation and characterisation of sources that emit the detected vibrations in circumstances of low wave attenuation characteristics and hence, strong propagation of waves around a mill liner/shell. This is important in order to gain a clear understanding of the location and nature of the source acoustic emission events in such mills, which in turn could be used as a means to monitor and optimise comminution in tumbling mills.

This paper focuses on the development of an automated signal analysis system for source event location and characterisation based on multiple sensor surface vibration monitoring of tumbling mills. Surface vibration data is simultaneously acquired from three piezoelectric accelerometers mounted in a triangular array on the liner/lifter bolts of the rotating mill outer shell, coupled to analogue radio transmitters/receivers. The vibrations are analysed in the context of seismological source location methodology and acoustic emission inspection techniques used in the testing of structures. Wave trains associated with individual source impact events are identified and the differences in arrival time at each accelerometer of such events are calculated. The location and characteristics of each large source event are estimated on the basis of an assumed propagation path and a fixed propagation speed for "Rayleigh-like" surface waves.

Formulation and solution of the inverse problem for multiple collision event location on a rotating cylindrical surface based on arrival time differences for each event at each transducer is discussed. Vibration event characterisation techniques, signal filtering and feature extraction for the accurate calculation of individual event arrival times, event matching for the identification of the arrival of a specific vibration event at each transducer and optimisation techniques for solution of the source location inverse problem are also briefly discussed. The techniques are demonstrated using multiple sensor surface vibration data collected at an industrial-scale AG/SAG mill, with single impact tests on the mill liner used to establish the characteristics of individual vibration events.

An automated signal analysis system for location of impact events in tumbling mills based on multiple-sensor surface vibration monitoring is established. Such a system provides insights into both the efficiency of the grinding process and the propensity for liner wear as a function of mill operating conditions.

INTRODUCTION

Tumbling mills are horizontal rotating drums used in the mineral processing industry for comminution (grinding) of large tonnages of rock in order to achieve liberation of valuable minerals from gangue minerals [1]. Mill construction includes an outer steel shell suitable for heavy loading and a wear resistant inner liner that is in direct contact with the mill contents (charge). The liner generally includes regularly spaced lifter bars that assist in lifting the charge with mill rotation (promoting ore breakage) and increase liner life by preventing excessive slipping. Mills are generally continuously fed and contain large numbers of loose grinding media such as steel balls or rods, hard rock or ore (for autogenous breakage), which are relatively generally much heavier than the ore particles being comminuted. Impact, attrition and abrasion mechanisms are responsible in varying degrees for breakage of ore particles in tumbling mills. Optimisation and control of the continuous grinding process in tumbling mills, particularly large Autogenous or Semi-Autogenous Grinding (AG/SAG) mills, are on-going and important issues in the mineral processing industry. The goals of a plant operator are generally to maximise mill throughput and maintain product particle size distribution to specification, while simultaneously minimising equipment wear, plant down-time for maintenance and power consumption for a process known to be highly energy-intensive with very low efficiency. Mill optimisation and control are often problematic due to substantial variations in the characteristics of the mill feed, time lags between changes in feed and charge process characteristics, uncertainties in the actual effects of such changes on the process and limited available on-line process state space information. On-line monitoring of process variables such as the mill volumetric filling, particle size distribution, grinding media content and slurry density is difficult for a number of reasons. These include the robust nature of the grinding environment within tumbling mills, the availability of appropriate monitoring technologies and problems in reliable sampling of feed and product streams.

Acoustic Emissions (AE) and surface vibration monitoring technologies are promising non-intrusive approaches to on-line monitoring of industrial tumbling mill process and machine condition. Collision events at the mill liner/charge interface and inside the charge generate transient mechanical stress waves that propagate as elastic waves through the inner liner to the shell, where they are manifested as surface vibration waves that in turn generate airborne sound waves. Surface vibration (commonly termed AE) measurements utilising accelerometers mounted on the stationary bearings of industrial ball mills have been used to provide software-based (soft-sensor) estimates of process variables (for example [2,3]). Measurements of AE by a microphone mounted on the rotating shell of an industrial SAG mill have also been used to estimate charge inclination, an important parameter in the optimisation and control of mill operation [4]. Point location of vibration source events using arrival time differences of disturbance(s) at multiple locations has been utilised in seismology [5], microseismic activity [6], materials characterisation and AE inspection of structures [7,8,9,10,11].

It has been shown that industrial (production-scale) AG/SAG mill surface vibration measurements by accelerometers mounted on the mill shell are highly sensitive to mill operating variables [12,13]. Statistical, spectral and discrete event signal analysis techniques have been used to derive soft-sensor estimates of charge characteristics. These include the positions of the "toe" region where cascading/cataracting media impact on the charge adjacent to the mill liner, "shoulder" region of dynamic equilibrium where rising charge begins to separate from the liner and cascade/cataract to positions of lower gravitational potential, mill loading, charge particle size and gross power draft ([12–15]. Recently, CSIRO Minerals has completed a large Australian Minerals Industry Research Association (AMIRA) project in industrial AG/SAG mill monitoring via surface vibrations [14,15]. In these studies the bulk characteristics of the signal over many mill revolutions are useful input variables for soft-sensor estimates of process characteristics and the scanning nature of the monitoring technique provides charge information at specific circumferential positions in a mill rotation cycle. These papers together address techniques and results associated with single-sensor analysis of AG/SAG mill AE monitoring data, based on the assumption of signal localisation to the region of the charge/liner interface adjacent to the point of attachment of the transducer on the shell. This impact zonal location approach is limited in applicability to situations where surface vibrations are highly damped during propagation around the mill shell.

The assumption of signal localisation used in AG/SAG mill single-sensor AE analysis is predicated on the following considerations:

1) Strong vibration attenuation mechanisms in the metallic structure of industrial AG/SAG mills.
2) The observation that a relatively "sharp" region of vibration signal is often associated with the passage of the "toe" of the charge past transducers mounted on an operating tumbling mill [12–14].

So-called zonal location of vibration events local to transducers (rather than calculation of point location for individual events), sometimes utilising data from a large number of transducers, has been effectively used in many areas of AE inspection of structures [16]. However, it remains unclear exactly how far surface waves due to large impact events inside an operating AG/SAG mill can propagate around a mill liner/shell and still be detected; hence the necessity to consider the point location problem for operating tumbling mills.

This paper reports on the development of an AE monitoring system for point (exact) location and characterisation of impact events on a mill liner. The inverse problem of vibration source event location is solved by considering the difference in arrival times of stress waves generated by each specific impact event at multiple transducers on the mill surface. This approach is essential to image charge characteristics, particularly estimates of the axial and circumferential position of source events in situations where vibration waves propagate large distances around the mill shell. Multiple sensor analysis of AE data for a wide range of industrial tumbling mills is important to determine if the surface vibration signal location assumption (and hence application of zonal or point location analysis techniques) is generally valid. Some of the technical details, assumptions and problems associated with solving the AE source location problem for industrial SAG mills are discussed. Results are presented for vibration source location based on both "tap tests" at known liner positions and full vibration data for an operating industrial SAG mill.

MULTIPLE SENSOR AE MONITORING HARDWARE SYSTEM

The multiple sensor-monitoring array consists of three transducers mounted in a triangular arrangement on the mill shell. The fact that the transducers are mounted on the rotating mill shell necessitates using a radio system to transmit the raw signal (sampled acceleration as a function of time) for storage and later analysis.

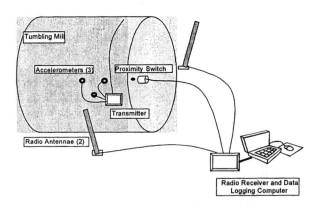

Figure 1. Schematic diagram of the multiple sensor AE tumbling mill monitoring system.

The transducers are small piezoelectric charge accelerometers mounted on mill liner/lifter bolts that secure the liner plates and lifter bars to the inside of the mill shell. This allows a direct path for AE waves from mill liner to transducer that does not include losses at medium boundaries such as the shell/liner interface. It is assumed that the dominant ray path for AE propagation as detected by transducers mounted on liner/lifter bolts is around the inner liner surface and then out to the shell through the liner/lifter bolt. The accelerometers register normal surface vibrations (acceleration) as an electrical charge via the piezoelectric effect. Charge amplifiers that convert the signal to a voltage are connected by coaxial cable to each transducer. The charge amplifiers are mounted in the same box as the on-board radio transmission system. The analog radio transmitter/receiver system consists of three transmitters mounted on the rotating mill shell (all in the same box) and three receivers mounted in an off-mill receiver box along with a rack-mounted computer for data acquisition and preliminary signal analysis. Vibrations sensed by the three transducers are continuously transmitted by radio from the mill. The on-board AE mill monitoring system is currently powered by sealed lead-acid batteries, although solar panels have also been successfully used in one outdoor application. Several transducer types have been tested, but the fundamental limitation on the upper frequency limit of the sensitivity of the monitoring system (and hence the accuracy of source location) is the frequency bandwidth of the analog radio transmission system. The effective bandwidth of the monitoring system is ~0.2–18 kHz. Data is typically sampled at 0.5-1×10^5 scans per second, depending on the nature of the test. It should be noted that data acquisition rates associated with AE source location are often much higher, with the use of transducers resonant at frequencies of ~150 kHz [16,17]. Sampling rates up to the order of 10^7 Hz have been reported for some applications [8,11]. Off-mill multiple sensor simultaneous data acquisition (using a data acquisition card for multiple analog channel simultaneous sampling and analog-to-digital conversion) is triggered by a proximity switch at a known mill rotation angle, with the time of data acquisition and sampling rate controlled by the user. Transducers are spaced close enough that they sample correlated vibration events, but far enough apart to result in detectable differences in event arrival times for the sensor system. These considerations are a function of the physical dimensions of a typical industrial mill (order of several metres); vibration wave speeds encountered (of the order of 3000 ms^{-1} for "Rayleigh-like" waves in steel); and the effective upper frequency limit of the sensor system. All these considerations lead to a sensor spacing interval of typically one metre resulting in a ~0.1 metre theoretical resolution based on the effective upper frequency limit of the system. Further details of the AE monitoring system hardware can be found in [14,15].

MULTIPLE SENSOR AE MONITORING SOFTWARE SYSTEM

The general goal of the AE analysis system is to characterise vibration sources using data collected by multiple transducers at different locations simultaneously. Specifically, the inverse problem of locating AE source events from transducer vibrations is of great interest to establish the location and strength of impacts on the mill liner and to test if the location assumption of single-sensor analysis is appropriate for industrial tumbling mills.

Vibration Event Characterisation

The AE signatures of single impact events were monitored during "tap tests" on a number of stationary tumbling mills. Each "tap test" generally consisted of a sequence of well-separated pulse impacts by rock or steel (sledge hammer) on the mill liner at a known position. Figure 2 shows the complex vibration signature typical of a single impact source event. This type of signature, representing an incoherent superposition of multiple, possibly reflected/refracted waves of different types with different speeds and propagation paths by a resonant sensor, is typical of that seen in non-destructive testing of engineering structures subject to stress pulse sources [9,16]. It is representative of a convolution in the time domain of source, propagation path (mill geometry and composition), and sensor effects. Analysis of individual vibration source events on an operating tumbling mill requires a way to decompose each transducer signal into sequences of multiple vibrations (emission "bursts") representative of discrete source events. This is achieved by imposing a threshold background "noise" energy level ("confusion limit") on the signal. In practice, separate energy threshold levels are often required for positive and negative excursions in accord with any DC bias in the electronics that produces a positive or negative signal. An automated search is made of the signal for consecutive nulls in the filtered signal with a sequence time length greater than a user-defined bound. Such sequences are taken as being between individual source events, ie, the individual events are assumed well-separated with a low intervening level of background "noise". A "typical" single impact in a "tap test" produces a transducer response of a few tens of milliseconds (see Figure 2). A separation of 1 ms between individual source events reliably allows individual tap impacts to be sensed without being decomposed into sub-event individual vibrations. However, for impacts during operation of an industrial tumbling mill, higher levels of background "noise" due to many events throughout the mill and higher energy impacts more closely spaced in time must be considered. This has so far led to an individual event separation time of ~100 µs (ten samples of a signal scanned at 10^5 scans/s).

Figure 3 shows two "typical" revolutions of AG/SAG mill surface vibration data. The periodicity of the signal over a mill rotation period (~5.4 sec. at ~11.1 rpm rotation rate) is readily apparent. However, as expected in an operating mill containing large amounts of grinding media, individual impact events are not readily apparent on a time scale of the order of the mill rotation period. Figure 4 shows a sub-set of the same data on a smaller time scale concentrated around the probable "toe" position of the charge. Individual impacts characterised by a rapid rise to a peak response followed by a "wake" of declining response are apparent. As well, clearly some impacts are overlaid due to being almost simultaneous in time of origin or at least simultaneous sensing by the transducer. However, for the most part individual events can still be discriminated with sufficiently fine separation in time. Thus individual AE source events can be identified even at positions in a mill rotation cycle where there are expected to be the largest number of almost simultaneous grinding media impacts.

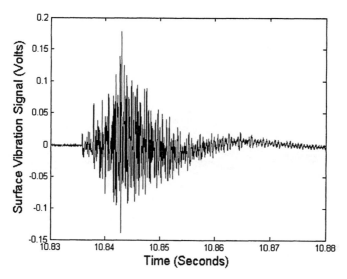

Figure 2. Raw signal from a single impact event on the mill liner as sensed at an individual transducer on the mill shell.

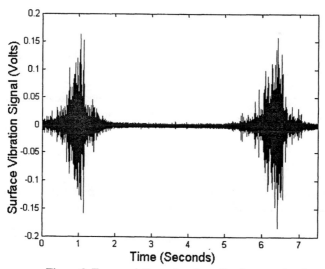

Figure 3. Two revolutions of surface vibration raw signal.

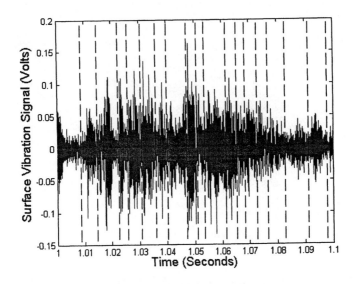

Figure 4. A "toe" region sub-set of a single revolution of surface vibration raw signal. Dashed lines indicate event boundaries.

Some Properties of Vibrations in Elastic Media

In seismology and AE inspection of structures, the dominant wave mode at sufficiently large distances from a vibration source is usually considered to be the Rayleigh surface acoustic wave [5, Ch. 5–7], [16], [18, Ch. 5] and [19, Ch. 8]. This is because geometric attenuation of a wavefront only occurs in two dimensions for surface waves, rather than in three dimensions for acoustic body waves. The energy associated with a surface wave decays because of geometric attenuation as r^{-1} (the inverse of the distance) from the source, body waves decay as r^{-2} and head waves as r^{-4} [5, Ch.6]. Attenuation of vibration waves occurs because of the following [5,9]:

1) Geometric spreading of the wavefront.
2) Internal friction (usually characterised by a damping coefficient with exponential decay of the wave with distance from the source).
3) Dissipation of energy into adjacent media.
4) Velocity dispersion with waves spreading due to variation of propagation speed with frequency.

Rayleigh surface waves are described by Rayleigh's solution to the wave equation for a semi-infinite medium i.e. propagation along a free surface [5,18]. The waves result from coupling of longitudinal and shear acoustic waves at a free surface. With Rayleigh waves, displacement decays exponentially with distance from the medium free surface, so the waves are confined to within a few wavelengths of the surface along which they propagate. Rayleigh waves are non-dispersive (propagation velocity is independent of frequency) for a medium with homogeneous (depth independent) properties. Hence the detection of the arrival of Rayleigh waves at multiple transducers can be used to accurately determine the position of a source event. Rayleigh wave velocity can readily be calculated as a function of the elastic properties of the medium [18, Ch. 5]. This leads to an estimate of the Rayleigh wave velocity in low-alloy cast steel (typical of tumbling mill liners) of $v_R \approx 2914$ ms^{-1} (3000 ms^{-1} is often quoted as the Rayleigh wave speed in steel [9]). Lamb (plate) waves propagate between two parallel surfaces and consist of a combination of symmetric and anti-symmetric modes with respect to the bounding surfaces that occur throughout the thickness of the material. These waves are dispersive such that the propagation velocity is a complex function of plate thickness and wave frequency. Accurate estimates of wave velocity and propagation path are essential to triangulate the source location of an individual impact event.

Vibration waves in real structures of finite dimensions are a combination of "pure" acoustic modes. For thick plates at high vibration frequencies, the propagation of Lamb waves tends to the speed of Rayleigh waves. A useful guide in this matter is wave group velocity as a function of the product of wave frequency and plate thickness (see [9] for a plot of the dispersion curve of the first four Lamb wave modes in steel). The zero anti-symmetric Lamb wave mode is usually dominant followed in importance by the zero symmetric mode. For a wave frequency/plate thickness product of ~1 mm MHz or greater, the zero order anti-symmetric Lamb mode can reasonably be treated as a "Rayleigh-like" wave. For industrial tumbling mills with a steel liner thickness of ~ 0.1 to 0.2 m., this implies that Rayleigh waves are only detectable at frequencies greater than ~10 to 5 kHz respectively. For the accelerometer, pre-amplifier and radio transmission system currently used in this project, this means that only the propagation of high frequency waves in the 10–20 kHz range can reasonably be considered as non-dispersive. Assuming the Rayleigh wave propagation speed, the calculated wavelength is ~0.30–0.15 m. for frequencies of ~10-20 kHz. Thus the AE waves in a mill liner can definitely only be thought of as "'Rayleigh-like" and there will be leakage of energy across the boundary with the backing layer or shell. Another consideration is that liner surface waves will also be dampened by contact with the charge slurry (termed Stonely waves as in [18, Ch. 5]) and the grinding media. In other words, at sufficiently high frequencies, it is reasonable to model vibration waves in tumbling mills as damped "Rayleigh-like" surface waves propagating at a constant velocity. Hence event arrival time difference can be used as the basis of a source location calculation. Application of a high pass filter may be useful in this regard as a pre-processing step to isolate high frequency wave components provided the filter accurately maintains phase information.

Vibration Event Arrival Time Estimation

AE source location relies on accurate determination of the arrival time of a wave train associated with an individual event at each transducer. Source location studies in AE inspection of structures have shown that the most reliable measure of arrival time is the rising edge of the wave train, which in some instances is associated with the arrival of a Primary (P) compressional body stress wave [8,9,11,16]. In this view only the leading edge of the waveform associated with an impact event is useful to calculate the source position. In this study it is assumed that the first detectable rising edge associated with a vibration event is due to a "Rayleigh-like" wave with a known (space and time invariant) propagation velocity.

A variety of algorithms have been tested to extract an AE event arrival time. The most accurate and unambiguous approach has been an adaptation of a method developed for AE testing of structures [11]. The following is a summary of the steps in this process subsequent to the identification of individual events in each transducer signal:

1. **Background removal** based on subtraction of the local mean signal associated with each event (reduction of signal bias).

2. **Smoothing** of the background subtracted signal associated with each individual event by a moving average method. Adaptive moving average filtering based on the entire signal has been advocated [11], but this is extremely computationally expensive (and probably unwarranted) in our application. A smoothing half-width of 5 data points has been found to be reliable for AG/SAG mill AE signals.

3. **Calculation of the second derivative** of the background subtracted, smoothed signal using a second order finite difference formula and subtraction of a weighted version of this second derivative from the processed signal (Laplacian filtering). This step is performed in order to emphasise regions in which the signal is changing from background "noise" to the advent of an event wave train. A weighting parameter has been determined which is independent of the transducer or test.

4. **Location of the leading edge of the first peak** in the filtered signal using a user-defined threshold level to define the leading edge. The value of this parameter is determined for each transducer from "tap tests".

5. **Arrival time identification** as the first filtered signal extremum point prior to the identified event leading edge.

Transducer Vibration Event Cross-Correlation

In order to calculate arrival time differences for specific AE events as seen at different transducers, it is necessary to determine which vibration sequences are in fact due to the same source event. This is achieved by time windowing based on the distance between the transducers and the estimated speed of Rayleigh waves on the surface of the elastic medium (group-velocity windowing [5, Ch. 11]. Hence, events at a reference transducer A can only be cross-correlated with events registered at transducers B and C within a certain event interval. The closest arrival time of an event at B and C to the reference event at A is deemed the matching counterpart signal. Event correlation is performed for every event registered at the reference transducer.

Formulation and Solution of the Inverse Problem for AE Source Event Location

Surface wave propagation around the mill liner/shell is modelled as circular waves spreading from each source location. The location of each source is effectively two-dimensional, confined to the liner surface that is modelled as a cylinder. The propagation path of the waves of interest is taken to be the shortest distance between source and transducer on a cylindrical shell. This model does not account for reflection/refraction of waves at boundaries of liner plates (between materials of different acoustic impedance) or for details of liner/lifter bar geometry. Multiple source events result in an incoherent superposition of waves on the liner surface at any given time. Figure 5 is a schematic diagram of wave propagation from a point source on a tumbling mill liner. Positions on the liner are defined in terms of coordinates (θ, Z), respectively representing angular and axial location on the cylinder.

The non-linear equations describing the relationship between each event arrival time difference, unknown source location and known transducer location (assuming three transducers in a triangular array) can be summarised as follows:

$$V_R \Delta t_{AB} = \sqrt{\Delta x_{AS}^2 + \Delta z_{AS}^2} - \sqrt{\Delta x_{BS}^2 + \Delta z_{BS}^2} \qquad (1)$$

and

$$V_R \Delta t_{AC} = \sqrt{\Delta x_{AS}^2 + \Delta z_{AS}^2} - \sqrt{\Delta x_{CS}^2 + \Delta z_{CS}^2}. \qquad (2)$$

The arrival time and position differences are defined by:

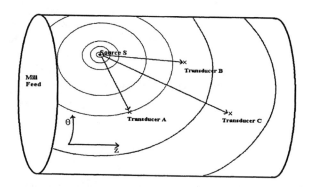

Figure 5. Schematic diagram of AE wave propagation from source to transducer on a tumbling mill.

$$\Delta t_{AB} = t_A - t_B \qquad \Delta t_{AC} = t_A - t_C \qquad (3)$$
$$\Delta x_{AS} = x_A - x_S \qquad \Delta x_{BS} = x_B - x_S$$

and

$$\Delta x_{CS} = x_C - x_S \qquad \Delta z_{AS} = z_A - z_S$$
$$\Delta z_{BS} = z_B - z_S \qquad \Delta z_{CS} = z_C - z_S.$$

Here v_R is the AE wave velocity (assumed independent of frequency for a "Rayleigh-like" wave), and t_A, t_B and t_C are the previously estimated known event arrival times at transducers A, B and C. The pairs (x_A, z_A), (x_B, z_B), (x_C, z_C) and (x_S, z_S) are the coordinates on the unrolled cylindrical surface representative of the mill liner/shell for transducers A, B and C, and for unknown source position S ($x_S = R_L \theta_S$ where R_L is the known inner radius of the mill liner). The system of non-linear equations (1) and (2) are solved for the unknown source location (x_S, z_S) of each AE event identified in the reference transducer raw signal and correlated with events sensed at the other two transducers. This is a well-posed inverse problem in that the available number of equations uniquely determines the solution. A minimum of at least three transducers is required to provide a unique solution for source position in two-dimensional geometry, based on a single arrival time measure for each vibration event. The solution procedure is the "fsolve" routine in the MATLAB Optimization Toolbox for finding the root (zero) of a system of non-linear equations. An analytical Jacobian associated with the objective function to be minimised (based on partial derivatives of Eq. 1 and Eq. 2 with respect to the unknowns x_s, z_s) is provided to the "fsolve" routine in order to speed convergence. The Gauss-Newton method forms the basis of the optimisation technique.

Only fully converged solutions are accepted for calculation of event statistics and graphical display. Solutions are filtered both for events with calculated source coordinates within the physical bounds of the unrolled cylindrical surface representative of the mill and for peak energy above a user-defined bound. Only relatively large events are considered to be able to be resolved accurately enough for source location to be undertaken. The event energy is estimated as the peak energy (peak sampled squared voltage of the event raw signal) associated with the event as sensed by the transducer closest to the source. More sophisticated models could be formulated that attempt to take into account geometric attenuation and damping during propagation of the signal from the source to the nearest transducer. For an operating mill, the rotation rate is also taken into account when calculating event location as seen by an external observer in a non-rotating (with respect to the earth) reference frame.

Tests of the AE Event Point Location Software System

Tests of the AE event location software were first conducted utilising an artificially generated set of data (modulated sawtooth waves analogous to a single revolution of tumbling mill AE data). A known time offset was introduced between three sets of waves representative of data collected at three transducers arranged in a triangular manner on a cylindrical surface. The peaks in the signal representative of vibration events were offset in time for each signal by the same amount for the entire wave train. This means that the signal could be interpreted as a sequence of source vibration events (here identified as a single oscillation of the AE signal) generated at a fixed (known) spatial position with respect to the sensors on a rotating mill. The artificial waveforms were sampled at a rate of 10^5 scans/second and a value of ~3000 ms^{-1} was assumed for the AE wave speed. The separation between each pseudo-event was ~0.1 seconds. The mill was taken to be of 5 metres radius and axial length. The event location routine calculated the position of several hundred source events to within ~1% for a specified source position 3 metres axially and azimuthally from the most distant sensor of a 1 metre vertex equilateral triangle of transducers. An interesting observation is that if the raw signal sampling rate is lowered below 10^5 scans/second, the source position calculation begins to fail for the previously stated wave propagation speed. Lowering the wave speed makes the problem easier and permits a lower signal sampling rate.

Event location analysis has been performed for "tap tests" undertaken on a number of stationary industrial AG/SAG mills. Impacts on the mill liners were at a variety of positions both internal and external to the boundaries of the transducer array. In all these tests the triangular transducer array on the shell had a vertex length of order one metre. As expected, vibrations are generally increasingly damped as the distance between the source and the transducer is increased. However, individual impact events on the liner of a stationary (quiet) mill can be detected at circumferential/axial distances of up to 3 metres by an accelerometer mounted on the shell end of a liner/lifter bolt. As expected, the rising edge of the wave train is the only reliable quantitative measure of vibration event arrival time for AE monitoring of impacts on a tumbling mill liner. Vibration arrival time differences (and source position location for a fixed wave propagation speed) are repeatable for both rock and steel impacts. The positions of "tap test" impacts within or near the transducer array (near-field) are reasonably accurately calculated, with an error of less than ~10–20 cm in comparison to the known source positions, by use of the arrival time extraction algorithm previously outlined. The same accuracy roughly applies for rock impacts at similar positions. This is in agreement with the expected error in source position calculation and a good result considering that industrial AG/SAG mills usually have a circumference of ~10–20 metres. In all cases, individual events were successfully identified, cross-correlated between transducers and source locations estimated. However, source location estimates for impacts well separated from the transducer array (far-field) are not as accurate although still reasonable. This reflects the difficulties of arrival time calculation for events with a small signal-to-noise ratio. Other considerations are the inherently increasing error in calculation of source locations at large distances with transducers located relatively close to one another and limited distance resolution due to relatively low effective and sampling rates and the high speed of AE wave propagation in metal liners. In general, "tap tests" show that event location can reliably identify well-separated near- and far-field vibration source events.

For "tap tests" on one SAG mill it was apparent that a systematic error existed in the estimation of AE source position from arrival time differences. This error appears to be due to a significant deviation of the AE wave propagation group velocity from the value for a Rayleigh wave in steel. It is hypothesised that this is due to the mill liner/shell construction being thin/complex enough to result in the first detected component of the AE wave train for each event (on which calculation of source location is based) deviating substantially from "Rayleigh-like" behaviour. The problem was addressed by finding an optimal wave speed to minimise errors between the calculated and actual difference in wave propagation distance between the source and the various transducers. Least-squares minimisation was used to optimise the wave group velocity for a sequence of "tap tests" at a variety of spatial positions with respect to the transducer array. More generally, "tap tests" at known impact positions can be utilised in a calibration procedure for calculation of the effective AE group velocity (associated with the first detected part of a vibration event wave train) on any particular tumbling mill. This procedure would be performed prior to prediction of the unknown positions of vibration sources during mill operation.

AE EVENT POINT LOCATION ON AN OPERATING TUMBLING MILL

AE event point source location was performed for vibration data collected on a plant trial of the monitoring system at an industrial tumbling mill. In each case ten consecutive revolutions of data from an array of three transducers were collected at a scanning rate of 10^5 scans/second. The relevant AG mill has a ~4 metre internal (liner) radius and a ~4.5 metre effective grinding (axial) length. The process of three-sensor data analysis enables large vibration event location, duration, rise-time and energy statistics to be collected and analysed in terms of information they can provide about the process and machine condition.

A three-dimensional perspective plot of the calculated position of AE source events ("hit map") over a revolution is shown in Figure 6, where source event positions are represented by crosses, marker size coded (small impact—small red cross, large impact—large dark blue cross). This schematic shows the location and intensity of AE source events on the mill liner, as viewed by an external observer in a non-rotating (with respect to the earth) reference frame. The positions of several hundred individual relatively large AE source events have been identified and plotted. In this instance the mill is operating at relatively high power draft (interpreted as being heavily loaded) and with a high feed rate. It is immediately apparent that the source events are largely local to the location of the transducers, being within ±1–2 m in the axial direction. As expected, large AE source events are within a restricted range of circumferential position angles, inferred to be associated with the "toe" position (the mill is rotating counter-clockwise viewed from the discharge end). Interestingly, the AE source event energy weighted average "toe" angle over ten revolutions is within a degree of that calculated by single-sensor analysis techniques.

This is also the case for data collected at substantially different (low power/charge load) operating conditions. This result supports the view that the vibration localisation assumption (at least in the circumferential direction) is valid for this particular mill. As expected, a slightly higher "toe" position occurs when the mill is more heavily loaded. Figure 6 also shows what appears to be a population of smaller events located above the "toe" position. This may indicate some direct impacts of cataracting grinding media on the liner rather than on the rest of the charge—a source of enhanced liner wear. The energy weighted average axial position of AE events over ten revolutions is near the centre of the scanning array of transducers (approximately 1–1.2 metres from the feed end of the mill) and is slightly further towards the discharge end of the mill at low power operating conditions. However, these results could be due to bias associated with preferred sampling of vibration events with sources near the small transducer array (total damping of axially more distant events) and a relatively crude model for estimation of source energy.

CONCLUSION

A multiple sensor surface vibration analysis system has been developed for investigation of mineral processing tumbling mill performance. High sampling rate simultaneous data acquisition from three transducers mounted in a triangular array on the rotating mill shell allow vibration arrival time differences to be used for point location of source AE events. The circumferential /axial location and strength of large AE source events on a mill cylindrical liner can be reliably estimated. Inferential (soft-sensor) estimates can be made of "hidden" process and performance variables such as charge toe and shoulder positions, mill volumetric loading, grinding medium content and liner wear as a function of operating conditions. The system has been successfully tested on a number of industrial AG/SAG mills.

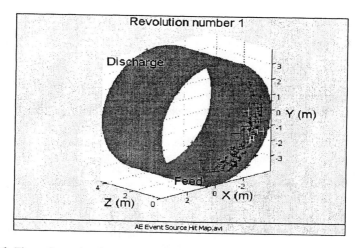

Figure 6. Three-dimensional perspective plot of calculated positions of large AE source events ("hit map") per revolution for an AG mill. Crosses—estimated impact positions (small energy impact—small red cross, large energy impact—large dark blue cross). Mill rotating counter-clockwise viewed from the discharge end. Double-click on picture to show a perspective plot movie of ten consecutive revolutions of AE source event positions.

An automated signal analysis system has been developed that includes identification of wave trains associated with AE discrete events, feature extraction of event arrival time, and cross-correlation of events detected at three different transducers. The inverse problem for individual AE event source location is formulated and solved utilising a model for propagation of circular ("Rayleigh-like") waves at constant speed on a cylindrical surface representative of the mill liner. This appears to be a reasonable first approximation to the behaviour of AE waves in a number of industrial AG/SAG mills.

Three-sensor surface vibration discrete event analysis has been developed as an adjunct to a single-sensor system, for situations in which AE waves propagate large distances around a tumbling mill. It is expected that multiple sensor vibration analysis will be further developed as part of an on-line AE analysis system for monitoring and control of tumbling mill operations. It is further anticipated that AE monitoring, coupled with the development of appropriate signal processing techniques and soft-sensor analysis models, will eventually have broad application in monitoring and control of a large range of mineral processing and materials manufacturing devices.

ACKNOWLEDGEMENT

The authors gratefully acknowledge the financial support of this work as part of AMIRA project P667 by Anglo Platinum Corporation Limited, Codelco-Chile, MIM Holdings Limited, Minera Los Pelambres, OK Tedi Mining Limited, Pasminco Limited, Phelps Dodge Corporation, Rio Tinto Limited and WMC Resources Limited. AMIRA is thanked for permission to publish this work.

REFERENCES

[1] B.A. Wills, 1992. Mineral Processing Technology. Pergamon Press.
[2] Y. Zeng and E. Forssberg, 1993. Monitoring grinding parameters by signal measurements for an industrial ball mill, International Journal of Mineral Processing, 40, 1–16.
[3] Y. Zeng, and E. Forssberg, 1995. "Multivariate statistical analysis of vibration signals from industrial scale ball milling". Minerals Engineering, 8, 389–399.
[4] F. Cartes, G. Rivas, F. Mujica, K. Schwarze, 1995, "On-line estimation of the inclination of the charge in a SAG mill". Copper-95, Intern. Conf., A. Casali, G. Dobby, M. Molina, W. Thorburn, (Eds.), 2, 225–232.
[5] K. Aki and P.G. Richards, 1980. Quantitative Seismology—Theory and Methods. W H Freeman—San Francisco.
[6] M. Kat and F.P. Hassani, 1989. "Application of Acoustic Emission for the Evaluation of Microseismic Source Location Techniques". Journal of Acoustic Emission, Vol. 8, 4, 99–106.
[7] Y.H. Pao, 1978. "Theory of Acoustic Emission" in Elastic Waves and Non-Destructive Testing of Materials, AMD-20, American Society of Mechanical Engineers, 107–128.

[8] C.B. Scruby, 1985. "Quantitative Acoustic Emission Techniques", Ch. 4 in Non-Destructive Testing, Vol. 8. Academic Press (London).
[9] A.A. Pollock, 1986. "Classical Wave Theory in Practical AE Testing" in Progress in Acoustic Emission III, The Japanese Society of NDI, 708–721.
[10] M.P. Collins and R.M. Belchamber, 1990. Acoustic Emission Source Location Using Simplex Optimization, Journal of Acoustic Emission, 9(4), 271–276.
[11] E. Landis, C. Ouyang, and S.P. Shah, 1991. Automated Determination of First P-Wave Arrival and Acoustic Emission Source Location, Journal of Acoustic Emission, 10(1), S97-S103.
[12] S.J. Spencer, J.J. Campbell, K.R. Weller and Y. Liu, 1999. Acoustic Emissions Monitoring of SAG Mill Performance, Proceedings of the Intelligent Processing and Manufacturing of Materials Conference, Hawaii, July 1999, 939–946.
[13] S.J. Spencer, J.J. Campbell, K.R. Weller and Y. Liu, 2000. Monitoring of SAG Mill Performance Using Acoustic Emissions, Proceedings of the XXI International Mineral Processing Congress, Rome, July 2000, A4-13-A4-20.
[14] J.J. Campbell, S.J. Spencer, D.N. Sutherland, T. Rowlands, K.R. Weller, P. Cleary and A. Hinde, 2001. SAG Mill Monitoring using Surface Vibrations, Proceedings of the International Conference on Autogenous and Semiautogenous Grinding Technology, 30 Aug – 3 Sept, 2001, Vancouver, Canada, 2, 373–385.
[15] J.J. Campbell, R.J. Holmes, S.J. Spencer, V. Sharp, K.J. Davey, D.G. Barker, P.L. Phillips, 2003. The Collection and Analysis of Single Sensor Surface Vibration Data to Estimate Operating Conditions in Pilot-Scale and Production-Scale AG/SAG Mills, Proceedings of the XXII International Mineral Processing Congress, Capetown, September–October 2003, accepted.
[16] A.A. Pollock, 1989. "Acoustic Emission Inspection" in ASM Handbook—Non-Destructive Testing and Quality Control, ASM International, A17, 278–294.
[17] D.J. Yoon, Y.H. Kim, and O.Y. Kwon, 1990. New Algorithm for Acoustic Emission Source Location in Cylindrical Structures, Journal of Acoustic Emission, 9(4), 237–242.
[18] J.D. Achenbach, 1973. Wave Propagation in Elastic Solids. North-Holland—Amsterdam.
[19] J.D.N. Cheeke, 2002. Fundamentals and Applications of Ultrasonic Waves. CRC Press—London.

Fracture Toughness and Surface Energies of Covalent Materials: Theoretical Estimates and Application to Comminution

D. TROMANS

Department of Metals and Materials Engineering, University of British Columbia, Vancouver V6T 1W5, British Columbia, Canada

J. A. MEECH

Department of Mining Engineering, University of British Columbia, Vancouver V6T 1W5, British Columbia, Canada

ABSTRACT

Theoretical estimates of the ideal brittle fracture toughness and surface energies of 15 single phase covalent minerals and materials have been obtained, using analyses based on a Morse-type model of bonding. The materials studied covered a toughness spectrum from ice to diamond, and included polymorphs of C, SiC, Si_3N_4, and SiO_2, plus BN (borazone) and MoS_2 (molybdenite). Quantitative development of the toughness model required calculation of the crystal binding enthalpy from thermodynamic data, and the use of published elastic compliance and stiffness constants of single crystals. The computed critical stress intensity values, K_{IC}, for propagation of a macroscopically planar transgranular crack, of random orientation, ranged from ~0.11 MPa $m^{1/2}$ (ice) to ~3.3 MPa $m^{1/2}$ (diamond); a difference factor of ~30 times. The corresponding critical energy release rates, G_{IC}, ranged from ~1.21 J m^{-2} to ~6.8 J m^{-2}. Application of the model to estimate the cleavage toughness, $(K_{IC})_{hkl}$, of commonly observed (hkl) cleavage planes in monocrystals indicated $(K_{IC})_{hkl} < K_{IC}$, consistent with expectations. In addition, the study showed that the critical stress intensity $(K_{IC})_{Gb}$ and energy release rate $(G_{IC})_{Gb}$ for fracture along high angle grain boundaries are lower than their K_{IC} and G_{IC} counterparts. Thus, intergranular cracking of suitably oriented grain boundaries is likely in polycrystalline material. However, recognising the limitations and assumptions in the model, it is considered that K_{IC} and G_{IC} values are sufficiently adequate for general application to mineral comminution processes (crushing and grinding). The results are discussed with relevance to comminution, giving particular attention to energy efficiency, factors affecting the likelihood of particle fracture during impact loading, and the influence of K_{IC} on average particle size during ultrafine grinding. In particular, using the Bond work index, it is concluded that the energy efficiency of comminution is very low, of the order of only ~1%, based on the energy require to generate new surface relative to the energy consumed.

INTRODUCTION

For economic and environmental reasons, there is a compelling need to improve the energy efficiency of crushing and grinding operations (comminution) during the industrial milling of minerals where it has been estimated that only 1–2% of the total energy consumed is used for particle fracture [1]. In order to understand the origins of this large energy inefficiency, and to improve the efficiency of comminution processes, a fundamental research program has been undertaken to examine the principal factors controlling fracture of mineral particles.

During comminution, particle fracture initiates at pre-existing flaws (cracks) which propagate in response to local tensile stress components acting normal to the crack plane. Such stresses are generated even when the external loading on the particle is predominantly compressive [2,3]. The strain energy released during crack propagation is accompanied by generation of new surface area (crack surfaces). Resistance to crack propagation under crack opening (Mode I) conditions, the usual situation, is termed the fracture toughness G_{IC}. It is defined as the critical energy release rate per unit area of crack plane (J m^{-2}) that is necessary for crack propagation and is related to the Mode I stress intensity factor for crack propagation (K_{IC}) via Eq. (1) [1,4]:

$$K_{IC}(1-v^2) = (EG_{IC})^{1/2} \approx K_{IC} \quad \text{Pa m}^{1/2} \tag{1}$$

where E is the tensile elastic modulus (Pa), v is Poisson's ratio, and K_{IC} (Pa m$^{1/2}$) is given by Eq. 2:

$$K_{IC} = Y\sigma_c(a)^{1/2} \quad \text{Pa m}^{1/2} \tag{2}$$

In Eq. 2, σ_c is the critical tensile stress (Pa) for crack propagation, a is the flaw size (m), and Y is a shape factor related to the crack geometry, e.g. Y has the value $(\pi)^{1/2}$ for a straight through internal crack of length $2a$ and the value $2(\pi)^{-1/2}$ for an internal penny-shaped (disc-shaped) crack of radius a [4].

The $(1-v^2)$ term in Eq. 1 implies plane strain conditions, which is the usual situation for brittle fracture. For ideal brittle fracture (negligible plastic deformation at the crack tip), G_{IC} is equivalent to 2γ, where γ is the surface energy per unit area (J m^{-2}). Consequently, a higher γ should lead to increased toughness of brittle materials (*e.g.* minerals). In the ensuing text it is implicit that γ is to be understood as the ideal fracture surface energy because of the manner in which its value is estimated. Frequently, K_{IC} and G_{IC} are used interchangeably as the measure of toughness, because 1) they are directly related via Eq. 1, and 2) experimental measurement of K_{IC} is less difficult than G_{IC}.

It is evident that development of quantitative comminution models, for purposes of power consumption and particle fracture, should include the fracture toughness of the minerals involved. Other workers have recognised the likely importance of fracture toughness in this regard [5-8], but there has been insufficient knowledge of the K_{IC} and G_{IC} values of individual mineral phases to allow satisfactory pursuit of this approach. Reviews by Rummel [9] and Zhang [10] provide useful but limited toughness data that are concerned principally with rock bodies (mixed or poorly defined mineral phases).

The dearth of toughness information on mineral phases is undoubtedly related to specimen preparation difficulties and problems encountered while attempting to conduct standardised toughness tests [10], plus the inability to obtain a mineral sample of sufficient size when it exists either as a dispersed phase or fine particulate. To avoid these problems, the authors adopted a theoretical approach whereby minerals are treated as essentially brittle materials and a basic bonding model is used to estimate their fracture toughness and surface energy when stretched to fracture. By applying an ionic bonding model (Born model) to applicable minerals, together with published thermodynamic data and relevant elastic constants and elastic moduli, the authors obtained the fracture toughness and surface energy of forty-eight minerals [1]. The current study is concerned with utilisation of a suitable covalent bonding model, together with thermodynamic and elastic property data, to predict the toughness and fracture surface energy of minerals that are more properly treated as covalent crystals.

COVALENT BONDING MODEL

In general, the bonding spectrum of minerals contains species whose chemical bonding characteristics may vary from being predominantly ionic to predominantly covalent. Examples of minerals suitable for treatment in terms of dominant ionic bonding have been considered previously [1]. Those that may be treated principally as covalent minerals are likely to be composed of atoms whose electronegativity differences are near or less than ~1.7 [11]. These include silicon nitride [12,14], boron nitride [12,15], silicon carbide [12], and the polymorphs of carbon [12] and silicon dioxide [12]. In addition, based on electronegativity considerations, it is expected that molybdenum sulphide and ice may be treated as covalent solids to a first approximation.

The Morse equation, based on an exact solution of the Schroedinger wave equation, provides a reasonable approximation of the effect of interatomic separation distance r on the bonding energy (potential energy) U_r between two covalently bonded atoms in a diatomic molecule [16]:

$$U_r = U_{r_o}[2\exp\alpha_a(r_o - r) - \exp 2\alpha_a(r_o - r)] \qquad (3)$$

where α_a is a constant for the particular atom pair, and U_{r_o} is the bonding energy (minimum potential energy) at the equilibrium separation distance r_o (note that $U_{r_o} < 0$).

Application of Eq. 3 to the toughness modelling of covalent crystals requires the equation to be recast in terms of the total number of atoms, N, contained in a unit volume (1 m^3) of crystal at equilibrium (unstrained state). The value of N is obtained from:

$$N = a_n (N_A / M_V) \quad (4)$$

where a_n is the number of atoms per molecule (*e.g.* a_n =2 for SiC), N_A is the Avagadro number (6.023 x10^{23} mol^{-1}) and M_V is the molar volume (obtained by dividing the molecular weight by the crystal density).

The general form of Eq. 3 is now applied to the collection of N atoms by replacing r_o and r by the average equilibrium inter-atomic distance R_o (m) and the non-equilibrium inter-atomic distance R (m) in the crystal, respectively, leading to:

$$U_R = U_e [2\exp\alpha_c (R_o - R) - \exp 2\alpha_c (R_o - R)] \quad \text{J per } N \text{ atoms} \quad (5)$$

where α_c (m^{-1}) is a constant for the particular crystal, U_e is the equilibrium crystal binding energy per unit volume of unstrained crystal (N atoms), and U_R is the binding energy of the strained (non-equilibrium) crystal ($R \neq R_o$).
The value of U_e in terms of N is obtained via Eq. 6;

$$U_e = \Delta H_f / M_V = N(\Delta H_f)/(N_A a_n) \quad \text{J m}^{-3} \quad (6)$$

where ΔH_f (Jmol^{-1}) is the molar enthalpy of formation of the crystal.

Crystal Binding Energy

The required values of ΔH_f for covalent crystals of interest at 298 K were calculated from the molar enthalpies (Jmol^{-1}) of the individual component atoms (H_a) in the gas phase and the molar enthalpy of the crystal (H_{cr}) at 298 K via Eq. 7:

Table 1. Molar enthalpies of gaseous atoms and crystalline minerals at 298 K.[†]

Atom enthalpies (kJ mol^{-1})[‡]					
Atom	H_a	Atom	H_a	Atom	H_a
C	716.682	N	472.679	O (873 K)	261.374
B	564.999	O	249.174	Si	450.002
H (270 K)	217.413	O (270 K)	248.555	Si (873 K)	462.236

Crystal enthalpies (kJ mol^{-1})[‡]					
Mineral	Formula	H_{cr}	Mineral	Formula	H_{cr}
Graphite-2H	C	0	α-Quartz	α-SiO$_2$	-910.702
Diamond	C	1.895	β-Quartz (873 K)	β-SiO$_2$	-874.008
Moissanite-6H	α-SiC	-71.902	α-Cristobalite	SiO$_2$	-906.903
Moissanite-3C	β-SiC	-73.22	Stishovite	SiO$_2$	-861.318
Nierite	β-Si$_3$N$_4$	-828.901	Coesite	SiO$_2$	-905.585
Cubic Si$_3$N$_4$	γ-Si$_3$N$_4$	-787.801	Silica (quartz) glass*	SiO$_2$	-903.200
Borazone	BN	-250.500	Ice (270 K)	H$_2$O	-293.834
Molybdenite-2H	MoS$_2$	-276.144			

[†]298 K unless stated otherwise: [‡]Values from database by Roine [17]: *Amorphous.

$$\Delta H_f = H_{cr} - \Sigma H_a \quad \text{J mol}^{-1} \qquad (7)$$

where Σ indicates the summation of the enthalpies of the different atom components (e.g. in SiO$_2$, $\Sigma H_a = H_{Si} + 2H_O$).

Values of H_a and H_{cr} for the atom and crystal species considered were obtained from the thermodynamic database of Roine [17] and listed in Table 1.

Note that data for ice and β-quartz are given at 270 K and 873 K, respectively. Also, Roine [17] listed two molar enthalpy values for Si$_3$N$_4$. We assigned the lower one (-828.901 J) to nierite (β-Si$_3$N$_4$) and the upper value (-787.801 J) to γ-Si$_3$N$_4$. When this was done, the molar chemical free energy data [17] for the two polymorphs of Si$_3$N$_4$ at 298 K indicated that the assigned β-phase was lower (more stable) than the assigned γ-phase by ~40 kJ mol^{-1}. This result is consistent with the general pressure/temperature conditions governing formation of the γ-phase, which indicate γ-Si$_3$N$_4$ has the lower stability under ambient conditions (e.g. Zerr et al. [18-19] and Soignard et al. [14]). Also, theoretical energy-volume calculations suggest that the energy of the γ-phase is of the order of ~100 kJ mol^{-1} higher than the β-phase [18].

The resulting crystal energy, U_e, obtained from Eq. 6 is listed in Table 2, together with ΔH_f and M_V. Each M_V was obtained from published crystallographic data in the PDF File [20], except for silica glass and β-quartz where data from Gebrande [21] and Dana [22], respectively, were used.

Table 2. Values of ΔH_f, M_V and U_e for covalent minerals and crystals at 298 K.[†]

Mineral	Formula	Structure[‡]	ΔH_f (kJ mol^{-1})	M_V (10^{-5}m^{-3})	U_e (10^{11}J m^{-3})
Graphite-2H	C	h P6$_3$/mmc	-716.68	0.53497	-1.3397
Diamond	C	c..Fd3m	-714.786	0.34148	-2.0932
Moissanite-6H	α-SiC	h P6$_3$mc	-1238.585	1.2380	-1.0004
Moissanite-3C	β-SiC	c F$\bar{4}$3m	-1239.903	1.2469	-0.99440
Nierite	β-Si$_3$N$_4$	h P6$_3$/m	-4069.62	4.3838	-0.92834
Cubic Si$_3$N$_4$	γ-Si$_3$N$_4$	c Fd3m	-4028.52	3.4887	-1.1547
Borazone	BN	c F$\bar{4}$3m	-1288.178	0.71199	-1.8093
Molybdenite-2H	MoS$_2$	h P6$_3$/mmc	-1488.123	3.2050	-0.46431
α-Quartz	α-SiO$_2$	trig P3$_2$21	-1859.05	2.2680	-0.81968
β-Quartz (873 K)	β-SiO$_2$	h P6$_4$22	-1859	2.3710*	-0.78407
α-Cristobalite	SiO$_2$	tet P4$_1$2$_1$2	-1855.253	2.5774	-0.71981
Stishovite	SiO$_2$	tet P4$_2$/mnm	-1809.67	1.4018	-1.2910
Coesite	SiO$_2$	m P2$_1$/a	-1853.93	2.0746	-0.89364
Silica glass	SiO$_2$	amorphous	-1851.55	2.3108**	-0.80127
Ice (270 K)	H$_2$O	h P6$_3$/mmc	-977.215	1.9356	-0.50488

[†]298 K unless stated otherwise: *Data from Dana [22]: **Data from Gerbrande [21]
[‡]c = cubic: h = hexagonal: trig = trigonal: tet = tetragonal: m = monoclinic.

Bulk Modulus of Elasticity

The N atoms of crystalline material are contained in a volume V, where $V = NR^3$ ($V \to 1$ m^3 as $R \to R_o$). The bulk modulus B (Pa) is given by the usual definition [1,23]:

$$B = -V\left(\frac{\partial \sigma_h}{\partial V}\right) = V\left(\frac{\partial^2 U_R}{\partial V^2}\right)_{R=R_o} = V\left[\frac{\partial^2 U_R}{\partial R^2}\left(\frac{\partial R}{\partial V}\right)^2\right]_{R=R_o} \quad \text{Pa} \quad (8)$$

where σ_h is the hydrostatic compression stress (pressure) and $\sigma_h = -\partial U_R/\partial V$.

The second differential of Eq. 5 yields $\partial^2 U_R/\partial R^2$, which is inserted in Eq. 8 together with $V = NR^3$, $\partial R/\partial V = (3R^2N)^{-1}$, and $R = R_o$ to give the final equation for B:

$$B = \left[\frac{-2(\alpha_c)^2 U_e}{9NR_o}\right] \quad \text{Pa} \quad (9)$$

Following previous work [1], the isotropic value of B (equivalent to the averaged modulus for a polycrystalline aggregate), is obtained from the corresponding isotropic tensile elastic modulus E and Poisson's ratio v via Eq. 10 [24,25]:

$$B = \frac{E}{3(1-2v)} \quad \text{Pa} \quad (10)$$

Rearrangement of Eq. 9, followed by insertion of Eq. 10, allows α_c to be obtained in terms of E:

$$\alpha_c = \left[\frac{-3ENR_o}{2U_e(1-2v)}\right]^{1/2} \quad \text{m}^{-1} \quad (11)$$

Computed values of α_c, based on Eq. 11 and U_e from Table 2, are listed in Table 3 for the minerals studied, together with the corresponding N, R_o, E, and v values.

At this point, all the necessary parameters controlling R-U_R behaviour in Eq. 5 have been determined. Examples of such behaviour, computed via Eq. 5, are shown for diamond and moissanite-6H in Figure 1, where the equilibrium (unstrained) values of R_o and U_e for each mineral are indicated on the diagram.

Table 3. Values of N, R_o, E, ν, α_c and σ_{max}/E ratio for covalent minerals and crystals at 298 K.[†]

Mineral	Formula	N (10^{28} m^{-3})	R_o (10^{-10}m)	E (GPa)	ν	α_c (10^{10}m^{-1})	σ_{max}/E
Graphite-2H	C	11.25865	2.07096	274.84[a]	0.202	1.0979	0.11
Diamond	C	17.63771	1.78315	1099.6[a]	0.100	1.7603	0.08
Moissanite-6H	α-SiC	9.72992	2.17419	454.54[a]	0.169	1.4757	0.078
Moissanite-3C	β-SiC	9.66083	2.17936	436.20[a]	0.170	1.4492	0.079
Nierite	β-Si$_3$N$_4$	9.61756	2.18263	306[f]	0.286	1.5562	0.074
Cubic Si$_3$N$_4$	γ-Si$_3$N$_4$	12.08507	2.02264	379[h]	0.280	1.6553	0.075
Borazone	BN	16.91876	1.80805	909.05[d]	0.121	1.7442	0.079
Molybdenite-2H	MoS$_2$	5.63767	2.60796	113.43[b]	0.1	0.82062	0.117
α-Quartz	α-SiO$_2$	7.96684	2.32401	95.5[a]	0.078	0.61936	0.174
β-Quartz (873K)	β-SiO$_2$	7.62099	2.35865	99.76[a]	0.205	0.76195	0.139
α-Cristobalite	SiO$_2$	7.01046	2.42523	65.17[c]	-0.166	0.41648	0.248
Stishovite	SiO$_2$	12.89010	1.97962	536.16[e]	0.217	1.6763	0.075
Coesite	SiO$_2$	8.70969	2.25597	157.06[b]	0.274	1.0695	0.104
Silica glass	SiO$_2$	6.6285	2.4709	72.845[e]	0.167	0.62935	0.117
Ice (270 K)	H$_2$O	9.33532	2.20440	9.03[b]	0.326	0.39880	0.284

298 K unless stated otherwise: [a]Hearmon [26]; [b]Hearmon [27]; [c]Yeganeh-Haeri et al. [29]; [d]Grimsditch et al. [30]; [e]Bass [28]; [f]Dodd et al. [31]; [h]Zerr et al. [19].

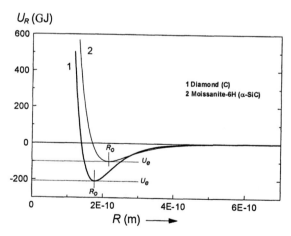

Figure 1. Computed R-U_R behaviour for diamond and moissanite-6H. Note that U_e is the equilibrium crystal binding energy per unit volume (1 m^3) when $R = R_o$.

Note that ν in Table 3 was obtained from the isotropic elastic modulus E and the isotropic elastic shear modulus μ_s for a randomly oriented polycrystalline aggregate via the relationship $\nu = (E/2\mu_s) - 1$ [24, 25]. Of the fourteen crystalline materials listed in Table 3, twelve pairs of E and μ_s moduli were computed from the anisotropic stiffness and compliance constants for single crystals. These include constants for nine minerals compiled by Hearmon [26,27]), constants for stishovite reported by Bass 28], a set of constants for α-cristobalite measured by Yeganeh-Haeri et al. [29] and a set for borazone (BN) measured by Grimsditch et al. [30]. Computations were conducted in two ways: 1) from stiffness constants (c_{mn}) assuming uniform stress in the polycrystalline aggregate (Voigt method), and 2) from compliance constants (s_{mn}) assuming uniform strain in the aggregate (Reuss method). The resulting values for each elastic modulus, E and μ_s, obtained by the Voigt and Reuss methods were then averaged (Harris procedure). The matrix and analytical procedures required for the Voigt and Reuss computations are outlined by Gebrande [21] and Wachtman [25].

Of the two remaining crystalline materials, β-Si_3N_4 and γ-Si_3N_4, experimental values of the E and μ_s moduli measured by Dodd et al. [31] and Zerr et al. [19], respectively, were used. The experimental E and μ_s moduli listed by Bass [28] were used for non-crystalline (amorphous) silica glass.

Note that the negative value for the Poisson ratio of α-cristobalite in Table 3 is a real effect arising from the values of the elastic constants c_{mn} and s_{mn}. This unusual behaviour has been explained rationally by Yeganeh-Haeri et al. [29] in terms of the detailed topology of the crystal structure of the unit cell.

Idealised Tensile Behaviour

Following previous procedures by the authors [1]), tensile behaviour is analysed for a cube of ideally brittle mineral (no plastic deformation) that is defect-free with no cracks, flaws or imperfections. The cube contains N atoms such that the initial length of the unstrained cube is $x = N^{1/3} R_o = 1$ m. Equation 12 defines the variation in uniaxial stress σ_x as the mineral aggregate is extended in the x direction [1]:

$$\sigma_x = \frac{1-2\nu}{3}\left(\frac{\partial U_R}{\partial x}\right) = \frac{1-2\nu}{3N^{1/3}}\left(\frac{\partial U_R}{\partial R}\right)_{R=R_x} \quad (12)$$

After substituting Eq. 5 for U_R, Eq. 12 becomes:

$$\sigma_x = \frac{2U_e(1-2\nu)\alpha_c}{3N^{1/3}}\{\exp[2\alpha_c(R_o - R_x)] - \exp[\alpha_c(R_o - R_x)]\} \quad \text{Pa} \quad (13)$$

where $\sigma_x \to 0$ as $R_x \to R_o$, as expected.

Based on Eq. 13, representative curves showing the R_x-σ_x behaviour for covalent materials examined as R_x increases from R_o are presented in Figs. 2 and 3.

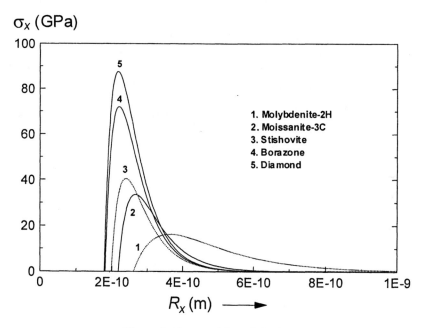

Figure 2. Computed R_x-σ_x behaviour.

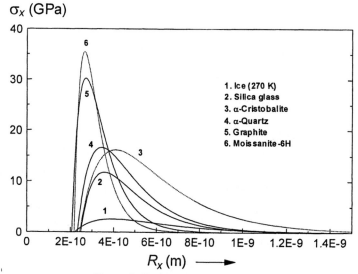

Figure 3. Computed R_x-σ_x behaviour.

Note that, initially, σ_x increases rapidly to a peak value, termed the maximum theoretical tensile fracture stress σ_{max}, after which it decreases asymptotically to zero as R_x increases. The σ_{max} values may be obtained from Eq. 13 by applying the condition $\partial\sigma_x/\partial x = 0$, to compute the value of R_x at σ_{max} (*i.e.* $R_{max} = R_o + (\ln 2)/(\alpha_c)$, and substituting R_{max} into Eq. 13. Results are listed as σ_{max}/E ratios in the last column of Table 2. The general shapes of the R_x-σ_x curves in Figs. 2 and 3 are similar to those predicted by the ionic bonding model developed previously [1]), except that the covalent model indicates that σ_x tends to decrease more rapidly after reaching σ_{max}.

Following previous analysis [1], the tensile elastic modulus E_{R_x} at any R_x is related to the differential of Eq. 13:

$$E_{R_x} = \frac{\partial \sigma_x}{\partial \varepsilon_x} = R_x \left(\frac{\partial \sigma_x}{\partial R_x}\right) = \frac{2R_x U_e(1-2\upsilon)\alpha_c^2}{3N^{1/3}}\{\exp[\alpha_c(R_o - R_x)] - 2\exp[2\alpha_c(R_o - R_x)]\} \quad \text{Pa} \tag{14}$$

where ε_x is the strain in the x-direction (*i.e.* $\partial x/x$).

If the substitutions $E_{R_x} = E$ when $R_x = R_o$ are made in Eq. 14, and the left hand side of the equation is divided by 1 m^2 while the right hand side is divided by the equivalent unit area $(N^{1/3}R_o)^2$ (recognising that $(N^{1/3}R_o)^2 = 1$ m^2), Eq. 15 is obtained:

$$E = \frac{-2U_e(1-2v)\alpha_c^2}{3NR_o} \quad \text{Pa} \tag{15}$$

After rearrangement, Eq. 15 is seen to be identical to Eq. 11, as required for analytical consistency.

FRACTURE TOUGHNESS

Transgranular Fracture

The area beneath each R_x-σ_x curve in Figure 2 is a measure of the brittle fracture toughness G_{IC} [1]. It corresponds to the average work (energy) per unit area of crack plane that is required for brittle fracture of flaw-free material. Formally, G_{IC} is expressed via the integration:

$$G_{IC} = \int_{R_x=R_o}^{R_x=R_{limit}} \sigma_x \partial R_x \quad \text{J m}^{-2} \tag{16}$$

where R_{Limit} is an upper bound for R_x beyond which the covalent model becomes invalid and σ_x becomes negligibly small.

Following the substitution of Eq. 13 for σ_x, and integration of Eq. 16:

$$G_{IC} = \left[\frac{U_e(1-2\upsilon)}{3N^{1/3}}\{2\exp[\alpha_c(R_o - R_x)] - \exp[2\alpha_c(R_o - R_x)]\}\right]_{R_x=R_o}^{R_x=R_{Limit}} \quad \text{Jm}^{-2} \quad (17)$$

Inspection of Figs. 2 and 3 indicates that R_{Limit} is reached (σ_x negligibly small) when the average separation between atoms in the x direction increases by ≤ 1.3 nm. However, for consistency and comparison with the authors ionic crystal model, where the increase is nearer to 2 nm [1]), integration is conducted for the condition $R_{Limit} = (R_o + 2 \times 10^{-9})$ m. Resulting G_{IC}, computations,

Table 4. Computed toughness values for transgranular and grain boundary cracking, plus surface and grain boundary energies of covalent minerals at 298 K (unless stated otherwwise).

Mineral	Formula	Transgranular crack		γ (J m^{-2})	Grain boundary crack		γ_{Gb} (J m^{-2})
		G_{IC} (J m^{-2})	K_{IC} (MPa m$^{1/2}$)		$(G_{IC})_{Gb}$ (J m^{-2})	$(K_{IC})_{Gb}$ (MPa m$^{1/2}$)	
Graphite-2H	C	5.5052	1.230	2.7526	4.2355	1.079	1.2697
Diamond	C	9.9503	3.308	4.9751	6.8308	2.741	3.1195
Moissanite-6H	α-SiC	4.80	1.477	2.40	3.2680	1.219	1.5320
Moissanite-3C	β-SiC	4.7654	1.442	2.3827	3.2639	1.193	1.5015
Nierite	β-Si$_3$N$_4$	2.8946	0.941	1.4473	1.9285	0.768	0.9661
Cubic Si$_3$N$_4$	γ-Si$_3$N$_4$	3.4193	1.138	1.7096	2.2906	0.932	1.1287
Borazone	BN	8.2631	2.741	4.1315	5.6625	2.269	2.6006
Molybdenite-2H	MoS$_2$	3.2294	0.605	1.6147	2.5324	0.536	0.6970
α-Quartz	α-SiO$_2$	5.3560	0.715	2.6780	4.6510	0.667	0.7050
β-Quartz (873K)	β-SiO$_2$	3.6427	0.603	1.8214	3.0026	0.547	0.6401
α-Cristobalite	SiO$_2$	7.7428	0.710	3.8714	7.1338	0.682	0.609
Stishovite	SiO$_2$	4.8191	1.607	2.4096	3.2393	1.318	1.5798
Coesite	SiO$_2$	3.0434	0.691	1.5217	2.2962	0.601	0.7472
Silica glass	SiO$_2$	3.7215	0.521	1.8608			
Ice (270 K)	H$_2$O	1.2866	0.108	0.6433	1.2052	0.104	0.0814

based on this limit and Eq. 17, are listed in Table 4, using the corresponding U_e, N, R_o, and ν in Tables 2 and 3. Critical stress intensities for brittle crack propagation K_{IC} were obtained from Eqs. 1 and 17, and the average fracture surface energy of each mineral was obtained from the condition $\gamma = G_{IC}/2$ for ideal brittle fracture. These are listed in Table 4. Essentially, the computed G_{IC}, K_{IC} and γ, are average values for transgranular cracking on randomly oriented planes in polycrystalline minerals. This follows from the use of an isotropic elastic modulus E in Eq. 10, obtained from the behaviour of polycrystalline aggregates.

Grain Boundary Fracture

Fracture along high angle grain boundaries is treated in an analogous manner to the model developed for ionic crystals [1]. Grain boundaries are considered to be composed of regions of atom coincidence and non-coincidence between neighbouring grains. In the coincident regions, the atom separation across the boundary is assumed to be the same as that within the crystal grain (*i.e.* R_o). Average atom separation across the non-coincident boundary region is mR_o, where m is >1. Assigning a fractional boundary area (*f*) to coincident sites, and a fractional area (1-*f*) to non-coincident sites, allows the grain boundary toughness $(G_{IC})_{Gb}$ to be expressed in a formal manner:

$$(G_{IC})_{Gb} = (f)\left[\int_{R_x=R_o}^{R_x=R_{Limit}} \sigma_x \partial R_x\right] + (1-f)\left[\int_{R_x=mR_o}^{R_x=R_{Limit}} \sigma_x \partial R_x\right] \quad J\,m^{-2} \quad (18)$$

where $R_{Limit} = (R_o + 2 \times 10^{-9})$ m.

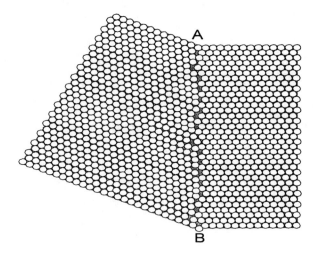

Figure 4. Schematic of a simple 25° tilt boundary showing regions of atom coincidence and non-coincidence.

After substituting Eq. 13 for σ_x, and integrating, Eq. 19 is obtained:

$$(G_{IC})_{Gb} = \left[\frac{U_e(1-2\upsilon)}{3N^{1/3}}\right] \times \begin{Bmatrix} f[\{2\exp[\alpha_c(R_o - R_x)] - \exp[2\alpha_c(R_o - R_x)]\}|_{R_x=R_o}^{R_x=R_{Limit}} \\ + \\ (1-f)[\{2\exp[\alpha_c(R_o - R_x)] - \exp[2\alpha_c(R_o - R_x)]\}|_{R_x=mR_o}^{R_x=R_{Limit}} \end{Bmatrix} \quad \text{Jm}^{-2}$$

(19)

In general, the value of f is dependent upon the characteristics of the crystal structure, the misorientation angle across the boundary and the type of boundary (e.g. twist, tilt or mixed). Thus, even in the same mineral phase, no two boundaries will be identical. However, assuming the average value of m is relatively constant and independent of f for high angle boundaries, Eq. 19 indicates that $(G_{IC})_{Gb}$ has a maximum value when $f = 0$ and a minimum value when $f = 1$, with an average condition represented by $f = 0.5$.

Examination of a simple 25° tilt boundary AB in Figure 4, based on close packing of equal sized spheres consistent with the appearance of boundaries in the bubble raft experiments of Bragg and Nye [32], indicates $m \approx 1.5$ and $f = 0.54$ (close to the average condition of 0.5). Hence, useful relative comparisons of $(G_{IC})_{Gb}$ for different minerals may be made by inserting $m \approx 1.5$ and $f = 0.5$ in Eq. 19, together with $R_{Limit} = (R_o + 2 \times 10^{-9})$ m. Resulting $(G_{IC})_{Gb}$ values are listed in Table 4 and should be considered estimates for high angle boundaries in pure, single-phase materials with "clean" boundaries (i.e. no segregation of impurity atoms at the boundary).

Interestingly, if the $(G_{IC})_{Gb}$ calculation for α-quartz is repeated with $f = 0$ and $m = 1.5$, corresponding to non-coincidence of any atoms across the boundary (amorphous boundary), the new $(G_{IC})_{Gb}$ is decreased to 3.946 J m^{-2}. This compares favourably with $G_{IC} = 3.7215$ J m^{-2} for silica (quartz) glass (amorphous material) in Table 4. The degree of agreement may be fortuitous, but it lends confidence to the analytical procedures used for estimating toughness.

Based on the $(G_{IC})_{Gb}$ estimates, the corresponding critical stress intensity factors for grain boundary cracking are listed in Table 4. They were obtained from Eq. 20 by analogy with Eq. 1:

$$(K_{IC})_{Gb}(1-\nu^2)^{1/2} = [E(G_{IC})_{Gb}]^{1/2} \approx (K_{IC})_{Gb} \quad \text{Pa m}^{1/2} \quad (20)$$

The grain boundary energy, γ_{Gb}, corresponding to the condition $f = 0$ and $m = 1.5$ for each mineral is listed in Table 4. It was obtained via the difference in toughness between transgranular and intergranular cracking:

$$\gamma_{Gb} = G_{IC} - (G_{IC})_{Gb} \quad \text{J m}^{-2} \quad (21)$$

Fracture of Monocrystals and Cleavage

For monocrystals, G_{IC} (and γ) is likely to be anisotropic in the same manner that elastic constants of single crystals are anisotropic, consistent with the fact that minerals often exhibit preferred cleavage planes [22]. Thus, suitably oriented monocrystals, where the applied tensile stress is oriented normal to the cleavage plane, are expected to exhibit lower toughness values than the averaged polycrystalline transgranular values listed in Table 4. Moreover, if the covalent model represented by Eq. 17 is a reasonable approximation for polycrystals, it should be possible to modify the equation to demonstrate that an observed cleavage plane has a lower toughness than the averaged transgranular value.

Consider cleavage on the (hkl) plane, where the tensile elastic modulus normal to the plane is E_{hkl} and the smallest equilibrium spacing between the cleavage planes is d. The first step is to rearrange Eq. 11 to produce Eq. 22, recognising that $NR_o = N^{2/3}$:

$$\frac{(1-2v)}{3N^{2/3}} = -\frac{E}{2U_e \alpha_c^2} \qquad (22)$$

After replacing E by E_{hkl}, and recognising that the units of N are m^{-3} (atoms per unit volume), $N^{2/3}$ (m^{-2}) is replaced by $N^{1/3}/d$ (m^{-2}), so that Eq 22 becomes:

$$\frac{(1-2v)}{3N^{1/3}} = -\frac{E_{hkl}}{2U_e d\alpha_c^2} \qquad (23)$$

It is proposed that substitution of Eq. 23 into Eq. 17 provides a general equation for estimating fracture toughness of the (hkl) plane:

Table 5. Computed toughness values for reported cleavage planes at 298 K.[†] Values compared with computed toughness for transgranular fracture (polycrystal).

Mineral	Polycrystal Transgranular		Cleavage Fracture[‡]				
	G_{IC} (J m^{-2})	K_{IC} (MPa m$^{1/2}$)	Cleavage Plane (hkl)	E_{hkl} (GPa)	d (10^{-10}m)	$(G_{IC})_{hkl}$ (J m^{-2})	$(K_{IC})_{hkl}$ (MPa m$^{1/2}$)
Graphite-2H	5.5052	1.230	(001)	36.364	3.3756*	0.447	0.1275
Diamond	9.9503	3.308	(111)	1171.9	2.060	9.179	3.28
Molybdenite-2H	3.2294	0.605	(001)	45.249	6.145*	0.5477	0.157
α-Quartz	5.3560	0.715	(101)	76.360	3.34321	2.977	0.477
β-Quartz (873K)	3.6427	0.603	(101)	95.6030	3.39157	2.428	0.482
Ice (270 K)	1.2866	0.108	(001)	11.628	3.660*	0.998	0.108

[†]298 K unless stated: [‡]Cleavage planes reported by Dana [22]: *d corresponds to (002) spacing.

$$(G_{IC})_{hkl} = \left[\frac{-E_{hkl}}{2d\alpha_c^2}\{2\exp[\alpha_c(d-d_x)] - \exp[2\alpha_c(d-d_x)]\}\right]_{d_x=d}^{d_x=d_{Limit}} \text{ Jm}^{-2} \quad (24)$$

where d has replaced R_o, d_x is the change in interplanar spacing in the x-direction (normal to the (hkl) plane) as a uniaxial tensile stress is applied in the x-direction, and $d_{Limit} = (d + 2 \times 10^{-9})$ m.

Based on Eq. 24, Table 5 shows the predicted toughness $(G_{IC})_{hkl}$ for cleavage planes of graphite-2H, diamond, molybdenite-2H, α-quartz and ice. The E_{hkl} modulus for each mineral was computed from the corresponding single crystal compliance constants (s_{mn}) reported by Hearmon [26,27]). This was achieved by transformation of axes to obtain the compliance s'_{11} normal to the (hkl) plane [24]), from which $E_{hkl} = 1/s'_{11}$. The corresponding d spacing was obtained from published crystallographic data [20]). Additionally, the critical stress intensity for cleavage crack propagation $(K_{IC})_{hkl}$ was obtained via Eq. 25, by analogy with Eq. 1:

$$(K_{IC})_{hkl} = [E_{hkl}(G_{IC})_{hkl}]^{1/2} \text{ MPa m}^{1/2} \quad (25)$$

In Table 5, the predicted toughness and critical stress intensity for cleavage are compared with corresponding values for random transgranular cracking in polycrystal aggregates obtained from Table 4.

In all cases, cleavage toughness is shown to be lower than that of the polycrystal, as required. Furthermore, the relatively small difference in toughness between G_{IC} and $(G_{IC})_{hkl}$ for diamond and for ice is consistent with the observation that in both cases cleavage is perfect, but difficult to achieve [22]. Similarly, the relatively large difference between G_{IC} and $(G_{IC})_{hkl}$ for graphite-2H and molybdenite-2H is consistent with the ease with which they undergo perfect cleavage [22].

Note that the fracture surface energy for cleavage γ_{hkl} of the minerals in Table 5 is readily obtained from the fracture toughness via $\gamma_{hkl} = (G_{IC})_{hkl}/2$.

DISCUSSION

Comments on Toughness Estimates

Of the fifteen covalent minerals/materials examined, the toughness estimates in Table 4 show that diamond has the highest G_{IC} and K_{IC} values, being higher than previous estimates for forty-eight different ionic minerals [1]. Experimental toughness data for diamond are unavailable for comparison with the theoretical results. However, the use of diamonds for hardness indentation testing of ceramics [12] and the widespread use of industrial quality diamonds for abrasives and cutting tools are entirely consistent with the higher toughness values predicted for diamond relative to other minerals.

Fracture toughness tests by Andersson and Salkovitz [33] on artificial polycrystalline graphite, using centre-notched specimens, showed that K_{IC} was ~

1.6 MPa m$^{1/2}$ for the most pore-free material. This compares favourably with a K_{IC} of 1.23 MPa. m$^{1/2}$ for polycrystalline graphite-2H in Table 4. Pabst [34]) reported an experimental K_{IC} of 0.7 MPa m$^{1/2}$ for graphite when using a notched four point bend specimen. This compares favourably with the $(K_{IC})_{Gb}$ estimate of 1.079 MPa m$^{1/2}$ in Table 4.

The experimental value of K_{IC} for polycrystalline SiC is ~3.4 MPa m$^{1/2}$ based on notched bend tests [34]. This is higher than the estimates of 1.477 MPa m$^{1/2}$ for α-SiC in Table 4. Similarly, the experimental K_{IC} for polycrystalline Si_3N_4 is ~2.6 MPa m$^{1/2}$ based on notched bend tests [34]. Again, it is higher than the estimate of 0.941 MPa m$^{1/2}$ in Table 4. The experimental values are likely to be on the high side, because of the difficulty in preparation of SiC and Si_3N_4. test material (usually prepared by hot pressing of powder) and difficulty in obtaining a sharp initial machined (diamond saw) notch from which to propagate a crack in the notched bend test [35]. However, even if this is not the case, it is still encouraging that the estimated and experimental values are of the same magnitude.

Experimental measurements of the fracture surface energy γ for amorphous silica (quartz) glass are reported by Kennedy et al. [35] to be ~3.5 J m^{-2}, when utilising an "atomically sharp crack" with the double cantilever beam method. This compares reasonably well with an estimated γ of 1.86 J m^{-2} in Table 4. Also, the estimated K_{IC} of 0.521 MPa m$^{1/2}$ in Table 4 compares favourably with the experimental value of 0.753 MPa m$^{1/2}$ listed by Gebrande [21] for the three point bend test.

Other experimental K_{IC} listed by Gebrande [21] for double torsion tests on α-quartz are 0.852 to 1.002 MPa m$^{1/2}$, which compare favourably with the estimated 0.715 MPa m$^{1/2}$ in Table 4. The review by Parks [36] gave a best estimate of 2 J m^{-2} for the surface free energy of quartz in vacuum, comparable with an estimated γ of 2.678 J m^{-2} for α-quartz in Table 4.

Experimental values of G_{IC} and K_{IC} of ice at 260 K, obtained by three point bend testing, are reported to be 1.5 J m^{-2} and 0.116 MPa m$^{1/2}$, respectively [21]. Again, these compare very favourably with G_{IC} and K_{IC} estimates of 1.287 J m^{-2} and 0.108 MPa m$^{1/2}$, respectively, for ice at 270 K in Table 4.

Overall, the covalent model estimates gave a reasonable correlation between predicted ideal brittle toughness behaviour and the available experimental data. The correlation is satisfactory in terms of both the magnitude and relative toughness of the different minerals. For crystalline materials, the G_{IC} and K_{IC} estimates are to be interpreted as being average values applicable to a macroscopic transgranular crack propagating normal to the applied tensile stress in a polycrystalline aggregate. At the microscopic level, such a crack usually exhibits a stepped topography [37] composed of crystalline facets (cleavage facets) of lower G_{IC} (lower γ) with higher energy steps. Consistent with this, application of the covalent model to cleavage fracture on some commonly observed cleavage planes gave toughness values lower than G_{IC} and K_{IC}, as shown in Table 5.

The predicted toughness for grain boundary cracking is lower than that for macroscopic transgranular cracking of a polycrystalline aggregate (see Table 4), indicating that failure along suitably oriented grain boundaries is a likely occurrence in polycrystals. The percentage difference between G_{IC} and $(G_{IC})_{Gb}$

ranged from ~6% for ice up to ~33% for nierite. The corresponding differences between K_{IC} and $(K_{IC})_{Gb}$ ranged from ~4% for ice to ~23% for nierite.

The relatively higher G_{IC} and $(G_{IC})_{Gb}$ toughness values for α-cristobalite, compared to other SiO_2 polymorphs, is attributed to its negative Poisson's ratio (see Eqs. 17, 19 and Table 3), consistent with expectations of enhanced toughness for minerals with a negative ratio [29].

Energy Efficiency of Comminution

For purposes of estimating upper values of the total energy required to generate new surface area during comminution of polycrystalline material, the fracture surface energy γ (i.e. $G_{IC}/2$) is sufficient. For simplicity of analysis, consider the comminution of an initial spherical particle of diameter D_i (m) that is fractured (milled) into very small particles with an average final diameter of D_f (m). The number of particles produced is $(D_i/D_f)^3$ and the resulting increase in surface area per unit volume ΔS_A becomes:

$$\Delta S_A = 6F_r \left[\frac{1}{D_f} - \frac{1}{D_i} \right] \quad m^{-1} \tag{26}$$

where F_r is a surface roughness factor (> 1) indicating that milled particles are not perfect spheres, but exhibit a roughened (stepped and faceted) topography with a surface area/diameter ratio that is F_r times that of spheres.

From Eq. 26, the increase in surface energy per unit mass ΔS_{En} is obtained:

$$\Delta S_{En} = \frac{\Delta S_A \gamma}{\rho} = \frac{6F_r \gamma}{\rho} \left[\frac{1}{D_f} - \frac{1}{D_i} \right] \approx \frac{6F_r \gamma}{\rho D_f} \quad Jkg^{-1}, \text{ when } D_i \gg D_f \tag{27}$$

where ρ is the mineral density (kg m^{-3}).

Bond [38] determined the average standard work index W_i for the crushing and grinding of numerous materials, where W_i is defined as the work input in kWh/short ton (1 short ton =2000 lb) that is required to reduce the feed from an infinitely large particle size to 80% passing 100 μm. [N.B. conversion of W_i to SI units, $(W_i)_{SI}$ requires that $(W_i)_{SI} = W_i(3.968 \times 10^3)$ J kg^{-1}]. Bond [38] cautioned the use of his W_i data, as large variations may be obtained between the same material. No doubt this is due to the fact that the '80% passing criterion' provides no information on the actual distribution of particle sizes, which may be expected to vary between >100 μm to <20 μm, with an averaged effective final diameter of D_{aef}. Nevertheless, recognising these limitations, Eq. 27 is useful for estimating the energy efficiency of particle fracture during comminution from the ratio $\Delta S_{En}/(W_i)_{SI}$:

$$\% \text{ efficiency} = \frac{\Delta S_{En}}{(W_i)_{SI}} \times 100 \approx \frac{6F_r\gamma}{\rho D_{aef}(W_i)_{SI}} \times 100 \qquad (28)$$

where D_{eaf} (< 100 μm) replaces D_f.

Based on Eq. 28, estimated energy efficiencies for crushing and grinding of six materials are presented in Table 6.

The required W_i for Table 6 were obtained from Bond [38], γ from Table 4, and ρ from X-ray diffraction data [20] of pore-free pure minerals. An approximate value of ~3 was assigned to F_r, together with an assumed value of ~40 μm for D_{aef}. The names of the materials listed in Table 6 are those reported by Bond [38]; selected covalent minerals likely to be their closest representatives are indicated in the table footnote. Within the limited data available, the results indicate that the energy efficiency during conventional crushing and grinding of covalent minerals/materials is very low, of the order of ≤ 1%, consistent with previous estimates for ionic minerals [1].

Table 6. Estimated energy efficiency of crushing and grinding based on the Bond work index.

Material	γ (J m^{-2})	ρ (10^3 kg m^{-3})	W_i (kWh/ton)*	$(W_i)_{SI}$ kJ kg^{-1}	Efficiency (%)
Graphite[a]	2.7526	2.245	45.03	178.7	0.31
Silicon carbide[b]	2.40	3.239	26.17	103.84	0.32
Quartz[c]	2.6780	2.649	12..77	50.67	0.9
Quartzite[c]	2.6780	2.649	12.18	48.33	0.94
Flint[c]	2.6780	2.649	26.16	103.8	0.44
Silica sand[b]	2.6780	2.649	16.46	65.31	0.7

*Bond's data [38]: [a]Graphite-2H; [b]Moissanite-6H; [c]α–Quartz.

Factors Affecting Energy Efficiency

Principal factors affecting energy efficiency during comminution are related to the conditions under which impact loading is insufficient to promote particle fracture. This may be assessed qualitatively in terms of fracture mechanics concepts, based on the presence of inherent particle flaws. Figure 5 is a schematic diagram of a mineral particle containing several penny-shaped flaws (cracks), some of which are surface flaws and others are internal flaws. The particle is subjected to opposing impact forces P, as occurs when particles are compressed between two balls in a ball mill, or two rods in a rod mill. Compressive stresses are generated in the particle during impact, their magnitudes being dependent upon P and location within the particle relative to the impact loading axis P-P. Maximum compressive stresses σ_P are generated along the loading axis. The latter situation is shown schematically in Figs. 6(a) and 6(b) for an internal penny shaped flaw, of diameter $2a$, lying within a particle of average diameter D. The flaw in Figure 6(a) is inclined at an angle θ with respect to the loading axis, whereas the flaw in Figure 6(b) is parallel to the axis. During impact, tensile stresses of magnitude $k\sigma_P$ ($k < 1$) are generated normal to the loading axis. This is a well-recognised behaviour of spherical and cylindrical particles, which has been treated in detail by Hu et al. [2]. Previously, the authors approximated k to Poisson's ratio ν [3]. A detailed stress analysis by Hu et al. [2] on the horizontal diametric plane indicates that k is ~0.28 when ν is 0.15 and ~0.25 when ν is 0.3.

Considering Figure 6(a), the component of induced tensile stress normal to the crack plane is $k\sigma_P\text{Cos}\theta$. Also, the component of compressive stress acting normal to the crack plane (tending to close the crack) is $\sigma_P\text{Sin}\theta$, so the combined crack opening stress is $\sigma_P(k\text{Cos}\theta - \text{Sin}\theta)$. The resulting stress intensity K_I acting on the flaw is obtained from Eq. 2:

$$K_I = Y\sigma_P(k\text{Cos}\theta - \text{Sin}\theta)a^{1/2} = 2\sigma_P(k\text{Cos}\theta - \text{Sin}\theta)\left(\frac{a}{\pi}\right)^{1/2} \quad \text{Pa m}^{1/2} \quad (29)$$

where Y is $2(\pi)^{-1/2}$ [4].

For a particle to be fractured $K_I = K_{IC}$. Hence, a larger K_{IC} (see Table 4) requires a higher σ_P (larger P) for particle fracture. If P is insufficient, fracture is absent despite repeated impacts (impact inefficiency). However, if the orientation of the particle changes during successive impacts, so that $\theta \rightarrow 0$ and $k(\text{Cos}\theta - \text{Sin}\theta) \rightarrow k$, K_I will increase at constant P and may reach K_{IC} (particle fracture). Thus, K_{IC}, P and flaw orientation (θ) determine impact efficiency. Impact without particle fracture imparts elastic deformation to the particle, the elastic strain energy being released as thermal energy (heat) after impact. Thus, impact inefficiency contributes directly to high-energy consumption during comminution.

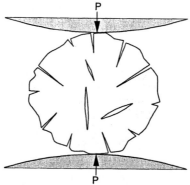

Figure 5. Schematic diagram of particle with many flaws under compressive force P.

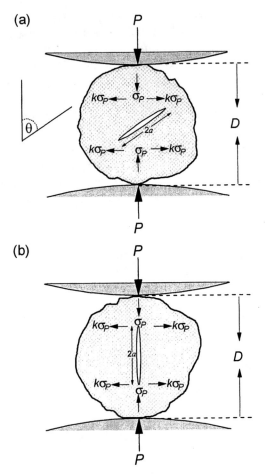

Figure 6. Small flawed particle under compressive loading. (a) Flaw inclined θ degrees to loading axis. (b) Flaw parallel to loading axis.

In ball mills and rod mills a distribution of P takes place, due to the random nature of the particle/ball interactions, leading to very inefficient particle fracture. An obvious way to narrow this distribution and increase (control) P is by high compression roller mill grinding, as proposed and developed by Schönert [39]. In several instances, such mills are reported to consume less energy [39, 40] and exhibit improved interparticle separation in mineral aggregates (*i.e.* liberation via interphase cracking), particularly in the processing of diamond ores [40]. Interestingly, it is now clear that liberation of diamonds without damage due to fracture is directly attributable to the high K_{IC} of diamond (see Table 4). Controlling P through compression rolling allows K_{IC} of the weaker host rock to be exceeded while remaining below K_{IC} for diamond.

Limiting Fine Particle Size

Inspection of Figure 6 shows the maximum flaw size must be less than the particle diameter (*i.e.* $2a < D$). Using the condition $K_I = K_{IC}$ and $\theta = 0$, Eq. 29 may be reset in terms of D by expressing $2a$ as a fraction Φ of D, and rearranging [3]:

$$\Phi D = \pi \left(\frac{K_{IC}}{2k\sigma_P} \right)^2 \text{ m} \tag{30}$$

where $\Phi < 0.5$ and $\theta = 0$.

Hence, for a constant σ_P, the lower limiting average D obtainable via particle fracture is expected to be strongly dependent upon $(K_{IC})^2$. This, of course, applies at very fine particle sizes when only one or two flaws are present. It will not be applicable at larger particle sizes where a significant population of different flaw sizes is likely (a population distribution that may be dependent on the type of mineral and its previous history). The influence of K_{IC} on fine particle fracture suggested by Eq. 30, together with the estimated K_{IC} values in Table 4 should prove useful for guiding the production of ultra-fine mineral powders via techniques such as stirred ball mills [41]. Production of ultrafine fine powders is often required for use as abrasives, or for starting material in the sintering of high quality products of intricate shape that require desirable combinations of thermal, chemical and hardness properties, *e.g.* sintered SiC [12].

Note that with ultra-fine particles ($D \sim 1$ μm or less), a brittle-ductile transition may be obtained at high compressive loading stresses (σ_P). This occurs when high induced tensile components (high $k\sigma_P$) remain insufficient to produce particle fracture, whereas the shear stress components ($\sim \sigma_P/2$) due to σ_P become sufficient to promote significant dislocation motion in the particle (plastic deformation). This has been described in previous detail by the authors [3]. It contributes to increased energy inefficiency (*i.e.* large energy consumption via plastic deformation with minimal particle fracture).

CONCLUSIONS

1. Theoretical analyses of the ideal fracture toughness of fifteen single phase minerals/materials, suitable for application of a covalent bonding model, gave toughness estimates for G_{IC}, K_{iC}, and fracture surface energy γ that appear to be of the correct order of magnitude and ranking with respect to each other.
2. The developed model allowed estimates of the grain boundary energy γ_{Gb} of high angle grain boundaries to be made, and showed that toughness values for intergranular cracking along suitably oriented high angle grain boundaries, $(G_{IC})_{Gb}$, are likely to be less than those of G_{IC} for macroscopically planar transgranular fracture of random orientation. Hence, a propagating crack may follow an intergranular path in situations where the grain boundary plane tends to be oriented normal to the local maximum tensile stress.

3. The model provided estimated cleavage toughness values, $(G_{IC})_{hkl}$ and $(K_{iC})_{hkl}$, for fracture of monocrystals on observed cleavage planes (hkl) that were lower than G_{IC} and K_{IC}, as required for a satisfactory model.
4. Comparisons between the standard Bond work index and the total ideal fracture surface energy required for generating new surfaces during comminution of covalent materials indicate that the energy efficiency of crushing and grinding operations is very low, of the order ~1% or less.
5. Energy efficiency during comminution is dependent upon impact efficiency (number of impacts before particle fracture), which in turn is dependent upon the particle loading force, the size and orientation of inherent flaws with respect to the loading axis, and K_{IC}.
6. The average limiting particle size achievable by ultrafine grinding is strongly influenced by K_{IC}.

ACKNOWLEDEMENTS

The authors wish to thank the Natural Sciences and Reesearch Council of Canada for financial support of the study.

REFERENCES

[1] D. Tromans. and J.A. Meech, 2002. Fracture toughness and surface energies of minerals: Estimates for oxides, sulphides, silicates and halides, Minerals Engineering, 15, 1027–1041.
[2] G. Hu, H. Otaki and M. Lin, 2001. An index of the tensile strength of brittle particles, Minerals Engineering, 14(10), 1199–1211.
[3] D. Tromans and J.A. Meech, 2001. Enhanced dissolution of minerals: Stored energy, amorphism and mechanical activation, Minerals Engineering, 14(11), 1359–1377.
[4] D. Broek, 1982. Elementary engineering fracture mechanics, 3rd Edition, Martinus Nijhoff Publishers, The Hague, Netherlands.
[5] K. Schoenert, 1972. Role of fracture physics in understanding communition phenomena, Transactions of Society of Mining Engineers-AIME, 252 (March), 21–26.
[6] G.L. Austin, 1984. Gaudin Lecture, Concepts in process design of mills, Mining Engineering, 36(6), SME-AIME, 628–635.
[7] R.A. Bearman, R.W. Barley and A. Hitchcock, 1991. Prediction of power consumption and product size in cone crushing, Minerals Engineering, 4(12), 1243–1256.
[8] T.J. Napier-Munn, S. Morrell, R.D. Morrison and T. Kojovic, 1999. Mineral comminution circuits: Their operation and optimisation, JKMRC Monograph Series in Mining and Mineral Processing 2, The University of Queensland, Australia, 53–54, 80.
[9] F. Rummel, 1982. Fracture and flow of rocks and minerals: Strength and deformability, Landolt-Börnstein tables, Group V, Vol. 1b, (Ed.) G. Angenheister, Springer-Verlag, Berlin, 141–194.

[10] Z.X. Zhang, 2002. An empirical relation between mode I fracture toughness and the tensile strength of rock, Int. J. Rock Mechanics and Mining Sciences, 39, 401–406.
[11] L. Pauling, 1960. The nature of the chemical bond, 3rd Ed., Cornell Univ. Press, Ithaca, NY, 97–102.
[12] I.J. McColm, 1990. Ceramic Hardness, Plenum Press, New York, 38-42, 209–251.
[13] R. Grün, 1979. The crystal structure of β-Si_3N_4; Structural and stability considerations between α and β- Si_3N_4, Acta Crystallographica, B35, 800–804.
[14] E. Soignard, M. Somayazulu, J. Dong, O.F. Sankey, P.F. McMillan, 2001. High pressure-high temperature synthesis and elasticity of the cubic nitride spinel γ-Si_3N_4, J. Physics: Condensed Matter, 13, 557–563.
[15] K. Eichorn, A. Kirfel, J. Grochowski and P. Serda, 1991. Accurate structure analysis with synchroton radiation. An application to Borazone, cubic BN, Acta Crystallographica, B47, 843–848.
[16] P.M. Morse, 1929. Atomic molecules according to the wave mechanics. II. Vibration levels, Physical Review, 34, 57–64.
[17] A. Roine, 2002. Outokumpu HSC Chemistry for Windows 5.11, Chemical Reaction and Equilibrium Software with Extensive Thermochemical Database, Outokumpu Research Oy, Pori, Finland.
[18] A. Zerr, G. Miehe, G. Serghiou, M. Schwarz, E. Kroke, R. Riedel, H. Fueß, P. Kroll, and R. Boehler 1999. Synthesis of cubic silicon nitride, Nature, 400, 340–342.
[19] A. Zerr, M. Kemf, M. Schwarz, E. Kroke, M. Göken and R. Riedel, 2002. Elastic moduli and hardness of cubic silicon nitride, J. Amer. Ceramic Society, 85(1), 86–90.
[20] PDF—Powder Diffraction File, 1995. PCPDFWIN, JCPDS-International Center for Diffraction Data, Swarthmore, PA, USA.
[21] H. Gebrande, 1982. Elastic wave velocities and constants of elasticity of rocks and rock forming minerals, Landolt-Börnstein Tables, Group V, Vol. 1b, (Ed.) G. Angenheister, Springer-Verlag, Berlin, 1–98.
[22] Dana's New Mineralogy, 1997. Eighth Edition, (Eds.) R.V. Gaines, H.C.W. Skinner, E.E. Ford, B. Mason, A. Rosenzweig, John Wiley & Sons Inc., New York, 1573–1575.
[23] J. Sherman, 1932. Crystal energies of ionic compounds and thermochemical applications, Chemical Reviews, XI (Aug.-Dec.), 93–169.
[24] J.F. Nye, 1985. Physical Properties of Crystals, Oxford University Press, Oxford, UK, 131–149.
[25] J.B. Wachtman, 1996. Mechanical Properties of Ceramics, J. Wiley and Sons, Inc., New York, NY, 24–35.
[26] R.F.S. Hearmon, 1979. The elastic constants of crystals and other anisotropic materials, Landolt-Börnstein Tables, Group III, Vol. 11, (Eds.) K-H. Hellwege and A.M. Hellwege, Springer-Verlag, Berlin, 1–244.
[27] R.F.S. Hearmon, 1984. The elastic constants of crystals and other anisotropic materials, Landolt-Börnstein Tables, Group III, Vol. 18, (Eds.) K-H. Hellwege and A.M. Hellwege, Springer-Verlag, Berlin, 1–154.

[28] J.D. Bass, 1995. Elasticity of minerals, glasses and melts, Mineral physics and crystallography: A handbook of physical constants, (Ed.) T.J. Ahrens, American Geophysical Union, Washington, DC, 45–63.

[29] A. Yeganeh-Haeri, D.J. Weidner and J.B. Parise, 1992. Elasticity of α-cristobalite: A silicon dioxide with a negative Poisson's ratio, Science, 257 (July), 650–652.

[30] M. Grimsditch, E.S. Zouboulis and A. Polian, 1994. Elastic constants of boron nitride, J. Applied Physics, 76(2), 832–834.

[31] S.P. Dodd, M. Cankurtaran, G.A. Saunders and B. James, 2001. Ultrasonic study of the temperature and pressure dependences of the elastic properties of β-silicon nitride ceramic, J. Materials Science 36, 2557–2563.

[32] L. Bragg and J.F. Nye, 1947. A dynamical model of a crystal structure, Proceedings of the Royal Society of London, 190, 474–481.

[33] C.A. Andersson and E.I. Salkovitz, 1974. Fracture of polycrystalline graphite, Fracture Mechanics of Ceramics, Vol. 2, (Eds.) R.C. Bradt, D.P.H. Hasselman and F.F. Lange, Plenum Press, New York, 509–526.

[34] R.F. Pabst, 1974. Determination of K_{IC}-factors with diamond saw-cuts in ceramic materials, Fracture Mechanics of Ceramics, Vol. 2, (Eds.) R.C. Bradt, D.P.H. Hasselman and F.F. Lange, Plenum Press, New York, 555–565.

[35] C.R. Kennedy, R.C. Bradt and G.E. Rindone, 1974. Fracture mechanics of binary sodium silicate glasses, Fracture Mechanics of Ceramics, Vol. 2, eds. R.C. Bradt, D.P.H. Hasselman and F.F. Lange, Plenum Press, New York, 883–893.

[36] G.A. Parks, 1984. Surface and interfacial free energies of quartz, J. Geophysical Research B, 89(B6), 3997–4008.

[37] D. Tromans and J.A. Meech, 2002. Enhanced dissolution of minerals: conjoint effects of particle size and microtopography, Minerals Engineering, 15, 263–269.

[38] F.C. Bond, 1961. Crushing and Grinding Calculations. Allis Chalmers Pub. No. 07R9235B, 1–14.

[39] K. Schönert, 1988. A first survey of grinding with high-compression roller mills. Int. J. Mineral Processing, 22, 401–412.

[40] R.E. McIvor, 1997. High pressure grinding rolls—A review, Comminution Practices, (Ed.) S. Kawatra, SME Inc., Littleton, CO, 95–97.

[41] Y. Wang and E. Forssberg, 1997. Ultra-fine grinding and classification of minerals, In Comminution Practices, ed. S. Kawatra, SME, Littleton, CO, 203–214.

Usable Heat from Mine Waters: Coproduction of Energy and Minerals from "Mother Earth"

M. M. GHOMSHEI and J. A. MEECH
The Centre for Environmental Research in Minerals, Metals, and Materials,
Department of Mining and Mineral Process Engineering,
University of British Columbia, 6350 Stores Road,
Vancouver, B.C., V6T 1Z4

ABSTRACT

In addition to minerals, all mines contain geothermal energy. The heat is provided by two sources: slow exothermic oxidization of minerals such as sulphides and coal; and access to higher heat flux at depth, resulting from the natural temperature gradient of the ground (20 to 40°C/km). In addition to these heat sources, mine workings (and waste dumps as well) provide the necessary permeability to allow extraction of this heat using naturally present ground waters.

Heat flux from the earth's interior to the mine wall surfaces can be as high as 1.5 kcal per square meter of mine opening per day. For coal and sulphide mines, in-mine mineral oxidation will add to this gradient heat. In sulphide-bearing mines, slow-burning of sulphide minerals (e.g. pyrite), caused by exposure to atmospheric oxygen, leads to extensive heat generation (about 12 kcal/g of oxidized pyrite). Some heat is therefore generated along an underground mine's walls where sulphide minerals are exposed to open air. Significantly more heat is generated inside waste dumps where higher mineral surfaces and easier air circulation permit higher rates of oxidation. While the temperature of underground mine waters rarely exceeds 20°C in shallow mines, the temperature inside many sulphide-bearing waste dumps can reach as high as 50°C. Besides these sources of heat, what really makes a mine a viable geothermal resource is the ease with which water flows within mine compartments (i.e. workings and waste dumps) allowing significant transfer of energy even in the absence of a high ground heat flux.

This paper discusses the quantity and value of recoverable heat from abandoned and operating mines. The Britannia Mine (an abandoned copper mine located 45 kilometres north of Vancouver) is presented as a commercially viable case. Using heat-pump technology together with intelligent control systems, the mine effluent can provide heat to a community of about 100 households that presently occupy this site.

INTRODUCTION

Earth is a hot planet. Geothermal energy resources are not however limited to the energy emanating from the earth's interior. Rather, a resource can be defined as any heat source stored or available within the ground. The origin of this heat can be 1- the internal heat of the planet; 2- chemical and mineralogical reactions; 3- radioactive decay; and 4- captured solar radiation near surface.

There is a net heat flux from inside the earth towards the surface. On land, an average of 50 to 75 mW/m^2 (1–1.5 kcal/day/m^2) of power is constantly emitted to the atmosphere. This heat is measurable, but is insignificant for a small area—for example, for the cross-section of an insulated coffee cup, it would take over a decade to warm up the coffee! However in a typical underground mine with as much as 10,000,000 m^2 of open surface, this total heat flux can approach 1–10 MW considering that the heat flux at depth in a mine is significantly higher than at surface. Note that over the long term (i.e., 100s to 1000s of years) the ground temperature around these openings must decline since total heat flux into the rock surrounding the mine from the earth's interior is constant. Eventually heat taken out through the walls must equal that coming into the mine rock (i.e. 50-70 mW/m^2 of perpendicular cross-section). The rock temperature depends naturally on the depth of the mine and the local geothermal gradient. In some areas (e.g., young volcanics), this heat flux can be much higher than normal leading to high temperatures at accessible depths.

Based on the ground water temperature and application, geothermal resources are classified into high-grade (above 150°C), medium-grade (150 to 30°C) and low-grade (below 30°C). The boundaries between these grades are fuzzy and depend on the particular application. High-grade resources are used for electric power generation. The medium-grade class is often used for direct-use energy applications while a low-grade area can become usable when the temperature is modified by a heat pump [1–3]. In Canada, high-grade resources are specific to small areas associated with west-coast young volcanic activity. Medium-grade resources are more widespread, mainly in the Western Provinces. The low-grade class (especially the lower end, i.e. below 15°C), is abundant and widespread all over Canada [4].

All geothermal systems perform like a heat exchanger, where the rock mass integrates the heat source and the heat exchanger. Increasing the surface with the earth's heat flux is therefore key in extracting heat from the ground. In geothermics, this factor is linked to permeability (natural open fractures or artificial mine openings). An equally important factor in a heat exchanger is the fluid that transfers the heat. In nature this fluid is generally either water or, to a lesser extent, air.

Enhancing ground permeability is not easy. The common process of hydrofracturing is expensive and therefore limited to cases of high-grade geothermal systems used to generate electricity. In medium and low-grade geothermal terrain, artificial fracturing is not cost-effective especially in cases of small-size projects. Nevertheless, permeability remains a key componenct of all types of geothermal resources.

THE MINE AS A HEAT EXCHANGER

In most Canadian underground mines, all three elements of a geothermal system (i.e. heat source, water and permeability) are readily available [5]. Permeability is extensively present due to mine workings (in addition to fracture zones) and water is abundant below the water table (relatively near surface in most regions of Canada). As for heat, the natural temperature gradient (~30 °C/km) guarantees at least a sizable low-grade heat flux into the mine openings. Thermal energy flux from mine workings can be expressed in terms of W/m^2 as conductive heat transfer. This value is enhanced by convective heat transfer through ground water. A mine's temperature gradient increases with depth leading to even higher temperatures. At the Con Mine in Yellowknife, North West Territories, mine waters reach temperatures well above 35°C at levels below 1500 m. Other less important heat sources are related to exothermic reactions triggered by exposure of minerals to oxygen (e.g. sulphide oxidation and slow-burning of coal). An underground mine can therefore be regarded as a significant heat deposit, where ground water does the mining.

OTHER MINE COMPONENTS

Besides underground workings, other mine components such as waste dumps and, to a lesser extent, tailings impoundments, can be considered as potential heat deposits. Most mines store large volumes of sulphide-bearing rock in piles known as waste dumps. Sulfide-bearing wastes generate acid rock drainage due to oxidation of sulfide minerals into sulfuric acid and consequent mobilization of deleterious heavy metals into ground waters. These waste dumps are considered a major environmental hazard that in most cases, require perpetual monitoring and treatment. Sulfide waste piles however, also contain significant amounts of heat that can potentially be used in situations where a heat market is available or created near the site. Contrary to underground workings, where the heat source is defined by low-temperature and high flow, in the waste dump the resource may provide relatively high temperature (above 40°C) and low flow. The heat source temperature, therefore, does not need to be enhanced for most applications.

HEAT GENERATED IN A WASTE DUMP

Assuming a typical 8,000,000 m^3 waste rock [6] with a pyrite content of 5% and an oxidation rate of 10^{-7} kg of oxygen per cubic meter per second [7] the total generated heat will be 10,000 kJ/s (10MW), assuming a typical pyrite reaction:

$$FeS_2 + 7/2O_2 + H2O \rightarrow Fe\,SO_4 + H_2SO_4 + 1440 kJ \text{ (energy)}$$

Note that the rate of oxidation in a waste dump is constrained by the diffusion of oxygen into the dump, as long as there is enough pyrite in the dump to consume the available oxygen. Based on a 5% sulphide content, this heat generation can last 35 years. If oxidation is hampered (by ARD mitigating measures), the heat generation will reduced to 1 MW (the furnace lasting for 350 years).

So, 10-MW thermal energy at medium grade (i.e. temperature above 35°C) is generated within a typical waste dump. This agrees well with measured heat flux in waste piles (e.g. White's waste rock dump at the Rum Jungle mine site [8]). A significant portion of this heat is transported by meteoric water infiltrating the dump. A smaller portion is carried by air, circulating within the rock pile. Considering the low specific heat of air and high specific heat of water, most of the generated heat tends to be stored in water. The temperature of ground water in some waste dumps can reach 50°C, making these piles an excellent source for medium-grade geothermal applications such as greenhouse, aquaculture, food processing, etc. [1].

Considering the extent and sustainability of the available energy, certain heat-intensive industries might be enticed to relocate to a mine site that might otherwise turn into a ghost town after mine closure.

A STEP TOWARDS SUSTAINABILITY

The possibility of using heat energy carried by mine waters is of special interest to those concerned with making the mining industry more sustainable and environmental friendly. In recent years the Canadian mining community has taken significant steps towards accommodating environmental and social concerns into the decision-making process. These steps are particularly remarkable during mine closure and rehabilitation. Costs associated with such undertakings can be especially high in the case of rehabilitating a site that requires perpetual monitoring and remediation after closure. In an attempt to account for perpetual operating costs, such rehabilitation programs would clearly benefit from sustainable funding sources. In many cases the only revenue available derives from a fund set up at the start of mining that would earn interest over the lifetime of the operation to provide sufficient resources for long-term rehabilitation requirements. But sustainable revenue may actually be derived from the site itself by restoring it to productive use after mining—by economic use of the geothermal heat present in the mine thus securing a sustainable revenue stream as a contribution to mine rehabilitation.

Heat-pump technology has already proven successful at recovering heat from mine waters at Springhill, Nova Scotia [4,5]. Yet recovering heat from a mine remains a marginal practice due to several drawbacks. Firstly contrary to that of Springhill, most mine waters are saline and/or acidic causing potential corrosion and scaling problems inside heat-extraction equipment. Secondly, the economics of heat extraction depends on a variety of complex issues [9] related to 1- resource temperature and flow, 2- climate conditions, 3- cost of energy extraction, and 4- availability of a market for the heat in the vicinity of the mine. With rising energy prices, emerging regulations on emission reduction, and energy security issues, utilities have started to explore new frontiers. The barriers are therefore lower than in the past and the time may be right to revisit marginal "green energy" technologies, such as extraction of low-grade geothermal heat.

FIRST CANADIAN SUCCESS STORY (Springhill Nova Scotia)

In the case of the Springhill Mine Water project [4,5], water from a coalmine (abandoned after a mining disaster in 1958) is extracted to provide heat to commercial and residential users in the town. Water in the mine is at a temperature of 18 to 20°C within a few meters of surface, well above normal shallow ground water temperature of 6 to 10°C in the area. The heat flux from this mine is likely related significantly to exothermic reactions at the exposed coal-faces and circulation from the deeper parts of the mine. The first stage of developing this resource came on stream in 1989 with a system installed within an 8000-m² expansion to an existing industrial building. Water at 18°C is pumped from the mine, passed through a heat exchanger, and then returned to the mine at 12°C. The system is designed to provide cooling in the summer when needed. It is estimated that the system is at least 70% more efficient than conventional heating cooling sources.

BRITANNIA MINE IN PERSPECTIVE

Considering the technical and market constrains, each mine site should be studied in the context of its unique site-specific conditions. An example of a costly mine rehabilitation case is the Britannia Mine located on Howe Sound about 45 kilometres north of Vancouver. Dubbed the worst ARD pollution problem in North America by Environment-Canada, about 1 tonne of copper and zinc combined along with an average of 14,000 m³ of acidic effluent continue to discharge into the Sound each and every day since 1976. With the presence of the Britannia Beach community (population of about 250), combined with exposure to the Vancouver-Whistler traffic along the Sea-to-Sky Highway, this site is appropriate for an urgent, yet sustainable rehabilitation program

In the Britannia Mine, the total heat flux to the mine workings can be calculated from an estimate of total surface area of the mine workings. This information is not directly available, as the mine was abandoned more than 30 years ago and existing underground maps are considerably out of date. Internal rock falls take place regularly which expose new surfaces and close off others from water and air flows. Based on the total weight of tailings produced (40 million tons), a conservative surface area estimation of 20,000,000 m² can be made. Assuming an average heat flux of 65 mW/m² [10] in this region, the total heat energy that can be captured through this surface area is about 1.3 MW thermal. The temperature of different mine levels will be averaged by the ground water circulating within the mine conduits. The total heat generated by sulphide oxidation is also significant. Based on the water chemistry (total anions and cations, including zinc and copper), the total ARD per litre of effluent is estimated to be less than 200 grams of equivalent pyrite oxidation per cubic metre of effluent. This amounts to about 570 cal/L of effluent (~0.6°C temperature increase). For a flow of 600 m³/h, the total ARD heat swept up by the effluent is estimated to be about 400 kW, i.e., about 24% of the total heat. In reality, more ARD may be generated than stated here as some of the metal content is precipitated in the mine as the ground waters are neutralized by non-acidic water flows.

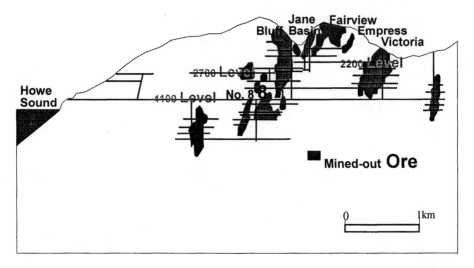

Figure 1. Cross-section of Britannia Mine. All mine effluent discharges to Howe Sound at the 4100 Level.

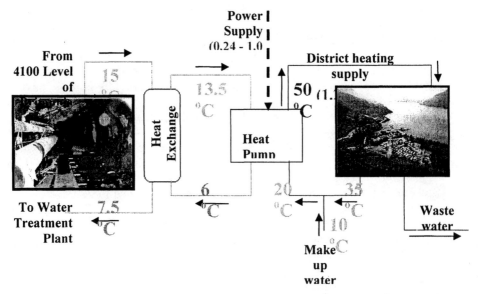

Figure 2. Schematic diagram of the proposed Geothermal Heat Pump system for Britannia Beach.

So, the heat from the mine is collected and conducted by the mine waters. This water is readily available for heat extraction as the effluent discharges at present from a single point at the 4100 Level of the mine (see Figure 1).

A glory hole in the Jane Basin area connects the underground workings to the open pit.

This effluent can be regarded as a geothermal resource that does not need any drilling as opposed to most conventional low-grade resources thus reducing the capital costs related to drilling, which can amount to 300 to 600 thousand dollars (assuming procurement of the same flow rate by boreholes). In addition there is no operating cost for pumping. The resource is therefore readily available for low-cost heat extraction through geothermal heat pump technology. Acid resistant and easy-to-clean plate heat exchangers [11–13] will be required to cope with the water chemistry and solids fouling the heat exchangers. Figure 2 shows a schematic diagram of the proposed system.

A summary of the main resource parameters at the Britannia mine site is presented in Table 1.

Heat pump technology can extract up to 4 MW net thermal energy from this single geothermal resource assuming an average flowrate of 600 m^3/h—far more than the heat requirements of the community. Note that the designed heat extraction is significantly greater than the calculated 1.7 MWt geothermal and chemical heat that enters the mine waters, considering that extra heat is recoverable by reducing the effluent to 7.5ºC (i.e. 4.5ºC below the average ground water temperature in the region). The heat pump receives electrical energy input at one fourth to one fifth of the energy output (in terms of heat) with the balance coming along for the ride from the ARD effluent.

With respect to flow fluctuations (low in January and high in June), the design capacity calculated from the average flow may not be available in low-flow episodes. At the lowest flow rate (200 m^3/h), the useful capacity may only be ~1.5 MWt. This means for an installed capacity of 4 MWt, the capacity factor may be episodically as low as 37% which is sufficient to meet current requirements at all times of the year. So to provide heat for a community of 100 households, capacity shortage is unlikely to be a problem, but as demand potentially increases in the future, episodic capacity variations may have to be moderated by:

1. using storage capacity within the mine to filter high-frequency flow fluctuations and
2. reducing the chill temperature of the heat pump to 3–4ºC which is reasonable in winter months.

Table 1. Britannia Mine Geothermal Parameters.

Parameter	Value
Average yearly peak flow (June)	1000
Average yearly low flow (January)	200
Yearly average flow	600
Effluent temperature	15
Discharge temperature (after heat recovery)	7.5
Available Heat Content (from average flow)	5.0
Heat required for 100 homes (on site)	1.2
Effluent pH	4.0 to 4.5

A fuzzy-based control system linked to statistical data on flow rates and local precipitation is being developed to optimize use of the mine's surge capacity and the dynamic performance range of the heat pumps. Curbing high-frequency flow fluctuations may also be advantageous to the performance of the effluent treatment plant, however this will depend on how fluctuations in storage levels in the mine workings affect metal levels in the effluent. We are currently evaluating an historical database to supply information to this intelligent flow-regulating system. The system envisaged will be composed of two rule-bases that account for seasonal fluctuations in demand and supply based on a projection of historical data over a planning horizon of 3 months duration. The rules can be depicted as Fuzzy Associative Memories (FAM maps) as shown in Figure 3(a) and Fig 3(b). The associated fuzzy set definitions are given in Figure 4.

For the heat extraction system, the performance of the heat pumps is tied to adjusting the Chill temperature set point to a high or low level as required while the flowrate is controlled to provide sufficient storage to meet the demand estimated for the next 3 months.

The capital costs (including heat exchangers, pumps, piping, and heat pumps) for a medium-scale facility at Britannia Beach are estimated at 2 and 2.5 million dollars, (for a 2.5 MWt facility expandable to 4 MWt capacity in the form of hot water at 45–50ºC on an annual basis). This can translate into an average savings in electrical consumption of $25,000 to $35,000 per month based on a 1.2 MWt net capacity demand (for a developed site) and assuming a power rate of $55/MWh.

Table 2. Main Economic Elements of the Britannia Mine GHP System.

Capital Cost (for full scale)	$	2,000,000–2,500,000
Savings on energy costs	$/month	25,000–35,000
Payback Period	years	5–8

The payback for such a project is 5–8 years [9]. These economic concepts are summarized in Table 2.

Geothermal development at the Britannia Mine site is presently under pilot study. Although the payback may not entice a conservative investor, the forecasted escalation in energy prices in British Columbia together with a new design for the distribution system in the town are expected to further reduce the payback period to the goal of 2 to 3 years.

(a) Rule base for optimization of heat pump controller set point (demand/flowrate are instantaneous)

Heat Demand Flowrate	Low	Medium	High
Low	Medium Chill Temperature	Medium-Low Chill Temperature	Low Chill Temperature
Medium	Medium-High Chill Temperature	Medium Chill Temperature	Medium-Low Chill Temperature
High	High Chill Temperature	Medium-High Chill Temperature	Medium Chill Temperature

(b) Rule base for optimization of mine storage (demand/flowrate are projected 3-month moving averages)

Heat Demand Flowrate	Low	Medium	High
Low	Medium Storage	High Storage	Very-High Storage
Medium	Low Storage	Medium Storage	High Storage
High	None	Low Storage	Medium Storage

Figure 3. FAM maps for the Britannia Beach GHP System.

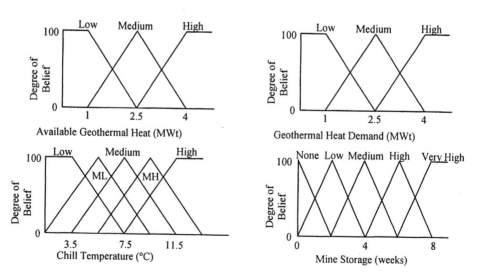

Figure 4. Fuzzy Set definitions for the FAM maps input and output variables.

ENVIRONMENTAL BENEFITS

Extraction of heat energy from mine effluent produces no greenhouses gases and generates no chemical waste. In addition, this energy source requires very little land (only for heap pumps and heat exchangers) and is very quiet, causing no harm to the local ecology. The electrical energy used to run the heat pumps, may however be less benign, depending on the source of electricity. In the British Columbia context, this is not an issue since the majority of BC's electrical power derives from hydroelectric generation. Considering that mine energy is used as an alternative to electricity, there is a net environmental benefit equivalent to the savings in electricity consumption (which can be as much as 70% in the case of Britannia Beach). In comparison with natural gas (which is the least harmful among all fossil fuels), the net savings in greenhouse gas emission can be as much as 4000 tons per year for each MWt geothermal capacity on site, with the added bonus of zero NOX and SOX emissions.

CONCLUDING REMARKS

1. Abandoned and operating mines contain significant amounts of extractable green heat. This energy can be of the order of 1 to 10 MW depending on the depth of the mine, the surface area of the mine workings, the local temperature gradient and the chemical reaction of exposed minerals.

2. Mine workings and natural fracture zones within the rock mass provide the necessary permeability to allow extraction of the significant heat flux through groundwater.

3. The ease by which mine heat is accessed provides an incentive for low to medium-grade geothermal development at mine sites where the market economy is right. Such development can generate funds to offset perpetual remediation costs associated with most sulphide-bearing mines.

4. All ingredients for a successful mine-site geothermal development are present at the Britannia Mine site. Development of this site (presently under feasibility study) is expected to provide heat to the entire Britannia Beach community and facilitate research for similar developments at other mine sites.

REFERENCES

[1] M.M. Ghomshei and Sadlier-Brown, 1996. Direct Use Energy from Hotsprings and Subsurface Geothermal Resources of British Columbia, Bitech Publishers, Ltd. Richmond, British Columbia.
[2] Lienau and Lunis, 1991. Geothermal direct use engineering and design guidebook, Geo-Heat Center. Oregon Institute of Technology. Klamath Falls, Oregon.
[3] P.J. Lienau, T. Boyd, R. Rogers, 1995. Ground-Source Heat Pump Case Studies and Utility Programs, Technical Report to USDOE, Geothermal Division, Oregon Institute of Technology, Geo-Heat Center, Klamath Falls, OR.

[4] M.A. Jessop, M.M. Ghomshei, M.J. Drury, 1991. Geothermal Energy in Canada, Geothermics, vol. 20, No. 5/6, pp. 369–385.
[5] Jessop, A.M., Macdonald, J.K, Spence, H., 1995., Clean Energy from Abandoned Mines at Springhill, Nova Scotia, Energy Sources, Volume 17. pp. 93–106
[6] M.M. Ghomshei, A. Holmes, R.L. Lawrence, T. Carriou, 1997. Acid Rock Drainage Prediction of Samatosum Mine, B.C., International Conference on Acid Rock Drainage, Vancouver. pp. 351–366.
[7] A.I.M. Ritchie, 1994. Sulphide Oxidation Mechanism: Controls and Rates of Oxygen Transport, In Jambor and Blowes (eds), Short course handbook on environmental geochemistry of sulfide mine-wastes. Mineralogical Association of Canada, Waterloo, Ontario, pp. 201–246.
[8] J.R. Harries and A.I.M. Ritchie, 1987. The effect of rehabilitation on the rate of oxidation of pyrite in a mine waste dump, Env. Geochem. Health 9, 27–36.
[9] M.M. Ghomshei, J.A. Meech, D.W. Fraser, R.A. Dakin, 2003. Geothermal Heat Pump Options: Fuzzy Arithmetic for a Bright Decision, in Intelligence in a Materials World, Selected Papers from IPMM-2001. 3rd International Conference on Intelligent Processing and Manufacturing of Materials. CRC Press, New York., 609–624.
[10] Lewis and Wang, 1992. Influence of Terrain on Bedrock Temperature, Paleogeography, Paleoclimatology, Paleoecology (Global and Planetary Change Section). 98, 87–100
[11] K. Rafferty, 1991. Heat Exchangers, In: Geothermal Direct Use Engineering and Design Guidebook, Eds. Lineau and Lunis, Geo-Heat Center. Oregon Institute of Technology.
[12] K. Rafferty, 1994. Groundwater and Ground-Coupled Heat Pump Systems, Geo-Heat Quarterly Bulletin, 15(3), 9–12.
[13] K. Rafferty, 1999. Scaling in Geothermal Heat Pump Systems, Geo-Heat Center, Oregon Institute of Technology, prepared for U.S. Department of Energy, 63.

Al Clusters as a Tool for Hydrogen Storage

A. GOLDBERG, M. MORI and A. BICK
Accelrys K.K. Nakarin-Auto Bldg. 5F. 2-8-4, Shinkawa Chuo-Ku, Tokyo 104-0033, Japan

ABSTRACT

Accelrys module $DMol^3$ was used to consider the Al_{13} cluster as an object for hydrogen storage. Density Functional Theory (DFT) and the Generalized Gradient Approximation (GGA) were used to obtain optimized and Transition State (TS) structures of $Al_{13}H_n$ (n=1-12) clusters. The binding energy of hydrogen atom to the surface of Al cluster is 2.4–2.5 eV and the hydrogen atom can be adsorbed on the cluster surface without crossing a potential barrier. However this binding energy is not enough in order to dissociate the hydrogen molecule. Different optimized structures were obtained with one hydrogen atom being displaced in different positions on the Al_{13} cluster surface. We considered TS between those positions. The optimized structures with one hydrogen atom on the Al_{13} cluster surface were used to obtain optimized structures with two hydrogen atoms displaced on the surface of Al_{13} cluster. The possibility to adsorb larger number of hydrogen atoms on the cluster surface was investigated. Our theoretical study is expected to provide predictions on the possibility of using Al clusters as a tool for hydrogen storage.

INTRODUCTION

As natural resources continue to deplete and environmental contamination continues to be a big issue, development of alternative fuel sources is one of the most crucial topics of the modern world industrial technologies. In this respect Fuel Cells (FC) seem to be one of the most prospective sources of energy for the future. FC being highly efficient offers an opportunity to achieve near zero emission. Although the FC technique has been known for many years it is still in a development stage [1].

The hydrogen storage device is the crucial issue in FC application. There are several problems that are waiting to be solved in order to make one step further in using FC in industry. Current status of hydrogen storage technologies is inappropriate in terms of size of the storage devices, safety and cost. This is why hydrogen storage systems are attracting much interest of researchers. A safe, compact and inexpensive system that is capable of storing large amount of hydrogen would be of great interest. In particular the light metal clusters such as Al are promising material for hydrogen storage devices. Hydrides of Al clusters offer the advantage of high storage densities and in addition they are characterized by their high safety potential as hydrogen release is endothermic.

Metal clusters possessing unique properties are difficult for experimental investigation [2]. Typically prepared by molecular beams their structure investigation is practically impossible due to highly reactivity. Although the problems encountered during the theoretical treatment are enormous the electronic structure calculations using ab initio methods can shed light on structure, evolution of energetic, thermodynamic, spectroscopic and dynamic properties of metal clusters. It is important to understand the cluster size and hydrogen site-specific effects in order to provide bridging between clusters, nanowires and bulk properties of the material to be used for hydrogen storage.

According to the jellium model [3] the Al clusters are organized in structures that reflect the existence of electronic shells with magic numbers similar to those in isolated atoms. This theoretical finding was confirmed experimentally by the population analysis of the cluster beams with magic electron numbers: 2, 8, 20, 40, 70, 112 ... [4]. It was shown [2] that clusters Al_7 and Al_{13} are stable due to the fact that these clusters have number of electrons that is close to the jellium closed shell [2]. However it appeared that larger clusters are energetically favored compared with smaller ones. The reason for that is that the jellium model fails for larger clusters as the spacing between shells vanishes [2]. It was shown that adding hydrogen atom to the Al_{13} cluster can stabilize the structure as the jellium electronic shell is complete in this case. Alonso et al. [5] showed this is the way to increase HOMO-LUMO gap.

In this study we simulated the interaction of hydrogen atom and hydrogen molecule with the surface of Al_{13} cluster. We obtained stable isomers of the Al_{13} cluster with different number of hydrogen atoms in it. The possibility of adsorbing greater amount of hydrogen atoms was considered. Our paper is intended to provide predictions on the usage of Al clusters as a tool for hydrogen storage in FC technology.

CALCULATION DETAILS

The calculations within the framework of this paper were performed using the DFT program DMol^3 provided by Accelrys Inc. [6]. DMol^3 is based on the numeric atomic functions basis set that contains exact solutions to the Kohn-Sham equations for the atoms [7]. In the present study, a double numeric polarized (DNP) basis set has been used that includes all occupied atomic orbitals plus a second set of valence atomic orbitals plus polarized d-valence orbitals. For exchange, the Becke model (B88) [8] in conjunction with the correlation Lee-

Yang-Parr [9] nonlocal functional (BLYP) equation was used. An unrestricted spin approach was applied with all electron being considered explicitly.

Atom centered grids were used for the numerical integration. The "Medium" option in DMol was used that includes about 1000 grid points for each atom. A numerical integration real space cut-off of 5 Å was imposed [10].

Self-Consistent-Field (SCF) convergence was set at the root-mean-square (rms) change in the electronic density to be less than 1×10^{-6} Ha. Geometries were optimized using efficient algorithm taking advantage of delocalized internal coordinates [11]. Although the time spent for the optimization in Cartesian and delocalized coordinates on each iteration is almost the same the number of iterations using delocalized orbitals is much less. The criteria applied for the convergence threshold in geometry optimization were 1×10^{-5} Ha for energy, 0.002 Ha/Å for force and 0.005 Å for displacement. For all the optimized structures we performed frequency analysis to prove the obtained structure is the minimum energy structure. For Transition State (TS) search we applied LST/QST Halgren, Libscomb method [12]. Consequently the TS optimization procedure has been used followed by the frequency analysis. As a result we obtained one negative frequency that was a proof for the correctly located TS. Subsequently we confirmed the obtained TS is between considered minima applying Nudge Elastic Band (NEB) method [13] implemented in $DMol^3$. This method allows finding all local minima between two considered optimized states.

As a starting configuration the icosahedral structure of Al_{13} cluster was always used that was reported [2] as the most stable structure for Al_{13} cluster. To add hydrogen atom we started from positioning it at distance of about 4 Å from the surface. Eventually the three minima of $Al_{13}H$ were found. We proceeded with hydrogen molecule and greater amount of hydrogen investigating the possibility of adsorption and distribution of hydrogen atoms on the surface of Al_{13} cluster. The results of this investigation are reported in the next section.

RESULTS AND DISCUSSION

Al_{13} Cluster

Aluminum is a typical metal with deficient metallic bonding. This should lead to the structure with large average number of first nearest neighbors and high symmetry. The highest possible cluster symmetry is icosahedral (I_h) one. It was shown by Ahlrichs et al [2] that Al_{13} cluster structure with I_h symmetry has the lowest energy (Figure 1).

It was confirmed experimentally [4] that small clusters of many metals are featuring the existence of electronic shells. The high stability of the clusters with closed electronic shell with electron numbers 2, 8, 10, 20, 34, 40, 58, 92, ... is a robust feature. Al_{13} having a compact icosahedral structure and 39 valence electrons is lacking just one electron to complete its outer electron shell [14]. Closed shell structure is supposed to have a large HOMO-LUMO gap, the electron affinity of the Al_{13} has to be low and the Al_{13}^- anion should be more stable than the neutral cluster. Our calculations showed that the anion cluster is 2.46 eV lower than the neutral one. Ahlrichs et al. [2] reported 3.5 eV. Although

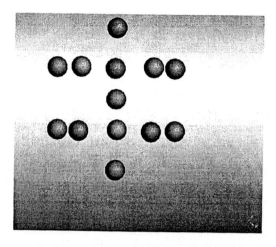

Figure 1. Al_{13} cluster with ideal I_h structure.

there is a difference in the number, it is confirmed that the anion cluster is more stable than the neutral one. Interestingly we also found that the Al_{13}^{2-} anion is more stable than the neutral one. Its ground state energy is only 1.56 eV higher than the Al_{13}^- anion. This finding shows that Al_{13} cluster can favor adsorption of more than one hydrogen atom on its surface from the electronic point of view. The obtained electron affinity for the neutral cluster is 2.76 eV which is rather low in comparison to the ionization potential that is 5.99 eV. The result expected for the HOMO-LUMO gap was not confirmed. It was expected that HOMO-LUMO gap for the Al_{13}^- anion would be larger than the Al_{13} neutral cluster since for an anion, it is a closed-shell electronic configuration. It appears that for an anion, the HOMO-LUMO gap is 1.73 eV while that for the neutral cluster is 1.83 eV. This feature can be attributed to the fact that a prediction for stable electronic levels in the jellium model is obtained under the assumption of an average spherical potential while for partially occupied HOMO, high symmetry is not preserved. The results discussed above are summarized in Table 1.

Table 1. Comparison between neutral and negative Al_{13} clusters.

Cluster	HOMO-LUMO gap (eV)	e-affinity (eV)	Ionization potential (eV)	Relative Ground State energy (eV)
Al_{13} neutral	1.83	2.76	5.99	2.46
Al_{13}^- anion	1.73	1.55	2.76	0

Al₁₃H

First we have investigated the adsorption of hydrogen atom on the surface of Al_{13} cluster. Initially hydrogen atom was positioned at a distance of 4 Å off the surface as it is shown on the Figure 2a. The geometry optimization run brought hydrogen atom close to the surface (Figure2b) indicating there is no potential barrier for the hydrogen atom to be adsorbed on the surface of the Al_{13} cluster. Furthermore when hydrogen atom was initially placed inside the cluster it moved out to the surface under the geometry optimization procedure. As a result we can assume that hydrogen atom prefers surface displacement within the cluster.

Three stable isomers were found with the hydrogen atom being adsorbed on the surface of Al_{13} cluster. Those isomers are shown on Figure 3. Hydrogen atom can occupy different positions in the Al_{13} cluster. It can be attached to Al atom (Figure 3a) or can be displaced between two Al atoms (Figure 3b) or between three Al atoms (Figure 3c). These three isomers hereafter will be referred to as **(1)**, **(2)** and **(3)**, respectively. The isomer **(3)** has energy that by 4.66 kcal/mol higher than the lowest energy isomer **(1)**. This energy difference is rather high that makes the probability of finding isomer **(3)** in the Boltzmann distribution negligibly small. On the contrary the energy difference between the lowest energy isomer **(1)** and the isomer **(2)** is only 0.42 kcal/mol. With this energy difference the number of lower energy isomers **(1)** would be twice as much as those with hydrogen atom between two Al atoms (isomer **(2)**) in the Boltzmann distribution at room temperature. This evaluation is based on the ratio

$$\frac{\rho_1}{\rho_2} = \frac{1}{\exp\left(\frac{-\Delta E}{kT}\right)}$$

where ρ_1 is the population of the isomer **(1)**, ρ_2 is the population of the isomer **(2)**, ΔE is the energy difference between two isomers, k is the Boltzmann constant and T is the temperature.

As can be seen in Figure 3a, adsorption of a hydrogen atom onto one Al atom (isomer **(1)**) induces distortion of the Al_{13} cluster on its opposite side while the Al_{13} cluster in isomer **(2)** experiences only minor changes in the vicinity of the hydrogen atom. As expected the Al-H distance is smallest in the case of H attached to one Al (1.585 Å). For the second lowest energy isomer **(2)**, the Al-H distances are 1.775 Å and 1.782 Å. These distances for the two Al-H bonds can swap and there is a small potential barrier between the two states. However they are practically indistinguishable.

We obtained a Transition State (TS) structure between the two lowest energy isomers of the $Al_{13}H$ cluster. The potential barrier for transition is about 5 kcal/mol. This rather small value points towards easy mobility of H atom along the surface of an Al_{13} cluster. As the binding energy of hydrogen to the surface is also not high (~ 2.4 to 2.5 eV) the release of hydrogen atoms from the surface

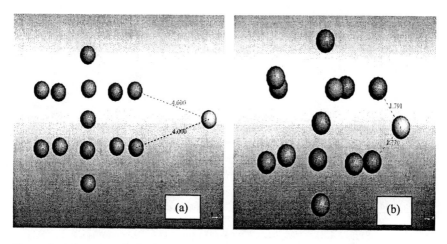

Figure 2. Adsorption of a hydrogen atom onto the Al_{13} cluster surface. (a) is the initial state and (b) is the final state.

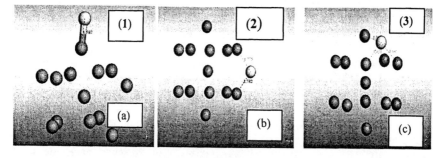

Figure 3. Three different isomers of $Al_{13}H$ cluster.

should not require a large expenditure of energy. The binding energy is calculated according to the following expression:

$$E_B = \frac{E_{Al_{13}} - E_{Al_{13}-H_n} - E_{H_n}}{n}$$

where $E_{Al_{13}}H_n$ is the energy of the cluster $Al_{13}H_n$, $E_{Al_{13}-H_n}$ is the energy of the same cluster without hydrogen atoms, E_{H_n} is the energy of the hydrogen atoms without Al atoms. For $Al_{13}H$ we used n=1.

From the electronic density map in Figure 4, we can infer that the bonding between hydrogen and Al atoms is ionic. In the case of a hydrogen atom being displaced between two Al atoms (isomer (**2**)), it must share its electron between

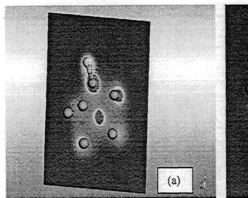 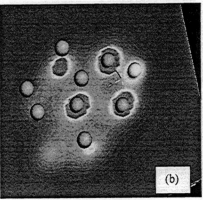

Figure 4. Electron density map. (a) isomer (**1**), (b) isomer (**2**).

Al atoms making the bonding weaker in comparison to the case where it shares its charge with only one Al atom (isomer (**1**)), although in both cases the bond is purely ionic. This confirms the idea of a closed electronic shell in an Al_{13}^- cluster. The hydrogen atom must donate its only electron to the cluster to complete its electronic shell resulting in ionic bonding between the cluster and the hydrogen atom.

We estimated the HOMO-LUMO band gap, electron affinity and ionization potential for $Al_{13}H$ neutral and $(Al_{13}H)^-$ anion. The results are shown in Table 2.

The HOMO-LUMO band gap obtained for the anion cluster is 1.43 eV. Measured by photoelectron spectroscopy the HOMO-LUMO gap for the same anion clusters is 1.4 eV [15] that is in complete agreement with our estimation. Corresponding sizeable HOMO-LUMO gap of 1.53 eV was obtained for the neutral clusters as well. Apparently Al_{13} clusters in presence of H atom get distorted changing its almost spherical symmetry. It results in decrease of the HOMO-LUMO gap form 1.83 eV for the neutral cluster without hydrogen to 1.53 eV with the hydrogen atom. In case of anion the HOMO-LUMO gap was decreased accordingly from 1.73 eV to 1.43 eV. Electron affinity of the neutral

Table 2. Comparison between neutral and negative $Al_{13}H$ clusters.

Cluster	HOMO-LUMO gap (eV)	e-affinity (eV)	Ionization potential (eV)	Relative Ground State energy (eV)
$Al_{13}H$ neutral	1.53	1.80	6.00	1.86
$(Al_{13}H)^-$ anion	1.43	1.67	1.80	0

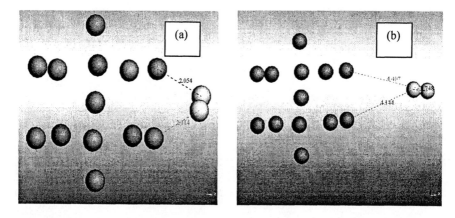

Figure 5. Repulsion of H_2 molecule from the Al_{13} cluster. (a) initial state, (b) final state.

cluster also reduced from 2.76 eV to 1.80 eV. The latter value is in good agreement with the experimental value of 2.0 eV [15]. The calculated ionization potential of the $Al_{13}H$ cluster is 6.00 eV. Relatively low electron affinity and high ionization potential points to the high stability of the neutral cluster. However we have to stress that the anion $(Al_{13}H)^-$ cluster still has lower ground state energy than the neutral one. This feature can not be explained by jellium model and has to be clarified further. Moreover the electron affinity of the second electron to the $(Al_{13}H)^-$ is even lower than that of the neutral cluster.

$Al_{13}H_2$

We investigated the interaction of the hydrogen molecule H_2 with the Al_{13} cluster. The molecule is repulsive to the cluster. Initially positioned at a distance of about 2 Å (Figure 5a) it was repelled to a longer distance of about 4 Å (Figure 5b). The dissociation energy of the hydrogen molecule is about 4.5 eV that is almost as twice as high as the binding energy of hydrogen atom to the Al_{13} cluster. As a result in order to adsorb hydrogen atoms onto the surface of Al_{13} cluster it is necessary first to use a catalyst to dissociate the hydrogen molecule. The main source of hydrogen atoms is hydrogen (H_2) molecule. So it is absolutely necessary to find the way of dissociation of the H_2 molecule into separate atoms. Work in this direction is still in progress. Three different isomers of $Al_{13}H$ cluster can contribute to a different combination of adsorption of two hydrogen atoms onto the Al_{13} cluster surface. The two lowest energy isomers are shown in Figure 6. The most stable isomer is the one with one hydrogen atom attached directly to an Al atom with the other being displaced between two Al atoms with one Al atom being shared. This is the combination of $Al_{13}H$ isomers (**1**) and (**2**). This isomer will be referred to from now on as: isomer (**a**). The second lowest energy isomer with two hydrogen atoms attached directly to two different Al atoms is about 1.5 kcal/mol higher than the first one (isomer (**b**)). This makes the population ratio a:b in the characteristic Boltzmann distribution to be 92:8. Although $Al_{13}H$

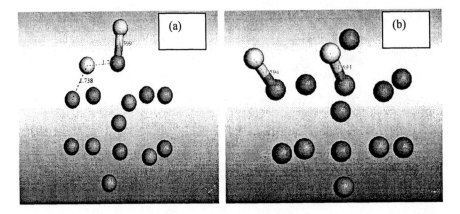

Figure 6. Two lowest energy isomers of $Al_{13}H_2$ cluster.

clusters have a noticeable contribution from the second lowest energy isomer (**2**) in the overall population opposite to the $Al_{13}H_2$ cluster, the population is dominated by the contribution from the lowest energy isomer (**a**).

$Al_{13}H_n$

For practical application of the Al clusters for hydrogen storage devices there is a need to adsorb as more hydrogen atoms as possible. We tried to follow the trend by increasing the number of atoms that can be adsorbed onto the Al_{13} cluster surface. We found out that with some fluctuations the binding energy of hydrogen atoms to the cluster is always about 2.4–2.5 eV and it does not depend on the number of hydrogen atoms. The calculations were done for n = 1, 2, 3, 4 and 12. The obtained structures for n = 1, 2 and 3 are shown on Figure 7 while that for n = 12 is shown on Figure 8. It makes us think that even greater amount of hydrogen can be attached to the surface. Although large amount of hydrogen atoms may distort and destroy the small cluster as Al_{13} it might be possible to adsorb larger amount of hydrogen atoms on the bigger clusters. The work on larger clusters is in progress.

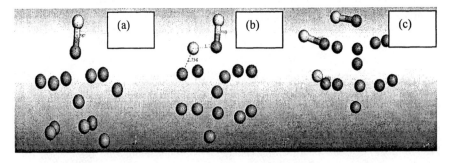

Figure 7. Isomers of $Al_{13}H_n$, n=1 (a), 2 (b), 3 (c). Distortion of the Al_{13} cluster reduces with increasing n.

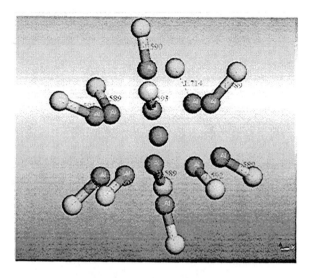

Figure 8. $Al_{13}H_{12}$ Cluster.

As a final remark we would like to draw attention to the fact that an increased number of hydrogen atoms adsorbed onto the Al_{13} cluster surface can stabilize the cluster. The lowest energy structure with one hydrogen atom attached to the Al_{13} cluster surface distorted the cluster. Additional hydrogen atoms have smaller effects on the cluster geometry. This fact supports the idea of lower energy of hydrogen atom adsorption on a cluster surface.

CONCLUSIONS

1. Hydrogen is adsorbed on the Al_{13} cluster surface without crossing a potential barrier.
2. Hydrogen molecule H_2 is repulsive to the surface of the Al_{13} cluster.
3. The $Al_{13}H$ cluster has three different stable minima with the hydrogen atom being attached to one Al atom, being displaced between two Al atoms or being displaced between three Al atoms. The most stable isomer is with one hydrogen atom attached to one Al atom. The most stable isomer leads to significant distortions of the cluster.
4. With an increase in the number of hydrogen atoms adsorbed on an Al_{13} cluster surface, the distortion is reduced and the binding energy with small fluctuations does not change.

ACKNOWLEDGEMENTS

We want to thank Prof. Tachibana for fruitful discussions that stimulated this work.

REFERENCES

[1] K. Glassford, 2000. Simulation Technology through Vertical Marketing, Accelrys Fuel Cell Technology Report.
[2] R. Ahlrichs, S.D. Elliott, 1999. Clusters of Aluminum, a Density Functional Study, Phys. Chem. Chem. Phys. 1, 13–21.
[3] W.A. de Heer, W.D. Knight, M.Y. Chou, 1987. Electronic shell structure and metal-clusters, Solid State Phys. 40, 93–181.
[4] M.F. Jarold, 1992. In Clusters of Atoms and Molecules, ed. H. Haberland, Springer, Berlin, Heidelberg.
[5] J.A. Alonso, M.J. Lopez, L.M. Molina, F. Duque, A. Mananes, 2002. Conditions for the self-assembling of cluster materials, Nanotechnology, 13, 253–257.
[6] B. Delley, 1990. An all-electron numerical method for solving the local density functional for polyatomic molecules, J. Chem Phys. 92(1), 508–517. DMol^3 is available as a part of Materials Studio by Accelrys Inc.
[7] B. Delley, 1995. In Density Functional Theory: A Tool for Chemitsry. J.M. Seminario, P. Politzer, Eds. Amsterdam, The Nerthelands.
[8] A.D. Becke, 1988. A multicenter numerical integration scheme for polyatomic molecules, J. Chem Phys. 88(4), 2547–2553.
[9] C. Lee, W. Yang, R.G. Parr, 1988. Development of the Colle-Salvetti correlation-energy formula into a functional of the electron density, Phys. Rev. B, 37(2), 785–789.
[10] Materials Studio DMol^3 Version 2.2
[11] J. Andzelm, King-Smith, G. Fitzgerald, 2001. Geometry optimization of solids using delocalized internal coordinates, Chem. Phys. Lett. 335(4), 321–326.
[12] T.A. Halgren, W.N. Lipscomb, 1977. The Synchronous-Transit Method for Determining Reaction Pathways and Locating Molecular Transition States, Chem. Phys. Lett. 49(2), 225–231.
[13] G. Henkelman, H. Jonsson, 2000. Improved tangent estimate in the nudged elastic band method for finding energy paths and saddle points, J. Chem. Phys. 113(22), 9978–9985.
[14] F. Dukue, A. Mananes, 1999. Stability and electronic properties of pure aluminum clusters, Eur. Phys. J. D 9, 223–227.
[15] S. Burkart, N. Blessing, B. Klipp, J. Muller, G. Gantefor, G. Seifert, 1999. Experimental verification of the high stability of $Al_{13}H$: a building block of a new type of cluster material? Chem. Phys. Lett. 301(6), 546–550.

Theoretical Study of Clathrate Hydrates with Multiple Occupation

V. R. BELOSLUDOV
Institute of Inorganic Chemistry, SB RAS, Novosibirsk, Russia and Center for Northeast Asia Studies of Tohoku University, Sendai, Japan

T. M. INERBAEV
Institute of Inorganic Chemistry, SB RAS, Novosibirsk, Russia

R. V. BELOSLUDOV, M. F. SLUITER and Y. KAWAZOE
Institute for Material Research, Tohoku University, Sendai, Japan

J. KUDOH
Center for Northeast Asia Studies of Tohoku University, Sendai, Japan

ABSTRACT

In this work, the electronic, structural, dynamic and thermodynamic properties of structure II, H and tetragonal Ar clathrate hydrates have been calculated and the effect of multiple occupancy on their stability has been examined using first-principles and lattice dynamics calculations. The dynamic properties of these clathrates have been investigated depending on the number of guest molecules in a clathrate cage. It has been found that selected hydrate structures are dynamically stable. The calculated cell parameters are in agreement with experimental data. We also report the results of a systematic investigation of cage-like water structures using first-principles calculations. It has been observed that Ar clusters can be stabilized in different water cages and the stability is strongly dependent on the number of argon atoms inside the cages.

INTRODUCTION

Clathrate hydrates comprise one type of crystalline inclusion compounds consisting of guest atoms or molecules and host framework of water molecules linked by hydrogen bonds. The clathrate hydrates are formed when water molecules arrange themselves in a cage-like structure around small molecules and hence many of their physical and chemical properties are different from ice [1]. These compounds are a potential source of energy in the future since natural gas (methane) hydrates occur under conditions of high pressure and low temperature in the permafrost regions or below the seafloor. Global estimates of methane in clathrate hydrates may exceed 10^{16} kg, which represents one of the largest sources of hydrocarbons on Earth. Speculations about large releases of methane from clathrate hydrates have raised serious, but unresolved questions about its possible role in climate change. Among many potential applications of clathrate hydrates, these compounds can also be used as gas (such as CO, CO_2, O_2 or H_2) storage materials. Therefore, a good understanding of the chemical and physical properties of clathrate hydrates such as structure, dynamics and stability is essential for practical manipulation of this class of inclusion compounds.

In the early 1950's, two types of gas hydrates known as cubic structures I (CS-I) and II (CS-II) were identified by Stackelberg, Pauling and Claussen [2,3,4]. More recently, the third type of gas hydrates crystal with hexagonal structure H (sH) was determined [5,6]. At the present time, most of recognized gas hydrates existing at moderate pressure (hundreds bars) belong to these types of structures. The gas hydrates are acutely sensitive to pressure variation by virtue of friable packing of host framework and relatively weak binding energy between water molecules. Experimental studies [7,8,9,10,11] on different gas-water systems at pressures up to 15 kbar showed that the sequential change of hydrate phase is observed by increasing pressure.

There is a well know general rule that CS-I hydrates are formed by molecules with van der Waals diameters of up to about 5.8 Å while CS-II hydrates are formed by large molecules up to about 7.0 Å in size. The exceptions of this rule are Ar, Kr, N_2, and O_2 molecules with small van der Waals diameters up to about 4.3 Å, which form CS-II hydrates [12,13]. At a low temperature Holder and Manganeillo [14] predicted using van der Waals and Platteeuws [15] model that these small molecules should favor CS-II. This result was verified by Davidson and co-workers [12].

At the present time, the possibility of double or more occupancy in the large cages of CS-II has been a subject of topical interest. Crystallographic studies of the structure and filling fractions of the small and large cages of N_2, O_2 and air hydrate by high-resolution neutron powder diffraction have shown that the large cages are partly doubly occupied for these hydrates at high pressures [16,17]. The degrees of filling 1.8 and 2.3 molecules per cage respectively were obtained by powder neutron diffraction at pressure of 3.4 and 4.3 kbar, respectively [11]. The doubly occupy the large cages of CS-II N_2 and Ar hydrates was also examined by molecular dynamics calculations [18,19,20,21]. Thus, it was shown that N_2 hydrate in double occupancy is stable at temperature 80 and 273 K below 2.5 kbar [18] and double occupied Ar hydrate with may be also stabilized by high external pressure [19]. Recently, the CS-II H_2 clathrate hydrate was synthesized and studied at high pressure [22]. It was proposed that the small cages are doubly occupied and the large cages are quadruply occupied by hydrogen molecules with a diameter of 2.72 Å.

Phase diagram of argon—water system was studied at high pressure and the formation of four hydrates was established [10,11,23]. Powder neutron diffraction study showed that in Ar—water system CS-II hydrate exists from ambient pressure up to 4.5 kbar. After increasing the pressure, the phase transition occurs and the Ar hydrate with hexagonal structure is formed up to 7.6 kbar. In the pressure range of 7.6 to 10 kbar, the Ar hydrate with unknown before type of structure was obtained and a new tetragonal crystal structure of hydrate with one type of doubly occupied cavity was proposed in [11].

In the present study, the electronic, structural, dynamic and thermodynamic properties of CS-II, sH and tetragonal Ar clathrate hydrates have been investigated and the effect of multiple occupancy on stability of these hydrates has been examined using first-principles and lattice dynamics calculations.

THEORETICAL METHODS

First-Principles Calculations

Full geometry optimization and vibrational analyses of selected cage-like structures of water clusters with and without enclathrated guests were performed at the Hartree-Fock (HF) level. A large yet computationally manageable basis set, 6-31+G(d) including polarization and diffuse functions, was used. The inclusion of diffusion functions in the basis set is necessary for a better description of the structure and energetic of hydrogen bonded complexes [24]. The optimizations were performed using the redundant internal coordinate procedure [25] and the vibrational frequencies were calculated from the second derivative of the total energy with respect to atomic displacement about the equilibrium geometry. If all of eigenvalues of Hessian matrix are positive the corresponding frequencies are real. This means that these structures are indeed (at least local) minima. The stabilization energy (SE) was considered as the difference between the total cluster energy and the energies of separated empty water cages and guest atoms at an infinite distance. All first-principles calculations were carried out using the Gaussian 98 package [26].

Dynamical and Thermodynamical Properties

The free energy F_{qh} of crystal in this model is calculated within the framework of lattice dynamics (LD) in the quasiharmonic approximation as

$$F_{qh} = U + F_{vib}$$

where U is the potential energy, F_{vib} is the vibrational contribution:

$$F_{vib} = \frac{1}{2}\sum_{j\vec{q}}\hbar\omega_j(\vec{q}) + k_B T \sum_{j\vec{q}} \ln\left(1 - e^{-\hbar\omega_j(\vec{q})/k_B T}\right)$$

where $\omega_j(\vec{q})$ are the frequencies of crystal vibrations. The eigenfrequencies $\omega_j(\vec{q})$ of molecular crystal vibrations are determined by solving numerically the following system of equations

$$m_k \omega^2(\vec{q}) U_\alpha^t(k,\vec{q}) = \sum_{k',\beta}[\tilde{D}_{\alpha\beta}^{tt}(\vec{q},kk')U_\beta^t(k',\vec{q}) + \tilde{D}_{\alpha\beta}^{tr}(\vec{q},kk')U_\beta^r(k',\vec{q})]$$

$$\sum_\beta I_{\alpha\beta}(k)\omega^2(\vec{q}) U_\beta^r(k,\vec{q}) = \sum_{k',\beta}[\tilde{D}_{\alpha\beta}^{rt}(\vec{q},kk')U_\beta^t(k',\vec{q}) + \tilde{D}_{\alpha\beta}^{rr}(\vec{q},kk')U_\beta^r(k',\vec{q})]$$

where $\widetilde{D}_{\alpha\beta}^{ii'}(\vec{q},kk')$ (α, β=x,y,z) are translational ($i,i'=t$), rotational ($i,i'=r$) and mixed ($i=t$, $i'=r$ or $i=r$, $i'=t$) elements of the molecular crystal's dynamical matrix, the expressions for which are presented in [27,28], $U_\alpha^j(k,\vec{q})$ (α, β=x,y,z) is the amplitude of vibration, m_k and $I_{\alpha\beta}(k)$ are the mass and inertia tensor of k-th molecule in the unit cell.

In the quasiharmonic approximation, the free energy of a crystal has the same form as in the harmonic approximation but the structural parameters at fixed volume depend on the temperature. This dependence is determined self-consistently at calculation of the system's free energy. To obtain the equation of state P(V) at fixed temperature, the expression

$$P = -\left(\frac{\partial F_{qh}}{\partial V}\right)_T$$

was used. For calculations of the free energy, the molecular coordinates (the centers of mass positions and orientations of molecules in the unit cell) were determined by the Newton-Raphson method. In this method new coordinates of molecules can be found from the minimum of potential energy of the expanded lattice.

The interactions between water molecules in hydrate were described by the modified empirical TIP4P potential [29]. The protons were placed according to the Bernal-Fowler rule and the water molecules were oriented so that total dipole moments of the unit cells of the hydrates are equal to zero. The long-range electrostatic interactions were computed by the Ewald method. The free energy and the derivatives of free energy were calculated using 3x3x3 k-points inside the Brillouin zone. The guests were considered as spherically symmetric Lennard-Jones particles. The potential parameters for the argon-argon interactions were taken from [15].

RESULTS AND DISCUSSION

Dynamic of CS-II, sH and Tetragonal Ar Clathrate Hydrates with Different Filling of Cavities

The phonon density of states (the number of vibrational modes in a given frequency interval) is the most widely discussed and best understood dynamical property of molecular crystal. To calculate this property for CS-II, sH and tetragonal Ar clathrate hydrates, the LD method was used. The clathrate hydrate of CS-II consists of two fundamental cages (small (5^{12}) and large ($5^{12}6^4$)) with radius of about 3.91 and 4.73 Å, respectively. The unit cell of the CS-II hydrate contains 136 water molecules, forming 16 small and 8 large cavities. For comparison, the calculations of phonon density of states (DOS) of the CS-II hydrate have been performed for various fillings: empty host lattice, single occupancy of the large and small cages and double occupancy of the large cages and single occupancy of the small cages. The results are shown in Figure 1. The peculiarity of this plot is a gap of about 240 cm^{-1} which divides the low- and high-frequency vibrations of lattice. For empty host lattice, the analysis of the eigenvectors derived from the LD method revealed that the low-frequency region (0-300 cm^{-1}) consists of translation modes of water and the high-frequency region (520–1000 cm^{-1}) consist of libration modes of water host framework. Argon atoms influence the vibrations of the host water framework only slightly and guest vibrations are located in the vicinity of the peaks of phonon density of states at 0-40 cm^{-1}. The peak in the negative region, in the case of the single occupancy of both types of cages, corresponds to the motions of argon in large cages with imaginary frequencies. This means that the argon atoms are not localized in potential minima and can be freely moved inside the large cages. However, the all frequencies of water framework are positive in all the cases and hence both the single and the double occupancies do not disrupt the dynamical stability of the host lattice.

The unit cell for sH hydrate contains 34 water molecules, forming 3 small, 2 medium and one large cavities. Radius of small (5^{12}), medium ($4^3 5^{12} 6^3$) and large ($5^{12} 6^8$) cages are of about 3.91, 4.06 and 5.71 Å, respectively [5]. We examine the dynamical properties only for two cases. There are dynamic of empty host lattice and dynamic of Ar hydrate of sH with maximum experimentally predicted [11] number of guest atoms (fivefold occupancy of the large cages and single occupancy of the small and medium cages). The DOS calculations have been done using the experimental values of cell parameters at T=293 K. The results are shown in Figure 2. The large intensive peak at 20 cm^{-1} corresponds to translation of the guest atoms. After inclusion of argon atoms, the vibrational spectrum of host lattice has practically same features as in the case of the empty hydrate structure and hence the dynamical stability of H hydrate is not significantly changed even for fivefold occupancy of the large cages. The unit cell for the argon hydrate with the tetragonal structure contains 12 water molecules, forming one cavity ($4^2 5^8 6^4$) [11]. The DOS calculations have been performed both for empty host lattice and double occupancy of the cages using the experimental values of cell parameters at T=293 K. The results are shown in Figure 3. The empty host lattice is dynamically stable because all frequencies of water framework are positive. Dynamical stability of water lattice preserves after inclusion two guest atoms in each cage. The density of vibrational states of the empty tetragonal hydrate has same features as the density of vibrational states of ice Ih [30], hydrates of CS-I [31], CS-II and sH. The frequency region of molecular vibrations is divided into two zones. In the lower zone (0–315 cm^{-1}) water molecules mainly undergo translational vibrations, whereas in the upper one (540–980 cm^{-1}) the vibrations are mostly librational. In comparison with hydrates of CS-I, the frequency spectrum of tetragonal argon hydrate is shifted towards higher frequencies, which may be explained by greater density of a new tetragonal crystal argon hydrate compared to hydrates of CS-I. Vibrational frequencies of argon atoms in the cavities lay in the region 20–45 cm^{-1} and 60–110 cm^{-1}. The guest atoms influence the phonon spectrum of host framework, diminishing the density of states in upper zone of translational vibrations and the density of states of librational vibrations.

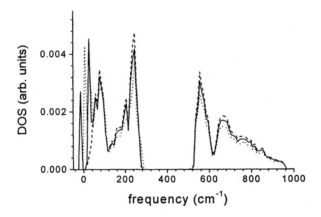

Figure 1. Phonon density of vibrational states (DOS) of CS-II Ar hydrates for various fillings: empty host lattice (solid line); double occupancy of the large cages (dashed line) and single occupancy of the large cages (dotted line). For the small cages, only single occupancy is considered.

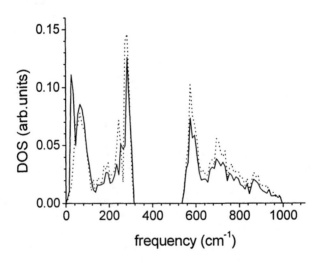

Figure 2. Phonon density of vibrational states (DOS) of structure sH Ar hydrates for various fillings: empty host lattice (dotted line), fivefold occupancy of the large cages and single occupancy of the small and medium cages (solid line).

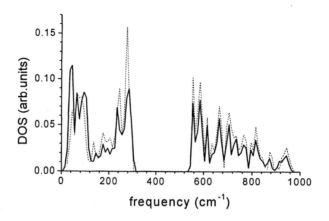

Figure 3. Phonon density of states of empty host lattice (dotted line) and doubly occupied tetragonal Ar clathrate (solid line).

Equation of State of the CS-II, sH and Tetragonal Ar Clathrate Hydrates

The thermodynamic functions P(V) of three types of hydrate structures can be estimate using obtained vibrational spectrum within the LD method. The equation of state P(V) has been calculated at 293 K. It has been found that the studied hydrates are thermodynamically stable in selected range of pressure. Moreover, these results can be compared with experimental P(V) data for Ar hydrate of three structural types [11]. In the case of Ar hydrate of CS-II type, the calculated P(V) for single occupancy of the large and small cages is most closely correlated with experimental points as shown in Figure 4. The largest difference between theory and experiment has been obtained for H hydrate using the experimental proposed multiply (5 Ar atoms) occupation of large cages [11] (see Figure 5). In this case, the P(V) function of sH Ar hydrate with Ar*4.87H_2O stoichiometry, for which the occupancy of the large cages is double is closer to experiment.

The good agreement with experimental data has been observed in the case of double occupancy of Ar atoms in the hydrate cages of tetragonal structure. Figure 6 shows that at the experimental determined lattice parameters (a=6.342 Å, c=10.610 Å) the calculated value of pressure is P=9.8kbar, which correlates well with the experimental value (P=9.2kbar) [11].

Stability of Ar Clusters in Large Cages of Clathrate Hydrates CS-II, sH and Tetragonal Structures

Since the multiple occupation for Ar hydrate has been proposed only for the large cages and the large cavities of selected clathrate hydrates have different

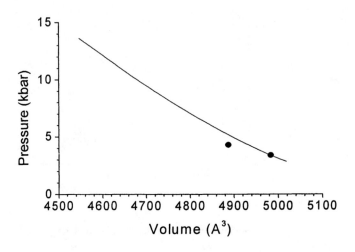

Figure 4. Equation of state of single occupied CS-II Ar hydrate at T=293 K (solid line) and • experiment [11].

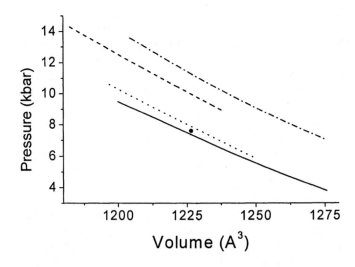

Figure 5. Equation of state of sH Ar hydrate: empty host lattice (solid line); fivefold (dotted and dashed line); threefold (dashed line) and double (dotted line) occupancy of the large cages (in all the cases, except empty host, the occupancy of the small and medium cages are single) at T=293 K and • experiment [11].

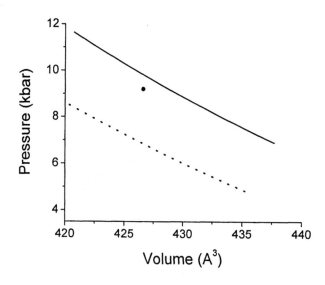

Figure 6. Equation of state of single (dash line) and double (solid line) occupied tetragonal Ar hydrate at T=293 K and • experiment [11].

number of water molecules, the hexakaidecahedron ($5^{12}6^4$), icosahedron ($5^{12}6^8$) and tetradecahedral ($4^25^86^4$) water cages with enclathrated Ar atoms have been optimized using first-principles calculations (see Figure 7). The energy values for all investigated structures are presented in Table 1. The interaction between one Ar and the hexakaidecahedron cage $(H_2O)_{28}$ is equal to –0.41 kcal/mol. In the case of the icosahedron cage $(H_2O)_{36}$, the interaction between one argon atom and the cage is equal to –0.29 kcal/mol. The negative value of SE means that Ar has positive stabilization effect on these cages and hence single occupations can be achieved without applying high external pressure.

The analysis of calculated frequencies shows that the translation of Ar atom in these two cages is characterized by imaginary frequency that is in agreement with LD calculations (see Figure 1). Moreover, other calculated vibrational frequency of an Ar atom in both cavities is ~20 cm^{-1}, which is also close to the value obtained by the LD method. To calculate the H-bond energy (HBE), we assumed the binding energy (difference between the total cluster energy without guest atoms and the separate H_2O monomers at infinite distance) for the water cluster is solely due to H-bonding. The value of HBE was determined as the binding energy divided by the number of H-bonds. In these cases, we found the HBE values for the respective empty cages. Moreover, the structural features of the water cavities (distances, angles, etc.) are very similar to those existing in the empty cages, and hence, they represent the cage structures with no distortion.

In the case of double occupancy, the negative value of SE was obtained only in the case of $5^{12}6^8$ cage. The positive values of SE are found for hexakaidecahedron and tetradecahedral cavities. However, these energies are very small and hence the double occupancy may be possible in the case of CS-II and tetragonal structures under applying the external pressure. Moreover, the HBE as well as shape of cages are not changes as compared with the HBE and the structure of the respective empty cavities. Addition one more Ar atom leads to significant increasing the SE and decreasing the HBE values for hexakaidecahedron and tetradecahedral cages because of the distortion of the water cages. Moreover, the imaginary frequencies are found. The analysis of these frequencies shows that the Ar clusters are strongly interacting with the water cages since there is a strong coupling between the vibrations of guest and host molecules.

In the case of sH hydrate, the Ar_n (n up to 5) clusters can be stabilized in the icosahedron cavity. Frequency calculations reveal that all frequencies are real and so, these structures are in local minima. The SE value is increased to 5.68 kcal/mol and HBE is not significant changed (see Table 1). Stabilization of the Ar_6 cluster inside this cage is energetically unfavorable since the value of SE is twice as larger as for the Ar_5 cluster.

The present calculations results can support to observed experimental data [11]. Thus, the double occupancy of the large cage of CS-II hydrates can be achieved under external pressure. The following increasing pressure with changing of stoichiometry form $Ar*4.25H_2O$ to $Ar*3.4H_2O$ leads to phase transition of hydrate structure and formation of sH hydrates with fivefold occupancy of icosahedron cages as observed in experiment. After further pressure increasing the formation of high dense tetragonal phase with stoichiometry $Ar*3H_2O$ is preferable. The disagreement between the LD results and experimental data on P(V) diagram can be explained by fact that in the LD calculations all the cages were filled. However, in real situation, it is difficult to realize the full occupation of hydrate cavities. Moreover, it is experimentally known that it is not necessary to occupy all cavities by guest molecules and even a small number of guests is enough to form the clathrate structure [1]. Therefore, the multiple occupation for sH which was predicted experimentally [11], may be realized by only in a limited number of large cavities.

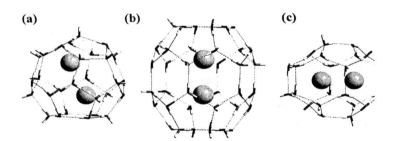

Figure 7. Structures of the water cages with double occupancy: (a) hexakaidecahedron ($5^{12}6^4$); (b) icosahedron ($5^{12}6^8$); and (c) tetradecahedral ($4^25^86^4$) cages.

Table 1. Stabilization energy for Ar clusters in the different large hydrate cages, H-bond energy (HBE) for the studied hydrate cages, number of Ar atoms and stoichiometry of the studied Ar hydrate ([a] the occupation in the small and medium cages in the cases of CS-II and sH hydrates are single).

Type of water cage	Number of Ar atoms	Stoichiometry[a]	SE(kcal/mol)	HBE(kcal/mol)
Cage ($5^{12}6^4$)	0			−6.06
	1	Ar*5.67H$_2$O	−0.41	−6.06
	2	Ar*4.25H$_2$O	2.14	−6.06
	3	Ar*3.4H$_2$O	9.25	−5.99
Cage ($5^{12}6^8$)	0			−5.97
	1	Ar*5.67H$_2$O	−0.29	−5.96
	2	Ar*4.87H$_2$O	−0.54	−5.96
	3	Ar*4.25H$_2$O	−0.04	−5.95
	4	Ar*3.78H$_2$O	3.54	−5.94
	5	Ar*3.4H$_2$O	5.68	−5.94
	6	Ar*3.1H$_2$O	11.85	−5.90
Cage ($4^2 5^8 6^4$)	0			6.17
	1	Ar*6H$_2$O	0.99	6.17
	2	Ar*3H$_2$O	2.50	6.16
	3	Ar*1.5H$_2$O	26.81	5.87

CONCLUSIONS

The electronic, structural, dynamic and thermodynamic properties of CS-II, sH and tetragonal Ar clathrate hydrates were investigated. The LD calculations showed that the selected hydrate structures with the different filling fractions are dynamically stable. The systematic investigation of the large cage of CS-II, sH and tetragonal structures with different number of Ar atoms was also performed using first principles calculations. It has been found that in the studied hydrates the multiple occupancies of the large cages is possible. Moreover, the stability of Ar clusters in the large cages is correlated well with experimental phase transition from CS-II to a new tetragonal hydrate structure [11].

ACKNOWLEDGMENT

V.R.B. thankfully acknowledges the kind hospitality at the CNAS of the Tohoku University. We would like to thank the Information Science Group of the Institute for Materials Research, Tohoku University for their continuous support of the SR8000 supercomputing system. In Russia, this work also supported by the Russian Foundation for Basic Research through Grant No. 00-03-32508 and by Presidium of Siberian Branch of Russian Academy of Sciences (Grant No. 76).

REFERENCES

[1] E.D. Sloan, Jr. 1998. Clathrate Hydrates of Natural Gases, 2nd ed. Marcel Dekker, NY, 1–705.
[2] M. von Stackelberg, H.R. Miller, 1954. Feste Gashydrate II. Strukture und Ruumchemie. Z. Elektrochem., 58, 25–39.
[3] W.F. Claussen, 1951. Suggested structures of water in inert gas hydrates, J. Chem.Phys., 19, 1425–26.
[4] L. Pauling, R.E. Marsh, 1952. Structure of chlorine hydrate, Proc. Natl. Acad. Sci. U. S., 38, 112–118.
[5] J. Ripmeester, J. Tse, C. Ratcliffe, B. Powell, 1987. A new clathrate structure, Nature, 325, 135–137.
[6] K.A. Udachin, C.I. Ratcliffe, G.D. Enright and J.A. Ripmeester, 1997. Structure H hydrate: a single crystal diffraction study of 2,2-dimethylpentane 5(Xe, H_2S)34H_2O, Supramol. Chem., 8, 173–176.
[7] Yu.A. Dyadin, E.G. Larionov, T.V. Mikina and L.I. Starostina. 1997. Clathrate formation in Kr-H_2O and Xe-H_2O systems under pressures up to 15 kbar, Mendeleev Commun., 74–76.
[8] Yu.A. Dyadin, E.G. Larionov, D.S. Mirinski, T.V. Mikina and L.I. Starostina, 1997. Clathrate formation in the Ar- H_2O systems under pressures up to 15 kbar, Mendeleev Commun., 32–34.
[9] Yu.A. Dyadin, E.Ya. Aladko and E. G. Larionov, 1997. Decomposition of methane hydrates up to 15 kbar, Mendeleev Commun., 34–35.
[10] Yu.A. Dyadin, E.G. Larionov, A.Yu. Manakov, F.V. Zhurko, E.Ya. Aladko, T.V. Mikina and V.Yu. Komarov, 1999. Clathrate hydrates of hydrogen and neon, Mendeleev Commun., 209–210.
[11] A.Yu Manakov, V.I. Voronin, A.E. Teplykh, A.V. Kurnosov, S.V. Goryaainov, A.L. Ancharov, A.Yu. Likhacheva, 2002. Structural and spectroscopic investigations of gas hydrates at high pressures, Proceedings of the 4th International Conference on Gas Hydrates, Yokohama, Japan, 630–635.
[12] D.W. Davidson, Y.P. Handa, C.I. Ratcliffe and J.S. Tse, 1984. The ability of small molecules to form clathrate hydrates of structure II, Nature, 311, 142–143.
[13] D.W. Davidson, Y.P. Handa, C.I. Ratcliffe, J.A. Ripmeester., J.S. Tse, B.J.R. Dahn, F. Lee, L.D. Calvet, 1986. Crystallographic Studies of Clathrate Hydrates. Pt. I, Mol. Cryst. Liq. Cryst., 141–149.

[14] G.D Holder and D.J. Manganiello, 1982. Hydrate dissociation pressure minima in multicomponent systems, Chem. Eng. Sci., 37, 9–16.
[15] J. van der Waals, J. Platteeuw, 1959. Thermodynamic properties of gas hydrates, Adv.Chem.Phys., 2, 1–57.
[16] W.F. Kuhs, B. Chazallon, P.G. Radaelli and F. Pauer, 1997. Cage occupancy and compressibility of deuterated N_2-clathrate hydrate by neutron diffraction, J. Incl. Phenom., 29, 65–77.
[17] B. Chazallon and W.F. Kuhs, 2002. In situ structural properties of N_2-, O_2-, and air-clathrates by neutron diffraction, J. Chem. Phys., 117, 308–320.
[18] E.P.van Klaveren, J.P.J. Michels, J.A. Schouten, D.D. Klug and J.S. Tse, 2001. Stability of doubly occupied N_2 clathrate hydrates investigated by molecular dynamics simulations, J. Chem. Phys., 114, 5745–5754.
[19] H. Itoh, J. S Tse and K. Kawamura, 2001. The structure and dynamics of doubly occupied Ar hydrate, J. Chem. Phys., 115, 9414–9420.
[20] E.P. van Klaveren, J.P.J. Michels, J.A. Schouten, D.D. Klug and J.S. Tse, 2001. Molecular dynamics simulation study of the properties of doubly occupied N_2 clathrate hydrates, J. Chem. Phys., 115, 10500–10508.
[21] E.P. van Klaveren, J.P.J. Michels, J.A. Schouten, D.D. Klug, J.S. Tse, 2001. Computer simulations of the dynamics of doubly occupied N_2 clathrate hydrates, J. Chem. Phys., 117, 6637–6645.
[22] W.L. Mao, Ho-kwang Mao, A.F. Goncharov, V.V. Struzhkin, Q. Guo, J. Hu, J. Shu, R. Hemley, M. Somayazulu, Y. Zhao, 2002. Hydrogen Clusters in Clathrate Hydrate, Science, 297, 2247–2249.
[23] H.T. Lots, J.A. Schouten, 1999. Clathrate hydrates in the system H_2O–Ar at pressures and temperatures up to 30 kbar and 140 °C, J. Chem. Phys., 111, 10242–10247.
[24] M.J. Frisch, J.E. Del Bene, J.S. Binkley, H.F. Schaefer III, 1986. Theoretical studies of the hydrogen-bonded complexes $(H_2O)_2$, $(H_2O)_2H^+$, $(HF)_2$, $(HF)_2H^+$, F_2H^-, and $(NH_3)_2$, J. Chem. Phys., 84, 2279–89.
[25] C. Peng, P.Y. Ayala, H.B. Schlegel and M.J. Frisch, 1996. Using redundant internal coordinates to optimize geometries and transition states, Journal of Computational Chemistry 17, 49–56.
[26] Frisch, M. J., et al., Gaussian 98, Revision A. 9; Gaussian, Inc.: Pittsburg, PA, (1998).
[27] R.V. Belosludov, I.K. Igumenov, V.R. Belosludov, V.P. Shpakov, 1994. Dynamical and thermo-dynamical properties of the acetylacetones of copper, aluminium and rhodium, Mol. Phys., 82, 51–66.
[28] V.R. Belosludov, M.Yu. Lavrentiev and S.A. Syskin, 1988. Dynamical Properties of Molecular Crystals with Electrostatic Interaction Taken into Account. Low Pressure Ice Phases (I_h and I_c), phys.stat.sol.(b), 149, 133–142.
[29] V. R. Belosludov, V. P. Shpakov, J. S. Tse, R.V. Belosludov, and Y. Kawazoe, 2000. Mechanical stability of gas hydrates under pressure, Ann. N.Y. Acad. Sci., 912, 993–1002.

[30] J. Tse, 1994. Dynamical properties and stability of clathrate hydrates, N.Y. Acad. Sci., 715, 187–206.

[31] V.R. Belosludov, M.Yu. Lavrentiev, Yu.A. Dyadin, S.A. Syskin, 1990. Dynamic and Thermo-dynamic Properties of Clathrate Hydrates, J. Inc. Phenom. Molec. Recog. Chem., 8, 59–69.

Chapter 7: Intelligent Instrumentation and Metrology

Image Mining of Evanescent Microwave Data for Nondestructive Material Inspection

S. OKA

Air Force Research Laboratories, Wright-Patterson Air Force Base, Dayton OH, USA

S. LeCLAIR

Missile Defense Agency, Washington, DC, USA

ABSTRACT

In this study, we propose an image-mining algorithm for non-destructive inspection using microwave imaging. Since the microwave imaging of a tiny feature on a metal surface is subject to the influence of surface roughness, surface coating and the evanescent spatial resolution, it is difficult to observe the target feature clearly. In order to extract a target feature from the microwave image, we employ a combination of two image-processing algorithms: Tilt Noise Removal by Neural Network, and Blind Deconvolution Deblurring. The outline of our strategy is to remove the global tilt noise by a back-propagation neural network, and to restore the Gaussian noise by Lucy-Richardson PSF estimation. We evaluated the performance of the proposed algorithm by extracting a 1-mm edge equilateral triangle etched on a metal surface and covered with 4.6 mils of coating.

INTRODUCTION

Microwave imaging is a novel approach to non-destructive inspection of electronic, semiconductor, superconductor, biological, chemical, and other materials. Microwaves are able to capture surface roughness features with very fine resolution without damaging or touching the sample material. The most significant advantage of microwave scanning is to penetrate the material surface deeply and elucidate the inside aspect which cannot be observed by other optical devices [1].

The microwave probe tip is set very close to a sample without touching it, and it then scans along a regular direction. While the probe scans the surface of the sample, the frequency shift value between the initial and current points is recorded

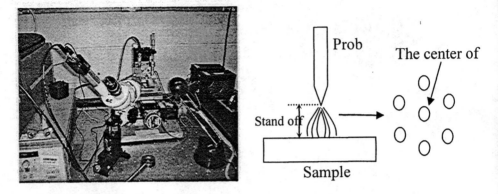

Figure 1. The microwave imaging device (left), and the schematics of evanescent microwave probe (right).

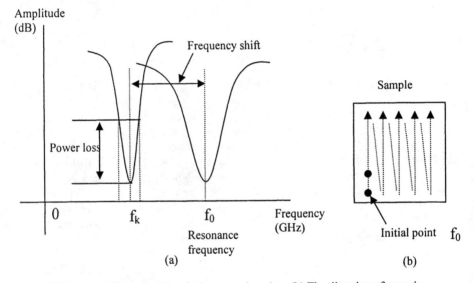

Figure 2. (a) The principle of microwave imaging. (b) The direction of scanning.

together with the power loss value at the current point. The frequency shift is the gap between the center frequency of two resonance curves while the power loss is the gain value of the resonance curve as shown in Figure 2. The probe tip emits several radial microwaves as shown in Figure1, and the average value of several emission reflectances is recorded as the frequency shift and the power loss.

BENCHMARK SAMPLE

Figure 4 shows a benchmark sample of microwave imaging used in our experiment. Figure 4(a) is the CCD image of the tested metal sample with an

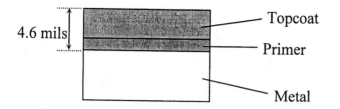

Figure 3. The coating of the experimental sample.

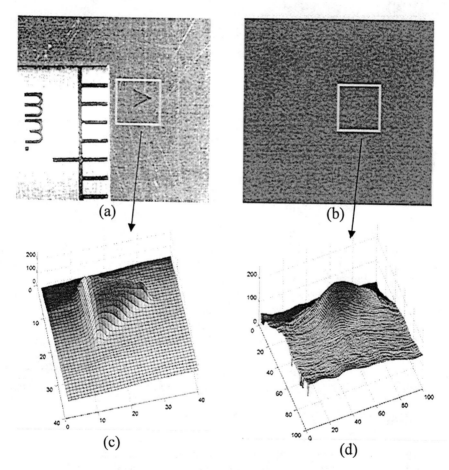

Figure 4. (a) CCD image of bare triangle etched on metal surface,
(b) CCD image of triangle covered with a 4.6 mil coating,
(c) 3D surface plot of the frequency shift data of (a),
(d) 3D surface plot of the frequency shift data of (b).

etched 1 mm triangle on the surface. Figure 4(b) is the CCD image of the same sample with an etched triangle covered with a coating of 4.6 mils thickness (see Figure 3). Figure 4(c) is the frequency shift data of the bare triangle of (a), scanned at 6.848 GHz with a 50 µm resolution and a 30 µm stand off. Figure 4(d) is the frequency shift data of the triangle under the 4.6mils coating of (b), scanned with 6.843 GHz with a 20 µm resolution and a 30 µm stand off. Stand off is the distance between the microwave probe and the material surface (see Figure 1).

PURPOSE OF STUDY

Our image-mining strategy consists of the following two algorithms:

- Tilt Noise Removal by Neural Network
- Blind Deconvolution Deblurring
- As shown in Figure 4(d), it is obvious that the shape of the triangle under the thick coating is distorted and entirely blurred in the microwave imaging. The purpose of this study is to extract the

TILT NOISE REMOVAL

As a first step of the image-mining strategy, we should consider noise removal to obtain precise object data. Microwave images contain various types of distortion. One of the critical factors is "tilt" as shown in Figure 5. Due to the high resolution used, microwave imaging is influenced with a slight tilt noise by the experimental equipment. Most studies on image noise removal have focused on filter processing, where uniform random noise (Gaussian noise, white noise) is mainly discussed. Well-known noise removal filters are High Pass and Low Pass filtering using FFT, two-derivative Laplacian filtering, Median filtering, Gaussian filtering, and combinations of these. However, these simple filtering processes are ineffective at removing "tilt" noise, because "tilt" is not a uniform local noise such as Gaussian noise, but rather a global noise spread over an entire image. To remove "tilt" noise from microwave images, we employ a back-propagation neural network since it is able to estimate a non-linear two-dimensional "tilt" noise field.

Procedure

1. Select several points (k_1, k_2, ..., k_n) surrounding a target object in an input image, and store these as a learning data set. (x_i, y_i) which is the x and y coordinate of point k_i, and feed them to a neural network as input data. t_i is the pixel density at (x_i, y_i) used as the teaching signal of k_i (see Figure 6). All data values are normalized between 0 and 1.
2. Learn n pieces of data (k_1, k_2, ..., k_n) using the back-propagation algorithm.
3. Estimate a 2-D "tilt" plane by interpolating each pixel value based on the learning result.
4. Subtract the input image from the estimated tilt plane image.

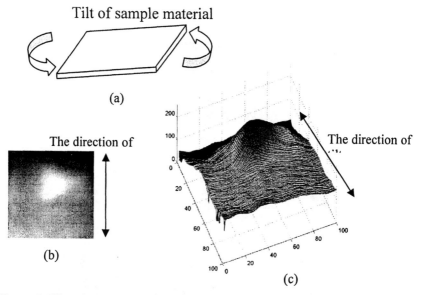

Figure 5. Tilt noise of microwave imaging. (a) Schematic of tilt noise. (b) Gray-scale image of the original data. (c) 3D surface plot of the original data.

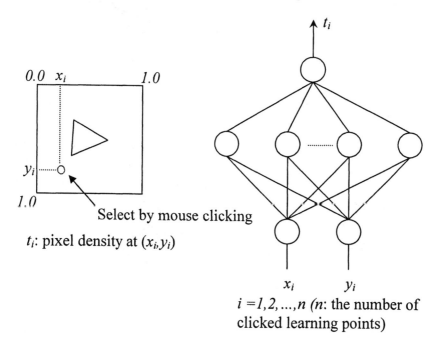

Figure 6. Selection of learning data (left) and the three-layered neural network used to estimate "tilt" noise (right).

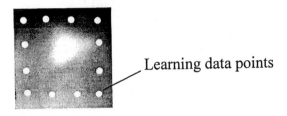

Figure 7. Twelve learning data points of "tilt" noise estimation.

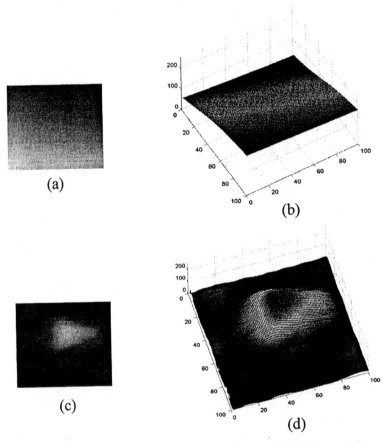

Figure 8. The result of tilt noise removal by the neural network. (a) Gray-scale image of the estimated tilt plane acquired by the neural network. (b) 3D surface plot of the estimated "tilt" noise acquired by the neural network. (c) Gray-scale image of the subtraction between the original image and the estimated tilt plane. (d) 3D surface plot of the subtraction between the original image and the estimated tilt plane.

Simulation Result

We used the frequency shift data, and selected 12 points as a learning data set as shown in Figure 7. In our simulation, the number of hidden neuron was set to seven, the learning constant was set to 0.1, and the learning iteration steps were set to 30,000. Figure 8 shows an estimated "tilt" plane obtained by the neural network and the subtraction image between the original image and the estimated tilt plane, where in, we can see that the "tilt" noise has been cleanly removed from the original image.

BLIND DECONVOLUTION DEBLURRING

The next step after removing "tilt" noise must consider image deblurring to extract a clear shape of the target object from the microwave image. Image deblurring is a common topic in most image processing papers on optical or other sensing devices. Generally, the basic deblurring model is as follow:

$$g = H*f + n$$

g : The blurred image
H: The distortion operator or PSF (Point Spread Function).
   Image deblurring occurs by convolving the matrix H and an ideal input image.
f: True image
n: Additive noise
$*$: Convolution

Based on this basic blurring model, deblurring is realized by deconvolution of the blurred image with the PSF that is a unique property of deblurring of each image-capturing device. The quality of deblurring is determined by knowledge about the particular PSF. However, in most practical cases including microwave imaging, the exact PSF is generally unknown *a priori*. To restore the blurred microwave image, we employ a blind deconvolution algorithm to estimate the true image and PSF simultaneously without any prior knowledge of the distortion function. The blind deconvolution function of MATLAB 6.1 Image Processing Toolbox was used in our simulation where the Richardson-Lucy algorithm is implemented [2].

Simulation Result

Figure 9(a) and Figure 9(b) show results of a deblurred image and the estimated PSF using Figure 8(c) as an input image. In this algorithm, we must adjust the appropriate size of the PSF to obtain a good restored image. In this simulation, a 17x17 PSF matrix size gave the best restoration result.

Improving Deblurring

As shown in Figure 9, the contour of the triangle is still distorted after deblurring. To improve the edge degradation, we used an edge weight array for re-deblurring. Figure 10(b) is the edge weight array of the triangle image obtained by the canny edge detector and the morphology dilation algorithm. Figure 10(a) and Figure 10(c) show the result of re-deblurring where the contour edge of the triangle is restored much better.

CONCLUSION AND FUTURE WORK

Evanescent microwave imaging can elucidate the inside aspect without damaging or touching the material with a resolution below 10 μm. Our image mining algorithm succeeded in extracting the target object from the microwave image by a combination of neural network and blind deconvolution deblurring. As future work, we would like to try other non-destructive inspection problems and finally realize the 3D reconstruction of a target object by capturing depth information by using various frequencies of microwave imaging.

ACKNOWLEDGEMENTS

Professor Yoshiyasu Takefuji of Keio University, and the National Research Council is gratefully appreciated for providing one of the authors (SO) with the opportunity to challenge this study. We also wish to thank William Fitzgerald for providing reliable experimental data, and Dr. John Jones and David Jacobs for many helpful suggestions.

REFERENCES

[1] M. Tabib-Azar, R. Muller and S.R. LeClair, 2000. Nondestructive imaging of grain boundaries in polycrystalline materials using evanescent microwave probes, Engineering Application of Artificial Intelligence, 13(5), 549–564.
[2] L.B. Lucy, 1994. The restoration of HST images and spectra. II., Space Telescope Institute, Hanisch R.J. and White R.L. (eds.), 79–85.

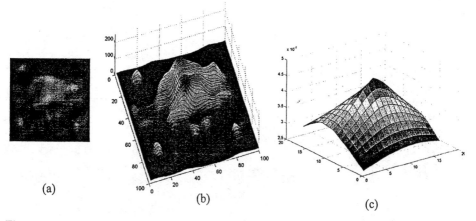

Figure 9. The result of deblurring. (a) Gray-scale image of the deblurred image (b) 3D surface plot of the deblurred image. (c) Estimated PSF (two-dimensional 17x17 matrix).

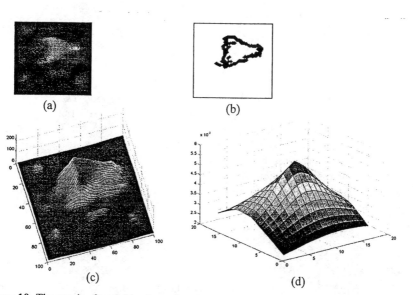

Figure 10. The result of re-deblurring using an edge weight array. (a) Gray-scale image of the re-deblurred image, (b) Edge weight array by canny edge detector. (c) 3D surface plotting of the re-deblurred image. (d) Estimated PSF (two-dimensional 17x17 matrix).

Magnetohydrodynamic Stability in Pulse EMC

K. FUJISAKI
Plant Engineering Technology Center, Nippon Steel Corp., 20-1, Shintomi, Futtsu, Chiba, 293-8511, Japan

ABSTRACT

This paper explains how pulse EMC (Electro-magnetic casting) works at improving the surface steel quality from the viewpoint of fluid dynamics by means of a magnetic-hydrodynamic calculation on a free surface. In order to obtain good surface quality, the free surface profile must follow the pulsed electromagnetic force and the fluid dynamics must be stable and uniform in the rectangular plane. By means of the pulse mode of an EMC operation, fluid dynamics become stable and uniform and any unstable flow in continuous EMC operation is eliminated allowing the free surface profile to be maintained as convex.

INTRODUCTION

Electromagnetic force applications in steel-making provide superior characteristics such as non-contact operation and quick response. The process has been well-researched [1–3] and used in practice [4] to obtain high quality steel at high productivity. However it is still an important problem for steelmakers to apply funding to research other improvement potential in this technique.

New applications of EMF have been expected in steel-making and a national project of the Japanese government has been in place for several years now. The major steel making plants and those that produce steel-making equipment and electricity including companies from outside of Japan have been collaborating in this project. This national research project ended in March 2001 achieving the prescribed results [5].

One of the National Project themes is Pulse EMC (Electro-Magnetic Casting) technology [6] which gives good surface quality in continuous casting. Since the Pulse EMC technology overcomes instabilities in molten steel flows, this paper evaluates the process stability using a magneto-hydrodynamic calculation.

SYSTEM CONSTRUCTION
Construction of EMC

Pulse EMC technology is used within the continuous casting mold in a steel-making plant. Continuous casting involves control of the solidification of molten steel. Before this process, the molten steel is already refined and so, the initial solidification in this mold determines most of the ultimate steel quality.

Construction of a Pulse EMC is shown in Figure 1. The black body at the center is a submerged nozzle though which the refined molten steel is poured from an upper vessel called a tundish. The amount of molten steel to be poured is controlled by an actuator such as a sliding nozzle or a stopper, so the position of the molten steel free surface can be held constant [7]. The actuator is controlled by information from a level sensor such as an electromagnetic type that detects the position of the free surface.

Around the submerged nozzle, the molten steel flows out into the mold. The mold is made of copper to cool down the molten steel by water with stainless steel used to support the copper mold. The molten steel solidifies at the contact surface with the copper mold and is pulled downward in the casting direction. Outside the mold, an electromagnetic coil is arranged and a single-phase alternate current is supplied by a power inverter at 200Hz frequency. Since the electromagnetic field is supplied to the mold, the mold is separated at 4 corners and isolated electrically to depress the increased formation of eddy currents within the mold and to promote the penetration of the electromagnetic field into the molten steel.

When the electromagnetic field is supplied to the molten steel, eddy currents are induced in the molten steel so that the electromagnetic force is supplied to the center of the submerged nozzle as in Figure 2. The molten steel flows toward the center and then upward at the center. So in this way the free surface moves upward and its shape becomes convex.

Lubrication powder is placed over the free surface. The powder flows down between the molten steel and the copper mold by means of mechanical oscillation of the mold in order to prevent burn out by the molten steel against the mold.

When the shape of the free surface becomes convex, insertion of the power is promoted which introduces a large thickness of power between the molten steel and the mold. Since the heat transfer from the molten steel to the mold is reduced by the powder, the cooling rate is reduced. This slower cooling provides good surface quality from a metallurgical sense.

In the above explanation, good surface quality depends on the convex shape of the free surface and the convex shape derives from the electromagnetic force by way of flow dynamics. In this process, since the Reynolds number is about 10^5–10^6, the flow dynamics are turbulent and have a kind of swing like a Karman flow in spite of the symmetric shape or the absence of disturbance forces. When the single phase AC current is supplied to the molten steel continuously, the free surface shape becomes unstable basically because of the turbulence disturbances. Therefore, continuous operation of the EMC causes unstable flow. Since unstable flow introduces poor surface quality, long time operation is impossible and EMC is difficult to use in a commercial operation [7].

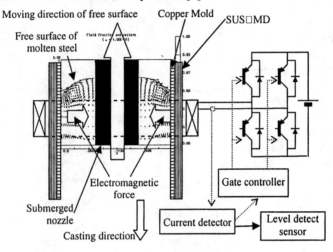

Figure 1. Construction of pulse-EMC (side view).

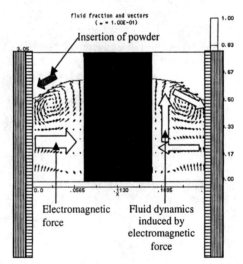

Figure 2. Basic concept of pulse-EMC.

Pulse EMC Operation

In order to solve the unstable flow problem, single-phase AC current must be used in a pulse mode operation - the so-called Pulse EMC [8]. In Pulse EMC, a 200Hz single phase AC current operates for only 50 ms and then no current is applied for the next 50 ms. This pulse operation continues repeatedly as in Figure 3. Operating for 50 ms on corresponds to 10 waves of 200 Hz single-phase AC current. Pulse EMC prevents instability in the fluid dynamics and maintains a convex shape on the free surface since the current cuts off before the fluid flow becomes unstable.

A 50ms-on-50-ms-off pulse corresponds to a 10 Hz electromagnetic force. The inertia of the molten steel is too large to follow this force do it is assumed that the molten steel receives only the average EMF as shown in Figure 3(b) which maintains the convex shape on the free surface.

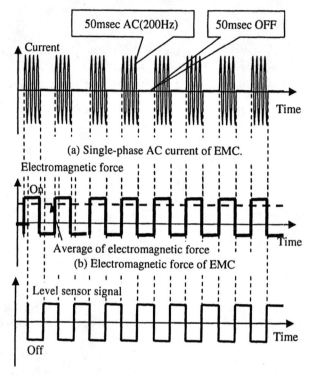

Figure 3. Current waveform of pulse-EMC.

Level Sensor

In Pulse-EMC operation, a problem occurs with the level detection sensor. Since an electromagnetic sensor is used, the electromagnetic field derived by Pulse-EMC becomes a serious noise. In case of pulse mode operation, since the electromagnetic field circumstance changes in every 50 ms, the level detect sensor can not supply the signal continuously.

To solve this problem, the level detection sensor operates only during the 50-ms-off operation of the pulse EMC as shown in Figure 3(c). The information about whether single phase AC current is supplied to the EMC coil or not is obtained from the power source such as the inverter shown in Figure 1. The 50-ms-off operation is long enough to detect the position of the free surface.

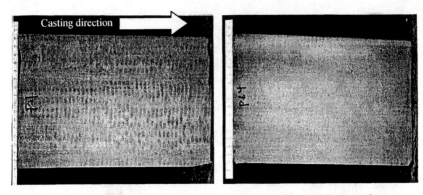

(a) No electromagnetic force.   (b) Pulse-EMC operation.

Figure 4. Operating results of Pulse-EMC.

Casting Results

The Pulse-EMC casting system was operated in practice at the Muroran works continuous billet casting process which is used commercially. As results of these tests, photos of the billet steel surface are shown in Figure 4. These surfaces are after casting before rolling. With no EMF operation, there are clear oscillation marks vertical to the casting direction on the billet surface. These marks are generated when the mold oscillates mechanically during operation. With Pulse-EMC, there are no oscillation marks on the surface. So it can be said that Pulse-EMC improves in surface quality since powder insertion between the steel and the copper mold is better.

MAGNETOHYDRODYNAMIC CALCULATION

The magneto-hydrodynamic analysis is calculated to understand how Pulse-EMC operates. Direct calculation of oscillation mark phenomena is currently too difficult to calculate. Then the stability of fluid dynamics is discussed here in order to evaluate the surface quality by means of magneto-hydrodynamic calculation. The magneto-hydrodynamic calculation used here is based on the analysis technology used in electromagnetic stirring [10] and the "shadow method" [11] developed for free surface magneto-hydro- dynamic phenomena.

Electromagnetic Field

Using the $\vec{A}-\psi$ method, the electromagnetic field based on Maxwell equations is expressed as:

$$\nabla \times \left([\mu]^{-1} \nabla \times \vec{A}\right) + \sigma \left(\frac{\partial \vec{A}}{\partial t} + \nabla \psi\right) = \vec{J_0} \qquad (1)$$

where, $\vec{A}$ is the vector potential [Wb/m],
ψ is the scalar potential [V/m],
$\vec{J_0}$ is the current density [A/m2],
• is the electromagnetic conductivity [S/m],
$[\mu]^{-1}$ is the inverse of the permeability [m/H].

The Finite Element Method is used in discrete 3-D space and the jω method is used where $\omega = 2\pi f$ (f = $\vec{A} - \psi$ frequency [Hz]). Using in Eq. 1, introduces the electromagnetic force density in the molten metal as follows:

$$F_{emf} = \sigma(\frac{\partial \vec{A}}{\partial t} + \nabla \psi) \times (\nabla \times \vec{A}) \qquad (2)$$

Fluid Dynamics

Fluid dynamics is expressed as the equation of mass continuity and the Navier-Stokes equation in incompressible conditions as follows:

$$\nabla \vec{u} = 0 \qquad (3)$$

$$\rho \frac{D\vec{u}}{Dt} = \mu \nabla^2 \vec{u} - \nabla P + \rho \vec{g} + \vec{F}_{em} + \rho K \vec{u} \qquad (4)$$

where, $\vec{u}$ is the velocity vector [m/s],
- is the density of the fluid material [kg/m3],
μ is the coefficient of viscosity [kg/ms],
P is pressure [N/m2],
$\vec{g}$ is the gravity acceleration vector [m/s2],
D/Dt is the total differential,
K is Darcy's flow coefficient [1/s].

The total differential equation is expressed as followed:

$$\frac{D}{Dt} = \frac{\partial}{\partial t} + u\frac{\partial}{\partial x} + v\frac{\partial}{\partial y} + w\frac{\partial}{\partial z} \qquad (5)$$

where, u, v, w are the velocities in the x, y, and z directions respectively.

The electromagnetic force expressed as Eq. 2 is installed in the Navier-Stokes equation (Eq. 4) as the disturbance force. Since the Reynolds number is around 10^5–10^6 in this process, turbulent phenomena occurs and is modeled by LES (Large Eddy Simulator). Three dimensions in space and time variation are modeled by the finite differential method.

To express the free surface boundary within a fixed mesh, a function F is defined whose value is unity at any point occupied by fluid and zero otherwise. This is called the VOF (Volume of Fluid) method [12]. The value of F in the mesh represents the fractional volume of the cell occupied by fluid. The time dependence of F is governed by Equation 6:

$$\frac{DF}{Dt} = 0 \qquad (6)$$

The combined magneto-hydrodynamic calculation with electromagnetic field and fluid dynamics of the free surface phenomena is expressed as in Figure 5. Here, the electromotive force, which is a cross term of the velocity and magnetic flux density is ignored, because this term is very small relatively speaking and the AC current has such high frequency.

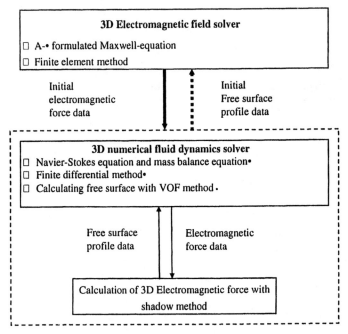

Figure 5. Magneto-hydrodynamics by the shadow method.

EVALUATION
Moving Height and Contact Angle

The magneto-hydrodynamics by the shadow method in Figure 5 is applied to the Pulse-EMC process. Figure 6 shows the calculation results. The white part is a schematic view of the free surface profile of the molten steel at one point in time. The square-shaped hole at the center of the free surface is the submerged nozzle. Though a cylindrical shape is used in practice as a submerged nozzle, a rectangle was assumed and used for convenience in the calculation.

Figure 6 shows that the free surface profile is non-uniform for the rectangular axis. The rectangular axis is a line which follows 4 corners of the copper mold. To evaluate insertion of the powder between the solidified steel and the mold, two evaluation indexes are introduced - "*moving height*" and "*contact angle*" - at the contact point of the molten steel and the copper mold.

A standard line (such as the average line) in the fluid at some point in time is assumed along the rectangular axis at the contact line between the molten steel and the copper mold. Then a "*moving height*" is defined as the height from the standard line to the molten steel surface at each contact point and at each point in time as shown in Figure 6. When the *moving height* is flat along the rectangular line, then the free surface of the molten steel is considered to be uniform. When the *moving height* is not flat along this rectangular line, the free surface of the molten steel is considered to be unstable.

Next, a "*contact angle*" is defined as the angle between the molten steel and the copper mold vertical to the rectangular line at each contact point and at each point in time as shown in Fig. 6. When the *contact angle* is 90°, the free surface of the molten steel is flat. When the *contact angle* is less than 90°, the angle becomes acute and powder insertion is promoted. When the *contact angle* is greater than 90°, the angle becomes obtuse and powder insertion is inhibited.

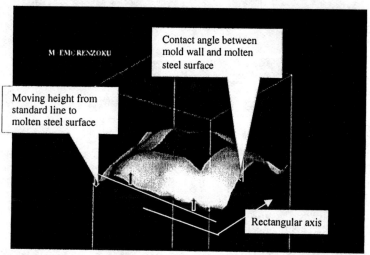

Figure 6. Free surface profile by magneto-hydrodynamics using the shadow method.

Calculation Results

The free surface profiles are shown in Figure 7. To clarify the Pulse-EMC characteristics, Continuous-EMC operation, in which a single-phase AC current was supplied, is also shown for comparison. A cylindrical-shaped nozzle as shown in Figure 7 was used in this case.

In Pulse-EMC operation, the free surface profile is clear and uniform. In the continuous operation, the free surface profile is unclear and non-uniform showing some small creases. The creases indicate that the free surface has small up and down movement in the vertical direction. So we can say that Pulse-EMC operation has a stable profile, while Continuous-EMC operation shows an unstable profile.

To show the dynamic and 3-dimensional characteristics, the *moving height* and *contact angle* characteristics are shown in Figure 8 and Figure 9 respectively. The vertical axis shows 1000 ms in time for which 10 pulses are supplied. The horizontal axis is the rectangular axis with 4 corners and 4 edges. The 5 dotted vertical lines in each figure correspond to the 4 corners respectively. Since the horizontal axis surrounds the mold, the starting and end points are the same corner. Each figure shows contours: Figure 8 has contours of the *moving height* from the standard line while Figure 9 contains contours of the *contact angle*.

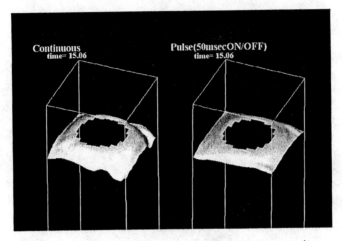

(a) Continuous operation.   (b) Pulse-EMC operation.

Figure 7. Free surface shape.

In the continuous operation, Figure 8(a) and Figure 9(a) show that the *moving height* and *contact angle* are not uniform for either the rectangular axis or over time. The un-uniformity along the rectangular axis means that the initial solidification does not start uniform and it is bad for surface quality. The un-uniformity along 1 s time series means that the free surface does not maintain stable profile. This means that the continuous operation is unstable.

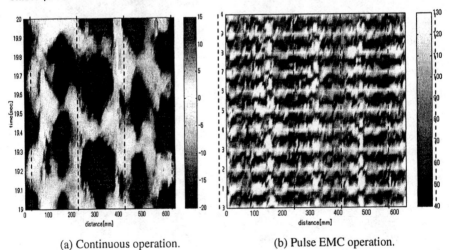

(a) Continuous operation.   (b) Pulse EMC operation.

Figure 8. Moving height from basic line to molten steel surface.

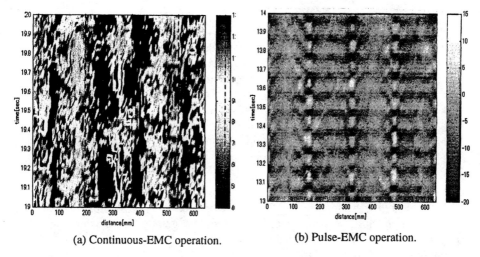

(a) Continuous-EMC operation.　　　(b) Pulse-EMC operation.

Figure 9. Contact angle between mold wall and molten steel surface.

Though the shape of mold is simple and rectangular and the electromagnetic force is continuous, the calculation result shows that the free surface profile is unstable. The reason is considered as that the fluid dynamics is turbulent and turbulent flow generates small disturbance and the convex profile of the free surface is unstable for small disturbance.

In the pulse EMC operation, Figure 8(b) and Figure 9(b) show that the *moving height* and *contact angle* are uniform for the rectangular axis and 10 times up and down in 1 second (vertical axis). The uniformity along the rectangular axis means the initial solidification starts uniformly which is good for surface quality. The 10 times up and down phenomena means that the moving height follows the pulse mode of the electromagnetic force as shown in Figure 3(b), since 1 second has 10 pulses of EMF. This means that the Pulse-EMC operation is stable. Table 1 shows the standard deviation of the moving height in Continuous-EMC and Pulse-EMC operation. The standard deviation in Continuous-EMC operation is almost twice that of Pulse-EMC operation.

From the above discussion, it is can stated that the free surface profile should follow the pulsative electromagnetic force and the fluid dynamics must be stable and uniform in the rectangular axis in order to obtain good surface quality. By means of the pulse mode of EMC operation, the fluid dynamics are stable and uniform in comparison with the unstable flow in continuous-EMC so that the free surface profile maintains convex.

Table 1. Standard Deviation of Moving Height.

	Continuous operation	Pulse EMC operation
Standard deviation of *moving height*	7.86 [mm]	3.79 [mm]

CONCLUSION

It is clear why Pulse-EMC makes for good steel surface quality from the viewpoint of understanding the fluid dynamics in the free surface by means of magnetic-hydrodynamic calculations. In order to obtain good surface quality, the free surface profile should follow a pulsed electromagnetic force so that fluid dynamics remain stable and uniform in the rectangular axis of the mold. By means of the pulse EMC operation, the fluid dynamics are made stable and uniform in comparison to the unstable flow in normal continuous-EMC operations and the free surface profile is maintained as convex.

ACKNOWLEDGEMENT

This research was supported by the national project of JRCM (Japan Research and Development Center for Metal). The author greatly appreciates Mr. T. Tomizawa at Taiheikougyou Corporation for his technical support of the computational part of this work.

REFERENCES

[1] K. Fujisaki, K. Wajima, K. Sawada, T. Ueyama, 1997. Application of electromagenic technology to steelmaking plant, Nippon Steel Technical Report, 74.
[2] K. Fujisaki, J. Nakagawa, H. Misumi, 1994. Fundamental characteristics of molten metal flow control by linear induction motor, IEEE Trans, Magn., 30(6), 4764–4766.
[3] K. Fujisaki, T. Ueyama, K. Takahashi, S. Satoh, 1997. Phase charactersistics of electromagnetic strirring, IEEE Trans, Magn., 33(5), 4245–4247.
[4] H. Yamane, Y. Ohtani, J. Fukuda, T. Kawase, J. Nakashima, A. Kiyose, 1997. High power in-mold electromagnetic stirrer—Improvement of surface defects and inclusion control, Steelmaking Conf. Proc., 80, 159–164.
[5] K. Ayata, K. Miyazawa, E. Takeuchi, N. Bessho, H. Mori, H. Tozawa, 2000. Outline of national project on application of electromagnetic force to continuous casting of steel, The 3rd International Symposium on Electromagnetic Processing of Materials, 376–380.
[6] M. Tani, K. Shio, K. Wajima, E. Takeuchi, J. Tanaka, K. Miyazawa, 2000. The control of initial solidification by the imposition of pulsative AC electromagnetic field, The 3rd International Symposium on Electromagnetic Processing of Materials, 381–384.
[7] K. Fujisaki, 2001. Dynamic Behavior of Level Flow Control by Numerical Physical Model, IFAC-MMM 2001, 10th IFAC Symposium on Automation in Mining, Mineral and Metal Processing (MMM2001), 164–169.
[8] K. Fujisaki, T. Ueyama, K. Okazawa, 1996. Magnetohydrodynamic calculation of in-mold electromagnetic stirring, IEEE Trans, Magn., 33(2), 1642–1645.
[9] K. Fujisaki, T. Ueyama, 1998. Magnetohydrodynamic calculation in free surface, Journal of Applied Physics, 83(11), 6356–6358.
[10] C.W. Hirt and B.D. Nichols, 1981. Volume of fluid (VOF) method for the dynamics of free boundaries, J. Comp. Phys., 39, 201.

Ductile Machining Phenomena of Nominally Brittle Materials at the Nanoscale

J. PATTEN and H. CHERUKURI
Center for Precision Metrology, University of North Carolina at Charlotte, Charlotte, NC 20223, USA

G. PHARR
University of Tennessee, Knoxville, TN, USA

R. SCATTERGOOD
North Carolina State University, Raleigh, NC, USA

W. GAO
Tohoku University, Sendai, Japan

J. YAN
Department of Mechanical Engineering, Kitami Institute of Technology, Kitami, Japan

ABSTRACT

Precision machining operations can be performed in the nanometer regime. Diamond turning machines (DTM) have resolution at less than 10 nm. Special purpose nanometric cutting machines, such as Nanocut, have controlled resolution approaching 1 nm. Nanometric cutting tests have been performed by two of the authors (Patten and Gao), using Nanocut II, at controlled depths of cut down to the 2 to 5 nm range. Many of the precision machining applications in the nanometer to micrometer level involve applications using hard and brittle materials such as semiconductors and ceramics. Components made from these materials include microelectronics, optics, electro-optics, optoelectronics, and structural (mechanical) devices, including MEMS and mechatronics applications. At such small size scales semiconductors and ceramics materials exhibit a dual nature. Generally at macroscopic scales (on the order of a few mm) these materials are ideal brittle solids, similar to glass at room temperature. However, at a small enough scale (nm to μm) these materials can experience ductile behavior similar to metals. Therefore, in order to machine these materials in a ductile mode the nanometric characteristics of the materials must be accounted for. This requires relying on the field of nanotechnology to explore this phenomenon. More precisely, the ductile response of these materials is based upon their nanoscale behavior. This phenomenon gives rise to a ductile to brittle transition with increasing size. The size or scale effect is characterized as competition between plastic deformation (a volume effect) and brittle fracture (a *surface area effect). At small scales the energy for plastic deformation is less than that required for fracture, while at larger scales the opposite is true, i.e. fracture is more energetically favorable. If "a" is the characteristic length scale, then the volume and is proportional to a^3 and the area is determined from a^2. Energy scales with volume for plastic deformation and scales with the area for brittle fracture. Finite element analysis and modeling (FEA/FEM) and molecular dynamics (MD) models fail to capture this dual nature or split personality of these nominally brittle materials, due to inadequate material models, i.e. constitutive models and potential energy functions. However, experimental counterparts such SEM, TEM, and AFM, clearly show this ductile and brittle behavior. Theoretical analyses, ab-intio quantum mechanics, depict this class of materials as exhibiting a high-pressure metallic phase that presumably provides the observed ductile behavior. Thus, at small size scales (nanometer) the ductile behavior of these materials is governed by the high-pressure metallic phase, and brittle fracture is controlled by the native covalent or ionic material phase. This paper will provide a unified theory that combines all of the above-mentioned experimental, analytical, modeling, and theory into a cohesive explanation of the ductile behavior of these nominally brittle materials.*

INTRODUCTION

The emphasis of this paper is on the role of high-pressure phase transformation (HPPT) in ductile machining of semiconductors and ceramics at the nanoscale. The metallic high-pressure phase of these materials is the origin of the materials' ductility. The focus of this work is on the ductile nature of these nominally hard and brittle materials. The work complements the larger body of knowledge concerning brittle or fracture characteristics of these materials. This current paper emphasizes the process of single point diamond turning (SPDT) and its corollaries: indentation, scratching, grinding, and polishing.

BACKGROUND

The metallic nature of the high-pressure phases of semiconductors and ceramics is responsible for their metal like ductility at room temperature. However, at present it has not been possible to directly observe this metallic ductility *in-situ*, therefore skeptics (non-believers) are justified in their continued negativism towards this explanation of the ductile behavior of these nominally hard and brittle materials. The metallic nature of the high-pressure phase (HPP) of semiconductors and ceramics provides the convincing knowledge as to why these nominally brittle materials can behave in a ductile fashion. The answer is really that the hard-brittle semiconductors and ceramics do not behave in a ductile manner, based upon traditional plasticity models with origins in dislocation generation and motion; rather it's their high-pressure metallic counterparts that are ductile and plastically deform. It is this dual nature, or split personality of these materials that is the origin of the ductile-brittle transition. In essence the high-pressure metallic phase can be ductile while the room condition covalent phase is brittle. Understanding that these material phases, (which exist simultaneously, and are really two different materials) one a brittle covalent material at room or atmospheric conditions and the other a metal at high-pressures, is central to the theme of this paper. In this section some background research performed by the authors will be presented for silicon, germanium and silicon nitride, which will help the reader to see the big picture and understand how many related behaviors are consistent with the above comments.

Nanocutting

Two of the authors (Gao and Patten) have performed extensive nanocutting tests of silicon and silicon nitride [1–4]. Figure 1 and Figure 2 show some representative results from Si and Si_3N_4 respectively.

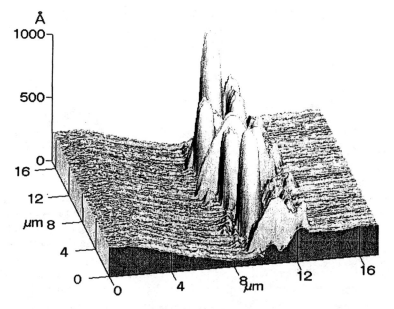

Figure 1. Nanocutting result for Ductile –Brittle Transition Silicon (AFM).

Nanocutting provides a simplified 2-D cutting condition and permits controlled depth of cuts down to a few nanometers. The device also provides concurrent force data. The device can be used to study the ductile to brittle transition of silicon [1] but not Si_3N_4, as the critical depth of cut is beyond the equipment's usable range for the materials studied. Results of the nanocut experiments confirm the ductile response of machining of Si and Si_3N_4 (Figure 1 and Figure 2) and through post process inspection have established a ductile-brittle transition in Silicon (Figure 1). Similar work has been performed on Germanium with corresponding results [5]. The beneficial result of a highly negative rake angle tool has been demonstrated in all of these materials (Si, Ge and Si_3N_4). A 45° rake angle appears to be the most advantageous, similar to results obtained in SPDT

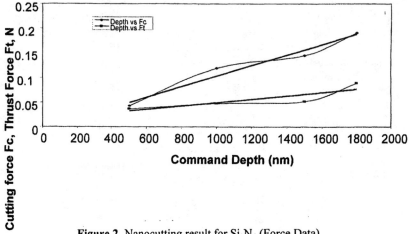

Figure 2. Nanocutting result for Si_3N_4 (Force Data).

[6,7]. While brittle behavior can be demonstrated in Si [1] and Ge [5] with nanocut devices, the total controlled depth of cut of 2 μm appears to be less than that required to initiate brittle fracture in Si_3N_4 (with a –45° rake angle for the materials tested thus far by the authors). The force data obtained from nanocutting experiments show two predominant results. First, for ductile cuts, the cutting force F_c is greater than the thrust force F_t (Fc>Ft) for rake angles up to and including –45°. For more negative rake angles (i.e. –80 and –85°) Ft>Fc [4,7], i.e. the thrust force exceeds the cutting force (Fc<Ft). The specific cutting energy (cutting energy/volume of material removed) increases as the depth of cut decreases, indicating less efficient cutting and increased influence of rubbing and plowing. At extremely small depths of cut, when these latter effects dominate, F_t>F_c as we approach the rubbing/sliding frictional behavior where $F_c=\mu F_t$, and $\mu<1$ is the characteristic of the coefficient of friction (apparent coefficient of friction). Additionally, by varying the rake angle, in addition to depth of cut, the ductile—brittle transition can be evaluated. The conclusion from these experiments is that the more negative the rake angle the more ductile the material's response. There appears to be little benefit to be achieved from rake angles beyond ~ - 45°. This issue is addressed in more detail later in the paper.

DUCTILE MODE MACHINING: THEORY AND EXPERIMENT

With the previous background, it is clear that ductile machining of semiconductors and ceramics is possible and feasible provided that the HPPT mechanism is sufficient to avoid brittle fracture. Numerous studies have shown ductile chip formation and fracture free surfaces generated by ductile machining [8,9]. Figure 3 shows an example of ductile chips formed from silicon nitride. Many other authors have demonstrated similar chip morphology for silicon and germanium [10,11].

Figure 3. SEM of Si_3N_4 Ductile Chip.

The key parameters to achieve ductile machining are: small chip cross-section (critical chip thickness or critical depth of cut), which involves a combination of feed (f), depth (d) or depth-of-cut (DoC), and tool nose radius (R), and large negative rake angle. (Note: the cutting edge radius (r) can provide an effective rake angle at small depths of cut (D < r) [6]. Less negative rake angle tools can produce ductile cuts, but require a smaller chip thickness to do so. These parameters all work in conjunction to control the extent of the HPPT and thus, the ductile response of the material.

It is also quite obvious that brittle fracture can readily occur with large chip dimensions (s) and less negative rake angle tools. The competition between plastic deformation, which is a volume effect (s^3), and brittle fracture, which is a surface area effect (s^2), is based upon the "size effect" (s) such as has been incorporated into Griffith's fracture criterion. The premise is that it takes a given amount of energy (specific cutting energy, Es) to cause plastic deformation and fracture (via crack initiation and propagation). At small scales, it is easier or energetically more favorable to create plastic deformation ($Es^3 < Es^2$) whereas at larger scales the reverse is true, i.e. it takes less energy (and thus is more energetically favorable) to produce brittle fracture ($Es^2 < Es^3$), i.e., when relying on principles of minimum energy or least work. Fracture will be discussed in more detail later relative to the ductile to brittle transition.

SINGLE POINT DIAMOND TURNING (SPDT)

The following sections provide details of a commonly-used machining process associated with ductile regime machining of semiconductors and ceramics. The process parameters and variables that directly affect the ability to ductile-machine these materials will be highlighted and discussed in detail.

Single point diamond turning (SPDT) is a well-established technology for fabricating non-ferrous metal mirrors for optical applications [12]. SPDT emerged in the 1960's with the demands for advanced science and technology in energy, computer, electronics, and defense. Development of this technology was a revolutionary advancement of conventional machining technology to sub-micrometer to nanometer precision. Some of the features that enable this technology are ultra-precision tool fabrication, air-bearing spindle, pneumatic/hydrostatic slides, feedback control, vibration isolation, and temperature control, etc. Compared to other precision machining technologies such as grinding, lapping, and polishing, SPDT has an excellent ability to precisely control the machining contour through high precision numerical control.

Since the late 1970's, SPDT has also been extended for precision machining of brittle materials [13-15]. Since then, significant experimental work has been reported [8,10,11,16–25]. These studies illustrate a fundamental fact that a variety of brittle materials are machinable by SPDT. These brittle materials can be removed by plastic flow as is the case with metals, leaving a crack-free surface when the machining scale is sufficiently small. Also, a common conclusion drawn from these studies is that in the diamond turning of brittle material, tool rake angle is one of the most important factors. Unlike in traditional metal cutting where a zero or a positive rake angle is beneficial, when cutting brittle material, a negative rake angle is advantageous.

The advantage of a negative rake angle in machining brittle material originates from the tool-induced compressive stress state (hydrostatic pressure). It is the internally generated hydrostatic pressure that behaves similarly to the external high-pressure and makes a nominally brittle material deform in a plastic manner and be removed in a ductile mode [26]. Such high-pressure conditions are similar to that occurring at light loads under an indenter in indentation testing as shown in Figure 4, where spherical symmetry of the bottom half of a spherical cavity is maintained in the deformation zone [27]. Immediately below the indenter, the material is considered as a radially expanding core, exerting a uniform hydrostatic pressure on its surroundings. Encasing the core is an ideally plastic region within which flow occurs according to some yield criterion. Beyond the plastic region lies the elastic matrix.

As known from the theory of plasticity, while the yield strength of the material is determined by the stress state, the magnitude of the hydrostatic stress determines the extent of plastic deformation prior to fracture [28]. In other words,

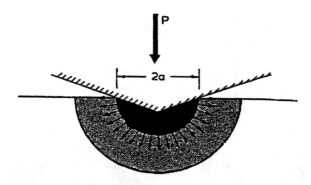

Figure 4. Model of elastic-plastic indentation. Dark region denotes a hydrostatic core, the shaded region plastic zone and the surrounding region elastic matrix [27].

hydrostatic pressure determines the strain at fracture that in turn determines the ductility or brittleness of the material under a state of stress. Some authors attribute this in part to a structural transformation from diamond cubic (Si-I) into a metallic state β-Sn (Si-II) that could occur under the indenter due to high hydrostatic pressure (10~13 GPa) [29,30]. They propose that material around the indenter would become sufficiently ductile (due to *in-situ* metallic transformation) to sustain plastic flow. Careful measurements of electrical conductivity during indentation close to the indenter on silicon shows a significant increase in conductivity (from semi-conducting to highly conducting) [31,32] strongly supporting a change to metallic state underneath the indenter in which case the material is deformed plastically or in a ductile mode.

Because diamond turning with a negative rake angle tool is geometrically akin to the indentation test with an indenter, the conditions of high hydrostatic pressure can be generated immediately underneath the cutting tool with a highly negative rake. Such a high hydrostatic pressure becomes a prerequisite for machining brittle materials by plastic flow at room temperature. Morris et al. [9] and Patten [8] investigated the ductile chips removed from single crystal germanium generated during diamond turning and confirmed the occurrence of high-pressure phase transformation (HPPT). Due to the *in-situ* metallic transformation, the initially brittle material becomes sufficiently ductile to sustain plastic flow.

DUCTILE TO BRITTLE TRANSITION

In these nominally brittle materials, ductile behavior comes from the metallic HPP transformation and the brittle behavior originates in the covalently-bonded phase. The covalent material can be considered an ideal brittle material, i.e. they can fracture in tension or shear (and compression if the pressure is less than the HPPT pressure). Therefore to model the transition from ductile to brittle behavior one needs to know the crack size distribution and the critical stress value (as a function of crack size and orientation). Or conversely, brittle behavior can be assumed if conditions are unfavorable for generation and sufficient extent of the HPPT to accommodate the imposed deformation, i.e., the material is going to be displaced due to the tool interacting with the workpiece. If the material cannot respond in a ductile fashion due to insufficient generation of HPPT, it will respond in a brittle fracture mode.

Much work has been published on the brittle-fracture of these materials [33,34]. Perhaps the leading work is that of Lawn [35]. Hiatt [36] has attempted to model fracture behavior using FEM by using a two-step procedure during machining simulations. First he assumes ductile behavior, and then he inserts a crack into the resultant stress field and evaluates the potential for brittle fracture (based on an analysis similar to that of Lawn). Of course the fallacy of these two methods of analysis (Lawn [35] and Hiatt [36]) is that neither includes the influence of the HPPT. Additionally, the model proposed by Hiatt does not include the fracture that occurs in front of the tool during chip formation and therefore, it models only one aspect of the brittle fracture during machining.

Traditional crack tip blunting, i.e., plastic behavior at the crack tip in tension is not a concern for these ideal brittle materials. They do not produce plastic deformation in tension, plasticity only occurs in compression due to HPPT. This simplifies the analysis somewhat since this special-case crack tip plasticity need not be included in the ductile to brittle transition analysis.

Traditional brittle fracture analysis and modeling can adequately predict fracture in these nominally brittle materials even in the absence of HPPT if the fracture is initiated and propagated in a purely brittle manner, i.e., in tension rather than in compression, which avoids the influence of HPPT. This situation occurs in the trailing tensile stress field, behind or in the wake of the tool, and it can apply to brittle fracture in front of the tool during chip formation if the process is appropriately configured to account for HPPT.

FUTURE WORK

This section will cover some current and planned research by the authors and others on the subject of ductile machining of semiconductors and ceramics with respect to the high-pressure surface phenomena in these materials.

Silicon Carbide and other nominally hard and brittle materials

The current authors and others have clearly demonstrated the ductile machining of silicon, germanium, and silicon nitride, along with the occurrence of a concomitant HPPT to a metallic state. Similarly, silicon carbide (SiC) has been experimentally shown to be ductile based on polishing [37] and machining tests [38]. Also, theoretical studies [39] have shown SiC to have a HPPT to a metallic phase. Recently, two of the authors (Pharr and Patten) have demonstrated ductile behavior of SiC during nanoindentation, i.e. purely plastic indents without fracture! Therefore, at this stage it seems reasonable to conjecture at least, that SiC may also possess a ductile machining mode which is achievable via HPPT to a metallic state. Further testing of this hypothesis is planned by one of the authors (Patten). One advantage of SiC over Si_3N_4 is that single crystal wafers are available for testing purposes, eliminating the added complication and analysis of the secondary binder phase associated with bulk Si_3N_4 materials. Also, crystal dependencies such as crystal planes and directions, along with various polytypes can be studied with single crystal SiC similar to work done with Si and Ge.

On the other hand, silicon carbide is much harder than the other materials studied and is reported on in this paper. [Ge, 9 GPa; Si, 12 GPa; and Si_3N_4, 14 to 22 GPa, compared to SiC, 20 to 80 GPa]. SiC will present additional challenges to ductile machining, especially with respect to tool wear. While a practical SPDT of SiC may be a long way off in reality, studying the basic process mechanisms associated with ductile deformation and behavior of this material is interesting and potentially of use to the scientific community. The existence of an HP ductile (metallic) phase of SiC may result in applications in other areas than machining such as semiconductor, microelectronic, or MEMs devices.

Silicon's affinity for carbon, resulting in excessive wear of diamond tools during machining of silicon (perhaps even the formation of SiC), may limit the ability to apply SPDT to SiC [40]. Cutting fluids that create a barrier to this chemical reaction, such as some experimental alcohol fluids, may be useful in limiting this negative effect.

SUMMARY

This paper was based on our current understanding of the role and necessity of HPPT to a metallic phase of semiconductors and ceramics required to achieve ductile machining. Our analysis is based on a covalent crystalline semiconductor (or ceramic) to metallic phase forward transformation (with application of pressure, i.e., a hydrostatic stress state) and the subsequent metallic HPP to covalent–amorphous semiconductor (or ceramic) back transformation upon release of this pressure. For the present discussion, we choose to ignore the influence of other process parameters such as shear [41], speed effects [42], etc. However, other authors have differing opinions and models on how ductile machining in general, and the role of HPPT in particular, produce the end result [43]. These issues (i.e., discrepancies) are still unresolved, and so a unified theory is still not generally accepted. It is the authors' hope and intention that the current work will promote additional research and understanding, advancing us towards this goal, i.e., an acceptable unified theory for the ductile regime machining of semiconductors and ceramics.

ACKNOWLEDGEMENT

John Patten and Harish Cherukuri wish to acknowledge the generous support of the National Science Foundation (NSF) through the Machining of Ceramics grant number EEC 9903838 (1999) Win Aung, Program Manager. Patten, Pharr and Scattergood also acknowledge the generous support from NSF for their FGR-High-Pressure Phase Transformations in Semiconductors and Ceramics grant DMR-0203552 (2002) Lynnette Madsen, Program Manager. These grants have helped researchers at the University of North Carolina at Charlotte, University of Tennessee, and North Carolina State University to extend their work on semiconductors and ceramic materials, and in particular to study the role and effect of high-pressure phase transformations. John Patten and Harish Cherukuri also wish to express their appreciation for the continued support of the Center for Precision Metrology at UNC Charlotte (R. Hocken, Director) that houses their research program.

REFERENCES

[1] W. Gao, R.J. Hocken, J.A. Patten, J. Lovingood, D.A. Lucca, 2000a. Precision Eng., 24, 320–328.
[2] W. Gao, R.J. Hocken, J.A. Patten, J. Lovingood, 2000b. Review of Sci. Instruments, 71, 4325–4329.
[3] J.A. Patten, W. Gao, R. Fesperman, 2000a. Proceedings ASPE 2000a Annual Meeting, 22, 106–109.
[4] J.A. Patten, W. Gao, 2001 Precision Engineering, 25, 165–167.
[5] D.A. Luccca, P. Chou, R.J. Hocken, 1998. Annals of the CIRP, 47, 475–478.
[6] J. Patten, 1998a. Proceedings ASPE Spring Topical Meeting on Silicon Machining, 17, 88–91.
[7] J. Yan, K. Syoji, T. Kuriyagawa, 1999b. Precision Science and Technology for Perfect Surface, Y. Furukawa, Y. Mori and T. Kataoka (Eds.), JSPE Publ. Series No.3 (Proc. 9th ICPE), Osaka, 92–97
[8] J.A. Patten, 1996. High-pressure phase transformations analysis and molecular dynamics simulations of single point diamond turning of germanium (PhD dissertation, North Carolina State University)
[9] J.C. Morris, D.L. Callahan, J. Kulik, J.A. Patten, R.O. Scattergood, 1995. J. Am. Ceram. Soc., 78, 2015–2020.
[10] J. Yan, K. Syoji, H. Suzuki, T. Kuriyagawa, 1998a. Journal of the Japan Society for Precision Engineering, 64, 1345–1349. (In Japanese)
[11] W.S. Blackley, R.O. Scattergood, 1991. Prec. Eng., 13, 95.
[12] K. Iwata, T. Moriwaki, K. Okuda, 1987. Journal of JSPS, 53, 1253–1258 (in Japanese)
[13] T.T. Saito, 1978. Optical Engineering, 17, 570–573.
[14] R.J. Benjamin, 1978. Optical Engineering, 17, 574.
[15] D.L. Decker, D.J. Grandjean, J.M. Bennett, 1979. Science of Ceramic Machining and Surface Finishing II, NBS Special Publication, 562, 293–303.
[16] B.A. Fuchs, P.P. Hed, P.C. Baker, 1986. Applied Optics, 25, 1733–1735.
[17] H. Matsunaga, S. Hara, 1986. Journal of the JSPE, 52, 26–29. (In Japanese)
[18] K.E. Puttick, M.R. Rudman, K.J. Smith, A. Franks, K. Lindsey, 1989. Proc. Roy. Soc., A, 426, 19.
[19] C.K. Syn, J.S. Taylor, 1989. ASPE/IPES Conference poster session
[20] T. Nakasuji, S. Kodera, S. Hara, H. Matsunaga, N. Ikawa, S. Shimada, 1990. Ann. CIRP, 39, 89.
[21] P.N. Blake, R.O. Scattergood, 1990. J. Amer. Ceram. Soc., 73, 949.
[22] R. Shore, 1995. Proc. SPIE, 2576, 426–431.
[23] T. Shibata, S. Fujii, E. Makino, M. Ikeda, 1996 Prec. Eng., 18 130.

[24] J. Yan, K. Syoji, T. Kuriyagawa, and H. Suzuki, 1998b. Proceedings of the Japan Society for Precision Engineering Autumn Meeting, 259.
[25] T.P. Leung, W.B. Lee, S.M. Lu, 1998. J. Mater. Proc. Tech., 73, 42.
[26] J. Yan, M. Yoshino, T. Kuriyagawa, T. Shirakashi, K. Syoji, R. Komanduri, 2001a. Materials Science and Engineering A, 297, 230–234.
[27] K.L. Johnson, 1970. Journal of the Mechanics and Physics of Solids, 18, 115-126.
[28] W. Johnson, P.B. Mellor, 1973. Engineering Plasticity, (Van Nostrand Reinhold Co., London)
[29] A.P. Gerk and D. Tabor, 1978. Nature, 271, 732-734.
[30] A. Kailer, Y.G. Gogotsi, and K.G. Nickel 1997. Journal of Applied Physics, 81, 3057–3063.
[31] I.V. Gridneva, Y.V. Milman, V.I. Trefilov, 1972 Physica Staus Solidi., 14, 177–182.
[32] D.R. Clarke, M.C. Kroll, P.D. Kirchner, R.F. Cook, and B.J. Hockey, 1988. Physical Review Letter, 60, 2156–2159.
[33] NIST, 1993. Machining of Advanced Materials (Proc. Int. Conf. Mach. Adv. Mat.) NIST Special Pub., 847.
[34] ASPE, 1998. ASPE Spring Topical Meeting on Silicon Machining, 17.
[35] B. Lawn, 1993. Fracture of Brittle Solids (Cambridge Univ. Press).
[36] G.D. Hiatt, 1992. A Fracture Mechanics Technique for Predicting the Ductile Regime in Single Point Diamond Turning of Brittle Materials (Ph.D. thesis, North Carolina State University, Raleigh NC)
[37] M. Kikuchi, Y. Takahashi, T. Suga, 1992. J. Am. Ceram. Soc., 75, 189–194.
[38] A. Zhong, 2002. J. Mat. Proc. Tech., 122, 173–178.
[39] J.A. VanVechten, 1973. Physical Review B, 7, 1479–1507
[40] R.O. Scattergood, 2000-2002. private conversations with J. Patten.
[41] J.J. Gilman, 1993b. Science, 261, 1436–1439.
[42] Y. Gogotsi, C. Baek, F. Kirscht, 1999. Semicond. Sci. Technol., 14, 936–944.
[43] F.Z. Fang, V.C. Venkatesh, G.X. Zhang, 2002. Int. J. Adv. Manf. Tech., 19, 9.

Fractal Dimension: A New Machining Decision-Making Parameter

A. M. M. SHARIF ULLAH
Department of Mechanical Engineering, College of Engineering, United Arab Emirates University, P.O. Box 17555, AL AIN, UAE

M. R. RAHMAN and V. KACHITVICHYANUKUL
ISE Program, AIT, P.O. Box 4, Klong Luang, Pathumthani 12120, Thailand

K. H. HARIB
Department of Mechanical Engineering, College of Engineering, United Arab Emirates University, P.O. Box 17555, AL AIN, UAE

ABSTRACT

While keeping surface finish within the required limits one should not rely on standard surface roughness parameters (e.g., Ra), only; fractal dimension should also come into consideration. Otherwise, the complexity of the machined surface remains undefined. Keeping this in mind, turning experiments have been carried out, wherein varying cutting conditions (depth of cut, feed rate, and cutting velocity) and tool insert, the values of the fractal dimension of roughness profiles have been determined along with the values of other widely used roughness parameters. Using a soft computing framework that combines both fuzzy computing and probabilistic computing the experimental results have been converted to two linguistic rules, one of them specifies the limits of cutting conditions for keeping surface roughness with the required limits, and the other specifies that if the same limits of cutting conditions are applied then fractal dimension is likely (not absolutely likely) to be within the required limits. In other words, keeping both fractal dimension and roughness within a specified limit is perhaps a difficult task. Further study should be carried out to see whether similar remarks could be made for other machining processes. Nevertheless, it is confirmed that fractal dimension is an important machining decision-making parameter in the sense that if we try to keep both fractal dimension and surface roughness (particularly, Ra) within the required limits then we get a set of recommended ranges of cutting conditions, if we omit the case of fractal dimension, we get another set of recommended ranges of cutting conditions.

INTRODUCTION

Fractal Dimension and Machining

Machining of materials exhibits highly irregular shapes and behaviors, which are supposed to be the results of chaotic phenomena—the phenomena that have similar properties at all levels of magnification [1–3]. The concept of self-similar dimension (i.e., fractal dimension) is often used to quantify the complexity of a chaotic behavior or shape [4,5]. Mandelbrot actually coined the word "fractal" in mid-1970s to describe objects, shapes, or behaviors that have similar properties at all level of magnification [6]. Mathematically, fractal dimension (FD) is expressed by the following expression:

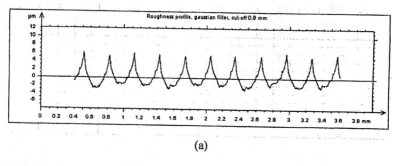

(a)

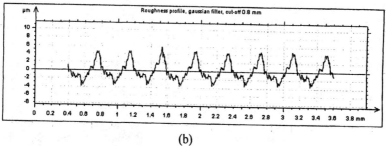

(b)

Figure 1. Two roughness profiles having same Ra but different fractal dimension.

$$FD = \operatorname*{Lt}_{\substack{r \to 0 \\ N(r) \to \infty}} \frac{\log(N(r))}{\log\left(\frac{1}{r}\right)}. \quad (1)$$

In (1), r denotes the scale of observation; N(r) denotes the number of similar objects (shape or behavior) at observation scale r.

Like many other engineering fields, the impact of fractal dimension has equally been felt in researching machining of materials [7–9]. To realize the main argument, refer, for example, to the roughness profiles (a) and (b) in Figure 1, which are two turned surfaces having same Ra[1] [9] value (1.78 μm) but different fractal dimension (1.21 and 1.26, respectively). Since profile (a) is less irregular than the other one they should have different values of FD. In fact, FD of profile (a) is 1.21 and profile (b) is 1.26. This means that profile (a) is less complex than profile (b).

Now, some of the commercial computer applications for analyzing surface texture now have a function to determine fractal dimension of roughness profile (e.g. Talyprofile™ of Digital Surf™), magnifying the profile up to 1 nm scale using box-counting method[2]. This means in machining decision-making, fractal dimension can be considered, along with other factors, as explained below.

Machining Decision-Making

Generally speaking, in most machining decision-making problems three relative motion components (depth of cut, feed, rate, cutting velocity) and cutting tool insert (i.e., tool material and geometry) are set, keeping productivity as high as possible, surface finish within the required limits, tool wear as low as possible. While doing this, the involvement of machinability data[3] is somewhat unavoidable, also the guidelines of cutting tool makers [13,14]. Now, regarding keeping surface finish within the required limits, one should not rely on standard surface roughness parameters [10] (e.g., Ra, Rt, etc.) only; fractal dimension should also come into consideration. Otherwise, the complexity of the machined surface remains undefined, which should not be neglected, as pointed out in the previous sub-section. On the other hand, regarding keeping tool wear low, one should follow the depth of cut and feed rate limits specified by the tool venders because tool venders provide a range of depth of cut and feed rate for a tool ensuring a definite tool life, say 10 minutes. If such recommendations are not followed, then low tool wear might not be guaranteed. Regarding keeping productivity high, one should keep depth of cut, feed rate, and cutting velocity as high as possible because product of these three relative motions determines productivity.

---

[1] Ra is arithmetic average of the points of a roughness profile measured according to ISO Standard, 4287:1997. It is the most frequently used surface finish indicator [11].
[2] This may not be a standard practice. For more details, see the proposal of ASME B46 SC9 [12].
[3] There is (most probably) no machinability databases exist including the values of fractal dimension. This means we would like to see the impact of fractal dimension in machining decision-making, experiments should be carried out to get the data recording fractal dimension of machined surface along with other roughness parameters at different cutting conditions varying cutting tools (geometry, material). These pieces of data should be used to show the effect of fractal dimension in machining decision-making. See the section called Initialization for the set of data used in this study.

Approach

Following the perspective stated in the above, if a machinist searches for optimal cutting conditions and cutting tools, the machinist is going to encounter heterogeneous information, i.e., a mixture of measurements (numerical or c-granular information) and human-perception (f-granular information) [15]. This implies that a computational framework capable of handling numerical, c-granular, and f-granular information (and even other forms of information) is needed. As such, a framework supporting soft computing[4] [16] should be chosen. Otherwise, it would be difficult to deal with the computational complexity underlying such a heterogeneous form of information [15—18]. The remainder of this paper thus describes the soft computing based computational framework used in this study and how it works or should work in providing machining decisions. Simultaneously, how fractal dimension influences the decision making process is also emphasized.

COMPUTATIONAL FRAMEWORK

Figure 2 shows the computational framework used in this study. It is actually a soft computing based framework that combines two methods of computing, fuzzy computing and probabilistic computing. Fuzzy computing is going to help encode requirements and interpret data, also to generate operating guidelines, as if a human being does it. On the other hand, probabilistic computing (which has been found to be very effective when one relies on a set of data representing a particular scenario [19]) is going to help find rules (knowledge). In the framework are four schemes, namely, (1) Perception Based Requirement-Expression Scheme, (2) Numerical Data Scheme, (3) Perception Based Data Interpretation Scheme, and (4) Probabilistic Reasoning Scheme.

Using scheme (1), one can express requirements linguistically. The scheme use fuzzy sets to put such requirements into formal computation. Using scheme (2) one can acquire necessary numerical data. The information in scheme (1) might or should interfere such data acquisition process. A set of numerical data (most probability machinability data if machining decision is going to be made) due to the action of schemes (1) and (2) is acquired, as indicated by using "A" in Figure 2. Once a set of data is available, the values of the parameters related to requirements should be interpreted as acceptable or not acceptable using fuzzy logical approach. As a result, numerical data becomes a modified set of data, termed as "Requirement Constrained Semi-Linguistic Data." The process for coming up to this point after A is indicated by "B" in Figure 2. The processes A and B are hereinafter referred to as *initialization* because of their role in the whole framework.

---

[4] Soft computing means any combination of fuzzy, neural, evolutionary, and probabilistic computing so that computing methods combined should support each other, rather than competing with each other. In real-life decision-making problems soft computing has been found to be very useful and its use is growing extensively [18].

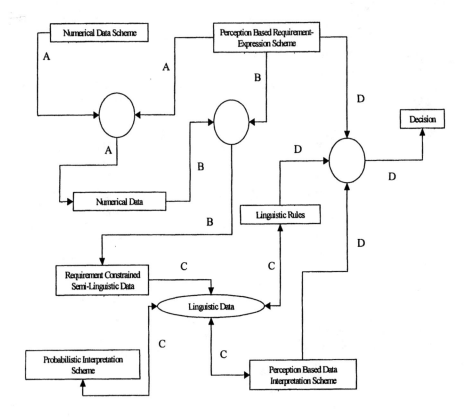

Figure 2. Computational Framework for Machining Decision.

When *initialization* is completed the framework is ready to perform a process of interactive computation hereinafter referred to as *dynamic computing*, as shown by C in Figure 2, in which schemes (3) and (4) are involved. First, dynamic computing acknowledges the set of requirement constrained semi-linguistic data and then generate linguistic data using the Perception Based Data Interpretation Scheme, i.e., scheme (3). Afterward, it analyzes the linguistic data generated using Probabilistic Interpretation Scheme, i.e. (4) and then try to generate some linguistic rule. If scheme (3) does not find any suitable rules, then information used in scheme (3), i.e., information used in interpreting data, should be changed. In this stage, new variable can be considered that should be calculate-able from the parameters involved in numerical data. This process continues until there is a set of easy-to-implement linguistic rules.

When *dynamic computing* is completed, the framework is ready to make decisions (set cutting conditions, cutting tools, etc.), hereinafter referred to as *decision-making*, which is indicated by "D." In decision-making, information finally used to generate linguistic rules in scheme (3) and information set at the beginning in scheme (1) play important role, as shown in Figure 2 by D.

INITIALIZATION

Having described the silent features of the framework in general, it is now time to describe it specifically—what results it produces in showing the implication of fractal dimension in selecting cutting conditions and inserts and how. This section describes *initialization* only; *dynamic computing* and *decision-making* are described in the following two sections.

Suppose that a machinist is going to turn a workpiece made of SAE 1045 steel. The surface roughness (Ra) should be kept near to 1.2 μm. Considering the fact that fractal dimension quantifies the complexity of a roughness profile, the machinist should also keep fractal dimension as low as possible say below 1.2 (here 1.2 is box-counting fractal dimension). This means that the machinist has to set cutting conditions (depth of cut, feed rate, and cutting velocity) as well as the insert in such a way so that surface finish remains near to 1.2 μm and fractal dimension below 1.2. Therefore, the requirements are "keep Ra near 1.2 μm" and keep "fractal dimension of roughness profile below 1.2." These two linguistic requirements can be put into formal computation by using fuzzy logical approach. As shown in Table 1, "keep Ra near to 1.2 μm" refers to a fuzzy set labeled "Acceptable" in the universe of discourse [0 μm, 2.5 μm][5]. Similarly, the other requirement "keep fractal dimension below 1.2" refers to a fuzzy set labeled "Acceptable" in the universe of discourse = [1,2]. The shape of the membership functions for acceptable requirement may vary from machinist to machinist.

The above requirement setting indicates that pieces of machinability data including the values of both fractal dimension and roughness for various combinations of cutting conditions and tool material and geometry are needed. Since fractal dimension is not yet a standard parameter for machined surface evaluation, most probably, there is no machinability database available that provides information of fractal dimension at various cutting conditions. This implies that experiments should be carried out to acquire such data. In other words, the Numerical Data Scheme for this study means carrying out turning experiments. Accordingly, experiments have been carried out to get machinability data for turning. Two cutting tool inserts have been used, one for general machining and the other for high feed machining, as shown in Table 2, both of which are from Sandvick™ Coromant™. Job material used is SAE 1045 steel (cylindrical bar), according to the requirement. Using Taylor-Hobson™ Surtronic™ 3+ Stylus Instrument and Surface Texture analyzing software called Talyprofile™ (Version 2.0) of Digital Surf™ have been used to find out the values of Ra and Fractal Dimension, as shown in Table 2. If the values of surface

---

[5] To note, the universe of discourse surface roughness, here, [0 μm, 2.5 μm], should be used to determine whether a piece of machinability data should be considered or should not be considered for the whole decision-making process. As such, for this particular case, a piece of machinability data having Ra value above 2.5 μm should not be considered.

Table 1. A Requirement Scheme.

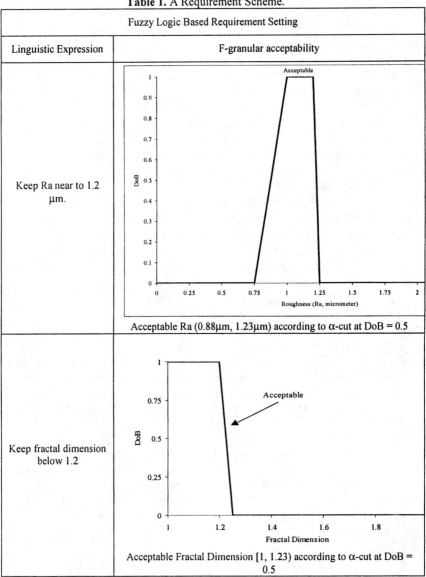

Fuzzy Logic Based Requirement Setting	
Linguistic Expression	F-granular acceptability
Keep Ra near to 1.2 µm.	*(graph showing Acceptable region peaking near Ra = 1.2)*
	Acceptable Ra (0.88µm, 1.23µm) according to α-cut at DoB = 0.5
Keep fractal dimension below 1.2	*(graph showing Acceptable region for Fractal Dimension up to ~1.2)*
	Acceptable Fractal Dimension [1, 1.23) according to α-cut at DoB = 0.5

roughness and fractal dimension comply with the numerical limits shown in Table 1, then they are interpreted as acceptable, denoted by AC, as shown in Table 2. If not, then they are interpreted as not acceptable, denoted by NA. This interpretation provides the requirement constrained semi-linguistic having following structure: <numerical depth of cut (say, 1.08 mm), numerical feed rate (0.2 mm/rev), numerical cutting velocity (180 m/min), linguistic roughness (NA), and linguistic fractal dimension (AC)>.

Table 2. Experimental data including roughness and fractal dimension.

Experiments	Depth of cut mm	Feed rate mm/rev	Cutting speed m/min	Surface Roughness (Ra) μm		Fractal Dimension (Box-counting)	
Cutting Tool Maker: Sandvick™ Coromant™ Insert Code: CNMG 120404 PF (Basic insert for general machining) Depth of cut range [0.30mm, 1.50 mm] Feed rate range [0.07mm/re , 0.30 mm/rev]							
1	0.84	0.17	80	0.988	AC	1.27	NA
2	0.84	0.21	100	1.83	NA	1.21	AC
3	0.84	0.25	120	2.38	NA	1.23	NA
4	0.84	0.17	160	0.915	AC	1.19	AC
5	0.84	0.21	180	0.962	AC	1.16	AC
6	0.84	0.25	200	1.17	AC	1.31	NA
7	1.08	0.17	160	1.45	NA	1.30	NA
8	1.08	0.21	180	1.00	AC	1.27	NA
9	1.08	0.25	200	0.855	NA	1.18	AC
10	1.08	0.17	240	0.702	NA	1.11	AC
11	1.08	0.21	260	0.770	NA	1.17	AC
12	1.08	0.25	280	1.49	NA	1.29	NA
13	1.32	0.17	240	1.00	AC	1.30	NA
14	1.32	0.21	260	1.96	NA	1.29	NA
15	1.32	0.25	280	2.43	NA	1.26	NA
Cutting Tool Maker: Sandvick™ Coromant™ Insert Code: CNMG 120404 WF (Wiper insert for high feed machining) Depth of cut range [0.25mm—1.50mm] Feed rate [0.10mm/rev—0.40 mm/rev]							
16	0.81	0.18	160	1.15	AC	1.26	NA
17	0.81	0.24	180	1.71	NA	1.27	NA
18	0.81	0.30	200	1.77	NA	1.21	AC
19	0.81	0.18	240	1.09	AC	1.15	AC
20	0.81	0.24	260	0.938	AC	1.30	NA
21	0.81	0.30	280	1.90	AC	1.21	AC
22	1.06	0.18	160	0.978	AC	1.29	NA
23	1.06	0.24	180	1.14	AC	1.10	AC
24	1.06	0.30	200	0.944	AC	1.27	NA
25	1.06	0.18	240	1.43	NA	1.30	NA
26	1.06	0.24	260	1.83	NA	1.27	NA
27	1.06	0.30	280	1.85	NA	1.23	NA
28	1.31	0.18	240	1.20	AC	1.31	NA
29	1.31	0.24	260	1.83	NA	1.29	NA
30	1.31	0.30	280	2.33	NA	1.19	AC

AC means acceptable. NA means not acceptable. Refer to the numerical range shown in Table 1.

DYNAMIC COMPUTING

Following dynamic computing is carried until getting a set of suitable rules (here two rules only, as shown at the end of this Section).

1. Acknowledge requirement constraint semi-linguistic data
   ↓
2. Define Variables
   ↓
3. Define f-granular recommended range
   ↓
4. Find c-granular recommended range using α-cut at DoB = 0.8
   ↓
5. Generate Linguistic Data
   ↓
6. Identify Linguistic Rule using Probabilistic Reasoning
   ↓
7. Suitable Linguistic Rules for Decision-Making
   ↓   ↓
   YES   NO
   Adopt Rules for Decision-Making   Go to Step 2

Table 3. A Perception Based Data Interpretation Scheme.

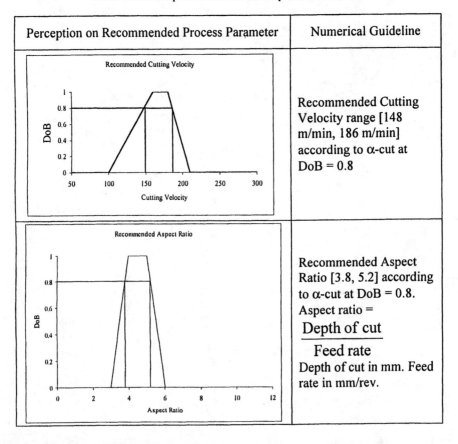

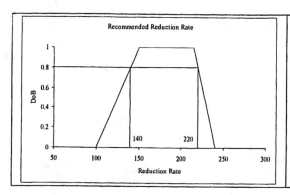

Recommended Reduction Rate [140, 220] according to α-cut at DoB = 0.8.
Reduction Rate = <u>Cutting Velocity</u> / Depth of cut
Cutting Velocity in m/min, Depth of cut in mm.

Table 3 shows the final version of the Perception Based Interpretation Scheme. As seen from Table 3, besides cutting velocity, two new derived variables are used—one of them is aspect ratio (the ratio of depth of cut and feed rate) and the other is reduction rate (the ratio of cutting velocity and depth of cut). Table 4 shows the linguistic data used to find out rules. It is needless to say that first three columns of Table 4 correspond to the interpretation scheme shown in Table 3 and last two columns correspond to the interpretation scheme shown in Table 1.

Table 5 shows the Probabilistic Reasoning Scheme. As seen from Table 5, probability of getting acceptable roughness is unit and acceptable both roughness and fractal dimension is 0.6 given that all cutting velocity, aspect ratio and reduction rate are kept within the recommended ranges. The probability of getting acceptable both roughness and fractal dimension given that all cutting velocity, aspect ratio, and reduction rate are kept out of the recommended range is 0.14. Note that Prob(X|Y) is much greater than Prob(X|¬Y). This ensures that recommended ranges are really helpful in ensuring the acceptable performance. This probabilistic analysis suggest that if cutting velocity, aspect ratio and reduction rate are kept within the recommend range, then the roughness is absolutely likely to be acceptable but fractal dimension is likely to be acceptable. This result can be expressed by the following two IF-THEN rules:

Rule 1: **IF** V_c = R AND a_p/f = R AND V_c/a_p = R **THEN** Ra is *absolutely likely* to be AC.

Rule 2: **IF** V_c = R AND a_p/f = R AND V_c/a_p = R **THEN** FD is *likely* to be AC.

DECISION-MAKING

In making decisions the linguistic rules found in dynamic computing as well as the numerical guidelines is needed. In addition, the f-granular information used in expressing requirements i.e., information in Table 1 is also needed. Furthermore, f-granular information expressing the perception of probability is needed because linguistic rules contain perception of probability (e.g., fractal dimension is "likely" acceptable). Otherwise, expected output (expected range of FD and Ra

Table 4. A Set of Linguistic Data.

Cutting Velocity	Aspect Ratio	Reduction Rate	Roughness	Fractal Dimension
NR	R	NR	AC	NA
NR	R	NR	NA	AC
NR	NR	R	NA	NA
R	R	R	AC	AC
R	R	R	AC	AC
NR	NR	NR	AC	NA
R	NR	R	NA	NA
R	R	R	AC	NA
NR	R	R	NA	AC
NR	NR	NR	NA	AC
NR	R	NR	NA	AC
NR	R	NR	NA	NA
NR	NR	R	AC	NA
NR	NR	R	NA	NA
NR	NR	R	NA	NA
R	R	R	AC	NA
R	NR	NR	NA	NA
NR	NR	NR	NA	AC
NR	R	NR	AC	AC
NR	NR	NR	AC	NA
NR	NR	NR	AC	AC
R	NR	R	AC	NA
R	R	R	AC	AC
NR	NR	R	NA	NA
NR	NR	NR	NA	NA
NR	R	NR	NA	NA
NR	NR	NR	NA	NA
NR	NR	R	AC	NA
NR	NR	R	NA	NA
NR	R	R	NA	AC

R and NR mean recommended and not recommended, according to the guidelines in Table 3.

Table 5. Results of Probabilistic Reasoning.

X	Y	Prob (X | Y)	Reasoning
Ra = AC	V_c = R AND a_p/f = R AND V_c/a_p = R (All recommended)	1	If cutting velocity, aspect ratio, and reduction rate are kept within the recommended range, then the roughness is absolutely likely to be acceptable but fractal dimension is likely to be acceptable.
Ra= AC AND FD = AC	V_c = R AND a_p/f = R AND V_c/a_p = R (All recommended)	0.6	
Ra = AC AND FD = AC	V_c = NR AND a_p/f = NR AND V_c/a_p = NR (All not recommended)	0.14	

V_c is cutting velocity, f is feed rate, a_p is depth of cut, Ra is arithmetic average roughness, and FD is fractal dimension.

Table 6. Decision-making.

Linguistic Rules	Constraints (Consult Table 3)	Recommendations	Expected Results	
			Ra	FD
Rule 1: IF V_c = R AND a_p/f = R AND V_c/a_p = R **THEN** Ra is *absolutely likely* AC **Rule 2: IF** V_c = R AND a_p/f = R AND V_c/a_p = R **THEN** FD is *likely* to be AC	V_c = [148, 186] V_c/a_p = [140,220] a_p/f = [3.8,5.2]	(For example) a_p = 0.9 mm f = 0.2 mm/rev V_c = 160 m/min (For example) Insert: CNMG120404 W or P	Expected Ra = [0.88μm, 1.23μm] according to Rule 1 and Perception of Probability	Expected FD = [1, 1.24] according to Rule 2, Table 1, and Perception of Probability.

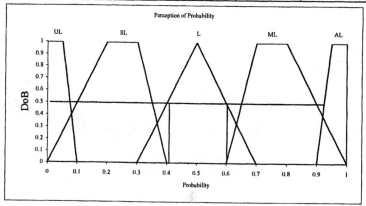

UL, SL, L, ML, AL mean unlikely, some likely, likely, most likely, and absolutely likely, respectively.

when antecedent is true) cannot be determined properly. Table 6 shows an example of machining decision-making—choose insert CNMG120404 W or P, and set cutting conditions $a_p = 0.9$ mm, $f = 0.2$ mm/rev, $V_c = 160$ m/min, in order to achieve surface finish: Ra = [0.88μm, 1.23μm], FD = [1, 1.24].

DISCUSSIONS

The rules found clearly indicate that it is somewhat difficult to keep roughness and fractal dimension together within the required limits. Even though surface roughness can be kept within the required limits, for fractal dimension it should be enlarged to 1.24 from 1.23; the computation involved is as follows.

>Get the expected value (average) of the granule "likely", E^6(likely) $\to$ 0.5
>(according to the f-granular information in Table 6)
>Reset $\alpha = 0.5 \times E$(likely) $\to 0.5 \times 0.5 = 0.25$
>Find new c-granular information using α-cut at DoB = $0.5 \times E$(likely) (=0.25) $\to$ [1, 1.24]

This computation underlies the fact that "if FD is AC" means a range [1, 1.23], then "FD is likely AC" is a range larger than [1, 1.23], and it should be determined by reducing the value of DoB for α-cut by the expected value of the probability granule labeled "likely."

This study is unique not only for considering the fractal dimension as a new machining decision-making parameter, but also for including all four types of decision-relevant information, as identified by Zadeh [15], numerical information (e.g., FD = 0.98), c-granular information (FD = [0.9,1.1]), f-granular information (FD is acceptable), fr-granular information[7] (FD is likely acceptable). To understand this aspect more clearly, note the following line diagrams. These diagrams represent how various forms of information are involved in the framework used in this study.

$$\text{Perception} \to \text{F - Granular Information} \xrightarrow{\alpha\text{-cut at DoB}=0.5} \text{C - Granular Information} \begin{array}{c} A \quad NA \\ \uparrow \quad \uparrow \end{array}$$

$$\uparrow$$
$$\text{Numerical Information} \leftarrow \text{Experiment}$$

(When numerical data is converted to requirement constrained semi-linguistic data.)

```
                                          R         NR
                                          ↑          ↑
Perception → F-Granular Information —α-cut at DoB = 0.8→ C-Granular Information
                                                      ↑
                                          Numerical Information ← Experiment
```

(When requirement constrained semi-linguistic data is converted to linguistic data.)

```
Prob. Reasoning → IF-THEN Rules → Fr-Granular Info. —α-cut at DoB = 0.8×E(granule)→ C-Granular Info.
                                        ↑
                            F-Granular Info. (for Probability)
```

(When setting expected outcomes, FD range, Ra range, from the rules.)

Another point should also be noted here is that unlike other widely used soft computing methods, the method (framework) combines fuzzy computing and probabilistic computing. In this regard further investigation should be carried out incorporating the context of Bayesian Belief Network [20], which is also a potential member of soft computing and suitable for finding out rules from a small set of data.

CONCLUSIONS

Fractal dimension should come into consideration, along with other standard machined surface roughness parameters, in order to ensure surface finish—otherwise, the complexity of the machined surface remains undefined. Using a soft computing framework that combines both fuzzy and probabilistic computing experimental results of turning (i.e., numerical information) have been converted into two linguistic rules. This rules show that fractal dimension is an important machining decision-making parameter in the sense that if we try to keep both fractal dimension and surface roughness (particularly, Ra) within the required limits then we get a set of recommended ranges of cutting conditions, if we omit the case of fractal dimension, we get another set of recommended ranges of cutting conditions. This implies that keeping both fractal dimension and roughness within a specified limit is perhaps a difficult task. Further study should be carried out to see whether similar remarks could be made for other machining processes. However, from the viewpoint of information, this study is a unique one simply because all four forms of information (numerical, c-granular, f-granular, fr-granular) are found to be in use in systematic manner.

REFERENCES

[1] Moon, F.C. and Kalmar-Nagy, T. (2001). Nonlinear models for complex dynamics in cutting materials. Royal Society Philosophical Transactions: Mathematical, Physical & Engineering Sciences, 359, 1781, 695–711.

[2] Moon, F.C.(1994). Chaotic Dynamics and Fractals in Material Removal Processes, In Chapter 2 of Nonlinearity and Chaos in Engineering Dynamics, Thompson, J.M.T. and Bishop, S.R. (Eds.), John Wiley and Sons, New York.

[3] Wiercigroch, M. and Budak, E. (2001) Sources of nonlinearities, chatter generation and suppression in metal cutting. Royal Society Philosophical Transactions: Mathematical, Physical & Engineering Sciences, Volume: 359, 1781, 663–693.

[4] Kantz, H. and Schreiber, T. (1997). Nonlinear Time Series Analysis. Cambridge Nonlinear Science Series 7. Cambridge University Press, Cambridge.

[5] Purintrapiban, U., Kachitvichyanukul, V. and Ullah, A.M.M.S. (2003). Neural Network for Detecting Cyclic Behavior on Process Mean, Proceedings of the 17th International Conference on Production Research, August 3–7, 2003, Blacksburg, Virginia, USA.

[6] Mandelbrot, B. B. (1967). How long is the coast of Britain? Statistical self-similarity and fractional dimension, Science, 156, 636–638.

[7] Brown, C.A.; Johnsen, W.A. and Hult, K.M. (1998) Scale-sensitivity, Fractal Analysis and Simulations, International Journal of Machine Tools and Manufacture, 38 (5–6), 633–637.

[8] Brown, C.A. and Siegmann, S. (2001). Fundamental scales of adhesion and area-scale fractal analysis, International Journal of Machine Tools and Manufacture, 41 (13015014), 1927–1933.

[9] Higuchi, M., Ullah, A. M. M. Sharif and Yano, A. (1997). The structural Design of Superfinishing Stone using Fractal Modelling. Proceedings of the First International Conference on Intelligent Processing and Manufacturing of Materials (IPMM'97), 14–17 July 1997, Gold-Coast, Australia.

[10] —.Geometrical Product Specifications—Surface texture: Profile method—Terms, definitions and surface texture parameters. ISO 4287:1997.

[11] Chiffre, D.L.; Lonardo, P.M.; Trumpold, H., Chemintz, T.U.; Lucca, D.A., Goch, G.; Brown, C.A.; Raja, J. and Hansen, H.N. Quantitative Characterisation of Surface Texture, Annals of CIRP, 49/2/2000, 635–651

[12] —.Brown, C.A. (Eds.). (2000). Terminology and Procedures for Evaluation of Surface Textures using Fractal Geometry. ASME B46 SC9 Proposal, URL: http://www.wpi.edu/~/asmeb46.pdf.

[13] Ullah, A.M.M.Sharif.; Yano, A. and Higuchi, M. (1997). Protein Synthesis Algorithm and a New Metaphor for Selecting Optimum Tools. International Journal of the Japan Society of Mechanical Engineers, Series C, 40, 540–546.

[14] Ullah, A.M.M.Sharif.; Yano, A. and Higuchi, M. (1998). Virtual Machinability data versus real machinability data and their use during manufacturing decision-making. Technical Report of Kansai University, 40, 15–26.

[15] Zadeh, LA. (2001). A New Direction of AI—Toward a Computational Theory of Perceptions, AI Magazine, 73–84.
[16] Zadeh, L.A. (1994). Fuzzy Logic, Neural Networks, and Soft Computing. Communications of ACM, 37(3), 77–84.
[17] Zadeh, L.A. (2000). From Computing with Numbers to Computing with Words—From Manipulation of Measurements to Manipulation of Perceptions, In Azvine, B.; Azarmi, N. and Nauck, D. (Eds.): Intelligent Systems and Soft Computing: Prospects, Tools and Applications. Lecture Notes in Computer Science, Springer, 1804, 3–4.1804.
[18] Negnevitsky, M. (2002). Artificial Intelligence: A Guide to Intelligent Systems. Addison-Wesley, England.
[19] Ochiai, K.; Ullah, A.M.M.Sharif, Higuchi, M.; Yano, A.; Matsumori, N. and Yoshizawa, I. (1997). Acquisition of Knowledge for Design of Grinding Wheels by the Machine Learning Methodology ID3. Journal of the Society of Grinding Engineers, 41, 5, 186–189. (In Japanese.)
[20] Pearl, J. (1988). Probabilistic Reasoning in Intelligent Systems: networks of plausible inference. Morgan Kaufmann, San Francisco.

APPENDIX A: BOX-COUNTING FRACTAL DIMENSION

In practice, Fractal Dimension (FD) of an arbitrary irregular shape or behavior that is supposed to be the result of chaotic phenomena is determined by the following numerical method.

- Fit the irregular shape using a simple liner- or area-object.
- Determine N(r), for all r = 1, ½, ¼,....
- Plot N(r) and 1/r on a log-log graph.
- Determine approximate FD from the slope of the plot.

Even though the basic steps are the same, as mentioned in the above, depending on the fitting-object (line, area, etc.), different names are used to specify the methods. For example, note box-counting method, as shown in Figure A.1, wherein area-objects (box) are used to determine the fractal dimension of an irregular shape. The (irregular) shape in Figure A.1 has fractal dimension approximately 1.75, as seen from the plot in Figure A.1. This value (1.75) indicates that the shape has deviated from line to an area but still it is not a perfect area-element (here, box). (According to the method shown in FigureA.1, a perfect area (box) has fractal dimension 2). In other words, for box-counting method, the value of the fractal dimension determines how complex the shape is compare to a line or an area.

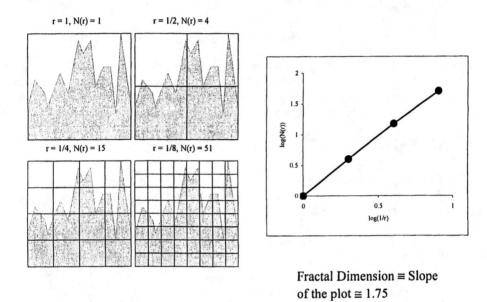

Figure A.1. Box-counting fractal dimension determination process.

EB-Reinforcement by Formation of Atomic Scale Defects in Inorganic Glasses

Y. NISHI, N. YAMAGUCHI, A. KADOWAKI, K. OGURI and A. TONEGAWA
Department of Materials Science, Tokai University, 1117 Kitakaname, Hiratsuka, 259-12, Japan

ABSTRACT

The effects of electron beam (EB) irradiation on fracture toughness were studied for silica and soda lime glasses. The EB irradiation process improved hardness, ductility and fracture stress and also enhanced fracture toughness. The EB-reinforcement mechanism was explained by molecule stress relaxation induced by dangling bond formation in the surface network structure of silica glass. Based on ESR and RDF results, the strengthening mechanism was deduced by relaxation-induced by free volume with dangling bonds that control the fracture toughness of the glass. This new treatment process is a promising concept to be applied to materials used in the fields of aerospace and rapid transit.

INTRODUCTION

The brittleness of ceramic materials is a serious problem yet to be solved. Once solidified, fracture toughness usually cannot be improved so, it is very difficult to control the properties homogeneously. From an engineering point of view, high fracture resistance is demanded for inorganic glasses and so methods to homogenize these materials are needed.

Both EB irradiation and heating are homogenizing processes. Heating close to equilibrium is a good way to normalize the mechanical properties of materials. On the other hand, EB irradiation is effective at activating surface atoms, breaking chemical bonds and promoting the migration of movable atoms in glasses to a depth of 0.1 mm of the surface [1-6]. Since EB-irradiation uniformly randomizes the glassy structure, it may be a possible tool to improve fracture toughness. On this basis, we have undertaken this research to investigate the effects of EB irradiation on stiffness and fracture toughness in silica and soda lime glasses.

EXPERIMENTAL PROCEDURE

The purity of the silica glass was 99.9999%. The mass composition of the soda lime glass used in this work was 73% SiO_2, 13% Na_2O, 8% CaO, 4% MgO and 2% Al_2O_3. The samples were manufactured into 30.05 mm square sheets with a thickness of 3 mm for silica and soda lime glass.

The sheets were irradiated using an electron-curtain processor (Type CB175/15/180L, Energy Science Inc., Woburn, MA, Iwasaki Electric Group Co. Ltd. Tokyo) [1–6]. The sheet beam was generated by a tungsten filament in a vacuum environment with an acceleration potential and irradiating current density of 170 kV and 4.0 mA respectively. The sheet electron beam is the most important feature of the experiment and to perform a homogeneous treatment of the samples, they were kept under the protection of nitrogen gas at one atmosphere with a residual concentration of oxygen below 400 ppm. EB irradiation was not applied continuously in order to control the surface temperature of the sample—a cycle time of 0.238 s was maintained by transporting the sample on a conveyor at a speed of 9.56 m/min. The temperature of the surface of each sample was kept below 323 K just after irradiation. Repeated applications were used to increase the total radiation dose with the dosage being proportional to the yield value determined from the irradiation current, I, (mA) the conveyer speed, S, (m/min) and the number of irradiations, N, according to the following equation:

$$\text{Dosage (MGy)} = 0.216*(I/S)*N \tag{1}$$

The yield value was calibrated by FWT nylon dosimeters. The length of the titanium window is 38.0mm. The surface electrical potential (128 kV) was estimated by electrical potential (170 kV) and titanium window thickness (13µm). The distance between sample and window is 35 mm. Based on the surface electrical potential of 128 kV and density and thickness of inorganic materials. The EB treatment transmitted the sample. Based on the density, ρ, (g/cm^3) and irradiation voltage (V: kV), EB-irradiation depth, D_{th}, (µm) was expressed by the following equation [7]:

$$D_{th} = 0.0667\, V^{5/3}/\rho \tag{2}$$

Here the density ρ (g/cm^3) was 2.22 g/cm^3 for silica glass. Thus, the EB-irradiation depth calculates out as 0.1 mm for silica glass.

To evaluate fracture toughness, K_{Ic}, critical values of indentation diameter, d_f (a measure of ductility), load, P_f (a measure of fracture stress), and toughness, E_f upon fracture were obtained, using a micro-Vickers' indentation tester [8].

To obtain the atomic scale glassy structural change precisely, the density of dangling bonds in each sample was obtained by using an electron spin resonance spectrometer (ESR, SA2000, Nippon Denshi Ltd., Tokyo) [3]. Based on the ESR signals, the apparent signals of dangling bonds were found. The microwave frequency range used in the ESR analysis was an X-band of 9.45 ± 0.05 GHz with a field modulation of 100 kHz. Spin density was calculated using an Mn^{2+} standard sample.

Formation of dangling bonds by EB irradiation should enhance mean atomic distance and decrease coordination number. The changes were calculated by a radial distribution function (RDF) of Si-O atom pairs of silica glass before and after EB-irradiation. By using a rotating type X-ray diffraction machine (Mo target, Ni filter, 50 kV and 70 mA) with step scanning (0.05°/step, 10 sec/step) [9], RDF curves were measured.

RESULTS AND DISCUSSION

EB-Reinforcement by Formation of Atomic Scale Defects in Silica Glass

Based on X-ray diffraction patterns of the silica glass before and after EB treatment, remarkable structure changes were not observed. On the other hand, it was possible to detect dangling bonds generated by the EB irradiation, resulting in reinforcement. The dangling bonds density in silica glass was measured by means of an electron spin resonance (ESR) spectrometer. Figure 1 shows ESR signals of silica glass before and after different dosages of electron beam irradiation. The ESR-signals were formed by 0.84 MGy EB-irradiation. A sharp single peak in the ESR signal occurs for dangling bonds of E-prime center in irradiated pure silica glass in which the chemical bonding energy is 150 kcal/mol for Si-O pairs in silica glass [10–11]. When the EB-irradiation causes dangling bonds to form at the atomic bonding of Si-O pairs, molecular stress relaxation in the network structure is induced. The relaxation was expected to prevent crack propagation and induce ductility. To evaluate the effects of EB irradiation on fracture toughness, the K_{Ic} value was obtained for the silica glass samples. Figure 2 shows how K_{Ic} changes as a function of EB irradiation dose. EB irradiation doses as low as 0.864 MGy enhanced the fracture toughness significantly.

Annealed glass generally shows a tight network structure as a dense random packing model of simple elements [12]. When the EB-irradiation forms free volume [13] with dangling bonds at the weaker atomic bonding pairs of Si-O inter-atomic force-distance, molecular stress relaxation is indicated. The stress relaxation prevents crack propagation and induces ductility. It also enhanced K_{Ic} values and enlarged fracture toughness.

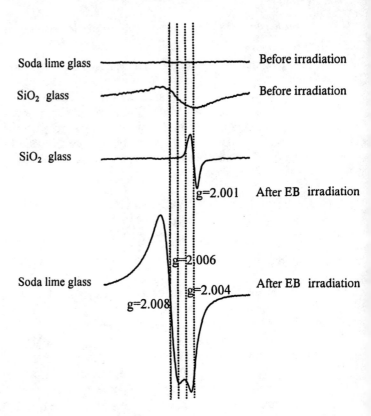

Figure 1. ESR signals of silica and soda glass before and after electron beam irradiation.

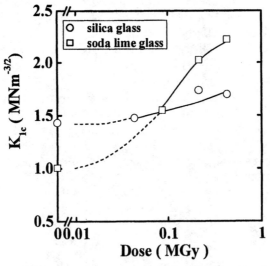

Figure 2. Changes in fracture toughness values of K_{IC} of silica and soda lime glasses versus EB-irradiation dose.

The EB irradiation enhanced the intensity of dangling bond signals, that is, the density of spins (dangling bonds), as shown in Figure 1. The EB treatment increased the density of dangling bonds related to free volume which mainly acted as relaxation sites, resulting in reinforcement below 0.864 MGy. With the broad peak of the Si-O pair obtained in the network structure with dangling bond, then EB-reinforcement is explained for silica glass.

To evaluate the formation of free volume [13], the mean atomic distance and coordination number was obtained by the radial distribution function (RDF) of the X-ray diffraction data [9] as shown in Figure3. The EB irradiation process slightly changed the RDF-shape of the silica glass. An EB-irradiation level of 0.864 MGy enhanced the reduced mean atomic distance (~3%) and decreased the reduced coordination number (13%). The results confirm the formation of free volume with dangling bonds in the network structure of silica glass. Namely, the broader peak of the force-distance curve can be found in the glass irradiated above 0.864 MGy.

Figure 4 shows schematic inter-atomic force-distance [IAFD] curves estimated by the potential curve of Si-O pairs. The crystal curve is sharp at the inter-atomic distance, whereas the glass curve is broad. Glass generally has both extensive and compressive retained strain. The IAFD curve of EB-treated glass should be broader than that before EB-irradiation because of relaxation of Si-O pairs induced by formation of free volume with dangling bonds. Thus, high ductility and fracture toughness are explained for the EB irradiated glass. Based on these discussions, fracture toughness can be controlled by EB irradiation.

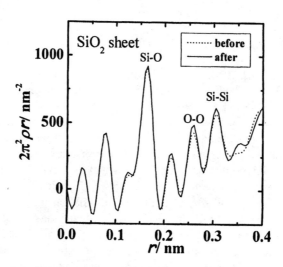

Figure 3. Radial distribution function of silica glass before and EB-irradiation.

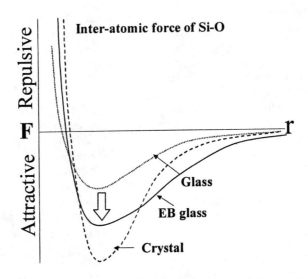

Figure 4. Schematic inter-atomic force-distance [IAFD] curves estimated to potential curve of Si-O pairs in crystal and glasses before and after EB-irradiation.

EB-Reinforcement by Formation of Atomic Scale Defects in Soda Lime Glass

Soda lime glass used in this work is made of SiO_2, Na_2O, CaO, MgO and Al_2O_3 with respective bond energies of 799.6 kJ/mol, 256.1 kJ/mol, 402.1 kJ/mol, 363.2 kJ/mol, 511kJ/mol [14]. EB-irradiation breaks the weak chemical bonds in Na-O, Ca-O and Mg-O pairs and relaxes the glassy structure so that the mechanical properties of the glass can be controlled.

Based on the X-ray diffraction patterns of soda lime glass before and after EB treatment, large structure changes could not be observed. However, it is possible to form free volume with dangling bonds, resulting in reinforcement. In order to evaluate the free volume, the dangling bond density in soda lime glass has been measured by means of an electron spin resonance (ESR) spectrometer. ESR signals of soda lime glass before and after electron beam irradiation are shown in Figure 1. The EB irradiation generated the dangling bond signals and thus the density of spins (dangling bonds). When the EB treatment generated free volume [13] induced by breaking weak bonding of metal (Na, Ca, Mg)-oxygen atoms, it reinforced strong metal (aluminum and silicon)-oxygen atoms bonding by relaxation to optimum atom sites from unstable atom sites. Thus, stress relaxation by EB irradiation is expected.

Change in fracture toughness of K_{1c} of soda lime glasses against EB irradiation dose is shown in Figure 2. The K_{1c} value of soda lime glass is lower than that of silica glass, however, the EB reinforcement process enhanced the K_{1c} value of soda lime glass to near that of silica glass. If the free volume formation relaxes the loading stress, then EB irradiation should enhance the K_{1c} value.

Based on these results and discussion, we conclude that the fracture toughness of soda lime glass can be controlled by the EB-irradiation process to a large extent.

CONCLUSION

In summary, EB-reinforcement enlarges the mean inter-atomic distance and decreases the coordination number for silica glass. The EB-reinforcement also enhanced the density of dangling bonds for inorganic glass. Based on the formation of atomic scale defects, the EB irradiation improved ductility, fracture stress and enhanced the fracture toughness of inorganic glass.

REFERENCES

[1] Y. Nishi, H. Izumi, J. Kawano, K. Oguri, Y. Kawaguchi, M. Ogata, A. Tonegawa, K. Takayama, T. Kawai, M. Ochi, 1997. Effect of electron beam irradiation on water wettability of hydroxyapatites for artificial bone, J. Mater. Science, 32, 3637–3639.

[2] K. Oguri, K. Fujita, M. Takahashi, Y. Omori, A. Tonegawa, N. Honda, M. Ochi, K. Takayama and Y. Nishi, 1998. Effects of electron beam irradiation on time to clear vision of misted dental mirror glass, J. Materials Research, 13, 3368–3371.

[3] K. Oguri, N. Iwataka, A. Tonegawa, Y. Hirose, K. Takayama and Y. Nishi, 2001. Misting-free diamond surface created by sheet electron beam irradiation, J. Materials Research 16, 553–557.

[4] Y. Nishi, T. Toriyama, K. Oguri, A. Tonegawa and K. Takayama, 2001. High fracture resistance of carbon fiber treated by electron beam irradiation, J. Materials Research, 16, 1632–1635.

[5] A. Kimura, A. Mizutani, Toriyama, K. Oguri, A. Tonegawa and Y. Nishi, 2001. Fracture stress enhancement by EB treatment of carbon fiber, Proc. 6th Applied Diamond Conference/Second Frontier Carbon Technology Joint Conference ([ADC/FCT 2001], Auburn University Hotel and Dixon Conference Center, Auburn, AL, USA, August 4–10, 2001), 779–784.

[6] Y. Nishi, S. Takagi, K. Yasuda and K. Itoh, 1991. Effect of aging on Tc of YBa2Cu3O7-y irradiated electron beam, J. Appl. Phys., 70, 367–370.

[7] J.B. Pedley and E.M. Marshall, 1983. Thermochemical data for gaseous monoxides, J. Phys. Chem. Ref. Data, 12, 967–1031.

[8] Y. Nishi, T. Katagiri, T. Yamano, F. Kanai, N. Ninomiya, S. Uchida, K. Oguri, T. Morishita, T. Endo and M. Kawakami, 1991. A new method for the evaluation of brittleness in ceramics, Journal of Applied Physics Letters, 58, 2084–2086.

[9] Y. Nishi, H. Harano, T. Fukunaga and K. Suzuki, 1988. Effect of peening on structure and volume in a liquid-quenched $Pd_{0.835}Si_{0.165}$ glass, Phys. Rev. B, 37 2855–2858.

[10] D.L. Griscom, E.J. Friebele, 1982. Effect of ionizing radiation on amorphous insulators, Rad. Effects, 65, 303–312.

[11] D.L. Griscom, 1991. Optical properties and structure of difects in silica glass. The Centenial Memorial Issue of the Ceramic Soc. Japan, 99, 923–942.
[12] J.D. Bernal, 1964. The Bakerian lecture, The structure of liquids, Proc. Royal Society London, A 280, 299–322.
[13] M.H.Cohen and D. Turnbull, 1961. Free-volume model of the amorphous phase: glass transition, J. Chem. Physics, 34, 120–125.

Texture Observations of Ferromagnetic Shape Memory of Nanostructured Fe-Pd Alloy by Laser and Electronic Microscope

T. OKAZAKI

Faculty of Science and Technology, Hirosaki University, Hirosaki 036-8561, Japan

T. KUBOTA

Institute for Materials Research, Tohoku University, Sendai 980-8577, Japan

H. NAKAJIMA and Y. FURUYA

Faculty of Science and Technology, Hirosaki University, Hirosaki 036-8561, Japan

S. KAJIWARA and T. KIKUCHI

National Institute for Materials Science, Tsukuba 305-0047, Japan

M. WUTTING

Department Material and Nuclear Engineering, University of Maryland, College Park, MD 20742-2115, USA

ABSTRACT

We have found a large magnetostriction of 1.1×10^{-3} for FSMA (ferromagnetic shape memory alloy) Fe-29.6at%Pd ribbon of about 60 µm thickness prepared by rapid solidification melt-spinning, when a magnetic field of 800 kA/m is applied normal to the ribbon surface at room temperature. The magnetostriction of this ribbon is 10 times as large as that of the polycrystalline bulk value. The magnetic field induces a strain caused by conversion of variants in the martensitic phase. This strain increases with temperature and reaches a maximum value at a phase transformation temperature of 380–400 K, though the austenite finishing temperature A_f of the single crystal is 320–330 K. On the other hand, the mechanical shape recovery effect of the ribbon has a two-step phase transformation temperature of 300–330 K and 380–420 K. In order to investigate the origin of this phenomenon, we studied the texture of the alloy by using a laser microscope and a high-resolution electronic microscope. The cross section of the ribbon shows a columnar structure of about 10µm in width. The ribbon consists of three parts: upper surface have small grains of 2–3 µm with strong [100] texture and bottom one is covered by roller traces. While, the inner part has a fine layer-structure of 30–40 nm sized grain thickness. We conclude that this nano-scale composite structure causes the phase transformation temperature to increase from 300 K at the surface to 380–400 K in the inner part of the ribbon.

INTRODUCTION

Ferromagnetic shape memory alloy (FSMA) Fe-Pd is expected to be useful as a magnetic-field-drive sensor/actuator material for a micro-machine and intelligent/smart material system. The FMSA exhibits a large strain caused by martensitic twin's initiations and its movements, i.e., a new type of mangetostriction [1]. Although Fe-31.2at%Pd single crystal [2] and polycrystalline bulk [3] samples exhibit large magnetostriction, phase transformation temperatures are low than room temperature. In previous study [4,5], we showed that Fe-29.6at%Pd alloy ribbon prepared by the rapidly solidification melt-spinning method has stronger crystal anisotropy and giant magnetostriction of 1.0×10^{-3} at room temperature with good shape memory effect. The magnetostriction of this ribbon is 10 times as large as polycrystalline bulk value. Its value increases with temperature and has a maximum at the phase transformation temperature of 380–400 K, though the austenite finishing temperature A_f of Fe-29.6at%Pd single crystal is 320–330 K [6].

In the present study, to investigate the origin of the temperature discrepancy, we observed the texture of ribbon sample by using laser microscope and high resolution electronic microscopes. We found that the martensite twin in the inner part of ribbon consists of fine layer-structures of 30–40 nm thickness.

EXPERIMENTAL DETAILS

The rapidly solidified Fe-29.6at%Pd ribbon sample with 60 μm in thickness, was prepared by originally designed electro-magnetic melt-spinning single-roll method from bulk alloy [4]. Some of ribbon was annealed at 1173 K for 1 h in vacuum atmosphere to study the effect of heat treatment on magnetostriction. The structure of ribbon was researched by XRD method with Cu Kα radiation. The microstructure of texture of ribbon was observed by using laser and high resolution electronic microscopes. The magnetization M vs. applied magnetic field H loop was measured by VSM.

The magnetostriction ε was measured by a strain-gauge attached on the sample (see Figure4(a)). The magnetic field was applied perpendicular to the rolling direction (RD) and strain changes were measured along RD. The shape memory effect was evaluated from the changes of shape recovery of the curled ribbon with temperature.

RESULTS AND DISCUSSION
Microstructure of Ribbon

Figure 1 shows laser microscope photographs of (a) the contact surface and (b) the free surface of as-spun ribbon. Microstructure of roll contacting surface, which is covered by roller traces, is observed in (a). As seen in (b), the free surface consists of microstructure of grains with about 3 μm size. The grain size can be controlled by changing the roll-speed. The microstructure of both surfaces annealed for 1 h at 1173 K was hardly changed.

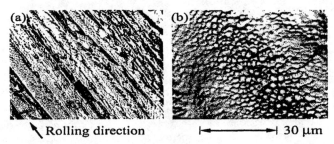

Figure 1 Laser microscope photographs:
(a) the contact surface, and
(b) the free surface of as-spun Fe-29.6at%Pd ribbon.

Figure 2 shows (a) the schematic diagram of the rapidly solidification apparatus and processed ribbon sample, and (b) the micrograph of as-spun Fe-29.6at%Pd ribbon. The cross section of ribbon shows unevenness of columnar texture of about 20 μm. Moreover, there is a fine chilled region of 10–15 μm on roll contacting side.

Figure 3 shows the XRD profiles of (a) the roll contact surface and (b) the free surface for as-spun and annealed ribbons. The fct martensite and fcc austenite phases coexist in the ribbon because fcc and fct {111} peak at near $2\theta = 42°$, fct (200), (020), fcc {200} and fct (002) peak at $47° < 2\theta$

<52°. The second and third peaks are larger than the first one for the free surface, indicating that the ribbon has a strong [100]-oriented texture. The [100] direction is an axis of easy magnetization. Moreover, the fct (200), (020) and fct (002) peaks are characteristic of martensite phase increases after annealing at 1173 K for 1 hour. With increasing temperature, the fct phase is completely transformed to fcc at about 333 K.

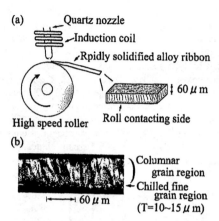

Figure 2. (a) Schematic diagram of the rapid solidification apparatus and processed ribbon sample, (b) Micrograph of the as-spun Fe-29.6 at%Pd ribbon.

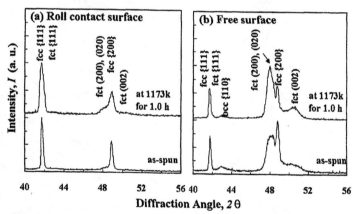

Figure 3. XRD profiles on (a) roll contact surface and (b) free surface of as-spun and annealed Fe-29.6 at%Pd ribbons.

Magnetic and Magnetostriction Properties

Figure 4(a) is the *M–H* loops of as-spun ribbon, where θ is the rotation angle between the transverse direction of ribbon and *H*, and θ=0° and 90° denote *H* parallel and perpendicular to plate of ribbon. When H=200 kA/m parallel to the plate, the saturation magnetization M_s of ribbon is 153 $4\pi \times 10^{-7}$ Wb·m/kg, indicating that it is ferromagnetic. When H of 400 kA/m applies to the θ=90° case, the magnetization of ribbon can not be saturated because of a large demagnetizing field, and the coercive force (Hc) is larger than one for θ=0°. Figure 4(b) shows θ-dependence of Hc for as-spun, 0.5 h, 1.0 h, 12.0 h ribbons and the bulk sample. The H_c of these ribbons has a maximum at θ = 85°, and decreases with annealing time because of a decrease in internal stress that finally reaches to the bulk value. The above results indicate that the ribbon has magnetic anisotoropy.

Figure 5 shows the magnetostriction ε vs. H curves of (a) the as-spun ribbon and (b) the ribbon annealed for 1 h at 1173 K, which measures change in strain in RD. The strain is minus, that is ε parallel to H is positive because the a-axis is a direction of easy magnetization (a>c in fct lattice). The value also depends remarkably on θ and has a maximum at θ=78□- 85□direction.

The reason that a maximum arises in the thickness direction is considered to be that the [100] direction of the texture of the ribbon are distributed around the center of the pole figure [7]. As shown in Figure5, the magnetostriction of the 1 h ribbon is larger than that of the as-spun sample.

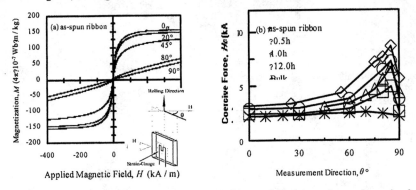

Figure 4. (a) *M–H* loops of as-spun Fe-29.6 at % Pd ribbon and (b) θ-dependency of coercive force for as-spun, 0.5h, 1.0h, 12.0h ribbons and the bulk sample.

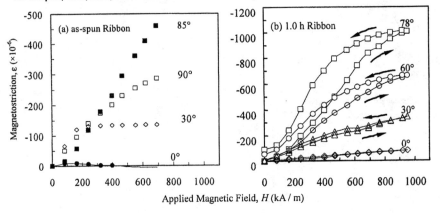

Figure 5. Magnetostriction ε vs. H curves for (a) the as-spun ribbon and (b) ribbon annealed for 1 h at 1173 K.

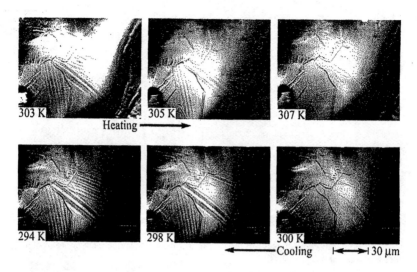

Figure 6. Laser microscope photographs on surface of coarse grain (about 30μm).

Two-step Phase Transition Temperature

Figure 6 shows the laser microscope photographs of surface of coarse grain (about 30 μm). On heating process, the martensite twin observed at 303K disappears at 307K. On cooling process, the same martensite twin is recovered at 298K. From these results, A_f and the martensite starting temperature M_s of the grains are determined to be 307K and 298K, respectively, in almost perfect agreement with values from XRD measurement for the same ribbon [4] and single crystals [6].

Dependence of the magnetostriction and shape recovery ratio of as-spun ribbon on temperature is shown in Figure 7(a). The magnetostriction increases from 3.2×10^{-4} at room temperature to 7.0×10^{-4} at 376 K. In the range of T>376 K, the strain decreases and suddenly falls to 1.6×10^{-4}. We consider that this phenomenon is closely related with the mobility of martensite twin, that is, the mobility of variants is activated by heating, and a maximum strain arises near the phase transformation temperature, A_s. A_s and A_f of the ribbon are found to be 376 K and 450 K, respectively. Moreover, the shape recovery ratio Φ_{T2}/Φ_{T1} increases with temperature, where Φ_{T1} and Φ_{T2} are diameters of curled ribbon at 300K and T K, respectively (see Figure 7(b)). The variation is rapid in a temperature range of 300–330K and of 380–420K. These results suggest that the ribbon has two-step phase transformations from martensite to austenite. Low A_s-A_f, at 300–330K is consistent with that from laser microscope observation of the surface in Figure 6 and by XRD [4]. Therefore, the low A_s-A_f is a net phase transformation temperature of Fe-29.6 at%Pd alloy. Note that high A_s-A_f, at 380–420K is peculiar to a rapidly solidified ribbon.

Nano-Scale Composite Structure

To investigate the origin of high A_s-A_f, we observed inner part of ribbon by using high resolution electronic microscope. The observed photograph is shown in Figure 8. There, it is seen that grains of about 1μm size consist of fine layer-structures of 30–40 nm thickness. That is, they are martensite twins because the widths of dark and light layers are in the ratio of about 2 to 1. These nano-scale layers are parallel to the columnar structure which is in the same direction of thermal diffusion on roll. This nano-scale composite structure is stress-induced martensite twin. Therefore, it can be considered that nano-scale martensite twins make phase-transformation temperature increase from 300–330 K in surface to 380–420 K in inner part.

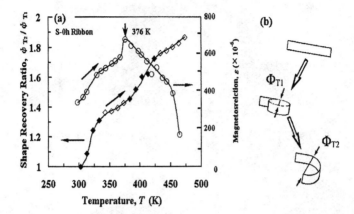

Figure 7. Temperature dependence of the magnetostriction and shape recovery ratio of as-spun Fe-29.6at%Pd ribbon.

Figure 8. High-resolution electron-microscope photograph of the inner part of a ribbon.

CONCLUSION

Rapidly solidified FSMA Fe-29.6at%Pd ribbon has a two-step phase transformation temperatures. To investigate the origin, we observed the texture by using laser and high resolution electron microscopes. The cross-section of ribbon has columnar structure of about 10 μm in width. The ribbon consists of three parts: upper surface have small grains of 2–3 μm with strong [100] texture and bottom one is covered by roller traces. While, the inner part has fine layer-structures of 30–40 nm thickness in grains. It can be concluded that this nano-scale structure makes phase-transformation temperature increase.

REFERENCES

[1] K. Ullakko, J.K. Huang, V.V.Kokorin and R.C.O'Handley, 1997. Magnetically cotrolled shape memory effect in Ni_2MnGa intermetallics, Scr. Matter, 36, 1133–1138.

[2] J. Koeda,Y. Nakamura, T.Fukuda, T.Kakeshita,T.Takeuchi and K.Kishio, 2001. Giant magnetostriction in Fe-Pd alloy single crystal exhibiting martensitic transformation, Trans. Mater. Res. Soc.Jpn., 26, 215–217.

[3] H.Y. Yasuda, N. Komoto, M. Ueda and Y.Umakoshi, Sci.Tec.Adv. Mater., 2002, in press.

[4] T. Kubota, T. Okazaki, Y. Furuya and T. Watanabe, 2002. Large magnetostriction in rapidly-solidified ferromagnetic shape memory Fe-Pd alloy, .J.Magn. Magn. Mater., 239, 551–553.
[5] T. Kubota, T. Okazaki, H. Kimura, T. Watanabe, M.Wuttig and Y.Furuya, 2002. Effect of rapid solidifycation on giant magnetostriction in ferromagnetic shape memory iron-based alloy, Sci.Tec.Adv. Mater., 2, 201–207.
[6] M. Sugiyama, R. Ohshima and F.E. Fujita, 1984. Martensitic transformation in the Fe-Pd alloy system, Trans. Mater. JIM, 25,585–592.
[7] Y. Furuya, N.W. Hagood, H. Kimura and T. Watanabe, 1998. Shape memory effect and magnetostriction in rapidly solidified Fe-29.6 at%Pd alloy, Trans. Mater. JIM, 39, 1248–1254.

New Advanced Diffusion Simulators for Boron Ultra-Shallow Junction

Y. SATO, M. MIYATA and M. UEHARA
Advanced Material Laboratory, Seiko Epson Corporation, 3-3-5, Owa, Suwa-shi, Japan 392-8502

H. NAKADATE
CRC Research Institute Inc., Tokyo, Japan

G. S. HWANG, E. HEIFETS, T. CAGIN and W. A. GODDARD III
California Institute of Technology, Pasadena, CA, USA

ABSTRACT

We have developed a new diffusion simulator that can be applied for any ion implantation condition, e.g., less than 1.0 keV. To develop it, we investigated underlying mechanisms of the diffusion of boron and the dynamics of defect-dopant clustering/dissolution during annealing using quantum mechanical calculations. We found that both clustering/dissolution and diffusion of a small boron cluster play important roles in diffusion. We have implemented such microscopic phenomena into a continuum equation. We have also developed an ion implantation simulator by implementing new functions in the SASAMAL code which can be applied for these ultra-low implantation energy systems.

INTRODUCTION

A process simulator has been used to design semiconductor devices. It is very important to predict the evolution of the doping profile of implanted ions during annealing because the distribution of implanted ions strongly affects the switching characteristics of a semiconductor device. It is possible to predict the evolution of boron implanted with a relatively high implantation energy using conventional diffusion simulators. However, this is not possible in the case of ultra-low energies such as 1keV or less because the governing low boron diffusion in such low implantation energy systems is different to that in the relatively high energy ones. Conventional systems have been developed without detailed knowledge of the diffusion mechanism. Our objective is to develop a new predictive diffusion simulator that can be applied to VLSI, i.e. to any implantation energy system. A second objective is to develop a new implantation simulator that can predict the doping profile of boron implanted with a low energy.

MICROSCOPIC CALCULATIONS

The microscopic calculations are summarized in Figure 1. We investigated microscopic processes that can occur during annealing [1].

Possible Clusters

We identified many kinds of clusters $Bs_lBi_mI_n$ (where Bs is substitutional boron, Bi is interstitial boron, and I or Sii is interstitial silicon). Some examples are shown in Figure 2 and Figure 3. For example, the cluster BsBi2 shown in Figure 2(a) has a triangular structure with one of the three boron atoms originally being Bs. The cluster BsSii2 in Figure 3(b) also has a triangular structure similar to BsBi2. We calculated the binding energy of each cluster. Our results give us the binding energy of Sii in BsSii2 as -1.65eV. We can estimate the dissolution rate using this value for BsSii2 ---> BsSii + Sii.

Clustering & Dissolution:

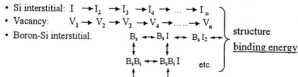

Diffusion:

- Si interstitial (I)
- B substitutional (B_s)
- B_s-I pair
- Vacancy
- B_s-B_i pair

} diffusion mechanism
  activation energy barrier

Microscopic Calculation:

- 64-atom supercell
- Local density approximation
- Plane wave basis set: E_{cut}=20 Ry
- k-point sampling: (2×2×2) M-P mesh

Figure 1. Summary of microscopic calculations.

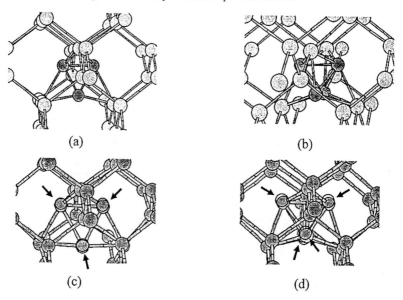

Figure 2. Structures of clusters. (a) BsBi2, (b) BsBi3, (c) I2, (d) I3.

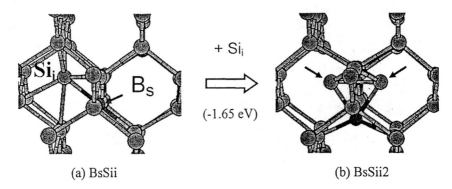

Figure 3. Structures of clusters. (a) BsSii, (b) BsSii2.

Diffusion Pathway

We identified the diffusion pathway of BsBi to be as depicted in Figure 4. The most stable structures of BsBi are shown in Figure 4(a) and Figure 4(g) while the transition state is in Figure 4(d). We can estimate the diffusion rate of BsBi by calculating the energy barrier of the pathway.

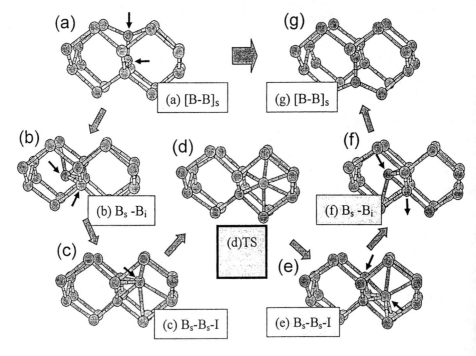

Figure 4. Diffusion pathway of BsBi.

Mesoscopic Calculation of Diffusion

We carried out kinetic Monte Carlo (KMC) simulations to see the behavior of defects that are generated during the ion implantation process. Figure 5 shows a snapshot of KMC simulation. Figure 5(a) shows the initial structure while Figure 5(b) shows the structure after 12 sec. We assumed that there exist Bs, Sii and vacancies in the initial structure and that the concentration of each is 10^{21} cm^{-3}. We further assumed that the temperature of the surface of the silicon wafer during ion implantation can rise to 500K. We carried out the KMC simulation under these conditions. We found that few vacancies survive after implantation. Therefore, we neglected the presence of vacancies in the diffusion simulation.

We calculated the diffusion of boron using KMC and our findings are as follows:

(1) Near surface: Evaporation of boron or BsI from the surface will generate the kink that appears near the surface (see Figure 6).

(2) High concentration region: The higher the concentration of boron is, the higher the possibility of clustering is. Relatively large cluster cannot diffuse, so the diffusion of boron is governed mainly by slow dissolution of the clusters.

(3) Low concentration region: Clustering is significantly suppressed, and diffusion of BsI prevails.

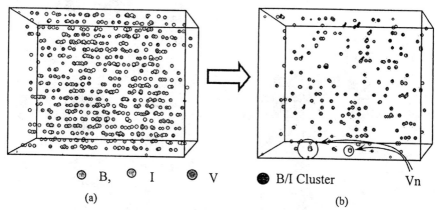

Figure 5. KMC calculation of B-I-V. (a) initial structure, (b) after 12 sec.

Calculation Method

As described above, KMC may be a powerful tool to predict diffusion, but it is not practical to make use of KMC for diffusion as it requires a huge simulation box to describe the wide range of concentration, e.g., from 10^{21} through to 10^{16}. Therefore, a more effective simulator is required.

We developed a new continuum equation to predict diffusion without losing the atomistic insights obtained both from the microscopic simulations and from the mesoscopic ones. For example, the equation for interstitial silicon I is given by Eq. 1:

$$\frac{\delta n}{\delta t} = -D\nabla^2 n + \underset{c1}{\sum K_{d,cl} n_{cl}} - \underset{c2}{D R_{cp} n \sum n_{c2}} \tag{1}$$

where n is the concentration of I, n_{cl} is the concentration of other clusters which contain a silicon atom, n_{c2} is the concentration of other clusters, I or boron, D is the diffusivity of I, $K_{d,cl}$ is the dissolution rate of cluster c1, R_{cp} is the capture radius. The three terms on the right side of the equation represent the gain in concentration of I by Fick's Law, by the dissolution of clusters, and by clustering respectively. The same equation must be evaluated for each cluster. If the cluster is immobile, the first term will disappear.

Results

Figure 6 shows a comparison between our simulations and experiments. In our calculations, we used the concentration of boron obtained by SIMS measurement as the initial structure. Our simulations reproduce the experimental data very well.

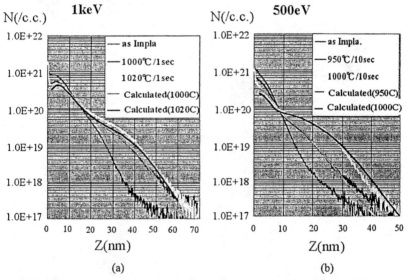

Figure 6. Comparison between calculation and experiment. Implantation energy of (a) and (b) is 1keV and 500eV, respectively.

MESOSCOPIC CALCULATIONS OF ION IMPLANTATION
Calculation Method

We modified the SASAMAL code to reproduce the doping profile of boron implanted with ultra-low energy. SASAMAL is based on a binary collision approximation. New functions implemented were as follows:

(1) Mean free path normal to the surface is longer than those in other directions.
(2) Energy loss due to elastic collision is smaller than that in other directions.

(3) Energy loss due to non-elastic collision is zero if the kinetic energy of the boron atom is less than the critical value.

The first two functions imitate channeling while the third imitates the tail of the doping profile.

Results

Figure 7 compares our simulation results with the experimental data. The implantation energy in Figure 7(a) is 1 keV and 200 eV in Figure 7(b). There is fairly good agreement between the calculations and the experiments in the case of 1 keV, but there is considerable discrepancy in the high concentration region for 200eV.

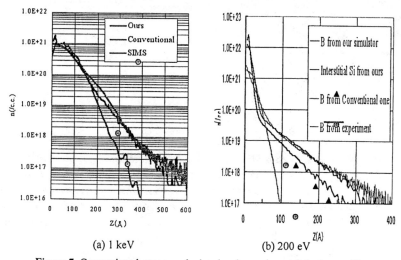

Figure 7. Comparison between calculated and experimental doping profiles.

CONCLUSIONS

We investigated the mechanism of diffusion of boron and the dynamics of defect-dopant clustering/ dissolution during annealing using quantum mechanics. We found both clustering/dissolution and diffusion of a small boron cluster play important roles in diffusion. We estimated the diffusion and dissolution rates of each cluster by finding the diffusion pathway and by calculating the cluster binding energy. We implemented such microscopic phenomena in a continuum equation. Using our equation, we can reproduce diffusion of boron implanted by ultra-low energy where experimental data of an as-implantation sample was used as the initial condition. Concerning the simulator, there is good agreement for low concentrations, but some improvement is needed to reproduce the high concentration region in case of 200eV.

ACKNOWLEDGEMENT

We thank Dr. Y. Miyagawa of the National Industrial Research Institute of Nagoya, Japan for providing us with the SASAMAL code.

REFERENCES

[1] G. S. Hwang et. al., 2001. MRS2001 Spring meeting J9.5.

Nano-Sensors for Gas Determination Based on Heterostructure SnO$_2$-Si

V. V. IL'CHENKO, A. I. KRAVCHENKO, V. P. CHEHUN, A. M. GASKOV and V. T. GRINCHENKO

Kiev National University, Radiophysical Faculty, Physical Electronics Department, 64 Volodimirska St., Kiev, 02033, Ukraine

ABSTRACT

Diode-like thin film heterostructures based on nanometer-scaled semiconductor oxides can be used as a gas-sensitive receptor and transducer system [1,2]. Instead of resistive oxide sensors which have a linear current-voltage characteristic, the nonlinearity current-voltage characteristics of similar diode hetero-structures can be used for the new approaches to the creation gas-sensitive structures.

EXPERIMENTAL DETAILS AND RESULTS

The heterostructure n-SnO$_2$(Ni)/p-Si wa manufactured by the technique described in [1]. Thicknesses of the n-SnO$_2$(Ni) layer varied from 25 to 80 A°.

A feature of the technique is that during measurement of DC current-voltage characteristics, we add an alternating voltage with an amplitude of about 0.1 V at a frequency range of 100–600 kHz. This research technique allows us to measure not only the DC current-voltage characteristic, but also d^2I/dU^2 of the current-voltage characteristics. In Figure 1 current-voltage characteristics are presented accordingly: curves *a* and *c* are obtained in air: *a* - at modulation frequencies of 100, 200, 400, 500 kHz, *c* - 310 kHz. Curve *b* is in alcohol vapor at modulation frequencies of 335 kHz and 300 K.

In Figure 2, the values of d^2I/dU^2 are shown for the same cases at a resonant frequency of 310 kHz (*c* is in air) and 335 kHz (*b* - is in alcohol vapor).

In Figure 3, the value of d^2I/dU^2 as a function of the resonance frequency in arbitrary units is presented for different gas environments: *c* - in air; *b* in alcohol vapor.

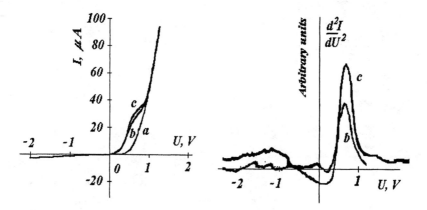

Figure 1. Current-voltage characteristic on the DC.   **Figure 2.** Second derivative of the I-V characteristic.

Sensitivity and selection of a gas sensor on a base of a thin heterostructure n-SnO2/Si-p supposes exposure of dominance peculiarities of adsorbed gas to changes in the functionally important parameters of the current-voltage characteristic of such transitions.

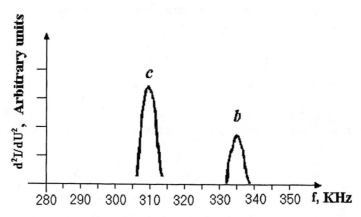

Figure 3. Second derivative of the I-V characteristics as a function of frequency.

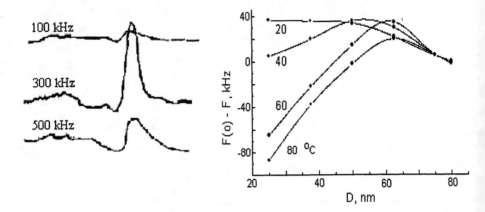

Figure 4. Second derivative of I-V vs. V.

Figure 5. Difference in modulation frequencies in air vapor vs. film and air with alcohol thickness at different temperatures.

In Figure 4, the second derivative of the current-voltage characteristics for the structure described above in air with alcohol vapor and for different modulation frequencies are presented. In Figure 5, the difference between modulation frequencies in pure air and air with alcohol vapor is presented from which we observe the maximum steepness of the current-voltage characteristic (maximum of second derivative), as a function of film thickness (d) and for different temperatures.

CONCLUSION

It can be seen that frequency modulation of the measuring signal can be used to analyse changes in the structural parameters depending on the gas environment.

REFERENCES

[1] R.B. Vasiliev, L.I. Ryabova, M.N. Rumyantseva, A.M. Gaskov, 2000. Proceedings Eurosensor 2000, 27-30.08, Copenhagen, Denmark, 365–368.
[2] O.I. Bomk, L.G. Il'chenko, V.V. Il'chenko, G.V. Kuznetsov, 2000. Proceedings of Eurosensor 2000, 27–30.08. Copenhagen, Denmark, 89–191.

Electrochemical Detection of Urea Using Self-Assembled Monolayers on a Porous Silicon Substrate

D.-H. YUN
Department of Biomicrosystem Technology, Korea University, Seoul 136-701, Korea

J.-H. JIN
Graduate School of Biotechnology, Korea University, Seoul 136-701, Korea

N.-K. MIN
Department of Control and Instrumentation Eng., Korea University, Chochiwon 339-700, Korea

S.-I. HONG
Department of Chemical Engineering, Korea University, Seoul 136-701, Korea

ABSTRACT

A new structure for an amperometric urea sensor was fabricated by electrochemical etching techniques. In this sensor, three thin-film electrodes are patterned on the porous silicon (PS) substrate by using platinum RF sputtering and gold, silver evaporation. The attachment of urease (Urs) to a carboxylic terminated self-assembled monolayers (SAMs) chemisorbed onto gold electrode was achieved using carbodiimide coupling. The formation of self-assembled chemisorbed layers of urease on gold has been studied by XPS (X-ray photoelectron spectroscopy). Their electrochemical properties are characterized and a result that the amperometry is preferable for the detection of urea to the potentiometry is obtained. Measured sensitivity of the Urs/SAMs/Au/PS electrode is ~11.21 $\mu A/mM \cdot cm^2$ and that of the Urs/SAMs/Au/Planar Si electrode is ~4.39 $\mu A/mM \cdot cm^2$. About three times of sensitivity increase is observed in the PS electrode.

INTRODUCTION

The metabolic function of kidney is reflected in the concentration of organic compounds such as urea in blood or urine. Therefore, the estimation of urea is frequently performed in the medical field. The use of urease as a biocatalyst for the development of urea biosensor has attracted continued interest from biochemical and clinical analysts [1]. The general principle for fabricating a urea biosensor is based on immobilization of urease onto a membrane or support in which urea is catalytically converted into ammonium and bicarbonate ions [2]:

$$(NH_2)_2 CO + H^+ + 2H_2O \xrightarrow{\text{UREASE}} 2NH_4^+ + HCO_3^- \qquad (1)$$

$$2H_2O \rightarrow 4H^+ + O_2 + 4e^- \qquad (2)$$

As can be seen in Eq. 1, the urea hydrolysis reaction induces proton consumption increasing the pH. At this condition, anodic current caused by water dissociation begins to flow as in Eq. 2 in which the H^+ produced compensates for the pH shift. The magnitude of this current can be used as a monitor of the urea concentration [3].

Enzyme sensors, i.e., analytical devices that incorporate enzyme components and transducers, have been widely applied in chemical and biological fields due to their high sensitivity and high selectivity [4–6]. Most of these enzyme sensors are based on the catalytic properties of the enzymes immobilized on the transducers, and the development of new immobilization methods for enzyme components is still an important trend in enzyme sensor research. Until now most enzyme biosensors have had a membrane fixed on top of the transducer. That method often leads to malfunction of the sensor, arising from problems such as inadequate membrane adhesion and insufficient mechanical stability. To solve this kind of problems, porous silicon (PS) biosensors have been represented [7–9]. In this work, urease (Urs) immobilized biosensor was fabricated on the PS substrate. For a stable and reproducible enzyme immobilization, self-assembled monolayers (SAMs) were coated on the Au/PS layers.

Materials and Methods
Reagents
Hydrofluoric acid (CMOS grade, 49%) and ethanol (HPLC solvent) from J. T. Baker were used as components of an electrochemical etchants without further purification. The deionized water was made from milli Q pure water system in laboratory conditions. 3-Mercaptopropionic acid was purchased from ACROS. Urease (EC 3.5.1.5, type β, from Jack Beans) was purchased from Sigma Chemical Co. (St. Louis, USA.). EDC (1-ethyl-3-[3-(dimethylamino)-propyl]carbodiimide), NHS (N-hydroxysulfosuccinimide) were used for the immobilization of urease.

Device Fabrication
The implementation of the proposed structure was shown schematically in Figure 1. A boron-doped, (100) oriented, p-type silicon wafers that have 1–10Ω·cm resistivity was used as electrode substrates to form electrochemically etched PS layers on their surfaces. The electrolyte solution for growing PS layer was composed of HF (49%) : C2H2OH (95%) : deionized water = 1 : 2 : 1 by the volume percent and $-7mA/cm^2$ current density was maintained in a specifically designed Teflon cell to form a uniform PS layer of 2 μm width and 10 μm depth pores. Samples were rinsed in running deionized water and blown dry by nitrogen gas immediately after electrochemical etching. The PS substrate was oxidized at 400°C in dry oxygen for 1 hour followed by annealing at the same temperature in nitrogen for 15 minutes. The Ti (200 Å) and Pt (2000 Å) layers which make up the counter electrode were deposited by RF-sputtering in vacuum at a base pressure of 2×10^{-5} mbar and a temperature of 100°C. The role of the Ti underlayer is important in acting as an adhesive between the Pt thin film and the silicon dioxide. An Ag thin film was deposited next by evaporation to become a reference

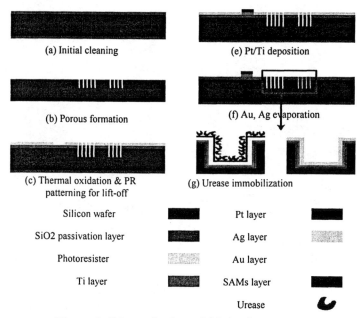

Figure 1. Schematic view of fabrication process.

electrode with a thickness of 2000 Å. Then Ti and Au were evaporated to achieve a thickness of 200 Å and 2000 Å respectively.

Preparation of Self-Assembled Monolayers of Alkanethiols on Gold

The gold surfaces were cleaned immediately before use by etching for 5min in freshly prepared, hot "piranha solution" comprising a mixture of 1:3 (30% H_2O_2, concentrated H_2SO_4) at 80°C. The gold electrode was then immersed in solutions of 3-mercaptopropionic acid and ethanol. The 3-mercaptopropionic acid was vacuum distilled to ensure a low level of disulfide in the loading solution. Immediately prior to the experimental testwork, 3-mercaptopropionic acid was dissolved at a concentration of 2mmol/L in an absolute ethanol. After an immersion time of 12 hour, the electrode was thoroughly rinsed with absolute ethanol and blown dry with nitrogen.

Immobilization of Urease

Urease was immobilized onto the SAMs-modified gold electrode by covalent attachment. For the covalent immobilization, the gold electrode modified with SAMs containing terminal carboxylic groups were treated with a stirred 0.1 mol/L sodium phosphate buffer solution at pH 7.4, containing NHS (3mmol/L, N-hydroxysulfosuccinimide) and EDC (100 mmol/L, 1-ethyl-3-[3-(dimethylamino)-propyl]carbodiimide) at room temperature. After a reaction time of 3 hour, they were rinsed with the phosphate buffer and immediately placed in a stirred

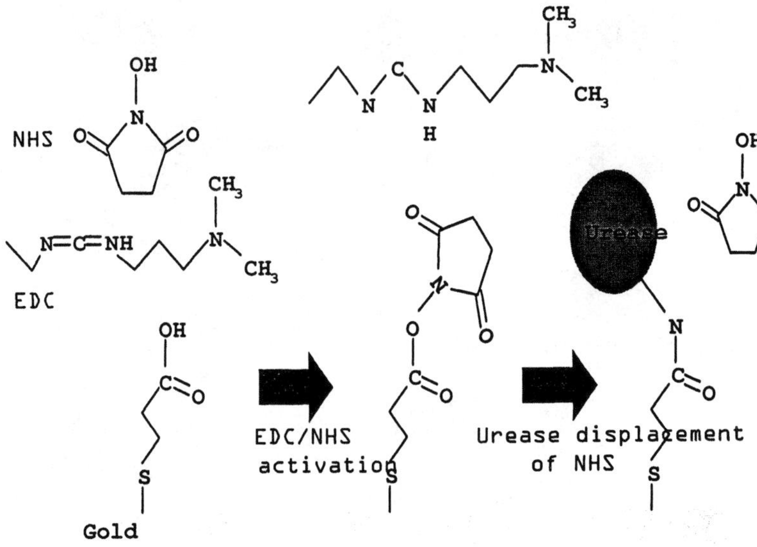

Figure 2. Reaction scheme for EDC/NHS mediated amine coupling to a carboxylate SAMs.

5mg/cm^3 Urs solution in 0.1mol/L sodium phosphate buffer a pH 7.0. This step was allowed to continue overnight at room temperature. The electrode was then thoroughly rinsed with the following sequence of liquid: pH 6.3 phosphate buffer solution, 1mol/L NaCl, Milli Q Plus water, and pH 7.0 sodium phosphate buffer solution. Fig. 2 below shows a schematic of the coupling chemistry for the carboxylate SAMs. Briefly, the carboxyl groups are activated to NHS ester, followed by displacement of the NHS by amine groups on the urease. This results in the covalent attachment of the urease via peptide bond.

Results and Discussion
Characterization of PS layer surface
Figure 2 shows SEM images of the p-type porous silicon formed at room temperature. About 2 µm diameter and 10 µm depth pores are formed in 5 minutes. P-type PS layer is more useful as electrode substrates in the development of PS-based urea sensors than n-type PS. The former shows larger diameter and inner space than the latter.

Characterization of the Immobilization of Urease by XPS
The presence of urease immobilized on the surface of the SAMs could be confirmed by XPS. Fig. 4 shows the XPS spectra obtained with containing urease, immobilized by covalent attachment. The C 1s region shows three well-resolved peaks. The peak at 285.8 eV was attributed to carbon atoms in a hydrocarbon-like environment, present in amino acid side chains of the polypeptide backbone of the enzyme and in carbohydrate residues of its polysaccharide shell. The peak at 286.8 eV was attributed to carbon linked to oxygen by single bond, present in the

polysaccharide shell and in amino acid side chain. The highest binding energy signal at 288.7 eV was attributed to carbon atoms linked to oxygen via double bond, present in the amino groups of the polypeptide backbone, in some amino acid side chains and in N-acetylglucosamine residues of the polysaccharide shell. The O 1s signal contained two components: (1) a signal at 533.3 eV, assigned to oxygen atoms in C=O groups, present in both the polypeptide backbone and in the N-acetylglucosamine residues and (2) a signal at 532.1 eV, assigned to oxygen atoms in C-O groups, present in amino acid side chains and carbohydrate residue. The N 1s peak at 400.1 eV was assigned to nitrogen atoms present in the amide bonds in the polypeptide chain, amino acid side chains, and N-acetylglucosamine residues [10,11].

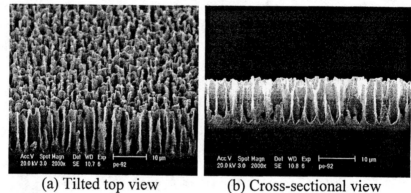

(a) Tilted top view   (b) Cross-sectional view

Figure 3. SEM images for the p-type porous silicon. Current density is -7 mA/cm^2.

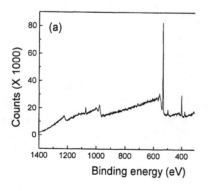
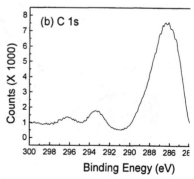

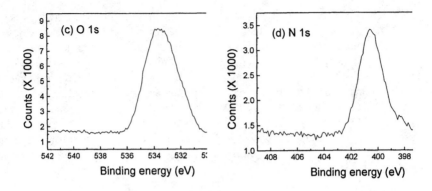

Figure 4. XPS spectrum of urease immobilized on SAMs/Au/PS layers; (b) C 1s region of (a); (c) O 1s region of (a); (d) N 1s region of (a). X-ray power, 2.5 kW; pass energy, 150 eV; TOA, 10°.

Dependence of amperometric sensitivity on the pH

Figure 5 shows optimum pH requirement of amperometric sensitivity at 35°C at a urea concentration of 10 mmol/L. The optimum pH is observed at pH 7.3. It can be noted that fall in activity in the fall in activity in the immobilized state. A broadening of profile towards both acidic and alkaline range was observed implying that the enzyme becomes less sensitive to pH changes.

Dependence of sensitivity on the operating temperature

Figure 6 shows the effect of operating temperature on the amperometric sensitivity of urea sensors. The applied potential is kept for 0.6 V *vs.* Ag/AgCl thin-film reference electrode. The maximum sensitivity was observed at ca. 35°C. About 68% sensitivity was recorded at 40°C after which a decrease was observed which is likely to be due to the denaturation of urease at higher temperatures.

Long-term storage stability

The long-term storage stability of the urea sensor was investigated. The current response of the urea sensor to a series of urea solutions of different concentrations in 1.0 mmol/L phosphate buffer solution at pH 5.0 was monitored every 5 days with storage at 4°C. As can be seen from Fig. 7, the urea sensor has still retained 90% of its response sensitivity after storage for 30 days. This indicates the good long-term storage stability.

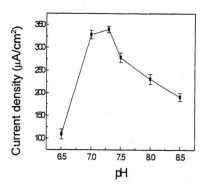
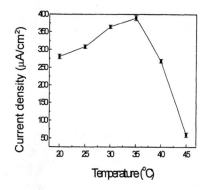

Figure 5. The effect of pH on the amperometric sensitivity of the urea sensor. The Maximum sensitivity obtained at pH 7.3.

Figure 6. The effect of operation temperature on the amperometric sensitivity of urea sensor. Maximum sensitivity obtained at 35°C.

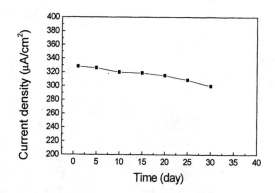

Figure 7. Quality degradation of amperometric urea sensor with time at 0.6V vs. Ag/AgCl and 35°C Urea concentranion is 10 mmol/L.

Amperometric determination of urea concentration

To obtain a calibration curve, the current responses of the urea sensor to urea solution of different concentrations were measured under the optimum conditions (pH, 7.3; temperature 35°C). As can be seen from Fig. 7, the current density increases with urea concentration over the concentration range 0–100 mmol/L. Measured sensitivity of the Urs/SAMs/Au/Porous Si electrode is ~11.21

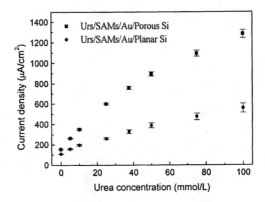

Figure 8. Calibration curves of the urea sensors. Sensitivity Urs/SAMs/Au/Porous Si electrode is ~11.21 µA/mMcm^2 and Urs/SAMs/Au/Planar Si electrode is ~4.39 µA/mMcm^2.

µA/mMcm^2 and that of the Urs/SAMs/Au/Planar Si electrode is ~4.39 µA/mMcm^2. About three times of sensitivity increase is observed in the porous silicon electrode.

CONCLUSION

The use of porous silicon as a sublayer of SAMs provides an opportunity for high sensitivity to a broad range of urea concentrations. Furthermore, measured sensitivity of the Urs/SAMs/Au/Porous Si electrode was ~11.21 µA/mMcm^2 while that of Urs/SAMs/Au/Planar Si electrode was ~4.39 µA/mMcm^2 in the linear range of 1–50 mmol/L urea concentrations. This self-assembly process has little effect on the activity of urease, but the sensor shows high sensitivity compared to that using conventional covalent attachment onto a planar electrode. Another advantage of this self-assembly urea sensor is that it can be repeatedly regenerated by washing with a high pH, high ionic strength solution. PS-based urea sensors offer promise for the use of PS layers as biosensor elements.

ACKNOWLEDGEMENT

This work was supported by Grant No. R01-2002-000-00591-0 from the Basic Research Program of the Korea Science & Engineering Foundation.

REFERENCES

[1] S.H. Park, J.H. Jin, N.K. Min, S.I. Hong, 2002. Poly(3-methylthiophene)-based urea sensors with planar Pt electrodes on silicon substrate, J. Korean Physical Society, 40(1), 17–21.

[2] R. Singhal, A. Gambhir, M.K. Pandey, S. Annapoorni and B.D. Malhotra, 2001. Immobilization of urease on poly(N-vinyl carbazole)/stearic acid Langmuir-Blodgett film, Biosensors and Bioelectronics, 17(8), 697–703.

[3] K. Yoneyama, Y. Fujino, T. Osaka and I. Satoh, 2001. Amperometric sensing system for detection of urea by a combination of the pH-stat method and flow injection analysis, Sensors and Actuators B, 76(1-3), 152–157.

[4] D. Ivnitski, I. Abdel-Hamid, P. Atanasov and E. Wilkins, 1999. Biosensors for detection of pathogenic bacteria, Biosensors and Bioelectronics, 14(7), 599.

[5] R. Koncki, A. Hulanicki and S. Glab, 1997. Biochemical modifications of membrane ion-selective sensors, Trends in analytical chemistry, 16(9), 528–536.

[6] Z. Wu, L. Guan, G. Shen and R. Yu, 2002. Renewable urea sensor based on a self-assembled polyelectrolyte layer, Analyst, 127(3), 391–395.

[7] R.W. Bogue, 1997. Novel porous silicon biosensor, Biosensor & Bioelectronics, 12(1), xxvii–xxix.

[8] D.v. Noort, S.Welin-Klintström, H. Arwin, S. Zangooie, I. Lundström and C.F. Mandenius, 1998. Monitoring specific interaction of low molecular weight biomolecules on oxidized porous silicon using ellipsometry, Biosensors and Bioelectronics, 13(3–4), 439–449.

[9] M.J. Schöning, F. Ronkel, M. Crott, M. Thust, J.W. Schultze, P. Kordos and H. Lüth, 1997. Miniaturization of potentiometric sensors using porous silicon microtechnology, Electrochemica acta, 42(20–22), 3185–3193.

[10] K.M.R. Kallury, W.E. Lee, and M. Thompson, 1992. Enhancement of the thermal and storage stability of urease by covalent attachment to phospholipid-bound silica, Analytical Chemistry, 64(9), 1062–1068.

[11] M. Wirde and U. Gelius, 1999. Self-assembled monolayers of cystamine and cysteamine on gold studied by XPS and voltammetry, Langmuir, 15(19), 6370–6378.

Chapter 8: Intelligence in Materials Science

Atomic Scale Simulations for Compositional Optimization

R. W. GRIMES and M. PIRZADA
Department of Materials, Imperial College London, London SW7 2BP, UK

P. K. SCHELLING and S. R. PHILLPOT
Materials Science Division, Argonne National Laboratory, Argonne, IL 60439, USA

K. E. SICKAFUS
MS-G755, Los Alamos National Laboratory, Los Alamos, NM 87545, USA

J. F. MAGUIRE
Materials and Manufacturing Directorate, Air Force Research Laboratory, OH, USA

ABSTRACT

Taking the pyrochlore system $A_2B_2O_7$ as the example, atomic scale computer simulation will be used to identify optimum compositions for two disparate applications. The first focuses on these materials as thermal barrier coatings (TCB's) so that thermo-mechanical properties must be predicted. The second considers pyrochlores as potential electrolytes for electrochemical devices. In both cases, results of calculations are presented as process value contour maps, which focus attention on compositional regions of significance.

Pyrochlore materials are already known to exhibit very low thermal conductivities (~1 W/mK). However, they have recently received significant attention as candidate materials for a new generation of thermal barrier coatings on jet engine turbine blades (the aim is to replace or provide alternatives for fluorite based TBS's). To elucidate the importance of the choice of A and B ions on the thermal conductivity, we have used molecular-dynamics simulation to study the thermal conductivity of pyrochlores for 50 different combinations of A and B cations. The two-dimensional contour maps show thermo-mechanical properties as a function of A and B cation radii. We find that some pyrochlores exhibit disorder in the oxygen sublattice, resulting in an even lower thermal conductivity. Thus, contour maps provide a means to analyze and convey a considerable amount of data that can itself be understood in terms of a basic physical model.

Given their close relationship to the fluorite structure, it is also not surprising that pyrochlore materials are being considered as alternatives in high temperature electrochemical devices. For such applications it is imperative to maximize oxygen transport. This is achieved by decreasing the activation energy for oxygen hopping migration. Initially we modelled the consequence of compositional variation on activation energy, although the actual migration mechanism itself changed as a function of composition. Again, contour map *analysis was able to unravel the complexity so that optimum compositions could be identified.*

INTRODUCTION

Materials with similar structures often have similar properties. For example, fluorites and fluorite related oxides are relatively good oxygen ion conductors while the transport properties of oxides with the rock salt structure are poor. Unfortunately property variations within a structure class of materials can be so significant as to render this general observation of only limited value. Furthermore, it does not aid in compositional optimization, that is, in selecting a composition that will have the best property within a class of materials.

In order to facilitate compositional optimization we must first identify a trend in a property as a function of some variable of the component ions. Staying with the example of oxygen ion conduction, the property could be the activation energy for migration and the component ion variable ionic radius. Indeed we shall provide results for this later within the pyrochlore class of materials. Such materials have formula $A_2B_2O_7$ so that both the A and B cations can be varied resulting is a large number of possible compositions. If we can identify a suitable property trend as a function of A and B cation radii it can be used to optimize composition selection. However, it is rare that sufficient data exists from which to generate such a relationship for a property of any real technological significance. Furthermore, in selecting a material for an application there is usually more than one property criterion that it is important to optimize, thereby imposing a compromise composition. It is for these reasons that we have chosen to compute property values rather than search for experimental data. The fact that we are only concerned with trends in properties rather than absolute values allows us to accept that degree of approximation inherent in a computational approach. On the other hand, we do make contact with experiment wherever possible as a means to assess the quality of the simulation results.

Two examples will be presented as a means to demonstrate the potential of this approach. The first involves the selection of a pyrochlore oxide composition suitable for use as an electrolyte in an electrochemical device (e.g. a solid oxide fuel cell or an oxygen generator). The second concerns composition selection for a pyrochlore thermal barrier coating material as would be used to protect a jet engine turbine blade. In both cases yttria-stabilized zirconia (YSZ) is presently used [1,2] but pyrochlore compositions are being investigated experimentally [3–6]. The aim is to provide guidance for experimental studies and thereby bring about a more focused effort.

PYRCHLORE CRYSTAL STRUCTURE

We will be concerned with so-called (3-4) pyrochlore compounds, i.e. A is nominally a 3+ valence cation and B is a 4+ valence cation. The pyrochlore structure (see Fig 1.) is closely related to fluorite and can be considered as an ordered defective fluorite. It exhibits space group $Fd\overline{3}m$ with eight formula units

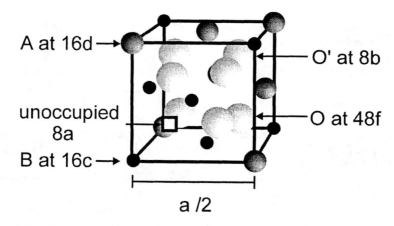

Figure 1. Partial unit cell of the pyrochlore structure.

within the cubic unit cell [7]. If we write the general formula as $A_2B_2O_6O'$, and choose to fix the origin on the B site, the ions occupy four crystallographically non-equivalent positions: A at 16d, B at 16c, O at 48f and O' at 8b. There is an interstitial site that, within this space group, is at 8a. This will become important when we consider migration of oxygen ions between 48f sites.

METHODOLOGY

Interionic Forces

Calculations presented here are based upon a classical Born-like description of an ionic crystal lattice [8]. The interatomic forces acting between ions are resolved into two terms: long-range columbic forces and short-range forces, which are modelled using parameterized pair potentials. We define the perfect lattice using a unit cell, which is repeated throughout space using periodic boundary conditions as defined by the usual crystallographic lattice vectors. The lattice energy (U_L) can then be expressed as follows:

$$U_L = \frac{1}{4\pi\varepsilon_0} \sum_{i \neq j} \frac{q_i q_j}{r_{ij}} + Ae^{\frac{-r_{ij}}{\rho}} - \frac{C}{r_{ij}^6} \qquad (1)$$

where A, ρ and C are adjustable parameters specific to ions i and j, r_{ij} is the interionic separation and q_i is the charge on ion i. A multi-structural parameter-fitting approach was applied as discussed previously [9].

Defect Energies—Static Simulations

The migration of oxygen ions is mediated by oxygen vacancies i.e. unoccupied oxygen lattice sites. Since the model must therefore involve the formation of defects, an approach is needed to simulate the effect of a defect on the surrounding lattice ions. Indeed, defect energies are not useful unless they include the contribution due to lattice relaxation. However, this relaxation is greatest in the vicinity of the defect. Therefore the lattice is partitioned into three regions [10]. In region I, centred around the defect, ions are treated explicitly and relaxed iteratively to zero strain via a Newton-Raphson procedure. Region IIa is an interfacial region in which the forces between ions are determined via the Mott-Littleton approximation [11] and ions are relaxed in one step. The interaction energies of the ions between region IIa and region I are calculated explicitly. Finally, the outer region IIb is effectively a point charge array whose relaxation energy is determined using the Mott-Littleton approximation, this provides the Madelung field of the remaining crystal. All calculations were carried out using the CASCADE code [12].

The electronic polarisability of ions is accounted for via the shell model [13]. This assumes a massless shell with charge $Y|e|$ coupled to a core of charge $X|e|$ via an isotropic harmonic force constant k. The net charge state of each ion equals $(X+Y)|e|$, assuming full formal charge states are with coulomb interactions summed using Ewalds' method [14]. Values for the charges and force constants were published previously [15].

Thermal Properties—Molecular Dynamics Simulations

Due to the considerably greater computational overhead compared to the static technique, these simulations were performed using rigid ion models (i.e. without a shell model). Furthermore a partial charge model was employed where $q_A = +2.55$, $q_B = +3.4$ and $q_O = -1.7$ and the short-ranged interactions parameters were taken from Minervini *et al.* [16]. The Coulomb energies and forces were calculated using a real-space direct summation method [17], which has previously been successful in describing zirconia and yttria-stabilized zirconia [18].

The thermal conductivity (κ) simulations were performed on <001>-oriented perfect crystals. The simulation cell had a cross section of 2x2 unit cells in the x-y plane and was 24 unit cells long in the z-direction. Since each crystallographic unit cell contains 8 formula units of $A_2B_2O_7$, the simulation cell contained a total of 8448 ions. For the present simulations, the average temperature was chosen to be 1473K, which is close to the current operating temperature for many TBCs. The lattice parameters were determined from zero-pressure simulations at T=1473K on 4x4x4 unit cells; reference calculations using larger unit cells showed no system size dependence. The thermal conductivity was determined in a manner analogous to experiment, with heat being added and removed in unit-cell thick slices 12 lattice parameters apart. Because the unit cell is symmetric, this sets up two identical heat fluxes. The two resulting temperature gradients between the heat source and sink were calculated, from which the thermal conductivity was determined by Fourier's law. Details of this method have been given elsewhere [19]. The simulation errors were determined by finding the variation in the computed value of κ for smaller intervals of simulation time, and are about ±10% of the simulated κ in each case.

Contour Map Generation

The results for each solution mechanism are presented in the form of energy contour maps. In these, the calculated normalized solution energies associated with a pyrochlore compound $A_2B_2O_7$ is a value on a two dimensional grid. The location of the energy value is defined by B^{4+} and A^{3+} cation radii, plotted along the x and y axes respectively. The energy contours are then generated using software [20] that interpolate and connect equal energies over the cation radii surface. The advantage of these maps is that they allow us to compare trends and identify compositional regions of interest.

OXYGEN MIGRATION IN PYROCHLORE OXIDES

Background

$A_2B_2O_7$ pyrochlore materials are being considered for use in a range of applications that require oxygen ion transport through the lattice. In addition to obvious technologies such as solid-oxide fuel cells and oxygen generators [5,6] oxygen migration is also important for use as a host lattice for the immobilisation of actinides in nuclear waste [21]. In the context of this work, the aim has been to identify A and B component cations that yield the lowest migration activation energy.

Migration simulations

Oxygen ion conduction in pyrochlores, like in fluorites, proceeds via an oxygen vacancy mechanism [22]. The migration mechanism consists of sequential jumps of oxygen ions into vacant sites. The activation energy for migration is then defined as the difference between the energy of the system when the migrating oxygen ion is at the saddle point and the energy of the oxygen vacancy at equilibrium. The saddle point energy is calculated by introducing a fixed oxygen ion at the saddle point location and then relaxing the surrounding lattice. Evaluating the potential energy surface both parallel and perpendicular to the diffusion path identifies the configuration of the saddle point.

In the pyrochlore lattice, the lowest energy contiguous pathways for oxygen ion migration are provided by jumps from a filled 48f to a vacant 48f site along <100> directions [23–25]. The first step in this study is therefore to model the structure of the 48f oxygen vacancy.

Clearly this is a well-defined problem. The atomic configuration (in this case the crystal structure) is known in detail, the model parameters can be selected and the simulation yields an energy that can be directly compared with known experimental data. However, this apparent simplicity conceals complex processes.

RESULTS AND DISCUSSION

The 48f Vacancy

The structure of the 48f vacancy was determined for 54 $A_2B_2O_7$ compounds: A = Lu to La, B = Ti, Ru, Mo, Sn, Zr and Pb. In many cases, the lowest energy configuration is a single vacancy with a symmetric relaxation of neighbouring ions: towards the vacancy for anions, away from the vacancy for cations. However, some compounds exhibit a different lowest energy configuration. This is formed when a 48f oxygen ion adjacent to the vacancy relaxes considerably towards the unoccupied 8a site (see Figure 2). This forms the so-called split

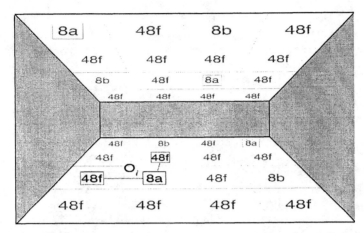

Figure 2. The structure of the split vacancy in $A_2B_2O_7$ compounds. A box, □, indicates an anion lattice site that is not occupied by an oxygen ion, O_i is the oxygen interstitial ion in the split vacancy. Two adjacent anion (001) sublattice planes are shown; $z = 3/8$ and $z = 5/8$.

vacancy [23] consisting of two 48f vacancies, orientated along <110>, and an interstitial oxygen ion. However, the interstitial ion never actually occupies the 8a site being symmetrically displaced away from 8a towards the two 48f sites (see Figure 2).

IMPLICATION FOR THE 48f-48f MIGRATION MECHANISM

For compounds that do not exhibit split vacancy formation, the 48f to 48f <100> mechanism will be a simple activated hopping process. All oxygen ion jumps will be along <100> directions. For those compounds that exhibit a split vacancy, the overall motion of the oxygen ions is not along <100>. In fact, what has occurred is the reorientation of a split vacancy associated with one 8a site to a new 8a site [15]. If we assume the oxygen interstitial occupies the 8a site, the overall mass transport is, on average, in a <111> direction.

Calculated Activation Energies

The activation energies for oxygen migration are presented in the form of an energy contour map in Figure 3. The ordinate shows A cation radius, the abscissa shows B cation radius. Equal activation energies are connected by contours. Thus compounds on the same contours are predicted to exhibit the same activation energies. The white points in Figure 3 are compounds for which a calculation was made and from which the contours were constructed.

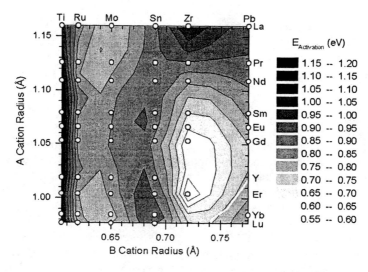

Figure 3. Contour map of oxygen migration activation energy (eV).

To assess the reliability of these predictions we must compare the calculated activation energies for specific compounds with available experimental data. Although data is limited, there is generally excellent agreement [15]. However, for $Gd_2Zr_2O_7$, the experimental values range from 0.73–0.90 eV [24,26,27] while the predicted activation energy is 0.58eV. At first sight it seems that this prediction is somewhat low. However, it is well established that $Gd_2Zr_2O_7$ exhibits significant disorder [24,26] which will increase the experimental activation energy [25]. Although the value presented in this work assumes a fully ordered $Gd_2Zr_2O_7$ lattice, it does provide an estimate of the activation energy for local oxygen rearrangements even though it doesn't mirror the long-range migration value of a significantly disordered material. So this highlights a limitation of using these types of computed property values—they are only as good as the mechanistic hypothesis on which they are based. Comparison with experimental data remains important.

Implications

Given the agreement with experiment described above, we may now use the activation energy contour plot (Figure 3) in a predictive capacity to select pyrochlore compositions, which exhibit low activation energies. However, the onset and effect of disorder makes prediction less straightforward. Compounds that exhibit higher degrees of disorder also display higher activation energies. Therefore, although $Y_2Zr_2O_7$ for example might seem to have an especially low activation energy, the extent of cation disorder in this compound and the difficulty in forming such a stoichiometry [28] causes a significant increase in activation energy. Conversely, $Sm_2Zr_2O_7$ exhibits a much lower disorder (<3% [16]), which is sufficiently low as to not unduly increase the activation energy. As such $Sm_2Zr_2O_7$ will exhibit the lowest activation energy although compounds just above (i.e. $Nd_2Zr_2O_7$) and to the right (e.g. $Gd_2Sn_2O_7$) should also be considered.

A guide to the extent of disorder expected in all these materials is provided in reference 16.

PYROCHLORES AS THERMAL BARRIER MATERIALS

Background

The efficiency of a turbine engine improves at higher operating temperature. Furthermore, a more complete combustion results in lower environmental impact. To increase the maximum operating temperature of the high-temperature alloys used to manufacture turbine blades, a micron-thick ceramic layer is bonded to the surface of the alloy to protect it from the intense heat. The present material of choice for TBCs is yttria-stabilized zirconia (YSZ), typically with a composition of $(Y_2O_3)_{0.08}$-$(ZrO_2)_{0.92}$ and a thermal conductivity of ~2-3W/mK [2]. Any potential replacement for YSZ should have a thermal conductivity significantly lower than this value [3]. On this basis, oxides with a pyrochlore structure have emerged as leading candidates for a new generation of TBCs. Indeed, $Gd_2Zr_2O_7$ with a columnar microstructure has been successfully grown as a barrier coating on a Ni-based superalloy with traditional bond coats [3,4].

In this second example we use molecular dynamics simulations to determine the thermal conductivity of 40 pyrochlore oxides (A= Lu to La, B=Ti, Mo, Sn, Zr and Pb). We identify the regions of composition space for which the thermal conductivity is lowest and on this basis could suggest a number of compositions as possible candidates for replacing YSZ as a prefered thermal barrier coating material. However, while low thermal conductivity is a prerequisite for any replacement material, there are other crucial requirements. These include high-temperature stability, chemical compatibility with the thermally grown oxide bond layer (typically Al_2O_3), a coefficient of thermal expansion as close as possible to that of the superalloy (for mechanical integrity), and a low bulk modulus (for good compliance). Ideally we should therefore also systematically explore the elastic properties and the thermal expansions of these materials.

RESULTS

In Figure 4 we show a contour plot of the computed thermal conductivity as a function of the ionic radii for each composition simulated.

We see that the thermal conductivity does not show a strong systematic dependence on the size of the A ion, but does tend to decrease with increasing size of the B ion. More particularly, while there is a gradual decrease in κ going from B=Ti to B=Zr, the values for B=Pb are significantly lower than for any other choice of B ion. Interestingly, analysis shows that there is a significant amount of oxygen disorder induced through the MD simulation in the Pb-containing systems, which is absent in other compositions; such diffusion is known to decrease the thermal conductivity [29] and may be developed as a strategy for further reducing the thermal conductivity of pyrochlores. In trying to identify ideal compositions for minimum thermal conductivity, we see that some of the Sn-based systems have slightly lower thermal conductivities than either the Mo.

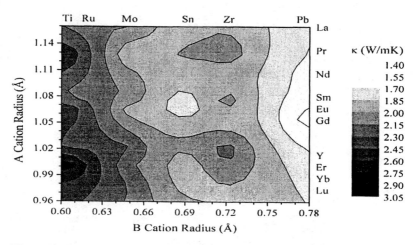

Figure 4. Contour map of thermal conductivity, κ (in W/mK), determined at T=1473K by simulation, as a function of ionic radii of the A and B ions for pyrochlores, $A_2B_2O_7$, or Zr-based systems. Unfortunately there is no experimental data on the stannates to compare against.

COMPARISON TO EXPERIMENTAL DATA

The thermal conductivity of only five pyrochlores, all zirconates, have been determined experimentally. Encouragingly these values are in the range of 1–1.6 W/mK (see Table I) [3,4,30–34]. For $La_2Zr_2O_7$ there are two different measurements that agree with each other rather well [3,4,33]. Conversely for $Gd_2Zr_2O_7$ there is a difference of more than 50% between the smallest and largest measured values [3,4,32]. Presumably this disagreement is a result of uncharacterized differences in microstructure/porosity [34] and impurity levels in the samples studied. For the other three zirconates, there is only one set of experimental data. In the absence of systematic experimental studies over the whole of the A-B composition space, it is not evident what the optimum materials choice might be and thus again illustrates the potential use of maps derived from computed values.

In Table 1 we compare simulation results to the available experimental data. We see that the simulation results are always larger than the experimental values, though not by a consistent amount. These larger values are consistent with our previous simulations of YSZ, where MD simulation results for the thermal conductivity at 1000K was about 30% larger than the corresponding experimental result [29]. While the wide range in the value of $\kappa_{simul}/\kappa_{exptl}$ could point to an inadequacy in our model, it is equally likely a result of the differing microstructures and differing degrees of composition control in the experiments.

Table 1. Comparison of simulated thermal conductivities at T=1473K with experimental values, for all pyrochlores that were investigated experimentally. In some cases the experimental values were obtained by extrapolation from results at somewhat lower temperatures.

	κ_{expt} (W/mK)	κ_{simul} (W/mK)	$\kappa_{simul}/\kappa_{exptl}$
$La_2Zr_2O_7$	1.56 [33] 1.5 [30] 1.5 [3, 4]	1.98	1.27–1.32
$Nd_2Zr_2O_7$	1.6 [32] 1.33[34]	1.83	1.14–1.38
$Sm_2Zr_2O_7$	1.5 [32] 1.5 [31]	2.09	1.39
$Eu_2Zr_2O_7$	1.6 [31]	1.99	1.24
$Gd_2Zr_2O_7$	1.6 [32] 1.0 [30] 1.1-1.4 [4]	1.91	1.19–1.91

SECONDARY CRITERIA

It is also important to match the coefficient of thermal expansion of the thermal barrier coating as close as possible to that of the superalloy. In this case, our aim is to make it as large as possible since the metal has a larger coefficient than do any of the ceramics. A contour map of the thermal expansion is shown in Figure 5 indicating that large B ions result in large thermal expansion coefficients. In Table 2, we also compare simulation results of the coefficient of thermal expansion with

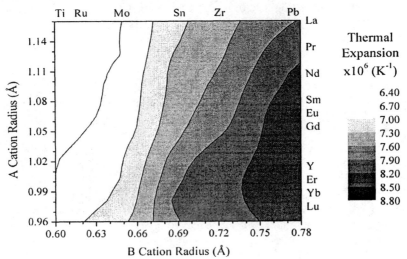

Figure 5. Contour map of the coefficient of thermal expansion, α, (in K^{-1}) at 1473K.

available experimental data [35,36]. Again the calculated values are approximately 30% larger than the corresponding experimental ones.

Finally, for maximum mechanical integrity, it is important that TBCs be as elastically compliant as possible. We have therefore also computed the zero temperature elastic properties using static simulations and standard lattice-dynamics techniques, the results of which are reported elsewhere.

COMPOSITION SELECTION FOR THE BARRIER COATING

Taken as a whole, the above data indicate that the best way to lower the thermal conductivity while increasing the thermal expansion coefficient and decreasing the bulk modulus, is to choose as large a B ion as possible. By contrast the choice of the A ion seems to be less important and may be chosen in order to meet other design criteria such as chemical compatibility with bond coats and corrosion/oxidation stability. Here we have examined only chemically pure materials; alloying of different pyrochlores may also decrease the thermal conductivity further, although further improvements may only be incremental.

Thus, it seems likely that the most effective way to reduce the thermal conductivity of pyrochlores is by increasing the concentration of grain boundaries. While grain growth at elevated temperature is a potential problem for single component systems, a nanocomposite of two pyrochlores may not be subject to grain growth, and thereby offer lower thermal conductivity.

CONCLUDING COMMENTS

The two examples presented here show how computed property values can be used to generate a property composition relationship that facilitates compositional optimization. The first example used a static simulation technique to predict activation energies for oxygen transport. As such it assumes a quasi-harmonic approximation for what is a dynamic process and consequently we were forced to impose a mechanistic model for the migration process. However, the model was complex in that it exhibited clear compositional dependence and good agreement was observed between computed values and experiment. Importantly, by comparing with known experimental structural data, the limitations of the model became apparent. Once again it is clear that in such complex systems it is crucial to engage with experiment.

The second example used MD simulations to predict thermo-mechanical properties of pyrochlores for thermal barrier applications. This example differs

Table 2. Comparison of thermal expansion and bulk modulus from experiment and simulation.

	$\alpha_{expt}(\times 10^{-6} K^{-1})$	$\alpha_{simul}(\times 10^{-6} K^{-1})$
$Er_2Ti_2O_7$		6.71
$La_2Zr_2O_7$	9.1 [7]	7.45
$Nd_2Zr_2O_7$		7.60
$Sm_2Zr_2O_7$	10.8 [21]	7.70
$Gd_2Zr_2O_7$	11.6 [20]	7.81

from the first in that no external mechanistic model was imposed. The physical processes responsible for the thermal conductivity result from the atomistic processes being modelled. Of course this does not mean we were able to model the complete system. In fact here we could only consider thermal conduction in the bulk material and ignored contributions from internal interfaces and interfaces with substrates such as the bond coat. Although absolute agreement with experimental data was less satisfactory, the property trends with composition were useful so that compositional selection was possible.

At the present time the same approach is being used to predict property trends in the compositionally extensive perovskite system. In these cases the issues are dielectric constant for microwave application and radiation tolerance. This underlines the potentially broad applicability of this approach to composition selection.

ACKNOWLEDGEMENTS

P.K.S. and S.R.P. are supported by the US Department of Energy, BES-Materials Science under contract W-31-109-Eng-38. R.W.G. gratefully acknowledges support through EOARD contract No. F61775-00-WE016. KES is supported by the Department of Energy, Office of Basic Energy Sciences, Division of Materials Sciences.

REFERENCES

[1] J. T. S. Irvine, A. J. Feighery, D. P. Fagg, S. Garcia-Martin, 2000. Structural studies on the optimisation of fast oxide ion transport, Solid State Ionics, 136–137, 879–885.
[2] D. P. Hasselman, L. F. Johnson, L. D. Betsen, R. Syed, H. L. Lee, and M. V. Swain, 1987. Thermal diffusivity and conductivity of dense polycrystalline ZrO_2 ceramics: a survey, Am. Ceram. Soc. Bull. 799–806.
[3] M. J. Maloney, in US Patent and Trademark Office (United Technologies Corporation, USA, 2001), No. 6, 177,200.
[4] M. J. Maloney, in US Patent and Trademark Office (United Technologies Corporation, US, 2001), No. 6,231,991.
[5] S. Kramer, M. Spears and H. L. Tuller, 1994, Conduction in titanate pyrochlores: role of dopants, Solid State Ionics, 72, 59–66.
[6] B.J. Wuensch, K.W. Eberman, C. Heremans, E.M. Ku, P. Onnerud, E.M.E. Yeo, S.M. Haile, J.K. Stalick, J.D. Jorgensen, 2000. Connection between oxygen-ion conductivity of pyrochlore fuel-cell materials and structural change with composition and temperature, Solid State Ionics, 129, 111–133.
[7] M. A. Subramanian, G. Aravamudan and G. V. Subba Rao, 1983. Oxide pyrochlores—a review, Prog. Solid State Chem., 15, 55–143.
[8] M. Born, 1923. Atomtheorie des Feten Zustandes, Teubner, Leipzig, Germany.
[9] L. Minervini, R.W. Grimes, K.E. Sickafus, 2000. Disorder in pyrochlore oxides, J. Am. Ceram. Soc., 83, 1873–78.

[10] C.R.A. Catlow, W.C. Mackrodt, 1982. Computer Simulation of Solids, Springer-Verlag, Berlin, Germany, 1982.
[11] N.F. Mott, M.J. Littleton, 1932. Conduction in polar solids. I. Electrolytic conduction in solid salts, Trans. Faraday Soc., 34, 485–99.
[12] CASCADE® M.Leslie. DL/SCI/TM31T. Technical Report SERC Daresbury Lab., 1982.
[13] P.P. Ewald, 1921. Die berechnung optischer und elektroststischer gitterpotentiale, Ann. Phys. (Leipzig), 64, 253–87.
[14] B.G. Dick, A.W. Overhauser, 1958. Theory of the dielectric constants of alkali halide crystals, Phys. Rev., 112, 90–103.
[15] M. Pirzada, R. Grimes, L. Minervini, J. Maguire, K. Sickafus, 2001. Oxygen migration in $A_2B_2O_7$ pyrochlores, Solid State Ionics, 140, 201–208.
[16] L. Minervini, R. Grimes, Y. Tabera, R. Withers, K. Sickafus, 2002, The oxygen positional parameter in pyrochlores and its dependence on disorder, Phil. Mag. A 82, 123–135.
[17] D. Wolf, P. Keblinski, S. R. Phillpot, J. Eggebrecht, 1999. Exact method for the simulation of Coulombic systems by spherically truncated, pairwise r(-1) summation, Journal of Chemical Physics 110, 8254–82.
[18] P.K. Schelling, S.R. Phillpot, D. Wolf, 2001. Mechanism of the cubic-to-tetragonal phase transition in zirconia and yttria-stabilized zirconia by molecular-dynamics simulation, J. Am. Ceram. Soc., 84, 1609–19.
[19] P.K. Schelling, S.R. Phillpot, K. Keblinski, 2002. Comparison of atomic-level simulation methods for computing thermal conductivity, Phys. Rev. B 65, 144306.
[20] Microcal Origin® Microcal Software Inc., MA 01060 USA.
[21] J. Lian, X.T. Zu, K.V.G. Kutty, J. Chen, L.M. Wang, R.C. Ewing, 2002. Ion-irradiation-induced amorphization of $La_2Zr_2O_7$ pyrochlore, Phys. Rev. B, 66, 054108.
[22] H.L. Tuller, 1994. In Defects and disorder in crystalline and amorphous solids, (Ed. C.R.A. Catlow), Kluwer, 189.
[23] M.P. van Dijk, A.J. Burggraaf, A.N. Cormack, C.R.A. Catlow, 1985. Defect structures and migration mechanisms in oxide pyrochlores, Solid State Ionics, 17, 159–67.
[24] M.P. van Dijk, K.J. de Vries, A.J. Burggraaf, 1983. Oxygen ion and mixed conductivity in compounds with the fluorite and pyrochlore structure, Solid State Ionics, 9&10, 913–20.
[25] R.E. Williford, W.J. Weber, R. Devanathan, J.D. Gale, 1999. Effects of cation disorder on oxygen vacancy migration in $Gd_2Ti_2O_7$, J. Electroceram., 3, 409–24.
[26] P.K. Moon, H.L. Tuller, 1989. Intrinsic fast oxygen ion conductivity in the $Gd_2(Zr_xTi_{1-x})_2O_7$ and $Y_2(Zr_xTi_{1-x})_2O_7$ pyrochlore systems, Mat. Res. Symp. Proc., 135, 149–163.
[27] A.J. Burggraaf, T. van Dijk, M.J. Veerkerk, 1981. Structure and conductivity of pyrochlore and fluorite type solid solutions, Solid State Ionics, 5, 519–22.
[28] V.S. Stubican, G.S. Corman, J.R. Hellmann, G. Senft, 1983, Adv. Ceram., 12, 96.

[29] P.K. Schelling, S.R. Phillpot, 2001. Mechanism of thermal transport in zirconia and yttria-stabilized zirconia by molecular-dynamics simulation, J. Am. Ceram. Soc. 84, 2997–3007.

[30] G. Suresh, G. Seenivasan, M.V. Krishnaiah, P.S. Murti, 1997, Investigation of the thermal conductivity of selected compounds of Gd and La, J. Nuclear Materials 249, 259–261.

[31] G. Suresh, G. Seenivasan, M.V. Krishnaiah, P.S. Murti, 1998. Investigation of the thermal conductivity of selected compounds of lanthanum, samarium and europium, Journal of Alloys and Compounds 269, L9–12.

[32] J. Wu, X. Wei, N.P. Padture, P.G. Klemens, M. Gell, E. Garcia, P. Miranzo, M.I. Osendi, 2002. Low-thermal-conductivity rare-earth zirconates for potential thermal-barrier-coating applications, J. Am. Ceram. Soc. 85, 3031–35.

[33] R. Vassen, X.Q. Cao, F. Tietz, D. Basu, D. Stover, 2000. Zirconates as new materials for thermal barrier coatings, J. Amer. Cer. Soc. 83, 2023–28.

[34] S. Lutique, R.J.M. Konings, V.V. Rondinella, J. Sommers, T. Wiss, 2003. The thermal conductivity of $Nd_2Zr_2O_7$ pyrochlore and the thermal behaviour of pyrochlore-based inert matrix fuel, J. Alloys Comp., 352, 1–5.

[35] Y.S. Toulokian, R.K. Kirby, R.E. Taylor, T.Y.R. Lee, 1977. Thermal Expansion-Non Metallic Solids, IFI/Plenum New York.

[36] G.L. Catchen, T.M. Rearick, 1995. O-anion transport measured in several $R_2M_2O_7$ pyrochlores using perturbed-angular-correlation spectroscopy, Phys. Rev. B 52, 9890-99.

[37] P.K. Schelling, S.R. Phillpot R.W. Grimes, 2003. Optimum pyrochlore compositions for low thermal conductivity by simulation, submitted to Phil. Mag. Lett.

Effects of Different Initial Conditions of Liquid Metals on Solidification Microstructures

J.-Y. LI

Chemistry and Chemical Engineering, Hunan University, Changsha, P.R. China

R.-S. LIU, K.-J. DONG, C.-X. ZHENG and F.-X. LIU

Department of Physics, Hunan University, Changsha, P.R. China

ABSTRACT

The effects of different initial conditions of liquid metal Al on the solidification microstructures have been studied with molecular dynamics simulation in a liquid metal Al system by rapid cooling. From the results, it has been found that under different initial conditions, the 1551 bond-type and the icosahedron (expressed by (12 0 12 0)) related to the 1551 bond-type play a remarkable role during microstructure transitions. Highly interesting is that though the total number of icosahedron in each system will be obviously increased with the decrease of temperature, the number of icosahedral cluster repeatable not increased with the same tendency and has a maximum during the middle section of temperature, this point is corresponded to the grass transition temperature T_g. It is also demonstrated that the 1771 bond-type only appears in the liquid and supercooled liquid states, and cannot be found in the solid state, and the critical temperature of the 1771 bond-type also corresponded to the grass transition temperature T_g. These results will give us a new way to understand the microstructure transitions of liquid metals from liquid to solid states.

INTRODUCTION

It is well known that amorphous and nano-crystal metals and alloys prepared by melt quenching possess various excellent physical and chemical properties, and these properties are mainly determined by their microstructures. The solid microstructures are mainly related to their initial microstructures in liquid state and their cooling rate from liquid state. As we know, the initial microstructures in liquid state are influenced by their thermal history. For understanding the physical origins of these features, it is important to clarify the concrete transition mechanisms of microstructure configurations of liquid metals during their rapid cooling under different initial conditions with different thermal history. However, up to now, under the present experimental conditions, the microscopic transition mechanisms cannot be exactly determined.

With the development of computer technique, the simulation studies on the microstructures of liquid and amorphous metals have been made by means of molecular dynamics (MD) method [1–6], and the method has been improved gradually.

For this purpose, based on the authors' previous work [4–6], the effects of different initial conditions of liquid metal Al on its solidification microstructures have been investigated by tracing rapid cooling processes using molecular dynamics simulation. Some new results were obtained.

SIMULATION CONDITIONS AND METHODS

We performed a simulation study on the microstructural transitions under different initial conditions in a liquid system consisting of 500 Al atoms during rapid cooling processes by using molecular dynamics. The conditions for simulating calculation are as follows: at first, the 500 Al atoms are placed in a cubic box and then the system runs under periodic boundary condition. The interacting interatomic potential adopted here is also the effective pair potential function of the generalized energy independent non-local model—pseudo potential theory developed by S. Wang et al.[7,8], and the function as following and its curve as shown in Figure 1.

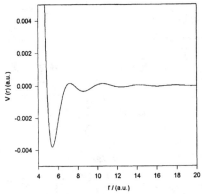

Figure 1. Effective pair potential $V(r)$ of liquid metal Al at 943K.

Where Z_{eff} and $F(q)$ are, respectively, the effective ionic valence and the normalized energy wave number characteristic, which were defined in detail in refs [7, 8]. This pair potential is cut off at 20.0 au (atomic unit). The time step of these runs is chosen at 5.95×10^{-15}s. The cooling rate is 33.5×10^{12}K/s.

$$V(r) = \left(Z_{eff}^2 / r\right)\left[1 - \left(\frac{2}{\pi}\right)\int_0^\infty dq F(q)\sin(rq)/q\right] \quad (1)$$

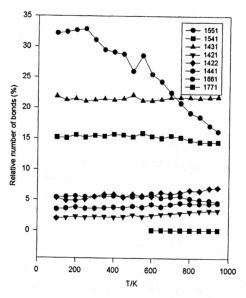

Figure 2. Relations of the relative number of various HA bond-types with temperature during rapid solidification process of liquid metal Al under initial condition of 15000 time steps.

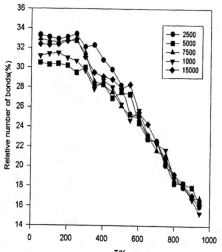

Figure 3. Relationship of relative number of 1551 bond-type with temperature under different initial conditions.

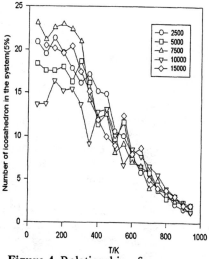

Figure 4. Relationship of average number of icosahedrons with temperature under different initial conditions

The simulating calculation is started at 943K (the melting point of Al is 933K). First of all, in order to obtain the different initial states let the system run 2500,

5000, 7500, 10000 and 15000 steps, respectively, at the same temperature (943K). Secondly, for each initial condition, the system temperature is decreased at given cooling rate to some given temperatures: 900, 850, 800, 750, 700, 650, 600, 550, 500, 450, 400, 350, 300, 250, 200, 150 and 100K. Then, let the system run 4000 time steps, and the structural configurations of this system, i.e. the space coordinates of each atom were recorded for each 200 time steps, and 20 recorded data obtained at each given temperature. Finally, the bond-types between related atoms are detected by means of the index method of Honeycutt-Andersen (HA)[9].

RESULTS AND DISCUSSION

As shown in previous work [3–6], to check the reliability of the pair-potential used in the simulations of disorder liquid or solid system, the pair distribution function g(r) of the system obtained from the above-mentioned simulation has been compared with the results from the experimental work of Waseda[10], and a good agreement obtained.

Similarly our previous work [3–6] has shown that the bonding relations between atoms in the system can be described using the bond-type index method of HA [9]. Applying these HA indices to the present results, the relative bonding numbers at each given temperature can be obtained. For example, the relations of the relative number of various HA bond-types with temperature during rapid solidification process of liquid metal Al under initial condition of 15000 time steps have been shown in Figure 2.

From Figure 2, it can be seen that among the various bond-types only the relative number of 1551 bonds changes sharply during rapid solidification, that is to say, the 1551 bond-type must play a remarkable role during microstructure transitions. For convenience of comparison, we only list the 1551 bond-type for each system under different initial conditions as shown in Figure 3. From Figure 3, it can be seen that a temperature range of ~550–600K is the important transition point corresponding to the glass transition point T_g of liquid metal Al determined by the Wendt-Abraham ratio method [11]. Above the glass transition point T_g, the 1551 bond-types and the icosahedra increase almost at the same rate for all different initial conditions. Below the transition point T_g, the 1551 bond-types and the icosahedra increase with different rates for different initial conditions and the variations are remarkable. At T_g, the effects of different initial conditions occur suddenly.

In Figure 2, it is seen the 1771 bond-type not only appears in the liquid and supercooled liquid states but is also found in the solid state with the critical temperature of the 1771 bond-type corresponded to the glass transition temperature T_g.

Based on the above, we can consider that the initial conditions seriously influence the solidification microstructures and the effects are mainly demonstrated at temperatures over the glass transition point.

Up to now, in order to describe the disorder systems of liquid and amorphous states, the HA bond-type index method is an important method. However, for the structures of various clusters formed by special conditions and numbers of bond-types, it is very difficult to describe them by this method. In this paper, we have used a new cluster bond-type index method (CBTIM) to describe some important types of clusters, especially larger clusters with more then 100 atoms [12]. The method is based on the work of D.W. Qi and S. Wang [13], who had adopted it to express Frank-Kasper polyhedron, Bernal polyhedron, and other defective icosahedra successfully using four integers. The four integers of CBYIM express in turn: the first integer represents the number of surrounding atoms which form a cluster with a central atom; the second, third and fourth integers represent the numbers of 1441, 1551 and 1661 bond-types respectively, by which the surrounding atoms connect to a central atom. For example, (12 0 12 0) expresses an icosahedron composed of 13 atoms (one is the central atom with coordination number Z=12) connected to twelve 1551 bond-types; (13 1 10 2) expresses the defective icosahedron composed of 14 atoms (with coordination number Z=13) connected tp one 1441, ten 1551 and two 1661 bond-types. Figure 5(a) and 5(b) show respectively the icosahedra and defective icosahedra.

According to the CBTIM, we obtain the average number of icosahedra in each system under different initial conditions during rapid solidification as shown in Figure 4 which also shows that the tendency of various curves also change at the same transition point T_g as observed in Figure 3.

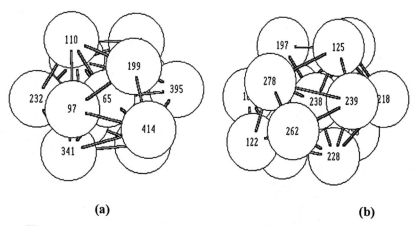

(a)   (b)

Figure 5. Schematics of icosahedron, defective icosahedron, at 550K.
(a) An icosahedron (12 0 12 0) with center atom 65.
(b) A defective icosahedron (13 1 10 2) with center atom 238.

To go further, in order to understand the effect of initial conditions on solidification microstructures, we have analyzed the numbers of icosahedron in each system under different initial conditions, and found that among these icosahedra some of which are repeatable and can be repeated times by times, and some of which can not be repeated during the 20 times in the thermal running at each given temperature. The total numbers of icosahedra in different system are shown in Figure 6. For convenience, the total numbers of icosahedra, the numbers of icosahedra repeatable and the numbers of icosahedra no repeatable of two initial conditions of 2500 and 15000 time steps, are shown in Figure 7, Figure 8, and Figure 9, respectively.

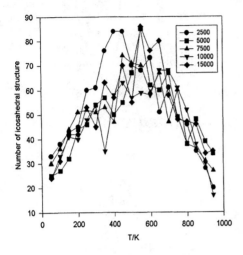

Figure 6. Relationship of the total number of icosahedral structures in each system with temperature under five different initial conditions.

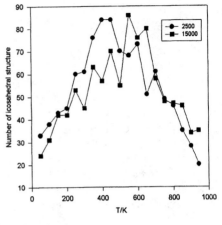

Figure 7. Relations of the total number of repeatable icosahedral structures with temperature at 2500 and 5000 time steps.

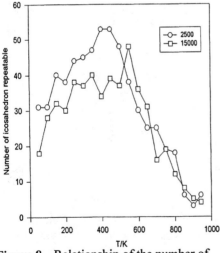

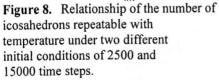

Figure 8. Relationship of the number of icosahedrons repeatable with temperature under two different initial conditions of 2500 and 15000 time steps.

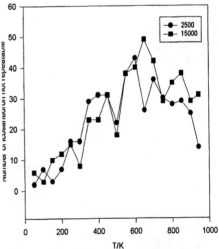

Figure 9. Relationship of the number of icosahedrons not repeatable with temperature under two different initial conditions of 2500 and 15000 time steps.

From Figure 6, it can be seen that the total numbers of icosahedra have both sharp and random changes in the middle section of the temperature range 300–700K for all the different initial conditions. From Figure 7, it is more obvious that the maximum number of icosahedron appears to move from a lower temperature range of 400–450K to a higher level of 550–650K by increasing the initial time steps from 2500 to 15000. From Figure 8, it can be seen that with decrease in temperature, the number of icosahedral structures repeatable at first increased from a low value up to a maximum level and then decreased to a value higher than the first. On the other hand, from Figure 9, the number of icosahedral structures not repeatable at first increased from a higher value up to a maximum and then decreased to a value lower than the first.

Based on the above, we consider that the initial condition, namely the thermal history of the system not only influences the solidification microstructure, but also can be used to control the degree of influence by changing the initial time steps.

CONCLUSIONS

From the results above, the following conclusions can be made:

1. Under different initial conditions, the 1551 bond-type and its related icosahedron (expressed by (12 0 12 0) play remarkable roles during microstructure transition. Of great interest is that although the total number of icosahedrons in each system obviously increases as temperature drops, the number of repeatable icosahedral clusters do not increase with the same tendency and show a maximum in the middle of the temperature range corresponding to the glass transition temperature T_g.
2. The data also demonstrate that 1771 bond-types only appear in the liquid and supercooled liquid state and are not found in the solid state. The critical temperature of the 1771 bond-type also corresponds to the glass transition temperature T_g.
3. The initial conditions seriously influences solidification microstructures and the effects are mainly seen above the glass transition point.
4. The initial conditions, namely the thermal history, not only influence solidification microstructures, but can be used to control the degree of influence by changing the initial time steps.

ACKNOWLEDGEMENT

This work was supported by The National Natural Science Foundation of China under Grant No. 50071021 and Grant No. 50271026.

REFERENCE

[1] H. Li, X. F. Bian, J. X. Zhang, 1999. Mater. Sci. Eng. A, 271, 167.
[2] J. Lu, J. A. Szpunar, 1993. Acta Metal. Mater., 41, 2291.
[3] V. S. Stepanyuk, A. Szasz, A. A. Katsnelson, et al, 1993. J. Non-Cryst. Solids, 159, 80.

[4] R. S. Liu, J. Y. Li, Z. Zhou, et al., 1999. Mater. Sci. Eng B, 57, 214.
[5] R. S. Liu, H. R. Liu, C. X. Zheng, et al., 2001. Chin. Phys. Lett. 18, 1383.
[6] R. S. Liu, D. W. Qi, S. Wang, 1992. Phys Rev. B, 45, 451.
[7] S. Wang, S. K. Lai, 1980. J. Phys., F, 10, 2717.
[8] D. H. Li, X. R. Li, S. Wang, 1986. J. Phys., F, 16, 309.
[9] J. D. Honeycutt, H. C. Andersen, 1987. J. Phys. Chem., 91, 4950.
[10] Y. Waseda, 1980. The Structure of Non-crystalline Materials, McGraw-Hill, NY, 270.
[11] H. R. Wendt, F. F. Abraham, 1978. Phys. Rev. Lett. 41, 1244.
[12] Liu R S, Li J Y, Dong K J et al., 2002. Mater. Sci. Eng B, 94, 141.
[13] D. W. Qi, S. Wang, 1991. Phys. Rev. B, 44, 884.

Combined Approach of Statistical Moment and Cluster Variation Methods for Calculation of Alloy Phase Diagrams

K. MASUDA-JINDO
Department of Materials Science and Engineering, Tokyo Institute of Technology, Yokohama, Japan

V. VAN HUNG
Department of Physics, Hanoi National Pedagogic University, Hanoi, Vietnam

R. KIKUCHI
Materials Science and Mineral Engineering, University of California, Berkeley, CA, USA

ABSTRACT

The thermodynamic quantities of metals and alloys are studied using the moment method in the quantum statistical mechanics, going beyond the quasi-harmonic approximations. Including the power moments of the atomic displacements up to the fourth order, the free energy, specific heats C_v and C_p, mean square atomic displacements and thermal lattice expansion coefficients are derived explicitly in terms of the second and fourth order vibrational coupling constants. The calculated thermodynamic quantities are favorably compared with the experimental results. Further applications are presented both for the phase separating and order-disordered binary alloys. For phase-separating binary alloys like CuAg and AgPd systems, the change of lattice constant with the composition and the reduction of the transition temperature are demonstrated. For the calculations of order-disordered phase transitions, we investigate the effects of thermal lattice vibration on the long range order (LRO) parameter and order-disorder transition temperatures of the ordered binary alloys. Some of the alloy phase diagrams (Ni-Al etc.) are also calculated and the theretical phase diagrams are compared with the experimental results.

INTRODUCTION

It is widely recognized that thermal lattice vibrations play an important role in determining the phase stabilities of various alloy systems [1–3]. It is important to account for the anharmonic effects of thermal lattice vibrations in the computations of thermodynamic quantities of metals and alloys. So far, most of the theoretical calculations of thermodynamic quantities of metals and alloys have been done on the basis of harmonic or quasi-harmonic (QH) approximations of thermal lattice vibrations [4–6].

A number of theoretical approaches have been proposed to overcome the limitations of the QH theories. The first calculation of the lowest-order anharmonic contributions to the atomic mean-square displacement $<u^2>$ or the Debye-Waller factor was done by Maradudin and Flinn [7] in the leading-term approximation for a nearest-neighbour central-force model. Since then, many anharmonic calculations including the lowest-order anharmonic contributions have been performed for metal systems [8,9]. The method requires acknowledge of a number of Brillouin-Zone sums [2] and the calculations are performed for the central force model crystals. Recently, some attempts have been made to take into account the bond length dependence of bond stiffness tensors in the calculations of the free energy of the substitutional alloys [1,3]. The anharmonic effects of lattice vibrations on the thermodynamic properties of the materials have also been studied by employing the first order quantum-statistical perturbation theory [10,11] as well as by the first order self consistent phonon theories [12–14]. However, the anharmonicity theories proposed so far are still incomplete and have some inherent drawbacks and limitations.

In this study, we use the finite temperature moment expansion technique to derive the Helmholtz free energies of metals and alloy systems, going beyond the QH approximations. The thermodynamic quantities, mean-square atomic displacements, specific heats and elastic moduli are derived from the explicit expressions of the Helmholtz free energies. The Helmholtz free energy of the system at a given temperature T will be determined self-consistently with the equilibrium thermal lattice expansions of the crystal.

THEORY

We derive the thermodynamic quantities of metals and alloys, taking into account the fourth order anharmonic contributions in the thermal lattice vibrations going beyond the harmonic or QH approximations. The basic equations for obtaining thermodynamic quantities of given crystals are described in our previous publications [15–18]: The equilibrium thermal lattice expansions are calculated by a force balance criterion and then the thermodynamic quantities are determined for equilibrium lattice spacings. The anharmonic contributions of the thermodynamic quantities are given explicitly in terms of the power moments of the thermal atomic displacements. The thermodynamic quantities of the harmonic crystal (harmonic Hamiltonian) will be treated in the Einstein approximation. In this respect, the present formulation is similar conceptually to the treatment of quantum Monte Carlo method by Frenkel [19,20].

We start with the Helmholtz free energy of the system, which is given by

$$\Psi_0 = U_0 + 3N\theta\left[X + \ell n\left(1-e^{-2X}\right)\right] + 3N\left\{\frac{\theta^2}{k^2}\left[\gamma_2 X^2 x \coth^2 X - \frac{2}{3}\gamma_1\left(1+\frac{X\coth X}{2}\right)\right]\right.$$
$$\left. + \frac{2\theta^3}{k^4}\left[\frac{4}{3}\gamma_2^2 x \coth X\left(1+\frac{X\coth X}{2}\right) - 2\gamma_1(\gamma_1+2\gamma_2)\left(1+\frac{X\coth X}{2}\right)(1+X\coth X)\right]\right\}$$

(1)

where $X = \hbar\omega/2k_BT$ (k_B being the Boltzman constant) and the second term denotes the harmonic contribution to the free energy. k, γ_1 and γ_2 are harmonic and unharmonic parameters originating from the second and fourth order derivatices of the atomic cohesive energy E_{ci}, respectively. The free energy expression of Eq. 1 is for fcc and bcc (cubic) systems. The Helmholtz free energy of hcp crystals is given in a similar fashion including second, third and fourth order moments [15–18]. In the present study, the internal energies of metals and alloys are evaluated both by Lennard-Jones type of potentials and the tight-binding (TB) total energy calculation scheme. For fcc metals and alloys, we use the later TB scheme since the reliable calculations are now possible for this crystalline system. In particular, it is useful to apply the bond order potential (BOP) formalism [21] to derive the pairwise bond energies between the constituent atoms taking into account the many atom interactions around the atom pairs. For hcp and bcc metals, we use the Lennard-Jones (L.-J.) type of potentials.

Within the BOP formalism [21], one can calculate the bond energy of individual sites in the following form:

$$E_{bond}^{i\alpha} = \frac{4}{\beta}\text{Re}\sum_{p=0}^{M-1} z_p(E_p - \varepsilon_{i\alpha})G_{i\alpha,i\alpha}(E_p), \quad (2)$$

where $G_{00}^{i\alpha}(Z)$ denotes the site ($i\alpha$) diagonal Green's function. E_p and Z_p are given respectively as

$$E_p = \mu + \frac{2M}{\beta}(z_p - 1), \quad (3)$$

$$z_p = \exp(i\pi(2p+1)/2M), \quad p=0, 1, \ldots, 2M-1 \quad (4)$$

with residues $R_p = -z_p/\beta$ (β being k_BT^{-1}). The number of valence electrons of the orbital α at site i is given by

$$N^{i\alpha} = \frac{4}{\beta}\text{Re}\sum_{p=0}^{M-1} z_p G_{00}^{i\alpha}(E_p). \quad (5)$$

One can also use the alternative expressions for the pairwise bond energy in the following form:

$$U_{ij} = -2\sum_{\alpha,\beta} H_{ij}(\sum_{n=0}^{\infty}\chi_{0n,n0}(E_F)\delta a_n^\lambda + \sum_{n=1}^{\infty}\chi_{0,n-1,n,0}(E_F)2\delta b_n^\lambda) \quad (6)$$

Eq. 6 allows us to calculate the bond energy between atomic sites i and j, taking into account the multi-atom interactions around the ij atom pair.

With the aid of the free energy formula $\Psi=E-TS$, one can find the thermodynamic quantities of metal and alloy systems, where the internal energy term E is evaluated by using either pair potentials or TB-BOP formalism. The specific heats and elastic moduli at temperature T are directly derived from the Helmholtz free energy Ψ of the system. For instance, the isothermal compressibility χ_T is given by

$$\chi_T = 3(a/a_0)^3 / \left[2P + \frac{1}{3N}\frac{\sqrt{2}}{a}\left(\frac{\partial^2\Psi}{\partial r^2}\right)_T\right], \quad (7)$$

On the other hand, the specific heats at constant volume C_v is calculated from

$$C_v = \frac{\partial E}{\partial T}\bigg|_V \quad (8)$$

The specific heat at constant pressure C_p is determined from the thermodynamic relations

$$C_p = C_v + \frac{9TV\alpha_T^2}{\chi_T}, \quad (9)$$

where α_T denotes the linear thermal expansion coefficient and V the atomic volume. The relationship between the isothermal and adiabatic compressibilities, χ_T and χ_s, is simply given by

$$\chi_s = \frac{C_v}{C_p}\chi_T. \quad (10)$$

One can also find the "thermodynamic" Grüneisen constant as

$$\gamma_G = \frac{V}{C_v}\left[\frac{\partial S}{\partial V}\right]_T = \frac{\alpha_T B_S V}{C_P}, \qquad (11)$$

where $B_S \equiv \chi_S^{-1}$ denotes the adiabatic bulk modulus.

RESULTS AND DISCUSSIONS
Thermodynamic Quantities of Cubic and HCP Metals

With the use of the analytic expressions of the Helmholtz free energy presented in the previous section, it is straightforward to calculate the thermodynamic quantities of metals and alloys at the thermal equilibrium. First, the equilibrium lattice spacings of cubic metals are determined, self-consistently by a force balance criterion including temperature dependent k, γ_1 and γ_2 values. The thermal lattice expansion can also be calculated by standard procedure of minimizing the Helmholtz energy of the system: We have checked that both calculations give almost identical results on the thermal lattice expansions. We have calculated the thermal lattice expansion and mean square atomic displacements of some fcc metals as well as for hcp metals, for which the reliable many body potentials [22,23] are available. So far, a number of the second moment approximation (SMA) base TB potentials have been proposed for fcc metals. Specifically, we use the SMA TB potentials by Rosato et al. [22] and by Cleri and Rosato [23] for fcc and hcp metals, whose parametera are presented in Table 1. They are known to give good descriptions of cohesive properties of fcc elements. In the TB scheme by Rosato et al. [22], the interaction range is limited to the first nearest neighbours, while in the TB scheme by Cleri and Rosato [23], it is extended to the fifth neighbours. The calculated thermodynamic quantities, bulk moduli B_T, thermal expansion coefficients α_T and Grüneissen constants γ_G of the fcc metals are presented in Table 2.

We have checked our thermodynamic calculations based on the SMA-TB scheme by also performing the calculations using the refined TB method of sp^3d^5-basis functions [26]. The thermal lattice expansions of Rh and Pd metals are calculated by using the sp^3d^5-basis functions (whose parameters are taken from Ref. 26), and they are found to be 7.8 x 10^{-6} degrees^{-1} and 12.6 x 10^{-6} degrees^{-1} respectively at room temperature. These values agree well with the experimental results 8.6 x 10^{-6} degrees^{-1} and 11.76 x 10^{-6} degrees^{-1} respectively.

The temperature dependence of the bulk moduli B_T of cubic metals is also calculated using the present statistical moment method. The ratios of bulk moduli of room temperatures, B_T/B_0 with respect to those of the absolute zero temperature, are calculated to be 0.85~0.90 (fcc metals), which compare favorably with the experimental results.

A similar expression of the Helmholtz free energy in Eq.1 is also obtained for non-cubic metals and alloys. For instance, for treating hcp metals, we introduce a new parameter β related to the third derivative of the potential energy

Table 1. Parameters of the second moment TB potentials for cubic metals.

	A (eV)	ξ (eV)	P	q	E_c (eV/atom)	a (Å)
Al	0.0334	0.7981	14.6147	1.112	-3.339	4.050
Ni	0.1368	1.7558	10.00	2.70	-4.435	3.523
Cu	0.0993	1.3543	10.08	2.56	-3.544	3.615
Rh	0.0629	1.660	18.450	1.867	-5.752	3.803
Pd	0.1746	1.718	10.867	3.742	-3.936	3.887
Ag	0.1231	1.2811	10.12	3.37	-2.960	4.085
Au	0.2061	1.790	10.229	4.036	-3.779	4.079
Pt	0.2975	2.695	10.612	4.004	-5.853	3.924
Ti	0.1519	1.8112	8.620	2.390	-4.853	2.492
Zr	0.1934	2.2792	8.250	2.249	-6.167	3.232
Co	0.0950	1.4880	11.604	2.286	-4.386	2.507
Cd	0.1420	0.8117	10.612	5.206	-1.166	2.959
Zn	0.1477	0.8900	9.689	4.602	-1.359	2.653
Mg	0.0290	0.4992	12.820	2.257	-1.519	3.176

from [22 and 23].

Table 2. Bulk modulus, linear thermal expansion, and Grüneisen constant calculated with the use of the SMA TB potentials.

Element	B_T(GPa) cal T=0	B_T(GPa) cal RT	B_T(GPa) exp	$\alpha(\times 10^{-6} K^{-1})$ cal	$\alpha(\times 10^{-6} K^{-1})$ Exp	γ_G cal	γ_G exp
Al	87	75	72	24.5	23.6	2.09	2.19
Cu	153	137	137	15.9	16.7	2.21	2.00
Ni	190	182	184	14.7	12.7	2.01	1.88
Ag	114	96	101	23.5	19.7	2.78	2.36
Rh	306	280	271	10.9	8.2	2.19	2.43
Pd	204	171	181	14.3	11.6	2.22	2.18
Au	185	164	173	17.2	14.2	3.21	3.04
Pt	301	259	278	11.2	8.9	3.06	2.56

$$\beta = \frac{\partial^3 E_{ci}}{\partial y_i \partial x_i^2}, \tag{12}$$

where E_{ci} denotes the atomic cohesive energy of site i. The thermodynamic quantities of hcp metals Ti, Cd and Zn are calculated by using the L.-J. type of potentials while hcp metals Zr and Mg by the second moment TB potentials [23], and they are presented in Table 3 together with the experimental results. One can see in Table 3 that the calculated thermal lattice expansion, specific heats C_v and C_p, and compressibility χ_T and χ_s are generally in good agreement with the corresponding experimental results [27]. The good agreement between the calculated and experimental results is rather surprising, in view of the phenomenological pairwise potentials used in the calculations.

Table 3. Comparison of calculated and experimental thermodynamic quantities of hcp metals.

	T(K)	a(Å)	χ_T (10^{-13}cm²/dyn)	χ_S (10^{-13}cm²/dyn)	$\alpha(10^{-5}$ K^{-1})	Exp [27]	C_v (cal/mol.K)	C_p (cal/mol.K)	Exp [27]
Ti	300	2.7980	7.72	7.55	1.27	0.85	5.67	5.80	5.97
	500	2.8049	8.12	7.85	1.26	0.98	5.92	6.12	6.51
	700	2.8119	8.54	8.16	1.27	1.05	6.02	6.29	7.01
	900	2.8189	8.98	8.50	1.29	—	6.08	6.42	—
Cd	100	3.0405	17.64	17.09	3.40	2.67	5.44	5.61	—
	200	3.0506	19.14	18.15	3.41	3.02	5.91	6.23	—
	300	3.0608	20.82	19.30	3.51	3.15	6.05	6.52	6.19
	400	3.0714	22.73	20.55	3.70	3.34	6.13	6.78	6.49
	500	3.0824	24.93	21.91	3.95	3.88	6.20	7.06	6.78
	600	3.0942	27.56	23.46	4.26	—	6.26	7.36	7.10
Zn	200	2.6889	13.69	13.12	3.05	2.84	5.74	5.98	—
	300	2.6969	14.65	13.82	3.08	2.97	5.95	6.31	6.07
	400	2.7051	15.71	14.58	3.17	3.05	6.06	6.53	6.31
	500	2.7134	16.89	15.38	3.29	3.16	6.14	6.74	6.55
	600	2.7220	18.20	16.26	3.46	3.42	6.20	6.94	6.79
	700	2.7310	19.70	17.22	3.66	—	6.26	7.16	7.50

The lattice specific heats C_v and C_p at constant volume and at constant pressure are calculated using Eq. 8 and Eq. 9 respectively. However, the evaluations by Eq. 8 and Eq. 9 are the lattice contributions, and their values may not be directly compared with the corresponding experimental values. We do not include the contributions of lattice vacancies and electronic part of the specific heats C_v^{ele}, which are known to give significant contributions in metals for higher temperature region near the melting temperature. In particular, it has been demonstrated that lattice vacancies make a large contribution to the specific heats for high temperature region [28]. The electronic contribution to the specific heat at constant volume C_v^{ele} is proportional to the temperature T and given by $C_v^{ele} = \gamma T$, γ being the electronic specific heat constant [28,29]. The electronic specific heats C_v^{ele} values are estimated to be 0.8%~13.4% of C_v^{lat} for metals considered here by free–electron model [29]. Therefore, the present formulae of the lattice contribution to the specific heats, both C_v and C_p, tend to underestimate the specific heats for higher temperature region, when compared with the experimental results. As expected from above mentioned reasoning, the calculated C_p values of hcp metals are smaller than the experimental values at high temperatures. Here, we also note that the SMM gives reasonable results of specific heats for wide temperature range, in contrast to those of MD simulations. In the MD simulations, the heat capacities per atom at constant pressure C_p can be obtained for metals by taking the numerical derivative of the internal energy with respect to temperature [30]. The MD simulations by Mei et al. [30] give reasonable values of C_p for higher temperature region when compared with the experimental data. However, it should be noted that the MD simulations are only adequate above the room temperature, and calculated C_p value deviates from the experimental data at low temperatures because quantum effects are not taken into account in the classical MD simulations.

As a final remark in this subsection, we point out that the calculated thermal lattice expansions by the SMA-TB potentials are different qualitatively from those by the L.-J. potentials. As can be seen in the results of Figure 1, the thermal lattice expansions become much smaller for higher temperature region when one uses the SMA-TB potentials for hcp metals, in contrast to the calculations of L.-J. potentials. This may indicate that the SMA-TB potentials of Ref. 23 are not well suited for treating the thermodynamic quantities of hcp metals, although it works well for fcc metals.

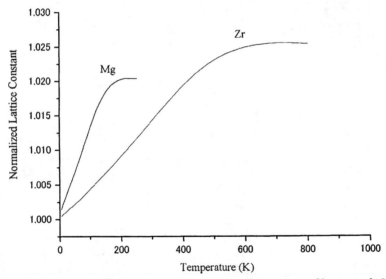

Figure 1. Temperature dependence of normalized constants of hcp metals Mg and Zr.

THERMODYNAMIC QUANTITIES OF BINARY ALLOYS

The statistical moment method outlined above can be applied straightforwardly to substitutional alloy systems, by simply taking into account the configurational entropy term S_{conf} in the free energies

$$\Psi = \Psi_0(c) - TS_{conf}. \tag{13}$$

In the present study, the configurational entropies are calculated by using the cluster variation method (CVM), in a tetrahedron cluster approximation [32,33]. With the use of the tetrahedron cluster approximation of CVM, the configurational entropy of the fcc binary alloys is expressed formally as

$$S_{conf}/N = -k_B \left[2\sum_{ijkl} L(z_{ijkl}) - \sum_{ij}\sum_{\alpha,\alpha'} L(y_{ij}^{\alpha\alpha'}) + \frac{5}{4}\sum_{i}\sum_{j} L(x_i^\alpha) \right], \quad (14)$$

where x_i^α is the concentration of i atoms in sublattice α, $y_{ij}^{\alpha\alpha'}$ is the nearest-neighbour pair probability of finding an i atom at α and a j atom at α', and z_{ijkl} the probabilities of tetrahedron cluster configuration ijkl. After a bit of algebra, one can find z_{ijkl} as

$$z_{ijkl} = \exp\left(\frac{\lambda}{2kT}\right) \cdot \exp\left(\frac{-e_{ijkl} + \bar{\mu}}{2k_B T}\right) \frac{\left[y_{ij}^{\alpha\beta} y_{il}^{\alpha\gamma} y_{ik}^{\alpha\delta} y_{jk}^{\beta\gamma} y_{jl}^{\beta\delta} y_{kl}^{\gamma\delta} \right]^{1/2}}{\left[x_i^\alpha x_j^\beta x_k^\gamma x_l^\delta \right]^{5/8}} \quad (15)$$

with $\bar{\mu} = (\mu_i + \mu_j + \mu_k + \mu_l)/4$.

We now describe the calculation results of thermodynamic quantities of binary alloys obtained by using the present combined SMM-CVM approach. Using the derivatives of the interatomic potentials and the vibrational parameters for the binary alloys, one can determine the thermal lattice expansion, LRO parameter η, and ODT temperature T_c of the order-disordered alloys.

First, we will apply the combined SMM-CVM approach to phase diagram calculations like CuAu alloys. In Figure 2, we show the calculated LRO parameters η of the Cu_3Au alloy as a function of the temperature T. The calculated LRO parameters η by the SMM coupled with the Bragg-Williams approximation and by the combined SMM-CVM (TTR) are presented by symbols ☐ and ☐, respectively. The arrow in the figure indicates the experimental order-disorder transition temperature and the results of the harmonic approximation are shown by symbols. One sees in Fig. 2 that there are big differences among the LRO parameters η calculated by the three different methods. The order-disorder transition temperature T_c becomes lower when we use the better approximation of the CVM and take into account the anharmonicity of thermal lattice vibrations. Therefore, for the quantitative calculations of the alloy phase diagrams, the anharmonicity of thermal lattice vibrations play an important role and can not be neglected. The temperature dependence of the nearest-neighbor distances of the Cu_3Au alloy are also calculated and compared with the experimental results. (Here, we have adjusted TB parameters for Cu_3Au alloy of Ref.23, by 7% for best-fitting of the nearest-neighbour distances.) The temperature dependence of the lattice constant of Cu_3Au alloys obtained by the present method is in good agreement with the experimental results, while that obtained by the QH approximation [31] shows strong nonlinear behavior at higher temperatures, in qualitative disagreement with experiments.

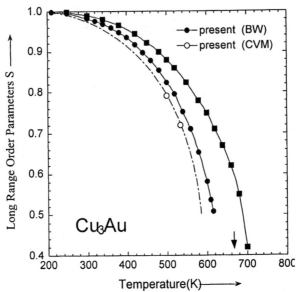

Figure 2. Temperature dependence of the LRO parameter η of Cu_3Au alloy: symbols □ , □ and □ represent the η values obtained by the present SMM combined with Bragg-Williams and tetrahedron CVM approach, and those by harmonic approximation, respectively. The arrow indicates the experimental transition temperature.

We also calculated bulk moduli B_T/B_0 of the Cu_3Au alloy, as a function of the temperature T, B_0 being the bulk modulus of the absolute zero temperature. We have found that the bulk modulus depends strongly on the temperature and is decreasing function of T. Specifically, the B_T of Cu_3Au alloy is decreased approximately ~10% at 600K compared to the B_0 value. The decrease of B_T with increasing temperature arises from the thermal lattice expansion and the effects of the vibration entropy, which also influence the transition temperature T_c of the order-disorder transformation.

Secondly, we will show the results of the phase diagrams of Ni-Al binary alloys. The Ni-Al alloys are, as is well known, technologically important materials because of their peculiar temperature dependence of the strength properties. In particular, the $\gamma + \gamma'$ two phase field which is the basis of the γ' particle strengthened nickel-base superalloys. The γ phase is the disordered fcc nickel solid solution and the γ' phase is the $L1_2$ ordered Ni_3Al. In order to allow a direct comparison between the present results using the combined SMM-CVM method and those obtained by the conventional CVM by Sanchez et al. [33], we have used the same 8-4 type of Lennard-Jones potentials for the Ni-Al alloys. Using, the L.-J. potentials and SMM approach, we have calculated the thermal lattice expansions of Ni_cAl_{1-c} alloys, and presented the results in Fig.3. The equilibrium lattice constants of the alloys depend sensitively on the concentration of the alloys as well as on the temperature. The phase equilibrium between the fcc disordered phase and the $L1_2$ ordered structure can be approximately described by an Ising-type model using constant pair and/or many body interactions, as demonstrated for Cu-Au phase diagram. However, in the case of the Ni-Al system this approximation is only qualitatively correct since it predicts a narrower γ/γ' two phase region than is experimentally observed. The inclusion of the lattice constant dependence of the interaction energies by using a Lennard-Jones pair potential is of great importance and it leads to the quantitative agreement between the CVM and the experimental observations. In Figs.4a and 4b, we show the γ-γ' phase boundaries of Ni-Al alloy calculated by using the conventional CVM by Sanchez et al. [33], and those calculated by using the combined SMM-CVM method One can see in Fig.4 that the γ / γ' two phase region becomes wider by including the thermal lattice expansions and the anharmonicity of thermal lattice vibrations. The agreement with the experimental observations is further improved in our SMM-CVM calculations.

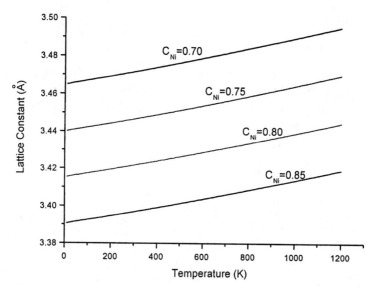

Figure 3. Temperature dependence of lattice constants of Ni_cAl_{1-c} alloys.

On the basis of the combined SMM-CVM approach, we have also investigated the phase diagrams of the phase separating binary alloys. For phase separating alloys, like CuAg and PdRu systems, we have calculated the phase separating diagrams using the electronic many body potentials [22,23]. The calculated phase separating diagrams of Cu-Ag alloy is shown in Fig.5. One can see in Fig.5 that the phase separating curve shows the asymmetric behavior, in contrast to the calculations based on the Ising-like Hamiltonian, and the transition temperatures T_c are reduced by ~20% compared to those of the quasi-harmonic calculations local volume relaxations [34,35].

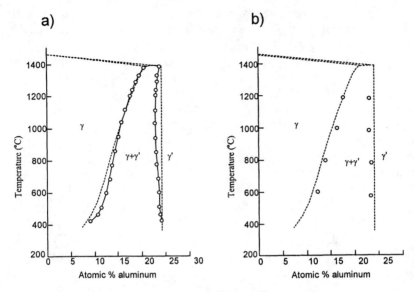

Figure 4. The calculated γ/γ' phase boundaries of Ni-Al alloys calculated with the conventional CVM (a) [Ref. 33], and those with the use of the combined CVM and SMM approach (b). The experimental boundaries are shown by dashed lines.

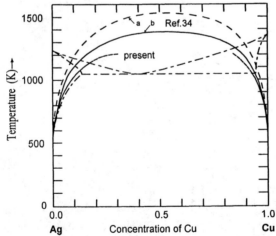

Figure 5. Phase separation diagrams of CuAg alloy calculated by the present SMM-TTR approach compared with the previous calculation by Sanchez et al. [34]; curves (a) and (b), without and with vibrational modes, respectively.

CONCLUSIONS

We have presented an analytic formulation to obtain the thermodynamic quantities of metals and alloys based on finite temperature moment expansion from statistical physics. The thermal lattice expansion coefficients are derived explicitly in terms of three characteristic parameters k, γ_1, and γ_2 for cubic crystals and four parameters, k, β, γ_1, and γ_2 for hcp crystals. This formalism accounts for the quantum mechanical zero-point vibrations as well as higher order anharmonic terms in the atomic displacements enabling us to calculate various thermodynamic quantities of metals and alloys over a wide temperature range. We can calculate the thermodynamic quantities efficiently and accurately by using these analytical formulae that account for the many-body electronic effects in metallic systems. The calculated thermodynamic quantities of metals are in good agreement with experimental results. We have also shown that the thermodynamic quantities of binary alloys, e.g., phase separation diagrams and order-disorder transition temperatures are significantly influenced by including the anharmonicity of thermal lattice vibrations.

REFERENCES

1. Van de Walle, G. Ceder, U.V. Waghmare, 1998. Phys. Rev. Lett., 80, 4911.
2. V. Ozolins, M. Asta, 2001. Phys. Rev. Lett., 86, 448.
3. Van de Walle, G. Ceder, 2002. Rev. Mod. Phys. 74, 11.
4. V.L. Moruzzi, J.F. Janak, K. Schwarz, 1988. Phys. Rev. B37, 790.
5. A.P. Sutton, 1989. Phil. Mag. 60A, 147.
6. M. Asta, R. McCormark, D. de Fontaine, 1993. Phys. Rev. B93, 748.
7. A.A. Maradudin, P.A. Flinn, R.A. Coldwell-Horsfall, 1961, Ann. Phys., 15, 337. A.A. Maradudin, P.A. Flinn, 1963. Phys. Rev. 129, 2529.
8. G. Leibfried, W. Ludwig, 1961. Theory of Anharmonic Effects in Crystals, Academic Press, New York.
9. J. Rosen, G. Grimvall, 1983. Phys. Rev. B27, 7199.
10. T. Yokoyama, Y. Yonamoto, T. Ohta, A. Ugawa, 1996. Phys. Rev. B54, 6921.
11. A. I. Frenkel, J. J. Rehr, 1993. Phys. Rev. B48, 585.
12. R.C. Shukla, E. Sternin, 1996. Phil. Mag. B74, 1.
13. R. Hardy, M.A. Day, R.C. Shukla, E.R. Cowley, 1994. Phys. Rev. B49, 8732.
14. R. C. Shukla, 1996. Phil. Mag. B74, 13, ibid., 1998. Phys. Stat. Sol. 205(b), 481.
15. N. Tang and V. V. Hung, 1988. Phys. Status Solidi B149, 511.
16. N. Tang, V.V. Hung, 1990. Phys. Status Solid., B162, 371.
17. Vu Van Hung, K. Masuda-Jindo, 2000. J. Phys. Soc. Jap., 69, 2067.
18. Vu Van Hung, H.V. Tich, K. Masuda-Jindo, 2000. J. Phys. Soc. Jap., 69, 2691.
19. D. Frenkel, A.J.C. Ladd, 1984. J. Chem. Phys. 81, 3188.
20. D. Frenkel, 1986. Phys. Rev. Lett., 56, 858.
21. A.P. Horsfield, A.M. Bratkovsky, M. Fearn, D.G. Pettifor, M. Aoki, 1996. Phys. Rev. B53, 12694.
22. V. Rosato, M. Guillope, B. Legrand, 1989. Phil. Mag. 59A, 321.
23. F. Cleri, V. Rosato, 1993. Phys. Rev. B48, 22.
24. R.C. Shukla, C.A. Plint, 1989. Phys. Rev. B40, 10337. (references therein).
25. N.I. Papanicolaou, G.C. Kallinteris, G.A. Evangelakis, D.A. Papaconstantopoulos, 2000. Comp. Mat. Sci., 17, 224.

26. P.E.A. Turchi, D. Mayou, J.P. Julien, 1997. Phys. Rev. B56, 1726.
27. Smithells, 1962. Metals Reference Book, Fourth Edition, 2.
28. M. Simersha, 1961. Acta Crystallogr. 14, 1259; 1962. Czech. J. Phys. B12, 858.
29. K.A. Gschneidner Jr., 1964. Solid. State. Phys. 16, 275. American Institute of Physics Handbook, 1982. 3rd ed., Eds. B.H. Billings et al., McGraw-Hill, NY.
30. J. Mei, J.W. Davenport, G.W. Fernando, 1991. Phys. Rev. B43, 4653.
31. D. Roy, A Manna, S.P. Sen Gupta, 1974. J. Phys. F: Metal Phys. 4, 2145, D.T. Keating and B.E. Warren, 1951. J. Appl. Phys. 22, 286.
32. F. Cleri, G. Mazzone, V. Rosato, 1993. Phys. Rev. B47, 14541.
33. J. M. Sanchez, J. R. Barefoot, R. N. Jarret, J.K. Tien, 1984. Acta Met., 32, 1519.
34. J. M. Sanchez, J. P. Stark, V. L. Moruzzi, 1991. Phys. Rev. B44, 5411.
35. T. Mohri, K. Terakura, T. Oguchi, K. Watanabe, 1988. Acta Metall. 36, 561.

Fabrication and Mechanical Properties of TiNi/Al Smart Composites

G. C. LEE
Research Institute of Industrial Science and Technology (RIST), Pohang, Korea

J. H. LEE
Department of Metallurgical Engineering, Dong-A University, Pusan, Korea

Y. C. PARK
Department of Mechanical Engineering, Dong-A University, Pusan, Korea

ABSTRACT

Al alloy matrix composite with TiNi shape memory fiber as reinforcement was fabricated by hot pressing to investigate microstructures and mechanical properties. The SEM and EPMA analyses showed that the composites had good interfacial bonding. The stress-strain behavior of the composites was evaluated between room temperature and 363K a function of pre-strain and we found the tensile stress at 363K was higher than that at room temperature. The tensile stress of the composite increased with increasing amounts of pre-strain and also depended on the volume fraction of fiber and heat treatment. The "smartness" of the composite is assigned because of the shape memory effect of the TiNi fibers which generate compressive residual stress in the matrix material when heated after being pre-strained. Finite element analysis was used to predict the mechanical properties of TiNi/Al6061 composite. The FEA results represented the experimental ones very well.

INTRODUCTION

A "smart" structural system is aimed at integrating functions, structures, and information in order to add judgement and learning functions to the material and structural system [1,2]. The process aims to produce self-repairing materials and structures, the ultimate goal being to apply the method to living materials and structural systems to achieve energy, resource savings and further safety improvements.

This study involved basic research to establish a "smart" structural system. Smart composites were fabricated and material properties evaluated. Smart composites with TiNi shape memory fibers (SMFs) as reinforcement were been fabricated by hot pressing [3]. 6061Al alloy was used for the composite matrix and the SMF was loaded at room temperature such that the fibers experienced phase transformation from austenitic to martensitic phases. Then they were heated to induce the reverse transformation. This shape memory effect (SME) in the matrix is the principal factor that enhances the tensile properties of a smart composite [4–6]. The martensite transformation temperature depends on the chemical composition of the alloy, the method of fabrication and the heat treatment.

The objectives of this work were to study:

1) microstructures and mechanical properties of 6061Al matrix composite containing TiNi fiber;
2) best processing conditions for "smart" composite by hot pressing;
3) effects of static strength of the composite with compressive stress formed by this SME; and
4) effects of pre-strain and volume fraction of fiber on the strength of the composite.

Finite element analysis (FEA) was used to predict the mechanical properties of the TiNi/Al6061 composite. ANSYS (v. 5.6) was used to carry out the finite element analysis. The FEA results were compared with experimental results to verify accuracy.

EXPERIMENTAL METHOD

To fabricate the smart composite, TiNi fibers (Ti-50.0 at.% Ni) of 500 μm diameter (made by Kantoc Ltd., Japan) and 6061 Al-alloy sheet metal were used. A 250-ton hydraulic hot press was designed to reinforce the Al-alloy matrix by placing a continuous TiNi shape memory alloy (SMA) fiber phase within the composite. A chamber was attached to the center of the hot press, to control the temperature. A fixture was designed to hold and array the TiNi shape memory alloy strings as shown in Figure 1. Without the fixture, it was difficult to manufacture a uniform TiNi/Al 6061 smart composite with various fiber volume fractions.

To investigate the effect of volume fraction and pre-strain, tensile test specimens (JIS 6) with several volume fractions were made by hot pressing under the optimum processing conditions. The TiNi fibers were fixed inside 150mm x 22mm x 1mm 6061 Al sheets at constant spacing and these sheets were stuck between the press dies as shown in Figure 2.

Figure 1. The fixture for fixing and arraying SMA fibers.

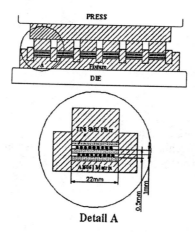

Figure 2. Schematic diagram of hot pressing.

To optimize the processing condition of composites, an orthogonal array was used. Three parameters that could affect the strength of TiNi/Al 6061 composite were considered. The parameters and levels were as follows:

Temperature : A1(783K), A2(803K), A3(823K)
Pressure : B1(40MPa), B2(50MPa), B3(60MPa)
Time : C1 (10min.), C2(20min.), C3(30min.)

The processing temperature was measured between the two Al sheets and the press pressure was calculated using the size of Al sheet before pressing. Since the surface of Al sheets can be oxidized easily in air, the hot press was preformed in argon. Tensile test specimen of flat bar type were cut out from freeform as shown in Figure 3. To determine the optimum conditions of the composite, bonding state between 6061 Al matrix and TiNi fibers was observed using SEM images. Also, EPMA analysis was used to evaluate chemical composition of diffusion layer. Under the optimum conditions, TiNi/Al6061 composites with three different volume fractions of TiNi fibers(3.2, 5.2 and 7pct.), were fabricated.

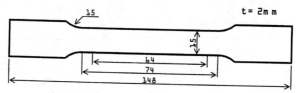

Figure 3. The shape and dimensions of tensile test specimen.

To enhance the mechanical properties of the matrix, T6 heat treatment was used. Solution treatments were conducted at 793K and 813K in air for 1 hour followed by quenching in ice-water. To enhance mechanical properties of the matrix, specimens were heat-treated at 448K in air for various times followed by a further water quenching procedure. Figure 4 shows Vickers hardness as a function of aging time. As can be seen, the hardness of materials heat treated at 813K is higher than that at 793K. For specimens heat-treated at 813K, prolonged heat treatments at 448K for more than 5 hours does not change the hardness significantly.

As a final processing step, the specimens were pre-strained (1, 3 and 5%) at room temperature in air. Mechanical testing was conducted on the pre-strained specimens at 363K. Specimen temperature was measured on the surface of the specimen using a thermocouple.

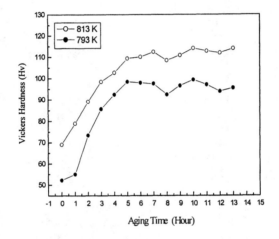

Figure. 4. Vickers hardness as a function of aging time.

FINITE ELEMENT ANALYSIS

Thermomechanical behavior and mechanical properties of Al 6061 matrix composite with shape memory alloy fiber were studied by finite element analysis using a commercial software application, ANSYS (v. 5.6). Figure 5 shows the design concept of the TiNi/Al6061 composite. To evaluate composite strength using ANSYS, the design concept of the smart composite was simulated. Step 1 is for pre-strain applied for displacement control. Displacement was changed to load, which was slowly removed for pre-strain. For finite element analysis, 1%, 3%, 5%, and 7% pre-strain was considered. To express the phase transformation (from the austenitic phase to the martensitic phase) at step 3, the temperature dependent elastic modulus and thermal expansion coefficient were used. Finally, stress-strain curves were obtained as displacement control in step 4.

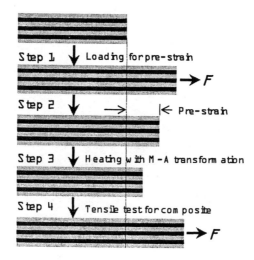

Figure 5. Design concept of TiNi/Al6061 composite.

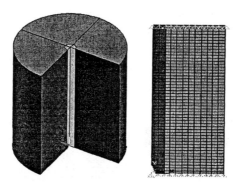

Figure 6. FE model (2D axisymmetric) and Boundary condition.

In this paper, the surface between the 6061 Al matrix and TiNi fiber was assumed perfectly bonded and the analytical model was assumed to be a 2D axisymmetric model of one fiber surrounded by the matrix. Figure 6 shows a 3D solid model and a 2D axisymmetric model while Table 1 shows the applied material properties of the TiNi fiber and the 6061 Al matrix for the finite element analysis.

Table 1. Mechanical properties of TiNi and Al6061.

Material Property	Al 6061		TiNi	
	R.T.	363K	R.T.	363K
Elastic Modulus [GPa]	70	70	41	83
Poisson's Ratio	0.33	0.33	0.43	0.43
Coefficient of Thermal Expansion [$\times 10^{-6}$ /K]	23.4	23.4	6.6	11
Yield Stress [MPa]	275	260	280	710

Results and Discussion
Optimum Processing Conditions

As shown in Table 2, the L9 experiment consists of nine rows and four columns where each row corresponds to a particular experiment and each column identifies settings of a design parameter. From SEM images and EPMA analysis, we can determine the optimum fabrication condition as test number 4 - $A_2 B_1 C_3$ (803K, 40MPa, 30min). Figure 7 shows a cross-sectional view of the composites under these conditions. The diagram shows strong bonding was formed between the fiber and the Al 6061 matrix under the optimum processing condition and the bond line between the each Al sheet has disappeared. Figure 8 shows chemical composition changes through an interface between matrix and fiber by EPMA analysis. According to this result, the diffusion layer was about 4μm thick.

Table 2. $L_9(3^4)$ orthogonal array.

Test number	column number				test condition	data
	1	2	3	4		
1	1	1	1	1	$A_1B_1C_1$	bad
2	1	2	2	2	$A_1B_2C_2$	bad
3	1	3	3	3	$A_1B_3C_3$	bad
4	2	1	2	3	$A_2B_1C_3$	Good
5	2	2	3	1	$A_2B_2C_1$	Good
6	2	3	1	2	$A_2B_3C_2$	bad
7	3	1	3	2	$A_3B_1C_2$	bad
8	3	2	1	3	$A_3B_2C_3$	bad
9	3	3	2	1	$A_3B_3C_1$	bad
	a	b	a b	a b^2		
location	A	B	E	C		

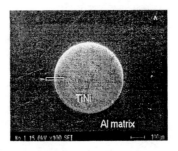

Figure 7. SEM micrograph of the TiNi/Al 6061 composite.

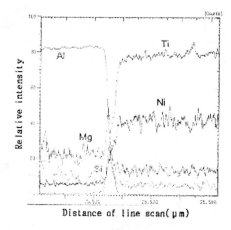

Figure 8. EPMA analysis of the interface.

Mechanical Properties of TiNi/Al6061 Composite

To evaluate the volume fraction effect on tensile strength of composites, tensile stress was measured for each volume fraction at two temperatures (room temperature between Ms and As temperature of TiNi and 363K above A_f temperature) as shown in Figure 9.

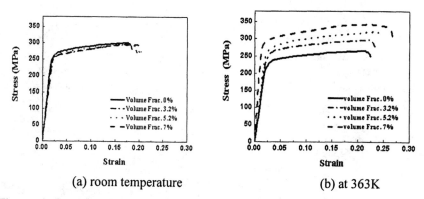

(a) room temperature   (b) at 363K

Figure. 9. Stress-strain curve of TiNI/Al6061 composite at room temperature and 363K.

At room temperature (298K), tensile stress does not increase with increasing volume fraction of the fiber. This is because, in pseudoelasticity region at room temperature, 6061 Al matrix has higher strength than TiNi SMA fiber. However, at 363K, composites with higher volume fraction of fiber has higher tensile stress than those with low volume fraction of fiber. This phenomena can be explained in terms of dependence of strength of SMA fiber on temperature. Since tensile strength of TiNi SMA fiber at 363K is twice higher than that of room temperature, it shows the effect of SMA fiber reinforcement.

As can be seen in Figure 10, the stress-strain behavior of the composites was evaluated at 363K as a function of pre-strain. The results show that the tensile stress of this composite increased as the amount of pre-strain increased. Tensile stress also depended on the volume fraction of fiber. The "smartness" of the composite is due to the shape memory effect of the TiNi SMA fiber which generates compressive residual stress in the matrix material when heated after being pre-strained.

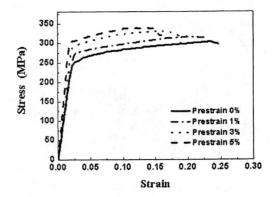

Figure 10. Stress-strain curve of TiNi/Al6061 composite as a function of pre-strain at 363K(V_f =5.2pct.)

Figure 11 shows the change in tensile strength as a function of pre-strain at room temperature and 363K. The difference between A and B is due to the effect of the fiber reinforcement while for B and C, the difference is the result of the shape memory effect of the SMA fiber. The differences among C, D and E are due to the volume fraction of fiber.

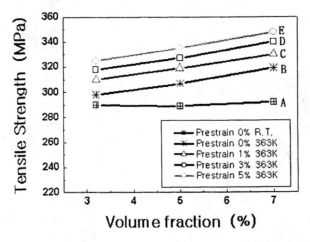

Figure 11. Relationship between volume fraction and tensile strength as a function of prestrain.

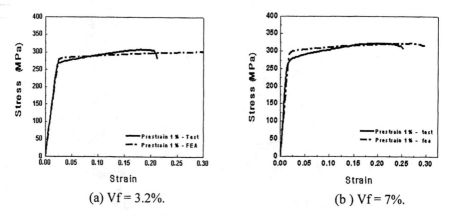

(a) Vf = 3.2%.   (b) Vf = 7%.

Figure 12. Stress-strain curves of TiNi/Al6061 composite obtained through experimental method and FEA.

Figure 13 shows the relationship between tensile strength and pre-strain evaluated by finite element analysis at 363K. The data show that the tensile stress of this composite increases as the amount of pre-strain increases and that composites with higher volume fraction of fiber have higher tensile stress than those with low volume fractions. As shown in Figure 13, the tensile strength of TiNi/Al6061 composite does not increase above 5 % pre-strain.

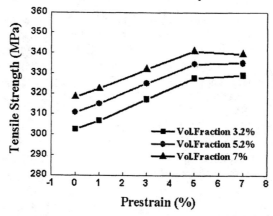

Figure 13. The relationship between pre-strain and tensile strength as a function of volume fraction at 363K using finite element analysis.

Figure 14 shows that the experimental results and FE results of TiNi/Al6061 composite with 3.2 % and 5.2 % volume fractions of fiber correspond to each other extremely well. This observation verifies that finite element analysis is a good method to use to predict the strength of smart composite materials.

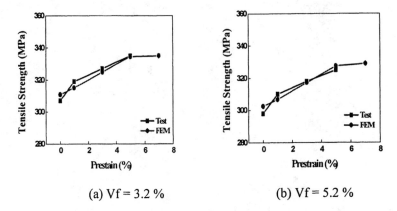

(a) Vf = 3.2 %   (b) Vf = 5.2 %

Figure 14. FE analysis and experimental results of the tensile strength in the composite as a function of pre-strain of TiNi fiber.

CONCLUSION

Optimum processing conditions for hot pressing were determined in order to fabricate TiNi shape memory alloy fiber-reinforced 6061 Al-matrix composites. It was found that the optimum conditions for hot pressing are at a temperature of 803K and a pressure of 40MPa for 30 minutes duration. To investigate tensile strength as a function of volume fraction of fiber, tensile tests were performed at room temperature and 363K. At room temperature, as volume fraction increased, a lower tensile strength was observed. On the other hand, as volume fraction increased a higher tensile strength was observed at 363K and the tensile strength at 363K was higher than that of room temperature. This can be explained in terms of fiber reinforcement. The tensile stress of the composite increased as the amount of pre-strain was increased at 363K. This seems to be due to the shape memory effect of the TiNi SMA fiber which generates compressive residual stress in the matrix material when heated after being pre-strained. In this study, the effect of fiber reinforcement and shape memory effect of fiber were verified quantitatively. The FEA results represented the experimental ones very well. We know that the FEA is a good method to predict the mechanical properties of TiNi/Al6061 composite.

REFERENCES

[1] Y. C. Park, Y. Furuya, 1992. Thermal Cyclic Deformation and Degradation of Shape Memory Effect in Ti-Ni Alloy, Nondestr. Test. Eval., 4(8), 541–554.

[2] Y. Furuya, 1995. Design and Experimental Verification of Intelligent Materials Using Shape Memory Alloy, Proceeding of the International Symposium on Microsystems, Intelligent Materials and Robots, Sendai, Japan, 313–318.

[3] K. Hamada, J.H. Lee, K. Mizuuchi, M. Taya, K. Inoue, 1998. Thermomechanical Behavior of TiNi Shape Memory Alloy Fiber Reinforced

6061 Aluminum Matrix Composite, Metallurgical and Materials Trans., 29A, 1127–1135.

[4] H. Ehren Stein, 1986. Production and Shape Memory Effect of Nickel-Titanium, Proceeding of the International Conference on Material Transformation, 1083–1086.

[5] Y. Hunag, G. Yang, P. He, 1985. The Investigation of Internal Friction Electric Resistance and Shape Change in NiTi Alloy During Phase Transformation, Scripta Metallurgica, 19, 1033–1038

[6] M Taya, Y. Furuya, Y. Yamada, R. Watanabe, S. Shibata, T. Mori, 1993. Proceedings Smart Material, Ed. V.K. Varadam, SIM. Vol. 1916, 373.

Mechanochemical Doping to Prepare a Visible-Light Active Titania Photocatalyst

Q. ZHANG, J. WANG, S. YIN, T. SATO and F. SAITO
IMRAM, *Tohoku University, 2-1-1, Katahira, Aoba-ku, Sendai 980-8577, Japan*

ABSTRACT

When titania is doped with a nonmetallic element such as sulphur and nitrogen, it has been reported to demonstrate visible-light activity. However, it is not always easy to carry out the doping operation. The authors have developed a new method to dope an element into a titania sample by grinding with sulphur/nitrogen-containing materials followed by calcination at low temperature. This is called mechanochemical doping which allows the titania sample to possess two absorption edges around 400 and 540 nm in visible light of wavelength. Under irradiation of visible light of wavelength over 510nm, titania doped with sulphur/nitrogen exhibits excellent photocatalytic ability for nitrogen monoxide destruction.

INTRODUCTION

Titania (TiO_2) nanocrystals have attracted increasing attention due to its wide applications in many fields, especially in photochemical research [1,2]. Under light irradiation, titania photocatalyst can decompose many polluting substances, such as poisonous nitrogen monoxide in the atmosphere and/or organic pollutants in water. Although three titania polymorphs, rutile, anatase and brookite occur in nature, anatase and rutile titanias are mainly used as photocatalysts. It is reported that anatase shows higher photocatalytic activity than rutile due to its higher chemical activity [3,4]. This enhancement is ascribed to differences in the Fermi level and the extent of surface hydroxylation of the solid [4]. In addition, rutile usually shows harder agglomeration and larger particle size than does anatase since it is normally prepared by calcining anatase at high temperature. A problem in the application of titania as a photocatalyst is the large band gap energy, for example anatase shows photocatalytic activity only under ultraviolet (UV) light irradiation of wavelength below 387 nm which corresponds to the band gap energy of 3.2 eV. It is well known that solar energy contains only about 5% of UV light and the large part of its energy lies in the visible light. In order to use solar energy efficiently, it is necessary to develop a visible-light reactive photocatalyst. Asahi et al. [5] reported that nitrogen-doped titania with high visible light photocatalytic activity could be prepared by sputtering N_2 (40%) into a TiO_2 target in Ar gas atmosphere followed by annealing in N_2 gas at 823 K for 4 h. It was observed that the color of such nitrogen-doped titania was yellow. Ihara et al. [6] reported that visible-light-active titania could also be prepared by a RF plasma treatment.

Recently, mechanochemical technology has shown promise to synthesize functional materials, by modifying solid particle surfaces, processing and recycling wastes, etc. [7,8]. In each case of mechanical stressing, the solid induced formation of a fresh oxygen-rich surface resulting in electron transfer from the O^{2-} ion on the oxide surface to other organic substances. As a result, this operation leads to destruction of weak bonding in an organic substance and formation of new bonds between oxide and nonmetallic element.

In this study, a rutile titania powder was subjected to mechanochemical-doping with nitrogen/sulphur to prepare a visible-light active titania photocatalyst. It was found that both nitrogen- and sulphur-doped photocatalysts show excellent photocatalytic ability for nitrogen monoxide destruction.

EXPERIMENTAL INFORMATION

A commercial titania powder, P-25 (titania standard sample), obtained from Nippon Aerosol Co., Japan and TiO_2 chemical agent (abbreviated as W-TiO_2) from Wako Pure Chemical Industries, Ltd., Japan were used as raw titania materials. Generally the P-25 sample is much more expensive among the titania samples due to its preparation process as compared to W-TiO_2. Hexamethylenetetramine (HMT) and urea (Kanto Chem. Co. Inc., Japan) served as nitrogen sources. Sulphur powder (Wako Pure Chemical Industries, Ltd., Japan) was the sulphur source.

Mechanochemical doping with nitrogen/sulphur to prepare a visible-light active titania photocatalyst was done as follows: The P-25 powder was mixed with 5 wt% HMT or 10 wt% urea before being introduced into a zirconia pot, described later. Sulphur at 10 wt% was mixed with the W-TiO_2. A planetary ball mill (Pulverisette-7, Fritsch, Germany) was used to grind these mixtures. Seven zirconia balls of 15 mm diameter and 4 grams of mixture were introduced into a zirconia pot of 45 cm^3 inner volume. The grinding was operated at 700 rpm for different periods of time. The ground samples with nitrogen-containing compounds were calcined in air at 673 K for 1h. The ground samples with sulphur were calcined at 673 K for 1h with an Ar-gas environment.

The phase constitution of the products was determined by X-ray diffraction analysis (XRD, Shimadzu XD-D1). The absorption edges and band gap energies of the products were determined from the onset of diffuse reflectance spectra of the sample measured using an UV-VIS spectrophotometer (Shmadzu UV-2000). The bonding state of sulphur/nitrogen to the titanium sample was investigated by X-ray photoelectron spectroscopic analysis (XPS, PHI 5600 ESCA system, Ulvac-Phi Inc., Japan). The photocatalytic activity for nitrogen monoxide oxidization was determined by measuring the concentration of NO gas at the outlet of the reactor (373 cm^3) during photo-irradiation at a constant flow of 1 ppm NO-50 vol% air mixed (balance N_2) gas (200 cm^3/min). The samples were placed in a hollow of size 20×15×0.5 mm on a glass holder plate and set in the center of the reactor. A 450-W high-pressure mercury lamp was used as the light source, where the light wavelength was controlled by selecting various filters, i.e., Pyrex glass for cutting off the light of wavelength < 290 nm, Kenko L41 Super Pro (W) filter < 400 nm and Fuji, triacetyl cellulose filter < 510 nm.

RESULTS AND DISCUSSION
N-doped Samples

Figure 1 shows the XRD patterns of the starting P-25 powder sample and the samples prepared by mechanochemical doping. The P-25 sample consists of ~70 wt% anatase phase and ~30 wt% rutile phase. After planetary ball milling with urea and HMT at 700 rpm for 1 h, the peak intensity of anatase greatly decreased and the samples mainly consisted of rutile. Usually, anatase transforms to rutile at high temperature such as 973 K in air. These results indicate that high mechanical energy accelerates the phase transformation of anatase to rutile. TG-DTA analysis indicated that residual organic substances burned out at about 673 K, thus the prepared samples were calcined at this temperature for 1 h. After calcination, the crystallinity of rutile increased.

Figure 2 shows the diffuse reflection spectra of the P-25 powder sample and the samples prepared by mechanochemical reaction. The P-25 sample possessed an absorption edge at 405 nm corresponding to a band gap of 3.1 eV (Fig. 2-a). No change was observed after calcination of the P-25 sample in air at 673 K. The color of the milled powder sample prepared by planetary milling of P-25 TiO_2-10% urea mixture was yellow-green, suggesting the formation of nitrogen doped titania during the milling. It is obvious that nitrogen-doped rutile titania prepared by the present study exhibits two absorption peaks with the onset of diffuse reflectance spectra of 402 nm (3.1 eV) and 543 nm (2.3 eV) (Fig. 2b)). While color of the powder sample after calcination was bright-yellow, and its absorption became large and the second absorption edge shifted to around 562 nm (2.2 eV) (Fig. 2-(c)). Instead of urea, when the P-25 powder was subjected to planetary milling with 5% HMT, the color of product was black-gray and it also showed two absorption edges at 400 nm (3.1 eV) and 520 nm (2.4 eV) (Fig. 2d). The color of the powder after calcination was dark yellow, and its absorption became large and the second absorption age shifted to around 542 nm (2.3 eV) (Figure 2-(e)). The black-grayish color is thought to be decomposition of HMT during grinding. It suggests that C-H and C-N bonds in HMT were destroyed to form concentrated carbon during the planetary milling. After calcination at 673 K, the carbon content was decreased, and the color of powder was dark yellow which suggests the formation of nitrogen doped TiO_2.

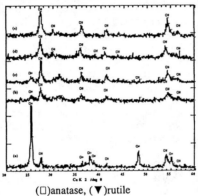

(□)anatase, (▼)rutile

Figure 1. XRD patterns of the prepared samples: (a) the P-25 titania sample; (b) the P-25 -10 wt% urea mixture ground at 700 rpm for 1 h; (c) the (b) sample calcined in air at 673 K for 1h; (d) the P-25 - 5 wt% HMT mixture ground at 700 rpm for 1 h; (e) the (d) sample calcined in air at 673 K for 1h

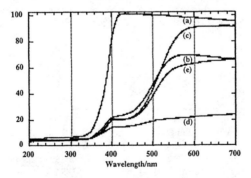

Figure 2. Diffusion reflectance spectra of the prepared samples: a) the P-25 titania sample; (b) the P-25 -10 wt% urea mixture ground at 700 rpm for 1 h; (c) the (b) sample calcined in air at 673 K for 1h; (d) the P-25 - 5 wt% HMT mixture ground at 700 rpm for 1 h; (e) the (d) sample calcined in air at 673 K for 1h.

Figure 3 shows the relationship between wavelength and photocatalytic ability of N-doped samples for oxidation of nitrogen monoxide. Note that the P-25 sample absorbs light energy at a level above the band gap energy to generate electron/hole pairs. In the presence of oxygen, electrons in the conduction band are trapped rapidly by molecular oxygen to form $\cdot O_2^-$, which then generate highly active $\cdot OOH$ radicals [9,10]. The NO reacts with these oxygen radicals, molecular oxygen, and water to produce HNO_2 or HNO_3. It is obvious that the yellow nitrogen-doped titania sample possesses visible-light photocatalytic activity. Under visible light of wavelength over 510 nm, ~27% of the NO is continuously removed by the nitrogen-doped titania sample prepared by planetary milling with 5% HMT followed by calcination in air at 673 K. The powder prepared by planetary milling with 10% urea showed lower activity than those with 5% HMT.

On the other hand, photocatalytic activity under visible light irradiation of the powders as-prepared without calcination was negligibly small. This may be due to depression of NO adsorption by the remaining reaction products such as CO_2, NH_3, carbon, organic molecules, etc. In addition, as expected no photocatalytic activity was observed for the P-25 sample because of its large band gap energy of 3.1 eV. In the case of light irradiation at wavelengths over 400 nm, similar results were seen, i.e., photocatalytic activity of the powder samples after calcination (Figure 3(c) and Figure 3(e): 30–33%) was higher than before calcination (Figure 3(b) and Figure 3(d): 10-12%). Under near-UV light irradiation at wavelengths over 290 nm, ~43-53% of nitrogen monoxide was removed. The powder sample activities after calcination showed almost the same level as that of the P-25 sample, but slightly higher than those before calcination. When the light was turned off, the nitrogen at the reactor outlet returned to an initial concentration of 1 ppm within 10 min., indicating light energy is necessary to oxidize nitrogen monoxide.

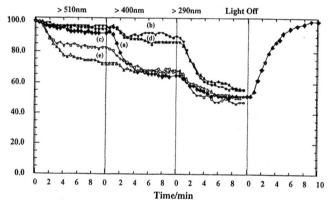

Figure 3. Relationship between wavelength and photocatalytic ability for the oxidation of nitrogen monoxide of the prepared samples: a) the P-25 titania sample; (b) the P-25 -10 wt% urea mixture ground at 700 rpm for 1 h; (c) the (b) sample calcined in air at 673 K for 1h; (d) the P-25 - 5 wt% HMT mixture ground at 700 rpm for 1 h; (e) the (d) sample calcined in air at 673 K for 1h.

S-doped Samples

Figure 4 shows the relationship between the wavelength of light and the photocatalytic ability of the S-doped titania samples for oxidation of nitrogen monoxide. In the present study, the 2h ground sample without addition of sulphur was compared to the S-doped samples. It is clear that the ground $W-TiO_2$ exhibits slightly photocatalytic activity after irradiation of light with a wavelength over 510 nm, i.e., it shows a slight decrement in NO concentration. When the titania ($W-TiO_2$) sample was ground with sulphur for 20 min., about 20% NO was been removed if the wavelength was over 510 nm. After prolonged grinding for up to 120 min, decomposition of NO gas was nearly 40%, demonstrating very high photocatalytic ability under visible-light irradiation. All the prepared samples exhibit similar high photocatalytic ability under ultraviolet light, similar to the typical property of titania samples.

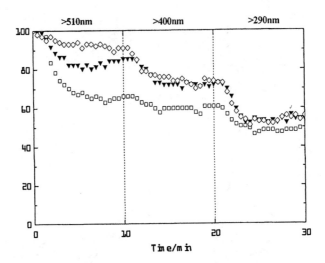

Figure 4. The relationship between light wavelength and the photocatalytic ability of S-doped TiO_2 for the oxidation of nitrogen monoxide. ◊: 2hrs ground W-TiO_2 ▼: S-doped sample prepared by 20 min grinding and calcinations □: S-doped sample prepared by 2 hrs grinding and calcination

Figure 5 shows the S2p spectrum of the S-doped sample prepared by 2-hours grinding and calcination. The S2p state has a broad peak because of the overlap of the split sublevels, the 2p3/2 and 2p1/2 states, with separation of 1.2 eV by spin-orbit coupling [11]. As observed in the spectrum, two peaks appear around 168 eV and 161 eV respectively. Several reports suggest that the peak positioned around 168 eV is due to absorbed SO_2 molecules on the TiO_2 surface while the peak around 163-160 eV probably results from bonding between S and Ti atoms. The XPS analysis has clearly shown the formation of sulphur-doping of a titania sample. It is also worth noticing that the co-existing of S-Ti bonding with SO_2 suggests a possible doping pathway, namely substitution of oxygen from TiO_2 by sulphur. Furthermore, the substituted oxygen tends to combine with other sulfur to form SO_2 absorbing on the sample surfaces.

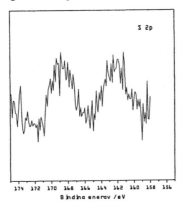

Figure 5. XPS spectrum of the S-doped sample prepared by 2 hrs grinding and calcination.

CONCLUSION

We have developed a new method to dope titania by grinding it with materials containing sulphur and nitrogen followed by calcination at low temperature. This mechanochemical doping with sulphur/nitrogen allows the titania sample to achieve two absorption edges around 400 and 540 nm within the visible light wavelength range. In addition, titania doped with sulphur/nitrogen exhibits excellent photocatalytic ability for nitrogen monoxide destruction under irradiation of visible light of wavelength above 510nm. Mechanochemical doping has the possibility to prepare other types of oxide particles doped with different kinds of nonmetallic elements besides nitrogen/sulphur.

REFERENCES

[1] A. Fujishima and K. Honda, 1972. Electrochemical photolysis of water at a semiconductor electrode Nature, 238, 37.
[2] B. O'Regan and M. Gratzel, 1991. A low-cost, high-efficiency solar-cell based on dye-sensitized colloidal TiO_2 films, Nature, 353, 737–740.
[3] J. Augustynski, 1993. The role of the surface intermediates in the photoelectrochemical behavior of anatase and rutile TiO_2, Electrochem. Acta, 38, 43–46.
[4] R. I. Bickley, T. Gonzalez-Carreno, J. S. Lees, L. Palmisano, and R. J. Tilley, 1991. A structural investigation of titanium-dioxide photocatalysts, J. Solid State Chem., 92, 178–190.
[5] R. Asahi, T. Morikawa, T. Ohwaki, K. Aoki, and Y. Taga, 2001. Visible-light photocatalysis in nitrogen-doped titanium oxides, Science, 293, 269–271.
[6] T. Ihara, M. Miyoshi, M. Ando, S. Sugihara, and Y. Iriyama, 2002. Preparation of a visible-light-active TiO2 photocatalyst by RF plasma treatment, J. Mater. Sci., 36, 4201–4207.
[7] T. Ikoma, Q. Zhang, F. Saito, K. Akiyama, S. Tero, and T. Kato, 2001. Radicals in the mechanochemical dechlorination of hazardous organochlorine using CaO nanoparticles, Bull. Chem. Soc. Jpn., 74, 2303–2309.
[8] J. Lee, Q. Zhang, and F. Saito, 2001. Mechanochemical synthesis of lanthanum oxyfluoride by grinding lanthanum oxide with poly-vinyliden-fluoride, Ind. Eng. Chem. Res., 40, 4785–4788.
[9] S. Yin and T. Sato, 2000. Synthesis and photocatalytic properties of fibrous titania prepared from protonic layered tetratitanate precursor in supercritical alcohols, Ind. Eng. Chem. Res., 39, 4526–4530.
[10] S. Yin, D. Maeda, M. Ishitsuka, J. Wu, and T. Sato, 2002. Synthesis of $HTaWO_6$/(Pt,TiO_2) nanocomposite with high photocatalytic activities for hydrogen evolution and nitrogen monoxide destruction, Solid State Ionics, 151, 377–383.
[11] T. Umebayashi, T. Yamaki, H. Itoh, and K. Asai, 2002. Band gap narrowing of titanium dioxide by sulfur doping, Appl. Phys. Lett., 81(3), 454–456.

Influence of Mean Interatomic Distance on Magnetostriction of Fe-Pd Alloy Films

H. YABE and Y. NISHI

Department of Materials Science, Tokai University, 1117 Kitakaname, Hiratsuka, 259-1292, Japan

ABSTRACT

Based on theoretical equation suggested by Kittel, the magnetostriction was deduced to be mainly dominated by inter-atomic distance. Thus, influences of inter-atomic distance on magnetostriction were investigated for Fe-Pd alloy films, which were prepared by DC magnetron sputtering process. As results, the magnetostriction depended on reduction ratio of inter-atomic distance.

INTRODUCTION

The giant magnetostrictive $Tb_{0.3}Dy_{0.7}Fe_2$ films have been expected to apply the sensitive smart remote actuators [1–3]. However, these films exhibit high electromagnetic noise and low corrosion resistance. To solve these problems, it is important to develop a new magnetostrictive alloy that shows high corrosion resistance. Recently, the giant magnetostriction of Fe-Pd alloy films which shows high magnetostrictive susceptibility and high corrosion resistance, has been studied [4–7]. To obtain large magnetostriction, a theoretical examination of magnetostriction was carried out [7]. Magnetostriction λ, uses the coercive force H_c, the saturation magnetization I_s, and the inter-atomic distance Δa ts parameters, as follows [8]:

$$\lambda = C (H_c I_s / \Delta a) \qquad (1)$$

The constant parameter C is expressed by the following equation.

$$C = (G/\pi)(\delta/l) \qquad (2)$$

Here, G is a proportional constant, δ is the width of the magnetic wall, and l is the length of the magnetic domain. In order to examine the effect of each factor on the magnetostriction λ, Eq. 1 is calculated by the total differential function at H_c, I_s, and Δa.

$$\Delta\lambda = (\partial\lambda/\partial I)_{H=Hc} + (\partial\lambda/\partial\Delta a)_{H=Hc, I=Is} = (\partial\lambda/\partial H)_{I=Is} \qquad (3)$$

Finally, $\Delta\lambda$ is expressed by the following equation.

$$\Delta\lambda = C'\{(1/\Delta a)H_c - (H_c I_s)/(\Delta a)^2 + (I_s/\Delta a)\} \qquad (4)$$

where C' is a constant.

Based on the theoretical Eq. 4, the magnetostriction λ was deduced to be dominated by inter-atomic distance Δa. In order to confirm this theory, the influences of inter-atomic distance on magnetostriction were investigated for Fe-Pd alloy films which were prepared by a DC magnetron sputtering process.

EXPERIMENTAL TESTWORK

Preparation and evaluation of the Fe-Pd alloy films were carried out as follows. Fe-Pd films were prepared using a DC magnetron sputtering process as shown in Figure 1.(a) [6,7]. Chemical composition of the Fe-Pd film was controlled by changing the amount of Pd on the Fe target (see Figure 1.(b)) [6,7]. The base pressure was less than 5.5×10^{-5} Pa and the leakage rate was 5.0×10^{-7} Pa·m³/s [6,7]. The sputtering conditions were 1.1×10^{-1} Pa Ar gas pressure with 200 W DC sputtering power and 3600 s sputtering time [6,7]. The Fe-Pd film was about 3-5 µm in thickness on an Si-substrate [6,7]. Film composition was analyzed by energy-dispersive X-ray spectroscopy (EDS; JSM-6301F, JEOL) to be Fe-43.8at%Pd, Fe-60.3at%Pd and Fe-66.9at%Pd. The highly (111)-oriented structure was analyzed by thin film X-ray diffraction (TF-XRD; X'Part-MRD, PHILIPS). The magnetic properties of the film were characterized at room temperature using a vibrating sample magnetometer (VSM; Model BHV-55, RIKEN). The in-plane magnetostriction measurements were made on an optical cantilever system with a sensitivity of about 1×10^{-7} in a maximum magnetic field of ± 3 kOe [6,7]. Young's modulus values for the films and substrates were obtained from load-indentation curves obtained by a nano indenter (nano indenter; ENT-1100a, ERIONIX) [6,7,9].

RESULTS AND DISCUSSION

From TF-XRD, the sample structure was determined as face centered cubic (FCC). The strong X-ray peak of the (111) plane was observed for each sample [6,7]. From the TF-XRD results, the lattice constants were calculated. Table 1 shows these values for the Fe-Pd alloy films, together with literature data for solution treated austenite iron [10], Fe-30at%Pd alloy film [11], Fe-50at%Pd alloy [12] and pure Pd [13].

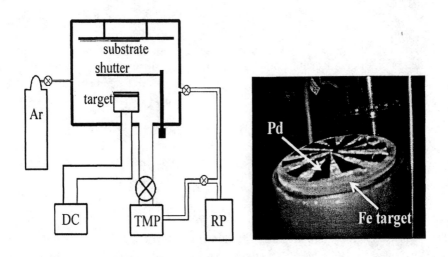

Figure 1. Schematic diagram of DC magnetron sputtering process.

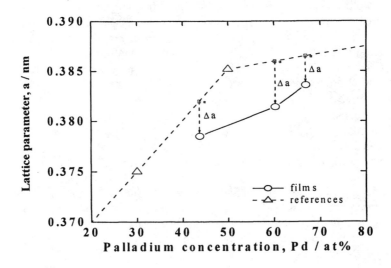

Figure 2. Lattice parameter plotted against content ratio of Pd to Fe in Fe-Pd alloy film and solution treated Fe-Pd alloy.

Table 1. Lattice constants of prepared Fe-Pd alloy films together with literature data of lattice constants of solution treated Fe-Pd alloys.

Composition	Shape	Preparetion	Lattice Constant	Ref.
Fe-43.8at%Pd	film	as depo.	0.3785 nm	-
Fe-60.3at%Pd	film	as depo.	0.3814 nm	-
Fe-66.9at%Pd	film	as depo.	0.3836 nm	-
Austenite iron	powder	*S.T.	0.360 nm	[10]
Fe-30at%Pd	film	*S.T.	0.375 nm	[11]
Fe-50at%Pd	powder	*S.T.	0.3852 nm	[12]
pure Pd	powder	*S.T.	0.38898 nm	[13]

* S.T. = Solution Treated

Table 2. Interatomic distance Δa of prepared Fe-Pd alloy films.

Composition	Inter-atomic Distance Δa
Fe-43.8at%Pd	- 0.0035 nm
Fe-60.3at%Pd	- 0.0045 nm
Fe-66.9at%Pd	- 0.0028 nm

Lattice parameter plotted against content ratio of Pd to Fe in Fe-Pd alloy film and solution treated Fe-Pd alloy are shown in Figure 2. The lattice constants of prepared film device are smaller than that of the literature data. The differences in lattice constants of sample and literature data are defined as Δa and this interatomic distance are summarized in Table 2.

The in-plane magnetostriction $\lambda_{//}$ of the film was obtained from measurements of the bending of a rectangular cantilever consisting of the film and substrate [5–7]. The $\lambda_{//}$-$H_{//}$ curve of the Fe-Pd alloy films show high magnetostrictive susceptibility from the earth's magnetic field of 0.5 Oe to $\pm$1 kOe [6,7]. Almost all film devices are saturated in the applied magnetic field of 1 kOe [6,7]. The Fe-60.3at%Pd alloy film has the largest magnetostriction above 500 ppm [6,7]. The saturated magnetostriction values of the Fe-Pd alloy films are summarized in Table 3. The saturated magnetostriction plotted against inter-atomic distance Δa of Fe-Pd alloy films are shown in Figure 3. A good fit was obtained between the interatomic distance Δa and the saturated magnetostriction of Fe-Pd alloy films.

Table 3. Saturated magnetostriction value of the Fe-Pd alloy films.

Composition	Saturated Magnetostriction
Fe-43.8at%Pd	307 ppm
Fe-60.3at%Pd	505 ppm
Fe-66.9at%Pd	75 ppm

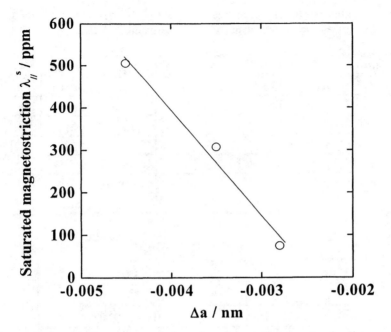

Figure 3. Saturated magnetostriction plotted against inter-atomic distance Δa of Fe-Pd alloy films.

CONCLUSION

The structures of Fe-Pd alloy films, prepared by DC magnetron sputtering method were shown to be a highly (111)-oriented fcc phase. The interatomic distance Δa were reduced in films on film formation. In addition, a good fit was obtained between interatomic distance Δa and saturated magnetostriction of these Fe-Pd alloy films.

ACKNOWLEDGEMENT

The authors wish to thank Mr. Toshio Kaneko, Saginomiya Seisakusho Inc., Engineering Management Department, Mr. Fumio. Kanazaki, Saginomiya Seisakusho Inc, Sayama R&D and Mr. Katsuyuki Takarazawa, Chief Researcher Engineering Department, Tanaka Kikinzoku Group for their useful advice and cooperative for the magnetron sputtering. In addition, parts of this study were carried out by Grant-in-Aid for Scientific Research (No. 06659) from the Japan Society for the Promotion of Science.

REFERENCES

[1] H. Uchida, M. Wada, A. Ichikawa, Y. Matsumura, H-H. Uchida, 1996. Effects of the preparation method and condition on the magnetic and giant magnetostrictive properties of the (Tb, Dy) Fe_2 thin films, Proc. 5th Int. Conf. New Actuators (ACTUATOR 96), Bremen, Germany, 275–282.

[2] M. Wada, H-H. Uchida, Y. Matsumura, H. Uchida, H. Kaneko, 1996. Preparation of films of (Tb, Dy) Fe_2 giant magnetostrictive alloy by ion

beam sputtering process and their characterization, Thin Solid Films 281-282 pp. 503–506.

[3] H. Uchida, Y. Matsumura, H-H. Uchida, H. Kaneko, 2002. Progress in thin films of giant magnetostrictive alloys, J. Mag. Mag. Mater. 239, 540–545.

[4] H. Yabe, K. Oguri, H-H. Uchida, Y. Matsumura, H. Uchida, Y. Nishi, 2000. Micrograph and crystal structure of high magnetostrictive susceptibility of Fe-Pd thin film, Inter. J. Appl. Electromagnetics and Mechanics, 12, 67–70.

[5] H. Yabe, Y. Nishi, 2003. High noise resistance of Fe-45at%Pd alloy film with high magnetostrictive susceptibility, Jpn. J. Appl. Phys., 42(1), in press.

[6] H. Yabe, Y. Nishi, 2003. Chemical composition dependence of magnetostrictive properties of Fe-Pd alloy films, Tetsu-to-Hagane, 89(2), 93–98.

[7] H. Yabe, 2003. Study in Fe-Pd alloy film as actuator material operated by magnetic field, Doctoral thesis, Tokai University, Japan, March 24, 2003, Ch.4, 48–50.

[8] C. Kittel, 1949. Reviews of Modern Physics, 21, 541–583.

[9] P.I. Williams, D.G. Load and P.J. Grundy, 1994. Magnetostriction in polycrystalline sputter-deposited TbDyFe films, J. Appl. Phys. 75, 5257–5261.

[10] H.J. Goldschmidt, 1949. Interplanar spacings of carbides in steels, Metallurgia, 40, 103–104.

[11] Z. Wang, T. Iijima, G. He, T. Takahashi, Y. Furuya, 2000. Structural characteristics and magnetic properties of Fe-Pd thin films, Inter. J. Appl. Electromagnetics and Mechanics, 12, 61–66.

[12] R. Hultgren, C.A. Zopffe, 1938. The crystal structures of the iron-palladium superlattices, Z. Krist., 99, 511–512.

[13] H.E. Swanson and E. Tatge, 1953. Standard x-ray diffraction powder patterns, Natl. Bur. Standards (U. S.) Circular 539, I, 21–22.

Chapter 9: Intelligent Materials Processing

Sub-Nanoscale Structure-Controlled Alloys Produced by Stabilization of Supercooled Liquid

A. INOUE

Institute for Materials Research, Tohoku University, Sendai, Japan

ABSTRACT

This paper presents recent results on high functional materials by utilization of subnanoscale ordered atomic configurations that can develop in glassy and crystalline bcc, hcp and fcc phases consisting of special elements. The glassy phase in Zr-based alloys consists mainly of sub-nanoscale icosahedral ordered atomic configurations including a high-density interface and exhibiting unique mechanical properties that cannot be obtained for conventional crystalline alloys. The Ti-based bcc alloys at nanograin sizes of ~5 nm are formed in alloy systems including atomic pairs with positive heats of mixing and exhibiting similar mechanical properties as those of bulk glassy alloys. The Mg-Zn-Y hcp solid solutions can have a novel long-periodic hexagonal structure through stabilization caused by atomic level segregation of Y and Zn elements. The long periodic structure has enabled us to obtain a nanogranular hcp solid solution exhibiting high strength and good ductility. The Al-Fe fcc solid solution is also formed as a sheet by vapor evaporation exhibiting high strength and good ductility. The high strength is mainly attributed to the solid solution strengthening caused by development of an atomic scale Al-Fe ordered atomic configuration. The novel concept of developing new functional materials by controlling sub-nanoscale ordered atomic configurations is useful and its extension is expected to lead to the fabrication of more varieties of functional materials.

INTRODUCTION

As one of the useful methods to reduce the consumption of energy and pollution on the earth, it is important to develop high-strength material with good ductility and low density. It is well known that the refinement of grain size in single-phase alloys causes simultaneous achievement of high strength and good ductility. The relation between yield strength and grain size is known as the Hall-Petch relation [1] and the validity of this relation has been recognized over a wide grain size range above ~50 nm [2]. For smaller grain sizes, no Hall-Petch relation is recognized and strength decreases with decreasing grain size [3]. Considering that the transition from the Hall-Petch relation to an inverse Hall-Petch relation occurs at a grain size of about 50 nm, the optimum grain size to achieve high strength and good ductility should lie in the vicinity of 50 nm. In addition to grain size refinement, it was recently noted that sub-nanoscale ordered atomic configurations in various metastable solid solutions such as glassy and crystalline (bcc, fcc and hcp) phases to develop metallic alloys with high strength and good ductility is also important. This

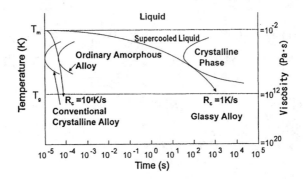

Figure 1. Schematic illustration of continuous cooling transformation curves of metallic liquids for different alloys in conventional crystalline, ordinary amorphous and bulk glassy types.

paper intends to present the formation and mechanical properties of glassy and crystalline solid solution phases consisting of sub-nanoscale ordered atomic configurations in Zr-, Fe-, Ti-, Mg- and Al-based alloys and to call attention to the importance of alloy design for development of these configurations.

SUB-NANOSCALE ORDERED ATOMIC CONFIGURATIONS IN GLASSY ALLOYS

Figure 1 shows a schematic illustration of continuous cooling transformation curves of some supercooled metallic liquids. The incubation time of a supercooled liquid is less than 10^{-5} seconds for conventional crystalline alloys and of the order of 10^{-4} seconds for ordinary amorphous alloys. This situation was the case for a long period before 1990 and the supercooled liquid of metallic alloys was too unstable to control for fabrication of useful materials. Since 1990, the stability of supercooled liquid has been dramatically enhanced and incubation times have increased to several hundreds and sometimes to several thousands of seconds [4–7]. The lowest critical cooling rate for glass formation now reaches the order of 0.01 K/s [8]. By stabilizing the supercooled metallic liquid, the various shapes of bulk glassy alloys can be produced such as massive ingots with diameters up to 75 mm, cylindrical rod with diameters of 25 mm and lengths of 30 cm and tubes with an outer diameter of 12 mm, an inner diameter of 10 mm and lengths of 1.5 to 1.8 m [9].

Table 1 summarizes typical bulk glassy alloy systems reported todate together with the calendar years when the first paper or patent was presented. Alloy systems can be divided into two types of nonferrous and ferrous groups. As the nonferrous alloy types, one can list Mg- [10], lanthanide (Ln)- [11], Zr- [12,13], Ti- [14], Pd- [15] and Cu- [16] based systems, while the ferrous alloy types consist of a variety of alloy components of Fe- [17], Co- [18] and Ni- [19] based systems. As examples, Fe-based bulk glassy alloys are obtained in alloy systems of Fe-(Al,Ga)-metalloids [17], Fe-(Nb,Mo)-(Al,Ga)-metalloids [20], Fe-(Zr,Hf,Nb)-B [21], Fe-Co-Ln-B [22], Fe-Ga-(Cr,Mo)-metalloids [23], Fe-(Nb,Cr,Mo)-metalloids [24], Fe-Ga-metalloids [25] and Fe-B-Si-Nb [26]. Considering that the Ln-metals consist of more than 10 elements, note that the total alloy systems now reach nearly one thousand kinds of ternary systems. In addition, more than half of these alloy systems were developed in the past five years and hence this research field is still in the growing phase even today.

Table 1. Typical bulk glassy alloy systems reported up to date together with the calendar years when the first paper or patent of each alloy system was published.

1. Nonferrous alloy systems		2. Ferrous alloy systems	
Mg-Ln-M (Ln=lanthanide metal, M=Ni,Cu,Zn)	1988	Fe-(Al,Ga)-(P,C,B,Si,Ge)	1995
Ln-Al-TM (TM=VI ~ VIII group transition metal)	1989	Fe-(Nb,Mo)-(Al,Ga)-(P,B,Si)	1995
Ln-Ga-TM	1989	Co-(Al,Ga)-(P,B,Si)	1996
Zr-Al-TM	1990	Fe-(Zr,Hf,Nb)-B	1996
Ti-Zr-TM	1993	Co-(Zr,Hf,Nb)-B	1996
Zr-Ti-TM-Be	1993	Ni-(Zr,Hf,Nb)-B	1996
Zr-(Ti,Nb,Pd)-Al-TM	1995	Fe-Co-Ln-B	1998
Pd-Cu-Ni-P	1996	Fe-(Nb,Cr,Mo)-(C,B)	1999
Pd-Ni-Fe-P	1996	Ni-(Nb,Cr,Mo)-(P,B)	1999
Pd-Cu-B-Si	1997	Co-Ta-B	1999
Ti-Ni-Cu-Sn	1998	Fe-Ga-(P,B)	2000
Cu-(Zr,Hf)-Ti	2001	Ni-Zr-Ti-Sn-Si	2001
Cu-(Zr,Hf)-Ti-(Y,Be)	2001		

It is important to point out the feature of alloy components leading to the formation of bulk glassy alloys through the stabilization of supercooled liquid. As summarized in Figure 2, all the bulk alloys found between 1988 and 1994 have simple three empirical component rules, i.e., (1) multi-component consisting of more than three elements, (2) significant atomic size mismatches above 12% among the three elements, and (3) negative heats of mixing among the three elements [4–6].

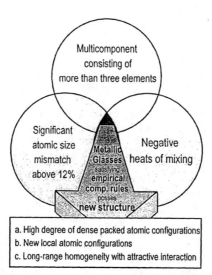

Figure 2. Features of alloy components for stabilization of supercooled liquid and high glass-forming ability.

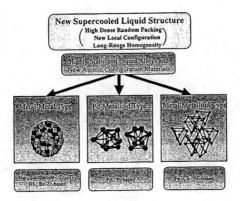

Figure 3. Schematic illustration of new supercooled liquid structure in metal-metal, metal-metalloid, and Pd-based.

Subsequently, this has clarified that alloys with the three component rules have high thermal stability of supercooled liquid against crystallization leading to the formation of bulk glassy alloys. Special attention was paid to local atomic configurations in the bulk glassy alloys. Therefore, it was examined that the feature of glassy structure in their multi-component alloys by various techniques such as anomalous X-ray scattering, neutron diffraction and high resolution TEM techniques and density measurement. As shown in Figure 3, it has been reported that the alloys with the three component rules have a new glassy structure with the features of a higher degree of dense packed atomic configurations, new local atomic configurations and long-range homogeneity with attractive interaction [4–6]. When the local atomic configurations are focused on more detail, the metal-metal type bulk glassy alloys as exemplified for Zr-Al-Ni-Cu, Hf-Al-Ni-Cu and Ti-Zr-Ni-Cu systems consist of an icosahedral atomic configuration with a size of approximately 1 nm [27]. On the other hand, metal-metalloid type glassy alloys in Fe-M-B, Co-M-B and Ni-M-B (M = Ln, Zr, Hf, Nb, Ta) systems have a network-like atomic configuration in which trigonal prisms are connected with each other in edge and corner sharing modes through glue atoms of lanthanide and transition metals [28]. In addition, Pd-based bulk glassy alloys in Pd-Cu-Ni-P system are composed of two kinds of polyhedrons consisting of Pd-Cu-P and Pd-Ni-P components [29].

After noticed the importance of the three component rules for stabilization of supercooled liquid region, the three parameters were transformed to thermodynamical parameters of misfit enthalpy for the difference in atomic size and mixing enthalpy for negative heats of mixing. By using the thermodynamical parameters, the following have been predicted: glass-forming ability, glass composition range and the best alloy composition with the highest thermal stability of supercooled liquid against crystallization in the multi-component alloys [30] by using of Miedema's semi-empirical model [31], quasi-chemical model [32] and Cowley's order parameter [33]. A rather good agreement between the predicted glass composition range and experimental data has been recognized for La-Al-Ni and Zr-Zl-Ni systems. The Cowley's order parameter represents the ease of the development of short-range order. It has also been reported that the Cowley's order parameter with the largest negative value corresponds well to the alloy composition with the largest supercooled liquid region before crystallization, indicating the close relation between the development of short-range ordered atomic configurations and the high stability of supercooled liquid against crystallization. Based on these data, it is concluded that bulk glassy alloys are mainly composed of sub-nanoscale ordered atomic configurations and hence can be regarded as a metallic material including very high volume fractions of sub-nanoscale interface boundary.

MECHANICAL PROPERTIES

Bulk Glassy Alloys

It has subsequently been shown that the bulk glassy alloys with sub-nanoscale ordered atomic configurations possess unique mechanical properties. Figure 4 shows the relation between tensile strength and Young's modulus for bulk glassy alloys, together with the data of conventional crystalline alloys [4–6]. The glassy alloys exhibit higher tensile strength, lower Young's modulus and larger elastic elongation, in comparison with those for the crystalline alloys. The difference is as large as about three times for their three properties. Very high tensile strength values exceeding 2000 MPa, which cannot be obtained for conventional crystalline alloys, are obtained for Cu- and Ni-based bulk glassy alloys. Very recently, it has also noticed that the addition of special elements with nearly zero and positive heats of mixing against the other constituent elements for Cu- and Ni-based alloys has enabled us to form bulk glassy alloys exhibiting simultaneously high strength exceeding 2000 MPa and distinct plastic elongation of 1 to 2.5% [34]. The simultaneous achievements of high strength above 2000 MPa and distinct plastic elongation have not been obtained for Zr-Al-Ni-Cu and Pd-Cu-Ni-P bulk glassy alloys.

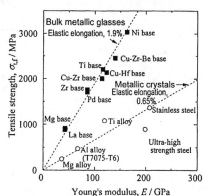

Figure 4. Relation between tensile strength and Young's modulus for bulk metallic glassy alloys. The data of conventional crystalline alloys are also shown for comparison.

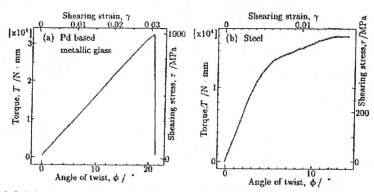

Figure 5. Relation between torque and twist angle for Pd-based bulk glassy alloys and 0.65 mass% carbon steel. The shearing stress and strain calculated from the torque and twist angle relation are also shown for comparison.

By choosing the bulk glassy alloy rods with a diameter of 10 mm which satisfy the JIS criterion, torsion deformation tests were performed at room temperature. Figure 5 shows the relation between torque and twist angle for a $Pd_{40}Cu_{30}Ni_{10}P_{20}$ bulk glassy alloy rod in comparison with that of conventional carbon steel rod [35]. The relation between shearing stress and shearing strain calculated from the experimental data on the torque and twist angle is also shown in the figure. It is noticed that the elastic twist angle of the glassy alloy rod is about 18 degrees which are about four times higher than that (about 4 degrees) for the steel rod and the shearing stress is also as high as about 980 MPa which is about three times higher than that (about 340 MPa) for the steel. Such a significant difference also seems to reflect the difference in the local atomic configurations between bulk glassy alloys and crystalline alloys.

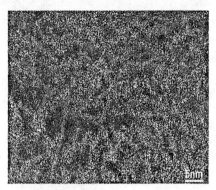

Figure 6. Bright-field TEM image of $Ti_{50}Zr_{10}Nb_{10}Ta_{30}$ alloy subjected to cold rolling to 99 % reduction in thickness and then annealing for 0.36 ks at 873 K.

Table 2. Mechanical properties of the $Ti_{50}Zr_{10}Nb_{10}Ta_{30}$ alloy subjected to cold rolling to 99% reduction in thickness and then annealed for 0.36 ks at 873 K. The data for other Ti-based alloys are shown for comparison.

Alloy	E / GPa	σ_f / MPa	σ_v / MPa	ε_E (%)	E_r / GPa	ε_T (%)	H_v
Pure-Ti	102	345	275	~0.3	4	20	100
Ti-6Al-4V	110	910	840	~0.5	84	18	260
Ti-Zr-Nb-Ta	42	1070	980	~1.7	29	15	350

E: Young's modulus, σ_f: Tensile strength, σ_v: Yield strength, ε_E: Elastic elongation, E_r: Elasticity, ε_T: Tensile elongation, H_v: Vickers hardness

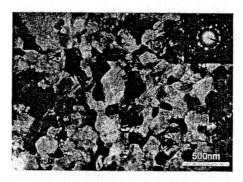

Figure 7. Bright-filed TEM image and selected-area electron diffraction pattern of $Mg_{97}Zn_1Y_2$ alloy produced by warm extrusion of atomized powder at 573 K.

Ti-Based BCC Solid Solution

In Ti-based alloys containing simultaneously Zr, Nb and Ta with zero and positive heats of mixing, a bcc solid solution with a grain size of about 5 nm including a high volume fraction of grain boundary shown in Figure 6, has been formed by thermo-mechanical treatment of cold rolling to 99% reduction in thickness, followed by annealing for 0.36 ks at 873 K [9]. The bcc Ti-Zr-Nb-Ta alloy exhibits unique mechanical properties, i.e., high tensile strength of 1070 MPa, low Young's modulus of 42 GPa and large elastic elongation of 1.7%, as shown in Table 2. The feature of mechanical properties is analogous to that for bulk glassy alloys and the analogy seems to reflect the existence of high volume fractions of interface (or grain) boundary in the glassy and bcc solid solution phases.

Mg-Based HCP Solid Solution

The Hall-Petch relation between tensile yield strength and grain size for various kinds of Mg-based alloys were examined [36]. Although Mg-based alloys with grain sizes of 350 to 1000 nm have been produced by using the rapid solidification/powder metallurgy (RS/PM) technique, there have been no data on the formation of Mg-based alloys with fine grains below 200 nm. The absence of such data seems to be due to the ease of recovery and recrystallization at low temperatures which is typical for the low melting temperature alloys. The Hall-Petch data suggest that refinement of grain size to 100 to 200 nm causes a high-strength Mg-based alloy. Recently, success was made in fabricating a Mg-based alloy with grain sizes of 100 to 150 nm by warm extrusion of atomized $Mg_{97}Zn_1Y_2$ alloy powder at 573 K. Figure 7 shows bright-field TEM image and selected-area electron diffraction pattern of the warm-extruded $Mg_{97}Zn_1Y_2$ alloy [37]. The alloy consists of fine grains with a size of 100 to 150 nm and each grain appears to include a high density of plane faults. The fine grain Mg-based alloy exhibits high tensile yield strength of 500 to 620 MPa combined with rather large elongation of 5 to 8%, as shown in Figure 8 [38]. The change in the mechanical properties can be controlled by changing warm extrusion temperature. The yield strength is about three times higher than that of the conventional Mg-based alloys and about 1.5 times higher than those for the RS/PM AZ91 and ZK61 alloys.

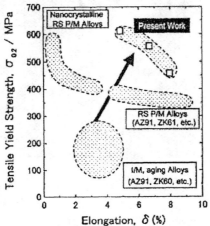

Figure 8. Relation between tensile yield strength and elongation for the RS/PM $Mg_{97}Zn_1Y_2$ alloy in comparison with conventional Mg-based alloys and RS/PM AZ91 and ZK61 alloys.

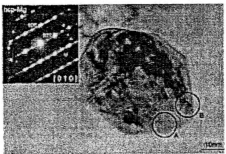

Figure 9. High-resolution TEM image and selected-area electron diffraction pattern of the RS/PM $Mg_{97}Zn_1Y_2$ alloy.

In order to examine in more details the faulted structure in the Mg-based alloys, the high-resolution TEM image and selected-area electron diffraction pattern are shown in Figure 9 [37]. In all the grains with an appropriate Bragg reflection contrast, one can observe the contrasts of a high density of faulted structure. The electron diffraction pattern reveals extra reflection spots at the position of $1/3(0001)_{Mg}$ along the $[0001]_{Mg}$ direction, indicating the formation of a long-periodic hexagonal phase in which the periodicity is three times longer than that for the ordinary hexagonal closed packed (hcp) Mg phase. The long periodicity is clearly demonstrated in the high-resolution TEM image of Figure 10 in which the atomic array of ABACAB can be clearly seen. One can also notice the existence of another type of long periodic hexagonal structure in the Mg-Y-Zn alloy [39]. As shown in Figure 11, the atomic array reveals the periodic hexagonal structure with periodicity which is seven times longer than that of the ordinary hcp Mg-phase.

It has subsequently been recognized that the segregation of Y and Zn elements occurs at the misfit sites of the atomic arrays, as seen in the HAADF image contrast shown in Figure 12 [40-42]. One can clearly see the periodic array of the bright contrast region corresponding to the enrichment of Y and Zn elements with a much larger negative heat of mixing against the other constituent elements [31]. The segregation of Y and Zn elements corresponds to the formation of ordered atomic configurations on an atomic scale caused by the difference in the atomic bonding nature. The atomic level segregation of Y and Zn elements enhances the thermal stability of the metastable long-periodic hexagonal structure, resulting in the formation of nanogranular hcp phase with higher strength and large elongation.

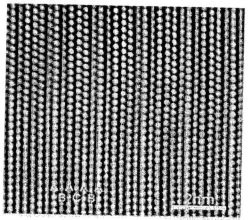

Figure 10. High-resolution TEM image of the RS/PM $Mg_{97}Zn_1Y_2$ alloy.

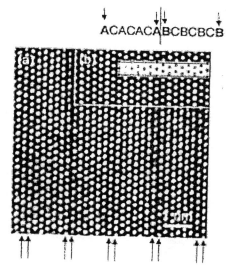

Figure 11. High-resolution TEM image of the RS/PM $Mg_{97}Zn_1Y_2$ alloy.

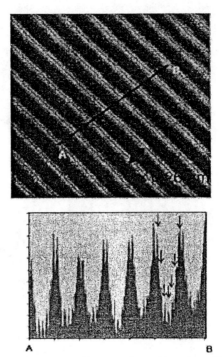

Figure 12. HAADF images of the RS/PM $Mg_{97}Zn_1Y_2$ alloy and intensity profile of the HAADF image contrast.

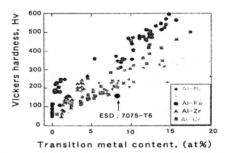

Figure 13. Vickers hardness as a function of solute content for Al-TM (TM=Ni,Ti,Fe,Zr,Cr) alloy sheets produced by the vapor deposition technique.

Al-Based FCC Solid Solution

By using the electron beam high-rate vapor deposition technique with two evaporation sources, Al-Fe supersaturated solid solution alloys containing 1–2 at%Fe have been formed in a sheet form of about 1 mm in thickness and about 12x12 cm^2 square [43]. Figure 13 shows the change in Vickers hardness number with Fe content for the Al-Fe deposited sheets together with data of other Al-TM (TM=Ni,Ti,Fe,Zr,Cr) deposited alloy sheets [43]. Notice that the Al-Fe sheets exhibit high Vickers hardness around 250 even at low Fe contents of 1–2 at%. The Al-Fe phase consists of an fcc Al phase with no secondary precipitates and grain size is between 50–200 nm. There is a clear tendency for grain size to decrease with increasing Fe content. Figure 14 shows the nominal tensile stress-elongation curves of the Al-Fe alloy sheets. The Al-Fe sheets with Fe contents of 1–2 at% exhibit high strength of 800–900 MPa with a plastic elongation of about 5%.

From the Hall-Petch relation of the present Al-Fe alloy sheets shown in Figure 15 in comparison with the data of the other Al-based alloys, the much higher strength of the Al-Fe deposited alloy sheets is due to the combination of grain size refinement and solid solution strengthening. Figure 15 also shows that the contribution to the strength value is larger for the solid solution strengthening. This is different from the tendency for other Al-based alloys in which the contribution of the grain size refinement is larger. With the aim of investigating the origin for the large contribution of the solid solution strengthening, the atomic configuration in the Al-Fe deposited alloy sheet was examined by the EXAF technique. Figure 16 shows the EXAF intensity profiles at Fe absorption edge corresponding to the local structure around Fe atom [44]. The peak position of the Al-Fe alloys with 1.0 to 2.5% Fe agrees well with that of Al_2Fe and Al_3Fe phases and is different from that of Fe in fcc-Al phase. However, the Al-Fe alloy sheets do not have any peak in a more longer distance region. These data suggest that Al-Fe local atomic ordering develops in the fcc-Al phase and the ordered atomic configuration is the origin for the large contribution to the high strength.

Based on the above-described data, it is concluded that the subnanoscale ordered atomic configurations in the glassy, bcc, hcp and fcc solid solutions may play an important role in the achievement of high functional properties.

CONCLUSIONS

The control of sub-nanoscale atomic configurations and microstructure in metastable solid solutions of glassy and crystalline (bcc, hcp and fcc) phases are very useful for developing advanced materials with good mechanical properties. The extension of this novel concept is expected to cause various novel materials with functional properties.

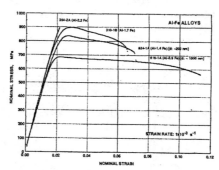

Figure 14. Tensile stress-elongation curves of the vapor-deposited Al-Fe alloy sheets with different Fe contents.

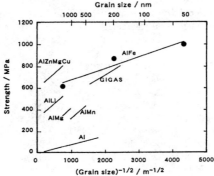

Figure 15. Hall-Petch relation of the vapor-deposited Al-Fe alloy sheets. The data for the other Al-based alloys are shown for comparison.

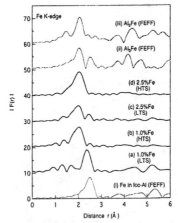

Figure 16. EXAFS intensity profiles of the Al-Fe alloy sheets produced by vapor deposition on a substrate at low (423 K) and high (523 K) temperatures. The data of other Al-Fe alloys are shown for comparison.

REFERENCES

[1] A.M. El-Sherik, U. Eeb, G. Palumbo and K.T. Aust, 1992. Deviations from Hall-Petch Behavior in As-Prepared Nanocrystalline Nickel. Scripta Metall. Mater., 27 (9), 1185–1188.

[2] K.A. Padmanabhan and H. Haln, 1996. Synthesis and Processing of Nanocrystalline Powder, ed. D.L. Bourell, TMS. Warrendale, 21.

[3] A. Inoue, H.M. Kimura, M. Watanabe and A. Kawabata, 1197. Work Softening of Al-based Alloys Containing Nanoscale Icosahedral Phase. Mater. Trans. JIM, 38(9), 756–760.

[4] A. Inoue, 1995. High Strength Bulk Amorphous Alloys with Low Critical Cooling Rates. Mater. Trans. JIM, 36(7) 866–875.

[5] A. Inoue, 2000. Stabilization of Metallic Supercooled Liquid and Bulk Amorphous Alloys. Acta Mater., 48(1), 279–306.

[6] A. Inoue, 2001. Bulk Amorphous and Nanocrystalline Alloys with High Functional Properties. Mater. Sci. Eng. A304-306, 1–10.

[7] A. Inoue, A. Takeuchi, 2002. Recent Progress in Bulk Glassy Alloys. Mater. Trans., 43(8), 1892–1906.

[8] N. Nishiyama, A. Inoue, 2002. Glass-forming ability of $Pd_{42.5}Cu_{30}Ni_{7.5}P_{20}$ alloy with a low critical cooling rate of 0.067 K/s. Appl. Phys. Lett. 80(4), 568–570.
[9] A. Inoue and A. Takeuchi, 2003. Recent Progress in Bulk Glassy, Nanoquasicrystalline and Nanocrystalline Alloys in Sendai Group. Mater. Sci. Eng., in press.
[10] A. Inoue, T. Nakamura, N. Nishiyama, T. Masumoto, 1992. Mg-Cu-Y Bulk Amorphous Alloys with High Tensile Strength Produced by a High-Pressure Die Casting Method. Mater. Trans. JIM, 33(10), 937–945.
[11] A. Inoue, T. Nakamura, T. Sugita, T. Zhang and T. Masumoto, 1993. Bulky La-Al-TM (TM=Transition Metal) Amorphous Alloys with High Tensile Strength Produced by a High-Pressure Die Casting Method, Mater. Trans. JIM 34(4), 351–358.
[12] T. Zhang, A. Inoue and T. Masumoto, 1991. Amorphous Zr-Al-TM(Tm=Co, Ni, Cu) Alloys with Significant Supercooled Liquid Region of Over 100K. Mater. Trans. JIM, 32(11), 1005–1010.
[13] A. Peker and W.L. Johnson, 1993. A Highly Processable Metallic-Glass $BeZr_{41.2}Ti_{13.8}Cu_{12.5}Ni_{10.0}Be_{22.5}$. Appl. Phys. Lett. 63(17), 2342–2344.
[14] A. Inoue, 1999. Synthesis and Properties of Ti-Based Bulk Amorphous ALloys with a Large Supercooled Liquid Region. Mater. Sci. Forum, 312–314, 307–314.
[15] A. Inoue, N. Nishiyama, H. Kimura, 1997. Preparation of a Thermally-Stable Bulk Amorphous $Pd_{40}Cu_{30}Ni_{10}P_{20}$ Cylinder of 72mm Diam., Mater. Trans., JIM, 38(2), 179–183.
[16] T. Zhang, K. Kurosaka and A. Inoue, 2001. Thermal and Mechanical Properties of Cu-Based Cu-Zr-Ti-Y Bulk Glassy Alloys. Mater. Trans., 42(10), 2042–2045.
[17] A. Inoue and J.S. Gook, 1995. Fe-Based Ferromagnetic Glassy Alloys with Wide Supercooled Liquid Region, Mater. Trans. JIM, 36(9), 1180–1183.
[18] T. Itoi and A. Inoue, 2000. Thermal Stability and Soft Magnetic Properties of Co-Fe-M-B(M=Nb, Zr) Amorphous Alloys with Large Supercooled Liquid Region, Mater. Trans. JIM, 41(9), 1256–1262.
[19] R. Akatsuka, T. Zhang, H. Koshiba, A. Inoue, 1999. New Ni-based Amorphous Alloys with Large Supercooled Liquid Region. Mater. Trans. JIM, 40(3), 258–261.
[20] A. Inoue, H.M. Kimura, G.S. Gook, 1996. Effect of Added Elements on Thermal Stability of Supercooled Liquid in $Fe_{72-x}Al_5Ga_2P_{11}C_6B_4M_x$ Glassy Alloys, Mater. Trans. JIM, 37(1), 32–38.
[21] A. Inoue, T. Zhang, H. Koshiba and A. Makino, 1998. New bulk amorphous Fe-(Co,Ni)-M-B (M=Zr,Hf,Nb,Ta,Mo,W) alloys with good soft magnetic properties, J. Appl. Phys. 83(11), 6326–6328.
[22] W. Zhang and A. Inoue, 1999. Thermal and Magnetic Properties of Fe-Co-Ln-B(Ln=Nd,Sm, Tb or Dy) Amorphous Alloys with High Magnetostriction, Mater. Trans. JIM, 40(1), 78–81.
[23] T.D. Shen and R.D. Schwarz, 1999. Bulk ferromagnetic glasses prepared by flux melting and water quenching, Appl. Phys. Lett., 75(1), 49–51.
[24] S. Pang, T. Zhang, K. Asami and A. Inoue, 2001. New Fe-Cr-Mo-(Nb, Ta)-C-B Alloys with High Glass-Forming Ability and Good Corrosion Resistance. Mater. Trans., 42(2), 376–379.
[25] B. Shen. H. Koshiba, T. Mizushima and A. Inoue, 2000. Bulk Amorphous Fe-Ga-P-B- C Alloys with a Large Supercooled Liquid Region. Mater. Trans., 41(7), 873–876.
[26] S.J. Pang, T. Zhang, K. Asami and A. Inoue, 2002. Formation of Bulk Glassy Ni-(Co-)Nb-Ti-Zr Alloys with High Corrosion Resistance. Mater. Trans., 43(7), 1771–1773.
[27] C.F. Li, J. Saida, M. Matsushita and A. Inoue, 2000. Precipitation of Icosahedral Quasicrystalline Phase in $Hf_{65}Al_{7.5}Ni_{10}Cu_{12.5}Pd_5$ Metallic Glass. App. Phys. Lett. 77(4), 528–530.
[28] T. Nakamura, E. Matsubara, M. Imafuku, H. Koshiba, A. Inoue and Y. Waseda, 2001. Structural Study of Amorphous $Fe_{70}M_{10}B_{20}$ (M=Cr, W, Nb, Zr and Hf) Alloys by X-ray Diffraction. Mater. Trans. 42(8), 1530–1534.

[29] C. Park, M. Saito, Y. Waseda, N. Nishiyama and A. Inoue, 1999. Structural Study of Pd-Based Amorphous Alloys wide Wide Supercooled Liquid Region by Anomalous X-ray Scattering. Mater. Trans., JIM, 40(6), 491–497.
[30] A. Takeuchi and A. Inoue, 2001. Quantitative Evaluation of Critical Cooling Rate for Metallic Glasses. Mater. Sci. Eng., A304-306, 446–451.
[31] Miedema, 1998. Cohsion in Metals, (eds F.R. de Boer, R. Boom, W.C.M. Mattens, A.R. Miedema and A.K. Nissen, Elsevier Science Publishers B.V., Netherlands), 1–758.
[32] P.J.Desre, 1999. Thermodynamics and Glass Forming Ability from the Liquid State. Mat. Res. Soc. Symp., 554, 51–62.
[33] J.M.Cowley: Phys. Rev. 77(5), 1950, 669.
[34] T. Zhang and A. Inoue, 2002. New Bulk Glassy Ni-Based Alloys with High Strength of 3000MPa. Mater Trans., 43(4), 708–711.
[35] K. Fujita, A. Inoue, T. Zhang and N. Nishiyama, 2002. Anelastic Behavior under Tensile and Shearing Stresses in Bulk Metallic Glasses. Mater. Trans., 43(8), 1957–1960.
[36] A. Kato, 1994.Doctor thesis, Tohoku University.
[37] Y. Kawamura, K. Hayashi, A. Inoue, T. Masumoto, 2001. Rapidly Solidified Powder Metallurgy $Mg_{97}Zn_1Y_2$ Alloys with Excellent Tensile Yield Strength above 600 MPa, Mat. Trans., 42(7), 1172–76.
[38] A. Inoue, M. Matsushita, Y. Kawamura, K. Amiya, K. Hayashi and J. Koike, 2001. Novel Hexagonal Structure and Ultrahigh Strength of Magnesium Solid Solution in the Mg-Zn-Y System. J. Mater. Res., 16(7), 1894–1900.
[39] A. Inoue, M. Matsushita, Y. Kawamura, K. Amiya, K. Hayashi and J. Koike, 2002. Novel Hexagonal Structure of Ultra-High Strength Magnesium -Based Alloys. Mater. Trans., 43(3), 580–584.
[40] E. Abe, Y. Kawamura, K. Hayashi and A. Inoue, 2002. Long-Period Ordered Structure in a High-Strength Nanocrystalline Mg-1 at% Zn-2 at% Y Alloy Studied by Atomic-Resolution Z-Contrast STEM. Acta Mater., 50(15), 3845–3857.
[41] D.H. Ping, K. Hono, Y. Kawamura and A. Inoue, 2002. Local Chemistry of a Nanocrystalline High-Strength $Mg_{97}Y_2Zn_1$ Alloy. Phil. Mag. Lett., 82(10), 543–551.
[42] K. Amiya, T. Ohsuna and A. Inoue, 2003. J. Mater. Res., in press.
[43] H. Sasaki, K. Kita, J. Nagahora and A. Inoue, 2001. Nanostructures and Mechanical Properties of Bulk Al-Fe Alloys Prepared by Electron-Beam Deposition. Mater. Trans., 42(8), 1561–1565.
[44] M. Sakurai, 2002. unpublished research, Tohoku University.

A Novel Method—Hot Quasi-Isostatic Pressing to Fabricate a Metallic Cellular Material Containing Polymer using an SPS System

Z. SONG, S. KISHIMOTO and N. SHINYA
Intelligent Materials Research Group, National Institutes for Materials Science, 1-2-1, Sengen, Tsukuba, Ibaraki 305-0047, Japan

ABSTRACT

During a pulse-current hot pressing, the direction of the pressure applied on the green body in a die is uni-axial in the sintering process. Therefore, an inhomogeneous deformation of the particles especially those soft particles occurs in the sintering process. In this report, a novel method, pulse-current quasi-isostatic pressing (PCHIP), was used to fabricate metallic cellular structures by using a spark plasma sintering (SPS) system. The metallic material with a very fine equiaxed cellular structure (cell size: in level of several tens of microns) was fabricated by this method. The method used phenol-resin-solid spheres coated by a Ni-P alloy layer. After sintering, the metallic layers were joined together, and the polymer materials inside the metallic cellular structure remained.

INTRODUCTION

Cellular metals are well known to have many interesting combinations of mechanical, physical and chemical properties. The scale of the cells produced by normal methods is in range of 0.1 mm to several centimetres [1]. Techniques to produce finer cellular structures are lacking. Hollow spheres made of Ni, Cu and steel can be used to fabricate a cellular structure by bonding, brazing or sintering [2,3]. Assembly of spheres in a special lattice structure can also be achieved. Although the size of the hollow spheres is controllable, the scale of the cellular structures fabricated by the mentioned references is rather large.

A novel metallic material with a very fine cellular structure (cell size of several tens of microns) has been developed by sintering alloy-coated polymer-solid spheres using normal sintering [4]. After sintering, the metallic coatings join together and the polymer materials inside the metallic cellular structure are consumed or carbonized at the high temperature. In this way a cellular structure was formed. Polymer material inside the cells benefits the mechanical properties of the material, such as damping and rigidity. So, the rapid sintering technique, called spark plasma sintering (SPS), was introduced to fabricate the cellular metal with the polymer remaining by using the same alloy-coated polymer-solid spheres [5].

SPS is a type of solid compression sintering method that is similar to hot pressing (HP) but faster. In using an SPS system, powder particles fill the die and are pressed by a pair of graphite punches while pulsed electric current flows through the particles allowing efficient localized heating at the contacts between the particles [6]. If the current is large enough, plasma may be generated at these contact points [7], but the process of how and when the plasma is generated has not yet been clarified. To avoid being entangled by the plasma, the technique is sometimes called pulse-current hot pressing (PCHP) [8]. It is a next-generation technique able to sinter at a lower temperature and in a shorter actuation time compared to the normal HP.

Figure 1 shows the processes and the cellular structures fabricated with the normal sintering and PCHP. In the PCHP process, the pressure applied on the particles in the die is in the axial directional. The pressure normal to the pressing axis is larger than that parallel to this axis. Actually, there is a distribution of radial pressure in the section normal to the axis. Several reviews of the compaction process were reported recently in a special issue of "MRS Bulletin" [9]. This pressure distribution always leads to inhomogeneous deformation of the particles, especially those that are softest. So the cellular structure fabricated by the SPS system is flattened by this pressure. In this report, a new method, pulse-current quasi-isostatic press (PCHIP) sintering [10] was used to fabricate a metallic cellular structure using the SPS system.

EXPERIMENTAL METHOD

Alloy-coated polymer-solid sphere particles were fabricated by electroless plating. The polymer is phenol resin and the alloy is nickel-phosphorus (2.5%wt). The average thickness of the alloy coat was about 3.5 µm; the diameter of the particle was distributed in a range between 10 and 80 µm. The particles were assembled in a flexible die and the green compact was formed by a cold isostatic pressing method.

Place the green compact in a ceramic die and then pour with fine hard spheres (ZrO_2, diameter 100µm). Graphite powder (5.0µm in diameter) was placed in the top and the bottom of the green compact. Therefore, the green compact was surrounded by ZrO_2 and graphite powders particles. Then the green was pressed by a pair of graphite punches (Figure 2). Pulse current was generated by an SPS system (SPS-515S, Sumitomo Coal Mining Co., Ltd.) and passed through the green compact from the graphite powder and punch. A high efficient heating was occurred. After a short time of sintering, by removing the punches and cleaning the particles surround the compact, a cellular material can be obtained.

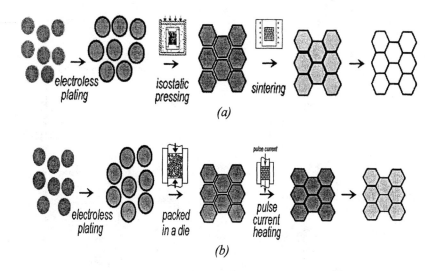

Figure 1. Former processes for the fabrication of cellular alloy by alloy-coated polymer-solid spheres. (a) normal sintering; (b) pulse current hot pressing sintering.

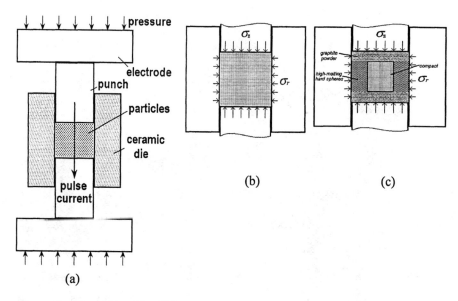

Figure 2. Schematics of the SPS system and the particles in the die. (a) Core part of the SPS system: A pulsed electric current passes through the particles from the graphite punch; (b) Former PCHP method: The particles in the die were pressed by the punches and the die wall ($\sigma_r < \sigma_z$). Obviously, the pressure in the die is axial directional; (c) Presented PCHIP method: The pressure field surrounding compact was homogenized by the hard spheres and graphite powder. The pulse current flows through the compact from the graphite powder.

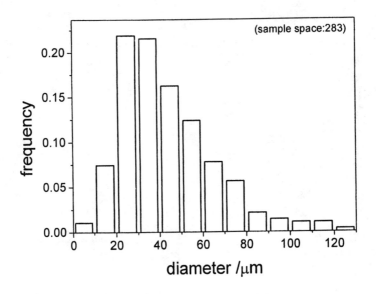

Figure 3. The size distribution of the phenol resin spheres coated with Ni-P.

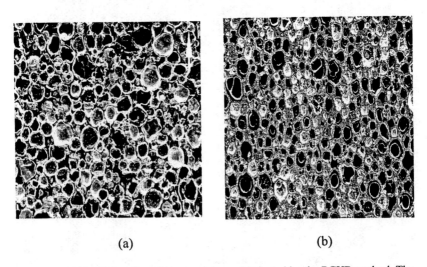

Figure 4. Images of cellular alloy. Structure (a) was fabricated by the PCHP method. The arrow indicates the direction of current and pressure. Structure (b) was fabricated by the PCHIP method.

RESULTS AND DISCUSSION

Figure 3 shows the size distribution of the spheres. Assuming that all particles after sintering were deformed into a polyhedral shape, the theoretical volume percent of the alloy in the sintered material is about 37%. Figure 4 shows the cellular structures of the specimens fabricated by the PCHP sintering method and the PCHIP method. Figure 5 shows the microstructure of a specimen fabricated by our PCHIP method. From these pictures, it can be seen that most of the cells in the image are nearly equiaxed. From the close-up image, it can also be seen that adjacent spheres are joined quite well and several spheres are deformed by the squeeze from their neighbours. Part of the polymer material inside the cell is carbonized or half-carbonized.

A quantitative analysis was done using an optical microscope. The area ratio of the alloy in the cross-section of the sample is about 44.3%, larger than the theoretical value. This result comes from two reasons. One is that the particles are compressed. After sintering the volume of space between the alloy coats is decreased while the volume of the alloy does not change. Another reason is that most of the large spheres are recessed by the squeeze from their neighbours.

Compared to normal PCHP, this new PCHIP method has two merits. One is that the amount of cracks in the sample by using this method is reduced; another one is that the cell shapes remain equiaxed. The alloy-coated polymer-solid spherical granules are soft comparing to the metal granules. Pressure gradients along the radial distance across the compact cause a serious density gradient resulting in radial cracking within the green body. For hard granules, the situation

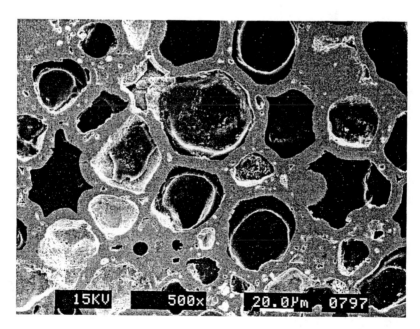

Figure 5. SEM image of the cellular structure fabricated by PCHIP method.

is much better. In this paper, pressure at the soft green body is homogenized by the surrounding hard granules and can be regarded as isostatic. The cracks due to density gradients in green body are reduced and so, compression of the spheres was isotropic.

To the particles in the specimen, there is a radial temperature distribution. When the current pulse passes through the particles, the temperature at the contact points can reaches a very high value for Joule heating while the centers of the particles simultaneously maintain a lower temperature. The temperature at the contact points is important for sintering the metal, but does not affect the polymer inside the cell. The center of the particle reaches the average temperature in a very short time. To avoid material inside the particles reaching a high temperature, the peak temperature should be controlled in the process. That is to say, a sharp temperature gradient from the circumference of the sphere to its centre is the result we need. In this work, a large current, short-actuation-time heating process was performed to obtain this effect.

CONCLUSIONS

This report provides a quasi-isostatic pressing method in an SPS system. By using this technique, a hollow sphere method for cellular structures was developed. A metallic cellular material with fine and equiaxed structure was obtained. After sintering, the metallic layers were joined together, and it appears that part of the polymer material inside the metallic cellular structure remained unchanged.

ACKNOWLEDGMENT

The authors gratefully acknowledge support of the Japan Society for the Promotion of Science.

REFERENCES

[1] J. Banhart, 2001. Prog. Mat. Sci., 46, 559–632.
[2] K.M. Hurysz, J.L. Clark, A.R. Nagel, C.U. Hardwikke, K.J. Lee, J.K. Cochran, T.H. Sanders, 1998. Mater. Res. Soc. Symp. Proc., Ed: D. Schwartz, D. Shih, A. Evans, H. Wadley, 521 191.
[3] O. Andersen, U. Waag, L. Schneider, G. Stephani, B. Kieback, 2000. Adv. Eng. Mater., 4, 192.
[4] S. Kishimoto, N. Shinya, 2000. Development of metallic closed cellular materials containing polymer, Materials Design 21, 575.
[5] Z. Song, S. Kishimoto, N. Shinya, M. Tokita, 1993. J.Soc.Powder Tech. (in Japanese) 30, 790.
[6] S.H. Risbud, J.R. Groza, 1994. Clean grain boundaries in Al-nitride ceramics densified with no additives by a plasma-activated sintering process, Philosophical Mag. B, (3)69 525–533.
[7] K. Mizuuchi, K. Inoue, M. Sugioka, M.Itami, Y. Okanda, M. Kawahara, 2002. Japan Society of Mechanics, 2002 Annual Conference, Tokyo, No. 936.

[8] I. Aydin, B. Briscoe, N. Ozkan, 1997. Modeling Powder Compaction: a review, MRS Bull., 45–51.
[9] S.J. Glass, K.G. Ewsuk, 1997. Ceramic Powder Compaction, MRS Bull., 24–28;
[10] J. Lannutti, 1997. Characterization and Control of Compact Microstructure, MRS Bull., 38–44.

Consideration of AC-Component Force of Electromagnetic Stirring in the MHD Calculation

S. SATOU
Ohita Setsubi Sekkei Corporation, Nisinosu-1, Ohita-city, Ohita 870-0902, Japan

K. FUJISAKI
Environmental & Process Technology Center, Nippon Steel Corporation, 20-1 Shintomi, Futtsu-city, Chiba 293-8511, Japan

T. FURUKAWA
ACE Laboratory, Saga University, 1 Honjo-machi, Saga-city, Saga 840-8502, Japan

ABSTRACT

In the continuous casting process, it is very important to understand the dynamic behavior of the free surface of the molten steel because the surface quality of the steel is almost decided by the conditions of the free surface which begins solidification. The conventional MHD calculation, which used only the DC component of the electromagnetic force, is enough to evaluate the time average flow. However, the time variations of the free surface and the AC component of the electromagnetic force must be argued, to evaluate the conditions of the free surface which decides the slab's surface quality. In this paper, the vibration of the free surface that could not be computed by the conventional MHD calculation using only the DC component of the electromagnetic force was computed by the new MHD calculation taking into account the full electromagnetic force which composed of the AC component and DC component of the electromagnetic force.

INTRODUCTION

The continuous casting mold produces "slab" that is the origin of sheet steel by cooling the molten metal and solidifying it. The quality of the slab is highly dependent on the fluid dynamic of the molten steel in the mold since it is so difficult to restore quality after solidification. The free surface shape and its dynamics are especially important because the surface quality of the slab is determined at the free surface position at the beginning of solidification. In recent years, electromagnetic stirring (EMS) was widely introduced to control the flow of molten metal because of its ability to produce high slab quality [1,2]. Both slab quality and productivity are extremely improved [3].

To evaluate the EMS operation and the fluid dynamics, 3-D magneto-hydrodynamic (MHD) calculations are generally used [4,5]. The conventional MHD calculation uses only the DC component of the electromagnetic force because the main purpose of the computation is to evaluate the effect of the EMF on the timed-average flow. The conventional

MHD calculation is sufficient to evaluate the effect of electromagnetic force. However, to improve the surface quality of the slab became of the higher demand of the steel marketplace, time variation of the free surface shape and the flow must be considered. Therefore, considering only the DC component of electromagnetic force is insufficient, the AC component of the EMF must be added to the conventional DC component calculation. In this paper, we research the MHD calculation taking into account both the DC and AC components of the EMF and evaluate the role of the AC component of the electromagnetic force at the free surface.

CASTING PROCESS USING THE EMS

Molten metal pours into the mold through a submerged nozzle with an obliquely downward flow. The liquid metal collides with the mold wall and divides into a descent flow and a rise flow which heads back up to the free surface. The liquid in the descent flow is solidified by cooling from the mold wall and continuously discharges in the casting direction.

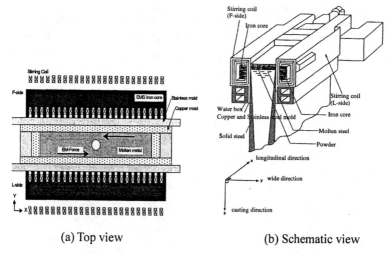

(a) Top view   (b) Schematic view

Figure 1. Continuous casting process using the EMS.

The molten metal in the rising flow reaches the free surface and flows towards the mold wall where it begins to solidify. The variation of the free surface shape and the uniformity of temperature distribution in the free surface are important because the molten metal that solidifies at the free surface forms the outer surface of the slab. A lubricant called "powder" floats on the free surface. The powder is used to prevent damage to the copper mold by contact between the molten metal and the copper mold around the free surface. However when the powder is taken up by rapid change in the free surface shape and becomes solidified within the cooling metal, the slab quality degrades. Stability of the free surface shape and maintenance of a moderate free surface velocity are important to reduce this effect.

To control the free surface shape and velocity, the EMS coil is installed outside the mold at the same height as the free surface as shown in Figure 1.

ELECTROMAGNETIC FORCE

The electromagnetic field by Maxwell equations is derived using the A-φ method as follows [6]. By using the jω method (ω = 2πf where f is frequency [Hz]) and a finite element method in the discrete 3-D field, the magnetic flux and eddy current densities together with the EMF are defined as follows:

$$\nabla \times \left(\left[\frac{1}{\mu}\right] \cdot \nabla \times \vec{A} \right) + \sigma \left(\frac{\partial \vec{A}}{\partial t} + \nabla \phi \right) = \vec{J}_0 \quad (1)$$

$$\vec{F} = \vec{J} \times \vec{B} = \sigma \left(\frac{\partial \vec{A}}{\partial t} + \nabla \phi \right) \times \left(\nabla \times \vec{A} \right) \quad (2)$$

Here, $\vec{A}$ is vector potential [Wb/m], ϕ is scalar potential [V/m], $\vec{J}_0$ is current density [A/m^2], σ is electromagnetic conductivity [S/m], μ is permeability [H/m], $\vec{J}$ is eddy current density [A/m^2], $\vec{B}$ is magnetic flux density [T], $\vec{F}$ is electromagnetic force [N/m^3].

The magnetic flux and eddy current densities in Equation 2 are shown in the jω complex field but they and the EMF can be redefined in the time domain as follows. Here, $B_{rx}, B_{ry}, B_{rz}, B_{ix}, B_{iy}, B_{iz}$ are the real and imaginary components of the magnetic flux density, while $J_{rx}, J_{ry}, J_{rz}, J_{ix}, J_{iy}, J_{iz}$ are the real and imaginary components of the eddy current density.

$$\vec{B} = \vec{B}_r + j\vec{B}_i = \vec{B}_r \cos\omega t + \vec{B}_i \sin\omega t = \begin{Bmatrix} B_x \\ B_y \\ B_z \end{Bmatrix} = \begin{Bmatrix} B_{rx} \cos\omega t - B_{ix} \sin\omega t \\ B_{ry} \cos\omega t - B_{iy} \sin\omega t \\ B_{rz} \cos\omega t - B_{iz} \sin\omega t \end{Bmatrix}$$

$$\vec{J} = \vec{J}_r + j\vec{J}_i = \vec{J}_r \cos\omega t + \vec{J}_i \sin\omega t = \begin{Bmatrix} J_x \\ J_y \\ J_z \end{Bmatrix} = \begin{Bmatrix} J_{rx} \cos\omega t - J_{ix} \sin\omega t \\ J_{ry} \cos\omega t - J_{iy} \sin\omega t \\ J_{rz} \cos\omega t - J_{iz} \sin\omega t \end{Bmatrix} \quad (3)$$

$$\vec{F} = \vec{J} \times \vec{B} = \begin{Bmatrix} F_x \\ F_y \\ F_z \end{Bmatrix} = \begin{Bmatrix} J_y B_z - J_z B_y \\ J_z B_x - J_x B_z \\ J_x B_y - J_y B_x \end{Bmatrix} \quad (4)$$

The electromagnetic force in the real domain is derived from Equations 2, 3 and 4. Only the x-directional component of the electromagnetic force is described below:

$$F_x = J_y B_z - J_z B_y$$
$$= (J_{ry} \cos\omega t - J_{iy} \sin\omega t)(B_{rz}\cos\omega t - B_{iz}\sin\omega t) - (J_{rz}\cos\omega t - J_{iz}\sin\omega t)(B_{ry}\cos\omega t - B_{iy}\sin\omega t)$$
$$= (J_{ry}B_{rz} - J_{rz}B_{ry})\cos^2\omega t + (J_{iy}B_{iz} - J_{iz}B_{iy})\sin^2\omega t - (J_{ry}B_{iz} + J_{iy}B_{rz} - J_{rz}B_{iy} - J_{iz}B_{ry})\sin\omega t \cos\omega t$$
$$= \frac{1}{2}(J_{ry}B_{rz} + J_{iy}B_{iz} - J_{rz}B_{ry} - J_{iz}B_{iy})$$
$$+ \frac{1}{2}(J_{ry}B_{rz} - J_{iy}B_{iz} - J_{rz}B_{ry} + J_{iz}B_{iy})\cos 2\omega t$$
$$- \frac{1}{2}(J_{ry}B_{iz} + J_{iy}B_{rz} - J_{rz}B_{iy} - J_{iz}B_{ry})\sin 2\omega t$$

(5)

Moreover, Equation 5 can be redefined as follows.

$$F_x = F_{DCx} + F_{ACx} \cos(2\omega t + \varphi_x) \qquad (6)$$

$$\begin{cases} F_{DCx} = \frac{1}{2}(J_{ry}B_{rz} + J_{iy}B_{iz} - J_{rz}B_{ry} - J_{iz}B_{iy}) \\ F_{ACx} = \frac{1}{2}\sqrt{(J_{ry}B_{rz} - J_{iy}B_{iz} - J_{rz}B_{ry} + J_{iz}B_{iy})^2 + (J_{ry}B_{iz} + J_{iy}B_{rz} - J_{rz}B_{iy} - J_{iz}B_{ry})^2} \\ \varphi_x = \tan^{-1}\left(\frac{J_{ry}B_{iz} + J_{iy}B_{rz} - J_{rz}B_{iy} - J_{iz}B_{ry}}{J_{ry}B_{rz} - J_{iy}B_{iz} - J_{rz}B_{ry} + J_{iz}B_{iy}}\right) \end{cases}$$

(7)

Then the complete electromagnetic force can be finally derived by redefining the y- and z-directional components in the same way:

$$\vec{F} = \begin{cases} F_{DCx} + F_{ACx} \cos(2\omega t + \varphi_x) \\ F_{DCy} + F_{ACy} \cos(2\omega t + \varphi_y) \\ F_{DCz} + F_{ACz} \cos(2\omega t + \varphi_z) \end{cases} = \vec{F}_{DC} + \vec{F}_{AC} \cos(2\omega t + \varphi) \qquad (8)$$

Here, $\vec{F}_{DC} = \{F_{DCx}, F_{DCy}, F_{DCz}\}^T$ is the DC component of the EMF, $\vec{F}_{AC} = \{F_{ACx}, F_{ACy}, F_{ACz}\}^T$ is the AC component of the EMF, $\varphi_x, \varphi_y, \varphi_z$ is the phase difference of the AC component of the EMF in the x, y, and z directions respectively, while $\{\ \}^T$ is transposed matrix.

The EMF is composed of a constant DC force component $\vec{F}_{DC}$ and a cyclic AC force component $\vec{F}_{AC}$, as shown in Equation 8. The AC force component has twice the cycling of the exciting input.

The conventional MHD calculation ignores the AC component of the EMF because the mass inertia in the fluid dynamic field is larger than the variation of the AC force component. When considering the timed-average flow of the molten metal this is not a problem but when considering the time variations of the free surface shape at low flows, a problem is created by ignoring this force.

MHD CALCULATION

Considering the molten metal as an incompressible fluid, its fluid dynamic field is defined by the equation of continuity and the Navier-Stokes equation as follows:

$$\begin{cases} \nabla \cdot \vec{\upsilon} = 0 \\ \rho \dfrac{D\vec{\upsilon}}{Dt} = \mu \nabla^2 \vec{\upsilon} - \nabla p + \rho \vec{g} + \vec{F} \\ \dfrac{D}{Dt} = \dfrac{\partial}{\partial t} + u \dfrac{\partial}{\partial x} + v \dfrac{\partial}{\partial y} + w \dfrac{\partial}{\partial z} \end{cases} \quad (9)$$

Here, $\vec{\upsilon} = \{u, v, w\}^T$ is velocity [m/s], μ is viscosity coefficient [Pa·s], p is pressure [Pa], ρ is density [kg/m^3], $\vec{g}$ is gravity [m/s^2], $\vec{F}$ is electromagnetic force [N/m^3].

The MHD calculation uses the LES (Large Eddy Simulation) which has excellent characteristics with high-Reynolds number fluid flows (turbulence model) together with the VOF (Volume of Fluid) method [7] that can analyze complicated surface shapes for the free surface model and the FDM (finite difference method) using a Staggered-Grid to discretize the three-dimensional field.

MHD CALCULATION RESULT

The distribution of the DC component of the electromagnetic force using the conventional method is shown in Figure 2(a), while the distribution of the full electromagnetic force (composed of the AC and DC components of the electromagnetic force) at $\omega t = 0$ using by the new method is shown in Figure 2(b). About point A in Figure 2, the time variation of the vertical electromagnetic force using each method is shown in Figure 3. As can be seen, the AC component of the vertical electromagnetic force using the conventional method is always constant and the full electromagnetic force composed of both the DC and AC components vibrates at twice the frequency of the exciting frequency.

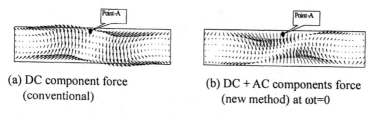

(a) DC component force
(conventional)

(b) DC + AC components force
(new method) at ωt=0

Figure 2. Distribution of the electromagnetic force (plane cross section).

Using only the conventional MHD calculation (i.e., considering only the DC component), the time variation of the vertical velocity at Point A is shown in Figure 4(a) while the free surface shape is shown in Figure 5(a). As can be seem in Figure 4(a), a long cyclic ripple flow is computed by the conventional method since the Reynolds Number coefficient of the molten metal is so large.

On the other hand, the results of the new MHD calculation using the full EMF are shown in Figure 4(b) and Figure 5(b) at Point A. The vertical velocity vibrates intensely compared with that predicted by the conventional MHD calculation while the vibration frequency is also twice the frequency of the exciting frequency as shown in Equation 8. This vibration of the vertical velocity allows for the computation of the disorder of the free surface shape as shown in Figure 5(b). As well, the timed-average of the vertical velocity is virtually the same for both calculation methods.

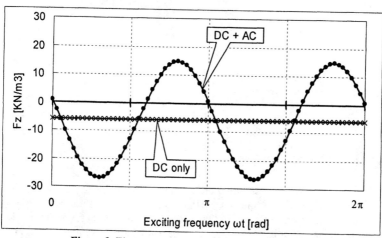

Figure 3. Time variation of the electromagnetic force.

We know that the free surface of the molten metal usually vibrates when using the EMS which is in agreement with this free surface vibration. This could not be predicted by the DC-only calculation.

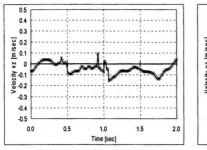 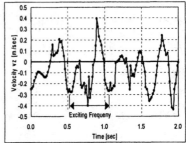

(a) By the conventional MHD calculation

(b) By the new MHD calculation

Figure 4. Time variation of the vertical velocity at Point-A.

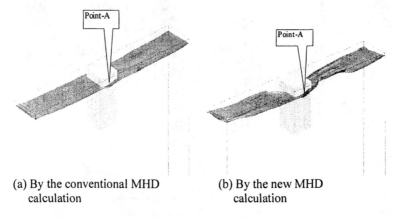

(a) By the conventional MHD calculation

(b) By the new MHD calculation

Figure 5. Free surface shape.

CONCLUSION

In this paper, we have reported on the characteristics of the EMF and the development and results of the new MHD calculation that accounts for the full electromagnetic force. As a result, the vibration of the free surface that could not be computed by the conventional MHD calculation using only the DC component was predicted by the new MHD method that includes both the AC and Dc components.

In continuous casting, the conditions of the free surface of the molten metal in the mold are very important since the conditions of the free surface are closely related to the quality and productivity of the slab. Therefore, the new MHD calculation that accounts for the full electromagnetic force as reported in this paper is an effective solution to this problem. In future work, we intend to research the design of the EMS coil to attempt to restrain the free surface vibrations by using this MHD calculation.

REFERENCES

[1] K. Fujisaki, K. Wajima, K. Sawada, and T. Ueyama, 1997. Application of electromagnetic technology to steel making plant, Nippon Steel Technical Report No. 74.

[2] K. Fujisaki, J. Nakagawa, and H. Misumi, 1994. Fundamental characteristics of molten metal flow control by linear induction motor, IEEE Transaction on Magnetics, 30(6), 4764–4766.

[3] H. Yamane, Y. Ohtani, J. Fukuda, T. Kawase, J. Nakashima, and A. Kiyose, 1997. High power in-mold electromagnetic stirring—Improvement of surface defects and inclusion control, Steelmaker Conf. Proc., 80, 159–164.

[4] K. Fujisaki, T. Ueyama, and K. Okazawa, 1996. Magnetohydrodynamic calculation of in-mold electromagnetic stirring, IEEE Transaction on Magnetics, 33(2), 1642–1645.

[5] K. Fujisaki and T. Ueyama, 1997. Magnetohydrodynamic calculation in free surface, Journal of Applied Physics, pt.2, 83(11), 6356–6358.

[6] T. Ueyama, K. Shinkura, and R. Ueda, 1989. Fundamental equation for analysis by using the A-φ method and 3-D analysis of a conducting liquid, IEEE Transaction on Magnetics, 25(5), 4153–4155.

[7] C.W. Hirt and B.D. Nichols, 1981. Volume of Fluid (VOF) method for the dynamics of free boundaries, Journal of Computational Physics, 39, 201.

Fabrication of Ceramic Hip Joint by Bingham Semi-Solid/Fluid Isostatic Pressing Method

F. TSUMORI, N. YASUDA and S. SHIMA
Department of Mechanical Engineering, Kyoto University, Yoshida-honmachi, Sakyo-ku, Kyoto, 606-8501, Japan

ABSTRACT

A new powder compaction method with a semi-solid mold is proposed and developed to fabricate a ball part for ceramic hip joints. To produce ball parts with a homogeneous density distribution, Cold Isostatic Pressing (CIP) followed by mechanical working has been employed in the past. However this conventional process has some problems about productivity and causes a big amount of material loss. In this study, Bingham Semi-solid/fluid Isostatic Pressing (BIP) process is proposed to overcome these problems. For demonstration, an attempt is made to fabricate near net-shape powder compacts of the ball part for hip joint by this method. Finally, productivity of the present method is discussed and compared with the conventional process.

INTRODUCTION

Currently there is a large demand for artificial joints, in particular hip joints, and it is anticipated that it would increase in future. As depicted in Figure 1, an artificial joint comprises a sliding pair of a ball and a socket; the former is connected to a stem that is inserted to the bone of the lower limb, while the latter is connected to the bottom of the thighbone. Regarding the clinically used artificial joints, the ball is metallic and the socket is made of polyethylene in most cases; they are called polyethylene-metal joints. These hip joints encounter a serious problem called "polyethylene disease", which is the reaction of a living body against fine particles generated because of the wear at the surface of the polyethylene socket.

To overcome this problem, metal-metal and ceramic-ceramic joints have been developed. For the former, the long term effect of metallic ions that are circulated with blood through the body is not yet known. On the other hand, balls made of alumina or zirconia have been attempted, but fracture or cracking occurs in an alumina ball due to low fracture toughness, while in the zirconia, transformations may cause a serious effect on the living body has been pointed out.

A hip joint of a pair of alumina ball and alumina socket is nontoxic, and very little wear occurs. However, there is quite a large technical issue in fabrication of alumina-alumina joints. They are ordinarily fabricated by the following procedure (see Figure 2): isostatic compaction followed by machining of the green body, sintering, coarse grinding and finish grinding. However, the whole process is time consuming and also the material yield is as low as 30%. The products are, therefore, quite expensive. An innovative process is thus required that provides artificial hip joint with a higher performance and high durability at lower cost.

In this study, an attempt is made to fabricate a near net-shape compact of the ball part. Candidate processes are cold isostatic pressing (CIP) and rubber isostatic pressing (RIP). The latter process is of a higher productivity than the former. However two problems are encountered owing to the rubber mold used in these two processes. The first is the stress state. The pressure exerted to the outer surface of the mold is isostatic, while the stress at the inner surface is not necessarily so, in particular, in the vicinity of the corner portions of the mold. The second is more serious. The elastic mold shrinks during compaction and expands greatly during unloading. The spring-back during unloading causes cracks and sometimes breakage in the powder compact. If the product shape has a recess like the ball part of the artificial joint, damage would occur more easily. To shape a recess to the ball part, unfavourable mechanical working is subsequently necessary.

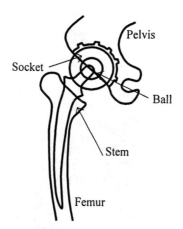

Figure 1. Schematic view of an artificial hip joint.

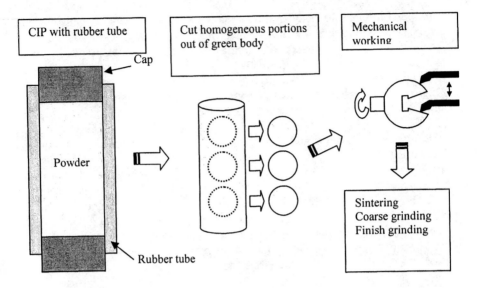

Figure 2. Conventional fabrication procedure of artificial hip joint.

In the present work, we propose using a "Bingham Semi-solid/fluid Pressing (BIP)" to fabricate ceramic hip joints. Bingham semi-solid material is employed for the mold material to realize ideal isostatic state and avoid cracks caused by the mold's spring-back during unloading. In the following section, details of the present BIP process are described.

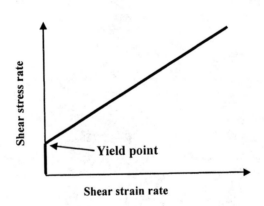

Figure 3. Typical strain rate-stress rate curve of Bingham semi-solid material.

BINGHAM SEMI-SOLID/FLUID ISOSTATIC PRESSING (BIP)

Mechanical Property of Bingham Semi-Solid Material

A typical strain-rate versus stress rate curve of the Bingham semi-solid material is shown in Figure 3. Two essential features for the present process are shown in this diagram:
1) The semi-solid material reveals a yield stress much smaller than the pressure commonly used for powder compaction processes.
2) The mold flows like liquid when the load exceeds the yield point.

In other words, the mold behaves like a solid below the yield point and like a liquid above the yield point. These characteristics are quite useful for the isostatic pressing process. The mold can form a cavity, in which the powder is filled, as a solid material before pressing, while it transmits pressure isostatically to the powder as a liquid material during pressing. The spring-back effect will be virtually neglected in this case, since the mold material is not a solid above the yield point.

In this study, semi-solid wax material called petrolatum or Vaseline is employed as a semi-solid mold material, which is cheap and will not remain in the sintered final product. It can easily be removed by heating and if this petrolatum remains in the green body, there is no problem because it is decomposed into CO_2 and H_2O during the subsequent sintering and no toxic ingredient is generated in the body.

Flow in the BIP Process

The present process consists of three steps; mold forming, pressing process and removal of the mold. At first, a master pattern is set in a metallic or rubber container, and heated semi-solid wax or liquid wax is poured into it as shown in Figure 4.

The container makes handling of the semi-solid material easier. After curing, the master pattern is removed. If it is difficult or impossible to remove the master without breakage of the mold due to its geometry, the mold should be designed to be split into two or more parts. Next, the mold is filled with powder and pressurized. The pressure is transmitted to the powder in the mold cavity. Semi-solid material works as a fluid medium which would realize the ideal isostatic state, so either isostatic pressing or other pressing method is favorably employed. For example, the semi-solid mold can be used as a capsule in CIP (See Figure 5). Another example is uni-axial pressing in a metallic container as in Figure 6. In the final step, the pressed powder compact is taken out of the mold. The compact is covered with semi-solid wax, however, it can easily be removed by heating. We are able to recycle the melted wax in the process. So, there is virtually no waste in the process. The powder compact will be sintered like other powder metallurgy processes.

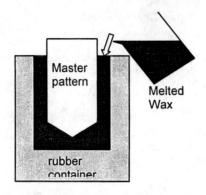

Figure 4. Forming process of semi-solid mold.

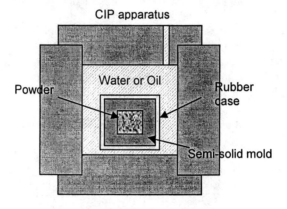

Figure 5. Schematic view of CIP with semi-solid mold.

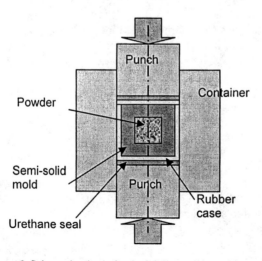

Figure 6. Schematic view of uniaxial die pressing with semi-solid mold.

EXPERIMENTS AND RESULTS

Homogeneity of Green Compact

If an ideal isostatic state is achieved by the present process, green compacts with a homogeneous density distribution would be obtained. To check this homogeneity, density distribution in a green compact should be observed, which is not available in the present work. Instead of the density distribution, shrinkage during sintering is measured. If the shrinkage ratio is constant in a sample, the density distribution would be homogeneous. As a sample, a green compact of Al_2O_3 powder with 0.2 µm mean particle size was made by BIP at 100 MPa, where the mold material was petrolatum with 50 penetration scale (ASTM D217 penetration test). In this process, the powder was filled in the cavity of petrolatum mold and pressed in a metallic container as shown in Figure 6. This compact was sintered under a temperature of 1350°C for 2 hours, and the shrinkage ratio due to sintering was measured. The shape of the compact is a circular cylinder as shown in Figure 7.

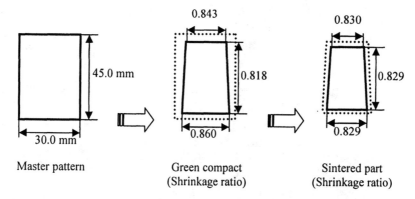

Figure 7. Shrinkage of BIPped sample.

The shape of the master pattern is shown to the left in the diagram. The shape of the green compact does not resemble the mold cavity because of inhomogeneities in the initial density distribution. Nevertheless, the shape of the sintered body is similar to that of the green compact. It can thus be concluded that the resultant sintered compact by the BIP process is of a homogeneous density distribution.

Near-Net-Shaping of Ball Part

As mentioned above, an artificial joint comprises a sliding pair of a ball and a socket. In this section, some samples of green compacts of this ball parts are shown for demonstration. The shape of the master pattern is a sphere with a recess and it is impossible to remove without breakage of the mold. The semi-solid mold is designed to be split into three parts. Figure 8 shows the flow of the experimental procedure. The same powder and semi-solid mold material were used as in the previous experiment. To press the powder in the semi-solid mold, die-pressing or CIP was employed. The powders were successfully compacted without breakage as shown in Figure 9.

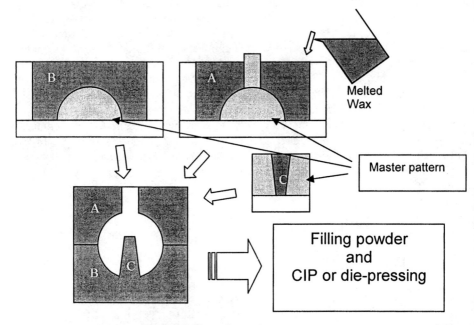

Figure 8. Flow of experiment.

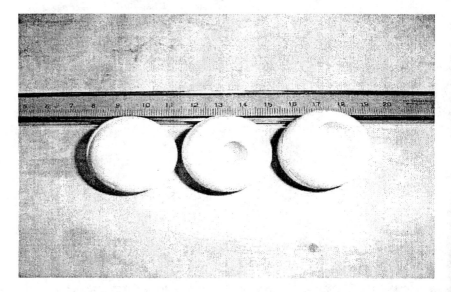

Figure 9. Green compacts fabricated by BIP process.

CONCLUSIONS

The present experiment shows that the newly developed "Bingham Semi-solid/Fluid Pressing (BIP)" has a large potential in fabricating ceramic hip joints by which homogeneous complicated green parts can be fabricated quite easily with a very high material yield (over 90%). With the aid of computer simulation, which we are developing now, the shape of the mold cavity would be optimized to produce a desired shape of the green body, thus the material yield would be more improved. In the present paper, the ball parts of hip joint are shown for demonstration, and there will be no problem in fabrication of socket parts.

REFERENCES

[1] ASM HANDBOOK, 1998. Vol. 7, Powder Metal Technologies and Applications.
[2] R. J. Henderson, 2000. Bag design in isostatic pressing, Materials and Design 21, 259–262.
[3] M. Sagawa et al, 2000. Rubber isostatic pressing (RIP) of powders for magnets and other materials, Materials and Design 21, 243–249.
[4] F. Tsumori et al, 2002. Development of Semi-solid Isostatic Pressing Method for Powder Compaction, Advanced Technology of Plasticity 2, 1237–1242.

Control of Surface Layer Formed in Fe-Si-0.4mass%Mn Alloys by Annealing under Low Partial Pressure of Oxygen

S. SUZUKI, H. HASEGAWA, S. MIZOGUCHI and Y. WASEDA

Institute of Multidisciplinary Research for Advanced Materials, Tohoku University, Aoba-ku, Sendai, Japan

ABSTRACT

Surface layer control is required for high-strength steel sheet to be used as advanced and practical metallic materials such as automobile bodies, since the sheets are hot-dip galvanized immediately upon production. This control is required because of characteristic layers that form on the surface of high-strength steel sheet containing a relatively high amount of silicon during annealing. In this work, secondary ion mass spectrometry (SIMS) and X-ray photoelectron spectroscopy (XPS) have been used to analyze surface layers formed in iron-silicon-manganese alloys with different amounts of silicon after annealing in hydrogen and argon gases with a low partial pressure of oxygen. SIMS depth profiles showed that silicon in these alloys reacts with oxygen that penetrates into the bulk to form silicon oxides. A characteristic distribution of silicon oxides is obtained in the surface layer depending on the bulk silicon concentration.

Manganese in these alloys is enriched to the outer side of the surface layer. On the other hand, XPS results show that silicon and manganese are enriched on the top surface in these alloys with the surface concentration of these elements depending on bulk silicon concentration. The experimental results are discussed on the basis of the thermodynamics of the elements. Factors that control surface layers formed in high-strength steel sheet during annealing are also discussed.

INTRODUCTION

High-strength steel sheets are used for structural materials such as automobile bodies, and their mechanical properties have been gradually improved by using modification of addition of alloying elements and process conditions [1]. However, a characteristic surface layer is known to form on the surface of high-strength low-alloyed steel sheets containing a few percents of alloying elements during annealing, and strongly influence the surface properties [2,3]. Thus, control of a surface layer formed in low-alloyed steel sheets, which are typically hot-dip galvanized, is required in order to establish manufacturing processes of galvanized high-strength steel sheets.

In general, steel sheets are annealed at high temperatures under a low partial pressure of oxygen, in which iron is not oxidized but some of reactive elements on the surface of steels are oxidized. For instance, a small amount of alloying elements such as silicon and manganese in steel sheets can be, more or less, oxidized on the steel surface during annealing. In such oxidation, that is selective oxidation, oxygen in an atmosphere of a low pressure of oxygen penetrates the steel to selectively react with a reactive alloying element [4,5]. The morphology of an oxide particles or layers formed in steels is influenced by the chemical characteristics of alloying elements and oxidation conditions. Among a variety of steels, selective oxidation has been studied in Fe-3mass%Si steel, that is the electrical steel, which is annealed at a high temperature under a medium partial pressure of oxygen in a decarburization process [6-8]. In these studies, the compositional and morphological changes of an internal oxidation zone formed in the Fe-3mass%Si sheet during annealing have been focused. Such selective oxidation is also considered to take place in low-alloyed steel sheets, of which the annealing atmosphere is not sufficiently controlled. The chemical composition and morphology of oxides formed in a surface layer of the low-alloyed steel by annealing are also likely to depend on the species and amount of alloying elements and oxidation conditions.

This background prompts us to study the effect of the silicon concentration on a surface layer formed in an iron-based alloy by annealing at a high temperature under a low partial pressure of oxygen. This work focuses on the depth distribution and surface concentration of alloying elements in surface layers formed in Fe-Si-Mn alloys, which is mostly fundamental system for the high-strength steel. Secondary ion mass spectrometry (SIMS) and X-ray photoelectron spectroscopy (XPS) were used for analyzing the depth profile and the chemical composition of a surface layer containing oxides of alloying elements. The experimental results are interpreted in conjunction with the thermodynamic characteristics of the alloying elements, and surface control of high-strength steel sheets is discussed on the basis of these results.

EXPERIMENTAL

Sample preparation

Buttons of Fe-Si-0.4mass%Mn alloys with 1, 2, 3 and 4.5mass% silicon, which are hereafter referred to as 1mass%Si, 2mass%Si, 3mass%Si and 4.5mass%Si, were prepared by arc melting. The method for sample preparation is the same as that adopted in the previous studies on *in-situ* observation of the microstructure [9]. The chemical compositions of these alloys are given in **Table 1**. Parts of the buttons were rolled to sheets of 0.5 mm in thickness, and the sheets were homogenized by annealing at 1273 K for 10^4s. After the sample sheets were mechanically polished, they were finally annealed at 1023 K for 1800 s in

9.8%H_2-Ar gas with the H_2O / H_2 ratio of 0.07. The present samples are fully recrystallized at this annealing temperature. This H_2O / H_2 ratio corresponds to the partial pressure of oxygen of 8x10^{-14} Pa at 1023 K, in which iron in alloys is not oxidized while silicon and manganese are oxidized [10].

MEASUREMENTS

Sputter depth profiles of constituent elements in surface layers of samples were measured using SIMS, PHI-6600 with quadrupole-type mass spectrometer. An incident beam of 5.0 keV Cs^+ ions was used for sputtering the sample surface, and positive secondary ions were analyzed. In particular, positive ions of $^{30}Si^+$, $^{55}Mn^+$, $^{144}(CsO)^+$ and $^{189}(CsFe)^+$ were selected in order to avoid interference of different kind ions. Sputter depth profiles of these ions are considered to reasonably correspond to depth distribution of silicon, manganese, oxygen and iron, since the matrix effect of CsM^+ ions is known to be relatively small, as reported in previous works [11].

Table 1. Chemical composition of samples used. (mass%)

Sample	Si	Mn	P	S	O
1mass%Si	1.01	0.41	0.003	0.015	0.0006
2mass%Si	1.91	0.38	<0.002	0.014	0.0009
3mass%Si	2.93	0.36	0.002	0.014	0.0012
4.5mass%Si	4.49	0.40	0.003	0.014	0.0023

Indeed, it has been confirmed that SIMS sputter depth profiles of iron-based alloys by these ions correspond to quantitative depth profiles obtained by glow discharge optical emission spectrometry, in which the matrix effect is small [8]. A stylus profiler was used for measuring the depth sputtered by SIMS. The sputtering rate was estimated from the sputtered depth and sputtering time.

X-ray photoelectron spectroscopy (XPS), PHI-5600, was used for characterizing the surface chemical composition and state samples. The operation conditions of XPS have already been described elsewhere [12]. The incident X-ray was monochromatized Al-Kα. XPS spectra were measured after the sample surface was slightly sputtered by argon ions, in order to remove a contaminated carbon layer from the sample surface. The surface chemical composition was evaluated from XPS spectra coupled with relative sensitive factors [13].

RESULTS AND DISCUSSION

Depth profiles by SIMS

Figures 1(a), (b), (c) and (d) show SIMS sputter depth profiles of $CsFe^+$, CsO^+ and Si^+, from surface layers of four steel samples annealed under the low partial pressure of oxygen. In these plots, secondary ion counts are shown in linear scale, and the level of ion counts of CsO^+ and Si^+ are adjusted by a factor for comparison. The ion counts of Si^+ in these depth profiles are fairly correlated with those for CsO^+, indicating that particles of silicon oxides, probably mainly SiO_2, are formed in the surface layer. However, it should be noted that the ion counts of CsO^+ do not decrease with increasing depth, and the shape of the depth profiles depends on the bulk silicon concentration. If oxygen penetrating into the alloy during annealing is selectively reacted with silicon dissolved in the alloys to form separate oxide particles, such shape is not expected. The shape of the depth profiles and dependence of bulk silicon concentration on the depth profiles imply that silicon oxide particles are interconnected in the surface layers, and the interconnection significantly occurs in the alloys with a large bulk amount of silicon. Indeed, such interconnection of silicon oxides has been observed in an Fe-3mass%Si alloy, in which oxidation is referred to as internal oxidation [8].

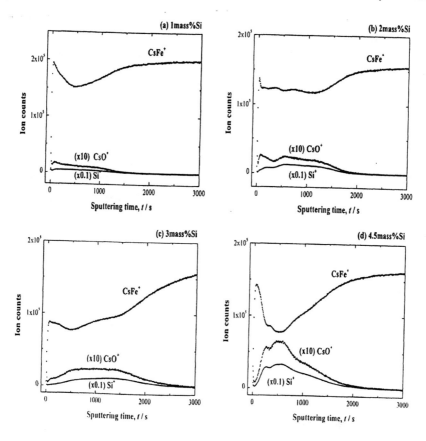

Figure 1. SIMS depth profiles of CsFe$^+$, CsO$^+$ and Si$^+$ for surface layers formed in (a) 1%Si, (b) 2%Si, (c) 3%Si and (d) 4.5%Si. The ion counts are plotted in linear scale.

In order to focus the distribution of alloying elements at low concentration level in surface layers formed in 1mass%Si and 2mass%Si, which are representative bulk silicon compositions in high-strength steels, secondary ion counts of CsFe$^+$, Mn$^+$, Si$^+$ and CsO$^+$ for 1mass%Si and 2mass%Si are plotted in logarithm scale as a function of sputter depth, as shown in **Figure 2(a) and (b)**, respectively. The profiles show that the ion counts of Mn$^+$ almost decrease with increasing depth in these alloys, although some step is observed in the surface side of depth profiles. This indicates that manganese is enriched to the surface side to form oxides, presumably manganese silicates. This profile may be induced by an oxygen potential gradient form the surface to bulk, and it is contrast to the depth profiles of silicon and oxygen, which are correlated in the surface layer.

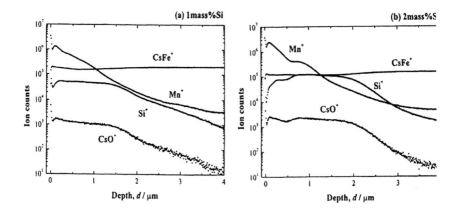

Figure 2. SIMS depth profiles of CsFe, Mn, CsO and Si for surface layers formed in (a) 1%Si and (b) 3%Si.

The ion counts are plotted in logarithm scale.

CHANGES IN THE AMOUNT OF ELEMENTS IN SURFACE LAYER

The SIMS results show that oxygen penetrates into the samples to form oxides of silicon and manganese in a surface layer by annealing under the low partial pressure of oxygen. Increases in the amounts of oxygen and these elements in the surface layer by annealing are semi-quantitatively evaluated by total secondary ion counts from surface layers of about five micrometers thick. **Figure 3** shows the total secondary ion counts of $CsFe^+$, Si^+ and CsO^+ as a function of the bulk silicon concentration in the alloy. Ion counts of Si^+ in these depth profiles correlate with those for CsO^+ indicating that the ions of Si^+ likely originate from silicon enrichment in surface oxide layers with the amount of silicon oxides increasing with increasing bulk silicon concentration. These results are consistent with the dependence of silicon concentration on total secondary ion counts in the sputter depth profiles as shown in Figure 1.

The secondary ion counts of $CsFe^+$, mainly attributed to metallic iron, appear to depend on the bulk silicon concentration in a more complicated manner. This is considered to include the matrix effect on occurrence of these secondary ions since oxide particles such as silicon oxides and silicate are finely distributed in metallic iron in the surface layer. Thus, quantitative discussion is difficult on the basis of only SIMS data. Nevertheless, the present sputter depth profiles suggest that oxide particles such as silicon oxides are distributed in the surface layer as interconnected grains while manganese distributes to the surface side to form manganese silicates.

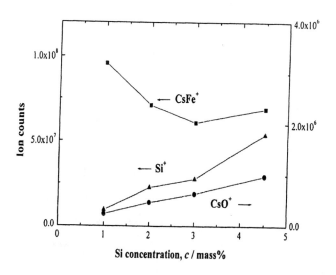

Figure 3. Total secondary ion counts from a surface layer formed by annealing as a function of the bulk silicon composition of the Fe-Si-mass%Mn manganese silicate [13].

SURFACE COMPOSITION AND STATE BY XPS

While SIMS provides information on relative compositional changes in the depth profiles, XPS is effective to analyses the chemical composition and state of the top surface layer thickness of a few nanometers. The XPS wide spectra using the results on the surface of 1%Si and 2%Si are presented in **Figure 4**. The spectra show that silicon and manganese are enriched in the top surface layer with the surface concentration of manganese increasing with increasing bulk silicon concentration. Silicon and manganese are oxidized at the sample surfaces during annealing to form silicon oxide and manganese silicate. O 1s peaks observed in these XPS spectra are attributed to the oxide formed during annealing and the native oxide formed by reaction of metallic iron on the sample surface with oxygen in air at room temperature after annealing.

In order to investigate the chemical state of oxygen, O 1s XPS spectra from the surface of 1mass%Si, 2mass%Si and 3mass%Si were measured as shown in **Figure 5**. Two peaks are seen at about 533.0 and 530.5 eV in the 1mass%Si sample. While the peak at 533.0 is allocated to SiO_2, the peak at 530.5 eV is attributed to native iron oxide, iron silicate and manganese silicate [13].

The Si 2p XPS spectra from the surface of 1mass%Si, 2mass%Si and 3mass%Si are shown in Figure 6, although the spectra appear to be noisy for the relatively low sensitivity of Si 2p. A peak between 103.5 eV and 104.0 eV in 1mass%Si and a peak at about 102.5 eV in 2mass%Si and 3mass%Si are observed in these spectra. These results also indicate dominant formation of silicon oxide on the surface of 1mass%Si and formation of manganese or iron silicate on the surface of 2%Si and 3%Si, which is also consistent with the results as shown in Figure 4. In addition, Mn 2p XPS spectra from the surface of the samples, which are not shown here, also suggest that manganese is mainly present as manganese silicate on the sample surface.

This peak is dominant in the spectra of 2%Si and 3%Si, which is consistent with the fact that manganese is considerably enriched to the top surface of these alloys, as shown in Figure 4.

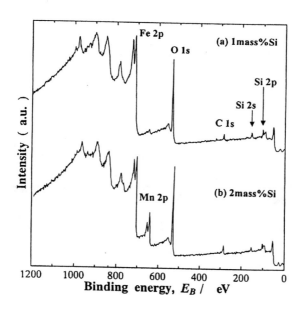

Figure 4. XPS wide spectra from the surface of (a) 1mass%Si and (b) 2mass%Si.

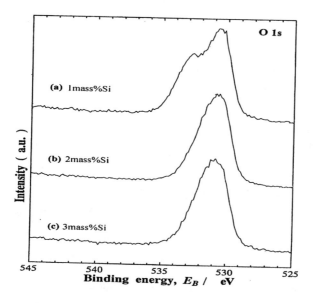

Figure 5. O 1s XPS spectra from the surface of (a) 1mass%Si, (b) 2mass%Si and (c) 3mass%Si.

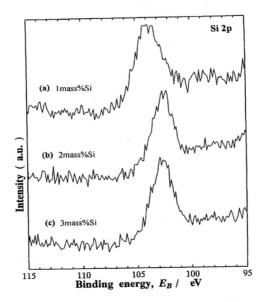

Figure 6. Si 2p XPS spectra from the surface of (a) 1mass%Si, (b) 2mass%Si and (c) 3mass%Si.

CHARACTERISTIC FEATURE OF FORMATION OF SUFACE LAYER

Let us consider the characteristic features of a surface layer formed in Fe-Si-Mn alloys by annealing under a low partial pressure. Figure 7 illustrates schematic diagrams for formation of surface layers in Fe-Si alloys containing (a) a small amount of silicon and (b) a large amount of silicon by annealing under a low partial pressure of oxygen, and in Fe-Si alloys containing a large amount of silicon by annealing under (c) a very low partial pressure and (d) a high partial pressure of oxygen. Oxygen originated from a low partial pressure of oxygen penetrates into the alloys to react with reactive elements such as silicon without oxidation of matrix iron. The frequency of interconnection between particles of silicon oxides increases with increasing bulk silicon concentration. Metallic iron formed by the interconnection of silicon oxides may be squeezed to outside of the surface layer, which results in the formation of and iron-rich layer on the surface side. In addition, the amount of manganese appears to be larger on the surface side of the surface layer, as shown in Figure 2. Since manganese can be reacted with silicon oxides to form manganese silicate [14], an increase in bulk silicon concentration is also likely to induce the enrichment of manganese, that is formation of manganese silicates, at the surface, as shown in Figure 4.

In addition, a continuous silicon oxide or manganese oxide layer was found to form during annealing under very low partial pressure of oxygen, as illustrated in Figure 7(c) [8]. On the other hand, an iron oxide layer forms by annealing under a high partial pressure of oxygen, as shown in Figure 7(d) [15]. The surface compositions of these two cases are considerably different from those of iron and steel.

Finally, processing factors for surface control of Fe-Si-Mn alloys annealed under a partial pressure of oxygen are discussed. Figure 8 presents a map of the processing factors of the bulk alloying elements and partial pressure of oxygen to control the distribution of oxides formed in the surface layer of Fe-Si-Mn alloys by annealing at high temperature. These results indicate that particles of silicon oxides dispersed in the surface layer form by annealing under a low partial pressure of oxygen with the amount of particles increases with increasing bulk Si concentration. On the other hand, a continuous layer of silicon oxide or manganese silicate is known to form in steels containing large amounts of silicon by annealing under very low partial pressures of oxygen [8], although it must be noted that reducing silicon oxides may be achieved by annealing in practical manufacturing processes. As well, iron oxides formed at high partial pressure of oxygen. The densely dispersed oxides in the surface layer and continuous oxide layer are likely inconvenient for surface reactions such as galvanization since the surface composition is considerably different from that of the alloy. Thus, the schematic diagram indicates there is an optimum condition to control the surface layer to obtain adequate surface properties of Fe-Si-Mn alloys. The condition become severe when the sample includes large amounts of Si and Mn.

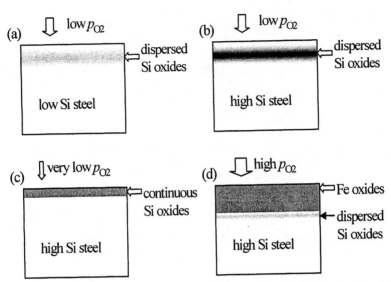

Figure 7. Schematic diagram for formation of surface layers in Fe-Si-Mn alloys containing (a) a small amount of silicon (low Si steel) and (b) a large amount of silicon (high Si steel) by annealing in a low partial pressure of oxygen, and in Fe-Si-Mn alloys containing a large amount of silicon by annealing under (c) a very low partial pressure and (d) a high partial pressure of oxygen.

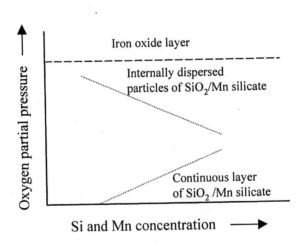

Figure 8. Map representing the morphology of oxides formed in Fe-Si-Mn alloys under conditions of different partial pressure of oxygen and different alloy composition.

CONCLUDING REMARKS

SIMS and XPS have been used to analyze surface layers formed in Fe-Si-0.4mass%Mn alloys with 1, 2, 3 and 4.5 mass% silicon that were annealed at a low partial pressure of oxygen. The main results obtained were as follows:

1) Oxygen originating from the atmosphere penetrates into the alloy to react with elements such as silicon and manganese without oxidation of matrix iron. Silicon oxide particles are interconnected with the frequency of interconnection between particles increasing with increasing bulk silicon concentration.
2) An iron-rich layer on the surface side is considered to form by the squeeze of metallic iron due to the interconnection of oxides.
3) Manganese is enriched to the outer side of the surface layer to form manganese silicate. Silicon and manganese are enriched on the top surface in these alloys.
4) A method was proposed on the basis of the alloying element map and partial pressure of oxygen to control the surface layer, i.e., the surface properties of Fe-Si-Mn alloys.

ACKNOWLEDGEMENT

The authors are grateful to Mr. T. Sato and Mr. M. Itoh for their help and maintenance of the XPS apparatus. A part of this work was supported by Nippon Steel Corporation.

REFERENCES

1. For example; K.Sugimoto, R.Kikuchi and S.Hashimoto, 2002, Steel Research, **73**, 253.
2. T.Usuki, A.Sakota, S.Wakano and M.Nishihara, Tetsu-to-Hagane, 1991, **77**, 84.

3. Y.Tsuchiya, S.Hashimoto, Y.Ishibashi, J.Inagaki and Y. Fukuda, Tetsu-to-Hagane, 2000, **86**, 396.
4. P.Kofstad: High Temperature Corrosion, Elsevier Applied Science, London, 1988, p.324.
5. D.L.Douglass, Oxid. Metals, 1995, **44**, 81.
6. W.Block and N.Jayaman, Mater. Sci. Tech., 1986, **2**, 22.
7. S.Suzuki, K.Yanagihara, S.Hayashi and S.Yamazaki, J.Surf.Anal.,, 1999, **5**, 274.
8. K.Yanagihara, S.Suzuki and S.Yamazaki, Oxid.Metals, 2002, **57**, 281.
9. H.Hasegawa, K.Nakajima and S.Mizoguchi, Tetsu-to-Hagane, 2001, **87**, 433.
10. I.Barin, Thermochemical Data of Pure Elements, VCH, Weinheim, 1989.
11. T.Gao, Y.Marie, F.Saldi and H.-N.Migeon, Int. J. Mass Spctrom. Ion Processes, 1995, **143**, 11.
12. S.Suzuki, T.Kosaka, H.Inoue, M.Isshiki and Y.Waseda, Appl. Surf. Sci., 1996, **103**, 495.
13. J.F. Moudler, W.F. Sticle, P.E. Sobol and K. Bomben, X-ray photoelectron spectroscopy, Physical Electronics: Minnesota, 1993, p.213.
14. S.K.Saxena, N.Chatterjee, Y.Fei and G.Shen: Thermodynamic data on oxides and silicates, Springer-Verlag, Heidelberg, 1993.
15. S.Suzuki, M.Wake, M.Abe and Y.Waseda, ISIJ Inter., 1996, **36**, 700.

Development of a Method to Fabricate Metallic Closed Cellular Materials Containing Organics

S. KISHIMOTO, Z. SONG and N. SHINYA
*Intelligent/Smart Materials Group, Materials Engineering Laboratory,
National Institute for Materials Science, Tsukuba, Ibaraki, Japan*

ABSTRACT

A new fabricating method for metallic closed cellular materials which have a high energy absorbability and a light weight, has been developed using a powder particle assembling technique. Powder particles of polymers coated with a nickel-phosphorus alloy layer using electroless plating were pressed into pellets and sintered at high temperature. A metallic closed cellular material containing polymer or organic materials was then fabricated.
The physical, mechanical and ultrasonic properties of this material were measured. The density of this material is same as that of aluminum alloy. The compressive tests of this material and the damping property measurements were carried out. The results of compressive tests showed that this material has a low Young's modulus and high-energy absorption. The result of the internal friction measurement showed that the internal friction of this material was same as that of pure aluminum. These results indicate that this metallic closed cellular material can be used for the passive damping and energy absorbing systems.

INTRODUCTION

Many current researches have been studied to develop many kinds of functional materials and systems. Particularly, passive and active damping functions are becoming increasingly important in terms of vibration control of the structures and energy absorbing system has been required to protect persons from injury during impact of accident. Therefore, these materials which has high-energy absorbability and high-damping functions are required.

Recently, cellular materials are receiving renewed attention as structural and functional materials. Cellular materials are thought to have unique thermal, acoustic and energy absorbing properties that can be combined with their structural efficiency [1]. Therefore, many kinds of cellular materials have been tested as energy absorbing and damping materials. Particularly, the closed cellular materials are thought to have many favorable properties and applications. However, there is a lack of technique to produce such fine closed cellular materials except for the gas forming [2–7], the sintering of hollow powder particles[8,9] and the two-dimensional honeycomb structures[1][1].

Therefore, authors have developed a fabrication process of metallic closed cellular material containing organic materials [10].

In this study, a metallic closed cellular material containing organic materials was fabricated and the physical, mechanical and ultrasonic properties of this closed cellular material containing organic materials are measured. And the effects of sintering temperature and thickness of the cell walls on the mechanical properties have been analyzed. The utility of this material is also discussed.

CONCEPTUAL PROCESS

The metallic closed cellular material fabricating process is shown in Figure 1. The process is as follows: 1) Powdered polymer particles are coated with a metal layer using electroless plating. 2) The powder particles are pressed into pellets (green pellets) by cold isostatic pressing. 3) After sintering at high temperature in a vacuum, the closed cellular material is produced.

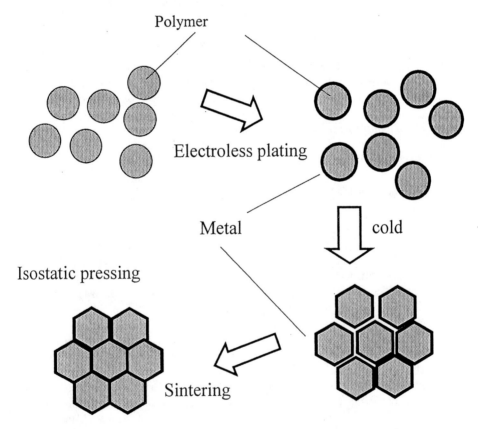

Figure 1. Flow diagram of metallic closed cellular material fabricating process.

EXPERIMENTAL PROCEDURE

Preparing the Metallic Closed Cellular Material

A thermal plastic polymer, polystyrene, particles of 10 μm diameter (Japan Synthetic Rubber Co., Ltd.) was selected for this study. These polystyrene particles were coated with 0.19 μm –0.53 μm thick nickel-phosphorus alloy layers using electroless plating. Figure 2 shows the schematic cross-sectional view of this polystyrene particle. These particles were pressed into pellets (green compacts) with about 8 and 16 mm diameters and 8 mm long by isostatic pressing at 200MPa and 363K. After this, these green compacts were sintered for 1 h at 1073K and 1123K in a vacuum.

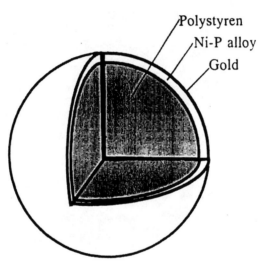

Figure 2. Schematic cross-sectional view of the polystyrene particles coated with Ni-P alloy.

Characterization

The microstructure of the powder particles, the green pellets before sintering, the cross-section after sintering and the fracture surface were observed using a scanning electron microscope (SEM). To prepare the specimen for observation of the cross-section, the specimen was cut and the cross-section surface was polished using emery paper (#600) and then 0.05 μm Al_2O_3 powders. To measure the mechanical properties, compressive tests were performed at room temperature. Ultrasonic measurements were carried out to estimate the attenuation coefficient of this material. The measurement was carried out with a 6.4 mm diameter probe generating a longitudinal wave of 10MHz at room temperature. In addition, damping tests were carried out to estimate the internal friction of this material. The measurement was carried out using an about 1.0mm thickness plate-type specimen and a free resonance vibration-type equipment (JE-RT, Nihon Techno-Plus Co., Ltd.). The internal friction was calculated by the half width method.

RESULTS

Micro-Structural Observation

Figure 3 shows an SEM image of the polystyrene powder particles coated with Ni-P alloy layers and Figure 4 shows an SEM image of the green compact after cold isostatic pressing at 363K and 200MPa. The polystyrene particles were deformed to polyhedrons by isostatic pressing. The surface of the polystyrene particles coated with the nickel-phosphorus alloy exhibited facets. An SEM image of the cross-section of this material after sintering at 1073K is shown in Figure 5. In this figure, the cell walls of the nickel-phosphorus alloy are observed as bright parts and the material inside the cell walls is observed as the darker parts.

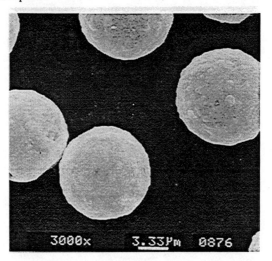

Figure 3. SEM image of polystyrene particles coated with Ni-P alloy layer.

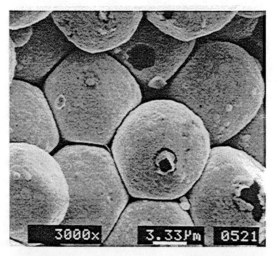

Figure 4. SEM image of green compact after isostatic pressing.

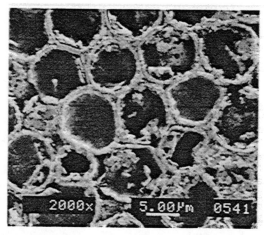

Figure 5. Cross-section of the metallic closed cellular material.

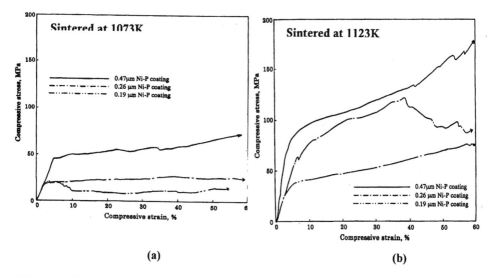

Figure 6. Compressive stress-strain curve for metallic closed cellular materials; (a) sintered at 1073K and (b) sintered at 1123K.

Compressive Tests

Compressive tests were carried out at room temperature. A typical example of the compressive test results is shown in Figure 6(a) and Figure 6(b). Figure 6(a) shows a stress strain curves of the specimens sintered at 1073K and Figure 6 (b) shows stress-strain curves of the specimens sintered at 1123K. The stress-strain curve shows a linear elastic region, a long plateau where the stress gradually increases and a wavy region where the stress repeatedly decreases and increases. A specimen sintered at 1123K has higher strength than that of specimens sintered at 1073K. Figure 7 shows the fracture surface of compressive tested specimens sintered at 1123K. The cells at fracture surface were crushed and broken and these fracture surfaces are almost parallel to the stress axis.

Figure 7. Fracture surface of metallic closed cellular material.

Ultrasonic Measurement

The ultrasonic attenuation of 16mm diameter specimen sintered at 800°C was measured at temperatures from room temperature to 100°C. The attenuation coefficient (about 3.8 - 4.8 dB/cm) is larger than that of pure aluminum (0.1–2.5dB/cm), but smaller than that of polystyrene (15.2dB/cm). In addition, as the temperature increases, the ultrasonic coefficient of this material gradually increases.

Damping Measurement

The internal friction of this material is measured by a free resonance vibration-type equipment. The internal friction of this material is 4.25×10^{-3} (sintered at 1123K) and 5.41×10^{-3} (sintered at 1103K). T compear with these results, the internal friction of pure aluminum was measured and its value is about 5.25×10^{-3}. These results suggest that the internal friction would decrease as the sintering temperature increases.

DISCUSSION

Metallic Closed Cellular Material

A metallic closed cellular material has been fabricated this study. The density of this material is from 2.2 g/cm^3 to 2.6g/cm^3, which is smaller than that of an aluminum alloy. As Figure 4 shows, the polystyrene particles were deformed to polyhedrons by isostatic pressing.

Figure 5 shows that cell walls of a nickel-phosphorus alloy are observed as bright parts and the material inside the cell walls is observed as darker in color. The amount of emitted secondary electrons per primary electron of the polymer or organic material is smaller than that of metals. During polishing, only 0.05μm Al_2O_3 powders were used. Therefore, the material inside cell walls is thought to be polystyrene or organic material. This indicates that the organic material remains inside of the cell walls after heat treatment and this metallic closed cellular material containing the organic material can be produced using this technique.

Energy Absorption

As shown in Figure 6, the stress-strain curve has a linear elastic region, a long plateau region and a wavy region. It seems that the presence of the plateau in the compressive stress-strain curve is responsible for the high energy absorption. Therefore, this metallic closed cellular material seems to have high-energy absorbing capacity. These results show that this material can be utilized as an energy absorbing material.

After the linear elastic region, a few cracks occur in the direction parallel to the stress axis. It is postulated that the fracture initiates from a defect in this material. Therefore, if this metallic closed cellular material has only few defects, the plateau area of the stress-strain curve will continue longer during the compressive test. It should be though that the specimen sintered at 1123K has fewer defects than specimens sintered at 1073K. Therefore, compressive stress of the specimens sintered at 1123K is higher than that of the specimens sintered at1073K.

Young's Modulus

Young's modulus of this material was measured using the data in linear elastic region for each specimen, and the relationship between Young's modulus and thickness of the cell walls of this material is shown in Figure 8. The relationship between Young's modulus and sintering temperature is also shown in this Figure. As the thickness of cell walls increases, the Young's modulus of the specimens increases. Therefore, it should be thought that Young's modulus of the specimens depended on the thickness of the cell walls of the specimen.

It should be thought that thickness of the cell walls of this material can be controlled by metal coating process. Therefore, it seems that Young's of this material can be controlled by changing the thickness of the cell walls and the sintering temperature.

Ultrasonic Attenuation Coefficient

The attenuation coefficient of this material (about 3.8-4.8 dB/cm) is larger than that of metallic materials (for example, the attenuation coefficient of pure aluminum is about 0.1–2.5 dB/cm), but smaller than that of polystyrene (15.2dB/cm). These results suggest that the material can be used as a passive damping material. As temperature is increased, the attenuation coefficient gradually increases and so, the attenuation coefficient can be controlled by changing the temperature.

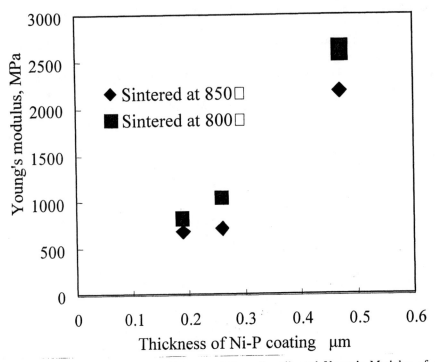

Figure 8. Relationship between thickness of the cell walls and Young's Modulus of closed cellular materials.

Internal Friction

The internal friction of this material, 4.25 ×10^{-3} (1123K) and 5.41x10^{-3} (1103K), is the same as pure aluminum (~5.25x10^{-3}) suggesting the material can be used as a passive damping material. In addition, as the sintering temperature increases, the internal friction would decrease. Therefore, it is believed that the internal friction of this material can be controlled by changing the sintering temperature. The results of the compressive tests show that the yield stress and Young's modulus are lower than that of another structural metal. As shown in Figure 6 and Figure 8, the yield stress and Young's modulus of this material can increase by changing the thickness of the cell walls and sintering temperature. Therefore, we believe that suitable conditions to fabricate a strong and stiff closed cellular material with a high damping capacity can be found.

CONCLUSION

A metallic closed cellular material containing organic materials has been developed. This metallic cellular material is light and has high-energy absorption and a large ultrasonic attenuation coefficient. In addition, Young's modulus depended on the thickness of the cell walls of the specimen and sintering temperature. The obtained results emphasize that this metallic closed cellular material can be utilized as the energy absorbing material and passive damping material.

REFERENCES

[1] L.J. Gibson, M.F. Ashby, 1988. Cellular solids - Structure and properties, Pergamon Press, Oxford, 41.
[2] J.T. Beals, and M.S. Thompson, 1997. Density gradient effects on aluminium foam compression behaviour, J. Mater., Sci. 32, 3595–3600.
[3] N. Chan and K.E. Evans, 1997. Fabrication methods for auxetic foams, J. Mater. Sci. 32, 5945–5953.
[4] Y. Sugimura, J. Meyer, M.Y. He, H. Bert-Smith, J.Grenstedt, A.G. Evans, 1997. On the mechanical performance of closed cell Al alloy foams, Acta Mater., 45(11), 5245–5259.
[5] G.J. Davies, S. Zhen, 1983. Review Metallic foams: their production, properties and applications, J. of Materials Science, 18, 1897–1911.
[6] S.K. Maiti, L.J. Gibson, M.F. Ashby, 1984. Deformation and energy absorption diagrams for cellular solid, Acta Metall 32(11), 1963–1975.
[7] J. Banhart, J. Baumeister, 1998. Deformation characteristics of metal foams, J. Material Sci. 33, 1431–1440.
[8] D.J. David, A.P. Phillip and H.N.G. Wadly, 1998. Porous and Cellular Materials for Structural Applications, Materials Research Society, Warrendale, PA, 205.
[9] U. Waag, L. Schneider, P. Lothman and G. Stephani, 2000.Metallic hollow spheres-materials for the future, Metal Powder Report, 55(1), 29–33.
[10] S. Kishimoto and N. Shinya, 2000. Development of metallic closed cellular materials containing polymer, Mater. Design 21, 575–578.

Bulk AxiSymmetric Forging of Magnesium Alloys

M. CHANDRASEKARAN, C. C. MUN and J. Y. M. SHYAN
Forming Technology Group, Singapore Institute of Manufacturing Technology, Singapore

ABSTRACT

Magnesium alloys are gaining prominence due to their lightweight. The market for magnesium alloys is fast expanding including automobiles and electronic applications such as PDA chassis. Three different materials were selected for axisymmetric extrusion tests namely; AZ31, AZ61 and the forging alloy, ZK 60. To establish the size and capacity of the press required to perform these forming trials and to know the formability, simulation using Finite Element Analysis was carried out with the known properties of the magnesium alloy. Two different die sets with a die shoe were designed to perform the forward and backward extrusion trials. The area reduction ratio for forward extrusion was fixed at 41% for the die design and simulation while it was 1.7 for backward can extrusion. The maximum strain is given as $\ln(A_o/A_f) \sim 0.88$ in the case of forward extrusion. The temperature was varied with a temperature controller built in-house from RT (room temperature) to 200°C. However, the results provided below only includes those tests carried out at 150°C to 200°C as the billets cracked while trying to form at temperatures below 150°C.

INTRODUCTION

Magnesium alloys are extensively used in various applications ranging from precision parts to large aerospace components. Automotive parts occupy a principal position in magnesium applications. Cast magnesium alloys are occupying the essential demands for magnesium automotive parts. Structural applications normally require energy absorption materials with reasonable elongation, high yield strength and high impact energy. Cast magnesium alloys fail to meet some of the requirements and it is here the wrought alloys of magnesium find extensive application by replacing parts made of aluminium or steel. The use of magnesium parts would cause a weight saving of around 30% and 70 % compared to aluminum and steel respectively. Low resistance to plastic flow of magnesium at elevated temperatures (close to hot working) increases the die life by more than 100% [1]. Hot working of Magnesium (Mg) alloys can be used to produce precise near net shape parts by forging or extruding the material for secondary processing. These products are easy to machine or weld. The hot working temperatures depend on the presses to be used and the alloys that require to be forged. Normal processing temperatures of most wrought magnesium alloys ranges from 250 to 450°C. One phenomena observed in the case of magnesium alloys are the DRX (Dynamic Recrystallization) and DRV(Dynamic Recovery), which improves the workability of the material at elevated temperatures. One of the disadvantages associated with magnesium is its flammability at temperatures above 478°C (alloy dependent). The workability of magnesium and specifically AZ31 and ZK60 are poor and fail during deformation below 250°C [1–5]. Hence, the present work is an attempt to evaluate the workability of three different magnesium alloys namely AZ31, ZK60 and AZ61 for formability.

EXPERIMENTAL METHODS

Theoretical Treatment

Three different materials were selected for axisymmetric extrusion tests namely; AZ31, AZ61 and the forging alloy ZK 60. To establish the size and capacity of the press required for performing the forming trials and to know the formability simulation using Finite Element analysis was carried out with the known properties of the magnesium alloy. The tonnage required for backward extrusion of magnesium was calculated using the elevated temperature properties using Schmitt method using the equation given below:

$$P_{B\,Max} = K * (\frac{A_0}{A_1} - 1)^{0.5} \tag{1}$$

where K is a factor dependent on the ultimate tensile strength of the material, S_u, A_0 and A_1 are the billet cross sectional area and the punch face area respectively. $P_{B\,Max}$ is the Maximum Punch pressure for the given reduction.

$$K = 2.4\ S_u + 50\ N/mm^2 \tag{2}$$

The maximum punch force can be calculated using the equation for forward extrusion

$$F_{max} = \frac{A_0 \varphi_{max} \sigma_{f.m}}{\eta_{def}\, m} \tag{3}$$

where A_0 is the original area, φ_{max} is the strain, $\sigma_{f.m}$ is the flow stress and η_{def} is the deformation efficiency and m is the force correction factor.

According to the equations, the estimated tonnage needed for the reduction is 10.2 tons without force correction factor. The most commonly used extrusion alloy, AZ31, was selected for various forming simulations to serve as a guideline. The simulated forging results of AZ31 alloy were compared with the actual experimental valued obtained. The stress values calculated using the theoretical method was compared with the simulation results.

EXPERIMENTATION

Two different die sets with a die shoe was designed for performing the forward and backward extrusion trials. The area reduction ratio for forward extrusion was fixed at 41% for the die design and simulation while for backward can extrusion the ratio was 1.7. The maximum strain is given as $\ln(A_o/A_f) \sim 0.88$ in the case of forward extrusion. The temperature was varied with the temperature controller built in - house from RT (room temperature)-200°C. However, the results provided below only include the tests carried out at 150°C to 200°C as the billets broke into pieces while trying to form at temperatures below 150°C. Few forward extrusion tests were done at temperatures higher than 200°C. Though the forming trials were successful, there was difficulty in removing the specimens from the die cavity. Secondly, the process of removing the samples in case of AZ31 and ZK 60 resulted in cracking because of which it was difficult to evaluate the samples and the process. Simulation results showed that magnesium is easily formed at elevated temperatures of 300°C as reported in some of the earlier findings [1,2,4]. However AZ61 samples did not show any evidence of crack formation during ejection of the formed sample.

The materials of AZ31, AZ61 and ZK60 in 1" rod form were procured from a commercial supplier. The rods were then sliced to make ϕ25 mm x 25 mm length billets with a tolerance of ±0.01 mm for forward extrusion and backward can extrusion. The die and the punch were assembled in a die shoe fabricated for the above tests and the extrusion trials were performed in a 50-ton hydraulic press. All the extrusions were conducted under isothermal conditions in which the die temperatures were varied depending on the extrusion temperature. However, the punch temperature was varied only for two extrusion conditions namely 100 and 150°C as further heating up of the punch was difficult due to the lower wattage of the heating assembly for the punch. Band heaters with power of 1 KW and 450W were used to heat the die and the punch respectively. The extruded samples were sectioned, polished and etched to reveal the microstructural features.

RESULTS

Theoretical analysis of the forming conditions using the equations above the approximate tonnage required for forming was calculated for a reduction of diameter from 25 mm to 16 mm inner diameter in backward can extrusion at room temperature.

Using the Schmitt method (Eq. 1 & Eq. 2) which estimates the forces to an accuracy of ± 5% and setting

F_{AZ31} = ~ 12.1 tons,
F_{AZ61} = ~ 17.7 tons,
F_{Zk60} = ~ 20 tons,

for forward extrusion, the force on the punch is given by:

$$F_{max} = F/m \qquad (4)$$

where F is a fictitious value and *m* is the force correction factor, a function of h/d_0.

Using Eq. 3 for the forward extrusion of AZ31B and substituting A_0 = cross sectional area of the billet (490.87), φ_{max} = Maximum strain (~0.89),

$\sigma_{f.m}$ = flow stress of the material (205 MPa).

Assuming a deformation efficiency (η_{def}) of 0.6 and *m*, force correction factor ~0.78 (based on a nomogram) [6], we get:

F_{max} = ~14 tons at room temperature.

FINITE ELEMENT ANALYSIS OF UPSETTING AND ECTRUCION OF MAGNESIUM

After theoretical estimation of the tonnage required for forming of the three materials namely AZ 31B, AZ61 and ZK 60, finite element analysis of the forming was done for AZ31 to know the deformation behaviour and the tonnage. Before the three basic axi-symmetric forming operations were performed, FEA calculations of these three processes were done. A commercially available finite element package ANSYS/LS-DYNA was used for the finite element analysis. 2-dimensional elements and a re-meshing algorithm were used. The main purpose of the FEA simulations in this work was to determine the press force required at various temperatures and the forming behavior of magnesium at three different temperatures. The material model used for the simulations can be described by the following equations:

$$\text{For plastic deformation: } \sigma = C \cdot \varepsilon^n \qquad (5)$$
$$\text{For elastic deformation: } \sigma = E \cdot \varepsilon \qquad (6)$$

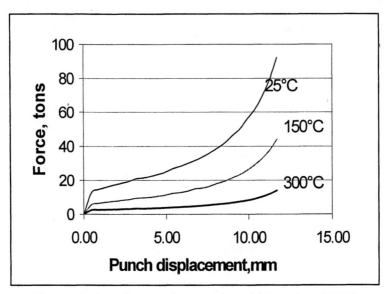

Figure 1. The punch force for upsetting extrusion of Magnesium AZ31B at various temperatures.

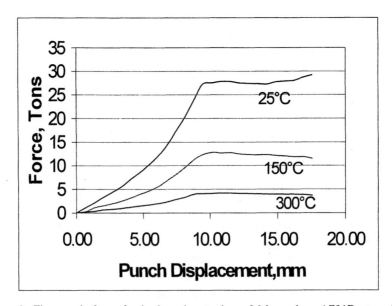

Figure 2. The punch force for backward extrusion of Magnesium AZ31B at various temperatures.

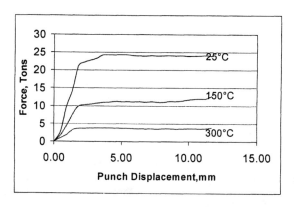

Figure 3. The punch force for forward extrusion of Magnesium AZ31B at various temperatures.

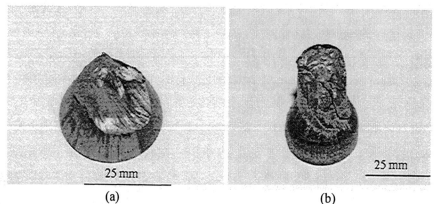

Figure 4. A typical failure observed on an AZ31 B billet extruded at room temperature with different reverse counter pressures applied: (a) 0 tons; (b) 5 tons.

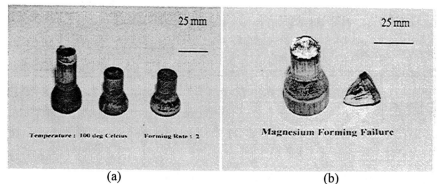

Figure 5. Typical forming heights achievable at low stress rates and the typical brittle failure of the magnesium specimens showing 45° cleavage fracture respectively.

Simulations for the three processes, namely upsetting, forward extrusion and backward can extrusion, were done at three temperatures, 25°C, 150°C and 300°C.

Figure 1 shows the required force for upsetting of a round magnesium billet with a diameter of 25 mm. The force for the deformations increases exponentially because of the constantly increasing area of the deforming blank. Figure 2 shows the required force for backward extrusion of a round magnesium billet with a diameter of 25 mm. The force reaches its maximum value very early on in the process. Figure 3 shows the required force for forward extrusion of a round magnesium billet with a diameter of 25 mm. The force reaches its maximum value around a punch travel of 10 mm with a complete blank height of 18 mm.

It can be seen from the simulation trials that backward can extrusion of AZ31B requires a high force of approximately 24 tons at room temperature for a displacement of 2 mm and above. The increase in test temperatures indicates a reduction in tonnage to approximately 12 tons at 150°C and to less than 5 tons at a temperature of 300°C. The simulation results of backward can extrusion indicate that very high forces are required for deformation. However, simulation of forward extrusion of AZ31B indicates a tonnage of about 10 tons for a displacement (extrusion length) of 5 mm increasing to as high as 26 tons for a displacement of 10 mm. Similar to backward can extrusion, the magnitude of reduction in force with an increase in test temperatures to 150°C is 10 tons and the approximate tonnage is close to 14 tons for a extrusion length of 10 mm. However, the force is stable after reaching the maximum values irrespective of extrusion length at all temperatures indicative of dynamic recovery on reaching a certain strain during deformation. Secondly the values predicted by the simulation and the theoretical calculations differ despite the fact that the strain value assumed for simulation was lower than those used for the theoretical calculation. This indicates that the equations used for the calculation of stress on the punch were invalid for magnesium alloys.

LABORATORY RESULTS

Based on the results obtained from the simulation and the theoretical estimate of the force required for forming, it was decided to use a 50-ton hydraulic press with a knock out load capacity of 5 tons. To verify the simulation results and those of the theoretical calculations, the AZ31B was initially deformed at room temperature with a counter pressure to generate hydrostatic stress state within the material and avoid any cracks during deformation. The billets cracked and a brittle failure was observed for the samples, that were extruded at room temperature. Forces as high as 40 tons were required for room temperature forging trials against a predicted load of ~25 tons.

The shape of the sample indicates that the specimens did undergo a deformation but the hydrostatic stress [7] generated by the ejector pin is not sufficient to arrest the crack generation in the sample. Since the number of slip planes that are active in room temperature is limited the billet fractured on the extruded face at approximately 45° angle.

The test temperature was increased to 100°C to see the effect of temperature on the formability of the alloy. The alloy could form at 100°C but to a limited extent, i.e., h/h_0 was low for complete formability without evidence of failure. Figure 5a and b below shows typical the failures of specimens with increase in forming length and the rate of application of force on the punch. With low-pressure rate on the ram of the press of 2 bar/sec, higher forming heights could be achieved.

At a forming temperature of 100°C, smaller billets were used and two billets were stacked in the die cavity to obtain a successful extrusion. AZ 31B formed completely at temperatures of 150° and above without any indication of failure on the specimen irrespective of the stress rates. Three specimens were formed for each test condition at 150°C with a stress rate of 2 bar/sec (~5.8 MPa/sec on the specimen) on the punch. The initial load on the specimen at which yielding starts is ~120 MPa which increases with the work hardening in the specimen. After extrusion, the specimens were allowed to cool under normal room temperature and also water quenched to check the presence of DRV (Dynamic Recovery) and DRX (Dynamic recrystallization during forming. Higher extrusion lengths were achievable with temperatures above 150°C with a rate of 2 bar/sec compared with extruding at a temperature of 100°C in case of AZ 31B. However, the AZ61 samples show evidence of failure at 150°C. The figure below shows the three specimens forward and backward extruded at 175°C, with a rate of 2 bar/sec. Backward can extrusion of both AZ31B and ZK60 at 175°C were possible but AZ 61 showed shear failure along the walls of the can. Reduction of the rate of forming to 1 bar/sec did improve the formability of the AZ61 alloy. This is due possibly to the higher hardness of AZ61 alloy.

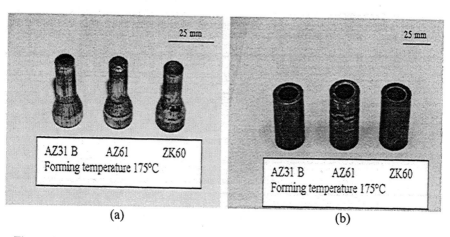

Figure 6. Magnesium alloy specimens (a) forward and (b) backward extruded at 175°C.

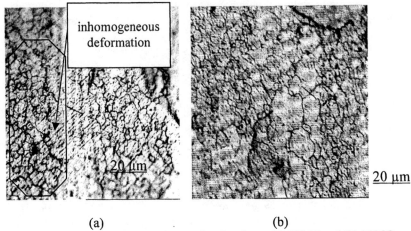

Figure 7. Microstructure of AZ31B specimens after forming at (a) 200 °C and (b) 175 °C and both air-cooled [500x].

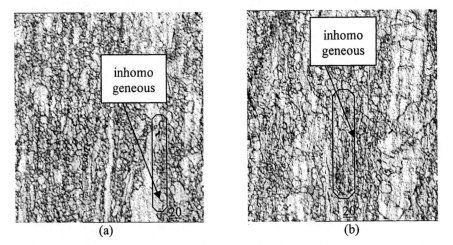

Figure 8. Microstructure of AZ31B specimens after forming at (a) 200 °C and (b) 175 °C and both water quenched [500x].

DISCUSSION

The microstructure of the specimen cooled at room temperature and water quenched after forming at 200 and 175°C are presented below. However, for the present study, a stress rate of 2 bar/sec was used to compare the formability at all temperatures and the an intermediate temperature of 175° was used to compare the effect of stress rate on formability and the force required for forming.

The specimens formed at 175°C showed fairly larger grains compared with those formed at 200°C in the case of air-cooled specimens. However, water-quenched specimens exhibited finer grains with sizes as small as a few micrometers in dimension. At lower temperatures of extrusion, strain-hardening and resistance to dimension. At lower temperatures of extrusion, strain-hardening and resistance to forming possibly cause a rise in local temperatures. The increase in resistance to deformation can be clearly seen from load vs. displacement curves in Figure 9. This possibly suggests that at lower forming temperatures, grain size is controlled by temperature rise associated with deformation. The microstructure was non-uniform in cross-section leaving bands separating the grains that suggest possible inhomogeneous deformation in the specimens. The band-like formation also suggests that a possible inhomogeneous deformation could have caused local increase in resistance to deformation. Thus the presence of smaller grains of ~2 μm can possibly be attributed to dynamic recrystallization. Both dynamic recrystallization and recovery processes are likely operative during forming of AZ31B.

Figure 9 below presents the load displacement curve for forward extrusion of AZ31B specimens.

It is clearly visible from the force displacement graph that the forming loads display an identical trend at 150 and 200°C while the loads were slightly higher initially for the specimens formed at 175°C. The microstructure also revealed dynamic recrystallization behavior at 175°C. Longer extrusion lengths were also possible with increase in the pressure rates for forming at 175°C. The higher force required at 175°C suggests that the work-hardening rate is possibly higher than the softening rate and so dynamic recovery is not pronounced at this temperature during initial stages of extrusion. Further increase in deformation results increase the local temperature because of an increase in resistance to deformation promoting DRV and DRX. At 150°C, this behavior is not pronounced possibly due to the softening from damage or temperature localization during forming. At 200°C, dynamic recovery possibly caused a reduction in forming force. Moreover the microstructure also suggests that DRV was predominant in the case of specimens extruded at 175°C. It has also been reported in earlier literature that over the temperature region of 180–240°C, the flow curves of AZ31B, AZ61 & ZK60 work-harden towards a peak and then failure occurs either before the peak value or immediately thereafter [8–10]. The observation of higher loads (hence higher stresses) is in conformity with the cited literature on the work hardening peak [8].

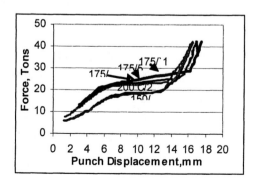

Figure 9. Load displacement curves of AZ31B material at different temperatures during forward extrusion.

Work hardening is limited by the occurrence of DRV, a thermally activated softening mechanism where the flow curves flatten to a plateau with gradual decline to failure, which happens with specimens, formed at 200°C. AZ61 and ZK 60 were also formed at the above temperatures and the microstructures revealed twinning to be the most predominant mode of deformation in the specimens at lower temperatures [8-10].

CONCLUSIONS

1. Magnesium alloys can be extruded at relatively low temperatures of 150°C.
2. Extrusion of AZ31B at a temperature of 100° is possible with application of counter pressure but has limited formability.
3. AZ61 and ZK60 have relatively low extrudability at lower temperatures as compared with AZ31B.
4. Both dynamic recovery and dynamic recrystallization are active at 175°C and above.
5. Typical grain sizes of 2–3 microns are identified for AZ31B specimens extruded at 175 and 200°C.
6. Internal energy due to deformation in AZ31B causes an increase in local temperatures above the recrystallization temperature of the sample.
7. The stress rate has a significant effect on the ability of the alloys to extrude regardless of the temperature.
8. Air cooled specimens have higher grain size compared with water quenched samples for the same conditions of forming suggesting grain growth during slow cooling of the sample.

ACKNOWLEDGMENT

The authors acknowledge with thanks the kind support of A Star Singapore Institute of Manufacturing Technology for the above work which is a part of the in-house project C01-P-145AR.

REFERENCES

[1] J. Becker, G. Fischer, K. Schemme, O. Fuchs, 1998. Light Weight Construction Using Extruded and Forged Semi-Finished Products Made of Magnesium Alloys, Magnesium Alloys and their Applications: Proceedings sponsored by VolksWagen, 15–28.
[2] A. Ben-Artzy, A. Shtechman, N. Ben-Ari and D. Dayan, 2000. Deformation Characteristics of Wrought Magnesium Alloys, AZ31, ZK60, Magnesium Technology 2000 Conference proceedings of the symposium held during the 2000 TMS Annual Meeting, Nashville, Tennessee, 363–374.

[3] P. Lukac, 1996. Strengthening and Softening of Some Magnesium-Base Alloys, Proceedings of Third International Magnesium Conference 10–12 April, 381–390.

[4] B.M. Closset, J.-F. Perey, C. Bonjour and P.-A. Moos, 1998. Microstructures and Properties of Wrought Magnesium Alloys, Magnesium Alloys and their Applications: Proceedings sponsored by VolksWagen, 195–200.

[5] E.A. El-Magd and H.P. Weisshaupt, 1998. Deformation Behaviour of a Mg-Al-Zn Alloy at high strain rate, Magnesium Alloys and their Applications: proceedings sponsored by VolksWagen, 189–194.

[6] P.S. Raghupathi,, 1998. Cold and Warm Extrusion, Handbook of Metal Forming, Ed. Kurt Lange et. al., McGraw-Hill, 15.1–15.65

[7] H.W. Wagener, 1998, Cold Extrusion of Magnesium Alloys and Magnesium-MMCs, Magnesium Alloys and their Applications: Proceedings sponsored by VolksWagen, 557–562.

[8] A. Mwembela, H.J. McQueen, E. Herba & M. Sauerborn, 1998. Hot Workability of Five Commercial Magnesium Alloys, Magnesium Alloys and their Applications: Proceedings sponsored by VolksWagen, 215–222.

[9] M. Kohzu, F. Yoshida, H. Somekawa, M. Yoshikawa, S. Tanabe & K. Higashi, 2001. Fracture Mechanism and Forming Limit in Deep Drawing of Magnesium Alloy AZ31, Materials Transactions, 42(7), 1273–1276.

[10] A. Galiyev and R. Kaibyshev, 2001. Microstructural Evolution in ZK60 Magnesium Alloy during Severe Plastic Deformation, Materials Transactions, 42, 1190–1199.

Chapter 10: Manufacturing Systems

A Knowledge Management System for Manufacturing Design

Y. NAGASAKA

Department of Business Administration, Konan University, Kobe, Hyogo, Japan

S. KISANUKI

Section of Product Lifecycle Management, QUALICA Ltd., Hirakata, Osaka, Japan

ABSTRACT

Most manufacturing design processes include non-routine tasks that depend on particular techniques of skilled workers. Not only explicit data, but also tacit knowledge is important. In this study, a knowledge management system for manufacturing design has been developed based on web computing technologies. The objectives of the software are to achieve more efficient and reasonable manufacturing design as well as inherit traditional know-how. Three modules have been developed. The first implements activity models for different manufacturing designs. Each activity model includes task models structured as a tree and related to background information. The second module obtains new information from a computer simulation of manufacturing processes or/and practical manufacturing data. The results of computer simulations are usually visualized by post-processing. Operators sometimes look at color contour lines of temperature and know which area has a problem. However it is not easy to accumulate know-how to analyze these simulation results. A methodology to extract and learn features of the contour lines has been studied. The third module performs data-mining and text-mining to rearrange the knowledge. The database includes specifications, activities, background information, computer simulation results as well as actual quality, costs, and customer claims after manufacturing. The C5.X engine was used to formulate rules to obtain good quality for a specific product. The rules help us understand what happens during manufacturing. The text-mining engine can search documents and memos quickly. The three models are integrated and applied in practical manufacturing situations.

INTRODUCTION

Most companies have recognized that intellectual capital is very important today. Especially, knowledge has become one of the most important management resources in business. It is said that there are two types of knowledge: explicit and tacit knowledge [1]. Operation manuals, documents on technical standards and drawings belong to explicit knowledge. Tacit knowledge such as know-how, paradigms and customs cannot be represented or shaped very easily. If tacit knowledge can be transferred to explicit digital data, everyone can understand what happened or how to perform some task. One of the most important objectives of knowledge management is to create a system which transforms tacit knowledge into explicit knowledge efficiently in daily work. This situation is then helpful to create new ideas within a company.

Information technologies are very helpful for knowledge management. In fact, many manufacturers are using information technologies to improve quality and productivity [2]. For instance, a database is very useful to retrieve suitable data to design a new product. Groupware is convenient in constructing a collaboration environment. CAD is used to achieve efficient drawings and CAM is popular for accurate manufacturing. Moreover, some computer simulations are used in practice to predict physical phenomena of production processes [3–5]. If these digital technologies are used well, it is possible to optimize pre-production processes, but how to use them depends on the company or its engineers. The know-how for the specific production process is also important. Nowadays, we naturally use information technologies for manufacturing routinely. This means that a lot of knowledge content is already in digital data form. However, a lot of information is discrete. Most of procedures for data processing are not stored routinely.

Background for decision-making is often not input clearly into the computer.

We need several modeling methodologies to compile knowledge systematically. Especially, the modeling of business activities is very important for knowledge management. In this study, a knowledge management tool for manufacturing design has been developed by considering an activity model [6]. Generally, manufacturing design includes a lot of non-routine work dependent upon the inherent technical abilities of the person in charge. It is also necessary to transform tacit knowledge into explicit knowledge by using a natural way to share and inherit important information. The objectives of this study are to achieve a more efficient and reasonable manufacturing design as well as to inherit important technologies.

KNOWLEDGE MANAGEMENT FOR MANUFACTURERS

Of course, knowledge has been one of the key elements for the business administration of manufacturing for some time. But today, we must consider knowledge as intellectual capital. If the intellectual capital increases, the value of the company increases. Generally speaking, the exchange cycle between tacit and explicit knowledge is well known in business. Daily communication spreads several kinds of tacit knowledge gradually and widely. Sometimes important tacit knowledge must be written down in documents and figures. Documents and figures are useful to be summarized and combined together but this takes time.

Namely, it is necessary to manage the transformation of tacit knowledge into explicit information efficiently as the first step. As a second step, we should consider how we can use the different types of explicit knowledge adequately in the business. In fact, explicit knowledge is very helpful in creating new ideas. A system to share such useful information is necessary. If tacit knowledge is continuously changed into explicit knowledge and saved for efficient utilization in a manufacturing facility, this will provide much helps for engineers to generate new ideas.

The problems of knowledge management in manufacturing processes can be summarized as follows:

(a) Know-how is not expressed explicitly.
(b) It is not easy to retrieve adequate information.
(c) It takes a lot of time to draw up a plan and make many documents.
(d) It is difficult to make sufficient technical standards.
(e) Inheritance of know-how is difficult.
(f) The methodology to evaluate knowledge value is not constructed yet.

Today, manufacturing engineers should engage in creative activities as much as possible. However, the time for creative activity is typically less than 20% of total working time in many companies. The employees spend more time searching for necessary information and they attend many meetings.

Generally speaking, manufacturing design depends on the peculiar characteristic know-how of the engineer because each engineer has different tacit knowledge. In other words, there is a problem in that the ideas for new manufacturing design depend on the experience of each specific engineer. If engineers communicate with each other, a lot of information may be exchanged based on their different experiences. Actually, it is not easy to extract enough explicit information only from a drawing and documents. We should make more useful documents including things such as process procedures and background information for each decision-making process. Moreover, communication is very important to get suitable knowledge whenever you need to know something. The intellectual environment will contribute a lot in thinking about new designs.

Modeling is a very useful technique in constructing an intellectual environment. It can change tacit knowledge into explicit knowledge. Figure 1 shows four models used for designing manufacturing processes: geometric, mathematical, data, and activity models.

Three-dimensional CAD can be used to generate a geometric model. It can include properties such as volume, thickness and machining specifications. We can consider the manner of production with a geometric model. Mathematical models of physical phenomena are used for computation. For example, a heat conduction model including latent heat during material processing is represented by finite difference equations that can be implemented on a computer. A geometric model is usually divided into many small finite elements that are applied to the mathematical model. Such a mathematical model is very useful for engineers who want to know what happens within a manufacturing process. Next, a data model is necessary in business to store and control data related to products. We must know a customer's name, delivery limit and other requirements. During pre-production processes, additional data such as cost estimation and manufacturing schedules are stored and controlled. Finally, an activity model is important in expressing purposes, tasks, entities, methods and so on [7]. The activity model can be built up as a business process model. A business process model is constructed with several elements including at least the business name and definition, attributes, behavior, relationships and constraints [8].

Considering the problems mentioned above, we can summarize the requirements for a knowledge management system for manufacturing design as shown in Figure 2. Enormous data in various forms such as figures, images and CAD drawings must be easily accumulated and retrieved. All activities of manufacturing design including background information should be clearly represented and linked to the database. Each engineer can make suitable activity models for himself/herself. Then successful activities should be unitarily stored as a company's intellectual capital. Useful information should be organized and distributed to group members. Of course, data processing must be performed with less operating load. Some mathematical models must be included in computer simulation software to predict manufacturing phenomena. Simulated stress distribution under loading is visualized by post-processors. Mold filling analysis of polymer injection processes is useful to improve the mold design. The operator judges which mold design is best based on an analysis of the data. These kinds of computer simulations are called CAE (computer-aided engineering).

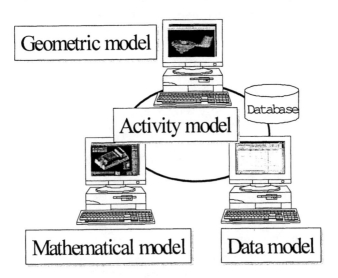

Figure 1. Models for manufacturing processes.

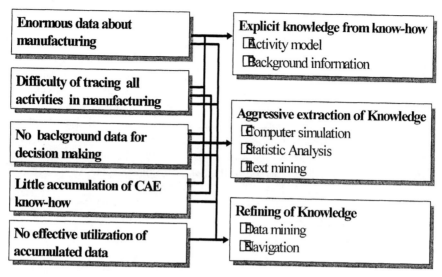

Figure 2. Problems and solutions by knowledge management.

Conventional statistical analysis is also important for manufacturing. SPC (statistical process control) in 6-Sigma, a management methodology for quality control is based on statistical data. The know-how to judge computer simulations and to use statistical analysis is also important. Not only is data saving important, but also data utilization. We need to refine and rearrange the accumulated data to reuse them. It is important to make clear the relationship between manufacturing procedures and customer satisfaction. All manufacturing processes should be improved to achieve customer satisfaction. Cost, quality and delivery are significant elements. Here, we need to know which activity is better or bad for these three elements. Namely, if the best practice is extracted from conventional activities, it becomes a part of the intellectual capital of the company. We should recognize significant requirements such as: how we can get explicit knowledge from our daily work; how we can extract knowledge more aggressively before manufacturing; and how we can refine knowledge for reuse. In this study, a framework for a knowledge management system has been constructed and several modules have been developed that consider these three key requirements.

FRAMEWORK FOR A KNOWLEDGE MANAGEMENT SYSTEM

The framework proposed for our knowledge management system is shown in Figure 3. The objectives of the system are to achieve more efficient and reasonable manufacturing design, as well as to inherit traditional technologies. Three types of knowledge are considered. Experiential knowledge is constructed by accumulating many activities. This includes tacit knowledge and background information such as design rationale – it is the most fundamental knowledge we need to save. Prediction and Analysis knowledge is also necessary to supplement the Experiential Knowledge. This can be obtained by in an aggressive manner such as computer simulation and statistical analysis. Moreover, fact knowledge is important. Fact knowledge is obtained as a result of all manufacturing activities. What happened? How is the quality and at what cost? Are the customers satisfied? Why was the quality not good? Why was the cost high? Causation between fact and processes should be investigated. This is very important knowledge, too.

The system is shown in Figure 4. Several modules have been developed and combined by considering the framework in Figure 3. The database includes product data and background information. Namely, the system can accept different data formats such as numbers, text, formulas, spreadsheets, graphs, images, CAD data and rule-bases. Data and document management tools can be used through a web browser as in Figure 5. It is important to achieve data consistency using a database management system. Moreover, some convenience tools are necessary to transform data written on paper into digital data and to store background information.

One of the most important modules is the "design palette" which can implement activity models for many kinds of manufacturing designs. A convenient reporting tool is also necessary to print out documents. Each activity model includes task models structured as a tree and related to background information.

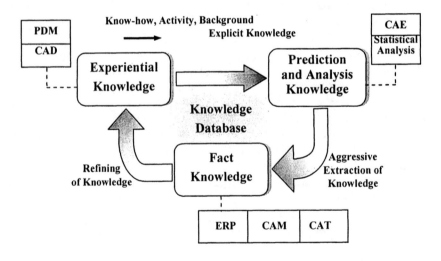

Figure 3. Relationships between three types of knowledge.

For instance, a drawing is not enough to know why and how the shape, dimensions and materials were determined. Here, design background information is linked to each item in the activity model. The activity model is built up and modified easily and interactively by users. The database is helpful for individual engineers. And also, it can be utilized to share information with other engineers in the same company. Common information such as design standards can be managed intensively. However, the knowledge base including the decision-making processes can be stored individually. After a while, effective data for individuals can be shared with other members of the group. This module helps to verify the design conditions and extract the best practice among conventional examples. Preparation of a network environment is indispensable for this purpose.

CAE Knowledge Tool	Knowledge map		Process management tool
Knowledge community I/F	Text mining tool	Data mining tool	
Design palette (Activity modeling tool)			Reporting tool
Data and document management tool			
Multimedia database			

Figure 4. Developed modules for knowledge management.

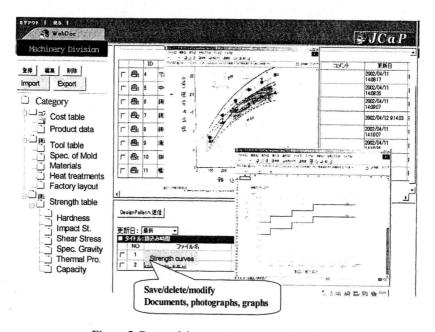

Figure 5. Data and document management tool.

A CAE knowledge tool is required to obtain any predicted information from computer simulation and statistical analysis of manufacturing processes. Usually, results of computer simulation are visualized by post-processing. Operators sometimes look at color contour lines of temperature etc. and know which area has problems. But, it is not easy to accumulate know-how to analyze computer simulation results. Here, representative data from computer simulation and statistic analysis is stored in the summary sheet, which we can find any differences between several cases easily. On the other hand, a methodology for extracting and learning features of contour lines has been studied. Then, important information from computer simulation is saved and utilized for manufacturing design.

Data mining and text mining tools are very important to make a knowledge map by rearranging the knowledge. The database includes specifications of products, activities, background information, results of computer simulation as well as real quality, cost and customer claims after manufacturing. The C5.X engine is used to see some rules to be used obtain good quality for a specific product. The rules are useful to understand what happened during manufacturing. Vextsearch, a text-mining engine is also available for use to retrieve suitable documents and memos. Words for several fields of manufacturing design are grouped and displayed with different colors in a word map. Then, important words of the document are heightened in the word map visually. We can check many documents quickly.

IMPLEMENTATION OF ACTIVITY MODEL

The activity model proposed in this study consists of task models that are structured as a tree. Each task has design items to be fixed, and some background information in the database is linked to each design item. Documents, drawings and computer files are created through several tasks. The values of design items are components of the documents and drawings. We can easily find the foreground information by looking at the results. However, it is usually very difficult to learn why or how a value was determined. The background information is important for determining how the engineers make a decision. If the design background information is stored and linked to each object in the drawing, it helps us to verify design conditions. The methodology in implementing the activity model on a computer should be discussed.

A GUI has been designed for the activity model and is named "Design Palette" as shown in Figure 6. There are three windows showing a task tree, design items and background information. If a task is selected, design items to be determined are shown in another window. Suitable values or comments must be input for each item. We can fix the values by navigating in the background information window. There may be several candidates for background information such as a technical standard table, a numerical equation, a drawing of a similar product and so on. The operator can see all candidates and choose one of them. If there is no good background information, he/she can fix the value as he/she wishes. Then, he/she can install the method as new background information and link it to the item. Even the names of design items can be changed by the operator. The operator can modify and arrange the activity model dynamically.

A general business object includes entities and methods as a capsule [8]. If you see a business object, some of the entities should be input to the operation. Others may only be referenced. Methods such as small calculation modules are often used to fix some values of these entities. The capsules must be designed and implemented beforehand. Only system engineers can modify these capsules. In this way, the background information includes entities and methods for reference. The entities and methods related to a design item can be determined automatically with the corresponding keywords held in this system. In addition, a natural language treatment using a concept-mining engine is used to select candidates for text data. However, the candidates may not be suitable information for the design item. If the operator opens and sees the background information, a "count" number increases. If the background information is actually used to fix the values of the design item, another "count" number increases. The "count" numbers show how useful a piece of data may be. Sorting the "count" numbers is useful to see which background information is most useful.

Let's look at the case of a die design. The die designer must fix geometric objects in a drawing. The geometric objects mean gates, overflows and runners. The class and attributes for each object are defined. Examples of attributes for the class of gate are: type; width; height; and so on. If a specific gate is designed, an instance is created. This kind of data generation for objects can be performed through the activity model mentioned above. The task is to fix an instance. If a parametric model of geometry can be defined for each class in CAD, a specific geometry is automatically drawn as a representation of the instance.

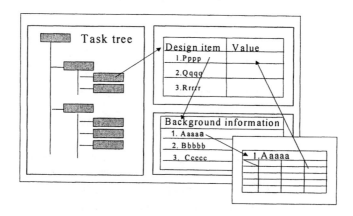

Figure 6. An implementation tool of the activity module "Design Palette".

The system developed here has already been applied to several practical cases. Figure 7 shows an example of a practical design palette. In practice, the first motive of this system is to save the cost of pre-production processes by reducing completion time. Different types of data can be stored on a personal computer and the necessary data can be found quickly making it useful for quality control. Some documents are automatically completed and printed out with adaptive forms through the operation of this system. Some data in the database can be related to report forms. It is easy to arrange the report form depending on requirements. Questions are asked of the designer and it must be recognized that even an engineer does not remember everything about why and how certain values were chosen. The system helps us understand more about the details of a drawing.

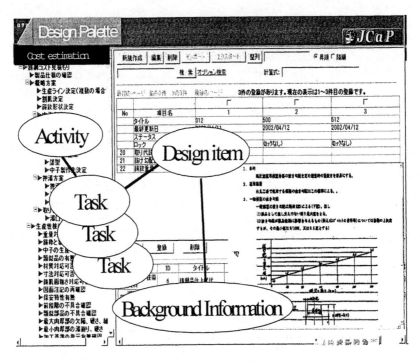

Figure 7. Developed GUI for design palette

KNOWLEDGE BY PREDICTION AND ANALYSIS

As everyone knows, experiential knowledge is very important in manufacturing, but when we don't have any experience for a new type of product, we must find out as much as we possibly can about similar products and obtain some predicted results. Statistical analyses such as time-series and regression analysis are generally used to predict phenomena quantitatively. We can check the correlation coefficient between an explanatory variable and certain dependent variables. If the correlation coefficient is sufficient, we can predict the explanatory variable for any dependent variable. X-R control charts are also used as a general tool to check on-site aberration. Representative statistical values such as mean, standard deviation, variance, and median are useful to know the situation of the production. These days, most of the machinery can be connected by computer network to import and export data automatically. This means we can obtain a lot of data continuously without cost because of information technology. SPC (Statistic process control) is a reasonable technique because we can judge based on facts. The situation builds up with input of daily data. It is possible to store these data in a database as shown in Figure 4.

Computer simulations of manufacturing processes based on stress, heat and fluid flow analysis have already been developed and used in practical work. However, it takes an expert to operate these computer simulations, i.e., knowledge about both mathematical models and manufacturing processes is necessary to judge the calculated results. As mentioned above, results of the computer simulation are usually visualized by post-processing, but it is not easy to accumulate the know-how to judge computer simulation results. With our system, some representative data from a computer simulation is stored in a summary sheet. Figure 8 shows an example of a computer simulation for casting design. Solidification time is presented with color contours three-dimensionally. Experts know which cross section is important. Then, a cross section is selected to show the details. Red areas are shown where the temperature gradient is not enough to achieve directional solidification. If these red areas are large, shrinkage defects possibly occur. We can compare the value of red area quantitatively between several casting designs. If much simulated data are stored and compared with the qualities of practical castings, then some criteria can be identified statistically.

Another method to accumulate know-how about computer simulation is as follows. If we can extract the characteristics of color contours numerically, this will allow the likeness of the characteristics to be compared quantitatively. Here, we consider that features of temperature contour lines are extracted as skeletons. Conventionally, this method is applied to geometry classification [9,10]. For instance, some temperature distribution is assumed three-dimensionally. At first, voxels (volume pixels) from the three-dimensional shape are generated. A voxel is a picture element in a three-dimensional coordinate system. Then, the temperature value of each voxel is calculated. This produces a three-dimensional temperature map. In the next step, skeletons of the three-dimensional temperature map are extracted as polygons from the mapped data. Voxels with the maximum temperature among neighboring voxels are selected as skeleton candidates. After selection, the candidates are converted to straight lines or circular rings. These are then represented by several vectors and stored as a tree structure. A standard tree involves, for example, four levels with each branch having four descendants. Each parent branch has the same number of descendants. The attributes include, for example, scaled volume, connection strength, and scaled X,Y,Z coordinates. These vectors are input to a skeleton classifier that is constructed on the basis of the back propagation neural network model. Some examples are shown in Figure 9. Extracted lines and rings are skeletons of the solidification time map. These are related to defects in the shrinkage cavity. These skeletons are understandable by anyone. Also, the pattern of skeletons can be learned by a neural network algorithm. Then, we can understand the likeness of skeleton patterns on a quantitative basis.

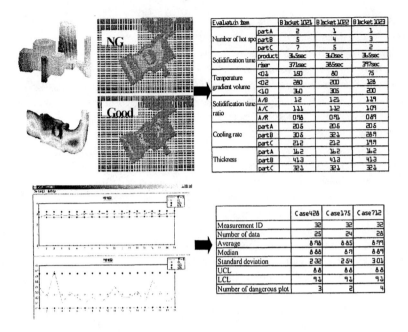

Figure 8. Summarized tables from computer simulation and statistics data.

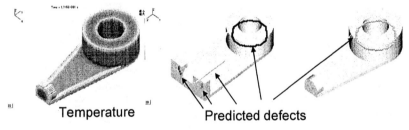

Figure 9. Examples of system output.

KNOWLEDGE MAPPING BY TEXT MINING

Documents written in a natural language are very important for communicating with each other. In this study, text-mining technology was used to retrieve and categorize text data of documents. Keywords are important for the retrieval. But, it is not necessary to include the keywords directly in the documents. The concept or meaning should be checked and evaluated quantitatively. The procedure to retrieve a suitable document is shown in Figure 10.

A large number of words can be located in a multi-dimensional vector space in this engine, Vextsearch. In the vector space, synonyms and relational words are mapped in the neighborhood by learning a lot of sentences. It is known that relational words co-occur very often in sentences. Then, the distance between two words or two sentences can be calculated within the vector space. Figure 11 shows the structure of the text mining process.

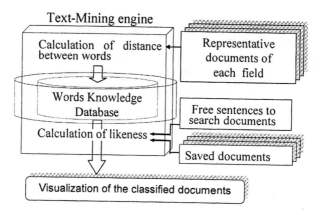

Figure 10. Procedure to retrieve a document.

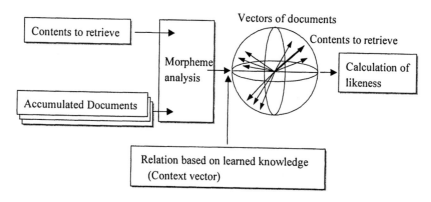

Figure 11. Structure of text mining.

The suitable vector space of words must be prepared depending on the field of manufacturing. Literature can be input through a user-friendly interface and classified in the directory of the thesaurus to build up several vector spaces. On the other hand, it is important to check the context of the writing. For example, sections such as background, observations, rationale and discussion should be included in a report. The engine can check if the document includes each of these parts quantitatively. Moreover, groups of marks in a word map can be used to visualize the types of fields written in a document. So, several word maps are created based on the word vector space. Words for several fields are grouped and displayed with different colors in a word map as shown in Figure 12. In this way, knowledge about the manufacturing design processes is visualized.

Many documents and literature are usually stored in a company. We can search for many types of information from these. But, it takes time to read whole documents. If several kinds of typical sentences are prepared, the text-mining engine can evaluate likeness for each sentence quantitatively. This is very helpful to know which field is contained in a particular document quicky. Radar charts are shown in Figure 13. These are examples to evaluate old documents about quality problems of castings. Which types of defects occurred: burr, hole, crack, surface defects, misrun, distortion or inclusion? We can see that Case V6509744 is about hole defects and misrun while Case V6505426 is about distortion problems and surface defects. Even without these keywords, the text-mining engine can classify documents depending on specific sentences in which a typical defect phenomenon is presented. This is also one of knowledge maps.

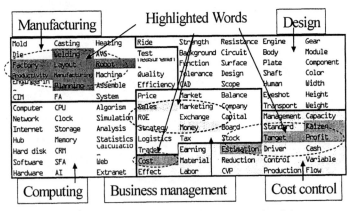

Figure 12. Knowledge map extracted from accumulated documents.

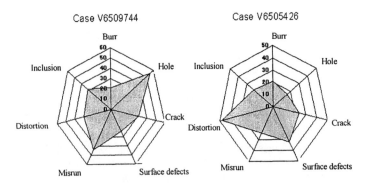

Figure 13. Radar charts to show the contents of a document (11).

REFINING OF KNOWLEDGE

An important objectives of knowledge management is to find the best practice. By evaluating activities, we can discover which parts of an activity need improvement. After performing practical manufacturing design using this system, the log data is stored. By analyzing this log, a workflow can be represented as in Figure 14. This is a typical PERT (Program Evaluation and Review Technique) diagram. Average clock time of each operation, stoppage clock time and output of each activity (documents and computer input data) are described in detail. This data type is useful for ABC (Activity Based Costing). As well, the critical path can be easily found.

The workflow must be analyzed to derive a reform bill after design. In recent years, ABM (Activity Based Management) has attracted attention as a key method for BPR (Business Process Reengineering). A reform bill can be derived using a 5-step approach:

<1> business process analysis,
<2> definition of process activity,
<3> cost investigation of each process activity,
<4> process value analysis and
<5> improvement planning in ABM.

In such an activity analysis, the cost of each activity is quantitatively understood in addition to processing time after the business process flow is subdivided into each detailed activity. In other words, we determine if some particular value is produced in each process examined. The value (V) is calculated by the equation of F (Function)/C (Cost) and compared with the allowance for cost reduction, (C-F). The activities are classified into two categories of added-value activity and non-added-value and divided into the three categories of core, support, and accompaniment activity. It is said that general activity consists of 30 % core, 35 % support, and 35 % accompaniment.

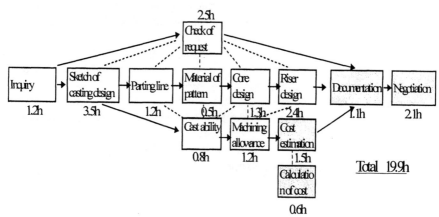

Figure 14. Workflow obtained from log data of activity.

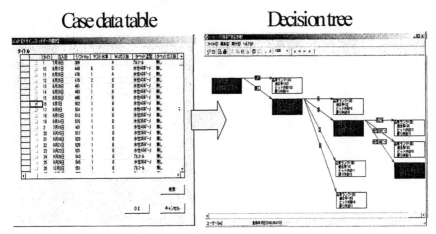

Figure 15. Decision tree made from case data.

It is difficult to estimate the value V of each activity in practice. Moreover it is troublesome to collect man-hours of every activity over a long period to know the true value of this indirect work. Analysis methods involve renewing the work flow chart by the following viewpoints:

(a) Classification between main processes and sub processes
(b) Extraction of repetitive processes
(c) Insertion of a new process

Workflow charts provide a lot of information. Time is important to evaluate the efficiency. But, more information is needed to measure the real value of each activity. What kind of causation does the activity have in production value? With our system, activity data can be stored automatically. Cost, quality, and delivery are facts to which we must pay attention, as are complaints from customers, sales and profit. For each instance, final cost is ranked with an integer by comparing to the standard cost. Rank is regarded as a goal of the activity. The data-mining engine helps to extract a workflow to arrive at the best or a better goal. More than a single attribute of the design and goal can be input from accumulated practical cases. Figure 15 shows an example of inputting some attributes and finding a decision tree to obtain good quality through the C5.0 software (12). We developed a GUI to select suitable attributes from the database and to execute the C5.0 engine.

A graphical presentation of the decision tree is useful to check for design problems before actual manufacture. For some products, adaptable cases may be too sparse to make a decision tree. In this case, skilful engineers can input trees based on their own ideas and experimental results. We can examine if the decision tree matches the real case. Hypothesis and inspection types of data mining are also useful to refine technical standards.

CONCLUSIONS

A framework of knowledge management for manufacturing design has been proposed and an integrated software system developed for this framework. Initially, the software can help individual engineers, but as knowledge is gained, the software is not needed routinely. If an engineer needs the software for all activities, the information for manufacturing design can be stored digitally. These data can be shared with other engineers. We consider activity-modeling to be one of the key methods although refining knowledge is also important. Moreover, predicted and analyzed knowledge can be added for later use. In this study, modules have been implemented as an integrated software system and several companies are using this software routinely.

ACKNOWLEDGEMENT

The authors wish to acknowledge all members of the Digital Meister Project of the Ministry of Economy, Industry and Trade in Japan for providing support for this study.

REFERENCES

[1] I. Nonaka and H. Takeuchi.1995. Knowledge-Creating Company: How Japanese Companies Creates the Dynamics of Innovation, Oxford University Press, Inc.
[2] Y. Nagasaka, 1999. Computer Simulation and Information Management Systems for Material Processing. Proc. IPMM99, 85–90.
[3] Y. Nagasaka, S. Kiguchi, M. Nachi and J.K. Brimacombe, 1989. Three Dimensional Computer Simulation of Casting Processes, AFS Transactions, 89-117, 553–564.
[4] Y. Nagasaka, J.K. Brimacombe, E.B. Hawbolt, I.V. Samarasekera, B. Hernandez-Morales and S.E. Chidiac, 1993. Mathematical Model of Phase Transformations and Elastoplastic Stress in the Water Spray Quenching of Steel Bars, Met. Trans., 24A, 795–808.
[5] I. Ohnaka, Y. Nagasaka and T. Murakami, 1996. A Computer Simulation System of Casting Processes for Concurrent Engineering Approach, Proc. MCS3-96, 46–51.
[6] Y. Nagasaka, 2000. Knowledge management for pre-production processes. Poster Proceedings of The Brimacombe Memorial Symposium, Vancouver, B.C., 255–264.
[7] H. Suzuki, F. Kimura, B. Moser and T. Yamada, 1996. Modeling information in design background for product development support, Annals of the CIRP 45(1), 141–144.
[8] J. Martin and J.J. Odell, 1998. Object-Oriented Methods, Prentice Hall PTR.
[9] M.C.Wu and S.R.Jen, 1996. A neural network approach to the classification of 3D prismatic parts. In. J. Manufacturing Technology, 11, 325–335.
[10] Y. Nagasaka, M. Nakamura and T. Murakami, 2001. Extracting and learning geometric features based on a voxel method for manufacturing design, Proc. IPMM2001, BT3, 1–10.

[11] M. Ate, Y. Nagasaka, I. Ohnaka, 2003, Arrangement of knowledge on the field of casting defects using natural language processing, J. Japan Foundry Eng. Soc., to be published.
[12] J.R. Quinlan, 1993. C4.5: Programs for Machine Learning, Morgan Kaufmann Pub. Inc.

Promising Information Technology in the Near Future and Current Obstacles: UWB/PLC and Cyber-Security in the Broadband Age

Y. TAKEFUJI
Faculty of Environmental Information, Keio University, Fujisawa, Japan

ABSTRACT

On Feb. 14 in 2002, the FCC (Federal Communications Commission) has authorized the ultra-wideband (UWB) technology [1]. On Sept. 30 in 2002 the first equipment authorization of UWB was awarded so that the UWB commercial products can be legally sold in US [2]. The UWB technology promises three areas including PAN (personal area network) applications with more than 100 Mbps, personal radar applications, and distance measurement applications. Ubiquitous computing or pervasive computing can be realized by the UWB technology. There is no network expert in the world who could predict the current legacy information infrastructure communication technology of ADSL and CATV modems instead of the fiber technology for the last one-mile network accessing. The powerline communication technology is able to provide us 200Mbps bi-directional communications using 30MHz bandwidth spectrum over an ordinary home power line [3]. Along with the spread of the broadband technology and ubiquitous computing, we will be more and more depending on the networking and digital information. In other words, we will use more and more networked digital data. Therefore cyber-security will play a key role in branding the products, the organizations, and the countries for securely handling networked digital data. ISO15408 and ISO17799 will play a security branding business. However the up-and-coming, promising, information technologies have been disturbed by bureaucratic opposition in several countries in order to protect their vested rights. In Japan, the national cyber-security committee organized by the bureaucrat may cause the hazardous crisis. A recent investigation of Japanese governmental web sites shows a number of high-risk vulnerabilities. In this paper, ideas for vitalization of the old/new industry will be given.

PROMISING TECHNOLOGY

Powerline Communications

In the current broadband age, legacy information infrastructure communication technology has been playing a key role for the last one-mile network accessing. Legacy information infrastructure communication technology includes ADSL technology using conventional telephone wire; cable TV modems using conventional cable TV wire; and wireless communication technology. The most promising legacy information infrastructure communications for the last one mile or ten meters network access can be achieved by powerline communication technology, because insignificant installation work using powerline communication technology is needed to establish a high speed information infrastructure communications in a building or house. Every powerline outlet will be a high-speed network access point while the conventional wireless technology cannot communicate through thick concrete walls. In this presentation, 200Mbps powerline communication devices will be demonstrated and explained in which adaptive OFDM technology is developed by DS2 (http://www.ds2.es/home/index_total.php). A summary of the current specifications are given below:

- 1280 OFDM carriers per 7MHz link
- Data rate per 7MHz link:
    - Up to 27 Mbps in downstream channel
    - Up to 18 Mbps in upstream channel
- Data rate per subcarrier adaptable according SNR detected
    - N° bits per carrier : 0, 2, 4, 6 and 8
    - Different carriers transmit at different data rates
- Modulation efficiency up to 7.25 bps/Hz
- Overlapped subchannels : efficient use of the spectrum saving bandwidth
- Total used bandwidth 1.7MHz~30MHz

A very interesting fact in this technology is that every OFDM carrier is independently programmable. In other words, undesirable signals/noises can be completely eliminated by this filtering function in order to accommodate any regulations in a country. The powerline communications with wireless devices might be a candidate for the next generation ubiquitous computing. Large scale powerline communication trials will be held in 2003 in Europe as shown in Figure 1. The powerline communication technology is useful in net home appliances where a power line can not only provide the electronic energy but also the high speed network accessing capability. The network function can give us VoIP, video streams, and internet access. All of home appliances are expected to have such powerline communication capability.

Figure 1. PLC in Europe (copyrighted by DS2).

Ultra-wideband Technology

FCC has deregulated the use of UWB technology as of Feb. 14, 2002, when 7.5GHz unlicensed spectrum from 3.1GHz can be used as shown in Figure 2. The short range UWB technology is currently standardized by 802.15.3a with an expected performance from TimeDomain devices (http://www.timedomain.com/) as follows:

110Mbps at 10 meters, 200Mbps at 2 to 3meters, or 480Mbps at less than 2 meters.

A USB2.0 cable provides 480Mbps communications in the range of 3 to 5 meters and a Firewire/1394 cable gives 400Mbps communications within 4.5 meters. Therefore, 802.15.3a UWB devices will play a key role in USB2.0 and 1394 as wireless cable replacements. The expected schedule for short range UWB (802.15.3a) is given in Figure 3.

FCC Regulatory Approval – 2/14/02

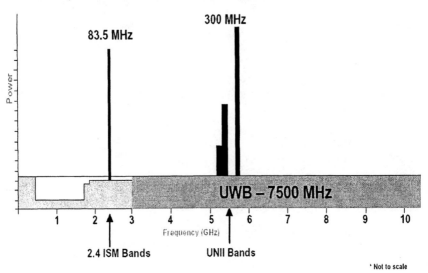

Figure 2. Deregulated UWB spectrum (copyrighted by TimeDomain).

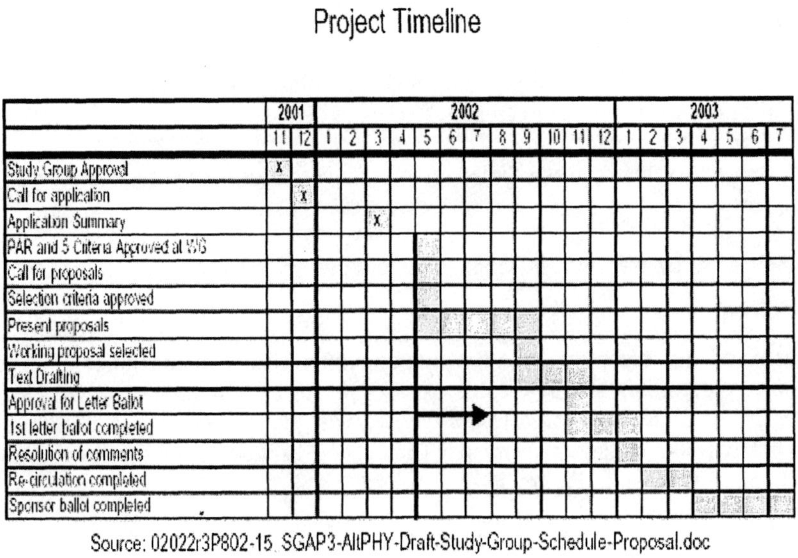

Project Timeline

	2001		2002												2003							
	11	12	1	2	3	4	5	6	7	8	9	10	11	12	1	2	3	4	5	6	7
Study Group Approval	X																				
Call for application		X																			
Application Summary					X																
PAR and 5 Criteria Approved at WG																					
Call for proposals																					
Selection criteria approved																					
Present proposals																					
Working proposal selected																					
Text Drafting																					
Approval for Letter Ballot																					
1st letter ballot completed																					
Resolution of comments																					
Re-circulation completed																					
Sponsor ballot completed																					

Source: 02022r3P802-15_SGAP3-AltPHY-Draft-Study-Group-Schedule-Proposal.doc

Figure 3. Timeline for 802.15.3a.

A variety of UWB applications including three major areas (communications, radar, distance measurement) is addressed in Figure 4. The UWB radar technology for ground investigation is called GPR (ground penetrating radar) technology for sensing/investigating the ground, bridges, or paved roads.

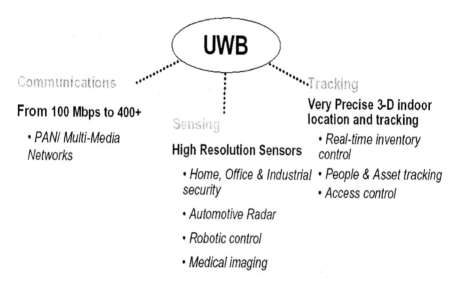

Figure 4. Applications of UWB (copyright Timedomain).

CYBERSECURITY RISKS IN THE BROADBAND AGE

The product liability act has never been applied to the software industry. Therefore software products including operating systems, applications software, device drivers or software programs generally have a large number of defects or bugs which are called "vulnerabilities" or "security holes." Currently about 60,000 security holes are known in the world where 20,000 vulnerabilities have been actively exploited by hackers or crackers. The largest official vulnerability database provided by US NIST is available at: http://icat.nist.gov/.

Despite the redoubled attention to security since the terrorist attacks of Sept. 11, 2001, 14 of 24 federal agencies flat out flunked their efforts to improve network safety, according to the Computer Security Report Card released last month by the House Subcommittee on Government Efficiency, Financial Management and Intergovernmental Relations [4]. Even worse significantly high risks have been observed in many countries including Japan. We do not have any data about quantification of trustworthy measurement on the existing e-commerce sites. All Internet users are interested in the security rating of a targeted organization through their web server. The security rating is like the financial rating by Moody's or S&P. The internal security rating can be evaluated by ISO security branding including ISO15408 (software and hardware) and ISO17799 (administration and management). In order to overcome the obstacles in the security problem, we have to face and understand the following facts:

1) there are unknown vulnerabilities in existing software programs,
2) there are known vulnerabilities without patch files,
3) there are many unpatched web sites although a decent security administrator can fix such vulnerabilities,
4) there are many security administrators who are not interested in patching vulnerabilities, and
5) there are many security administrators who do not understand that unpatched web sites are harmful for spreading viruses and DOS attacks.

Although hackers or crackers tend to do knowledge-sharing about attacking-skill in real-time, security administrators in an organization do not share their knowledge about how badly their web sites are attacked. In order to improve this circumstance, we have to provide the incentive or penalty to security administrators for knowledge-sharing.

REFERENCES

1. http://www.fcc.gov/Bureaus/Engineering_Technology/News_Releases/2002/nret0203.html
2. http://www.timedomain.com/Files/HTML/pressreleases/FCCCertification.htm
3. http://cgi.ds2.es/home/index_total.php
4. http://www.eweek.com/article2/0,3959,742734,00.asp

3D Interactive Input System

H. ELDEEB, H. ELSADEK and E. ABDALLAH
Electronics Research Institute, Dokki, Giza, Egypt, 12622

ABSTRACT

Personal computers now come with high performance 3D graphics hardware and rendering systems capable of offering applications that are truly 3D interactive. Virtual Reality (VR) applications are already recognized for their value in specialized domains. It is also convenient to have VR applications that run on the desktop that a wide number of users can acknowledge and use. Input devices such as the mouse, trackball and graphics tablet are often used for pointing and selection in conjunction with the present generation of 2D user interfaces [1]. However, current development in interactive computer graphics and user interface requires the user to navigate, select and rotate in 3D. Working toward this target, we use the idea of microwave holograms to design a dynamic system for 3D inputting (editing) [2] that combines elements of microwave holography with those of microstrip antenna arrays. Using this new system, an operator can create and draw, point by point or in a multipoint basis, any 3D graph or object that exists physically or even virtually and save it in computer memory. This paper includes a brief discussion of different 3D input device techniques and a detailed overview for our system, describing all its parts and measurements. Finally evaluation of our system is presented.

INTRODUCTION

Commodity priced PCs now come with high performance 3D graphics hardware and rendering systems capable of offering applications truly 3D interactive. Virtual Reality (VR) applications are already recognized for their value in specialized domains. However, it is also possible to have VR applications that run on the desktop, i.e., a non-immersive VR. For desktop systems, there is a need to match development in input devices suited to 3D applications beyond the simple 2D Mouse [1]. The applications for 3D input devices range over the non-immersive VR systems already mentioned as well as new user interfaces, web browsers, computer-aided design and manufacture, computer simulation, telerobotics, scientific data visualization, and areas such as molecular modeling.

Input devices such as the mouse, trackball and graphics tablet are often used for pointing and selection in conjunction with the present generation of 2D user interface. However, current development in interactive computer graphics and user interfaces require the user to navigate, select and rotate in 3D. This increase in power entails an increase in complexity in the user interface and it may be that the 3D system will fail to live up to its potential unless appropriate input devices are available to users [2-5]. Our 3D input system [6] is based on microwave and computer holographic techniques.

The organization of the paper is as follows: after a brief discussion of different 3D input device techniques, we introduce an overview of our system. Then, we describe the hologram recording part with an array antenna design and measurements. We then describe the software driver which includes the reconstruction process, and we compare our proposed reconstruction technique with the conventional FFT technique. We evaluate our system in comparison with some existing 3D systems by comparing important characteristics for 3D input devices. Finally, we discuss design implications of this study and direction of future research.

3D INPUT TECHNOLOGIES

Computer interfaces are subject to continuous change due to advancements in the underlying base technologies as shown in Figure 1.

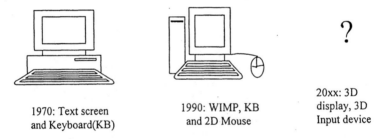

1970: Text screen and Keyboard(KB)

1990: WIMP, KB and 2D Mouse

20xx: 3D display, 3D Input device

Figure 1. Generations of Computer User Interfaces.

To implement the outlined solution for a technical system that can be used daily for hours at work, is a demanding task. [7, 8, 9].

(1) A crucial challenge is to develop a 3D display that can be viewed as comfortably as today's 2D displays. For instance cumbersome helmet-mounted displays will only be accepted in special applications for a limited period of time. Instead, free viewing display is needed (using holography or auto stereoscopic techniques), which allow the user to interact with the scenery by moving to appropriate perspective views.

(2) To support the cognitive aspect of human information processing, the interface should offer an adequate (spatial) model of the current computational environment. It is not yet clear how to optimally design these interfaces in detail.

(3) To support natural interaction with 3D display objects, reading of, e.g., hand gestures and voice control are straightforward approaches. However, interactions must not necessarily be limited to what is possible in communication with natural 3D environments. For example, a different virtually non-fatiguing interaction technique is gaze reading which allows one to visually control synthetic objects in the display space.

Many of these technologies have been adapted to measure body part locations in 3D, motivated by military cockpit and virtual reality (VR) applications [7] but such systems have significant limitations [10].

Mechanical arm-type tracking devices typically use potentiometers or optical encoders to measure the rotation of a joint between rigid linkages. These have a small working volume and mechanical trackers may wear out after a period of time. Electric Field (EF) sensors produce small signals to be measured by expensive electronic apparatus. Also, the non-uniform electric field nature of these sensors make it difficult to transform these signals into linear position coordinates. [7]. Acoustic methods are line-of-sight and are affected by echoes, multi-paths, air currents, temperature, and humidity. Optical systems are also line-of-sight, they require controlled lighting but are saturated by bright lights, and they can become confused by shadows. Infrared systems require significant power to cover large areas.

An electromagnetic tracker comprises a transmitter and a receiver. A fluctuating magnetic field generated in the coils of the transmitter is picked up by corresponding coils in the receiver. Variations in the received signal can be used to calculate the relative position and orientation of the receiver and transmitter with six degrees of freedom. Microwave holography is an electromagnetic tracking technique that consists of two steps, namely recording the object's 3D information (amplitude and phase) and then reconstructing these data to get an object's virtual image (position and orientation). Our research using this technique, addresses these electromagnetic sensing issues to help make it more accessible to interface designers. It will be shown that using holography technology with electromagnetics technology, can provide great improvement for input devices with more accuracy, better resolution and small latency time.

OVERVIEW OF THE SYSTEM SIMULATION AND ARCHITECTURE

As is well known, the main advantage of simulation and modeling of a physical problem is to try to reach the best design for the lowest cost. To build our system, we first had to decide what are the required accuracies, resolutions, and working volumes. Depending on these requirements, we then needed to develop a whole system simulation to adjust different system parameters such as wavelength, hologram size (the size of the array antenna and number of elements), hologram shape (the structure of array antenna) and sampling rate or distance between hologram points (inter-element distance between antennas). Figure 2 shows an overview for the system simulation and the output from each stage.

We chose the accuracy to be equal to or less than 1 cm in this work and attempted to achieve 1 mm. By inputting these requirements to our established simulation code, we found that the optimum parameters are: wavelength = 57 mm ((f=5.2GHz)-802.11b), the distance between hologram points (antenna elements) = 15 mm or less and the best hologram shape is square. A comparison is given in Table 1 of different hologram shapes tested to achieve this purpose. In the table the word "distance" refers to the array antenna inter-element distance. From Table 1, we see that sparse geometries (cross, frame) are useful when you need only moderate accuracy while complete shapes (full square) are required for high accuracy applications.

Figure 3 shows the 3D inputting system as a block diagram. The User hand unit itself works as a hand agent that contains a short passive reflecting dipole antenna with a modulating diode at low frequency and its power supply feed [6]. This unit is attached to the operator's hand to locate its movement. The illuminator is fixed in a certain corner within the reconstruction volume. The transmitter illuminates the whole volume under investigation. When the user clicks on the hand agent at the required position in space, the switch is closed causing the illumination signal to be modulated. The modulated scattered field is proportional to the local field at the corresponding mouse location. The receiver (microstrip antenna array) receives and stores the corresponding modulated scattering signal at every position of the array. The object signal is demodulated and then is allowed to interfere with the reference signal in the panel unit to form the interference pattern (hologram). The hologram data is digitized through an analog to digital converter in the same unit. Digital data is interfaced to the computer and retrieved through a software hologram

688 MANUFACTURING SYSTEMS

reconstruction driver that can be called a 3D-holographic card. Finally the retrieved point is drawn on the screen if necessary with any particular graphics software package. The proposed reconstruction algorithm allows us to reconstruct volume size equal to hologram size, which is double the volume if reconstructed by conventional FFT. More details about this reconstruction algorithm are provided later.

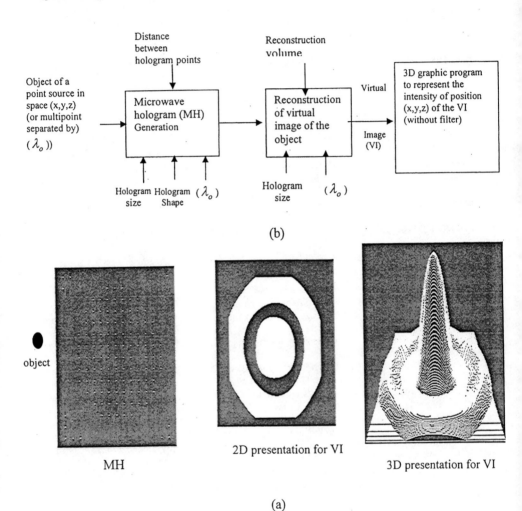

Figure 2. Overview of system simulation (a) Block diagram (b) Output of each stage.

Table 1. A comparison of different hologram shapes.

	Array Antenna Shape	Description	Design Advantage	Design Disadvantage
1	(4×4 square array, 10mm)	Square with 10mm distance between array antenna elements	Can achieve 1mm ($\lambda_o/50$) accuracy with low noise	Very difficult to achieve this distance practically at 5.2 GHZ with uncorrelated array elements
2	(4×4 square array, 15mm)	Square with 15mm distance between array antenna elements	Can achieve 1mm ($\lambda_o/50$) accuracy with cubic working volume	Difficult to achieve this distance but we did it practically
3	(Double Cross pattern)	Double Cross with 15mm distance between array antenna elements	Simple and more practical	Can not achieve 1mm ($\lambda_o/50$) accuracy with same cubic working volumes as (1) (Working volume ~ ellipsoid)
4	(Frame pattern)	Frame with 15mm distance between array antenna elements	Simple and more practical It takes the shape of frame around Monitor	Can not achieve 1mm ($\lambda_o/50$) accuracy

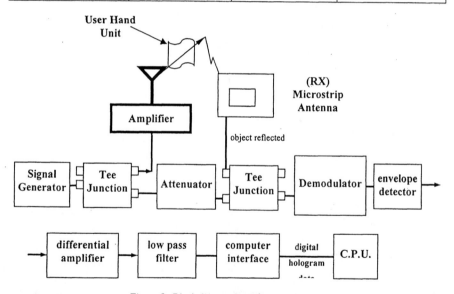

Figure 3. Block diagram for 3D inputting system.

MICROWAVE HOLOGRAM GENERATION

The first microwave hologram was produced at Bell Laboratories in 1950 [10]. We use a similar idea to design a dynamic electromagnetic system for our 3D input device. The hologram is constructed through an array of microstrip antennae. As a working frequency, we use S-band or C-band operation as this gives good compromise between positioning accuracy and the number of array elements. Using a conventional imaging system with FFT, a hologram plate size of double the working reconstruction volume is considered, making the antenna array size impractical for most applications [6]. A sampling grid of half wavelength reduces accuracy unless one uses very high operating frequencies. Instead, by using the reduced system proposed here, a hologram plate equal to the reconstruction volume can be used and location capability is improved to $\lambda/50$.

CORRELATION AMONG PLANAR ARRAY ANTENNA ELEMENTS

In the holographic 3D system, diversity reception is one of the significant and effective techniques to increase system accuracy for positioning purposes since we must sample the data as accurately as possible without any neighboring interference effects [5]. Small diversity antennas that provide high gain and low correlation coefficient must be developed. The correlation factor between array elements is a function of the element radiation pattern and the mutual coupling factor. The antenna can act separately if the correlation coefficient (ρ) in Eq. 1. is in the range 0.16 to 0.22. This corresponds to a cross-polar ratio greater than 9 dB in the element radiation pattern and to mutual scattering coefficients (< -15 dB) between elements [5, 10]. Assuming a Gaussian power distribution on an antenna element at elevation direction and uniform distribution at the Azimuth direction, the correlation factor becomes.

$$\rho = \frac{\left| \iint_{\phi\theta} (E_1^*(\theta,\phi).E_2(\theta,\phi).P(\theta,\phi)\sin\theta d\theta d\phi \right|^2}{\iint_{\phi\theta} (E_1^*(\theta,\phi).E_1(\theta,\phi).P(\theta,\phi)\sin\theta d\theta d\phi . \iint_{\phi\theta} (E_2^*(\theta,\phi).E_2(\theta,\phi).P(\theta,\phi)\sin\theta d\theta d\phi} \quad (1)$$

where: E_1 = the field pattern when element 1 fed and 2 terminated.
$P(\theta,\phi)$ = the receiving power distribution defined by:

$$P(\theta,\phi) = \left| \frac{1}{\sqrt{2\pi s^2}} \frac{\exp(-(\theta - \pi/2 - \theta_i))}{2s^2} \right|$$

θ_i is the direction of maximum incoming wave;
s is the standard deviation.

RECEIVING MICROSTRIP ANTENNA SYSTEM

A microstrip antenna array is the best candidate to satisfy the hologram recording process requirements because of its light weight, small size, low cost, high gain and directive beam that is suited for mono-frequency applications. Its planar or conformal configuration provides it with strong capabilities to be mounted on any device or wall surface to cover any required experimental area. A rectangular element is suitable for this application due to its simple structure allowing integration with any feed or matching network. It has low scattering field cross-section compared to other common shapes such as circular and triangular patches [9]. Also it has enough receiving area to handle the required hologram information.

Several treatments for different elements are done. Table 2 summarizes these processes. The antenna elements and arrays are analyzed by finite element method (HFSS software package) and fabricated using a photolithographic technique. Scattering parameters are measured using a network analyzer (HP8510c).

VIRTUAL IMAGE RECONSTRUCTION

The digital hologram data is retrieved by a new computer simulation model [7] called raytracing technique, which is based on the following equation:

$$\phi_i = \sum_\alpha \phi_{i\alpha} = \sum_\alpha I_\alpha \frac{\cos(kR_{i\alpha})}{R_{i\alpha}} \qquad (2)$$

Where ϕ_i is the retrieved image, I_α is the intensity of a grid point on hologram plane and $R_{i\alpha}$ is the distance between the hologram plane's grid point (α) and reconstruction point (i). The procedure in Eq. 2 merely inspects whether point (i) is a part of the original object or not. This is accomplished by using the correlation between the hologram intensity and the factor: $P_{i\alpha}$ = cos(k $R_{i\alpha}$)/ $R_{i\alpha}$ represents the hologram of a point source. If the hologram I_α coincides with the pattern $P_{i\alpha}$ generated by the point source at (i), the correlation ϕ_i becomes large and hence point (i) has a large probability to be included in the original object. With changing (i) over the 3D space and drawing the structure of ϕ_i, the image should coincide with the original object pattern. This method doesn't suffer from the limitations of FFT analysis [7] so the hologram sampling grid separation can be less than $\lambda_0/2$ and the holographic plate size can be in the range of $\beta < 45°$. Using the same reconstruction volume (axial), the hologram plate can be reduced to about one forth of its original FFT surface while keeping the same information content number of points L_N x W_M but with sampling grid size $\lambda_0/4$.

RESULTS

To verify our ideas, the hologram is generated at different points within the operator's hand reconstruction volume with use of the microstrip reduced size antenna array described above. Figure 4 shows the 3D representation of the retrieved image of points at positions (4 cm, 4 cm, 12 cm) with 1cm accuracy and at (30 mm, 30 mm, 45 mm) with 1mm accuracy respectively. From the diagram, it is evident that we were able to catch the 3D coordinates of a point within the object reconstruction volume with absolutely no error. We use a software filter to clarify the point position accurately. It is worth noting that our holographic system can identify multiple points at the same time hence there will be a modulating diode at each interesting point with different modulation frequencies. As we found empirically the resolution is about λ_0. We simulated a two-point reconstruction with a hologram with 16x16 antenna array elements to increase the reconstruction volume. Figure 5 is an example of 2-point reconstruction with a hologram size of 1 m x 1 m and point positions at (50 cm, 50 cm, 9 cm) and (70 cm, 70 cm, 9 cm) respectively.

Table 2. Comparison of different characteristics required for different holographic plate antenna systems. (O-plane coupling technique is to feed every adjacent element in the array orthogonally).

Element	Shape	Design technique	Bandwidth	Element size	Inter-element in 2x2 sub-array	Mutual coupling	Disadvantage
Reference rectangular patch		Aperture proximity coupling feed	1.5%	0.28λ x 0.29λ	$\lambda/2$ in x $\lambda/2$ in y	H=-33dB E=-31dB O=-39dB	impractical size
Reduced size rectangular patch		Inverted H slit for size reduction	0.55%	0.15λ x 0.18λ (75% reduction)	$\lambda/5$ in x $\lambda/2.5$ in y	H=-21dB E=-17dB O=-22dB	Reduced BW
Stacked reduced size patch		Additional stacking patch with bias between 2 patches for BW enhancement	1.7%	0.24λ x 0.10λ	$\lambda/4$ in x $\lambda/2.5$ in y		Difficult Fabrication
U-shape slot antenna		microstrip line fed through multilayer proximity coupling	2%	0.20λ x 0.12λ	0.24λ in x 0.24λ in y	E=-14dB O=-26dB	Bidirectional radiation (solved by using a $\lambda/4$ reflector plate under the slot)

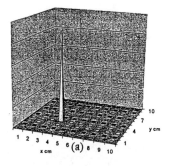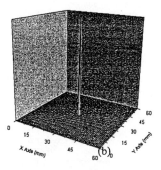

Figure 4. (a) Retrieved image of one point with hologram size of (10 cm x10 cm) and grid sampling of $\lambda_0/3$ and point at coordinates (4 cm, 4 cm, 12 cm) with 1cm accuracy;
(b) retrieved image of one point with hologram size of (60 mm x 60 mm) and grid sampling of $\lambda_0/4$ and point at coordinates (30 mm, 30 mm, 45 mm) with 1mm accuracy.

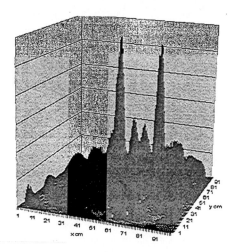

Figure 5. Reconstructed image of two-point hologram with points at positions (50cm,50cm,9cm) and (70cm,70cm,9cm) respectively and hologram size 1mx1m.

System Evaluation and Comparison with Other Existing Devices

There are many choices in designing or selecting an input device. The choice of each design dimension may have implications on the users' performance. Aside from application specific requirements, there are at least 11 aspects to the usability of any 3D input device, namely [5, 10]:

- type (sensing technology),
- speed (latency),
- accuracy,
- resolution (for multipoint input at the same time),
- ease of learning,
- fatigue,
- absolute or relative measurement of position,
- working volume,
- susceptibility,
- coordination,
- device persistence and
- ease of acquisition.

The first nine items are common to all input devices and their meaning is obvious. The tenth aspect, coordination, is unique to multiple degrees of freedom input control. There are many ways to measure the degree of coordination. One effective way to quantify this is based on the ratio of the length of actual trajectory and that of the most efficient trajectory in the coordination space, including translation space, rotation space and the 2D space between translation and rotation. By such a measure, in order to produce the most coordinated path, one must simultaneously move all degrees of freedom involved at the same space towards their respective goal status [10].

The eleventh item is ease of device acquisition. This is often overlooked in rating input device usability. Although a mouse is less dexterous than a pen-like input device (a stylus), the fact that a mouse is more easy to acquire is an important reason that makes it the dominant 2DOF input device. Many factors, such as distance to the computer keyboard home row (ASDFGHJKL keys), contribute to the ease of device acquisition. One of these is the device location persistence when released. With a mouse or a trackball, when released by the hand (in order to type something, for example) it stays in the last position. This is not true with a stylus.

System Performance and Future Work

Our system is developed to demonstrate its performance in a desktop environment:

* Frequency: S-C band for easy integration with personal wireless network standard communication.

* Robustness: testing the system with different point positions–results show our system is robust.

* Accuracy: 1mm ($\lambda_0/50$) point position accuracy. Accuracy improves when the wavelength and/or the inter-element distance are reduced. For example, we built a prototype for this system with an accuracy of 1cm at an operating frequency of 2.23 GHz. The reconstruction volume was 10 cm x 10 cm x10 cm and the holographic plate was a 2 x 2 reduced size rectangular array with inter-element distance ($\lambda_0/3$) [6].

* Resolution: The system has been tested with two points separated by λ_0 - the position of two points can be determined simultaneously.

* Working volume: Extension of our system has no limitations either in array surface or reconstruction volume or operating frequency. There is a linear relation between array size and the required reconstruction volume (array size α working volume).

* Lag: Since holography consists of two steps, the lag in our system is the sum of the lag in each step. For the first step, the lag time is measured in fractions of msec which can be

considered real time. The second lag occurs from the application of the reconstruction algorithm (software driver) used to reconstruct the virtual image of the object point in space. This lag depends on the CPU speed. For example, in this example, the lag time is 0.8 sec. for a hologram grid of 4 x 4, a reconstruction volume of 60 x 60 x 60 points running on a Pentium III 450Mhz machine.

CONCLUSION

We have proposed a new 3D input system to achieve accurate editing of 3D positions in space. The working volume can be increased by increasing the hologram size (array antenna elements). This will lead to the detection of more point positions at the same time, and also opens the application for motion detection and mobile robotics research. The results show that our system satisfies the requirement of a robust and friendly 3D user input device. The system can achieve real time operation by replacing the software reconstruction driver with a hardware device or by using a parallel architecture.

REFERENCES

[1] I. Poupyrev, 1995. Research in 3D user interfaces, IEEE Computer Society Student Newsletter, 3(2), 3–5.
[2] I. Poupyrev, 1998. Egocentric Object Manipulation in Virtual Environments: Empirical Evaluation of Interaction Techniques, Proceeding of Eurographics'98, 17(3).
[3] B. Frohlich, J. Plate, 2000. The Cubic Mouse: A new device for three-dimension input, CHI'2000, 2(1), 526–531.
[4] C. Hand, 1997. A Survey of 3-D interaction techniques, Computer Graphics Forum, 1615, 269–281.
[5] Hala Elsadek, Hesham Eldeeb, Franco Deflaviis, Luis Jofre, Esmat Abdallah, Essam Hshish, 2001. Microwave holographic 3D rendering system using a reduced size planar array antenna, Microwave and Optical Technology Letters, 29(6).
[6] Robert Skerjanc and Siegmund Pastoor, 1997. New generation of 3-D desktop computer interfaces, Proceeding of SPIE, 3012, 439–447.
[7] Hesham Eldeeb and Takashi Yabe, 1995. A fast method for an efficient reconstruction in computer holography, J. Plasma Fusion Res., 71, 331–335.
[8] Carl B. Dietrich, Jr., Warren Dietze, Kai Dietze, 2000. Smart antennas in wireless communications: Base station diversity, IEEE Transactions on Antenna and Propagation, 42(5).
[9] Ramesh Garg, Prakash Bharita and Inder Bahl, 2000. Microstrip Antenna Design Handbook, Artech House Press.
[10] I. Scott Mackenzie, Tatu Kauppinen, Niika Silfverberz, 2001. Accuracy Measures for Evaluating Computer Pointing Devices, Proceeding of CHI2001, 9–16

PART 2: NANOTECHNOLOGY

Chapter 11: Fullerene Materials Science

Novel Self-Assembled Material: Fullerenes Self-Intermixed with Phthalocyanines

M. DE WILD and S. BERNER
National Center of Competence in Research on Nanoscale Science, Department of Physics and Astronomy, Basel, Switzerland

H. SUZUKI
National Center of Competence in Research on Nanoscale Science, Department of Physics and Astronomy, Basel, Switzerland and Kansai Advanced Research Centre, Nishi-ku, Kobe, Japan

A. BARATOFF and H.-J. GUENTHERODT
National Center of Competence in Research on Nanoscale Science, Department of Physics and Astronomy, Basel, Switzerland

T. A. JUNG
Paul Scherrer Institut, Villigen, Switzerland

ABSTRACT

Self-assembling systems [1] are required to efficiently produce nanoscale molecular electronic devices. In this work, a novel route to highly-perfect molecular self-assembly through competing interactions on a metallic surface is presented. Depending on the relative surface coverage of two adsorbed species, Subphthalocyanine and C_{60}, periodic intermixed monolayers consisting of 1D chains with 1nm width or 2D hexagonal patterns are formed on atomically-clean Ag(111). The structural parameters and binary phase-diagram of this system were deduced from detailed room-temperature Scanning Tunneling Microscopy studies [2]. These novel intermixed patterns are different from previously known self-assembled molecular monolayers in that they form intermixed patterns at room temperature on uniform, unreconstructed and atomically-clean terraces. For the first time, control of the ordering of intermixed arrays of functionalizable molecules created by self-assembly has been demonstrated over hundreds of nanometers.

EXPERIMENTAL DETAILS

The first component, SubPc, is a polar molecule with a characteristic triangular symmetry as in Fig. 1.a. The second component is the well-known carbon fullerene (C_{60}), as in Fig. 1.b. In typical STM images (Fig. 1.c), individual molecules within these patterns can be clearly identified by their characteristic triangular or round shapes as determined in previous STM studies [3, 4]. The arrangement on the right hand side of the STM image in Fig. 1.d consists of zigzag rows of SubPc molecules separated by quasi-linear chains of C_{60} molecules. Within these rows, the C_{60} molecules are grouped into aligned triplets. In this phase, the molecules intermix at a fixed ratio of SubPc:C_{60} = 2:3. The C_{60} rows are parallel to the close-packed C_{60} rows of an island of the hexagonal phase of C_{60} on Ag(111) [4] appearing on the left side of Fig. 1.d. Such coexisting islands form when the C_{60} content of the deposited film exceeds a critical ratio.

A higher fraction of SubPc in the co-deposition process leads to a distinctly different pattern. Fig. 1e shows the chain-phase (I) co-existing with a new hexagonal phase (II). The triangular sub-unit of the hexagonal pattern consists of a C_{60} trimer which is surrounded by three SubPc molecules. The mixing ratio of SubPc:C_{60} in this pattern is 1:1.

RESULTS AND DISCUSSION

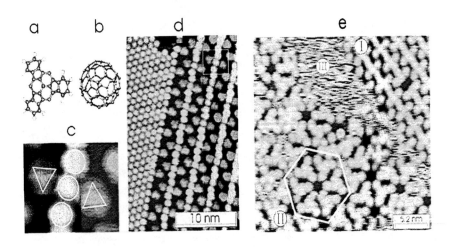

Figure 1. Self-intermixed phases **a**: Structure of SubPc. **b**: Structure of C_{60}. **c**: Artificially coloured STM image of the basic unit of the molecular chain pattern with superimposed schematic shapes (SubPc green, C_{60} yellow). **d**: Artificially coloured STM image of a monolayer of intermixed molecules on Ag(111) (imaged area, 17 x 25 nm). The bare substrate areas appear dark. **e**: Coexistence of two different intermixed patterns, labelled I and II (image area, 34 x 34 nm). On the right hand side, the chain phase (I) and on the left hand side, the hexagonal phase (II) is observed. Random tip excursions of single molecular height (III) are visible between the two ordered regions and are identified as mobile molecules in a 2D gas phase [3].

The multi-phase behaviour of this binary system on Ag(111) was studied as a function of composition in a series of deposition experiments. For the mixed phases, we note that the packing density of SubPc molecules in the chain and hexagonal phases is about 7% and 20% higher than in the honeycomb phase of pure SubPc even if we assign the same area to each C_{60} as in the pure hexagonal phase [2]. From Molecular Modelling and Photoelectron Spectroscopy Measurements we conclude that the observed mode of self-assembly and intermixing emerges from a delicate balance of interactions between static and fluctuating charge distributions of mobile and polarizable molecules that are bound non-covalently onto polarizable metallic substrates.

CONCLUSIONS

The self-assembly of such intermixed molecular systems is a critical step on the route to the efficient production of nanoscale molecular electronics devices and will likely also be important in a broad range of other applications such as in molecular data storage.

REFERENCES

[1] G. M. Whitesides, J. P. Mathias and C. T. Seto. Science 254, 1312 (1991).
[2] M. de Wild et al. ChemPhysChem 10 881 (2002).
[3] S. Berner et. al. Chem. Phys. Lett. 348, 175 (2001).
[4] T. Sakurai et. al. Progress in Surface Science 51, 263 (1996).

Electronic Structure of Self-Localized Excitons in a One Dimensional C_{60} Crystal

V. R. BELOSLUDOV
Institute of Inorganic Chemistry, SB RAS, Novosibirsk, Russia and Center for Northeast Asia Studies of Tohoku University, Sendai, Japan

T. M. INERBAEV
Institute of Inorganic Chemistry, SB RAS, Novosibirsk, Russia

R. V. BELOSLUDOV and Y. KAWAZOE
Institute for Material Research, Tohoku University, Sendai, Japan

J. KUDOH
Center for Northeast Asia Studies of Tohoku University, Sendai, Japan

ABSTRACT

Frenkel and charge-transfer excitons in a linear chain of fullerenes have been investigated using the Su-Shriffer-Heeger model supplemented by the Hubbard on-site Coulomb energy. The model is solved numerically within the adiabatic approximation. In this study, the electron charge distribution over the molecular surface, the Jahn-Teller distortions of the carbon atoms, the density of electron states and the total energy have been calculated self-consistently. It is shown that in the case of the electron excitation from the highest occupied state in the upper valence band to the lowest unoccupied state in the lower conductance band, a Frenkel exciton is formed on the central C_{60} molecule in the supercell at small conjugation parameter ($T/t < 0.25$) and with increasing conjugation a redistribution of the exciton to the other molecules in the supercell occurs. The formation of the charge-transfer exciton is observed in the case of the electron excitation from the highest occupied state in the upper valence band to the third lowest unoccupied state in the lower conduction band. For small T/t the positive and negative charges (hole and electron polarons, respectively) are located on two different molecules in the fullerene chain. With increasing conjugation, the polarons disintegrate and become delocalized over the whole chain. For $U=0$ the disintegration of the polaron states occurs simultaneously with increasing T/t. In the case of inclusion of the Hubbard interaction ($U=t$), the hole polaron continues to exist up to $T/t=0.0975$ whereas the electron polaron is already broken up.

INTRODUCTION

The outstanding discovery [1] of the fullerenes has created a new field of research at the boundary of chemistry and condensed matter physics, and the properties of the C_{60} molecule and C_{60} solids have been the subject of much attention from both experimental and theoretical viewpoints. The research interests, which are closely related to applications of fullerene-based compounds in nanotechnology, lie in investigating the bonding character and stability of the isolated fullerene and heterofullerene molecules and C_{60} solids, the reactivity of the C_{60} fullerene, and insertion of different atoms in C_{60} fullerene cages. Solid C_{60} is a molecular crystal bound by the van der Waals forces and its properties reflect both the properties of individual C_{60} molecules as well as the solid. However, the electronic configuration of crystal C_{60} is fundamentally different from usual molecular crystal, because of the conjugated character of the bonds in solid C_{60}. In a C_{60} molecule, as well as in conjugated polymers, the chemical bonding inside the molecule leads to one unpaired electron (π-electron) per carbon atom. These π-electrons delocalize on the quasi-two-dimensional surface of the C_{60} molecule and can move between the molecules of solid C_{60}. In spite of the fact that π-conjugation between C_{60} molecules is much weaker than inside the C_{60} molecule, solid C_{60} as a crystal has the remarkable physical properties.

The properties of the excited states in the pure and doped C_{60} solids have been a subject of many experimental and theoretical investigations because they have not only the fundamental interest but also direct practical applications, for example in molecular electronic devices. The fundamental interest is connected with idea that the pure and doped C_{60} solids are strongly correlated π-electron systems [2] and that π-electrons interact with the intramolecular distortions of the C_{60} molecules [3,4,5,6]. The reason is in the narrow bandwidth compared with the strong intramolecular Coulomb repulsion and the presence of Jahn-Teller (JT) distortions in the C_{60} molecule.

Experimental studies of low-energy fundamental excitations of the C_{60} molecule [7,8] and C_{60} solids [9–12] have shown the formation of self-trapped excitons where the excited electron and the remaining hole create a local deformation in the lattice and thus localize themselves into states across the energy gap in C_{60} below the conduction band and above the valence band in C_{60} solids, respectively. A number of attempts have been made to understand the nature of these excitonic states [10,13–20]. The existence of Frenkel excitons (where electron and hole are located in the same molecule) and also the charge-transfer (CT) excitons (where the two charges are located in different molecules) have also been demonstrated in solid C_{60}. The energies of the lowest levels of excitons and the optical-absorption spectra have been calculated.

Direct practical applications connected with the transmission signal and electron mobility depend on the properties of the excited electronic state. Therefore, the relaxation and trapping phenomena for excited carriers is important for the design of novel high speed electronic devices [12,21].

The investigations of the self-trapped excited states in fullerene crystals are also of much interest in the context of the photo-polymerization of C_{60} [22], polymerization that occurs spontaneously during cooling of doped AC_{60} compounds [23], and the synthesis of C_{60} dimers using a mechano-chemical technique [24]. During the polymerization the electronic structure between neighboring molecules significantly changes. Thus, the intermolecular bond changes from the van der Waals type to covalent. In order to understand this polymerization process, it is important to have an idea about the conditions under which the self-trapped excited state can form the covalent bond.

Since π-electrons in C_{60} solids are strongly correlated, the interplay between electron-phonon interaction and electron-electron repulsion may be important for understanding the nature of the excitonic states. For weak π-conjugations between C_{60} molecules one could suggest that the excited electrons form the excitons on single molecules. Will the molecular excitons remain such with increasing strength of conjugation? Do electron correlation and conjugation of the π-electrons have a strong effect on the conditions under which excitons can form? These questions are important to understand the nature of the unusual physical properties of C_{60} and the C_{60} solids.

We investigate the role of electron correlation and conjugation of π-electrons in the formation of excitons in C_{60} solids. For description of the exciton we use a generalized model of Su-Shrieffer-Heeger (SSH) [25] for the intermolecular and intramolecular degrees of freedom including a Hubbard-type on-site interaction. The method is based on the Hartree-Fock formalism and allows self-consistent calculations of the atomic and electronic structure of a crystal with the self-trapped exciton. Our previous results show that this model satisfactorily describes the physical picture of the phenomena and the properties of the fullerene compounds [6,26,27,28].

THEORETICAL METHOD

Model

The model Hamiltonian for the C_{60} crystal including electron correlation is written as the sum of the extended Su-Schrieffer-Heeger (SSH) Hamiltonian [14] and the Hubbard interaction term:

$$H = H_{ph} + H_{el-ph} + H_{el-el} \quad (1)$$

$$H_{ph} = \frac{K}{2}\sum_n\sum_{<ij>}(d_{ij}^n)^2 + \frac{K'}{2}\sum_n\sum_{<ij>}(d_{ij}^{n,n'})^2 \quad (2)$$

$$H_{el-ph} = -\frac{1}{2}\sum_{n,n'<ij>s} t_{ij}^{n,n'} a_{nis}^\dagger a_{n'is} + H.c. \quad (3)$$

$$H_{el-el} = U\sum_{ni} a_{ni\uparrow}^\dagger a_{ni\uparrow} a_{ni\downarrow}^\dagger a_{ni\downarrow} \quad (4)$$

Here $a^\dagger_{nis}$ creates a π-electron of spin s on the ith carbon atom of nth molecule; intramolecular hopping integrals for the nth molecule are represented by $t_{ij}^{nn}=t-\alpha d_{ji}^n$ for nearest neighbors and 0 otherwise; d_{ji}^n is the change in bond lengths between sites i and j of the molecule n; intermolecular hopping integrals from molecule n to n' are given by $t_{ij}^{nn'}=T-\alpha d_{ji}^{nn'}$ for nearest neighbors and 0 otherwise; $d_{ji}^{nn'}$ is the change in bond lengths between sites i in the molecule n and j in the molecule n'; α,α' are the electron-lattice coupling constants, U is the on-site Hubbard electron-electron interaction. In the approximation where displacements of carbon atoms caused by intramolecular and intermolecular vibrations are ignored, H_{ph} is the bond-stretching energy of the crystal with spring constants K and K'. For simplicity, we have not considered electron—electron interaction among electrons located at different sites. It can be supposed that this interaction will merely "renormalize" t and U [22]. As usual we treat the above Hamiltonian in the adiabatic approximation.

In the one-electron approximation; the wave function ψ_{is}^n of a π-electron with spin s on carbon atom i of molecule n can be presented as $\psi_{is}^n = a^\dagger_{nis}|0\rangle$ ($|0\rangle$ is the ground state of the system without π-electrons). The electron eigenfunctions $\psi_{\mu is}^n$ at site i corresponding to the energy level ε_μ can be found from the Schrödinger equation:

$$H\psi_{\mu is}^n = \varepsilon_\mu \psi_{\mu is}^n \tag{5}$$

The model is treated in the Hartree-Fock approximation

$$-\sum_j (t-\alpha d_{kj}^n)\psi_{\mu j\uparrow}^n - \frac{1}{2}\sum_j [(T-\alpha' d_{kj}^{n,n-1}) + (T-\alpha' d_{kj}^{n,n+1})](\psi_{\mu j\uparrow}^{n+1} + \psi_{\mu j\uparrow}^{n-1}) + U n_{\mu k\downarrow}^n \psi_{\mu k\uparrow}^n = \varepsilon_\mu \psi_{\mu k\uparrow}^n$$

$$-\sum_j (t-\alpha d_{kj}^n)\psi_{\mu j\downarrow}^n - \frac{1}{2}\sum_j [(T-\alpha' d_{kj}^{n,n-1}) + (T-\alpha' d_{kj}^{n,n-1})](\psi_{\mu j\downarrow}^{n+1} + \psi_{\mu j\downarrow}^{n-1}) + U n_{\mu k\uparrow}^n \psi_{\mu k\downarrow}^n = \varepsilon_\mu \psi_{\mu k\uparrow}^n \tag{6}$$

Here $n^n_{i\uparrow(\downarrow)}$ is the electron density on the ith carbon atom of the nth molecule with the spin up (down).

The total energy of the system is

$$E(\{d_{ij}^n\},\{d_{ij}^{n,n'}\},\{n_{js}^n\}) = \sum_\mu^{occ} \varepsilon_\mu (\{d_{ij}^n\},\{d_{ij}^{n,n'}\},\{n_{js}^n\}) + \frac{K}{2}\sum_{n<ij>}(d_{ij}^n)^2 + \frac{K'}{2}\sum_{n<ij>}(d_{ij}^{n,n'})^2 + U\sum_{ni} n_{i\uparrow}^n n_{i\downarrow}^n \tag{7}$$

We determine the lowest energy state using the variational theorem:

$$\partial E(\{d_{ij}^n\},\{d_{ij}^{n,n'}\},\{n_{js}^n\})/\partial d_{ij}^n = 0, \partial E(\{d_{ij}^n\},\{d_{ij}^{n,n'}\},\{n_{js}^n\})/\partial d_{ij}^{n,n'} = 0,$$
$$\partial E(\{d_{ij}^n\},\{n_{js}^n\})/\partial n_{is}^n = 0 \tag{8}$$

under the condition $\sum_{(ij)} d_{ij}^m = 0$, one can obtain d_{ji}^n, $d_{ji}^{n,n'}$ and n_{is}^n:

$$d_{ij}^n = -\frac{2\alpha}{K}\sum_{\mu s}^{occ}\psi_{\mu is}^{*n}\psi_{\mu js}^n + \frac{2\alpha}{90K}\sum_{(ij)}^{occ}\sum_{\mu s}\psi_{\mu is}^{*n}\psi_{\mu s\uparrow}^n, \; d_{ij}^{nn'} = -\frac{2\alpha'}{K'}\sum_{\mu s}^{occ}\psi_{\mu is}^{*n}\psi_{\mu js}^{n'},$$

$$n_{js}^n = \sum_{\mu}^{occ}\psi_{\mu is}^{*n}\psi_{\mu is}^n \qquad (9)$$

The nonlinear system of Eq. 5 and Eq. 8 is closed and its solution allows self-consistent determination of the electron structure of a one-dimensional C_{60} crystal, the distributions of bond lengths between carbon atoms and the charge densities with spins up and down on each site. For an ideal crystal one can use translational symmetry to reduce the system of *120N* (*N* - the number of C_{60} molecules in the crystal) equations to a system of 120 equations because the unit cell contains one C_{60} molecule. In the case of the formation of the self-trapped exciton, where the excited electron and the remaining hole create a local deformation of the lattice, the translational symmetry is broken and, in general, in order to find the solution the system of 120*N* equations has to be solved. To study the formation of the self-trapped exciton, it is necessary to consider a larger cell than the basic unit cell. Therefore we have constructed a supercell of *m* unit cells consisting of *m* C_{60} molecules. In this case, the system of *120N* equations is reduced to a system of 120*m* equations. The value of *m* is determined from the condition that the resulting system of equations would permit description of the creation of exciton in the C_{60} crystal to a reasonable degree of approximation. It is convenient to introduce in this case two indices *n* and *l* for molecules in the crystal supercell instead of one index *n*. Thus ψ_{is}^{nl} is the eigenfunction of an electron at site *i* of the *l*th ($l=1,2,...m$) molecule in the supercell *n* ($n=1,2,...N_C$) of the crystal. Using the Fourier transformation for a one-dimensional crystal

$$\psi_{js}^{nl} = \frac{1}{\sqrt{N_c}}\sum_q \psi_{js}^l(q)e^{iqa_c n} \qquad (10)$$

where N_c is the number of supercells, $a_C = ma$ is the length of the supercell, and *a* is the length of the unit cell, *q* is the modulus of the reciprocal lattice vector. Applying the translational symmetry $\psi_{is}^{nl} = \psi_{is}^l$, $d_{ij}^{nl} = d_{ij}^l$, $d_{ij}^{nl,nl+1} = d_{ij}^{n,n+1} = d_{ij}^{l,l+1}$, the system of equations (6) can be presented in matrix form as

$$\begin{bmatrix} A+C_\downarrow & 0 \\ 0 & C_\uparrow + A \end{bmatrix}\begin{bmatrix} \psi_{\mu\uparrow} \\ \psi_{\mu\downarrow} \end{bmatrix} = \varepsilon_\mu(q)\begin{bmatrix} \psi_{\mu\uparrow} \\ \psi_{\mu\downarrow} \end{bmatrix} \qquad (11)$$

where

$$\psi_{\mu s} = \begin{bmatrix} \psi_{\mu s}^1(q) \\ \psi_{\mu s}^2(q) \\ \psi_{\mu s}^3(q) \\ \vdots \\ \psi_{\mu s}^{m-1}(q) \\ \psi_{\mu s}^m(q) \end{bmatrix}, \quad A = \begin{bmatrix} A^1 & B^{12} & 0 & \cdots & 0 & 2\cos(qa_c)B^{1m} \\ B^{21} & A^2 & B^{23} & 0 & \cdots & 0 \\ 0 & B^{32} & A^3 & B^{34} & \cdots & 0 \\ \vdots & \vdots & \vdots & \ddots & \vdots & \vdots \\ 0 & \cdots & \cdots & \cdots & A^{m-1} & B^{m\,m-1} \\ B^{m1}2\cos(qa_c) & 0 & \cdots & \cdots & B^{m-1m} & A^m \end{bmatrix} \quad (12)$$

$$C_s = \begin{bmatrix} C_s^1 & 0 & \cdots & \cdots & \cdots & 0 \\ 0 & C_s^2 & 0 & \cdots & \cdots & 0 \\ 0 & 0 & C_s^3 & 0 & \cdots & 0 \\ \vdots & \vdots & \vdots & \ddots & \vdots & \vdots \\ 0 & \cdots & \cdots & 0 & C_s^{m-1} & 0 \\ 0 & \cdots & \cdots & \cdots & 0 & C_s^m \end{bmatrix}, \quad C_s^k = \begin{bmatrix} n_{1s}^k & 0 & \cdots & \cdots & \cdots & 0 \\ 0 & n_{2s}^k & 0 & \cdots & \cdots & 0 \\ 0 & 0 & n_{3s}^k & 0 & \cdots & 0 \\ \vdots & \vdots & \vdots & \ddots & \vdots & \vdots \\ 0 & \cdots & \cdots & 0 & n_{59s}^k & 0 \\ 0 & \cdots & \cdots & \cdots & 0 & n_{60s}^k \end{bmatrix} \quad (13)$$

Here N is the number of molecules in the crystal, A^l is the 60×60 matrix corresponding to the hopping of electrons between carbon atoms inside the lth fullerene molecule: $A_{ij}^l = t - \alpha d_{ji}^l$ for nearest neighbors of the ith and jth carbon atoms of the lth molecule, $B^{l\,l+1}$ is a 60×60 matrix that describes electron hopping between nearest neighbors of fullerene molecules l and $l+1$: $B_{ij}^{l,l+1} = T - \alpha d_{ji}^{l,l+1}$ for nearest neighbors of the ith carbon atom of the lth molecule and the jth carbon atom of the $l+1$th molecule and 0 otherwise. From Eqns. (8) one can obtain d_{ji}^n, $d_{ji}^{n,n'}$ and n_{is}^n:

$$d_{ij}^l = -\frac{2\alpha}{KN_C} \sum_{\mu sq}^{occ} \psi_{\mu is}^{*l}(q)\psi_{\mu js}^l(q) + \frac{2\alpha}{90KN_C} \sum_{\langle ij \rangle} \sum_{\mu sq}^{occ} \psi_{\mu is}^{*l}(q)\psi_{\mu js}^l(q),$$

$$d_{ij}^{l\,l'} = -\frac{2\alpha'}{K'N_C} \sum_{\mu sq}^{occ} \psi_{\mu is}^{*l}(q)\psi_{\mu js}^{l'}(q) \quad (14)$$

$$n_{js}^l = \frac{1}{N_C} \sum_{\mu q}^{occ} \psi_{\mu is}^{*l}(q)\psi_{\mu is}^l(q) \quad (15)$$

Simulation Details

As described above, in order to solve the problem of the self-trapped exciton in the C_{60} chain we have used the approximation in which the possibility of forming an exciton in a supercell was considered. The system of Eq. 5 and Eq. 8 of self-consistent nonlinear equations was solved numerically using ordinary iteration techniques for the single C_{60} molecule and Eq. 10, Eq. 13 and Eq. 14 for the C_{60} chain.

The ground state structure and wavefunction of single C_{60} (8) were determined assuming double occupancy (spin up and down) of the electronic level. All distances between carbon atoms on a neutral molecule were initially taken as equal. The wave function of the singlet excited state of the C_{60} molecule was determined using the same procedure as for the wave function of the ground state except one electron from the highest occupied orbital was transferred to the lowest unoccupied electronic level. In the case of the wave function of a triplet excited state, the spin of the transferred electron is changed. For the excited C_{60} molecule, distances between carbon atoms calculated for the neutral molecule were taken as an initial condition.

The ground and excited states of the C_{60} chain were found using the same method as for the C_{60} molecules. The sum over the occupied bands defined by the indices μ and q in Eq. 13 and Eq. 14 was performed in order to obtain the structure and wave function of the C_{60} chain at the ground state. For the excited state, half of the electrons with spin up (down) were transferred from the highest valence band to one of the conduction bands, and a similar summation was done in order to obtain the electronic structure and geometry of the C_{60} chain under excitation.

Considering the exciton on the C_{60} chain, the one-dimensional chain was modeled by three fullerene molecules in a supercell. The distances between atoms on the central molecule and their charges were initially defined as for an isolated excited molecule. In the case of the charge-transfer exciton on the C_{60} chain, the one-dimensional chain was modeled by eight fullerene molecules in a supercell. In this case, the initial configuration of one C_{60} molecule in the supercell is the same as for an isolated charged molecule with an electron polaron. The initial distances between the atoms and charges on other molecule in the supercell were set to those of an isolated charged molecule with a hole polaron. On the other molecules of the supercell, bond lengths and charges were defined for an isolated neutral C_{60} molecule in both cases.

In our calculations, the parameter T describing intermolecular electron-hopping was varied in the interval from $0.01t$ to $0.4t$ (where t is the value of the intramolecular hopping integral) and $\alpha=\alpha'$. The calculations were conducted in 61 points in the Brillouin zone in reciprocal space.

For the neutral C_{60} molecule in the case where electron-electron interaction is neglected, $U=0$ and with the parameters $t=2.5$ eV and $\alpha=6.31$ eV/Å [5,29,30], $K=49.7$ eV/Å [29], $K'=K$, and two distinct bond lengths exist. Bonds separating the pentagons and hexagons become longer than the bonds separating the hexagons, and changes in the bond lengths are identical to previous results [5,29,13,31]. The calculated width of energy bandgap in the electron spectrum of the neutral C_{60} molecule is equal to 2.25 eV, which agrees well with the experimental photoelectron and inverse-photoelectron spectra of solid C_{60} [2] as well as earlier calculations [5,30,32]. In the case of $U \neq 0$ for $0 \leq U \leq 2t$ in the neutral molecule, the bond lengths and energy gap in the electronic spectrum are the same as in the case of $U=0$. With increasing U, the behavior of the HOMO and

LUMO energy levels in the electronic spectrum is linear and identical to previous results [32]. For the C_{60} molecule with one extra charge (hole), when $U=0$ the same bond lengths between carbon atoms and electronic spectrum as in earlier reports [12,14,28] have been obtained.

RESULTS AND DISCUSSION

Following reference [30] we have numbered the 60 carbon atoms per C_{60} molecule as indicated in Figure 1. Bonds between carbon atoms are numbered as shown in Table 1. Numeration of bonds is chosen so the first 30 bonds separate the pentagons from the hexagons, and the remaining 60 bonds separate the hexagons.

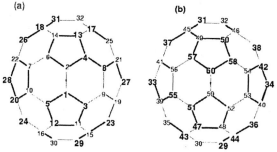

Figure 1. Numeration of carbon atoms in the C_{60} molecule. (a) and (b) show the front and back views of an exciton on single C_{60}, respectively. Bold numbers show atoms where negative charge is concentrated after formation of an exciton. The positive charge is concentrated on other atoms. The heavy curves indicate paths where dimerization is suppressed for the exciton.

In the notation accepted in this paper for the carbon atoms (Figure 1) and bonds between them (Table 1), the charge distribution of the exciton and relative changes in bond lengths on a single molecule C_{60} are shown in Figure 2. It can be seen that the non-homogeneous distributions of charge and relative changes in the bond lengths occur for the excited state.

Table 1. Numbering of bonds in the linear chain.

K	1	2	3	4	5	6	7	8	9	10	11	12	13	14	15
i	1	3	4	5	6	11	12	13	14	19	20	21	22	27	28
j	2	9	8	10	7	15	16	17	18	23	24	25	26	34	33
K	16	17	18	19	20	21	22	23	24	25	26	27	28	29	30
i	29	31	35	36	37	38	43	44	45	46	51	52	54	56	59
j	30	32	39	40	41	42	47	48	49	50	55	53	58	57	60
K	31	32	33	34	35	36	37	38	39	40	41	42	43	44	45
i	1	1	2	2	3	4	5	6	7	7	8	8	9	10	11
j	3	5	4	6	11	13	12	14	10	22	9	21	19	20	12
K	46	47	48	49	50	51	52	53	54	55	56	57	58	59	60
i	13	15	15	16	16	17	17	18	18	19	20	21	22	23	24
j	14	23	29	24	30	25	32	26	31	27	28	27	28	36	35
K	61	62	63	64	65	66	67	68	69	70	71	72	73	74	75
i	25	26	29	30	31	32	33	33	34	34	35	36	37	38	39
j	38	37	44	43	45	46	39	41	40	42	43	44	45	46	55
K	76	77	78	79	80	81	82	83	84	85	86	87	88	89	90
i	40	41	42	47	47	48	49	49	50	51	52	53	55	57	58
j	53	56	54	48	51	52	50	57	58	59	59	54	56	60	60

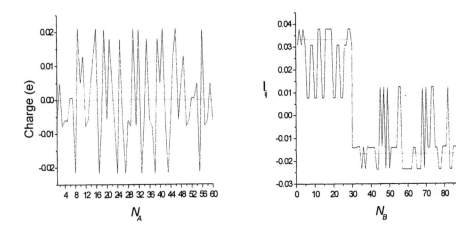

Figure 2. Charge distribution per atom and distortions in bond lengths d_{ij} on a single C_{60} molecule in the first excited state at $U = 0$. The numeration of the atoms (N_A) and bonds (N_B) correspond to the numberings indicated in Figure 1 and Table 1, respectively. Dotted lines indicate results for the neutral molecule in the ground state.

Frenkel Exciton

For weak π-conjugation ($T/t \ll 1$), in the case of singlet electron excitation from the highest occupied state in the upper valence band to the lowest unoccupied state in the lower conductance band, a Frenkel exciton is formed in a one-dimensional C_{60} crystal and is localized on the central molecule in the supercell. In Figure 3 and Figure 4, we show the results of calculating the charge distribution and changes in bond length in a supercell consisting of three C_{60} molecules for different values of the parameters T/t and U. Exciton formation can be characterized by the fact that non-uniform distribution of charge is realized on this molecule after electron excitation, in spite of the general electroneutrality of the central molecule. Moreover, distortions in the bond lengths for C_{60} in the first excited and ground states differ significantly.

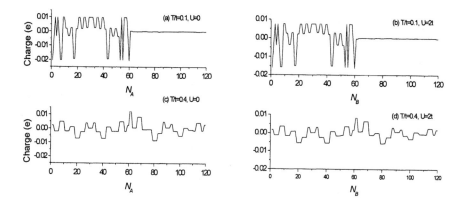

Figure 3. Charge distribution on carbon atoms in the chain modeled by a system of three periodically repeated C_{60} molecules in the case of the lowest excitation: (a) $U=0$, $T/t=0.1$, (b) $U=2t$, $T/t=0.1$, (c) $U=0$, $T/t=0.4$, (d) $U=2t$, $T/t=0.4$. The results are shown for the central (atom numbers 1-60) and one end molecule (atom numbers 61-120). Results for both end molecules are identical.

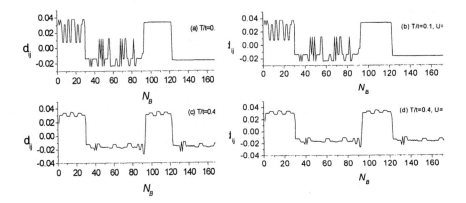

Figure 4. Change in bond length d_{ij} in the chain modeled by a system of three periodically repeated C_{60} molecules in the case of lowest excitation: (a) $U=0$, $T/t=0.1$, (b) $U=0$, $T/t=0.4$, (c) $U=2t$, $T/t=0.1$, (d) $U=0$, $T/t=0.4$. The results are shown only for the central (atom numbers 1-90) and one end molecule (atom numbers 93-184). The bonds 91 and 92 link the central and end molecules, while bonds 183 and 184 connect the end molecules of neighboring supercells in the chain. The results for both end molecules are identical.

For small values of the parameter T/t the non-uniform distribution of a charge is localized on the central molecule. For $U=0$ the charge distribution (Figure 3a) and distortions in the bond lengths (Figure 4.a) in this molecule differ slightly from the case of the isolated fullerene (Figure 2). Taking the Hubbard interaction into account produces an insignificant influence on the charge distribution in the system and has very little effect on the bond lengths (Figure 3.c and Figure 4.c). Increasing the value of the parameter T/t leads to an atomic charge redistribution on the ends molecules of the supercell (Figure 3.b) and hence the exciton is spread out. This is accompanied by changes in the bond lengths between carbon atoms (Figure 4.b) and in the electronic structure of the fullerene chain. Account of the Hubbard interaction has practically no influence on the charges and bond lengths (Figure 3.d and Figure 4.d).

Charge Transfer Exciton

In the case of excitation, electrons with spins up (down) from the highest occupied state in the upper valence band to the third lowest unoccupied state in the lower conduction band without changing the spin direction in a one-dimensional C_{60} crystal under weak conjugation ($T/t <<1$), the charge-transfer exciton, is formed on the two molecule in the supercell. This is accompanied, as in the case of the Frenkel exciton, by changes in the bond lengths between carbon atoms and in the electronic structure of the fullerene chain.

In Figures 5 and 6, the results of the charge distribution calculation and changes in the bond lengths in supercell consisting of eight C_{60} molecules for different values of the parameters T/t and U are given. For small values of T/t, a non-uniform charge distribution is found on the two molecules. Moreover, the charges are localized on separate molecules, forming hole and electron polarons respectively. For $U=0$ the charge distribution (Figure 4a) and distortions of bond lengths (Figure 5a) do not differ from that of a chain doped by only hole or electron respectively [28].

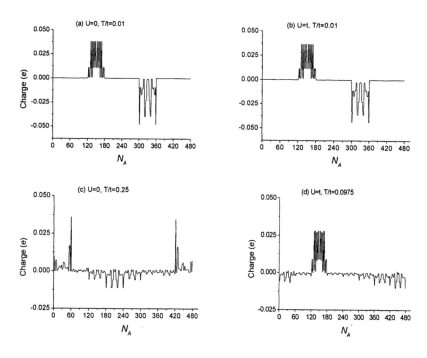

Figure 5. Charge distribution on carbon atoms in the chain modeled by a system of eight periodically repeated C_{60} molecules in the case of the charge transfer exciton: (a) $U=0$, $T/t=0.1$, (b) $U=t$, $T/t=0.01$, (c) $U=0$, $T/t=0.25$, (d) $U=t$, $T/t=0.0975$.

Accounting for the Hubbard interaction has only a small effect on the system charge distribution as well as on bond lengths (Figure 4b and 5b). Increasing T/t leads to disappearance of the hole and electronic polarons and a non-uniform charge distribution on the other molecules of the supercell (Figure 5c). Changes in bond lengths between carbon atoms are also observed (Figure 6c). For $U=0$, disintegration of the polaron states occurs with increasing T/t. For $U=t$ the hole polaron exists up to $T/t=0.0975$ while the electron polaron is disintegrated well before this point.

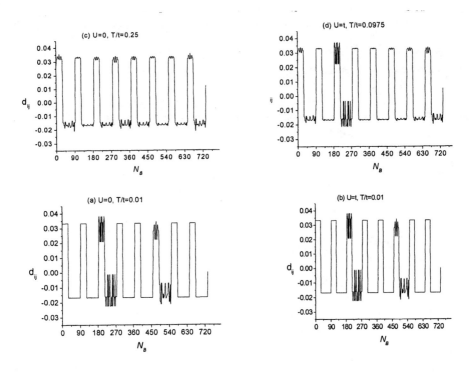

Figure 6. Change in bond length d_{ij} in the chain modeled by a system of eight periodically repeated C_{60} molecules in the case of the charge transfer exciton: (a) $U=0$, $T/t=0.1$, (b) $U=t$, $T/t=0.01$, (c) $U=0$, $T/t=0.25$, (d) $U=t$, $T/t=0.0975$.

Figure 7 shows electron state densities for one Frenkel exciton per supercell (Figure 7b) and one charge-transfer exciton per supercell (Figure 7d) with $U = 0$, in comparison with the neutral chain. The electronic structure of the chain is reconstructed. The self-trapped excitons create local lattice deformation and localize themselves into states across the energy gap in C_{60} and below the conduction band and above the valence band in C_{60} solids. Inclusion of the Hubbard interaction gives an analogous picture of the electron DOS, but all system energy levels shift to higher energies. Figure 8 shows the charge values on each molecule in the chain as a function of T/t for $U = 0$ in the case of one charge-transfer exciton per supercell. Note that for small values of T/t the charge is localized on the two molecules and starting from the threshold value $T^*/t \approx 0.12$, redistribution occurs. Changes in U within the considered limits do not change this picture.

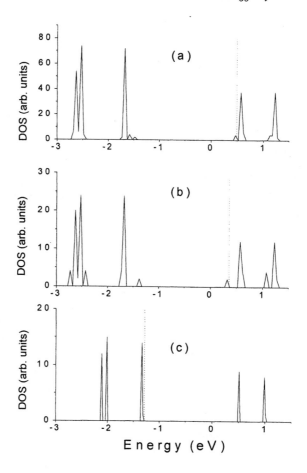

Figure 7. Electron DOS (in arbitrary units) of the chain for various cases ($U=0$, $T/t=0.1$): (a) one charge-transfer exciton in the supercell; (b) one Frenkel exciton in the supercell; (c) at the ground state.

CONCLUSION

The electronic structure of Frenkel exciton and associated Jahn-Teller distortions of carbon atoms in an excited chain of C_{60} were studied with the use of a generalized version of the Su-Shrieffer-Heeger model for the intermolecular and intramolecular degrees of freedom. Inclusion of electron correlation effects (Hubbard repulsion on-site) was also considered.

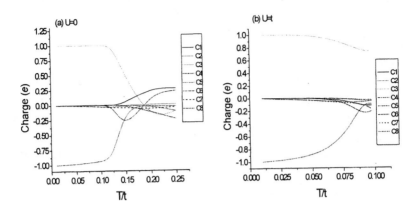

Figure 8. Distribution of the mean value of charge on each molecule of supercell of the chain modeled by a system of eight periodically repeated C_{60} molecules in the case of the charge transfer exciton: (a) $U=0$, (b) $U=t$.

When a molecule is excited to the lowest electron excitation, a self-localized exciton is generated. In the case of a linear chain, for weak π-conjugation, Frenkel and charge-transfer excitons are also generated. In these cases the electronic structures and distortions in exciton bond lengths differ from those of a single excited C_{60} molecule. These distinctions are caused by conjugation of π-electrons and electron-electron interactions. Inclusion of the Hubbard interaction does not influence the form of the exciton. The electronic states of the Frenkel and charge-transfer excitons localize themselves into states across the energy gap in C_{60} and below the conduction band and above the valence band in C_{60} solids. Further strengthening of π-electron conjugation leads to the disappearance of the excitons in the linear chain. Hubbard on-site repulsion does not change the physical picture of the formation and disappearance of excitons.

ACKNOWLEDGMENT

We would like to thank the Information Science Group of the Institute for Materials Research, Tohoku University for their continuous support of the SR8000 supercomputing system. One of the authors, V.R.B., has been supported by the Japan Society for the Promotion of Science (JSPS) Fellowship for Research in Japan (Grant ID No. RC30126005). This work at Russia was supported by "High T_c superconductivity" (Grant 98009).

REFERENCES

H.W. Kroto, J.R. Heath, S.C. O'Brien, R.F. Curl and R.E. Smalley, 1985. C_{60}: Buckminsterfullerene, Nature, 318, 162–163.

R.W. Lof, M.A. Veenendaal, B. Koopmans, H.T. Jokman and G.A. Sawatsky, 1992. Band gap, excitons, and Coulomb interaction in solid C_{60}, Phys. Rev. Lett., 68, 3924–3927.

F.C. Zhang, M. Ogata and T.M. Rice, 1991. Attractive interaction and superconductivity for K_3C_{60}, Phys. Rev. Lett., 67, 3452–3455.

K. Harigaya, 1992. Lattice distortion and energy-level structure in doped C_{60} and C_{70} molecules studied with the extended Su-Schrieffer-Heeger model: Polaron excitations and optical absorption, Phys. Rev. B, 45, 13676–13684.

B. Friedman, 1992. Polarons in C_{60}, Phys. Rev. B, 45, 1454–1457.

A. A. Remova, V. P. Shpakov, U-H. Paek and V. R. Belosludov, 1995 Band reconstruction of K_xC_{60} caused by the cooperative Jahn-Teller effect, Phys. Rev. B, 52, 13715–13717.

M. Matus, H. Kuzmany and E. Sohmen, 1992. Self-trapped polaron exsiton in neutral fullerence C_{60}, Phys. Rev. Lett., 68, 2822–2825.

P.A. Lane, L.S. Swanson, Q.-X. Ni, J. Shinar, J.P. Engel, Y.J. Barton and L. Jones, 1992. Dynamics of photoexicited states in C_{60}: An optically detected magnetic resonance, EPR, and light-induced ESR study, Phys. Rev. Lett., 68, 887–890.

W. Gass, J. Feldmann, E.O. Gobel, C. Taliani, H. Mohn, W. Muller, P.Haussler and H.-U. ter Meer, 1994. Fluorrescence from X traps in C_{60} single crystals, Phys. Rev. Lett., 72, 2644–2647.

S. Kazaoui, N. Minami, Y. Tanabe, H.J. Byrne, A. Eilmes and P. Petelenz. 1998. Comprehensive analysis of intermolecular charge-transfer excited states in C_{60} and C_{70} films, Phys. Rev. B, 58, 7689-7700, and references therein.

M. Knupfer and J. Fink, 1999. Frenkel and charge-transfer exitons in C_{60}, Phys. Rev. B, 60, 10731–10734, and references therein.

J.P. Long, S.J. Chase and M.N. Kabler, 2001. Excited-state photoelectron spectroscopy of excitons in C_{60} and photopolymerized C_{60} films, Phys. Rev. B, 64, 205415, and references therein.

13.S. Suzuki, D. Inomata, N. Sashide and K. Nakao, 1993. Jahn-Teller distortion in the lowest excited singlet state of C_{60}, Phys. Rev. B, 48, 14615–14622.

J. Fagerston and S. Stafstron, 1993. "Fundamental excitations in C_{60}, Phys. Rev. B, 48, 11367–11374.

J. Kim, Wu-Pei Su, 1994, Optical absorption of C_{60}: Singlet singlet-excitation calculations". Phys. Rev. B, 50, 8832-8837, and references therein.

R. Eder, A.-M. Janner, G.A.Sawatzky, 1996. Theory of nonlinear optical response of excitons in solid C_{60}, Phys. Rev. B, 53, 12786–12793.

E.L. Shirley, L.X. Benedict, S.G. Lourie, 1996. Excitons in solid C_{60}, Phys. Rev. B, 54, 10970–1977.

X. Jiang, Z. Gan, 1996, Exitonic spectra of solid C_{60}, Phys. Rev. B, 54, 4504–4506.

K. Harigaya, 1996. Charge-transfer excitons in optical-absorption spectra of C_{60} dimers and polymers, Phys. Rev. B, 54, 12087–12092.

R.W. Munn, B. Pac, P. Petelenz, 1998. Charge-transfer-induced Frenkel exciton splitting in crystalline fullerene, Phys. Rev. B, 57, 1328–1331.

R. Jacquemin, S. Raus, W. Eberhardt, 1998. Direct observation of the dynamics of excited electronic state in solids: F-sec time resolved photoemission of C_{60}, Solid State Commun., 105, 449–453.

A.M. Rao, P. Zhou, K.A. Wang, G.T. Hager, J.M. Holden, Y. Wang, W.T. Lee, X.X. Bi, P.C. Eklund, D.S. Cornett, M.A. Duncan, I.J. Amster, 1993. Photoinduced polymerization of solid C-60 films, Science, 259, 955–957.

P.W. Stephens, G. Bortel, G. Faigel, M. Tegze, A. Janossy, S. Pekker, G. Oszlanyi, L. Forro, 1994. Polymeric fullerene chains in RbC_{60} and KC_{60}, Nature, 370, 636–639.

G.W. Wang, K. Komatsu, Y. Murata, M. Shiro, 1997. Synthesis and X-ray structure of dumb-bell-shaped C-120, Nature, 389, 412–412

W.P. Su, J.R. Schrieffer, A.J. Heeger, 1980. Soliton excitations in polycetylene, Phys. Rev. B, 22, 2099–2011.

V.A. Levashov, A.A .Remova, V.R. Belosludov, 1997. Electronic structure of linear chain of fullerenes, JETP Lett. 65, 683–686.

V.A. Levashov, A.A. Remova, V.R. Belosludov, 1998. Linear Chain of Fullerenes under Pressure, Mol. Mat., 10, 197–200.

V.R. Belosludov, T.M. Inerbaev, R.V. Belosludov, Y. Kawazoe, 2003. Polaron in one dimensional C_{60} crystal, Phys. Rev B accepted for publication.

K. Harigaya, 1993. Polaron excitations in doped C_{60}: Effects of disorder, Phys. Rev. B, 48, 2765–72.

M. Springborg, 1995. Structural and electronic properties of polymeric fullerene chains, Phys. Rev. B, 52, 2935–2940.

J. Hawkins, A. Meyer, T.A. Lewis, S. Loren and F.J. Hollander, 1991. Crystal-structure of osmylated C_{60} – confirmation of the soccer ball framework, Science, 252, 312–313 .

J. Dong, Z.D. Wang, D.Y. Xing, Z. Domański, P. Edrös, P. Santini, 1996. Correlation effects on electronic and optical properties of a C_{60} molecule: A variational Monte Carlo study, Phys. Rev. B, 54, 13611–13615.

A Simple Route to New 1D Nanostructures

A. HUCZKO, H. LANGE, J. PASZEK and M. BYSTRZEJEWSKI
Department of Chemistry, Warsaw University, Warsaw, Poland

S. CUDZILO
Faculty of Armament and Aviation Technology, Military University of Technology, Warsaw, Poland

S. GACHET and M. MONTHIOUX
CEMES—UPR A-8011 CNRS, 31055 Toulouse Cedex 4, France

Y. Q. ZHU, H. W. KROTO and D. R. M. WALTON
School of Chemistry, Physics and Environmental Science, University of Sussex, Brighton, United Kingdom

ABSTRACT

Novel nanostructures have been created by thermolysis of mixtures containing PTFE, PVC, C_2Cl_6, C_6Cl_6 and different reductants like pure metals, metal alloys, metal silicides, B and Si, in a calorimetric bomb or in a reaction vessel. The stoichiometric effect of the Si-containing reactants was studied, in order to determine which material produced the highest nanofibre yield. Elemental analyses, XRD, SEM, HRTEM and EDX were used to characterize the combustion products. Elongated well-crystallized Si-related nano- and micro-fibers, coated with a thin SiO_2 amorphous or carbon layer were obtained, and were found to be dominated along with crystalline ball-like carbon nanoparticles. The highest yields of 1D nanostructures were obtained from F- and Si-containing precursors, due to the volatility of SiF_4.

INTRODUCTION

Self-sustaining reactions are initiated in a variety of highly exothermic powder mixtures. They offer direct reaction routes and solvent-free technologies that can produce unique metastable materials, which are unavailable by conventional techniques. Different names are used to identify this type of reaction, *i.e.* SHS (self-propagating high-temperature synthesis) [1] or thermolysis [2]. The first self-propagating solid state reaction was proposed by Goldschmidt in 1885 [3]. It involved metal oxide reduction with aluminium in the so-called thermite reaction. The process is usually initiated in a mixture of powder reactants by an electrically

heated wire. In this entirely solid state reaction a combustion front develops and propagates through the sample; no external source of oxygen is needed.
Systematic research on SHS was initiated in 1967 [4] and since then not only numerous experimental studies have been discussed in excellent reviews [4,5] but the processes have also been modeled theoretically [6,7]. One of the interesting condensed systems is represented by metal-polytetrafluoroethene (M-PTFE) mixtures in which vigorous exothermic reactions occur according to the idealized scheme [2]:

$$-(CF_2-CF_2)- + 4nM \rightarrow 2nC + 4nMF$$

In practice, these mixtures are widely used, *e.g.* in flares, as igniters for solid fuel rocket engines and in other pyrotechnics [8]. The process begins with a short activation period, during which a local temperature increase causes the reactants to reach a well-defined critical state of ignition. Once started, the reaction propagates through the powder charge as a combustion process. This step is almost concurrent with the release of reaction heat resulting in an abrupt temperature increase (approaching 10^6 Ks^{-1}). These specific reaction conditions provide a route to intriguing products as shown elsewhere for the M/PTFE system [2].

In this paper, the formation of novel nanostructures was achieved by thermolysis of mixtures of different oxidants and reductants in highly heterogeneous condensed systems. The composition and morphology of the solid products were studied.

EXPERIMENTAL DETAILS

Thermolysis was carried out in a calorimetric bomb using the procedure reported earlier [2]. Preliminary runs were also performed in an arc plasma generator [9] as shown in Figure1.

The pellet, made from the starting two-component mixture, was wrapped with kanthal wire and connected to the electrodes, then placed on a quartz substrate. The wire served as an ohmically-heated thermolysis promoter to initiate combustion. This new system allows better control of thermolysis by direct observation of the reaction zone through the view port (temperature and reaction time measurements) and pressure recording. Solid reaction products were analyzed using various techniques (DTA/TG, XRD, elemental analysis) [2], and were also examined by high resolution SEM and TEM at low and high resolution.

RESULTS AND DISCUSSION

Exploratory testing of mixtures containing polytetrafluoroethene (PTFE) and a pure metal or metal alloy (Zn, Cr, B, Si, Al_3Mg_4, ZrTi, FeSi, $CaSi_2$) showed [2] that not only do these highly exothermic reactions proceed vigorously, but also novel nanostructures are formed. Si-containing nanofibers/nanowires, together with uniform carbon nanoparticles, were efficiently synthesized, especially when $CaSi_2$ was used as reductant. Thus, in the following studies the influence of

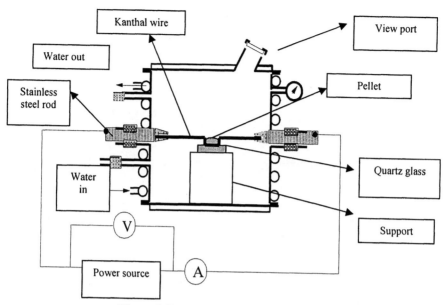

Figure 1. Experimental thermolysis set-up.

varying stoichiometry was studied for $CaSi_2$/PTFE and Si/PTFE mixtures. Also, other oxidants and reductants were evaluated, *e.g.* including chloroaromatic compounds and NaN_3. There are at least 3 advantages in this latter system: (i) chlorine can be more easily removed from its compound compared to fluorine; (ii) some ring structures—if preserved—can help the potential nucleation of aromatic nanocarbons and (iii) the product can be easily purified (water leaching of NaCl).

In practice, the calorimetrically measured heats of reaction ranged between 2800-3000 kJ/kg for NaN_3/C_6Cl_6 and NaN_3/C_2Cl_6 mixtures, relatively close to the value for NaN_3/PTFE mixture (3570 kJ/kg). The initial compositions of the tested mixtures, together with the results of elemental and XRD analyzes, are presented in Table 1.

The carbon content of the solid products varied between 10 and 80 wt %. A substantial proportion of the carbon was in an elemental form since the calcinations of products in air at 900 K drastically reduced the C content. The residual carbon can be related to carbide (mostly SiC) phases. Both crystalline graphite and turbostratic carbon were detected in the products resulting from Si-containing/PTFE mixtures. A relatively high amorphous carbon content was found in products resulting from thermolysis of NaN_3/C_nCl_m mixtures, which were leached with water to remove NaCl. Various types of novel nanostructure, including uniform carbon nanoparticles, huge spherical metal microparticles of

Table 1. Elemental and XRD analyses.

Reactants	Elemental analysis (wt %)		XRD analysis	
	C	H	Phases identified	C intensity
Si(26.0)/PTFE(74.0)	79.6	0.2	Si, SiC	srong
Si(26.0)/PTFE(74.0)[a]	26.1	0.4	Si, SiC	
Si(56.0)/PTFE(44.0)	21.8		Si, SiC	
FeSi(35.7)/PTFE(64.3)	44.2		Si, Fe_3Si, Fe_5Si_3, SiC	strong
$CaSi_2$(22.8)/PTFE(77.2)	43.3		Si, CaF_2, SiC	strong
$CaSi_2$(27.8)/PTFE(72.2)	42.2		Si, CaF_2, SiC	srtong
$CaSi_2$(37.8)/PTFE(62.2)	32.1		Si, CaF_2, SiC	medium
$CaSi_2$(47.8)/PTFE(52.2)	25.5		Si, CaF_2, SiC	medium
$CaSi_2$(57.8)/PTFE(42.2)	22.1		Si, CaF_2, SiC	medium
$CaSi_2$(57.8)/PTFE(42.2)[a]	8.8		Si, CaF_2, SiC	
$CaSi_2$(16.9)/C_6Cl_6(83.1)[b]	64.3	0.9	SiC, Fe_2C	medium
$CaSi_2$(46.9)/C_6Cl_6(53.1)[b]	15.4		Si, SiC, $FeSi_2$	
$CaSi_2$(39.6)/C_2Cl_6(60.4)[b]	17.2		Si, SiC, $FeSi_2$	
$CaSi_2$(43.2)/SPVC(56.8)[b]	22.6		Si, SiC, $FeSi_2$	
NaN_3(62.2)/PTFE(37.8)	60.3	0.5	NaF	strong
NaN_3(52.2)/C_2Cl_6(47.8)[b]	77.4	1.2		
NaN_3(47.8)/C_6Cl_6(52.2)[b]	79.7	1.1		medium
NaN_3(47.8)/C_6Cl_6(52.2)	28.1	0.4	NaCl	
NaN_3(60.0)/C_6Cl_6(21.1)/C_2Cl_6(18.9)[b]	74.5	2.0		
NaN_3(55.0)/SPVC(45.0)[b]	79.0	1.0		medium

[a] Product calcinated in air [b] Product leached with H_2O

unreacted reductants, and 1D ceramic nano- and microfibers were detected during SEM observations of the products. The yield of nanofibers, influenced by the stoichiometry of reactants, is significantly high for starting mixtures containing Si and PTFE. A typical morphology of these nanostructures is shown in Figure 2.

The fibers are ~5–150 nm (with a statistical mode at 20–50 nm) and up to 10 μm in length. TEM examinations revealed further detailed structural features of the thermolysis samples (Figure 3).

The products exhibit similar structural characteristics, namely well-crystallized, elongated single crystals 1D SiC (XEDS analysis) nanowires with a frequent display of a "herringbone" structure, corresponding to the association of twinned, faulted crystals (Figure 4).

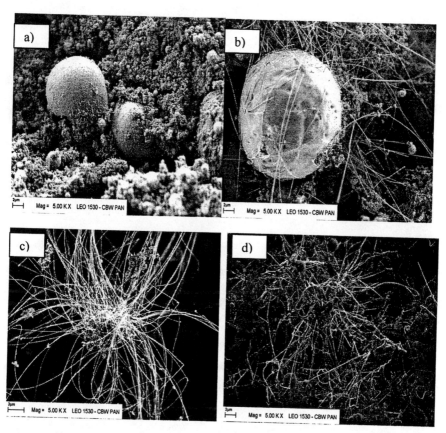

Figure 2. SEM images: CaSi$_2$/PTFE mixture. (a) 22.8/77.2; (b) 27.8/72.2; (c) 27.8/72.2; (d) 57.8/42.2.

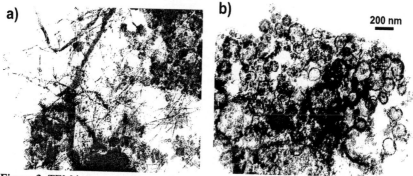

Figure 3. TEM images: (a) Si-containing precursors resulting mainly in nanofibers; (b) typical hollow carbon particles.

Figure 4. HRTEM images of elongated well-crystallized 1D SiC nanostructures from Si-containing samples.

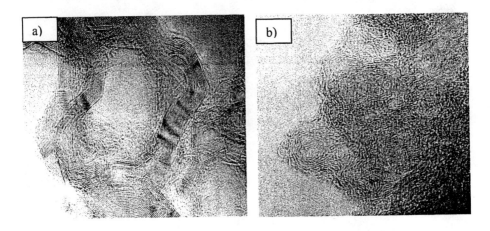

Figure 5. HRTEM images of Si-containing nanocarbon samples.

There are also "screw" type morphologies, similar to a common aspect associated with vapour-grown SiC whiskers. All nanofibers are coated with an amorphous carbon or silicon oxide layer, a few nm thick. Polyaromatic nanocarbons (Figure 5) are found as a microporous material.

The morphology is systematically turbostratic (Fig 5a). The uniform carbon particles (*ca.* 20-40 nm in diameter) are empty, often polyhedral shells, with an inner open cavity of *ca.* 10 nm diameter. Since graphenes exhibit few distortions, such a nanotexture requires either high temperature of formation (*ca.* 2000°C) or can originate from carbide decomposition. The polyaromatic hydrocarbon may also exhibit a lower microporosity and much lower nanotexture with many graphene distortions (Figure 5b). Such material resembles a low-grade carbon black. Huge spherical particles, 0.3–2 µm diameter, are also frequently found with Si present as a major component (Figure 6).

The products are possibly catalytically related to nanofiber growth, since the latter often emerge from the ball surface. Furthermore, large fibers, ca. 200-300 nm in diameter containing Si as a major element are seldom identified. They may be considered to be less crystallized bunches of SiC nanowires. Not only PTFE, but also other oxidants were used in the thermolysis of Si-containing mixtures (Table 1). Figure 7 shows SEM images of products resulting from $CaSi_2/C_2Cl_6$, $CaSi_2/SPVC$ and $CaSi_2/C_6Cl_6$.

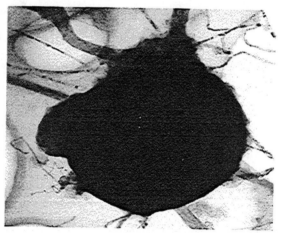

Figure 6. HRTEM image of product sample (unreacted reductant).

As expected, the Cl-containing organic precursors also yield Si-related nanofibers during thermolysis. However, their content is lower compared to PTFE. The use of NaN_3 as a reductant in mixtures with different oxidants led mainly to carbon particles (Figure 7d).

A preliminary experiment with a $CaSi_2(57.8)/PTFE(42.2)$ mixture was carried out in the new system shown in Figure 1. The combustion was successful and a black powder containing up to 35 wt % of elemental carbon was formed in the reactor. The XRD analysis showed the presence of CaF_2, Si, SiC, $FeSi_2$ and $CaSi_2$ as well as turbostratic carbon among the reaction products. SEM examinations (Figure 8) revealed the presence of Si-related nano- and microfibers in the thermolysis samples.

Thus, the desired reaction was at least partly carried out, and provides a promising application of this experimental system for more detailed studies; these are currently in progress.

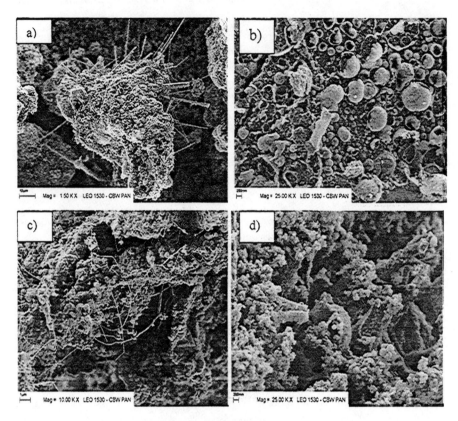

Figure 7. SEM images: (a) $CaSi_2(39.6)/C_2Cl_6(60.4)$; (b) $CaSi_2(43.2)/SPVC$ (56.8); (c) $CaSi_2(46.8)/C_6Cl_6(53.2)$; (d) $NaN_3(55.0)/SPVC(45.0)$.

Figure 8. SEM images from $CaSi_2/PTFE$ thermolysis (new system).

CONCLUSION

Thermolysis offers great potential for self-induced generation of 1D Si-related and carbon nanostructures.

ACKNOWLEDGEMENT

This work was supported by the Committee for Scientific Research (KBN) through the Department of Chemistry, Warsaw University, under Grant No. 7 T09A 020 20.

REFERENCES

[1] L. Takacs, 2002. Progr. Mat. Sci., 47, 355–414.
[2] A. Huczko, H. Lange, G. Chojecki, S. Cudziło, Y.Q. Zhu, H.W. Kroto, D.R.M. Walton, J. Phys. Chem., in print.
[3] V. Hlavacek, 1991. Ceramic Bull., 70, 240–245.
[4] A.G. Merzhanov. In: Z.A. Munir, J.B. Holt, editors. 1990. Combustion and plasma synthesis of high-temperature materials. Weinheim: VCH, 1–25.
[5] J.J. Moore, H.J. Feng, 1995. Progr. Mat. Sci., 39, 243–247.
[6] M. Eslarnoo – Grarni, Z.A. Munir, 1989. Mater., Sci. Report, 3, 227–232.
[7] F.B. Schaffer, P.G. McCormick, 1991. Metall. Trans. A, 22A, 3019-3024.
[8] E.-C. Koch, A. Dochnahl, Propellants 2002. Explosives, Pyrotechnics, 25, 37–40.
[9] H. Lange, P. Baranowski, P. Byszewski, A. Huczko, 1997. Rev. Sci. Instr., 68, Nr 10, 3723–27.

Structures and Intermolecular Interactions of Cocrystallites Consisting of Metal Octaethylporphyrins with Fullerene C_{70}

T. ISHII

Department of Advanced Materials Science, Kagawa University, Takamatsu, Kagawa, Japan

R. KANEHAMA, N. AIZAWA, M. YAMASHITA, K-I SUGIURA, H. MIYASAKA, T. KODAMA, K. KIKUCHI and I. IKEMOTO

Department of Chemistry, Tokyo Metropolitan University, PRESTO (JST), Hachioji, Tokyo, Japan

ABSTRACT

The new cocrystallites that contain fullerene C_{70} with the metal complexes of octaethylporphyrins (oeps) are reported. Metal complexes of porphyrin cocrystallites with fullerene C_{70} form solids with expected close contact between the curved □ surface of a fullerene and the planar π surface of a porphyrin, without the need for matching convex with concave surfaces. The structures of metal oeps in $C_{70} \cdot Pd(II)(oep) \cdot 2CHCl_3$ and $C_{70} \cdot Ag(II)(oep) \cdot 2CHCl_3$ reveal the remarkable syn-formed oep configurations, with the eight ethyl groups of the metal oep portions lying on the same sides of the porphyrin plane toward the C_{70}, suggesting that there is a strong porphyrin/porphyrin face-to-face interaction between two adjacent oep planes. The new cocrystallite $C_{70} \cdot V(IV)O(oep) \cdot C_6H_6 \cdot H_2O$ having also the syn-formed oep configuration, which exhibits a 1-D chain structure of C_{70} molecules with short intermolecular distances and strong fullerene/porphyrin intermolecular interaction, is also discussed.

INTRODUCTION

For more than 10 years the syntheses, structural analyses and the physical properties of fullerenes have been extensively investigated for the unique properties caused by their unique 3-D shapes.[1,2] It has been reported that the round-shaped fullerenes such as C_{60} and C_{70} are not appropriate to cocrystallize with planar molecules, and curving of the planar molecule to match the concave structure is required in order to fit to the round-shaped fullerenes.[3,4] There have been many reports of researchers obtaining curved surfaces that are able to encircle a fullerene by building complex structures from planar aromatic hydrocarbon units and other flat moieties, such as calixarenes,[5] oxacalix[3]arenes,[6] resorcina[4]arenes,[7] □-cyclodextrin,[8] azacrown ethers,[9] porphyrazine,[10,11] cyclotriveratrylene,[12,13] bis(ethylenedithio)–tetrathia fulvalene,[14] tetramethylenedithiodimethyl-tetrathiafulvalene,[15] Ni(omtaa),[16] Ni(tmtaa),[17] and Cu(tmtaa),[18] (omtaa = octamethyldibenzo-tetraazaannulenato, tmtaa = tetramethyldibenzo-tetraazaannulenato), etc.. According to the metalloporphyrin compounds which consist of fullerene, the first observation of a porphyrin-fullerene close approach was reported by Sun et al.[19] Recently, metal complexes of octaethylporphyrin (oep,), the dianion of octaethylporphyrin, cocrystallites with C_{60} have been reported [20] to form solids with remarkably close contact, that is, an interaction takes place between the curved π surface of a fullerene with the planar ☐ surface of a porphyrin, without the need for matching convex with concave surfaces, *i.e.* planar versus highly curved.[21]. In addition, unique cocrystallites of fullerene C_{60} with tpps (tpp is a dianion of tetraphenylporphyrin) have also been reported.[21,22]

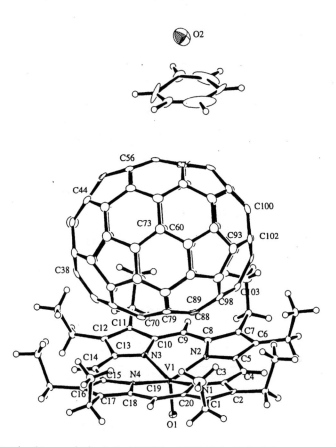

Figure 1. Atom numbering in C_{70}·V(IV)O(oep)·C_6H_6·H_2O with 30 % thermal ellipsoids.

Concerning the cocrystallites in the case of C_{70} with metal macrocycle, a 1:1 complex of Ni(omtaa) with C_{70}, *i.e.* zigzag alternating C_{70}-Ni(omtaa) chain structure, has been reported with corrugated sheets of close contact C_{70} molecules.[23]   Then, very few examples of cocrystallites consisting of C_{70} with metalloporphyrins have been reported up to now,[20,21] compared with the reported many examples of compounds in the case of fullerene C_{60}.[24]   Therefore, we report on the structures of the new metal complexes of octaethylporphyrins cocrystallized with fullerene C_{70}, such as C_{70}·V(IV)O(oep)·C_6H_6·H_2O, C_{70}·Pd(II)(oep)·$2CHCl_3$ and C_{70}·Ag(II)(oep)·$2CHCl_3$.   It is also expected that the anomalous structure and crystal packing can be observed in such a cocrystallite with C_{70}, reflecting its elliptical egg-shaped cage of C_{70} molecule. Comparison of the orientation of the 8 terminal ethyl groups on the metal oep between the C_{60} and C_{70} cocrystallites is very important in investigating mechanisms of fullerene/porphyrin and porphyrin/porphyrin intermolecular interactions in the cocrystallites with fullerenes. Significant structure of the 1-D chain ordering of the fullerene C_{70} molecules in the $C_{70} \cdot V(IV)O(oep) \cdot C_6H_6 \cdot H_2O$ cocrystallite, suggesting strong fullerene/porphyrin intermolecular interaction, will be described.

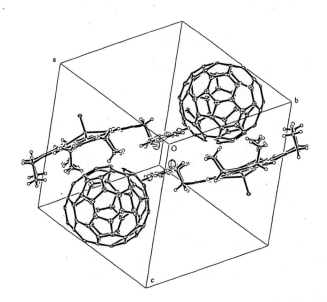

Figure 2. A drawing that shows the molecular packing in $C_{70} \cdot V(IV)O(oep) \cdot C_6H_6 \cdot H_2O$.

Table 1. Crystallographic and refinement data for cocrystallites $C_{70} \cdot VO(oep) \cdot C_6H_6 \cdot H_2O$, $C_{70} \cdot Pd(oep) \cdot 2CHCl_3$ and $C_{70} \cdot Ag(oep) \cdot 2CHCl_3$.

	$C_{70} \cdot VO(oep) \cdot C_6H_6 \cdot H_2O$	$C_{70} \cdot Pd(oep) \cdot 2CHCl_3$	$C_{70} \cdot Ag(oep) \cdot 2CHCl_3$
Formula	$C_{112}H_{52}N_4O_2V$	$C_{108}H_{46}Cl_6N_4Pd$	$C_{108}H_{46}AgCl_6N_4$
Formula weight	1536.61	1718.70	1720.16
Crystal system	Triclinic	Triclinic	Triclinic
Space group	$P\bar{1}$ (no. 2)	$P\bar{1}$ (no. 2)	$P\bar{1}$ (no. 2)
a/ Å	14.4894(8)	14.4723(1)	14.4461(2)
b/ Å	16.056(1)	14.7835(2)	14.7561(4)
c/ Å	14.3694(9)	18.8703(3)	18.9411(1)
α/°	90.178(3)	89.910(2)	90.4099(4)
β/°	95.363(2)	87.024(1)	87.4209(4)
γ/°	88.768(4)	60.879(1)	61.459(2)
V/ Å^3	3327.6(3)	3520.94(9)	3541.5(1)
Z	2	2	2
μ(Mo-Kα)/ cm^{-1}	2.20	5.55	5.74
T, °C	–190	–190	–190
No. of reflections measured (total, unique)	29791, 14857	33223, 15970	31461, 15615
Observed reflects ($I > 3\sigma(I)$)	8774	6329	7351
R1 (obs. data)	0.079	0.066	0.082
wR2	0.249	0.146	0.181

EXPERIMENTAL PROCEDURE

Preparation of Crystals

$C_{70} \cdot V(IV)O(oep) \cdot C_6H_6 \cdot H_2O$. A 0.003 g (0.003 mmol) sample of C_{70} was dissolved in 5 ml of benzene. The solution was filtered and then mixed with a filtered solution of 0.002 g (0.003 mmol) of V(IV)O(oep) dissolved in 3 ml of chloroform. The resultant mixture was allowed to stand for 5-7 days, during which dark crystals were formed. These were collected by decanting the solvent to yield 0.002 g (50 %) of product.

$C_{70} \cdot Pd(II)(oep) \cdot 2CHCl_3$. A 0.003 g (0.003 mmol) sample of C_{70} was dissolved in 5 ml of benzene. The solution was filtered and then mixed with a filtered solution of 0.002 g (0.003 mmol) of Pd(II)(oep) dissolved in 3 ml of benzene. The resultant mixture was allowed to stand for a 5-7 days, during which dark crystals were formed. Again, collection was made by decanting the solvent to yield 0.003 g (50 %) of product.

$C_{70} \cdot Ag(II)(oep) \cdot 2CHCl_3$. A 0.003 g (0.003 mmol) sample of C_{70} was dissolved in 5 ml of benzene. The solution was filtered and then mixed with a filtered solution of 0.002 g (0.003 mmol) of Ag(II)(oep) dissolved in 3 ml of benzene. The resultant mixture was allowed to stand for a 10-14 days, during which dark crystals were formed. Again, collection was made by decanting the solvent to yield 0.002 g (30 %) of product. IR (KBr): 1462, 1429, 1373, 1269, 1147, 1134, 1107, 1057, 1016, 958, 839, 796, 673, 642, 578, 536, 459 cm^{-1}.

X-ray DATA Collection. All the black block and platelet crystals having approximate dimensions of 0.10 x 0.10 x 0.10 for $C_{70} \cdot V(IV)O(oep) \cdot C_6H_6$, 0.35 x 0.30 x 0.10 for $C_{70} \cdot Pd(II)(oep) \cdot 2CHCl_3$, and 0.20 x 0.15 x 0.02 mm^3 for $C_{70} \cdot Ag(II)(oep) \cdot 2CHCl_3$ were coated with a light hydrocarbon oil and mounted on a glass fiber. Data for $C_{70} \cdot V(IV)O(oep) \cdot C_6H_6$, $C_{70} \cdot Pd(II)(oep) \cdot 2CHCl_3$ and $C_{70} \cdot Ag(II)(oep) \cdot 2CHCl_3$ were collected on a Rigaku RAXIS-RAPID 2 Imaging Plate diffractometer with graphite monochromated Mo-Kα radiation. Indexing was performed from two oscillations, which were exposed for 2.5, 0.3 and 0.6 minutes, respectively, for $C_{70} \cdot V(IV)O(oep) \cdot C_6H_6$, $C_{70} \cdot Pd(II)(oep) \cdot 2CHCl_3$ and $C_{70} \cdot Ag(II)(oep) \cdot 2CHCl_3$ compounds. The camera radius was 127.40 mm. Readout was performed in the 0.100 mm pixel mode. The cell constants and the orientation matrix for data collection corresponded to a primitive triclinic cell. The data were collected at a temperature of 83 ± 1 K in a cold dinitrogen steam to a maximum 2θ value of 55.0°. In total, 74 images corresponding to a 222.0° oscillation angle were collected with two different goniometer settings. The exposure time was 3.30, 1.00 and 6.00 minutes per degree, respectively. Data were processed by a PROCESS-AUTO program package. A symmetry-related absorption correction using the program ABSCOR [25] was applied, which resulted in transmission factors ranging from 0.76 to 0.96. The data were corrected for Lorentz and polarization effects. Check reflections were stable throughout data collection. The crystal data, including the distances and angles data are summarized in Tables 1-3.

Solution and Structure Refinement. The structure was solved by direct methods [26] and expanded using Fourier techniques[27]. The non-hydrogen atoms were refined anisotropically, and hydrogen atoms were refined isotropically. Hydrogen atoms were included through the use of a riding model. For $C_{70} \cdot V(IV)O(oep) \cdot C_6H_6$, $C_{70} \cdot Pd(II)(oep) \cdot 2CHCl_3$ and $C_{70} \cdot Ag(II)(oep) \cdot 2CHCl_3$ compounds, a numerical absorption correction was also employed. Neutral atom scattering factors were taken from Cromer and Waber[28]. The values for the mass attenuation coefficients are those of Creagh and Hubbel[29]. All calculations were performed using the teXsan [30] crystallographic software package from Molecular Structure Corporation.

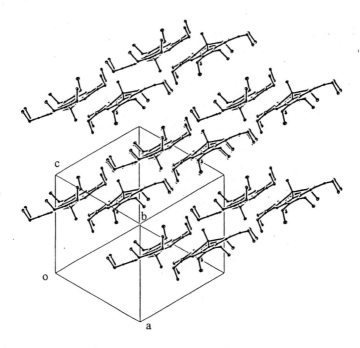

Figure 3. A drawing that shows that porphyrin V(IV)O(oep) molecular packing in $C_{70} \cdot V(IV)O(oep) C_6H_6 \cdot H_2O$. The fullerene C_{70} and solvents of C_6H_6 and H_2O molecules are omitted to guide to the eye.

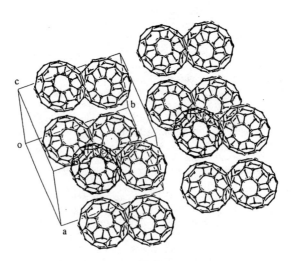

Figure 4. A drawing that shows that fullerene C_{70} molecular packing in $C_{70} \cdot V(IV)O(oep) \cdot C_6H_6 \cdot H_2O$. The V(IV)O(oep) and solvents of C_6H_6 and H_2O molecules are omitted to guide to the eye.

Table 2. Selected bond distances (Å) and angles (°) with estimated standard deviations in parentheses for cocrystallite $C_{70} \cdot VO(oep) \cdot C_6H_6 \cdot H_2O$.

V(1)-O(1)	1.589(4)	V(1)-N(1)	2.083(4)
V(1)-N(2)	2.084(4)	V(1)-N(3)	2.085(4)
V(1)-N(4)	2.077(4)		
O(1)-V(1)-N(1)	104.7(2)	O(1)-V(1)-N(2)	104.3(2)
O(1)-V(1)-N(3)	104.9(2)	O(1)-V(1)-N(4)	105.0(2)
N(1)-V(1)-N(2)	86.7(2)	N(1)-V(1)-N(3)	150.4(2)
N(1)-V(1)-N(4)	85.7(2)	N(2)-V(1)-N(3)	86.0(2)
N(2)-V(1)-N(4)	150.7(2)	N(3)-V(1)-N(4)	86.8(2)

Intermolecular distances

V(1)⋯C(79)	3.306(5)	O(1)⋯O(2)[1]	3.242(9)
N(1)⋯C(89)	3.154(7)	N(4)⋯C(70)	3.057(6)
C(73)⋯C(73)[2]	3.05(1)	C(76)⋯C(76)[3]	2.942(10)

Symmetry codes: (1) x+1, y, z-1; (2) -x+1, -y+1, -z; (3) -x+1, -y, -z

Dihedral angles between planes

plane(1)⋯plane(2)	89.4(2)	plane(1)⋯plane(3)	89.4(2)
plane(2)⋯plane(3)	0.1(3)		

Atoms defining plane: plane(1)

N(1)⋯N(2)⋯N(3)⋯N(4)⋯C(1)⋯C(2)⋯C(3)⋯C(4)⋯C(5)⋯C(6)⋯C(7)⋯C(8)⋯C(9)⋯C(10)⋯C(11)⋯C(12)⋯C(13)⋯C(14)⋯C(15)⋯C(16)⋯C(17)⋯C(18)⋯C(19)⋯C(20), plane(2) C(38)⋯C(44)⋯C(88)⋯C(100)⋯C(103), plane(3) C(56)⋯C(60)⋯C(93)⋯C(98)⋯C(102)

RESULTS AND DISCUSSION

The cocrystallites reported here were obtained in a form suitable for single-crystal X-ray diffraction by diffusion of a solution of the fullerene in benzene into a solution of the metal octaethylporphyrin in chloroform.

The 1-D C_{70} Chain Structure with the Short $C_{70}\cdots C_{70}$ Intermolecular Distance in Cocrystallite $C_{70} \cdot V(IV)O(oep) \cdot C_6H_6 \cdot H_2O$

The centric unit consists of one fully ordered C_{70} molecule, one vanadium(IV) oxide octaethylporphyrin, V(IV)O(oep), and solvents of benzene and water, as shown in Figure 1. There is no covalent bonding between the vanadium(IV) ions and the fullerene C_{70}, i.e. V(1)⋯C(79), 3.306(5). The porphyrin/fullerene interaction, which involves only a single V(IV)O(oep) molecule, is shown in Figure 2. The fullerene C_{70} is positioned so that the C_5 axis of the fullerene makes an angle of less than 0.6° with the porphyrin least-squared plane and almost the parallel to the porphyrin plane in $C_{70} \cdot V(IV)O(oep) \cdot C_6H_6 \cdot H_2O$. Such a small angle of the C_5 axis of the fullerene to the porphyrin plane has not been reported up to now, in comparison with the previous reports such as 16.4° in $C_{70} \cdot Co(II)(oep) \cdot C_6H_6 \cdot CHCl_3$, 16.2° in $C_{70} \cdot Ni(II)(oep) \cdot C_6H_6 \cdot CHCl_3$ and 16.6° in $C_{70} \cdot Cu(II)(oep) \cdot C_6H_6 \cdot CHCl_3$[20]. In the structure, the closest approaching fullerene carbon atoms are those at the most electron-rich sites, i.e. the intersection of three fused six-membered rings [20,21,31]. This suggests that van der Waals dispersion forces are the critical component of the interaction. The greater fullerene/porphyrin contact provided by the side-on rather than end-on orientation of C_{70} is also consistent with the importance of dispersion forces. These short, but nonbonding N(por)⋯C(fullerene) distances are N(1)⋯C(89), 3.154(7), N(3)⋯C(74), 3.279(7) and N(4)⋯C(70), 3.057(6) Å. Other selected bond distances and angles are summarized in Table 2. While these distances are too long to co-ordination bond, they are shorter than the normal van der Waals contact seen between neighboring fullerenes (greater than 3.2 Å) [32].

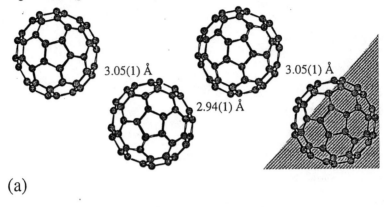

(a)

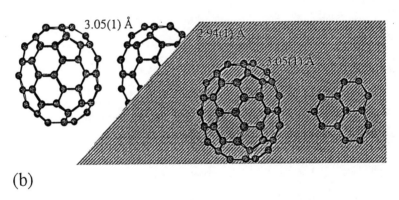

(b)

Figure 5. The continuous 1-D chain ordering of fullerene C_{70} molecules in $C_{70} \cdot V(IV)O(oep) \cdot C_6H_6 \cdot H_2O$, (a) top view and (b) side view.

In the cocrystallite, the *syn*-formed configuration of V(IV)O(oep) can be observed, though the two adjacent V(IV)O(oep) planes are separated enough to disappear the porphyrin/porphyrin face-to-face contact. The V(IV)O(oep) molecules play as a role of the template (Figure 3) in order to line up fullerene C_{70} molecules in a row, as shown in Figures. 4 and 5. The combination of the fullerene/porphyrin contact produces, so-called, a "placed eggs in a package-like" structure. The intermolecular distances between the nearest C_{70} molecules are significantly short, C(73)⋯C(73), 3.05(1); C(76)⋯C(76), 2.942(10) Å (Figure 5), in comparison with the C_{70}⋯C_{70} intermolecular distance in the ordinary porphyrin cocrystallites with C_{70}. Such a short distance between the C_{70} molecules has not been observed in the previous reports, *i.e.* 3.486 Å in $C_{70} \cdot Co(II)(oep) \cdot C_6H_6 \cdot CHCl_3$, 3.438 Å in $C_{70} \cdot Ni(II)(oep) \cdot C_6H_6 \cdot CHCl_3$ and 3.395 Å in $C_{70} \cdot Cu(II)(oep) \cdot C_6H_6 \cdot CHCl_3$ [20].

In addition to these fullerene/porphyrin interactions, there are significant porphyrin/porphyrin contacts instead of the pair-wise or face-to-face contact. The combination of porphyrin/porphyrin contact produces a "two slipped potlids-like" structure. The two slipped potlids arrangement is facilitated by the positioning all of the ethyl groups on the same side of the porphyrin from the adjacent porphyrin. In $C_{70}\cdot V(IV)O(oep)\cdot C_6H_6\cdot H_2O$, the *syn*-formed configuration of the V(IV)O(oep) has been observed, in contrast that the *anti*-formed or random configurations have been reported up to now in several pristine V(IV)O(oep) complexes.[33,34] In this compound, an oxygen atom is located at the opposite side of the porphyrin plane from the eight ethyl groups. The vanadium atom of V(IV)O(oep) is not located on the same plane of the least-squared porphyrin plane, *i.e.* N(1)-V(1)-N(3), 150.4(2); N(2)-V(1)-N(4), 150.7(2)°. This non-linear structure of the central metal is consistent with that observed in the pristine V(IV)O(oep) complexes (N-V-N, 149.9-150.2°)[34]. In the $C_{70}\cdot V(IV)O(oep)\cdot C_6H_6\cdot H_2O$ cocrystallite, it is supposed that the face-to-face interaction between two adjacent V(IV)O(oep) molecules, because the intermolecular distance between them is very long, *i.e.* V⋯V, 8.152 Å. The face-to-face interaction between adjacent porphyrins is strengthened by these V=O groups, with the result that the *syn*-formed configuration of oep of the "two slipped pod lids-like" structure, which has also been reported in $C_{60}\cdot Ru(CO)(oep)\cdot C_6H_5CH_3$ compounds,[24] is observed.

The Isomorphous *Syn*-Formed Series, $C_{70}\cdot Pd(oep)\cdot 2CHCl_3$ and $C_{70}\cdot Ag(oep)\cdot 2CHCl_3$

The isomorphous two additional new cocrystallites which consist of the fullerene C_{70} and M(oep) can be obtained, *i.e.* $C_{70}\cdot Pd(oep)\cdot 2CHCl_3$ and $C_{70}\cdot Ag(oep)\cdot 2CHCl_3$. The difference of the structure between them is quite small. Therefore, only the lattice structure of the cocrystallite $C_{70}\cdot Pd(oep)\cdot 2CHCl_3$ will be discussed mainly in this section.

The asymmetric unit in each consists of one C_{70} molecule, one metalloporphyrin and two solvent molecules of chloroform. The porphyrin/fullerene interaction, which involves only a single Pd(II)(oep) molecule, is shown in Figure 6. The intra-molecular Pd-N distances (Pd(1)-N(1), 2.030(7); Pd(1)-N(2), 2.029(7); Pd(1)-N(3), 2.024(7); Pd(1)-N(4), 2.016(6) Å) in $C_{70}\cdot Pd(II)(oep)\cdot 2CHCl_3$ are considerably longer than those in $C_{70}\cdot Co(II)(oep)\cdot C_6H_6\cdot CHCl_3$ (Co-N, 1.964(5)-1.967(5) Å),[20] reflecting the difference of their ionic radiuses. There is no covalent bonding between the palladium ion and the fullerene C_{70}. In the structure, the closest approaching fullerene carbon atoms are, of course, those at the intersection of three fused six-membered rings. The greater fullerene/porphyrin contact provided by the side-on rather than end-on orientation of C_{70} is also consistent with the importance of dispersion forces. The fullerene C_{70} is positioned so that the C_5 axis of the fullerene makes an angle of 17.4 ° with the porphyrin plane in $C_{70}\cdot Pd(II)(oep)\cdot 2CHCl_3$. This value of the angle is almost as same as those of the previous reports, mentioned above (16.2-16.6 °)[20]. The fullerene is positioned so that one of the carbon atoms (an atom in layer "c" of the fullerene) [35] is positioned almost directly over the metal at the center of the porphyrin. These short, but nonbonding distances are Pd(1)⋯C(40), 2.91(1); Pd(1)⋯C(41), 3.26(1); N(1)⋯C(41), 3.23(1); N(4)⋯C(39), 3.12(1) Å for porphyrin/fullerene distances, and Pd(1)⋯N(2), 3.190(7); Pd(1)⋯Pd(1), 3.470(2) Å for porphyrin/porphyrin distances. Other selected bond distances and angles are summarized in Table 3.

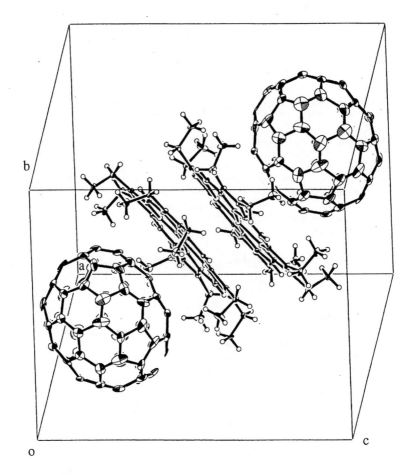

Figure 6. A drawing that shows the molecular packing in $C_{70} \cdot Pd(oep) \cdot 2CHCl_3$.

In our previous work, the unique *anti*-formed [36] configuration of the metal oeps (*anti*-M(oep)s) can be observed in the cocrystallites with C_{60}, *i.e.* $C_{60} \cdot Cu(II)(oep) \cdot 2C_6H_6$,[24] $C_{60} \cdot Ag(II)(oep) \cdot 2C_6H_6$,[37] and $C_{60} \cdot 2Ni(II)(oep) \cdot 2C_6H_5Cl$.[38]   Especially in the case of the Pd(II)(oep) and Ag(II)(oep) cocrystallites, the *anti*-formed configurations of the Pd(II)(oep) and Ag(II)(oep) can be observed in the cocrystallites with C_{60}, $C_{60} \cdot Pd(II)(oep) \cdot 1.5C_6H_6$ and $C_{60} \cdot Ag(II)(oep) \cdot 2C_6H_6$,[24] reflecting that the *anti*-formed configuration is also observed in the pristine M(oep) complexes. However, the significant *syn*-formed configuration (Figure 7) of the metal oeps are observed in this work in the cocrystallites with C_{70}, $C_{70} \cdot Pd(II)(oep) \cdot 2CHCl_3$ and $C_{70} \cdot Ag(II)(oep) \cdot 2CHCl_3$. The core molecular geometry of the Pd(oep) in $C_{70} \cdot Pd(II)(oep) \cdot 2CHCl_3$ is quite different from that is found in the pristine Pd(II)(oep) complex[39]. In the $C_{60} \cdot Pd(II)(oep) \cdot 1.5C_6H_6$ cocrystallite, the intermolecular distance between two adjacent Pd(II)(oep) molecules is large enough to remain into the *anti*-formed configuration as well that observed in the pristine Pd(II)(oep) complex. On the other hand, the intermolecular porphyrin/porphyrin distance between them in the $C_{70} \cdot Pd(II)(oep) \cdot 2CHCl_3$ cocrystallite is quite shorter (Figure 7), suggesting that there is the strong porphyrin/porphyrin face-to-face interaction between two adjacent Pd(II)(oep) molecules. As a result, the unique *syn*-formed configuration of the Pd(II)(oep) can be observed in the $C_{70} \cdot Pd(II)(oep) \cdot 2CHCl_3$ cocrystallite. This result suggests that the intermolecular interactions of the fullerene/porphyrin and porphyrin/porphyrin are very important in order to determine the configuration of the eight ethyl groups of the metal oeps in the cocrystallites with fullerenes. Notice from the result that the pair-wise porphyrin/porphyrin contact is greater in the fullerene cocrystallite than it is in pristine Pd(II)(oep) compound.

Table 3. Selected bond distances (Å) and angles (°) with estimated standard deviations in parentheses for cocrystallite $C_{70} \cdot Pd(oep) \cdot 2CHCl_3$.

Pd(1)-N(1)	2.030(7)	Pd(1)-N(2)	2.029(7)
Pd(1)-N(3)	2.024(7)	Pd(1)-N(4)	2.016(6)
N(1)-Pd(1)-N(2)	90.0(3)	N(1)-Pd(1)-N(3)	177.7(3)
N(1)-Pd(1)-N(4)	89.9(2)	N(2)-Pd(1)-N(3)	90.3(3)
N(2)-Pd(1)-N(4)	177.4(3)	N(3)-Pd(1)-N(4)	89.6(2)
Intermolecular distances			
Pd(1)···C(40)	2.91(1)	Pd(1)···N(2)[1]	3.190(7)
Pd(1)···C(41)	3.26(1)	Pd(1)···Pd(1)[1]	3.470(2)
N(1)···C(41)	3.23(1)	N(4)···C(39)	3.12(1)
Symmetry codes: (1) -x+1, -y+1, -z-1			

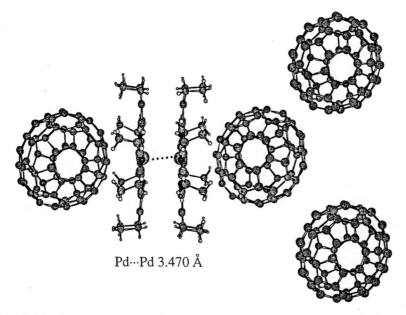

Pd···Pd 3.470 Å

Figure 7. A drawing that shows the molecular packing in $C_{70} \cdot Pd(II)(oep) \cdot 2CHCl_3$. Dashed line indicates the intermolecular Pd···Pd distance between two adjacent V(IV)O(oep) molecules.

CONCLUSION

Solutions of C_{70} and metal complexes of octaethylporphyrin yield new crystals that contain both the fullerene C_{70} and porphyrin. Naturally assembling cocrystallite of C_{70} fullerene with $V(IV)O(oep)$, $C_{70} \cdot V(IV)O(oep) \cdot C_6H_6 \cdot H_2O$, shows unusually short fullerene/fullerene contacts ($C_{70} \cdots C_{70}$, 2.94-3.05 Å) compared with typical □-□ interactions (3.0-3.5 Å). The structures of $C_{70} \cdot Pd(II)(oep) \cdot 2CHCl_3$ and $C_{70} \cdot Ag(II)(oep) \cdot 2CHCl_3$ are isomorphous and contain an ordered C_{70} cage surrounded by the $M(II)(oep)$ unit. In the C_{70} structures, the ellipsoidal fullerene makes porphyrin contact at its equator rather than its poles; a carbon atom from three fused six-membered rings lies closest to the center of the porphyrin.

ACKNOWLEDGEMENT

The authors are grateful to Profs. H. Imahori and Y. Sakata (Osaka Univ.) for their kind advices as to the syntheses of metal oeps. This research was supported partly by a Grant-in-Aid for Scientific Research on Priority Area (Nos. 10149104 and 11165235), "Fullerenes and Nanotubes Networks" from the Ministry of Education, Science and Culture, Japan.

REFERENCES

[1] C. A. Reed and R. Bolskar, 2000, Chem. Rev., **100**, 1075.
[2] D. Konarev and R. Lyubovskaya, 1999, Russ. Chem. Rev., **68**, 23.
[3] O. A. Dyachenko and A. Graja, 1999, Fullerene Sci. Tech., **7(3)**, 317.
[4] M. J. Hardie and C. L. Raston, 1999, Chem. Commun., 1153-1163.
[5] S. A. Olsen, A. M. Bond, R. G. Compton, G. Lazarev, P. J. Mahon, F. Marken, C. L. Raston, V. Tedesco and R. Webster, 1998, J. Phys. Chem., **102**, 2641.
[6] K. Tsubaki, K. Tanaka, T. Kinoshita and K. Fuji, 1998, Chem. Commun., 895.
[7] K. N. Rose, L. J. Barbour, G. W. Orr and J. L. Atwood, 1998, Chem. Commun., 407.
[8] Z. Yoshida, H. Takekuma, S. Takekuma and Y. Matsubara, 1994, Angew. Chem. Int. Ed. Engl., **33**, 1597.
[9] F. Diedrich, J. Effing, U. Jonas, L. Jullien, T. Plesnivy, H. Ringsdorf, C. Thilgen and D. Weinstein, 1992, Angew. Chem. Int. Ed. Engl., **31** 1599.
[10] D. M. Eichhorn, S. Yang, W. Jarrell, T. F. Baumann, L. S. Beall, A. J. P. White, D. J. Williams, A. G. M. Barrett and B. M. Hoffman, 1995, J. Chem. Soc., Chem. Commun., 1703.
[11] D. M. Hochmuth, S. L. J. Michel, A. J. P. White, D. J. Williams, A. G. M. Barrett and B. M. Hoffman, 2000, Eur. J. Inorg. Chem., 593-596.
[12] J. W. Steed, P. C. Junk, J. L. Atwood, M. J. Barnes, C. L. Raston and R. S. Burkhalter, 1994, J. Am. Chem. Soc., **116**, 10346.
[13] A. M. Bond, W. Miao, C. L. Raston, T. J. Ness, M. J. Barnes and J. L. Atwood, 2001, J. Phys. Chem. B, **105**, 1687-1695.
[14] A. Izuoka, T. Tachikawa, T. Sugawara, Y. Suzuki, M. Konno, Y. Saito and H. Shinohara, 1992, J. Chem. Soc., Chem. Commun., 1472.
[15] D. V. Konarev, E. F. Valeev, Y. L. Slovokhotov, Y. M. Shul'ga, O. S. Roschupkina and R. N. Lyubovskaya, 1997, Synth. Met., **88**, 85.
[16] P. D. Croucher, P. J. Nichols and C. L. Raston, 1999, J. Chem. Soc., Dalton Trans., 279.
[17] P. C. Andrews, J. L. Atwood, L. J. Barbour, P. J. Nichols and C. L. Raston, 1998, Chem. Eur. J., **4(8)**, 1382.

[18] P. C. Andrews, J. L. Atwood, L. J. Barbour, P. D. Croucher, P. J. Nichols, N. O. Smith, B. W. Skelton, A. H. White and C. L. Raston, 1999, J. Chem. Soc., Dalton Trans., 2927.
[19] Y.-P, Sun, T. Drovetskaya, R. D. Bolskar, R. Bau, P. D. W. Boyd and C. A. Reed, 1997, J. Org. Chem., **62**, 3642.
[20] M. M. Olmstead, D. A. Costa, K. M. Maitra, B. C. Noll, L. Phillips, P. M. Van Calcar and A. L. Balch, 1999, J. Am. Chem. Soc., **121**, 7090-7097.
[21] P. D. W. Boyd, M. C. Hodgson, C. E. F. Rickard, A. G. Oliver. L. Chaker, P. J. Brothers, R. D. Bolskar, F. S. Tham and C. A. Reed, 1999, J. Am. Chem. Soc., **121**, 10487-10495.
[22] T. Ishii, R. Kanehama, N. Aizawa, M. Yamashita, H. Matsuzaka, K. Sugiura, H. Miyasaka, T. Kodama, K. Kikuchi, I. Ikemoto, H. Tanaka, K. Marumoto and S.-I. Kuroda, 2001, J. Chem. Soc., Dalton Trans., **20**, 2975-2980.
[23] J. M. Marshall, P. D. Croucher, P. J. Nichols and C. L. Raston, 1999, Chem. Commun., 193.
[24] T. Ishii, N. Aizawa, M. Yamashita, H. Matsuzaka, T. Kodama, K. Kikuchi, I. Ikemoto and Y. Iwasa, 2000, J. Chem. Soc., Dalton Trans., **23**, 4407-4412.
[25] ABSCOR: T. Higashi (1995). Program for Absorption Correction, Rigaku Corporation, Tokyo, Japan.
[26] SIR92: A. Altomare, M. C. Burla, M. Camalli, M. Cascarano, C. Giacovazzo, A. Guagliardi and G. Polidori, 1994, J. Appl. Cryst., **27**, 435.
[27] DIRDIF94: P. T. Beurskens, G. Admiraal, G. Beurskens, W. P. Bosman, R. de Gelder, R. Israel and J. M. M. Smits. The DIRDIF-94 program system, Technical Report of the Crystallography Laboratory, University of Nijimegen, The Netherlands (1994).
[28] D. T. Cromer and J. T. Waber, "International Tables for X-ray Crystallography", Vol. IV, The Kynoch Press, Birmingham, England, Table 2.2 A (1974).
[29] D. C. Creagh and W. J. McAuley, "International Tables for Crystallography", Vol C, (A. J. C. Wilson, ed.), Kluwer Academic Publishers, Boston, Tables 4.2.4.3 pages 200-206 (1992).
[30] teXsan: Crystal Structure Analysis Package, Molecular Structure Corporation (1985 & 1999).
[31] D. Sun, F. S. Tham, C. A. Reed, L. Chaker, M. Burgess and P. D. W. Boyd, 2000, J. Am. Chem. Soc., **122**, 10704-10705.
[32] H. B. Burgi, R. Restori, D. Schwarzenbach, A. L. Balch, J. W. Lee, B. C. Noll and M. M. Olmstead, 1994, Chem. Mater., **6**, 1325.
[33] C. E. Schulz, H.-S. Song, Y. J. Lee, J. U. Mondal, K. Mohanrao, C. A. Reed, F. A. Walker and W. R. Scheidt, 1994, J. Am. Chem. Soc., **116**, 7196.
[34] F. S. Molinaro and J. A. Ibers, 1976, Inorg. Chem., **15**, 2278.
[35] A. L. Balch and M. M. Olmstead, 1998, Chem. Rev., **98**, 2123-2165.
[36] The term "anti" in this paper denotes the anti-symmetry of the terminal eight ethyl groups on the porphyrin plane according to the E,Z-nomenclature. That is a different meaning from a so-called anti-compound mentioned in a review article by Balch et al., ref no. [35].
[37] T. Ishii, N. Aizawa, R. Kanehama, M. Yamashita, H. Matsuzaka, T. Kodama, K. Kikuchi and I. Ikemoto, 2001, Inorg. Chim. Acta, **317(1-2)**, 81-90.
[38] T. Ishii, N. Aizawa, M. Yamashita, H. Matsuzaka, I. Ikemoto, K. Kikuchi, T. Kodama and Y. Iwasa, 2001, Synth. Met., **121**, 1165.
[39] A. M. Stolzenberg, L. J. Schussel, J. S. Summers, B. M. Foxman and J. L. Petersen, 1992, Inorg. Chem., **31**, 1678-1686.

Ab Initio Molecular Dynamics Simulation of Foreign Atom Insertion into C_{60}

K. OHNO and K. SHIGA
Department of Physics, Yokohama National University, Yokohama, Japan

T. MORISATO, S. ISHII, M. F. SLUITER and Y. KAWAZOE
Institute for Materials Research, Tohoku University, Sendai, Japan

T. OHTSUKI
Laboratory of Nuclear Science, Tohoku University, Sendai, Japan

ABSTRACT

In order to show the possibility of creating endohedral fullerenes by inserting a foreign atom with a suitably high kinetic energy into C_{60}, we have carried out large-scale ab initio molecular dynamics simulations on the basis of the all-electron mixed-basis approach that uses both atomic orbitals and plane waves as basis functions. The values inside the parentheses indicate the kinetic energy in units of eV needed for the penetration of the respective atoms through the center of a 6-membered ring of C_{60}: Li, Be (5); N (80), O (100), Na (no), S(40), Cl(60), K (no), V (no), Cu (no), As (40), Se (40), Sb (80), Te (40), Kr (120), Xe (160). The term, (no), indicates that penetration was not possible. The values for O, S and Cl are presented as new data.

INTRODUCTION

Endohedral fullerenes that have one or more atoms inside the C_{60} cage is attracting great interest these days. If their mass production becomes possible, they would have many interesting applications such as stable molecular devices at the nanometer scale. Although, endohedral complexes can be created simultaneously as well as ordinary fullerenes by using arc-discharge vaporization of composite rods made of graphite and the metal oxide, the production rate of endohedral C_{60} is quite low compared to that of ordinary C_{60}. An alternative way to produce endohedral C_{60} is to insert an atom inside the cage of the preexisting C_{60} afterwards. Creation of endohedral [Li@C_{60}]$^+$ and [Na@C_{60}]$^+$ species during the collisions of alkali-metal ions with C_{60} vapor molecules has been reported [1]. More recently, Sato *et al.* used a K$^+$ and C_{60} plasma to create K@C_{60} by collision [2]. One may also use the recoil of nuclear reactions to accelerate isotope elements such as ^{7}Be in collision with C_{60}. In fact, using high performance liquid chromatography (HPLC) combined with a UV detector and a Ge γ-ray detector, Ohtsuki confirmed that ^{7}Be is trapped inside the C_{60} cage after such a collision. Carrying on from these two recent experiments, we have conducted *ab initio* molecular dynamics (MD) simulations of the collision between Li, Na and Be cations and a C_{60} anion, and investigated the possibility of encapsulation [3,4].

Similarly, insertion of noble gas atoms into C_{60} has been tested experimentally [5-8]. Saunders et al. [6] showed that noble gas atom endohedral complexes are created from C_{60} crystals intercalated with noble gas atoms under pressure at a particular high temperature. Using the recoil of γ-ray emissions from nuclear reactions, Gadd et al. [7] created noble gas atom endohedral C_{60} from similar materials irradiated by a neutron source. Ohtsuki [8] also succeeded in detecting similar radioactive Xe@C_{60} and Kr@C_{60} by using the HPLC technique combined with the UV and γ-ray detectors.

The aim of the present paper is to show the possibility of creating endohedral C_{60} by inserting another atom with a suitably high kinetic energy (K.E.) into C_{60}. For this purpose, large-scale *ab initio* molecular dynamics simulations are performed in this study. In particular, we will present new results for O, S and Cl.

SIMULATION

We use an *ab initio* molecular dynamics method based on the all-electron mixed basis approach which uses both atomic orbitals (AO) and plane waves (PW) as a basis set within the framework of the local density approximation (LDA). So far, we have carried out similar simulations of Li- and Be-insertions into C_{60} [3,4] by using analytic Slater-type atomic orbitals. Our recent calculations involving the insertion of Kr, Xe [8], K, Cu, As [9], Se [10], Sb, Te [11] have used atomic orbitals determined numerically by a standard atomic code based on Herman-Skillman's framework with logarithmic radial meshes. For example, we adopt 318 numerical atomic orbitals for the C_{60} + Se system as well as 4,169 plane waves corresponding to 7 Ry cutoff energy. For dynamics, we assumed adiabatic approximations with the electronic structure being always in the ground state. In a supercell composed of 64x64x64 meshes, in which 1 a.u. = 0.52918 Å corresponds to 2.7 meshes, we put one stationary C_{60} molecule at $t = 0$ and one atom moving vertically with a given initial velocity toward the center of a six-membered ring (hereafter we will call this six-membered ring, u-C_6) of C_{60}. The initial distance from the center of u-C_6 is 1.5 Å. The basic time step is typically set to x $t = 4$ a.u. (~0.1 fs) and we perform five to six steepest descent (SD) iterations after each update of the atomic positions in order to converge the electronic states.

RESULTS

Here we will briefly describe some of the results of these simulations.

(A) Li, Be:
In the simulation where Li (or Be) atom with kinetic energy (K.E.) = 5 eV hits the

center of the u-C_6 of C_{60} or C_{70} perpendicularly, the Li atom penetrates into the cage through the center of the u-C_6 without difficulty [3,4]. It goes 1.6 Å deep into the cage and comes back to be trapped at a distance of 1.0 Å from the center of the u-C_6. In cases where the hit position is off-center (or at the center of a five-membered ring) or the initial K.E. is lower (~ 1 eV), both Li and Be cannot pass through the u-c_6.

(B) Na:
In the case of Na atom, simple insertion through the six-membered ring does not occur even when the initial K.E. is 50 eV [3], although penetration with ~ 70 eV is possible into carbon nanotube [12].

(C) K:
A potassium atom is trapped outside the cage when the initial K.E. is 50 eV, and passes through the whole cage when K.E. is 160 eV. It is trapped just below the open u-c_6 when the initial K.E. is 100 eV [13,14]. Penetration into carbon nanotube is possible with ~ 120 eV [12].

(D) N, O, S, Cl:
Initial kinetic energies of 80 eV, 100 eV, 40 eV and 60 eV are required respectively, for N [15], O, S and Cl atoms to be trapped inside of the C_{60} cage by insertion through u-c_6.

(E) Kr, Xe:
In the case of noble gas atoms, Kr with an initial K.E. greater than 80 eV or Xe with an initial K.E. greater than 120 eV penetrates into the cage of C_{60} through the center of u-C_6 without difficulty [8]. For kinetic energies above 150 eV for Kr or 200 eV for Xe, the foreign atom passes through the center of the six-membered ring at the opposite side of the cage [8]. For relatively low initial kinetic energies (typically x 300 eV for Xe and Kr), C_{60} shows a tendency to recover its original shape within the simulation period. The results of the simulations change of course, according to the impact energy, impact point and angle. For higher initial kinetic energies (x 300 eV for Xe and Kr), six C_2 losses occur simultaneously from the upper side of C_{60}. If a noble-gas atom is inserted toward off-center positions of a six- or five-membered ring, the damage suffered on C_{60} increases significantly.

(F) V:
In the case of a V atom hitting C_{60} with 100–150 eV, the V atom reacts with the u-C_6 and does not penetrate into the cage [13].

(G) Cu:
In the case of a Cu atom hitting C_{60}, the Cu atom is either trapped outside of C_{60} (for 50 eV) or trapped at the center of the u-C_6 (for 100 eV). For a higher initial K.E. ~ 150 eV, the Cu atom goes through the u-C_6 but is trapped just below the open u-C_6 [13,14].

(H) Se, Te:
In the case of Se and Te atoms hitting C_{60}, Se@C_{60} and Te@C_{60} are created with a 40 eV initial K.E. [10,11]. However, for higher energies such as 80 eV, both atoms pass through the other side of C_{60}. Figure 1 represents several snapshots of the Te simulation with an initial K.E of 40 eV.

(I) As, Sb:
In the case of As and Sb atoms, heterofullerenes such as C_{59}As and C_{59}Sb are possible as well as As@C_{60} and Sb@C_{60} [9,11] Penetration into the C_{60} cage is possible with 40 eV and 80 eV, respectively, for As and Sb [9,11].

(J) Ca, Si into C_{74}
In the case of an Si atom hitting C_{74}, the endohedral complex, Si@C_{74} is created with a 40 eV initial kinetic energy, although Si is reflected outside the cage when the initial kinetic energy is 20 eV [16]. On the other hand, in the case of a Ca atom, it is not encapsulated in C_{74} even with a 120 eV initial kinetic energy [16].

The results obtained so far are summarized in the periodic table given in Table 1. We find that Se and Te atoms can penetrate rather easily compared to the other fourth-row elements such as Cu, although we have not yet performed complete simulations involving all the fourth-row elements. From this table, one may find that penetration is possible when the electronegativity of the foreign atom is between 1.8–3.5 except for very small atoms such as Li and Be and noble gas atoms such as Kr and Xe. This feasibility tendency of insertions of these atoms is consistent with the experimental findings by Ohtsuki, et al. [4,8–11,14].

Table 1. Summary of the result of simulated atomic collisions at the center of a six-membered ring of C_{60} (or C_{74} if there is an indication). The initial kinetic energy is given in units of eV for each case. The term, -no-, indicates that the encapsulation did not succeed. The values inside the parenthesis are the electronegativity in Pauling's definition.

Li (1.0) 5 eV	Be (1.5) 5 eV	B (2.0)	C (2.5)	N (3.0) 80 eV	O (3.5) 100 eV	F (4.0)	Ne
Na (1.0) -no-	Mg (1.2)	Al (1.5)	Si (1.8) 40 eV (C_{74})	P (2.1)	S (2.5) 40 eV	Cl (3.0) 60 eV	Ar
K (0.8) -no-	Ca (1.0) -no- (C_{74})	V-Cu -no-	Ge (1.8)	As (2.0) 40 eV	Se (2.4) 40 eV	Br (2.8)	Kr 120 eV
Rb (0.8) -no-	Sr (1.0)		Sn (1.8)	Sb (1.9) 80 eV	Te (2.1) 40 eV	I (2.5)	Xe 160 eV

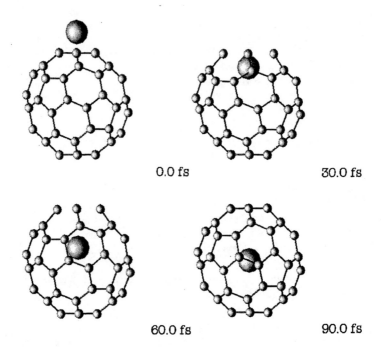

Figure 1. Snapshots of the simulation in which Te with a 40 eV initial K.E. is incorporated inside C_{60}.

CONCLUSION

The results of the present simulation reveal that the atoms like Se and Te which are much heavier than Na can go through the six-membered ring without difficulty when the K.E. is in a certain range. We found that the feasibility tendency of the insertion is related to the electro-negativity of atoms.

The results of the present simulation may tell us a story which is somewhat different from the open-window mechanism discussed in [6]. According to the open-window mechanism, one of the bonds of C_{60} should break off to create a widely opened hole in the cage, before atomic insertion. This may occur either by heating up C_{60} as was discussed in [6] or by a collision of another atom, as was discussed independently in [3] for the Na^+ collision against C_{60}. The possibility of realizing stable endohedral fullerenes by this open-window mechanism is much smaller compared to the present direct insertion process. That is, the creation rate of stable Se@C_{60} and Te@C_{60} by the direct insertion process can be much higher than, for example, that of stable Na@C_{60} by the open-window process. This is certainly consistent with the experimental evidence reported recently by Ohtsuki, et al. [4,8–11,14].

ACKNOWLEDGEMENT

The authors are grateful to the Center for Computational Materials Science of the Institute for Materials Research (IMR) at Tohoku University for the use of the HITAC SR8000 and to the Computer Center of the Institute for Molecular Science (IMS) for the use of the Fujitsu VPP5000.

REFERENCES

[1] Z. Wan, J. F. Christian and S. L. Anderson, 1992. Collision of Li^+ and Na^+ with C_{60}: Insertion, Fragmentation, and Thermionic Emission, Phys. Rev. Lett., 69(9), 1352–1355.
[2] T. Hirata, R. Hatakeyama, T. Mien and N. Sato, 1996. Production and control of $K-C_{60}$ plasma for material processing, J. Vac. Sci. Technol. A14(2), 615–618.
[3] K. Ohno, Y. Maruyama, K. Esfarjani, Y. Kawazoe, N. Sato, R. Hatakeyama, T. Hirata, M. Niwano, 1996. All-electron mixed-basis molecular dynamics simulations for collision between C_{60} and alkali-metal ions: possibility of $Li@C_{60}$, Phy. Rev. Lett., 76(19), 3590–3593.
[4] T. Ohtsuki, K. Masumoto, K. Ohno, Y. Maruyama, Y. Kawazoe, K. Sueki K. Kikuchi, 1996. Insertion of Be Atoms in C_{60} Fullerene Cages: $Be@C_{60}$, Phys. Rev. Lett. 77(17), 3522–24.
[5] T. Braun and H. Rausch, 1995. Endohedral incorporation of argon atoms into C_{60} by neutron irradiation, Chem. Phys. Lett. 237(5-6), 443–447.
[6] M. Saunders, R. J. Cross, H. A. Jimenez-Vazquez, R. Shimshi and A. Khong, 1996. Noble Gas Atoms Inside Fullerenes, Science, 271(5256), 1693–1697.
[7] G. Gadd, P. Evans, D. Hurwood, P. Morgan, S. Moricca, N. Webb, J. Holmes, G. McOrist, T. Wall, M. Blackford, D. Cassidy, M. Elcombe, J. Norman, P. Johnson, P. Prasad, 1997. Endohedral fullerene formation via prompt gamma recoil, Chem. Phys. Lett., 279, 108–14.
[8] T. Ohtsuki, K. Ohno, K. Shiga, Y. Kawazoe, Y. Maruyama, K. Masumoto, 1988. Insertion of Xe and Kr Atoms in C_{60}, C_{70} Fullerenes, Phys. Rev. Lett. 81(5), 967–970.
[9] T. Ohtsuki, K. Ohno, K. Shiga, Y. Kawazoe, Y. Maruyama and K. Masumoto, 1999. Formation of As- and Ge-doped heterofullerenes, Phys. Rev. B, 60(3), 1531–1534.
[10] T. Ohtsuki, K. Ohno, K. Shiga, Y. Kawazoe, H. Yuki, 2002. Se atom incorporation in fullerenes using nuclear recoil and *ab initio* molecular dynamics simulations, Phys. Rev. B, 65(7), 73402:1–5.
[11] T. Ohtsuki, K. Ohno, K. Shiga, Y. Kawazoe, Y. Maruyama, K. Shikano and K. Masumoto, 2001. Formation of Sb- and Te-doped fullerenes by using nuclear recoil and molecular dynamics simulations, Phys. Rev. B, 64(12), 125402:1–5.
[12] A. A. Farajian, K. Ohno, K. Esfarjani, Y. Maruyama and Y. Kawazoe, 1999. *Ab initio* study of dopant insertion into carbon nanotubes, J. Chem. Phys., 111(5), 2164–2168.
[13] K. Shiga, K. Ohno, Y. Maruyama, Y. Kawazoe and T. Ohtsuki, 1999. *Ab*

Initio Molecular Dynamics in an All-Electron Mixed Basis Approach: Application to Atomic Insertions to C_{60}, Modeling Simulation, Material Science Engineering, 7, 621–630.

[14] T. Ohtsuki, K. Ohno. K. Shiga, Y. Kawazoe, Y. Maruyama and K. Masumoto, 2000. Systematic study of foreign-atom-doped fullerenes by using a nuclear recoil method and their MD simulation, J. Chem. Phys., 112(6), 2834–2842.

[15] K. Shiga, K. Ohno, T. Ohtsuki and Y. Kawazoe, 2001. Formation of N-doped C_{60} Studied by *Ab Initio* Molecular Dynamics Simulations, Mater. Trans. 42(11), 2189–2193.

[16] K. Shiga, K. Ohno, Y. Kawazoe, Y. Maruyama, T. Hirata, R. Hatakeyama and N. Sato, 2000. Ab initio molecular dynamics simulation for the insertion process of Si and Ca atoms into C_{74}, Mater. Sci. and Eng., A290(1), 6–10.

Formation of Radioactive Fullerenes by Using Nuclear Recoil

T. OHTSUKI
Laboratory of Nuclear Science, Tohoku University, Mikamine, Taihaku, Sendai, Japan

K. OHNO and K. SHIGA
Department of Physics, Yokohama National University, Hodogaya, Yokohama, Japan

T. MORISATO, S. ISHII and M. F. SLUITER
Institute for Materials Research, Tohoku University, Aoba, Sendai, Japan

H. YUKI
Laboratory of Nuclear Science, Tohoku University, Mikamine, Taihaku, Sendai, Japan

Y. KAWAZOE
Institute for Materials Research, Tohoku University, Aoba, Sendai, Japan

ABSTRACT

Formation of atom-doped fullerenes has been investigated using several types of radionuclides produced from nuclear reactions. From the trace of the radioactivities using High Performance Liquid Chromatography (HPLC), it was found that formation of endohedral fullerenes (or heterofullerene) with small atoms (Be, Li), noble-gas atoms (Kr, Xe) and 4B-6B elements (Ge, As, Se, Sb, Te etc.) is possible by a recoil process following the nuclear reaction. We also found that heavier nuclides such as Po can penetrate into C_{60} by nuclear recoil. Other elements such as alkali, alkali-earth and transitional metals (Na, Ca, Sc, etc.), destroy most of the fullerene cage during entry. These experimental results indicate that chemical reactivity or inactivity seems to play an important role in forming foreign-atom-doped fullerenes.

INTRODUCTION

From the early days of the discovery of C_{60} and the successful production of a large number of fullerenes [1,2], much experimental data have been reported on various kinds of physical and chemical quantities associated with endohedrally doped [3–5], exohedrally doped [6,7] fullerenes and heterofullerenes [8–10]. In the formation of endohedral fullerenes for already created C_{60}, Saunders et al. [11] have demonstrated the possibility of incorporating noble-gas atoms into the fullerene cage under high-pressure and high-temperature conditions. Braun et al. [12] and Gadd et al. [13] have produced an atom-doped C_{60} by using prompt-gamma recoil induced by neutron irradiation. However, only partial knowledge about the formation process and the produced materials has been revealed about the nature of the chemical interaction between a foreign atom and a fullerene cage. Therefore, it is important and intriguing to synthesize such new complexes, i.e., foreign-atom incorporated fullerenes, and their properties need to be investigated due to understand how to produce large quantities of these materials.

In the present study, we examine the production of fullerene derivatives created when alkali and alkali-earth, transition metals, 3B-6B elements and noble-gas elements are produced by nuclear reactions induced by irradiation of samples with high-energy "bremsstrahlung" or charged particles. We found that Be [14], C, N [15], noble-gas elements [16] and 3B-6B [17–19] elements can be incorporated into fullerene cages, while alkali, alkali-earth and transitional metals will not remain inside the fullerene molecule. In order to check the possibility of direct incorporation, Ohno et al. [20-22] carried out *ab initio* molecular-dynamics (MD) simulation using our all-electron mixed-basis code. We found the experimental results are consistent with the results of MD simulations for several elements in the periodic table. Here, we report on the experimental framework and our accumulated results including newly obtained results such as Po, etc.

EXPERIMENTAL PROCEDURE

To produce a source of radioactive nuclides, several materials, e.g., Li_2CO_3, Ga_2O_3, etc. as listed in Table I, were used in powder form. The grain size of the materials was below 300 mesh. Purified fullerenes (C_{60}) were well-mixed with each material (at a weight ratio of 1:1) in an agate mortar, adding a few ml of carbon disulfide (CS_2). After drying, about 10 mg of the mixture was wrapped in pure aluminum foil of 10 μm in thickness for irradiation.

CHARGED-PARTICLE AND HIGH-ENERGY "BREMSSTRAHLUNG" IRRADIATION

According to the source nuclide used, either charged-particle or high-energy "bremsstrahlung" irradiation was used. In Table 1, the radionuclide produced, its characteristic γ-ray, half-life, and nuclear reaction are listed for each target material. Proton and deuteron irradiation with a beam energy of 16 MeV was performed at the Cyclotron Radio-Isotope Center at Tohoku University. The beam current was typically 5 μA and the irradiation time was about 1–2 hours. The samples were cooled with He-gas during irradiation. For production of a doping nuclide such as ^{11}C, ^{13}N, ^{22}Na, ^{47}Ca or ^{57}Ni, ^{120}Sb, a sample was set in a quartz tube and irradiated with "bremsstrahlung" of E_{max} = 50 MeV that originated from the bombardment of a Pt plate of 2 mm in thickness with an electron beam provided by a 300 MeV electron linac at the Laboratory of Nuclear Science at Tohoku University. The beam current was typically 120 μA and the irradiation time was about 6–8 hours. The samples were cooled in a water bath during irradiation. After irradiation in both cases (charged-particle and "bremsstrahlung" irradiation), the samples were left for one day to allow several of the short-lived radioactive by-products to decay.

Table 1. Nuclear data and experimental condition for the radioactive fullerenes.

Nuclide produced	γ–ray (α-ray)	Half-life ($T_{1/2}$)	Reaction	Material
^{7}Be	478 keV	53 d	^{7}Li(p,n)^{7}Be	C_{60}+Li_2CO_3
^{11}C	511 keV	20 m	^{12}C(γ,n)^{11}C	C_{60}
^{13}N	511 keV	10 m	^{12}C(d,n)^{13}N	C_{60}
^{69}Ge	511 keV	39.6 h	^{69}Ga(d,2n)^{69}Ge	C_{60}+Ga_2O_3
^{72}As	511 keV	26 h	^{72}Ge(d,2n)^{72}As	C_{60}+GeO
^{75}Se	136 keV	120 d	^{75}As(d,2n)^{75}Se	C_{60}+As_2S_3
^{79}Kr	261 keV	34.9 h	^{79}Br(d,2n)^{79}Kr	C_{60}+KBr
^{120}Sb	197 keV	5.76 d	^{121}Sb(γ,n)^{120}Sb	C_{60}+Sb_2O_3
^{121}Te	573 keV	16.8 d	^{121}Sb(d,2n)^{121}Te	C_{60}+Sb_2O_3
^{127}Xe	203 keV	34.6 h	^{127}I(d,2n)^{127}Xe	C_{60}+KI
^{210}Po	5.3 MeV(α)	138 d	^{209}Bi(d,n)^{210}Po	C_{60}+Bi_2O_3

CHEMICAL SEPARATION AND RADIOACTIVITY MESUREMENTS

The sample was dissolved in o-dichlorobenzene and filtered through a 0.2-millipore filter (0.45 μm) to remove insoluble material. The soluble portion was injected into an HPLC device equipped with a 5PBB (Cosmosil) (silica-bonded with a pentabromobenzyl group) column of 10 mm (in inner diameter) and 250 mm (in length) at a flow rate of 3 or 4 ml/min in each run. For confirmation of the fullerenes and their derivatives, a UV detector was installed with a wavelength of 290 nm. Downstream of the UV detector, two γ-ray detectors consisting of a bismuth-germanate scintillator and a photomultiplier (BGO-PM) were used to count the 511-keV annihilation γ-rays emanating from ^{11}C, ^{13}N, ^{69}Ge and ^{72}As in coincidence. Data from the radio-chromatogram were accumulated by means of a multichannel scaler system (MCS), using a personal computer. In order to measure the γ-rays emanating from other radionuclides from ^{7}Be, ^{75}Se, ^{79}Kr, ^{120}Sb, ^{121}Te, ^{127}Xe, eluent fractions were collected for 30 sec intervals (0–30, 30–60, 60–90,...sec), and the γ-rays activities of each fraction were measured with a Ge detector coupled to a 4096-channel pulse-height analyzer, the conversion gain of which was set at 0.5 keV per channel. The energy resolution of the Ge detector was 1.8 keV in FWHM with a photopeak of 1332 keV for the ^{60}Co source. In the case of ^{210}Po, α-spectrometry was used to detect the α-ray (5.3 MeV). In this way, each nuclide listed in Table 1 could be uniquely detected by means of its characteristic γ-rays (or α-ray), and any other sources could be ruled out. A schematic view of the radiochromatograph system is shown in Figure 1.

RESULTS AND DISCUSSION

Panels (a)~(f) of Figure 2 show the radiochromatogram for several materials mixed with the C_{60} samples, respectively, measured with the UV- and Ge-detector. The horizontal and vertical axes respectively indicate the retention time after injection into the HPLC and the absorption intensity of the UV as well as the α (or γ) counting rate of the radionuclide being produced. From a correlation of elution behavior between the UV chromatogram and the radioactivities of each objective atom, we found that several atom-doped fullerenes, namely $^{7}BeC_{60}$, $^{75}SeC_{60}$, $^{79}KrC_{60}$, $^{120}SbC_{60}$, $^{121}TeC_{60}$, $^{210}PoC_{60}$ can be produced by nuclear recoil implantation. Similar results were observed in the elution behavior between the UV detector and the ^{11}C, ^{13}N, ^{69}Ge, ^{72}As, ^{127}Xe populations. Formation of atom-doped fullerenes has also been investigated in several radioisotopes (Na, Ca, Sc, V, Cr, Mn, Co, Ni, Zn) using the same method. It was found that most of the elements (Na...Zn) could not be detected in the C_{60} portion although a small number of ^{48}V and ^{52}Mn atoms were retained within the C_{60} population.

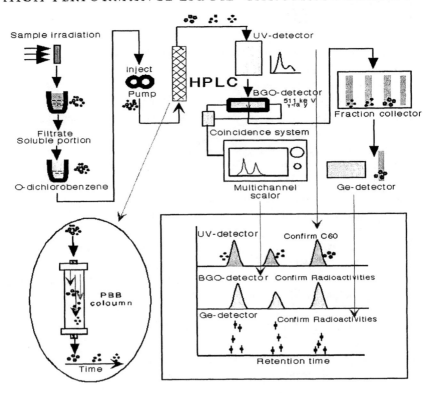

Figure 1. Schematic view of chemical separation and radioactivity measurement (radiochromatograph system). Dissolved fullerene was injected into the radio-chromatograph system (combined HPLC and radioactivity measurement).

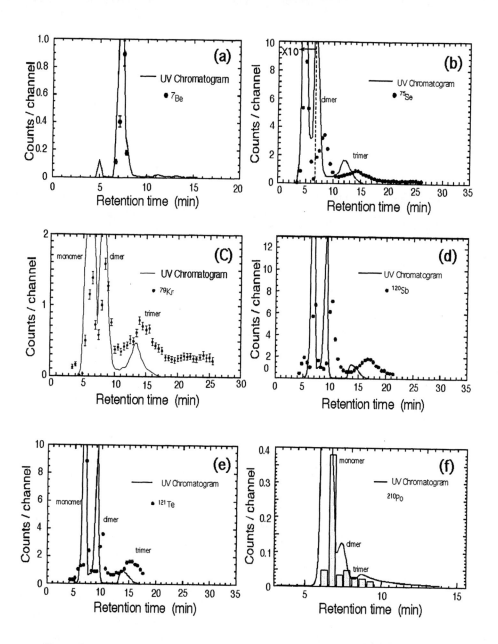

Figure 2. (a) HPLC elution curves of the soluble portion of the crude extracted in the proton irradiated sample of C_{60} that was mixed with Li_2CO_3. The horizontal axis indicates retention time, while the vertical axis represents counting rate of the radioactivity of ^7Be (closed circle), in each fraction as well as the absorbance in a UV chromatogram of C_{60} (solid line, arbitrary units). (b)–(f) same as (a), but for the deuteron or γ-irradiated sample of C_{60} mixed with As_2S_3 etc. Second and third peaks in (b)–(f) respectively indicate the formation of dimer and trimer.

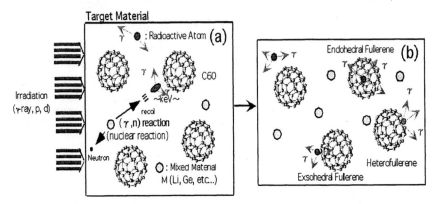

Figure 3. Schematic view of production mechanism for radioactive fullenenes. (a): Several radioactive nuclides are produced by (p, n), (d, n), (d, 2n) or (γ, n) reactions. Here, (γ, n) reaction (such as $^{12}C\,(\gamma, n)^{11}C$, $^{121}Sb\,(\gamma, n)^{120}Sb$) is used to show the production of radioactivities. (b): The kinetic energies of radionuclides are reduced in the sample to a magnitude that is appropriate for the fusion. Finally, the radionuclides hit the C_{60} cages and stop within the cage (formation of endohedral fullerene or heterofullerene).

Such clear correlation between the elution behavior of the C_{60} populations and the radionuclide populations indicates that the radioactive atoms (^{7}Be, ^{75}Se, ^{79}Kr, ^{120}Sb, ^{121}Te, ^{210}Po) in Figure 2 may be encapsulated inside the C_{60} cages, even though the ionic radius is larger than that of either a five- or six-membered carbon ring. Therefore, the first peaks in Figs. 1(a)-(f) can be respectively assigned to $^{7}Be@C_{60}$, $^{75}Se@C_{60}$, $^{79}Kr@C_{60}$, $^{120}Sb@C_{60}$, $^{121}Te@C_{60}$, and $^{210}Po@C_{60}$. It should be noted that no evidence for exohedral molecules has been presented so far by extraction of the soluble portions for the samples. Such molecules would be removed during solvation if they were exohedral. Therefore, the possibility should be considered in the present results that we have formed endohedral fullerene (or heterofullerene). The observation of the second and third peaks in Figs. 2(b)-(f) respectively corroborates formation of endohedral fullerene dimers and trimers with encapsulated radionuclides; e.g., in Figure 2(d) - $^{79}Kr@C_{60}$-C_{60} and $^{79}Kr@C_{60}$-$(C_{60})^2$. It seems that the shock of the collisions produces fullerene polymers by interacting with a neighbouring fullerene cage. The number of endohedral fullerenes (heterofullerenes) including such radioactivities is estimated to be about 10^{10} molecules.

In order to rationalize the experimental results, *ab initio* molecular dynamics (MD) simulations for the collision between several atom types and a C_{60} cage were carried out and the possibility of encapsulation was investigated. Ohno et al. have carried out simulations for Li, Be, K, Cu, As, Se, Kr, Xe insertions into C_{60} by an *ab initio* MD method based on the all-electron mixed-basis approach. Here, atomic orbitals (AO) and plane waves (PW) are used as the basis set within the framework of the local density approximation (LDA). The results of the simulation reveal that atoms such as Xe, Te, even in heavier elements, can go through the six-membered ring and form an endohedral fullerene without difficulty when the K.E. is in a certain range. This is certainly consistent with the experimental evidence shown in Figure 2.

There is a slight shift between the eluent peaks of the UV chromatogram and in the radioactive populations in Figure 2. That is, peaks in the objective radioactivities are shifted to later times from the peaks in the UV chromatogram (fullerene populations). The same trend was observed in the elution behavior of the metallofullerene extraction, such as ^{159}Gd@C_{82} [23]. Why does the delay occur in the retention time between the absorbance of the UV chromatogram and the γ-counting rate of the radioactivities? In these cases, it seems that the electron density of the endohedral fullerene (or heterofullerene) is distorted by the atom trapped in the cage. The distorted electrons may change the magnitude of interaction between the endohedral fullerene (or heterofullerene) and the resin inside the column.

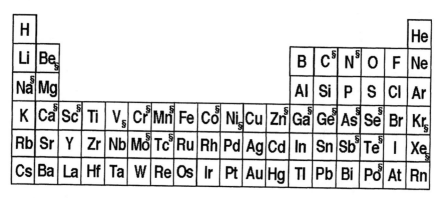

Figure 4. The Periodic Table. The symbol § indicates elements investigated in this work. Formation of foreign-atom-doped fullerenes (C_{60}) is possible with the elements shown in yellow (or shaded).

All the experimental results presented here support the following scenario: several radioactive nuclides are produced by (p, n), (d, n), (d, 2n) or (γ, n) reactions. The kinetic energies of the radionuclides are almost all of the same order of magnitude. As the form of the emitted neutron spectrum is approximately Maxwellian, the average neutron K.E. is about 2–3 MeV while the initial K.E. of the recoiled nuclides is estimated to be a few hundred keV (even if the reaction is accompanied by two neutron emissions). The energetic nuclides should destroy the fullerene cages because the K.E. is estimated to be of a different order of magnitude than the molecular bonding energies (eVs). Thus, the atoms escape from their own material due to the K.E. of a few hundred kiloelectron volts. Then, the kinetic energies are reduced in the sample to a magnitude that is appropriate for fusion. Finally, the radionuclides hit the C_{60} cages and stop within the cage (formation of endohedral fullerene or heterofullerene) and, the shock produces fullerene polymers by interaction with a neighbouring fullerene cage. A schematic representation of the production mechanism for atom-incorporated fullenene is shown in Figure 3.

In the present study, several elements in the Periodic Table have been investigated by recoil implantation following nuclear reaction. Elements shown by the symbol § in Figure 4 were investigated in this work. In the diagram, elements that were confirmed to combine with fullerene cages are shown in yellow (or shaded). It is interesting to note that group elements of small ion-radius (such as Be), groups 4B~6B and the noble gases are possible candidates to form complex materials such as atom-doping fullerenes more so than heavier elements such as the Po atom.

CONCLUSION

In this study, formation of atom-doped fullerenes has been investigated by using several types of radioactivity that are produced by nuclear reactions. It was found that noble-gas elements, such as Kr and Xe, as well as 3B-6B group elements, like C, N, Ge, As, Se, Sb, Te and Po remained in the C_{60} portion after a HPLC process. Most of the other elements, such as alkali, alkali-earth and transitional metals, could not be detected in the C_{60} portion. These facts suggest that the formation of endohedral fullerenes (or heterofullerenes) can be possible in 3B-6B group and noble-gas elements by a recoil process following nuclear reactions, while other elements (Na, Ca, Sc. etc.) may destroy most of fullerene cages due to strong chemical reactivity between atoms and fullerenes. By carrying out *ab initio* molecular-dynamics (MD) simulations on the basis of the all-electron mixed-basis approach, we confirmed that the insertion of several atoms into C_{60} through six-membered rings. These experimental and theoretical results indicate that chemical reactivity or inactivity seems to play an important role in the formation of foreign-atom-doped fullerenes.

ACKNOWLEDGEMENT

The authors are grateful to the technical staff of the Laboratory of Nuclear Science and the Cyclotron Radio-Isotope Center at Tohoku University for beam-handling, and to the technical staff working at IMR, Tohoku University for continuous support of the supercomputing facilities.

REFERENCES

[1] H. Kroto, J.R.Heath, S.C.Obrien, R.F. Curl, R.E. Smalley, 1985. C_{60} buckminsterfullerene, Nature, 318 (6042): 162–163.
[2] W. Kratschmer, L.D. Lamb, K. Fostiropoulos, D.R. Huffman, 1990. Solid C_{60} – a new form of carbon, Nature, 347, 354–358.
[3] Y. Chai, T. Guo, C. Jin, R.E. Haufler, L.P.F. Chibante, J. Fune, L. Wang, J.M. Alford, R.E. Smalley, 1991. Fullerenes with metals inside" J. Phys. Chem., 95, 7564–7568.
[4] R. D. Johnson, M.S. de Vries, J. Salem, D.S. Bethune and C. Yannoni, 1992. Electron-paramagnetic resonance studies of lanthanum-containing C_{82}, Nature (London), 355, 239–40.

[5] M. Takata, B. Umeda, E. Nishibori, M. Sakata, Y. Saito, M. Ohno, H. Shinohara, 1995. Confirmation by X-ray-diffraction of the endohedral nature of the metallofullerene Y-AT-C82, Nature, 377, 46–49.
[6] L.M. Roth, Y. Huang, J.T. Schwedker, C.J. Cassady, D. Ben-Amotz, B. Kahr, B.S. Freiser, 1991. Evidence for an externally bound Fe^+-buckminsterfullerene complex, FeC_{60}^+, in the gas-phase, J. Am. Chem. Soc., 113, 6298–6299.
[7] Y. Huang, B.S. Freiser., 1991. Synthesis of bis(buckminsterfullerene)nickel cation, $ni(C_{60})^{2+}$, in the gas-phase, J. Am. Chem. Soc., 113, 8186–8187.
[8] T. Guo, C. Jin and R.E. Smalley, 1991. Doping bucky – formation and properties of boron-doped buckminsterfullerene, J. Phys. Chem. 95, 4948–4950.
[9] H. Muhr, R. Nesper, B. Schnyder, R. Kotz, 1996. The boron heterofullerenes C59B and C69B: generation, extraction, mass spectrometric and XPS characterization, Chem. Phys. Lett. 249, 399–405.
[10] T. Pradeep, V. Vijayakrishnan, A.K. Santra and C.N.R. Rao, 1991. Interaction of nitrogen with fullerenes - nitrogen derivatives of C_{60} and C_{70}, J. Phys.Chem. 95, 10564–10565.
[11] M. Saunders, R. J. Cross, H. A. Jimenez-Vazquez, R. Shimshi and A. Khong, 1996. Noble Gas Atoms Inside Fullerenes, Science, 271, 1693–1697.
[12] T. Braun and H. Rausch, 1995. Endohedral incorporation of argon atoms into C_{60} by neutron irradiation, Chem. Phys. Lett. 237, 443–447.
[13] G. Gadd, P. Evans, D. Hurwood, P. Morgan, S. Moricca, N. Webb, J. Holmes, G. McOrist, T. Wall, M. Blackford, D. Cassidy, M. Elcombe, J. Norman, P. Johnson, P. Prasad, 1997. Endohedral fullerene formation via prompt gamma recoil, Chem. Phys. Lett., 279, 108–114.
[14] T. Ohtsuki, K. Masumoto, K. Ohno, Y. Maruyama, Y. Kawazoe, K. Sueki and K. Kikuchi, 1996. Insertion of Be Atoms in C_{60} Fullerene Cages: Be@C_{60}, Phys. Rev. Lett. 77, 3522–24.
[15] T.Ohtsuki, K.Masutomo, K.Sueki, K.Kobayashi, K.Kikuchi, 1995. Observation of radioactive fullerenes labeled with ^{11}C, J. Am. Chem. Soc.,117, 12869–12870.
[16] T. Ohtsuki, K. Ohno, K. Shiga, Y. Kawazoe, Y. Maruyama and K. Masumoto, 1988. Insertion of Xe and Kr Atoms in C_{60}, C_{70} Fullerenes and Formation of Dimers, Phys. Rev. Lett. 81, 967–970.
[17] T. Ohtsuki, K. Ohno, K. Shiga, Y. Kawazoe, Y. Maruyama and K. Masumoto, 1999. Formation of As- and Ge-doped heterofullerenes, Phys. Rev. B, 60, 1531–1534.
[18] T. Ohtsuki, K. Ohno, K. Shiga, Y. Kawazoe, H. Yuki, 2002. Se atom incorporation in fullerenes using nuclear recoil and *ab initio* molecular dynamics simulation, Phys. Rev. B, 65, 73402:1–5.
[19] T. Ohtsuki, K. Ohno, K. Shiga, Y. Kawazoe, Y. Maruyama, K. Shikano and K. Masumoto, 2001. Formation of Sb- and Te-doped fullerenes by using nuclear recoil and molecular dynamics simulations, Phys. Rev. B, 64, 125402:1–5.
[20] K. Ohno, Y. Maruyama, K. Esfarjani, Y. Kawazoe, N. Sato, R. Hatakeyama, T. Hirata, M. Niwano, 1996. All-electron mixed-basis molecular dynamics

simulations for collision between C_{60}^- and alkali-metal ions: the possibility of Li@C_{60}, Phy. Rev. Lett., 76, 3590–93.

[21] K. Shiga, K. Ohno, T. Ohtsuki and Y. Kawazoe, 2001. Formation of N-doped C_{60} studied by *Ab Initio* Molecular Dynamics Simulations, Mater. Trans. 42, 2189–2193.

[22] K. Shiga, K. Ohno, Y. Kawazoe, Y. Maruyama, T. Hirata, R. Hatakeyama and N. Sato, 2000. *Ab initio* molecular dynamics simulation for the insertion process of Si and Ca atoms into C_{74}, Mater. Sci. & Eng., A290, 6–10.

[23] K. Kikuchi, K. Kobayashi, K. Sueki, S. Suzuki, H. Nakahara, Y. Achiba, K. Tomura, M. Katada, 1994. Encapsulation of radioactive Gd-159 and Tb-161 atoms in Fullerene cages, J. Am. Chem. Soc. 116, 9775–9776.

Chapter 12: Nano-Machines and Biological Nanotechnology

Engineering with the Engines of Creation

C. D. MONTEMAGNO and J. J. SCHMIDT
Department of Bioengineering, University of California, Los Angeles, CA

ABSTRACT

Advances in single molecule manipulation coupled with a more complete understanding of biomolecular processes have enabled development of a new class of materials. These materials have the potential to emulate much of the functionality associated with living systems such as the active transport and transformation of matter and information and the transduction of energy into different forms. We will present the details of the technological demands and the results of efforts associated with the production of functional materials. Elements of the discussion will include the genetic engineering of active biological molecules into engineering building blocks, the precision assembly of these molecules into a stable, "active" material and, the potential of embedding intelligent behavior into the matrix of the assembled matter.

INTRODUCTION

The rapid pace of miniaturization in the semiconductor industry has resulted in ever faster and more powerful computing and instrumentation that have begun to revolutionize medical diagnosis and therapy. The lithographic techniques used to fashion electronic circuits in microprocessors and memory chips can also produce mechanical structures, commonly known as micro- and nanoelectromechanical systems (MEMS and NEMS). MEMS devices currently range in size from one to hundreds of micrometers (10^{-6} m), and can be as simple as the singly supported cantilever beams used in atomic force microscopy, or as complicated as a video projector with thousands of electronically controllable microscopic mirrors. NEMS devices exist correspondingly in the nanometer realm (10^{-9} m). The concept of using externally controllable MEMS devices to measure and manipulate biological matter (BioMEMS) on the cellular and subcellular levels has attracted much attention recently, as initial work has shown the ability to detect single base pair mismatches of DNA using cantilever systems [1], as well as the ability to control the grabbing and manipulation of individual cells and subsequently release them unharmed. (http://www.sandia.gov/media/NewsRel/NR2001/gobbler.htm).

The attraction of using ever smaller systems to analyze and alter biological systems *in vivo* and *in vitro* is evident: smaller systems are less invasive, require smaller amounts of analytes (and therefore are more sensitive), and have smaller volumes over which they are effective—creating the ability to localize diagnosis and therapy, confining them only to targeted regions. Although current technology has the limited ability to detect femtomolar concentrations of analytes from microliter volumes of solution in the laboratory, the grand vision is that of implantable or even autonomous mobile systems operating *in vivo* which are able to detect the naturally occurring minute quantities of biochemicals in their native environments and respond through signaling to external instruments, dispensation of drugs, or direct physical manipulation of their surroundings.

Of course, since BioMEMS involves the interface of MEMS with biological environments, the biological components are critically important. To date, they have mostly been nucleic acids, antibodies, and receptors involved in passive aspects of detection and measurement. These molecules are chemically attached to the surfaces of MEMS structures while retaining biological activity and their interactions are monitored through mechanical, electrical, or optical measurements. Because the biological components are nanometer size or smaller, the size of these systems is limited by the minimum feature sizes achievable using the fabrication techniques of the inorganic structures, currently under 1 micrometer. Commercially available products resulting from further miniaturization may be problematic due to the expanding cost and complexity of optical lithography equipment and the inherent slowness of electron beam techniques. In addition to size limitations, the effects of friction, which can excessively wear or seize moving parts, have plagued multiple moving parts in inorganic MEMS, limiting device speeds and useful lifetimes. Finally, MEMS are currently powered exclusively by electricity; an electrical power supply complicates any autonomous mobile system.

MECHANICAL-BIOLOGICAL PROTEIN DEVECES

Because of these constraints together with sheer scientific curiosity, recent discoveries of mechanical biological proteins have led to first generation proof-of-concept biological inorganic hybrid devices. These devices have relevant length scales under 50 nm, are powered by optical or chemical energy, and are capable of operation *in vivo*. Depicted in Figure 1 is a hybrid device made by our group using the rotary motor F_1-ATPase attached on a patterned inorganic substrate capable of rotating a micromachined nickel rod attached to its rotor [2].

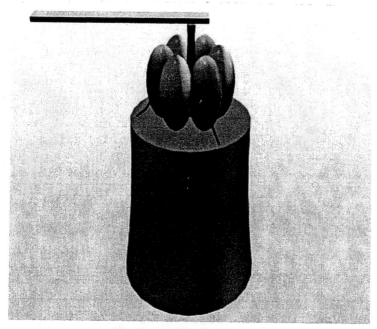

Figure 1. A depiction of an assembled hybrid device powered by F_1-ATPase. The motors were mounted on Ni capped SiO_2 posts defined by electron beam lithography. Ni rods (75 x 15 x 750 nm), also made using electron beam lithography, were attached to the rotor of the each motor. Rotary motion of the rods was observed using conventional bright field optical microscopy.

F_1-ATPase is a multiunit enzymatic protein under 12 nm in diameter, with a 3 nm rotor that rotates as the enzyme consumes the universal biological fuel, adenosine triphosphate (ATP). The nickel rods attached to these motors, made using electron beam lithography and standard material deposition and etching techniques, were 75 nm x 15 nm x 750 nm in size—their large length necessary to facilitate observation of the rotation using optical microscopy. Even with their large size, the rods were observed to rotate at approximately 8 rotations per second. Work by Yoshida *et al.* have suggested a high efficiency for F1-ATPase (> 80%) and have directly shown that the motors can rotate over 130 rotations per second when minimally loaded [3,4]. Other biological molecular motor systems have also been engineered. For example, kinesin, an ATP powered linear stepping motor that unidirectionally "walks" hand over hand on tracks of 25 nm wide microtubules, has natural functions which include intracellular cargo transport, ciliary movement, and cellular architectural modification. Microtubules and kinesin have both been deposited in patterns, and the resulting directed transport of microtubules or other cargo by the kinesin has been achieved in a limited manner [5,6].

Devices using motor proteins are not limited by the choice of molecules- the palette of motors and mechanical proteins is huge. Rotary and linear motors, valves, and screws can be found and even the ubiquitous ligand-receptor binding induces an electronically detectable hinging motion [7]. In summary, these mechanical molecules are tens of nanometers in size, can operate at high speeds (some at over 50 micrometers/sec), and are motile by their very nature, using chemical fuel and interacting primarily with their intended substrates. In addition, there are hundreds of different mechanical proteins in each organism, and thousands among the various species of plants and animals, and are even found in the simplest prokaryote. Each motor protein in each species of organism has been adapted to certain specific tasks over millions of years of evolution. In device design, specific mechanical proteins may be optimal for certain externally directed artificial tasks, using speed, power, or processivity of a particular motor. An incredible economic advantage in device production is realized in the synthesis of these proteins. The genes encoding these proteins can be inserted into bacterial expression systems using standard genetic engineering techniques and made in large quantities in relatively short periods of time: a typical production run in our laboratory produces 1016 motors in three days from start to finish, over 1014 motors produced per hour. These are systems which are ripe for adaptation in hybrid devices: organic and inorganic components combined in a system capable of performing tasks impossible by either alone.

Because of the small sizes of proteins, devices that can operate inside a cell are possible. Ultimately, it will be possible for micrometer-sized or smaller devices to be injected into the human body with specific intended targets (*e.g.*, cancer cells). As these devices reach their targets, a signal is released (such as that resulting from antibody-antigen or ligand-receptor binding), which then activates the device to release its contents to the environment surrounding the target or to inject its contents into the target itself (see Figure 2). The contents of the devices can be chemicals, drugs, or still smaller devices. These devices, once released, then interact with their new environments, inhibiting deleterious activities or restoring lost functions. Does this sound like fantasy? It is perhaps as little as a decade away. Many of the parts of this theoretical device now exist. As feature sizes of transistors fall below 100 nm, the number of transistors and other semiconductor circuitry able to be packed inside of a microscopic robot is increasing to the point of feasibility of an onboard computer. In addition, as circuit sizes shrink, their power requirements fall, allowing for smaller and lighter electrical power sources. Alternatively, these services may be entirely biological in nature—the motility provided by motor proteins, and computational power resulting as an emergent property [8] of many individual units acting cooperatively. Injection of drugs or devices into cells by these devices may be performed using syringes and needles of nanometer sharpness, easily made with semiconductor fabrication techniques.

Before such fantastic devices are constructed, control of synthesis, manufacture, and operation of individual system elements must be enhanced and refined. In our previous work, the number of functioning hybrid devices produced was limited by quantities of inorganic components. Future work focuses on fabricating pieces in amounts commensurate with their organic counterparts. This will enable the production of micromole amounts and more of hybrid nanomachines. Other efforts are directed toward the control of the nanomachines. Biological motors relying solely on chemical fuel such as ATP for power also have an additional shortcoming—control of catalytic activity. In their most primitive form, these

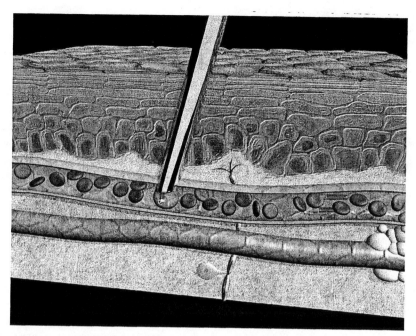

Figure 2. "Mechanical Drugs"—in this artist's rendering, robotic capsules are released in the bloodstream. Upon the capsules' rendezvous with their targets, their contents are discharged into the surrounding medium or the targets themselves.

motors will hydrolyze ATP until the supply is exhausted. Of course, nature is not so wasteful that she would let these motors consume ATP unnecessarily. Many proteins have natural built-in regulatory mechanisms controlling their activity. Examples of these in motor proteins include the δ and ϵ subunits in F_0F_1-ATPase [9] as well as the phosphorylation of the tail domain in Myosin-V [10]. In some applications, these natural controls will not be applicable or available. Our group has recently completed work in which an artificial control has been introduced into F_1-ATPase. Metal ion binding sites genetically engineered into the protein control its motion by forcing different moving parts to bind to an ion in solution, immobilizing it. Introduction of a chelator, such as 1,10-phenanthroline, removes the bound ion from the motor, restoring motility. This can be repeated many times, demonstrating repeatable control over these hybrid machines at the level of a single molecule [11].

BIOMIMETIC MATERIALS

Near term successes in integrating proteins into hybrid devices may result from membrane protein engineering. By substituting biomimetic polymers for the lipid environment, natural self-assembly processes can result in hybrid materials functionalized by membrane proteins. Use of polymer membranes is desirable since: they have a longer lifetime than lipid membranes, they are more rugged, and properties such as electronic and ionic conductivity and permeability, can be tailored to suit each application. The interior of these membranes must be hydrophobic and elastic so the natural protein environment is simulated as close as possible. A large number of materials demonstrate biocompatibility and incorporate proteins while maintaining their functionality, such as gelatin, hydrated polymers, sol-gels, poly(vinyl alcohol), and poly(acrylamide) [12]. For devices in which the protein environment must replicate the two dimensional nature of natural membranes, amphiphilic polymers can used, such as the triblock copolymer poly(2-methyloxazoline)-poly(dimethylsiloxane)-poly(2-methyloxazoline) (PMOXA-PDMS-PMOXA). The PMOXA blocks are hydrophilic, while the PDMS block is hydrophobic. The copolymer spontaneously forms a rich variety of phases from laminar to vesicular, similar to natural lipid systems. The porin OmpF has been shown to retain its natural functionality in this polymer [13].

CONCLUSIONS

Bionanotechnology is a nascent field, the locus of physics, chemistry, biology, and materials science at the nanoscale. Up to now, most of the engineering at these length scales has been non-biological inorganic micro- and nanomachined pieces, often studied in high vacuum or low temperatures. Using the machinery of life, Nature has created highly efficient manufacturing systems capable of producing exquisite machines nanometers in size that can operate at room temperature in aqueous environments. Our current understanding of these processes and the products made by them is not yet to the point where we are able to fully design or customize any but the simplest proteins; however, our knowledge is rapidly improving and integration of these biological molecules in crucial functional roles in NEMS devices has already begun. Advances in technology are quickening the pace of discoveries emerging from academic and industrial laboratories, which in turn lead to more technical innovation. This feedback loop will result in future devices and techniques presently unimaginable: autonomous mobile robots *in vivo* that sense and respond to their environments and dispense drugs that are mechanical in nature.

REFERENCES

[1] J. Fritz, M. Baller, H. Lang, H. Rothuizen, P. Vettiger, E. Meyer, H. Guntherodt, C. Gerber, and J. Gimzewski, 2000. Science 288, 316–318.
[2] R.K. Soong, G.D. Bachand, H.P. Neves, A.G. Olkhovets, H.G. Craighead, and C.D. Montemagno, 2000. Science 290, 1555–1558.
[3] R. Yasuda, H, Noji, K. Kinosita, and M. Yoshida, 1998. Cell 93, 1117–1124.
[4] R. Yasuda, H. Noji, M. Yoshida, K. Kinosita, H. Itoh, 2001. Nature 410, 898–904.
[5] H. Hess, J. Clemmens, D. Qin, J. Howard, V. Vogel, 2001. Nanoletters 1, 235–239.
[6] L. Limberis, & R.J. Stewart, 2000. Nanotechnology 11, 47–51 (2000).

[7] D. Benson, D. Conrad, R. de Lorimer, S. Trammell, H. Hellinga, 2001. Science 293, 1641–44.
[8] T. Nakagaki, H. Yamada, A. Toth, 2000. Nature 407, 470.
[9] S.P. Tsunoda, A.J. Rodgers, R. Aggeler, M.C. Wilce, M. Yoshida, R.A. Capaldi, 2001. Proc. of the National Academy of Sciences of the United States of America 98, 6560–6564.
[10] R.L. Karcher, J.T. Roland, F. Zappacosta, M.J. Huddleston, R.S. Annan, S.A. Carr, V.I. Gelfand, 2001. Science 293, 1317–1320.
[11] H. Liu, J.J. Schmidt, G.D. Bachand, S.S. Rizk, L.L. Looger, H.W. Hellinga, C.D. Montemagno, 2002. Nature Materials 1, 173–177.
[12] R. Birge, N. Gillespie, E. Izaguirre, A. Kusnetzow, A. Lawrence, D. Singh, W. Song, E. Schmidt, J. Stuart, S. Seetharaman, K. Wise, 1999. J. Phys. Chem. B 103, 10746–10766.
[13] C. Nardin, J. Widmer, M. Winterhalter, W. Meier, 2001. Eur. Phys. J. E4, 403–410.

Calculation of the Characteristics of a Molecular Single-Electron Transistor with Discrete Energy Spectrum

V. V. SHOROKHOV and E. S. SOLDATOV
Faculty of Physics, Moscow State University, Moscow, Russia

ABSTRACT

The main aim of this work is to calculate the I-V and control curves for a molecular single-electron transistor with a discrete energy spectrum. An analytical solution to the kinetic equations is obtained for the case of either fast or slow electron energy relaxation for a system with a discrete energy spectrum containing an arbitrary number of non-degenerated energy levels. The analytical expressions for the I-V curves and control curves are obtained. The effective recursive method for fast calculation of the Gibbs canonical distribution of electrons, as well as a method for fast calculation of the distribution function for slow relaxation are elaborated. The calculation of the tunnel characteristics of molecular SET with discrete energy spectrum was carried out for the case of slow relaxation and discrete equidistant energy spectrum at low temperature ($T=0$).

INTRODUCTION

Investigation of electron tunneling transport in devices based on single molecules is attracting heightened interest. Special attention must be paid to the study of electron tunneling transport in devices with characteristic size less than 2 nm since for such molecules, the effects of dimensional and charge quantization play significant roles. Therefore, during the characteristic calculation of such devices it is necessary to take into consideration the influence of quantum effects. In this paper, we consider the Coulomb effects in such quantum systems.

All approaches to consideration of such systems can be conditionally divided into two classes. The first method is based on statistical theory. This approach was previously described in [1, 2] for systems on the base of small metallic granules and quantum dots. The second approach is to calculate by microscopic self-consistent [3] and quantum chemistry methods. A detailed review of these methods can be found in [4]. Both techniques supplement each other to create a full picture of electron transport in the devices based on the single molecules. In this work we follow the first approach.

The goal of this work was to obtain an analytical solution for the system of the kinetic equations for cases of fast and slow relaxation of the electrons - their comparison and numerical calculation of characteristics of single-electron transistors on the base of the single molecule for different types of energy spectrum.

First, we consider a model of molecular single-electron transistors and the system of kinetic equations for such structures which describe the tunneling processes in the cases of slow and fast energy relaxation of electrons in the molecule. Secondly we describe the recursion solution of these kinetic equations for the case of slow relaxation of the electrons. We also describe a recursive equation to calculate Gibbs canonical distribution [5] which is necessary to solve the system of kinetic equations for the case of fast relaxation. Thirdly, we consider a solution of the system of kinetic equations for the case of slow relaxation. Finally, we present the results of numerical simulation and their discussion.

MANI EQUATIONS

We consider a system to consist of two flat electrodes on the dielectric substrate with a small molecule between them. The cluster molecules with metallic core and organic shell (number of atoms ~ a few dozen) are considered as the molecular part of the system. Examples of such molecules can be found in [6].

We suggest that electrons in the system can transport from the electrodes to the molecule and back only by tunneling. As well, we suppose that conditions for small thermal fluctuations and resonant width in comparison to Coulomb energy and mean level spacing are fulfilled:

$$<\Delta\varepsilon>|_{\varepsilon=\varepsilon_{HOMO}} \sim e^2/C_\Sigma \gg h\Gamma, kT \qquad (1)$$

where $<\Delta\varepsilon>|_{\varepsilon=\varepsilon_{HOMO}}$ - average characteristic spacing between energy levels of the molecule;

e^2/C_Σ - characteristic Coulomb system energy determined by capacitance of the system

C_Σ; $h\Gamma$ - characteristic resonant tunneling width which is determined by parameters of tunnel barriers between the molecule and electrodes [5];

kT - the thermostat temperature in energy units.

In Figure 1 the 1D model of the considered system is shown.

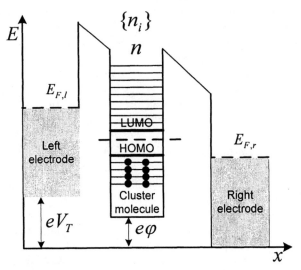

Figure 1 1D energetic model of the single-electron transistor on the base of the single molecule.

We suggest that energy spectrum $\{\varepsilon_i\}$ of the levels for considered molecule is known, $i = 1...N$, where N is number of energy levels. (The calculations of energy spectra one can find for example in [7]). The structure of the spectra is a variable parameter. For the convenience we will number the energy levels in order of increasing energy.

In this work we have used an approach of constant level spacing at creation of potential φ in the point of molecule localization. Such potential arises when tunneling voltage is applied to the electrodes:

$$\varphi = V_l = \frac{-(ne - Q_g) + \eta V_T}{C_\Sigma} \qquad (2)$$

In Eq. 2, n is the number of additional electrons in the molecule, this number can take only integer values; e is the electron charge; Q_i is additional induced charge on the molecule (this charge can be fraction of the electron charge and arises as a result of control voltage influence or as a result of the charge redistribution in the considered system); η is a coefficient which determines a relation of tunnel voltage V_T drops[1] on two tunnel barriers.

We have to use a probability distribution function $P(\{n_i\}, t)$ to describe the processes in the system. Such a function describes the probability to find the system in one of the state $\{n_i\}$

---

[1] In general, this coefficient can be determined using the geometry of the tunneling barriers between molecule and electrodes.

$$\sum_{\{n_i\}} P(\{n_i\}, t) = 1 \qquad (3)$$

where $n_p = 0,1$ is an occupation number of the p-th valent energy level by electron.

One can use the Fermi Golden rule to write the system of kinetic equations which describes the tunneling and relaxation processes in the system:

$$\frac{\partial}{\partial t} P(\{n_i\}, t) = -\sum_p P(\{n_i\}, t) \delta_{n_p,0} \left[\Gamma_p^l f\left(E^{init,l}(n) - E_F\right) + \Gamma_p^r f\left(E^{init,r}(n) - E_F\right) \right]$$

$$- \sum_p P(\{n_i\}, t) \delta_{n_p,1} \left\{ \Gamma_p^l \left[1 - f\left(E^{final,l}(n) - E_F\right)\right] + \Gamma_p^r \left[1 - f\left(E^{final,r}(n) - E_F\right)\right] \right\}$$

$$+ \sum_p P(n_1, \ldots, n_{p-1}, 1, n_{p+1} \ldots, t) \delta_{n_p,0} \left\{ \Gamma_p^l \left[1 - f\left(E^{final,l}(n+1) - E_F\right)\right] + \Gamma_p^r \left[1 - f\left(E^{final,r}(n+1) - E_F\right)\right] \right\}$$

$$+ \sum_p P(n_1, \ldots, n_{p-1}, 0, n_{p+1} \ldots, t) \delta_{n_p,1} \left[\Gamma_p^l f\left(E^{init,l}(n-1) - E_F\right) + \Gamma_p^r f\left(E^{init,r}(n-1) - E_F\right) \right]$$

$$+ R(\{n_i\}, t) \qquad (4)$$

where

$$E^{init,\binom{l}{r}}(n) = E_p + U(n+1) - U(n) + \begin{pmatrix} \eta \\ -(1-\eta) \end{pmatrix} eV \qquad (5)$$

is the electron energy in the left or right electrode before a tunneling event in the case of tunneling from electrode to molecule.

$$E^{final,\binom{l}{r}}(n) = E_p + U(n) - U(n-1) + \begin{pmatrix} \eta \\ -(1-\eta) \end{pmatrix} eV \qquad (6)$$

is the electron energy in the left or right electrode after a tunneling event in the case of tunneling from molecule to electrode.

where $U(n)$ is electrostatic energy of the system if n additional electrons are localized on the molecule [8]; $f(\varepsilon)$ is Fermi distribution function; $\delta_{n_p,0(1)}$ is the Kroneker symbol; E_F is the electrode Fermi energy; $\Gamma_p^{l(r)}$ - tunneling conductivity of left (right) barrier on the energy level p. The last expression in Eq. 4, $R(\{n_i\}, t)$, describes the excitation and relaxation of the electrons in the molecule[2]. The form of Eq. 4 was suggested in [1]. A similar system of equations was presented also in [2].

---

[2] The part which describes the relaxation processes can be expressed as

The equation for tunnel current flowing through the system can be written from the solution of Eq. 4 for the distribution function $P(\{n_i\})$:

$$I = -e \sum_{p=1}^{\infty} \sum_{\{n_i\}} \Gamma_p^l P(\{n_i\},t) \{\delta_{n_p,0} f(E^{init,l}(n) - E_F) - \delta_{n_p,1}[1 - f(E^{final,l}(n) - E_F)]\}. \tag{7}$$

The analytical solution of Eq. 4 is not possible for the general case. Therefore it is necessary to use certain boundary condition approaches to obtain a particular analytical solution to this system of kinetic equations.

First of all, we consider only stationary processes in this work, and so:

$$\frac{\partial}{\partial t} P(\{n_i\},t) = 0. \tag{8}$$

Let us try to simplify the distribution function $P(\{n_i\})$. For it let us consider a case when probability function $P(\{n_i\})$ of finding system in the state $\{n_i\}$ does not depend on the structure of occupation numbers configuration $\{n_i\}$ but depend only on the number of additional electrons on the molecule n, $P(\{n'_i\}) = P(\{n''_i\})$, where $\sum_p n'_p = \sum_p n''_p = n$. In this case one can produce the analytical solution of the system (4). In such approach the probability to find an electron on the level p can be described as:

$$P(\{n_i\})\delta_{n_p,1} = \sigma(n) F(p|n)/C_N^n \quad (P(\{n_i\})\delta_{n_p,0} = \sigma(n)(1 - F(p|n))/C_N^{N-n}), \tag{9}$$

where $\sigma(n)$ is the molecule charge state distribution function. We define the "charge state" every state $\{n_i\}$ with n additional electrons. $F(p|n)$ is a one-particle distribution function of the electrons that defines a conditional probability of finding an electron in the molecule on level p in the case of n additional electrons. The normalization conditions of the charge distribution function and partial distribution function can be written as:

$$R(\{n_i\},t) = \sum_{p,\{m_j\}} [P(\{m_j\},t)\delta_{m_p=0} P(\{n_i\},t)\delta_{n_p=1} + P(\{m_j\},t)\delta_{m_p=1} P(\{n_i\},t)\delta_{n_p=0}] \times [r(\{m_j\},\{n_i\},t) - r(\{n_i\},\{m_j\},t)]$$

where $r(\{m_j\},\{n_i\},t)$ describes the probability of changing of molecule state per time unit via relaxation or excitation of one or several electrons in the molecule.

$$\sum_p F(p|n) = n, \quad F(p|0) = 0$$

$$\sum_{l=0}^{N} \sigma(l) = 1 \tag{10}$$

One can write the equation for tunnel current based on Eq. 9:

$$I = -e \sum_{n=0}^{N} \sum_p \Gamma_p^l \sigma(n) \begin{cases} (1 - F(p|n)) f(E^{init,l}(n) - E_F)/C_N^{N-n} \\ - F(p|n)[1 - f(E^{final,l}(n) - E_F)]/C_N^n \end{cases} \tag{11}$$

It is possible to use this suggested approach in 3 cases:

- transport of the electrons through degenerated level in the molecule;
- fast relaxation in the molecule;
- slow relaxation of the electrons in the molecule.

The case of tunneling through degenerated energy level is well studied in [9]. In this case, the relaxation processes in the molecule are absent and so, $R(\{n_i\},t) = 0$. The partial distribution function can be written as $F(p|n) = n/N$.

In the case of fast relaxation the electron distribution function has sufficient time to relax to the equilibrium form between two sequential tunneling events. We consider the system to be in equilibrium at all times since the relaxation time is much smaller than the average time between separate tunneling events. One can use the Gibbs canonical distribution to describe the partial function in this case [5]:

$$F_{eq}(p|n) = \frac{\sum_{\{n_p=1,n\}} \exp\left(-\beta \sum_{k=1}^{N} n_k \varepsilon_k\right)}{\sum_{\{n\}} \exp\left(-\beta \sum_{k=1}^{N} n_k \varepsilon_k\right)}. \tag{12}$$

We have previously suggested a method for fast recursive calculation of Gibbs canonical distribution [10]:

$$F_{eq}(p|n) = \frac{n \exp(-\beta \varepsilon_p) \cdot (1 - F_{eq}(p|n-1))}{\sum_i \exp(-\beta \varepsilon_i) \cdot (1 - F_{eq}(i|n-1))} \tag{13}$$

Estimation of the characteristic relaxation time of electrons [10] in the cluster molecules with several dozens atoms shows that in such systems, the average time between separate tunneling events is much smaller than relaxation to the equilibrium state. This estimation allows the electron interaction part of the

Schrödinger equation to be disregarded. In our opinion, slow relaxation is the most probable regime of processes in the molecule. Therefore we place emphasis in this work to the study of stationary kinetic processes in a regime of slow relaxation. It is more convenient to consider a limit case of slow relaxation as extremely slow relaxation ($t_{rel} \to \infty$).

In the case of extremely slow relaxation the charge configuration $\{n_i\}$ remains unchanged between the separate tunneling events. The time scale on which the interactions between electrons in the molecule take place is much greater than average time between tunneling events so the profile of the partial distribution function $F(p|n)$ depends only on external factors to the molecule such as applied tunnel voltage, tunnel barrier parameters, electrode temperatures, etc. Inasmuch as the system remains unchanged between separate tunneling events, the last part of Eq. 4 can be written as:

$$R(\{n_i\},t) = 0 \tag{14}$$

Using all of the above for each of the three approaches, one can write the system of kinetic equations (i.e., Eq. 4) in simplified form as:

$$0 = -\sigma(n)\sum_p (1 - F(p|n))\left[\Gamma_p^l f(E^{init,l}(n) - E_F) + \Gamma_p^r f(E^{init,r}(n) - E_F)\right]/C_N^{N-n}$$

$$-\sigma(n)\sum_p F(p|n)\left\{\Gamma_p^l \left[1 - f(E^{final,l}(n) - E_F)\right] + \Gamma_p^r \left[1 - f(E^{final,r}(n) - E_F)\right]\right\}/C_N^n$$

$$+\sigma(n+1)\sum_p F(p|n+1)\left\{\Gamma_p^l \left[1 - f(E^{final,l}(n+1) - E_F)\right] + \Gamma_p^r \left[1 - f(E^{final,r}(n+1) - E_F)\right]\right\}/C_N^{n+1}$$

$$+\sigma(n-1)\sum_p (1 - F(p|n-1))\left[\Gamma_p^l f(E^{init,l}(n-1) - E_F) + \Gamma_p^r f(E^{init,r}(n-1) - E_F)\right]/C_N^{N-n+1} \tag{15}$$

In the case of tunneling through a degenerated energy level and in the case of extremely fast relaxation, the system described by Eq. 15 can be substituted by a system of kinetic balance equations:

$$\sigma(n)\sum_p F(p|n)\left\{\Gamma_p^l \left[1 - f(E^{final,l}(n) - E_F)\right] + \Gamma_p^r \left[1 - f(E^{final,r}(n) - E_F)\right]\right\}/C_N^n =$$

$$\sigma(n-1)\sum_p (1 - F(p|n-1))\left[\Gamma_p^l f(E^{init,l}(n-1) - E_F) + \Gamma_p^r f(E^{init,r}(n-1) - E_F)\right]/C_N^{N-n+1} \tag{16}$$

The solution of such equations can be written as:

$$\sigma(n) = \prod_{l=1}^{n} \left[\frac{(N-l+1)\sum_p (1-F(p|l-1))\left[\Gamma_p^l f(E^{init,l}(l-1)-E_F) + \Gamma_p^r f(E^{init,r}(l-1)-E_F)\right]}{l\sum_p F(p|l)\left\{\Gamma_p^l\left[1-f(E^{final,l}(l)-E_F)\right]+\Gamma_p^r\left[1-f(E^{final,r}(l)-E_F)\right]\right\}} \right] \Big/ Z \tag{17}$$

where the statistical sum Z can be expressed as:

$$Z = \sum_{k=0}^{N} \prod_{l=1}^{k} \left[\frac{(N-l+1)\sum_p (1-F(p|l-1))\left[\Gamma_p^l f(E^{init,l}(l-1)-E_F) + \Gamma_p^r f(E^{init,r}(l-1)-E_F)\right]}{l\sum_p F(p|l)\left\{\Gamma_p^l\left[1-f(E^{final,l}(l)-E_F)\right]+\Gamma_p^r\left[1-f(E^{final,r}(l)-E_F)\right]\right\}} \right] \tag{18}$$

In the case of extremely slow relaxation one can substitutes the system of equations (15) by the system of equations:

$$n\sigma(n)F(p|n)\left\{\Gamma_p^l\left[1-f(E^{final,l}(n)-E_F)\right]+\Gamma_p^r\left[1-f(E^{final,r}(n)-E_F)\right]\right\} =$$
$$(N-n+1)\sigma(n-1)(1-F(p|n-1))\left[\Gamma_p^l f(E^{init,l}(n-1)-E_F)+\Gamma_p^r f(E^{init,r}(n-1)-E_F)\right] \tag{19}$$

We have solved the system (19) in a recursive form. The solution in the case of extremely slow relaxation is:

$$F(p|n) = \frac{n \cdot K(p, n-1) \cdot (1-F(p|n-1))}{\sum_r K(r, n-1) \cdot (1-F(r|n-1))} \tag{20}$$

where the recursion function is

$$K(p|n) = \frac{\Gamma_p^l f(E^{init,l}(n-1)-E_F) + \Gamma_p^r f(E^{init,r}(n-1)-E_F)}{\Gamma_p^l\left[1-f(E^{final,l}(n)-E_F)\right]+\Gamma_p^r\left[1-f(E^{final,r}(n)-E_F)\right]} \tag{21}$$

Note that the structure of Eq. 20 is quite similar to that of Eq. 13, differing only by the recursion function.

Based on the solution to Eq. 20, one can write the recursive equation for the charge distribution function in the case of extremely slow relaxation:

$$\sigma(n) = \frac{1}{n} \sum_{l=0}^{n-1} (-1)^{n-l-1} \sigma(l) \sum_p \prod_{m=l}^{n-1} K(p|m) \tag{22}$$

LOW TEMPERATURE AND EXTREMELY SLOW RELAXATION APPROACH

The recurrent equations, Eq. 20 and Eq. 22, can be used for fast simulation of the characteristics of the considered system. However, these equations are not suitable for analytical consideration of the system. In a more direct form, the solutions can be obtained in a low-temperature ($kT \to 0$) approach. In this case the solution of the equation for the distribution functions $\sigma(n)$ and $F(p|n)$ will be different than «0» and «1» when the number of additional electrons on the molecule n and the level number p satisfies the following condition:

$$E_F - MAX(\eta eV, (\eta-1)eV) < E_p + U(n+1) - U(n) < E_F - MIN(\eta eV, (\eta-1)eV). \quad (23)$$

Let us also take into account the fact that the state with the lowest occupied levels will be the most energetically favorable when values of $\{p,n\}$ don't satisfy the inequality in Eq. 23. The additional conditions for $\{p,n\}$ can be obtained from this fact. Uniting these conditions with Eq. 23, one can write minimum (p_{min}, n_{min}) and maximum (p_{max}, n_{max}) values of $\{p,n\}$ for the tunneling events:

$$(p_{min}, n_{min}) = (MIN(p), MIN(n)), \text{ где}$$

$$\{p,n\}: \begin{cases} E_F - MAX(\eta eV, (\eta-1)eV) < E_p + U(n+1) - U(n) \\ \forall p > n, K(p|n)^{-1} > 0 \end{cases} \quad (24)$$

$$(p_{max}, n_{max}) = (MAX(p), MAX(n)), \text{ где}$$

$$\{p,n\}: \begin{cases} E_p + U(n+1) - U(n) < E_F - MIN(\eta eV, (\eta-1)eV) \\ \forall p > n, K(p|n) > 0 \end{cases}$$

For any value which satisfies $n < n_{min}$ or $n > n_{max}$, then $\sigma(n) = 0$. Using Eq. 18 and Eq. 24 and limiting the case of Eq. 19 to $kT \to 0$, one can obtain an analytical expression for the partial distribution function:

$$\forall n, n_{min} \leq n \leq n_{max}, F(p|n) \xrightarrow[kT \to 0]{} \begin{cases} 1, & p < p_{min} \\ (n - p_{min} + 1)/(n_{max} - p_{min} + 1), & p_{min} \leq p \leq p_{max} \\ 0, & p \geq p_{max} \end{cases}$$

$$\forall n, n < n_{min} \text{ или } n > n_{max}, F(p|n) \xrightarrow[kT \to 0]{} \begin{cases} 1, & p \leq n \\ 0, & p > n \end{cases}$$

$$(25)$$

and for the charge distribution function

$$\sigma(n) \xrightarrow{kT \to 0} \begin{cases} 0, n < n_{min} \; \text{или} \; n > n_{max} \\ \prod_{l=n_{min}+1}^{n} \left\{ K(p_{min}|l) \left(\frac{N-l+1}{l} \right) \left(\frac{n_{max}-l+1}{1-p_{min}+1} \right) \right\} \Big/ Z_\sigma, n_{min} \leq n \leq n_{max} \end{cases} \quad (26)$$

where the statistical sum is expressed as:

$$Z_\sigma = \sum_{k=n_{min}}^{n_{max}} \prod_{l=n_{min}+1}^{k} \left\{ K(p_{min}|l) \left(\frac{N-l+1}{l} \right) \left(\frac{n_{max}-l+1}{1-p_{min}+1} \right) \right\} \quad (27)$$

The equations for the charge distribution function and for the partial distribution in the cases of the degenerated level and that for fast relaxation can be found in [9] and [1] respectively.

RESULTS AND DISCUSSIONS

We have performed the numerical simulations of the I-V curves and the control curves for various sets of parameters. We present here those results obtained for the low-temperature limit in the case of equidistant energy spectra and slow relaxation. A typical I-V curve of the molecular SET with symmetrical barriers is presented in Figure 2.

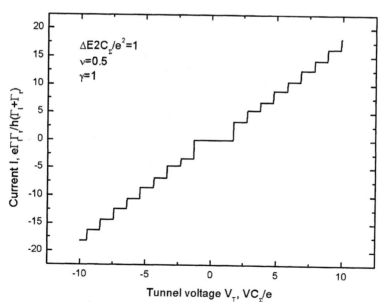

Figure 2. I-V curve of molecular single-electron transistor with identical tunnel junctions. The curve was calculated for the case of equidistant energy spectrum and slow relaxation. The spacing between energy levels equals the Coulomb characteristic system energy.

In this diagram, the characteristic staircase caused by changing values of (p_{min}, n_{min}) and (p_{max}, n_{max}) during tunnel voltage V_T changes can be seen. This corresponds to the energy level «entering» into the energy gap between Fermi levels of the transistor electrodes. The characteristic Coulomb blockade of the tunneling current can also be seen in this figure. The deviation of the width of the blockade region from e/C_Σ can be explained by interplay of charging effects with discrete energy electrons.

In Figure 3, one can see the dependence of the shape of I-V curves on tunnel barriers asymmetry. Note the increase in the width of steps on the I-V curves with increasing tunneling voltage. At the same time the height of the steps remains constant. One can also see the parabolic regions of the curves in the $\{V_T, \eta\}$ planes. Such parabolic sections can be explained by quadratic dependence of the electrostatic energy on the tunneling voltage. As well, one can see that the asymmetry of the tunnel barriers can increase the Coulomb blockade of tunneling current.

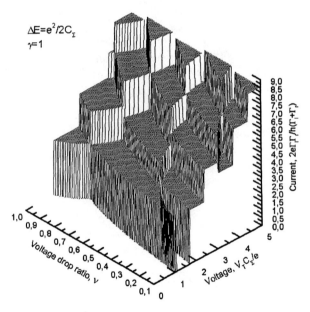

Figure 3. The series of I-V curves of the molecular SET transistor with discrete energy spectrum. The series are presented in relation to the tunneling voltage drop coefficient η. The value of $\eta = 0.5$ corresponds to a symmetrical transistor.

In Figure 4, the series of I-V curves are shown as a function of spacing between energy levels of the molecule. In this diagram, one can see that the conductivity of the transistor increases with an increase in the number of levels in the energy gap between Fermi levels of the electrodes. One can also see that the step density rises with an increasing number of energy levels in this gap. This provides conclusive proof that steps in the case of slow relaxation are related to peculiarities in the discrete energy spectra and not to Coulomb effects.

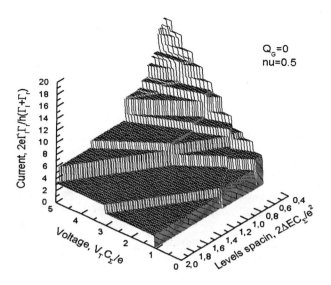

Figure 4. The series of I-V curves of the molecular transistor with discrete energy spectrum. The series are presented as a function of level spacing ΔE in the molecule.

Figure 5 shows the control curves. The characteristic rhombic profiles are parallel to the $\{V_T, Q_G\}$ planes, each corresponding to an average number of additional electrons in the molecule. Such rhombic profiles are analogous to those obtained in the theory of single-electronics.

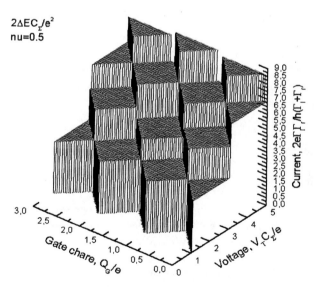

Figure 5. The control curve series for the molecular SET for the case of slow relaxation and discrete equidistant energy spectrum. The series are presented in relation to the applied tunnel voltage (operating voltage).

CONCLUSION

Thus we have presented an approach to calculation of molecular SET characteristics for the case of fast and slow electron relaxation which allows a simulation of various characteristics of molecular SET with discrete energy spectrum. The analytical basis for numerical calculations is elaborated. The calculation of tunnel characteristics of molecular SET with discrete energy spectrum was carried out as the first stage of study for the case of slow relaxation and discrete equidistant energy spectrum at low temperature (T=0). It is shown that the tunnel current blockade depend not only on the Coulomb characteristic energy but also on the average spacing between energy levels in the molecule. We have found also that the I-V curve staircase is a consequence of the discrete energy spectrum of the molecule: the regular step appears when the number of energy levels in the gap between Fermi levels of electrodes increases by one. Calculated control curves were qualitatively the same as obtained in the framework of orthodox theory.

ACKNOWLEDGEMENT

This work was supported in part by ISTC (Pr. No. 1991), INTAS (Pr. No. 99-864), and RFBR (No. 99-03-32218, No. 01-02-16580).

REFERENCES

[1] C.W.J.Beenakker, 1991. Phys. Rev. B, 44, 1646.
[2] D.V. Averin, A.N. Korotkov , 1990. JETP, 97, 1661.
[3] T. Tanamoto and M. Ueda, 1998. Interplay between the Coulomb blockade and resonant tunneling studied by the Keldysh Green's-function method, Phys. Rev. B, 14638–14641.
[4] K.K. Likharev, 1999. Proceedings of the IEEE, 87, 606.
[5] L.D. Landau, E.M. Lifshits, 1965. Quantum Mechanics, Addison-Wesley, Reading MA.
[6] G.B. Khomutov, E.S. Soldatov, S.P. Gubin, S.A. Yakovenko, A.S. Trifonov, A. Yu. Obidenov, V.V. Khanin, 1998. Thin Solid Films, 327-329, 550–555.
[7] S.P. Gubin, 1987, Chimiya klasterov, (Moscow. "Nauka").
[8] D.V. Averin, K.K. Likharev, 1991. In Mesoscopic Phenomena in Solids, Ed.: B.L. Altshuler, P.A. Lee, R.A. Webb, Elsevier, 173.
[9] Y.O. Klymenko, 2002. The Display of Coulonb Blockade Effects at Arbitrary Degeneracy of Molecular Junction Levels, Fizika nizkikh temperatur, 28(6), 558.
[10] V. Shorokhov, P. Johansson, E. Soldatov, 2002. Simulation of Characteristics of Molecular Set Transistor with Discrete Energy Spectrum of the Central Electrode, J. Appl. Phys. 91, 3049.

Electronic Transport through Benzene Molecule and DNA Base Pairs

A. A. FARAJIAN, R. V. BELOSLUDOV, H. MIZUSEKI and Y. KAWAZOE
Institute for Materials Research, Tohoku University, Sendai 980-8577, Japan

ABSTRACT

We have studied electronic transport through a benzene molecule attached to two gold electrodes via sulfur clips. The conductance characteristics of the device is derived for two different atomistic models of the gold contacts: a chain model and part of the Au(111) surface. The results show that the conductance characteristics strongly depend on the model used for the electrodes, therefore this factor must be taken into consideration when comparing different theoretical/experimental results. The relevance of the accurate determination of the relative position of sulfur clips and surface gold atoms are clarified. Transport through DNA base pairs attached to Au(111) electrodes will also be discussed.

INTRODUCTION

Individual molecules in general and organic compounds in particular, have been assigned the possible role of the basic building blocks of the next generation of electronic devices for quite some time. The current trend in miniaturization of electronic devices has resulted in intense interest in molecular devices which basically consist of a group of a few atoms in contrast to the previously-used 'bulk' materials. Among others, we can give two reasons for this attention on devices based on individual molecules: they are self-assembled and rather abundant.

As the size of the molecules to be used as the main functional part of the device have a dimension of a few angstroms, a natural question arises, namely—when these small building blocks are attached to electrodes, how does their main characteristics of interest depend on the actual arrangement of the atoms in the electrode. This is an issue of concern for both experimentalists and theoreticians: in experiments, it is necessary to know which surface in general, and which part of the surface in particular, are more suited for use as an electrode, while for theoretical calculations, we must know which model to use to describe the electrodes.

Two different coupling regimes can be considered when studying transport through individual molecules attached to metallic (e.g., gold) contacts: weak and strong. In the weak-coupling regime, the functional part of the device, i.e., the molecule, is separated from the electrodes by a 'spacer' that can be as small as the functional molecule itself. The role of the spacer is to screen out the effect of the details of the electrode structure (such as surface roughness) so as to prevent them from influencing the main characteristics of interest in the device. On the other hand, in the strong-coupling regime, the functional part of the device is usually attached to the electrodes via chemical bonds. This makes the coupling between the functional molecule and the atomic arrangement of the electrodes in the vicinity of the contact point strong enough to affect the transport characteristics of the device. In a sense, discarding the details of the spacer, the weak-coupling regime correspond to physical adsorption, i.e., van der Waals coupling, while the strong-coupling may be considered chemical bonding.

It is the purpose of this study to investigate these issues for the case of a simple organic molecule, i.e., benzene, attached to two gold electrodes via sulfur clips. Due to the presence of chemical bonds in attaching the functional part to the gold contacts, the coupling can be considered to be in the strong regime. As we shall see shortly, this strong coupling results in the dependence of the transport characteristics (conductance) on the atomistic model of the electrodes.

FIRST-PRINCIPLES MODEL

In calculating transport properties we must describe the system on a localized basis and since we are considering a relatively small number of atoms in this study we chose *ab initio* modeling for both the functional part of the device and the electrodes. For the functional part of the device, a benzene-1,4-dithiolate was considered. The sulfur atoms act like clips to attach the benzene molecule to the gold contacts. As for the specific atomic arrangements of the gold contacts, we considered two different cases: first, a simple chain arrangement of the gold atoms is considered; secondly, an arrangement of gold atoms corresponding to part of an Au(111) substrate was considered. These two atomic configurations with the organic molecule are depicted in Figure 1.

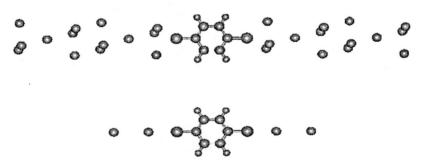

Figure 1. Two different arrangements of the first two unit cells of the electrodes.

In using *ab initio* modeling to describe the system, the Hamiltonian of the system was deduced. Moreover, for a non-orthogonal system, the overlap matrix is also needed. The data corresponding to the Hamiltonian and overlap matrices were obtained using the Gaussian 98 program [1]. The basis set involves selecting 13 vectors for Au (s&s&p&p&d), 5 vectors for C (s&sp), 4 vectors for S (s&p), and 1 vector for H (s).

CONDUCTANCE

In calculating transport properties, we adopted the Green's function approach which is quite general and able to handle a wide category of systems. In this approach, the system under consideration was divided into three parts: a semi-infinite part to the left, a finite part in the middle, and a semi-infinite part to the right. The semi-infinite parts resemble the 'leads' that are used to connect the middle finite part (the functional part) to an external bias. Using the Hamiltonian and the overlap matrices corresponding to the gold contacts, the surface Green's functions that describe the two semi-infinite electrodes attached to the left and right of the molecule were derived. These functions, together with the Hamiltonian and overlap of the molecule, as well as the Hamiltonian and overlap of the molecule-electrode part, were then used to determine the conductance of the system [2-4].

The same Green's function procedure was then used to calculate the electronic transport properties of DNA base pairs. These results are useful in describing the scanning tunneling microscope (STM) images of DNA, and in exploring the possible use of DNA base pairs as part of an electronic molecular device. The DNA base pair in this study is depicted in Figure 2.

Figure 2. DNA A-T base pairs attached to Au(111) electrodes.

RESULTS

In the strong coupling regime considered here, the atomistic model of gold contacts resemble the local environment in the vicinity of the position of the sulfur clips. The results of the calculations are depicted in Figure 3. As is apparent from this diagram, the conductance of the device strongly depends on the atomistic model used to represent the electrodes. More specifically, the inter-atomic distances within the gold electrodes and the relative position of the gold atoms that enter the transport calculation via the surface Green's function strongly affect the conductance characteristics. The decisive dependence of the conductance on the actual atomic arrangement of the gold electrodes shows that although the sulfur clips are thought to block the static charge transfer from the electrodes to the organic molecule, these clips cannot block the effects of the structure of the electrodes. Based on these results, we observe that the type of model used for the contacts of a molecular device plays an important role in transport calculations. This effect must be accounted for in order to be able to perform meaningful comparisons between different calculations/experiments based on the same organic molecule.

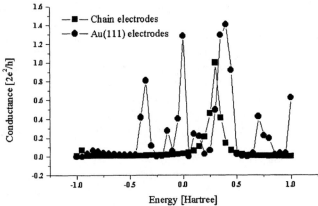

Figure 3. Conductance of the benzene-1,4-dithiolate (units of $2e^2/h$) vs energy.

CONCLUSIONS

In summary, we investigated the effects of gold-contact modeling in strong coupling with the functional molecular part of a nano-device. This resembles the atomic arrangement of the electrodes in the vicinity of chemical bonds that attach the functional part to the electrodes. We observe that the local configuration of the electrode atoms indeed influence the conductance characteristics of the molecular device.

ACKNOWLEDGEMENT

The authors would like to express their sincere thanks to the crew of the Center for Computational Materials Science at the Institute for Materials Research, Tohoku University, for their continuous support of the computing facilities. This study is supported by the Special Coordination Funds of the Ministry of Education, Culture, Sports, Science and Technology of Japan.

REFERENCES

[1] M. J. Frisch et al., 2001. Gaussian 98, Revision A.11.1, Gaussian Inc., Pittsburgh PA.
[2] A. Rochefort, P. Avouris, F. Lesage, D.R. Salahub, 1999. Phys. Rev. B60, 13824.
[3] W. Tian, S. Datta, S. Hong, R. Reifenberger, J. Henderson, C. Kubiak, 1998. J. Chem. Phys. 109, 2874.
[4] S. Datta, 1995. Electronic Transport in Mesoscopic Systems, Cambridge U. P., Cambridge.

Realization of "Molecular Enamel Wire" Concept for Molecular Electronics

R. V. BELOSLUDOV, H. SATO, A. A. FARAJIAN, H. MIZUSEKI,
K. ICHINOSEKI and Y. KAWAZOE
Institute for Material Research, Tohoku University, Sendai, Japan

ABSTRACT

The structural and electronic properties of conducting polymer covered with cyclodextrin molecules have been investigated using quantum mechanical simulations. Thus, the results of calculations showed that the structure of polyaniline, in the cases of β-CDs and molecular nanotube of cross-linking α-CDs has near-planar geometry, with the electronic configuration of the optimized structure being practically same as the one in the planar conformation. It has also been found that in these cases, there are no charge transfer between polymer fragment and frameworks of CDs. The doping effect on the geometric and electronic properties of polyaniline encapsulated by CDs has been also investigated. It has been shown that the doped polymer chain can be stabilized inside the molecular nanotube of cross-linking α-CDs. These results support the realization of molecular electronic device based on this complex.

INTRODUCTION

Despite a remarkable miniaturization trend in the semiconductor industry, in the next 10-15 years, conventional Si-based microelectronics is likely faced with fundamental limitations when feature lengths shrink below 100 nm [1]. Therefore fundamentally, new approaches for realization of electronic parts are required. In 1974, the possibility of an organic molecule functioning as a molecular rectifier was theoretically demonstrated by Aviram and Ratner [2] and later this was confirmed experimentally [3]. After that and specifically over the last several years, there have been many scientific efforts and significant advances in the development of electronic devices (such as the wire, diode, or transistor) integrated on a molecular scale [4,5]. Since the size of these molecular devices is a few nanometers, in parallel with the progress of more effective fabrication technologies, theoretical study of promising molecular structures has also been a key factor in designing new electronic devices with desired physical characteristics.

For applications with molecular electronics, the wire is a very important component because it can be used as a connection between a metal electrode and other functional molecules such as a molecular diode or transistor to create complex molecular circuits. It is important that the conducting part of a molecular wire must have metallic characteristics. For these reasons, conducting polymers are very attractive materials among the different candidates for molecular wires. Their electrical conductivity can be controlled over the full range from insulator to metal by chemical or electrochemical doping [6,7,8]. These can be synthesized with highly controlled lengths and can be integrated into complex circuits by chemical-bonding with other functional molecules without changing their electronic properties. To prevent possible interaction between different molecular wires, each polymer chain should be encapsulated into a bulky insulated structure and so, forming an insulated molecular enamel wire. According to the "molecular enamel wire" concept [4], the insulators are placed around a conducting center. It has been also suggested that the "molecular enamel wire" would be a key concept to realize a high performance molecular supercomputer.

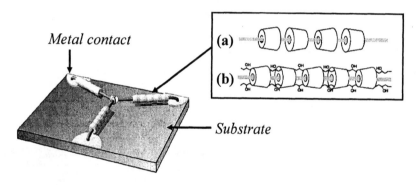

Figure 1. Schematic diagram of molecular circuit. (a) Inclusion complex formation of CDs and conducting polymer. (b) Polymer chain into molecular nanotube of cross-linking α-CDs.

Here, we report the structural and electronic properties of different CD-conducting polymer inclusion complexes. The aim of this study is to demonstrate the possibility of the formation of single molecular enamel wires, using quantum mechanical simulations.

A possible approach to realize this concept is the formation of an inclusion complex between the conducting polymer and cyclic cyclodextrin (CD) molecules shown in Figure 1. The cavity size of the CD component can be regulated by the number of D-glucose units in each CD molecule (6, 7, and 8 for α-, β-, and γ-CD, respectively) and a molecular tube can be created by cross-linking adjacent α-CD units using a hydroxypropylene bridge [9]. Recently, atomic force microscopy (AFM) and scanning tunneling microscopy (STM) observations have indicated the formation of an inclusion complex in which the polymer is fully covered by β-CD molecules [10] as well as a molecular nanotube of cross-linking α-CD molecules [11]. Moreover, theoretical studies indicate that β-CD molecules can be used as an insulated molecular structure to stabilize the isolated near-planar configuration of polymers with the electronic configuration of the optimized structure being almost the same as that in the planar conformation [12, 13].

THEORETICAL METHODS

Since the treated inclusion complexes consist of a huge number of atoms, the combined quantum mechanics and molecular mechanics calculations have been applied to optimize the polymer structure into CD molecular nanotubes. The two-layered "own N-layered integrated molecular orbital and molecular mechanics" (ONIOM) method [14] was applied. In this hybrid method, the structure of the polymer fragment was treated using quantum mechanics while the remainder of the system (CD molecules) was treated using a molecular mechanics force field. In this way, the structure of the selected cluster model was optimized. In order to examine possible charge transfer as well as interactions between polymer fragments and CDs, a single-point energy calculation was performed for optimal configuration of this inclusive complex. These calculations were performed using the Gaussian 98 set of programs [15].

RESULTS AND DISCUSSION

Structure of Polymer Chain in CDs

It is important to know the configuration of the polyaniline (PANI) fragment formed in the CDs since the source of conductivity for a conjugated conducting polymer is a set of π-type molecular orbitals that lie above and below the plane of the molecule when it is in a planar or near-planar configuration. The long polymer fragment (nine monomer units) was optimized in free space using both full optimization to find the lowest energy structure and partial optimization while maintaining the planar configuration of the PANI fragment. Figures 2a and 2d show the respective optimized structures. In the case of full optimization, the imaginary frequencies are absent, which means we are in a local minimum. In this configuration, adjacent benzene rings have a dihedral angle of 90^0 which will reduce the extent of π-orbital overlap between adjacent rings, break up the electron channels, and decrease the conductivity of the molecular wire. The planar structure is higher in energy by 38.65 kcal/mol at the HF/6-31G* level compared to the most stable structure. Moreover, the imaginary frequencies are found for the planar configuration to correspond to a combination of out-of-plane (bent) vibrations of benzene rings. Therefore it is necessary to apply external forces to stabilize the planar configuration.

The geometry of the PANI was optimized in β-CDs and the cross-linked α-CDs. These structures are shown in Figure 2b and 2c, respectively. The structure of the PANI in β-CDs lies higher in energy by 7.10 kcal/mol as compared to the most stable configuration of the same PANI fragment in free space. In the case of the cross-linked α-CD molecular nanotube, the energy difference is found to be 20.91 kcal/mole. Moreover, the length of polymer chain in this case (45.50 Å) is very close to the length of the planar conformation of the polymer chain (46.92 Å) as shown in Figure 2. These results indicate that the configurations of PANI in the cross-linked α-CD molecular nanotube are closer to the planar structure of PANI in free space than the configuration of PANI in β-CDs. It has also been found in both cases that there is no charge transfer between polymer fragment and CDs, and so, interaction between these molecules has a non-covalent characteristic. Thus, the near-planar configuration of the polymer chain is formed in a molecular nanotube of cross-linked α-CDs due to weak interactions such as Coulomb and van der Waals interactions between the host framework of the CDs and the PANI.

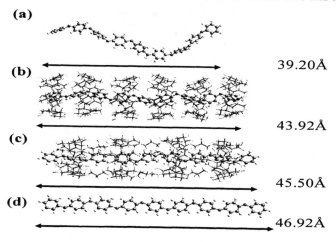

Figure 2. Structural analysis of PANI fragments (a) the most stable configuration in free space; (b) in β-CDs; (c) in molecular nanotube of cross-linking α-CDs, and (d) the planar configuration in free space.

In order to understand the electron transport through polymer chain in CDs, we have analyzed the spatial extent of the frontier orbital to provide a strategy by which the transport properties of these systems can be understood. Analysis of the molecular orbital energy diagrams (Figure 3) for the configurations of PANI in the CDs host framework shows that the lowest unoccupied (LUMO and LUMO+1) orbitals (as well as LUMO+2) are located on the polymer fragment and their contours are similar to those in the case of the planar configuration of PANI in free space.

Moreover, in these cases, there is no overlap of electron density between CDs and the PANI fragment. This reveals that the inclusion-complex based on a cross-linked α-CD nanotube and β-CD host framework can be used as a molecular enamel wire. This result may be important from a practical viewpoint as electrons will transport only through the polymer chain and no current leakage takes place across the CD molecules.

Doping of Polymer Chain in CDs

In order to realize the concept of a molecular enamel wire, it is also necessary to understand the stability and electronic properties of the conducting polymers in the metallic state when they are encapsulated within molecular nanotubes. For this purpose, the PANI fragment in the metallic (emeraldine salt) state has been optimized inside the cross-linked α-CD molecular nanotube. It is well known through experiment that protonation by acid-base chemistry leads to an internal redox reaction and the conversion from semiconductor (the emeraldine base, EB) to metal (the emeraldine salt, ES) [8]. The five monomers in free space were optimized using both full optimization to find the lowest energy structure and then, by partial optimization while maintaining the planar configuration of PANI with ES. It was found that the lowest energy structure of ES with five monomer units has a total spin of S=1 indicating the existence of two unpaired spins. The HOMO-LUMO energy difference is significantly reduced compared to the same energy difference for EB which has the same number of benzene rings. This indicates the transition of PANI from semi-conducting to metallic state. It was also found that by using the cross-linked CD molecular nanotubes, one can stabilize the near-planar configuration of the metallic form of PANI.

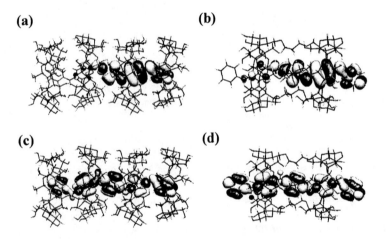

Figure 3. Contour of the lowest unoccupied orbitals: (a) LUMO of PANI-β-CDs; (b) LUMO of PANI – cross-linking α-CDs; (c) LUMO+1 of PANI-β-CDs and (c) LUMO+1 of PANI – cross-linking α-CDs.

Analysis of molecular orbital energy diagrams for the configuration of EB in a molecular nanotube shows that the single occupied molecular (SOMO, SOMO+1) as well the lowest unoccupied (LUMO and LUMO+1) orbitals (Figure 4) are located on the polymer fragment and their contours are similar to those in the case of the planar configuration of ES. These orbitals are located on polymer chains

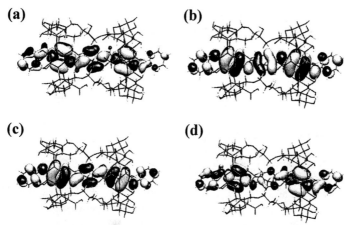

Figure 4. Contour of the selected molecular orbitals of ES of PANI fragment in molecular nanotube of cross-linking α-CDs: (a) SOMO; (b) SOMO+1; (c) LUMO and (c) LUMO+1.

and hence the CDs can be used as insulator between different single molecular wires. The present theoretical results provide support for the concept of "molecular enamel wires" [4] and hence indicate a high possibility of constructing such molecular parts.

CONCLUSION

The structure of PANI fragments in various inclusion complexes based on CD molecules was optimized using a combined quantum mechanics and molecular mechanics method. The results of calculations show that the structures of the PANI in molecular nanotubes of cross-linked α-CDs have near-planar geometry with the electronic configuration of the optimized structure being almost the same as that in the planar conformation. Moreover, the single chain of ES PANI (metallic form) can be covered with the insulator CD molecular nanotube. The theoretical results are in agreement with experimental data [10, 11] and may advance the application of such inclusion complexes in molecular circuits with more complex functionality.

ACKNOWLEDGMENT

The authors would like to express their sincere thanks to the staff of the Center for Computational Materials Science of the Institute for Materials Research, Tohoku University for their continuous support of the SR8000 supercomputing facilities. This study was performed through the Special Coordination Funds for

Promoting Science and Technology from the Ministry of Education, Culture, Sports, Science and Technology of the Japanese Government.

REFERENCES

C. Joachim, J. K. Gimzewski, A. Aviram, 2000. Electronics using hybrid-molecular and mono-molecular devices, Nature, 408, 541–548.

A. Aviram and M. A. Ratner, 1974. Molecular rectifiers, Chem. Phys. Lett., 29, 277–283.

M. Fujihira, N. Ohishi, T. Osa, 1977. Photocell using covalently-bonded dyes on semiconductor surface, Nature, 268, 226–228.

Y. Wada, M. Tsukada, M. Fujihira, K. Matsushige, T. Ogawa, M. Haga, S. Tanaka, 2000, Prospects and problems of single molecule information devices, Jpn. J. Appl. Phys., 39, 3835–49, and references therein.

J.C. Ellenbogen, J.C. Love, 2000. Architectures for molecular electronic computers: 1. Logic structures and an adder designed from molecular electronic diodes, Proc. IEEE, 88 386–426, and references therein.

H. Shirakawa, 2001. The discovery of polyacetylene film: the dawning of an era of conducting polymers (Nobel Lecture), Angew. Chem. Int. Ed., 40, 2574–2580.

A.G. MacDiarmid, 2001. Synthetic metals: A novel role for organic polymers (Nobel Lecture), Angew. Chem. Int. Ed., 40, 2581–2590.

Alan J. Heeger, 2001. Semiconducting and metallic polymers: The fourth generation of polymeric materials (Nobel lecture), Angew. Chem. Int. Ed., 40, 2591–2611.

A. Harada, J. Li, M. Kamachi, 1994. Double-stranded inclusion complexes of cyclodextrin threaded on poly(ethylene glycol), Nature, 364, 516–518.

K. Yoshida, T. Shimomura, K. Ito, R. Hayakawa, 1999. Inclusion complex formation of cyclodextrin and polyaniline, Langmuir, 15, 910–913.

T. Shimomura, T. Akai, T. Abe, K. Ito, 2002. Atomic force microscopy observation of insulated molecular wire formed by conducting polymer and molecular nanotube, J. Chem. Phys., 116, 1753–55.

R.V. Belosludov, H. Mizuseki, K. Ichinoseki, Y. Kawazoe, 2002. Theoretical study of inclusion complex of polyaniline covered by cyclodextrins for molecular device, Jpn. J. Appl. Phys., 41, 2739–41.

R.V. Belosludov, H. Sato, A.A. Farajian, H. Mizuseki, K. Ichinoseki, Y. Kawazoe, 2003. Molecular enamel wires for electronic devices: Theoretical study, Jpn. J. Appl. Phys., accepted for publication.

S. Humbel, S. Sieber and K. Morokuma, 1996. The IMOMO method: Integration of different levels of molecular orbital approximations for geometry optimization of large systems: Test for n-butane conformation and S(N)2 reaction: RCl+Cl-, J. Chem. Phys., 105, 1959–1967.

M. J. Frisch, et al., 1998. Gaussian 98, revision A.1; Gaussian, Inc.: Pittsburg, PA.

Mechanical Properties of a Thermoelectric SMA Manipulator

K. YAKUWA, Y. LUO and T. TAKAGI
Institute of Fluid Science, Tohoku University, Aoba-ku, Sendai, Japan

ABSTRACT

This paper presents the mechanical properties of a thermoelectric shape memory alloy (SMA) manipulator proposed by the authors. As actuator materials, SMAs have the advantage of large ratio of output energy to weight and can lead to compact design of actuators. Thermoelectric modules are applied both for temperature control and improving the responses of the SMA actuator. The use of SMA actuators as a manipulator requires the knowledge about the relation between their mechanical output and thermal behaviour. In this work, the free deformation and recovery forces of an SMA cantilever have been evaluated. Mechanical behaviour of the manipulator is discussed based on a proposed theoretical model of recovery forces.

INTRODUCTION

Shape memory alloys (SMAs) are functional materials revealing large deformation and recovery forces. As actuator materials, they have the advantage of large ratio of output energy to weight, and are adopted in a number of applications. The shape memory effect (SME) is induced by transformations from martensitic phase to austenitic phase and requires a temperature change covering the transformation temperature range. SMAs are usually electrically heated and cooled by natural convection. The natural cooling is one of the reasons of their poor response property. In the past decade, using thermoelectric elements to activate SMA actuators has been proposed [1,2]. Recently a series of thermoelectric SMA actuators have been developed for artificial heart muscles [3,4] and catheters [5].

In this work, a manipulator, consisted of an SMA cantilever and Peltier modules, was developed. As SMAs can play the roles of both the arms and motors in conventional manipulators, facilitating a design of manipulator with simple structures and lightweight. Thermoelectric elements were adopted here for the temperature control in order to obtain a desired mechanical output. Since the thermomechanical behaviour is closely related to the transformation phenomena of SMAs, the mechanical output of the manipulator depends not only on the temperatures of SMAs, but also on the situation of phase transformation, which has a hysteretic dependency on the temperatures. Therefore, the mechanical behaviour of the SMA manipulator has to be evaluated systematically in order to provide quantitative information for the further design.

THERMOELECTRIC SMA MANIPULATOR

Figure 1 shows a schematic drawing of a thermoelectric SMA manipulator. A prototype consists of an SMA ribbon for the manipulation, thermoelectric elements for the temperature control, and heat sinks. If electric current is applied in thermoelectric elements, heat transfer will be induced by the Peltier effect on boundaries between the elements and contacting solids. The heat transfer has a dependency on the applied current in the direction. The temperature change induces phase transformation in SMAs, and consequently generates large deformation of SMAs. If the shape of SMAs is restrained, recovery forces will be generated when a phase transformation occurs. The basic concept of the SMA manipulator is to use the recovery forces to perform a series of manipulation such as the holding, pressing, and lifting. The recovery forces depend not only on the restrained shapes but also the internal state of the hysteretic transformation.

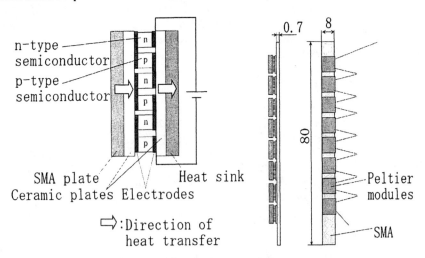

Figure 1. Schematic drawing of an SMA manipulator.

Basically simple temperature control of an SMA can realize control of its mechanical output, but a systematic investigation on the relationship between the temperatures of an SMA and the restrained shapes together with the state of phase transformation is needed.

SMA material adopted in the present work is Ti51at%Ni. Its transformation temperatures were determined by a differential scanning calorimeter. The start and finish temperatures are $A_s = 52°C$ $A_f = 64°C$ for A-phase and $M'_s = 53°C$ $M'_f = 44°C$ for R-phase. A prototype of SMA manipulator was designed to have an SMA ribbon with the dimension of 0.7mm in thickness, 8mm in width and 80mm in length. Eight Peltier modules (6.2mm × 8mm × 2.3mm) were attached on its surface and electrically connected in series. Heat sinks were bounded on the other side of Peltier modules for improving the efficiency of heat transfer. The SMA ribbon was subjected to a solution treatment with a flat shape followed by an aging treatment with a restrained arc shape (curvature = 35mm).

EXPERIMENTAL SETUP

The SMA material, Ti51at%Ni has a two-way shape memory effect. With the heat treatment described in the previous section, an Ti51at%Ni ribbon is expected to deform from a flat shape to an arc shape upon heating, and vice versa when subjected to a cooling. The free deformation of the SMA ribbon subjected to thermal cycles covering the transformation temperature range has been evaluated. The strain and temperature on the surface of the SMA ribbon were measured by strain gauges and thermocouples. The measurement of recovery forces was performed with the following procedure. Firstly the SMA ribbon was fixed at one end and heated up to a certain temperature, and its deflection of the free end was restrained. The recovery forces generated at the restrained end were then measured using force gauges when a further temperature change was applied. Figure 2 shows the experimental setup. All experiments were carried out in a thermostatic oven. If the ambient temperature is set to be lower than the start temperature of reverse transformation A_s, a further temperature change throughout the transformation temperature range can induce a complete transformation. However, when the ambient temperature has a value in the transformation temperature range, only a partial transformation could be obtained. As a consequence, the further heating can induce only a partial recovery force. The finish temperature of the reverse transformation A_f is a critical value, above which neither deformation nor recovery forces can be induced. Therefore, the evaluation was conducted in a temperature range lower than A_f.

RESULTS AND DISCUSSIONS

The experimental results of the free deformation are shown in Figure 3. Strains on the surface of the SMA ribbon increase rapidly as the temperature is raised up to A_s and saturate at A_f, then decrease from $M's$ on cooling. The large deformation was observed in the transformation temperature range. However, at the temperatures below A_s, slight deformation was observed. One possibility of this phenomenon is that there is an additional transformation occurred in this

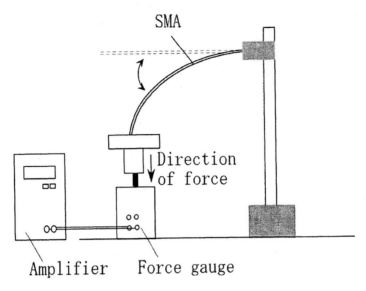

Figure 2. Experimental setup.

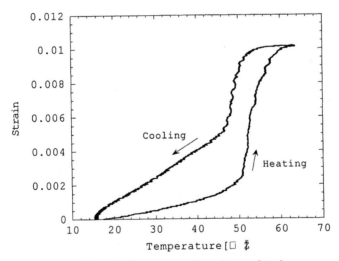

Figure 3. Temperature dependency of strain.

temperature range. Nevertheless, the maximum strain was found to be equal to the restrained strain of the aging treatment, 1%.

Figure 4 shows the experimental results of recovery forces of two representative cases: one with the ambient temperature of 20°C and the other of 55°C. In the first case, the SMA ribbon was in its R-phase. Therefore it experienced a complete reverse transformation upon heating to A_f. Since the vertical displacement of the free end of the ribbon was restrained, a maximum recovery force of 2.3N was obtained. In the second case, the SMA ribbon has a partially transformed state, therefore only a part of maximum recovery force was obtained.

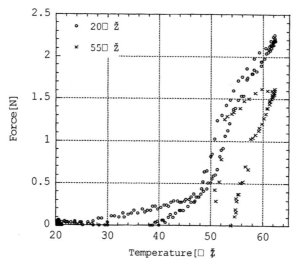

Figure 4. Experimental results of recovery forces.

Since the final goal is to control the mechanical output through a simple temperature control, a phenomenological model describing both the free deformation and the recovery forces is needed. In this paper, modeling of the mechanical behaviour of the SMA manipulator has been conducted. The basic idea is to express the transformation induced shape memory effect by a stress filed distributed homogeneously in SMAs. This field is determined by the condition of the heat treatment that SMAs experienced. For instance, a heat treatment under a uniaxial compression is assumed to induce a compressive stress field along the axial direction. In our case, the tension and compression stress fields are induced respectively in two layers with the neutral plane in the direction of thickness as a boundary, forming a bending moment. With a cosine function using characteristic parameters such as the transformation temperatures, Young's moduli, and the curvature at various temperatures, the moment is assumed to have the following expression

$$M(T) = \frac{E_A I}{2\rho_A}\left\{1-\cos\pi\left(\frac{T-A_s}{A_f-A_s}\right)\right\} \quad (1)$$

where ρ_A is the curvature of a fully deformed SMA ribbon, E_A is the Young's modulus of the SMA at temperatures above A_f. For a cantilever beam illustrated in Figure 5, we consider an initial state at a temperature T_l, where the beam has a curvature $\rho(T_1)$. If the vertical displacement of the free end is restrained, then a further temperature change to T_2 will induce a recovery force at the restrained end. The equilibriums of moments and forces are

$$M_A + M(x) + R_A x = 0, \quad R_A + R_B = 0 \qquad (2)$$

where M_A, R_A are the moment and reaction force at the fixed end, R_B is the recovery force, and $M(x)$ is the moment at x position. A deflection curve of the beam can be described by

$$\frac{d^2 y}{dx^2} = -\frac{1}{E(T)I}(-M_A - R_A x) \qquad (3)$$

with I being the moment of inertia of the beam. With the boundary conditions of this differential equation

$$\left\{\frac{dy}{dx}\right\}_{x=0} = 0, \quad y(0) = 0 \qquad (4)$$

The deflection of the beam at x position has the solution

$$y = -\frac{x^2}{6EI}(-3M_A + R_A x) \qquad (5)$$

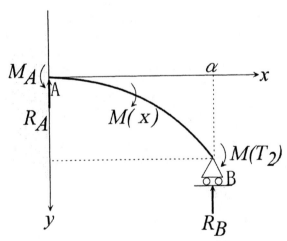

Figure 5. A cantilever beam model.

As the vertical deflection of the free end is restrained, then we have

$$y(\alpha) = y_1 = \frac{\alpha^2}{2\rho(T_1)}, \quad \alpha = L\cos\left(\frac{L}{\rho(T_1)}\right) \quad (6)$$

Finally, the reaction force, i.e., the recovery force at the free end becomes

$$R_B = \frac{\dfrac{3E(T_2)I}{\rho(T_1)} + 3M(T_2)}{3l + \alpha} \quad (7)$$

where E_R is the Young's modulus of the R-phase. $E(T)$ is defined

$$E(T) = \frac{1}{2}\left\{(E_R + E_A) + (E_R - E_A)\cos\pi\left(\frac{T - A_s}{A_f - A_s}\right)\right\} \quad (8)$$

Recovery forces were calculated with the above model. The Young's modulus was measured by the three-point bending method (E_R=40GPa, E_A=60GPa). Simulation results are shown in Figure 6. The initial temperatures are set to be 20°C and 55°C. The calculated maximum recovery force is 3.7N for the case of 20°C and 2.8N for 55°C. Although there are discrepancies comparing with the experimental results, numerical simulation predicts the recovery forces qualitatively. Possible reasons of the discrepancies are considered to be the difficulties of detecting the vertical forces by the force gauge. Because the free end of the cantilever beam has actually a small horizontal deflection, which may cause a friction on the surface of the force gauge and therefore reduce the accuracy of measurement. In addition, the linear model may have difficulties in capturing the non-linearity of the recovery forces.

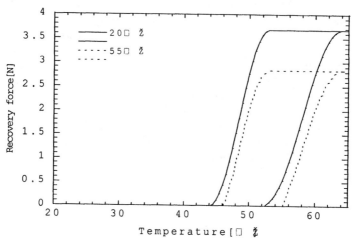

Figure 6. Analytical results of recovery forces.

CONCLUSION

In this work, a thermoelectric SMA manipulator is proposed. As a preliminary stage for the control of the mechanical output of the manipulator through the temperature control, the free deformation and recovery forces of an SMA cantilever beam have been evaluated. When the free end of the beam was restrained at temperatures below A_s, a complete phase transformation in the SMA induced a maximum recovery force of 2.3N with the present configuration. The obtainable maximum recovery forces decreased at higher temperatures in the phase transformation range. As A_s has a value of 52°C, the manipulator can be applied in a wide temperature range. A linear model of a cantilever beam is proposed to describe the mechanical behaviour of the SMA manipulator. The theoretical model is proved to be able to qualitatively predict the behaviour of recovery forces, with the improvement of the accuracy of experimental measurements and numerical modeling as remaining tasks in the future work.

ACKNOWLEDGEMENT

This work is partially supported by the Grant-in-Aid Scientific Research (No. 14350115) from the Japan Society for the Promotion of Science (JSPS).

REFERENCES

[1] A. Bhattacharya, D.C. Lagoudas, Y. Wang and V.K. Kinra, 1995. On the role of thermoelectric heat transfer in the design of SMA actuators: theoretical modelling and experiments, Smart Materials and Structures, 4, 252–263.
[2] P.L. Potapov, 1998. Thermoelectric trigging of phase transformation in Ni-Ti shape memory alloy, Materials Science and Engineering, B52, 195–201.
[3] S. Maruyama, R. Ibuki, S. Sakai, M. Makoto, M. Sato, T. Yambe, T. Takagi, Y. Luo and M. Behnia, 2001. Development of a novel artificial heart muscle using thermoelectric actuators, International Journal of Heat and Technology, 19, 75–80.
[4] Y. Luo, T. Toshiyuki, S. Maruyama, M. Yamada, 2000. Shape Memory Alloy Actuator Using Peltier Modules and R-Phase Transition, J. Intelligent Material Systems and Structures, 11(7), 503–511.
[5] S. Maruyama, M. Kawase, S. Sakai, T. Takagi, Y. Kohama, Y. Luo, Y. Tanahashi, S. Zama, M. Yamada, M. Sato, 2000. Proposal of Thermoelectric Actuator and Development of Active Catheter, JSME International Journal, Series B, 43(4), 712–718.

Single-Electron Tunneling in Planar Molecular Nanosystems

E. S. SOLDATOV
Faculty of Physics, Moscow State University, Moscow, Russia

S. P. GUBIN
Institute of General Inorganic Chemistry RAS, Moscow, Russia

V. V. KHANIN and G. B. KHOMUTOV
Faculty of Physics, Moscow State University, Moscow, Russia

V. V. KISLOV
Institute of Radio-Engineering and Electronics RAS, Moscow, Russia

I. A. MAXIMOV, L. MONTELIUS and L. SAMUELSON
Division of Solid State Physics and Nanometer Structure Consortium, Lund University, Lund, Sweden

A. N. SERGEYEV-CHERENKOV, M. V. SMETANIN, O. V. SNIGIREV and D. B. SUYATIN
Faculty of Physics, Moscow State University, Moscow, Russia

ABSTRACT

In the present work electron transport through monomolecular film of metal-organic molecular clusters $Pt_5(CO)_7(P(C_6H_5)_3)_4$ incorporated in matrix of PVP-20 polymer molecules was investigated. The molecular films were deposited by Langmuir–Shaefer technique on the metal nanoelecrodes formed by electron beam lithography. I-V curves of studied systems have shown that the nature of the electron current through the molecular system is a correlated electron tunnelling. It was shown that control characteristics I (Vg) of studied molecular systems have a periodic character typical for correlated electron tunnelling regime.

INTRODUCTION

A radical reduction of characteristic sizes of electronic devices has taken place in the last few decades. It brings with it rapid development of several areas of physics. One of the most remarkable areas is single-electronics (correlated-electron-tunneling) [1]. Single-electronics is an area of electronics that investigates the effects occurring in tunnel junctions with extremely small capacitances and, accordingly, sizes in which the energy of their recharging even by a single electron $e^2/2C$ (C is the junction capacitance, e is the electron charge) is larger than the thermal fluctuation energy kT.

The effect of single-electron tunneling (SET) offers extremely interesting possibilities in creating an elemental base for a new generation of nano-electronic devices in which information will be coded and processed by single electrons instead of a current as in present devices [1]. It provides a huge gain in the integration density of elements and allows the minimization of power consumption for single elements. Based on this phenomenon, new devices with interesting properties have been proposed and demonstrated [1].

In principle such systems can be made on the basis of super-small conducting granules. But in this case, it is impossible to get more or less suitable reproducibility of such objects which is vitally important for any application. Unfortunately, "classic" single-electron tunneling (in the system of thin-film metallic tunnel junctions prepared by electron beam lithography) is possible only at very low temperatures (about 100 mK) despite the use of the modern achievements such as nano-lithography for junction implementation. This essentially limits the possibilities of the phenomenon for both fundamental investigations and applications. This obstacle can be surmounted by reducing the junction sizes of SET transistors to the size of single molecules. A promising way to form small metal granules with reproducible properties is chemical synthesis of molecules with an appropriate intrinsic structure [2]. The small sizes of such systems allow the sufficiently higher operating temperatures of SET transistors [1, 3].

The use of single molecules as a basis of SET systems is a very promising way to solve this problem due to molecules being naturally invariable. Besides this limitation, molecules tend to self-organize into two-dimensional arrays on a surface [4]. This offers the possibility of patterning of electronic elements just on the surface by methods of surface chemistry. But the nature of the processes in such molecular electronic elements requires detailed study. Partly motivated by this, the study of molecular systems has been growing rapidly during the last few years [1]. Molecular clusters are one of the most promising molecules for future molecular electronics. The molecules have sizes from several angstroms to several nanometers [2]. It has been demonstrated by scanning tunneling microscopy (STM) [3, 4] that such molecules are suitable for molecular electronics. However, investigating electron transport through a molecular system in planar topology is very challenging from a practical point of view.

The main idea behind this work is to create the simplest single-electron device– a single electron transistor–on the basis of chemically-synthesized molecules using thin-film technology.

FABRICATION AND EXPERIMENT

Metal nano-electrodes were formed by electron beam lithography (EBL) on Si substrate covered by 300 nm layer of SiO_2. Poly-methylmethacrylate (PMMA) 950 KDa was used as electron resist. After EBL the resist was developed in MIBIK/IPA developer and metal film was formed by thermal evaporation of 5 nm

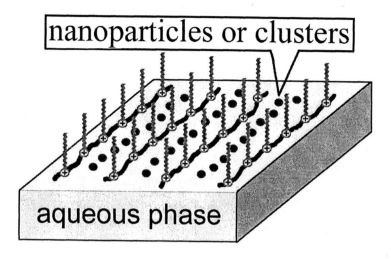

Figure 1. Schematic view of monolayer of studied cluster molecules in the matrix of inert polymer molecules of PVP-20 on a water surface.

Ti and 20 nm Au. Lift off was done in hot (60°C) acetone. The technique allowed the gaps between electrodes from 5 to 50 nm.

Molecular system of the samples is heterogeneous monomolecular film consisting of $Pt_5(CO)_7(P(C_6H_5)_3)_4$ molecules incorporated in matrix of inert polymer molecules of PVP-20. Such films were deposited by Langmuir – Shaefer technique [4] on the substrates with system of metal nano-electrodes. Schematic view of film structure is shown in Figure 1.

The setup for electron transport observations was optimized for current measurements in a range of $10^{-7} - 10^{-14}$ A. Applied source-drain voltage was up to 2 V. Gate potential was applied to Si substrate and varied in a range of ± 5 V. All measurements were done at room temperature.

RESULTS AND DISCUSSION

Use of PVP-20 polymer matrix is a key point in formation of samples. This provides stability and reproducibility of the electron transport measurements. After PVP-20 molecules polymerization $Pt_5(CO)_7(P(C_6H_5)_3)_4$ molecules appear to be incorporated in a rigid polymer film. Such film can be transferred on metal electrodes without significant breakage.

Figure 2 shows the I-V curves of such molecular film at several values of gate voltage. The I-V curves were stable and reproducible at different measuring regimes. It is opposite to I-V curves of homogenous $Pt_5(CO)_7(P(C_6H_5)_3)_4$ molecular films or I-V curves of heterogeneous films consisting of $Pt_5(CO)_7(P(C_6H_5)_3)_4$ in the matrix of stearic acid molecules.

$Pt_5(CO)_7(P(C_6H_5)_3)_4$ is a metal-organic molecular cluster. A typical structure of a metal-organic cluster is several metal atoms surrounded by a light organic shell

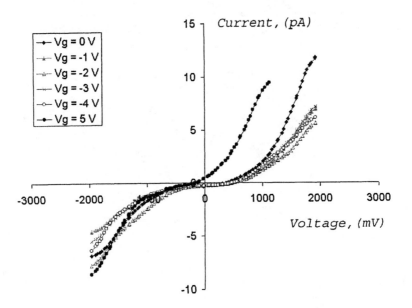

Figure 2. The room-temperature I-V characteristics of molecular films between nano-electrodes at various gate voltages.

[2]. The $Pt_5(CO)_7(P(C_6H_5)_3)_4$ molecular cluster has a core containing five Pt atoms. The diameter is 1 nm. Such molecules incorporated in the polymer matrix are like a grid of equal quantum dots separated from each other by tunnel barriers. Correlated electron tunneling through such grid is the most probable mechanism of electron transport in studied systems.

On the I-V curves presented in Figure 2 one can see the suppressed conductance regions. The sizes of the regions vary in range 0 – 0.5 V depending on gate voltage. Asymptotic resistances of the I-V curves are in range of 10^{11} - 10^{12} Ohm. Such resistances are much larger than the resistance of singe-electron transistor based on a single molecule (as measured in STM experiment [3]). The higher resistance indicates a large number of molecules in between the electrodes and it is in agreement on comparison of the gaps width between nanoelectrodes (5–50 nm) and the diameter of single molecule (1 nm).

The high rate of increase in resistance compared to STM experiment leads to a significant decrease in current. That is why the investigation of the current in such molecular system at non-destructive voltages appears to be possible only at much higher current sensitivity than in STM experiment [3]. Figure 3 illustrates the current dependence on gate voltage at fixed tunnel voltage.

One can see a non-monotonous shape of the current on gate voltage dependence. Such feature is characteristic of only correlated electron tunneling. The shape of I-V curves and the shape of current on gate voltage dependence provide a conclusion that the electron transport through the investigated molecular system has correlated electron tunneling nature.

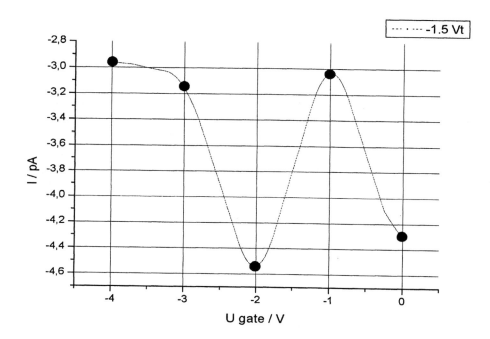

Figure 3. Dependence of current through molecular system on gate voltage at room temperature. The source-drain voltage is 1.5 V.

CONCLUSION

In this work, an electron transport through monomolecular film of $Pt_5(CO)_7(P(C_6H_5)_3)_4$ molecular clusters incorporated in matrix of PVP-20 polymer molecules was investigated. Transistor effect in such molecular system was observed, i.e. it was shown that the current through such system depended on external electric field and this dependence has non-monotonous shape. The shape of the current dependence on external electric field and the shape of I-V curves prove the realization of correlated electron tunneling in studied molecular systems.

ACKNOWLEDGEMENT

This work was supported in part by INTAS (Pr. No. 99-864), ISTC (Pr. No 1991), RFBR (№ 02-03-33158, № 01-02-16580).

REFERENCES

[1] K. Likharev, 1999, Single-electron devices and their applications, *Proc. IEEE*, 87(4), 606–32.
[2] S. P. Gubin, 1987, Chimiya Klasterov, (Moscow: Nauka).

[3] E.S. Soldatov, V.V. Khanin, A.S. Trifonov, D.E. Presnov, S.A. Yakovenko, G.B. Khomutov, S.P. Gubin, V.V.Kolesov, 1996, Single-electron transistor based on a single cluster molecule at room temperature, JETP Lett., 64(7), 556–560.

[4] S. Yakovenko, S. Gubin, E. Soldatov, A. Trifonov, V. Khanin, G. Khomutov, 1996, Clusters in Langmuir monolayer on the graphite surface, Inorganic Materials, 32(10), 1272–1277.

Integration of Smart Materials in Si to Create Opto- and Micro-Electronic Hybrid Devices

S. LIBERTINO and M. FICHERA
CNR—IMM sez. Catania, Catania, Italy

A. LA MANTIA
STMicroelectronics, Catania, Italy

ABSTRACT

In this work we studied the possibility of integrating biological materials within Si using processes compatible with VLSI technology and final front-end processing. In particular, we characterised the Purple Membrane (PM) of the Halobacterium salinarium, that contains bacteriorhodopsin, a photosensitive protein. When irradiated a complex photochemical cycle begins consisting of intermediate states with different absorption spectra and lifetimes. The PM was characterised in both solutions and in dehydrated nano- and micrometric (thicknesses ranged from 2 to 45 µm) films on Si-based surfaces. Various substrates were used as deposition surfaces such as glass, Si, and SiO_2. The films were structurally-characterized by scanning electron microscopy and optically-characterized by spectrophotometry and optical absorption. Todate the data collected demonstrate that the PM is suitable for hybrid integration since it keeps its optical properties both in solution and in the dehydrated films. Moreover, the film is very resistant to thermal degradation – a denaturation temperature above 180°C for the films deposited on Si-based surfaces was determined. Finally, the PM patterning achieved a micro-structuring of the Si surface. Stripes 0.6-2 µm wide and from 0.2 µm up to 10 µm deep were fabricated. The PM was subsequently deposited on structured wafer surfaces either by drop deposition or by dipping the sample into the PM solution.

INTRODUCTION

The future generations of microelectronic devices will need the introduction of smart materials as active layers, [1] to perform very complex functions in reduced space. Efforts to create such devices are driven by the need to shrink devices and the ever-growing performance-demands of Moore's Law [2]. Biological materials are good candidates for smart materials since Nature has implemented through evolution, many complex and specialised systems that can accomplish functions such as light emission, electrical signal emission, energy conversion and collection, movement, etc. The material of choice to achieve integration of organic materials with a microelectronic device is Si. In fact, it is nowadays considered the leading material in the microelectronic industry thanks to low-cost processing in fabrication the devices as well as the advent of VLSI technology. Many efforts are devoted to use it as the platform for monolithic integration of new functions and devices (e.g. optoelectronic devices). The possibility to integrate Si and biological materials having a signal transduction between them in hybrid devices, signals the birth of bioelectronics [3]. The first step is to develop know-how on biological material manipulation directed at hybrid device integration. The biological material must be bonded onto Si-based materials to avoid denaturation using processes compatible with VLSI. Moreover, other non-trivial problems must be faced: Si is not considered a bio-compatible material hence new technological solutions must be advanced to achieve such integration. Finally, the organic material must be patterned on the surface of the final device.

Our main goal is to realize hybrid (organic-inorganic) devices in which the biological part accomplishes complex functions to act as a smart material. In this work we characterised the properties of the Purple Membrane (PM) of the *Halobacterium salinarium*, a bacterium living in salty waters, containing a photosensitive protein, bacterio-rhodopsin (bR) [4]. The bR is produced to form the PM when the oxygen concentration in the water, necessary for metabolic processes, is too low. The bR is arranged in the PM to form a bi-dimensional hexagonal lattice. Each lattice point is given by three proteins organised in trimers. Space between the proteins are filled with lipids that provide compactness to the structure and a high stability against external agents. The PM resists to pH variation from 1 to 12 and to temperatures as high as 96 °C in water solution [4].

The bR within the PM is photosensitive and when irradiated at 570nm, a complex photochemical cycle begins characterized by a series of conformational changes in the internal structure (intermediate states). The final result is movement of a proton from outside to inside the bacterium membrane. The voltage difference thus created provides the energy needed to accomplish the metabolic function. The voltage returns to the "ground state" in less than 1 second [4]. The intermediates exhibit absorption spectra different from the ground state and have lifetimes ranging from a few picoseconds to 10 ms *in-vivo*. The bR could be used to implement optical functions, e.g., in optical switches, as a sensor for movement or light [5], or as an optical memory chip [6]. The aim of this work is to acquire know-how on protein manipulation, deposition and patterning on Si-based surfaces in an environment compatible with the VLSI technology.

EXPERIMENTAL DETAILS

We used PM fragments commercially available (MIB Munich) as a lyophilised purple powder. The PM solutions were prepared using deionized water with a resistivity of ~18 MΩ. The PM concentration was in the range of 1 to 10 mg/ml.

The solutions were deposited on different samples: Si, after a dip in HF; SiO_2 thermally grown on Si (to a thickness of 900 nm); and glass. Two different deposition methods were used: sample immersion in the PM solution and PM solution drops deposited on the sample. For the last deposition method, the drop volume was 10 µl with a variation in concentration of ~1%.

The treated samples were put in a chamber and a partial vacuum was generated by a membrane pump to facilitate the release of air bubbles lying on the film/substrate interface. The internal humidity of the chamber was controlled and raised to ~96% to slow down the drying process. In this way, uniform films were fabricated. Optical measurements were carried out with two pieces of equipments: a photo-spectrometer Perkin-Elmer λ4500 with a tungsten lamp at a power of 200 mW; and a photomultiplier with a lock-in method to detect light dispersed by a monochromator. The data were collected by computer. The absorbance (A) spectra were determined using the equation:

$$A = -\log \frac{I}{I_{ref}} \quad (1)$$

where I and I_{ref} are the light intensity of the lamp after the sample and after the reference, respectively. In particular, the reference for the solutions was a couvette containing the same solution but without the PM. All the PM solutions were contained in 1-cm square couvettes. In this way the measured absorbance is only due to the PM/photon interaction. Scanning electron microscopy (SEM) measurements were carried out using a Cambridge Autoscan MK250 and a Leo 1550 field emission unit. A thin Pt-layer (~100 Å) was deposited on the samples to reduce the charge on insulating layers during the SEM measurement. The film thickness was measured on cleft samples (cross sections).

RESULTS AND DISCUSSION

Optical Characterization of Solutions

We characterized the PM in solution as a function of both solution temperature and the PM concentration. As an example, in Figure 1, the absorption spectra are plotted for solutions with 0.1 mg/ml of PM (solid line A) and 1 mg/ml of PM (solid line B), calculated according to Eq. 1. The absorption peak at 570 nm is clearly detectable and shows that the photocycle has started confirming that the protein keeps its optical properties in solution. The absorption peak is quite broad since the photocycle must start with the widest wavelength band compatible with emissions from the sun as being fundamental to bacterium life. The absorbance value increases linearly with the PM concentration in solution as expected increasing the molar concentration (c) according to the Lambert–Beer law:

$$A = \varepsilon \, l \, c \quad (2)$$

where l indicates the optical path (1 cm for the solutions contained in the couvettes) The extinction coefficient (ε) derived is ~25000, in agreement with the literature data [7]. The couvettes were subsequently immersed in an oil bath and the temperature was increased under continuum mixing to avoid temperature gradients, up to 96 °C, the literature value [4] for bR denaturation. The absorption spectra after annealing are shown in the same figure. The PM in the low concentration sample is fully denatured ad observed by the dashed line (C). The protein lost its optical properties, the absorption peak at 570 is no more present and the solution assumed a yellowish color, visible also with naked eye. On the other hand, after 96 °C 20 min the PM in the highest concentration sample is still active, as clearly visible by the absorption spectrum (solid line E) in figure. An annealing at 99 °C is necessary to achieve the full PM denaturation in this sample (dashed line, D).

Similar measurements were performed as a function of the PM concentration and the resulting denaturation temperatures are summarised in Figure 2 for solutions with PM concentrations from 0.1 mg/ml up to 2 mg/ml. The solid line in the figure is only a guide for the eye.

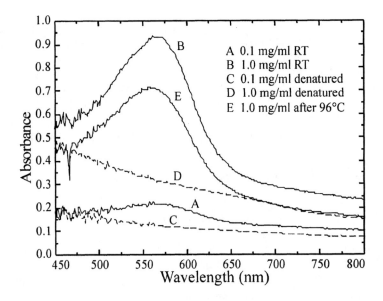

Figure 1. Absorbance spectra of solutions with 0.1 mg/ml and 1 mg/ml of PM before (A and B, resp.) and after annealing: at 96°C for 20 min (C and E, resp.); at 99°C for 20 min (D, 1 mg/ml PM solution).

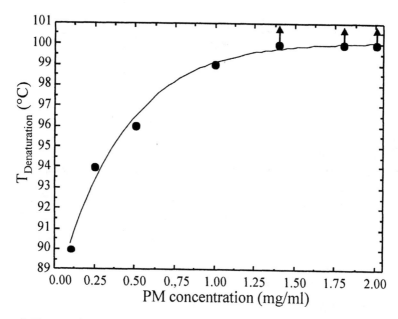

Figure 2. Denaturation temperature as a function of the PM concentration in solution. The arrows indicate that full denaturation is not achieved at the highest temperature (100 °C).

Fabrication of PM Films on Si-Based Surfaces

Different from literature data [8], PM in water does not denature at the same temperature for all concentration levels. In fact, the denaturation temperature is 90 °C for PM concentrations of 0.1 mg/ml, while it increases to 99 °C for a PM concentration of 1 mg/ml. Moreover, we have shown [9] that the PM thermal stability increases if a percentage of polyvinyl alcohol is added to the solution (from 15 to 25%). The arrows in the figure indicate that a higher temperature is necessary to achieve full denaturation of the protein in water at a PM concentration above 1.3 mg/ml. We believe the increased PM stability is due to PM–PM interactions. In fact, in order to denature, the seven helix that "protect" the chromophore must unfold. If they are embedded in a matrix through formation of large PM agglomerates or by encapsulation in a polyvinyl alcohol matrix, a higher energy is necessary to unfold. Our hypothesis is confirmed by measurements performed on large PM agglomerates (not shown).

Once we had characterized the PM properties in solution, we fabricated thick films on Si-based materials, following the procedure described in the experimental details section. A typical SEM image of a PM thick film on SiO_2 obtained from deposition of 10 μl of PM solution (1 mg/ml) is shown in the insert to Figure 3. The marker in the figure is 1 μm in length. The image also shows a flat surface without visible cracks or air bubbles. The roughness observed in the SEM is due to Pt deposition necessary to carry out measurements. In fact, the sample flatness was evaluated using Atomic Force Microscopy (AFM) measurements (not shown), performed on the same samples before Pt deposition and a roughness value of ~5% of the thickness was determined. The PM film is strongly bonded to the SiO$_2$ underlying and cleavage does not peel the PM from the surface.

SEM measurements were also performed on PM films deposited on Si and glass (not shown) under the same deposition and drying conditions to compare the film thickness and the macroscopic properties on the different substrates. To vary the film thickness two different approaches can be followed. One can either change the PM concentration in solution or the deposition method. The results obtained changing the PM in solution are summarized in Figure 3 where the film thickness as a function of the PM concentration in solution is plotted for drops of 10 µl deposited on Si (O), SiO$_2$ (o) and glass (Δ) surfaces, as measured from SEM cross sections of the samples. The graph shows that the film obtained on a Si surface from the 10 mg/ml solution is 5 µm tick, while on glass the thickness is below 3 µm.

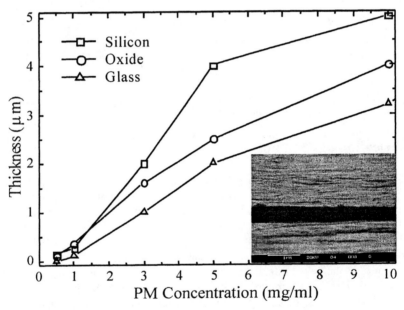

Figure 3. Film thickness as a function of PM concentration in solution for a 10 µl drop deposition on Si (□), SiO$_2$ (o) and glass (Δ) surfaces. In the inset: SEM image of 1 mg/ml of PM deposited on SiO$_2$.

The measurements do not show visible differences in the film quality for the various samples: both the surface flatness and the film uniformity are the same in all of the measured samples. The only visible difference is the film thickness. In fact, the solution concentration and surface hydrophilicity determine the film thickness while the drying process is the only variable to account for the final flatness and uniformity. The PM film are thickest on the Si samples while the SiO$_2$ samples have the thinnest film deposition. Since Si is hydrophobic, the solution drop does not expand on its surface. On the other hand, the SiO$_2$ and the glass are more hydrophilic which causes the water drop to expand. If the drop volume doubles, the final deposited area doubles but the film thickness remains unchanged. The data plotted in Figure 3 clearly confirm that, once the PM concentration in solution is defined, for levels ≥ 3 mg/ml, the substrate hydrophilicity determines the final thickness. Finally, very thin films (< 1mg/ml of PM in solution) are not uniform exhibiting long-range roughness. The data points plotted in the figure are the average thicknesses with a variability of ~50% of the average. The results suggest that one deposition method can be used to obtain thick films (≥ 0.5 μm) while a different one must be used to deposit thin films. The approach we followed and the results obtained are described later.

Once the PM films were deposited on the Si-based surfaces, we tested their optical properties in order to determine if the deposition procedure or the interaction with the Si-based surfaces denatured the PM or made it unable to begin its photocycle. To test the film, we performed absorption and diffused reflectance measurements. If the protein did not lose its optical properties, a strong absorption peak at 570 nm is observed. The absorbance spectra are shown in Figure 4.

In particular, the absorbance spectrum of a PM film deposited on glass using a solution with a PM concentration of 5 mg/ml is plotted in Figure 4.a. The sample exhibits a well-defined absorption peak at 566 nm. The water solution absorbance is higher than the film absorbance (see Figure 1) since the optical path, l, (see Eq. 2) is longer in this case. In fact, the film on glass thickness is in this case only a few microns, much smaller than the 1 cm of the couvette thickness. Finally, the PM films deposited on Si also exhibit a strong absorption peak at 570 nm. Since Si is opaque in the visible range, we performed diffused reflectance measurements. The light incident on the sample in the visible range is reflected and diffused. We collected this light and the results are summarised in Figure 4.b. In particular, the absorbance was obtained from the ratio between the reflectance (diffused and mirrored) spectrum of a Si surface (our reference) and that of a PM film deposited on Si. In this case a quite strong absorption peak is also observed, as would be expected if the protein maintains its optical properties after film formation. It should be mentioned that the peak is centred at 540 nm, hence shifted with respect to the other spectrum. Further measurements are in progress to understand whether the shift is due to the film deposition on the Si substrate and/or to the dehydration process as previously suggested in the literature [10] or to the different measurements apparatus used for the two sets of measurements. In fact, in this latter case the measure was performed using a spectrophotometer, hence the sample was lit with monochromatic light, while in the first case the sample was lit with the W lamp. This resulted in different sample warming during measurement.

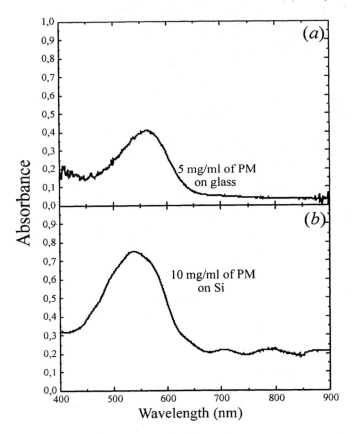

Figure 4. Absorbance in the visible range from (*a*) transmittance measurements of a glass sample with a PM film deposited from a solution of 5 mg /ml and (*b*) diffused and mirrored reflectance measurements of a Si sample with a PM film deposited from a solution 10 mg/ml.

Finally, we tested the thermal resistance of the deposited films to determine compatibility with the front-end processes necessary to finish an electronic device. We monitored absorption at 570 nm of PM films deposited on glass in-situ, as a function of sample temperature and the results, normalised to absorption at room temperature are plotted in Figure 5. The film does not show any variation in absorption up to 160 °C, while denaturation clearly occurs at 170 °C. Qualitative measurements of the film deposited on Si were performed and they show (see the insert of Figure 5) that the film maintains its colour (due to the photocycle start (*a*) up to 190 °C. Above 190 °C, the colour changes suggesting protein denaturation (*b*). The film denaturation temperatures determined in this work are much higher than the best previously reported value of 140 °C [11]. The results shows that film thermal stability is compatible with the final front-end processing, hence integration of hybrid devices is achievable if film patterning and the transduction of the information can be achieved.

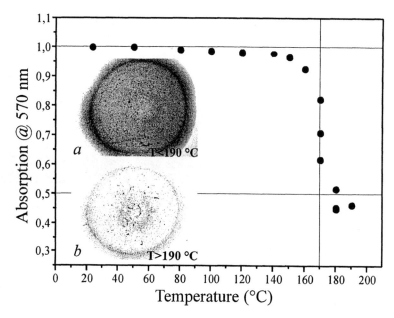

Figure 5. In-situ absorption measurements of a PM film on glass at 570 nm as a function of temperature. In the insert, optical images of a PM film on Si: (*a*) room temperature; (*b*) after 10 min at 190°C.

Patterning of PM Films

In the previous section we showed it is possible to deposit thick films of PM on Si-based surfaces without causing denaturation and we obtained compact and uniform films. To pattern the films, we changed the microstructure of the Si surface by defining 2μm wide square areas plasma-etched to a depth of 2μm. The structure is shown in Figure 6.*a*. On these samples, a drop of 10 μl of a solution with a PM concentration of 5 mg/ml was deposited. The samples were observed by SEM and the cross section and plan view are shown in Figure 6.*b* and Figure 6.*c*, respectively (the marker in each Figure is 1μm). The images show the structure and good quality of the etching performed to fabricate the trenches. The PM fills the microstructures taking on a compact shape. Some holes, visible in Figure 6*b* between the wall and the PM film, are due to imperfect drying. Slowing the drying and improving the chamber vacuum at the start of the process can remove these defects as air bubbles come away from the Si/PM interface while the PM is still in solution.

Another interesting observation is provided by the SEM image of the sample surface. As well as the regular holes corresponding to the microstructure, some cracks are evident also due to non-optimum drying. This result indicates the membrane has constant thickness and exhibits uniform coverage of the sample. More importantly, even small holes are filled by this procedure. Unfortunately, after drying we could not remove the PM from the Si surface. The PM was very resistant to chemical and plasma etches and can't be removed by selective etching of the sample surface. Even after denaturation, it remains on the surface. Perhaps mechanical etching or sample rinsing just after solution deposition may solve this problem to allow thick film patterning on Si.

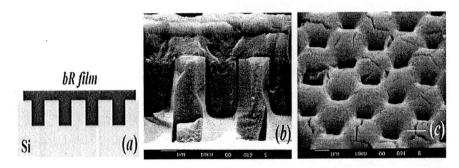

Figure 6. Microstructures for thick film deposition: (*a*) schematic of the section; (*b*) SEM image of the cross section (the marker is 1µm); (*c*) SEM of the surface (the marker is 1 µm).

In the literature a different approach has been followed: the bR is ablated using an excimer laser and a hard metal mask fashioned to cover the sample during ablation [12]. We followed a different approach. We structured the sample surface and identified the deposition procedure that allows selective bonding of the protein to different materials. The idea was to identify both the deposition procedure and the materials that allow the PM to bond only on selected sample areas. For this purpose, a different deposition method was carried out. The samples were immersed in the PM solution for 10 min (the optimum time to have maximum surface coverage), subsequently rinsed in deionized water, and then dried using a N_2 flow. Using this procedure, only the covalently-bonded PM remains on the sample surface. Si and SiO_2 surfaces were prepared for comparison by immersed in the PM solutions and reference samples, immersed in water without PM and then dried following the same procedure. A typical SEM image of a PM thin film deposited on SiO_2 is shown in Figure 7*a* (the marker is 0.2 µm), while a Si sample that underwent the same deposition and drying procedure is shown in Figure 7*b* (the marker is 10 µm). A nanometric PM film is deposited on SiO_2 even though the surface coverage is non-uniform.

We believe that non-uniform coverage depends on surface charging mechanisms. The PM does not bond to the Si surface when this deposition method is used. The reference samples exhibit a flat surface (not shown) confirming the deposited material on SiO_2 is PM. The data suggest that properly patterning the Si surface with SiO_2 and then using the deposition and drying procedures described, it should be possible to achieve protein localization in some areas of the sample surface.

 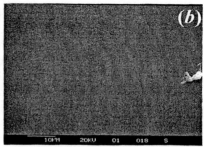

Figure 7. SEM images of (a) SiO$_2$ (the marker is 0.2 μm) and (b) Si (the marker is 10 μm) surfaces after dipping in PM solution and drying under N$_2$ flow.

We fabricated the structures as shown schematically in Figure 8a. A thin layer of 25 nm of SiO$_2$ was thermally grown on Si. On such a layer, 200 nm of polycrystalline Si was deposited and subsequently plasma-etched. The mask was designed to open up both a "squares" and "stripes" geometry. The "squares" have 600 nm on a side while the "stripes" are 600 nm wide and 2 mm long. The plasma-etching was carried out to completely etch the polycrystalline Si opening up SiO$_2$ areas. The PM was deposited according to the procedure described before and, after drying, measurement was done by SEM. An SEM cross-section image is shown in Figure 8b. The protein has filled the stripes leaving the polycrystalline Si uncovered. The side walls (against the polycrystalline Si) don't allow protein bonding. Their apparent coverage is only due to the protein film thickness. In fact, the protein film thickness after deposition on SiO$_2$ is ~200 nm, as monitored by SEM on flat SiO$_2$ surfaces. The maker in the figure is 200 nm and the "stripes" geometry is clearly visible. The dimensions are the expected ones and the walls are perfectly perpendicular confirming a correct etching process of the structure. The results demonstrate that by choosing the right structure and deposition procedure, it is possible to pattern PM thin films.

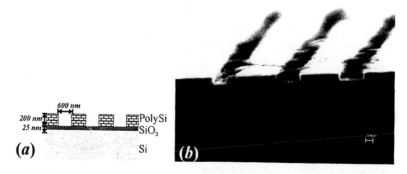

Figure 8. Microstructures for thin film deposition: (a) schematic of the section; (b) SEM image of the sample after dipping in PM solution and drying under N$_2$ flow (the marker is 0.2 μm).

Previously we showed that the PM films deposited on Si-based surfaces maintain their optically properties and denature at temperature ≥ 170°C, allowing the front-end final processes for fabrication of microelectronic devices. The optical properties of the thin PM patterned films were tested for monitoring the diffused reflectance spectrum. Even though only a weak signal was collected (not shown), an absorption peak at 540 nm, typical of the ground state absorption of the bR was observed. All these results clearly show it is possible to fabricate hybrid organic-inorganic devices using the bR as a "smart" material and the well-established Si fabrication technology.

CONCLUSION

We have shown how to fabricate thick and thin PM films on Si-based surfaces. In particular, thermal stability can be increased in solution to 100 °C by proper selection of PM concentration. The deposition and drying procedure assure good quality films, uniform and without macroscopic defects. Film thickness can be designed by properly changing the PM concentration. All fabricated samples exhibit a typical PM absorption spectrum indicating the protein maintains its optical properties in the dried film. Increased thermal stability was found: 170 °C for films on glass and 190 °C for films on Si. Finally, we can pattern thin PM films on Si-based surfaces by micromachining and using SiO_2 as the deposition surface. SiO_2 is a good anchoring site while Si is unsuitable. If thick PM films are required, selected areas can be filled with the protein and so, solutions are available to pattern the surface structure. Since the PM deposited on Si-based surfaces maintains its optical properties and allows the use of front-end final thermal processes, the results demonstrate that it is possible to fabricate hybrid organic-inorganic microelectronic devices using the bR as a "smart" material and the well-established Si fabrication technology.

ACKNOWLEDGEMENT

The authors wish to thank Dr. G. Strazzulla and Dr. G. Baratta for the spectrophotometry measurements on Si samples, Mr. A. La Mantia, Dr. P. Santoro and Mr. A. Vazzana for the SEM analysis, Mr. A. Spada and Mr. N. Parasole for the expert technical assistance. This work was partially funded by STMicroelectronics within the advanced research project MURST 8 "*Dispositivi elettro-ottici per l'immagazzinamento ed elaborazione di informazioni basati sul sistema Euglena Rodopsina-Si*" funded by the Italian government (law 488).

REFERENCES

[1] D. A. Loy, 2001. MRS Bulletin 26, May 2001.
[2] International Technology Roadmap for Semiconductors (ITRS) 2001. Available at the website http://public.itrs.net/Files/2001ITRS/Home.htm. Accessed February 2003.

[3] R. R. Birge, 1990. Photophysics and molecular electronic applications of the rhodopsins, Annu. Rev. Phys. Chem. 41, 683–733.
[4] N. Hampp, 2000. Bacteriorhodopsin as a Photochromic Retinal Protein for Optical Memories, Chem. Rev. 100, 1755–1776.
[5] T. Miyasaka, K. Koyama, 1993. Image sensing and processing by a bacteriorhodopsin-based artificial photoreceptor, Appl. Optics 32, 6371–6379.
[6] R. R. Birge, 1994. Protein-Based Three-Dimensional Memory, Am. Scientist 82, 348–355.
[7] http://www.shu.ac.uk/schools/sci/chem/tutorials/molspec/beers1.htm. Feb. 2003.
[8] A. A. Khodonov, O. V. Demina, L. V. Khitrina, A. D. Kaulen, P. Silften, S. Parkkinen, J. Parkkinen, T. Jaaskelainen, 1997. Modified bacteriorhodopsins as a basis for new optical devices, Sensors and actuators B 38-39, 218–221.
[9] S. Libertino, M. Fichera, A. La Mantia, D. Ricceri, 2003. Optical and structural characterization of bacteriorhodopsin films on Si-based materials, accepted for publication in Synthetic Metals.
[10] J. Tallent, J. Stuart, Q. Song, E. Schmidt, C. Martin, R. Birge, 1998. Photochemistry in dried films incorporating deionized blue membrane bacteriorhodopsin, Biophysical J. 75, 1619–1634.
[11] Y. Shen, C. Safinya, K. Liang, A. Ruppert, K. Rothschild, 1993. Stabilization of the membrane protein bacteriorhodipsin to 140 °C in two-dimensional films, Nature 366, 48–50.
[12] D. Haronian, A. Lewis, 1992. Microfabricating bacteriorhodopsin films for imaging and computing, Appl. Phys. Lett. 61, 2237–2239.

High Performance Thermo-Elastic Metallic Materials by Rapid Solidification and their Applications in "Smart" Medical Systems

Y. FURUYA
Faculty of Science and Technology, Hirosaki University, Hirosaki, Japan

Y. SHINYA
Graduate Student, Hirosaki University, Hirosaki, Japan

M. YOKOYAMA
Faculty of Science and Technology, Hirosaki University, Hirosaki, Japan

T. YAMAHIRA
Graduate Student, Hirosaki University, Hirosaki, Japan

S. TAMOTO
NEDO Fellow, Advanced Institute of Science and Technology, Tsukuba, Japan

T. OKAZAKI
Faculty of Science and Technology, Hirosaki University, Hirosaki, Japan

Y. TANAHASHI
Tohoku Kosai Hospital, Aobaku, Sendai, Japan

ABSTRACT

The purposes of this study is to develop fine fiber and foil-type sensor/actuator material elements by using the originally designed rapid solidification apparatus, and to investigate how to control the microstructures to prepare new multi-functional solid-state metallic sensor/actuator materials in order to apply to smart medical systems in the body. First, by changing the rotating speed of rapid-solidification roller of either plane surface type or triangle-tip type, the thin foil or very fine fiber sensor/actuator material elements could be produced. They had strong crystalline texture with fine columnar grains which were uniquely formed by rapid-solidification process. As the thermoelastic shape memory alloy (TSMA), sharp and non-linear TiNiCu alloy foil and fiber with very narrow transformation temperature range below 10K which works at human body temperature range, were successfully developed. As the ferromagnetic shape memory alloy (FSMA), FePd and FePt alloy foils with improvement of the ductility as well as striction could be developed. By using these developed SMA actuator material element, we design and developed the proto-type micro-gripper at the front of endoscope for medical devices.

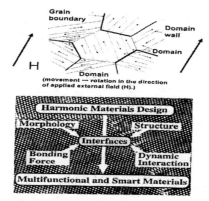

Figure 1. Harmonic material design and importance of microscopic interface structure design.

INTRODUCTION

Many metallic sensor/actuator materials with multi-functional characterizations (shape memory, magnetostriction and piezo-electricity, etc.) have phase transformations accompanied with mechanical, electrical, magnetic, and thermal energy conversion. Since these characteristics are strongly related to microstructural grain boundary morphology as shown in Figure 1 especially for the interaction of domain and grain boundaries, then the use of harmonic material design is important to improve performance [1].

In this paper, the improvement of the material functions of sensor/actuator materials (fiber and thin foils) is studied by controlling the metallurgical microstructure by using the originally designed rapid-solidification apparatus. Then, we evaluate and discuss the high performance of multi-functional materials from a microcrystalline viewpoint. Finally, we try to design and develop prototype micro-grippers at the front of the endoscope as a smart medical device by using the developed SMA actuator.

EXPERIMENTS

The rapid-solidification method used to develop foil- or fiber-type metallic sensor /actuator materials is shown in Figure 2.

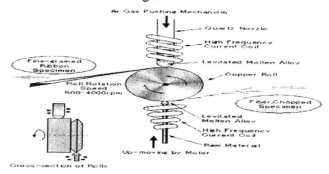

Figure 2. Rapid-solidification technique to develop foil/fiber type metallic sensor/actuator materials.

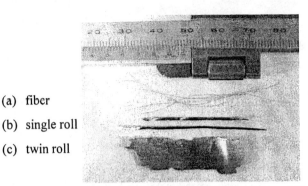

(a) fiber
(b) single roll
(c) twin roll

Figure 3. The example of the produced rapid-solidified ribbon and fiber samples.

The apparatus continuously produces thin foil (30~120μm thickness) or fine fiber (20~80μm diameter) by combining electromagnetic float-melting with a rotating roller of plane surface or triangle-tip type. The arc-melted button of alloy cut into the shape of a rectangular block was used as raw material. The block was placed at the bottom of a quartz nozzle surrounded by a high frequency electromagnetic coil to provide the downward flow of melted metal. This method enables us to control metallurgical microstructures delicately by changing rolling speed (i.e., changing the metal cooling rate). By using this new material processing method, we produce thin foil and fine fiber of TSMA ($Ti_{50}Ni_{50}, Ti_{50}Ni_{40-42}Cu_{8-10}$) and FSMA (Fe-29.6at%Pd, Fe-22.7at%Pt). Examples of the rapidly solidified materials are shown in Figure 3.

RESULTS AND DISCUSSION

Shape Memory Effects of Thin Foil and Fine Fiber of TiNiCu Alloys

Figure 4 shows shape memory function of fine fibers for $Ti_{50}Ni_{50}$, $Ti_{50}Ni_{42}Cu_8$ and $Ti_{50}Ni_{48}Co_2$ alloys prepared by using a tri-angle tip type roller in our new material processing method shown in Figure2. Shape memory strains of 3~4.5 at% for these fibers and foils are greater than for other actuator materials.

Moreover, the thermo-elastic shape memory of Ti50Ni37Cu13 alloy foil samples show non-linear hysteresis as well as very sharp crooked points at each phase transformation temperature compared to conventional $Ti_{50}Ni_{50}$ as shown in Figure 5. On the other hand, the shape memory strain of $Ti_{50}Ni_{48}Co_2$, which undergoes a low-temperature martensite-to-austenite phase transformation (As☐ 270 K), is slightly smaller than others, but has a small, linear hysteresis, indicating it is a good sensor/actuator element for cyclic use. We found that the developed thin foils and fine fibers have stronger and sharper longitudinal shape memory properties. Although its origin is not yet clear, fibers with a longitudinal crystalline anisotropy is useful in intelligent/smart actuators that are applicable to micro-device design for medical micro-machines in the human body.

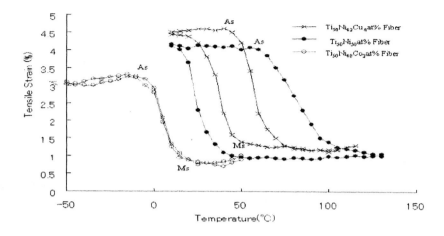

Figure 4. Thermoelastic shape memory for $Ti_{50}Ni_{50}$, $Ti_{50}Ni_{42}Cu_8$ and $Ti_{50}Ni_{48}Co_2$ fine fibers under. Loading (stress = 100 MPa)

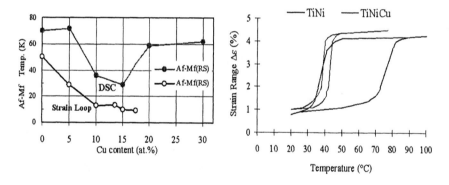

Figure 5. Transformation temperature range per cycle by DSC and shape memory strain, and Thermal strain loops in rapid-solidified TiNi and Ti50Ni37Cu13at% ribbons.

GREAT MAGNETOSTRICTION OF THIN FOIL FSMAs

Ferromagnetic shape memory alloy Fe-29.6at%Pd thin foil exhibits great magnetostriction as seen in Figure 6 which shows the dependence of strain on applied magnetic field direction (θ) referred to transverse direction [2]. The magnetostriction with a small hysteresis loop depends remarkably on θ and has a maximum value of -1200 ppm at θ = 87°, i.e., the direction almost perpendicular to the foil surface. The magnetostriction is about 40 times larger than that of randomly oriented polycrystalline material.

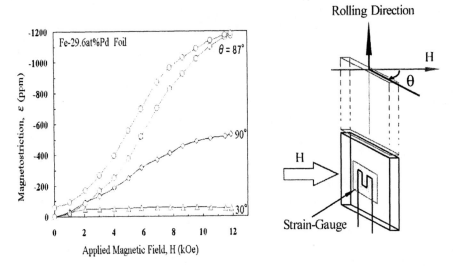

Figure 6. Magnetostriction of Fe-29.6at%Pd thin foil. The value of θ is the rotation angle between magnetic field and transverse direction of foil and schematic diagram of measurement method.

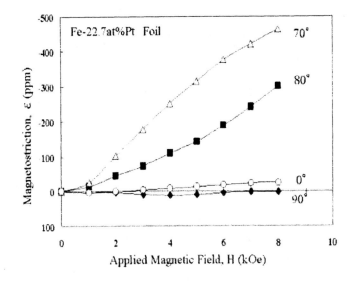

Figure 7. The same as Figure 4, except the sample is Fe-22.7at%Pt thin foil.

We have also succeeded in making thin foils from Fe-22.7at%Pt FSMA with large magnetostriction using rapid solidification. Figure 7 shows a maximum magnetostriction of -470 ppm arising at θ = 70° that is slightly off-normal to the foil surface. Thus, the maximum strain arises in a direction almost perpendicular to the foil surface. This directionality may be attributed to the fact that textures have strong crystal anisotropy of fine columnar grains with low-energy grain boundaries uniquely formed by rapid-solidification [3].

MICRO-GRIPPER OF ENDOSCOPE

Finally, as an application of the developed TSMA, TiNi alloy foil and FSMA, PePd alloy foil, we tried to design and fabricate a small gripper actuator at the front face of an endoscope for less-damage endoscopic surgery. The gripper consists of a composite material consisting of two layers (main matrix - TiNi foil (0.2mm thick and length 20mm) over which ferromagnetic iron Fe foil is attached. The gripper is set within the endoscope wall and small magnets are embedded on both sidewalls as shown in Figure 8. To achieve shape recovery effects for gripping, direct electric heating is done above the inverse transformation temperature (Af). The inverse opening motion of the gripper is done by bias-force from the embedded magnets. In the case of ferromagnetic SMA (FePd foil), it is not necessary to use Fe foil as FePd has ferromagnetic features that respond well to the magnet. We verified the cyclic motion of the gripper by switching the electrical direct heating on and off with a small battery (1.5V, 0.2A) as in Figure 8. This type of ferromagnetic type SMA gripper actuator is worthy for use in medical micro-surgery procedures in the human body.

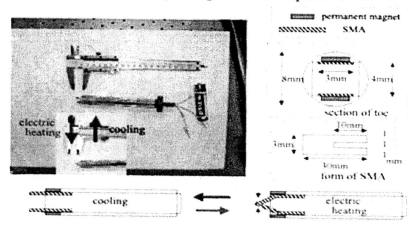

Figure 8. Endoscope (gastro camera) toe gripper.

CONCLUSION

By a rapid-solidification technique, drastic improvements in the functional characteristics of thermoelastic and ferromagnetic shape memory alloys can be achieved by introducing strong textures having heavy crystal anisotropy of fine columnar grains with low-energy grain boundaries. Thermo-elastic shape memory TiNiCu alloy foil showed very narrow hysteresis and sharp shape recovery motion.

On the other hand, the ferromagnetic shape memory FePd, FePt alloy foils showed larger magnetostriction than those of randomly crystalline-oriented bulk-materials of the same origin. Development of rapid-solidified fine fiber-type sensor/actuator materials seems very interesting and exciting, since it is possible that their texture have a quasi-one-dimensional microstructure along the surface direction, i.e., rolling direction, which is now under survey by our team at Hirosaki University, Japan. These sensor/actuator material elements become useful to the micro-electronic mechanical systems, and one trial of the developed actuator materials, we could succeed in making and actuating the prototype micro-gripper at the front face of the endoscope.

REFERENCES

[1] Y. Furuya, 1998. New type of shape memory alloys developed by electromagnetic nozzleless melt-spinning method, Proc. 1st Japan-France Intelligent Materials and Structures Seminar, Sendai, 113–122.
[2] T. Kubota, T. Okazaki, Y. Furuya, T. Watanabe, 2002. Large magnetostriction in rapid-solidified ferromagnetic shape memory Fe-Pd alloy, J. Magn. Mater., 239, 551–553.
[3] Y. Furuya, N. Hagood, H. Kimura T. Watanabe, 1998. Shape memory effect and magneto- striction in rapidly solidified Fe-29.6at%Pd alloy, Mater. Trans. JIM, 39, 1248–1254.

Systemic Evaluation for Replication Processing of DNA Double Strand as a Zippering Micro-Machine

H. HIRAYAMA

Department of Public Health Asahikawa Medical College, Midorigaoka, Asahikawa city, Japan

ABSTRACT

Digital replication processing of DNA molecule has been modeled on the basis of zippering machine. We described the probability that there are k successive pairs of unbound separated base pair s on the template DNA sites by a set of linear differential equations. We have applied the H-infinity control principle to evaluate noise filtering function of this DNA replication process.

INTRODUCTION

The ultimate mechanisms for genetic information transmission are transcription of genetic information from DNA and translation of the codes through RNA to produce protein [1]. The genetic information is preserved in the next generation by replication of the parent double-stranded DNA in terms of RNA. The genetic information is coded by a specific sequence of base molecules. In double-stranded DNA, each of the base pairs couples with its complementary base molecule through biochemical bridge formation. These molecular arrangements of the bases are replicated by a micro-molecular mechanism. The unwinding of double-stranded DNA and the complementary replication of each base molecule can be understood as a zippering machine [2]. It is useful to describe those phenomena in engineering aspects to predict temporal changes in the replication process. In conjunction with hereditary diseases which originate in miss-replication, it is also important to evaluate the noise filtering function of DNA replication [3]. This paper introduces H-∞ control theory [4] to a DNA replication model as a zippering micro-machine to show temporal changes in several probabilistic base-paired unbounded template DNA states.

THE BASIC MOLECULAR MECHANISMS

DNA Polymeraze

First we explain the elongation process of replicated DNA [5]. The central role of elongating DNA replication is done by DNA polymeraze which catalyzes polymerization among deoxyribonucleotides. The reaction can be summarized as

$$(dNMP)_m + n\, NTP \rightarrow (dNMP)_{m+n} + nPPi$$

where dNTP is deoxynucleotide triphosphate (N denotes an unspecified base molecule); $(dNMP)_m$ is one DNA chain in which m deoxynucleotides are polymerized to act as a primer; PPi is a phosphate molecule. The reaction is achieved on the basis of template molecules. The template carries information about arranging base molecules and is usually one DNA molecule (DNA dependent DNA polymeraze) or an RNA molecule (reverse transcriptase). By this template, the DNA polymeraze can catalyze the polymerizing of $dNMP_m$. The replicated DNA has a complementary arrangement of base molecules to those of the template such as A (Adenine)-T(Thymine) and G (Guanine)-C (Cytosine).

The DNA polymeraze copies the base sequence on the parent DNA chain from the 5' end to the 3' end (Figure 1). The DNA polymeraze requires two substrates–the DNA molecule that acts as a template and a double-stranded primer molecule whose base molecule arrangement is complementary to that of the template DNA. The DNA polymeraze adds a dNMP molecule to the 3' end of the template DNA chain one base at a time [1]. First, the DNA polymeraze binds to a dNTP as a primer carrying a specific base molecule that is complementary to the base molecule on the template DNA. Specifically, the DNA polymeraze catalyzes an ester bond formation between the 3'-OH of the primer RNA and the α-phosphate of the dNTP of the template DNA [1].

```
                        Template DNA
     --------------------------------------------------------------
3'      P     P     P     P     P     P     P              5'
        I     I     I     I     I     I     I
        T     G     C     A     T     G     C
        *     *     *     *     *     *     *
        A     C     G     T     A     C     G
        I     I     I     I     I     I     I
5'      P     P     P     P     P     P     P              3'
     --------------------------------------------------------------
            Primer   RNA     ------------Replication---------->
```

Figure 1. Schematic illustration of the nano aspect of DNA replication process for the leading chain. P is phosphate; * is a biochemical bridge between the complementary base pairs, one in the template DNA and its counter-part in the primer RNA. Bridge formation is achieved by DNA polymerase. The replication advances from the 5' end of the primer RNA to its 3' end.

Helicaze and Leading Chain

The enzyme that unwinds the double stranded DNA into two single DNA stands is called Helicase. It is a ring doughnut structure that encloses a given cross-sectional plane of the template DNA. The Helicaze moves over the DNA molecule as it opens the double-strand of DNA. This locomotion is driven by the energy derived from hydrolysis of ATP [1]. Since each daughter strand of the template DNA has a complementary base arrangement, when they are separated by the Helicaze, one single strand has an arrangement from the 3' end to the 5' end which is named the "leading chain". While the other single strand has its complementary arrangement from the 5' end to the 3' end and is named the "lagging chain". The DNA polymeraze acts only from the 5' end to the 3' end of the template DNA. So, replication of the leading chain is normally achieved from the 5' end to the 3' end on the template single DNA. Replication of the lagging chain can be explained as Okazaki fragments [6]. The complementary molecules to the lagging chain are produced by several short pieces of base molecules of 1500 bases. These Okazaki fragments are at first discontinuous as contrasted to the complementary leading chain. The small pieces of these fragments are associated by DNA Ligase. As a result, the complementary chain of the lagging chain is completed. The Primaze catalyzes the enzymatic reaction to produce primer RNA to synthesize the lagging chain. This paper confines itself to replication of the leading chain.

Clamping Factors

There are several molecular factors that keep the DNA polymeraze on the template DNA in order to complete the replication process. These are called "clamping factors" [7]. The clamping factor can bind to the DNA polymerase with the help of ATPase – a clamp loader. PCNA (proliferating cell nuclear antigen) is one of the more typical clamping factors. Clamping factors have ring structures that envelopes the DNA. It slides on the DNA as the DNA polymerase and the Helicase move along the template DNA molecule. The number and size of the template is assumed to be constant during replication, ignoring the possibility that a newly-formed daughter helix can serve as a new template for further replication.

ACCESS OF PRIMER RNA TO THE SITE ON POLYMERIZATION OF THE TEMPLATE DNA [2]

The RNA for the replication primer as a substrate is delivered to the template DNA by diffusion with a rate constant k_{dif}. The molecule is deposited on finding an empty site on the DNA template. When the deposited primer RNA is near to the growing end of the chain, replication occurs. The term k_{pol} is the rate constant for this replication while k_{des} is the rate of degradation of the RNA primer.

Replication Process

Replication occurs by means of single primer RNA attached to the leading chain of the two growing chains at the base of the Y form of the template DNA where Y is the center of the branching of one DNA to two [2] and proceeds from the 5' end to the 3' end. The parent helix unwinds ahead of these growing chains, leaving a short accessible region of separated base molecular pairs. The DNA Helicase has an effective region of catalytic action along the template DNA. The length of such a region extends for L sites. This region begins at all times at the center of branching of the parent helix. This effective region extends through the unbound portion of the parent helix into the bound portion as depicted in Figure 2. Unwinding of the parent helix takes place only within this effective region. The unwinding occurs for a given base pair with an effective rate constant k_o. Since the enzyme region moves along the template DNA with replication, the effective region drives the unwinding process by continuously permitting new portions of the bound helix to react with DNA Helicase. We set the following parameters: k_g is effective rate of rewinding of the parent helix in the region of separated base pairs; k_e is the rate of arrival of the Helicase; and finally, $H(t) = k_g \exp(-k_e t)$ is the modeling of chain growth by means of a first order rate process.

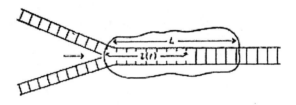

Figure 2. Schematic illustration of Y separation for replication of double stranded DNA. The Helicase active region extends L base pairs. l(t) is the instantaneous number of accessible template sites.

Base Pair Opening

All the base pairs between the active region and the end of the bound helix open instantly. This is achieved when the following condition is satisfied [1, 2]. When in the region of Helicase activity, a base pair opens spontaneously at a location less than or equal to some critical number "c" of the template DNA sites from the end of the bound helix. All of these template sites are then available to receive diffusing primer RNA. Thus, they can contribute to replicate a daughter helix. A base pair that opens at a distance longer than c base pairs from the unbound end causes no further openings. These are isolated from the replicating chains.

Mathematical Description

The time dependent competition between unwinding and replicating mechanisms can be described as a stochastic process. We introduce several sets of probabilities [2] to accomplish this.

1. $l(t)$: the expected number of successive separated and empty template sites at time t. For large $l(t)$, the replication proceeds unhindered. For $l(t)$, 1, replication is at least temporarily halted. $l(t)$ is a function of k_{pol}, k_o, k_g, k_e and k_{dif}.

2. $\lambda j\ (t)$: Each of the two growing chains has reached the jth template site at time t. The index j starts at 1.

3. We assume chain growth can be modeled as formation of a dimer brought along by the arriving enzyme.

4. $\Phi_k\ (t)$ where $k = 0,\ 1,\ 2,\ ...,\ L$ is the probability that there are k successive pairs of unbound template sites at time t. L is the maximum value of k. Since spontaneous opening of a base pair can occur only within the region of action of the enzyme, index k has an initial value of 0 which changes over time relative to the fixed index j. If chain growth has proceeded at time t to template position j_o, then relative to index j, $k = 0$ is equivalent to j_o+1. The probabilities $\Phi_k\ (t)$ and the function $l(t)$ are related by

$$l(t) = \sum_{k=1}^{L} k\ \Phi_k(t) \tag{1}$$

5. $p\ (t)$ is the probability that a given empty site of the template DNA is occupied by primer RNA,

6. $\psi\ (t)$ is the transition probability from daughter helixes of length j to those of length $j +1$ in the time interval $[\ t,\ t+dt\ ]$.

$$\psi(t)dt = k_{pol}\ p(t)(1 - \Phi_0(t))\ dt \tag{2}$$

7.
$$p(t) = k_{pol}(1 - exp(-k_{dif}\ t)) \tag{3}$$

8. The temporal change of the jth template arrival is:

$$d\lambda_2(t)\ /dt = -\ \psi(t)\ \lambda_2(t) \tag{4.a}$$

$$d\lambda_{j+1}(t)/dt = -\ \psi(t)(\lambda_j(t) - \lambda_{j+1}(t)) \tag{4.b}$$

$$d\lambda_N(t)/dt = \psi(t)\lambda_{N-1}(t) \tag{4.c}$$

We must distinguish several intervals for the index k on the basis of a zippering micro-machine [2]. For the first we have taken k values that satisfy the following condition:

$$c + 1 \leq k \leq L - c \tag{5}$$

$k_o\ \Delta t$ is the probability that a single bound base pair opens spontaneously and $k_g \Delta t$ is the probability that the helix rewinds one base pair in the interval $[t,\ dt]$. Then, we have on the basis of the digital nature of the opening of the double strand [2], the following expression:

$$\Phi_k(t+\Delta t) = \Phi_k(t) + (1 - k_{pol} p(t)\Delta t) [k_o\Delta t \sum_{p=1}^{c} \Phi_{k-p}(t)(1 - k_o\Delta t)^{c-1} + k_g\Delta t \Phi_{k+1}(t) - ck_o\Delta t \Phi_k(t)(1 - k_o\Delta t)^{c-1}$$

$$- k_g\Delta t \Phi_k(t)] + k_{pol} p(t)\Delta t [\Phi_k(t)(1 - k_o\Delta t)^c - \Phi_k(t)(1 - k_o\Delta t)c] \tag{6}$$

The first expression in brackets pertains to changes in $\Phi_k(t)$ when there is no polymerization in the interval $[t, t+dt]$ and the expression $(1 - k_{pol}p(t)\Delta t)$ is its probability. The four terms in the bracket have the following physiological meaning:

1. $\Phi_{k-1}(t)$ $(p=1)$ is the probability that there are $k-1$ open sites. $k_o\Delta t$ is the probability that one additional opening during the interval $[t, t+dt]$ of the first bound base pair occurs which is counted from the end of the kth opened base pair. The term $(1 - k_o\Delta t)^{c-1}$ describes the condition where no other base pairs open.

Figure 3(a) shows this unwinding process for the case where there are $k-1$ open sites. The symbol "I" describes one base molecule. Each of two rows of these I symbols in Figure 3(a)-i. shows an arrangement of bases that are separated from the upper and lower ones to express separated daughter DNA. To the right (Figure 3(a)-ii.) describes the base pair complementary sets. The middle row of I's expresses the chemical bonds between each complementary pair of bases such as A-T and G-C. A mark I* describes the first (counting starts from the ^ mark) base pair after which the bases in the double helix DNA are bound with their complementary pairs. Before the base designated by the I* mark, there are $k-1$ separated bases separated from their complementary pairs. Figure 3(a)-ii. shows that the first bound base pair denoted by I* has been unwound by the Helicase. The I* mark in the middle row has disappeared and two I* remain separated in the upper and lower I rows.

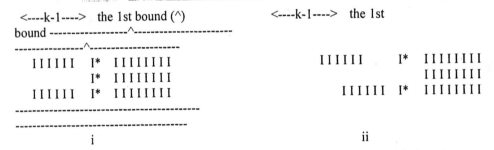

Figure 3(a). Winding (the left part of each Figure) and unwinding (the right part of each Figure) of the template DNA. I* is the opening base pair to which the Helicase acts most effectively. It is denoted by the ^ mark; 3(a)-ii. shows the first bond base pair has been separated.

```
<--k-2--> the 2nd bound                    <-- k-2--> the 2nd
bound
-----------^---------------------          
-----------^---------------------          
I I I I I   I I*    I I I I I I I          I I I I I   I I*   I I I I I I I
            I I*    I I I I I I I                              I I I I I I I
I I I I I   I I*    I I I I I I I          I I I I I   I I*   I I I I I I I
---------------------------------          
---------------------------------          

        i                                           ii
```

Figure 3(b). This Figure emphasizes that the base pair starts to open from the second bond pair (denoted by **I***) which is counted from the end (denoted by the ^ mark) of the separated base pair $[k-2]^{th}$ pair. When the Helicase has finished, two base pairs open simultaneously.

```
<---k-3---> the 3rd bound                  <---k-3---> the
3rd bound
-----------^---------------------
-----------^---------------------
I I I I I   I  I  I*  I I I I I I          I I I I I   I  I  I*  I I I I
I I              I*  I I I I I I                                 I I I I
I I I I I   I  I  I*  I I I I I I          I I I I I   I  I  I*  I I I I
---------------------------------------
---------------------------------------

i                                                   ii
```

Figure 3(c). This Figure emphasizes that the base pair starts to open from the third bond pair (denoted by **I***) which is counted from the end (denoted by the ^ mark) of the separated base pair $[k-3]^{th}$ pair. When the Helicase has finished, three base pairs open simultaneously.

2. There is another possibility that contributes to this issue: the probability that there were k-2 open sites at time t that the second bound base pair opened in the interval [t, t+dt] thus giving k open sites and no other base pairs opened. Figure 3(b) shows the case where k-2 base pairs have been separated. The Helicase catalyze the opening reaction from the second (the count start is denoted by ^) bound pair of the base pairs denoted by **I***. There are two bound base pairs beyond the end of the $[k-2]^{th}$ separated base pairs. Figure 3(b)-ii. shows the separated structure of the DNA where two base pairs have been separated simultaneously to achieve a total of k opened base pairs.

Figure 3(c) shows the case where k-3 base pairs have been opened and the enzymatic separation has started at the third base pair denoted by **I*** counted from the end (denoted by ^) of the $[k-3]^{th}$ separated base pairs.

3. Figure 3(d)-i. shows the case where k-q base pairs have been dissociated. The enzymatic separation starts from the end of this $[k-q]^{th}$ separated base pair until the qth base pair (denoted by **I***) which is counted from the end (counting start denoted

by the ^ mark) of the $[k-q]^{th}$ separated base pairs. Figure 3(d)-ii. shows q opened base pairs, $k-q$ previously opened and q additional opened pairs.

```
<--k-q-->   the qth bound                    <--k-q -->   the qth
bound
-----------^----------------------           
-----------^----------------------
IIII    I I I  I*  IIIII                     IIII    I I I  I*  IIIIII
        I I I  I*  IIIII                                          IIIIII
IIII    I I I  I*  IIIII                     IIII    I I I  I*  IIIIII
----------------------------------
----------------------------------
              i                                          ii
```

Figure 3(d). This Figure emphasizes that the base pair starts to open from the qth bond pair (denoted by I*) which is counted from the end (denoted by the ^ mark) of the separated base pair k-q th pair. When the Helicase has acted, q base pairs open simultaneously.

4. Finally, the probability that there is $k-c$ open sites at time t when the cth bound base pair opened, thus giving c more or k open sites at last and no other base pairs opened. The probability that there are $k+1$ open sites at time t is $\Phi_{k+1}(t)$ and that the helix has rewound one base pair with a probability of $k_g \Delta t$. The probability that there are k open sites at time t when any one of the first c bound base pairs opened, i.e, the case where there is opening to at least the $[k+1]^{th}$ site and that no other base pair opened.

5. The probability that there are k open base pairs at t and the helix has one base pair rewound.

6. The second bracket expression describes changes in $\Phi_k(t)$ when polymerization occurs over the interval $[t, t+dt]$ with probability $k_{pol}p(t)\Delta t$. 1- $\Phi_o(t)$ is not needed because it is sufficient to analyze only:

$$c + 1 \leq k \leq L - c \tag{7}$$

There must be at least one open site in considering all events relevant to the previous four possibilities:

1. The probability that there are $k+1$ open sites at time t and no new sites opened during $[t, t+dt]$. The single polymerization step results in k open base pairs at time $t + dt$.

2. The possibility that there are k open sites at time t and no new sites opened in $[t, t+dt]$. The single polymerization step results in only $k-1$ open base pairs at time $t + dt$. Then the total differential equations are given by the following expressions [2]:

$$d\Phi_o(t)/dt = - ck_o \Phi_o(t) + (k_g + k_{pol} p(t))\Phi_1(t) \tag{8}$$

$$d\Phi_k(t)/dt = k_o \sum_{p=0}^{k-1} \Phi_p(t) - (ck_o + k_g + k_{pol}p(t))\Phi_k(t) + (k_g + k_{pol}p(t))\Phi_{k+1}(t) \text{ for } k = 1, 2, 3, \ldots, c \quad (9)$$

$$d\Phi_k(t)/dt = k_o \sum_{p=0}^{c} \Phi_{k-p}(t) - (ck_o + k_g + k_{pol}p(t))\Phi_k(t) + (k_g + k_{pol}p(t))\Phi_{k+1}(t) \text{ for } k = c+1, \ldots, L-c \quad (10)$$

$$d\Phi_k(t)/dt = k_o \sum_{p=0}^{c} \Phi_{k-p}(t) - ([L-k]k_o + k_g + k_{pol}p(t))\Phi_k(t) + (k_g + k_{pol}p(t))\Phi_{k+1}(t) \text{ for } k = L-c+1, \ldots, L-1 \quad (11)$$

$$d\Phi_L(t)/dt = k_o \sum_{p=0}^{c} \Phi_{L-p}(t) - (k_g + k_{pol}p(t))\Phi_L(t) \quad (12)$$

These equations hold for $c \geq 2$ and $L \geq 2c + 1$ \quad (13)

H-INFINITY CONTROL

We assume that some genetic disturbances will occur during replication. The disturbances can be regarded as noise for signal input to controlled output. A well-controlled output is the normal physiological replication process goal and the influence of noise must be minimized. This can be interpreted that the magnitude of the transfer function from noise to output is minimized. Hence, we apply H-Infinity control to minimize the infinite norm of the system equation for noise. The mathematics are outlined in the Appendix.

CONTROL INPUTS

To drive the replication system, some control input is required. First, advancement of Helicase requires energy that may derive from hydrolysis of ATP. Secondly, the action of DNA polymerase needs some energy because the enzyme catalyzes ester bond formation between the 3'-OH of the primer RNA and the α-phosphate of the dNTP of the template DNA. Thirdly, the action of the clamping factor can be effective as long as the ATPase works well. Thus, ATP must be supplied in sufficient amounts to keep the clamping factor around the template DNA. The DNA polymerase and Helicase work effectively and synergistically, only when the clamping factor attaches to the template DNA. Thus the clamping factors can be regarded as control inputs. As a result, to achieve a given state probability Φ_k, at least three types of input are required. They must contribute to the advance of Helicase, the biochemical action of DNA polymerase, and to clamping factors. We know however, that these 3 inputs interact mutually or are functions of each other. However, since there is no functional relationship available for them, the control input can be expressed as a linear sum of these three inputs. We thus associate them to represent each state probability. Such linearization of three kinds of control inputs is treatable for application of control system analysis as in the present H-infinity control case. In a system operating under H-infinity control principles, the inputs are defined as the operation that minimizes the H-infinity norm of the system transfer function from noise to output. An outline of the mathematical expression for H-infinity control is given in the Appendix.

The present work shows computed results for temporal changes in the state probability Φ_k, in the observer, in the control inputs and in the worst case disturbances for the case of c=4 and L=9. The system parameters were set as $k_o = 2.0$, $k_g = 0.1$, $k_{pol} = 1.0$ and $p(t) = 0.1$.

COMPUTED RESULTS

Figure 4(a) shows temporal changes in state probabilities Φ_k for $k = 0$ to 8. As the state proceeds from Φ_o to Φ_5, probabilities increase gradually. For more advanced states, ($k = 6, 7, 8$), probabilities increase to a peak and then decrease. Figure 4(b) shows results of the corresponding observers. The time courses of the observers were about the same as the state probabilities. The observer peak amplitudes however, were larger than those of the state probabilities.

Figure 5(a) shows temporal changes in control inputs for corresponding state probabilities. The amounts of all the control inputs increased as the replication time passed. As the replication process advances to k=6,7,8, the time courses of the control inputs have been reduced to be gradual. Figure 5(b) shows time courses of the worst case disturbances on the corresponding state probabilities. The worst disturbances decreased continuously as time passed. Such reductions could be observed at all the replication states.

DISCUSSION

The key point of this modeling is to determine the state probability Φ_k that predicts an existence of k successive pairs of separated template sites and the right side of Eq. 6. The probability of k separated pairs was assumed to occur by linear sum among t probabilities from Φ_{k-1} to Φ_1. The probabilities less than the pairs of k are determined by the location of the active site of the Helicase. When the cleavage site of the Helicase acts just near the base of the separated DNA at the cross of the Y, only one or two bound base pairs are opened (Figure 3-a and Figure 3-b). When the active site of the Helicase occasionally acts on the middle portion of the right side of the Y, the enzyme catalyzes more than three or q base pairs as shown in Figure 3-c and Figure 3-d. We don't know however, how extensive the Helicase operates on the bound base pairs, whether the extent is small or large. For convenience we tentatively set the critical number of template sites from the end of the bound helix. In addition, we still do not know the mechanism of how these bound base pairs open spontaneously at a location less than this critical template site c, because the energy required to open a small number of base pairs must be different from those to open larger numbers of base pairs.

We have tentatively set the control inputs to achieve the corresponding state probabilities of DNA replication. Rigorously speaking, each control input must take a different functional form because as the state probabilities change, the amount of base pairs that open simultaneously are different because the energy required to cut the biochemical energy bond between the complementary bases on the template DNA are different from site to site and depend on the numbers of base pairs. Thus, control should be modified to accommodate the influence of the number of winding base pair molecules.

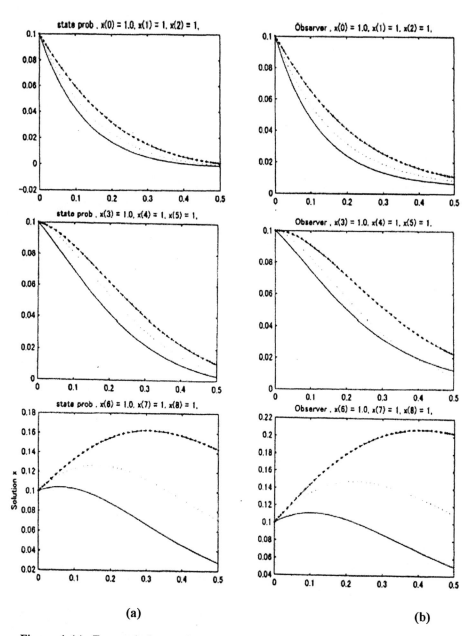

Figure 4 (a). Temporal changes in state probabilities Φ_o(continuous line), Φ_1(-*-*-*-*-*), Φ_2(dotted line), Φ_3(continuous line), Φ_4(-*-*-*-*-*), Φ_5(dotted line), Φ_6(continuous line), Φ_7(-*-*-*-*-*), Φ_8(dotted line). All initial conditions were set to 0.1. (b) The corresponding changes in the observers.

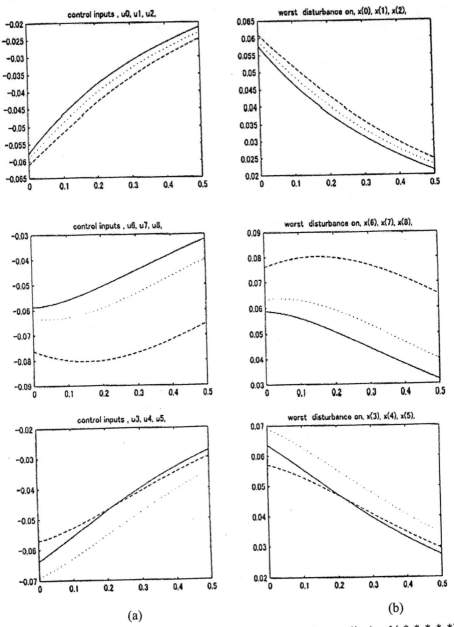

Figure 5(a). Temporal changes in control inputs u0,(continuous line), u1(-*-*-*-*-*), u2(dotted line), u3(continuous line), u4 (-*-*-*-*-*), u5(dotted line), u6,(continuous line), u7 (-*-*-*-*-*), u8(dotted line). The initial conditions were all set to 0.1;(b) Temporal changes in the worst case disturbances on the corresponding state probabilities.

H-infinity control is applied on the basis of intuitive speculation that the normal replication system works well even under the influence of natural biophysical noise and the system must operate to eliminate noise or to minimize the influence of noise. When this eliminating control does not work, noise will disturb the replication system. As a result, the genetic information is not copied well and so, wrong information is transferred producing inadequate proteins as a biophysical reaction output. In the normal physiological states, such erroneous genetic transformations do not occur and so, we suppose that replication of the genetic information must have operated in a way to minimize the disturbing effects of noise.

The present work is available to evaluate genetic diseases derived from incorrect replication of the template DNA.

CONCLUSION

The replication of DNA by primer RNA must operate so as to minimize noise that disturbs the replication of normal genetic information. H-infinity control will be available to help evaluate the minimization of noise in the replication process of copying genetic information.

REFERENCES

[1] J. Darnell, H. Lodish, D. Baltimore, 1990. Molecular cell biology, W.H. Freeman & Co., NY.
[2] J.M. Zimmerman, 1966. The kinetics of Cooperative unwinding and template replication, J. Theor. Biology, 13, 106–130.
[3] H. Hirayama, Y. Okita, 2001 Method for evaluating noise supressive function of Ach channel gating system by engineering H2 control, Machine intelligence & Robotics control, 3(1), 35–49.
[4] R.S. Sanchez-Pena, M. Sznaier, 1998. Robust systems. Theory and applications, Wiley & Sons, NY.
[5] J. Watson, M. Gilman, J. Witkowski, M. Zoller, 1983. Recombinant DNA, W.H. Freeman & Co., USA.
[6] R. Okazaki, 1968. Proc. Natl. Acad. Sci. USA, 59, 598.
[7] S. Waga, G. Hannon, 1994. The p21 inhibitor of cyclin dependent linase controls DNA replication by interaction with PCNA, Nature, 369. 574–578.

Appendix I

For L = 9, we have following differential equations

I. $c = 4; L = 9 \geq 2$ $4(=c) + 1$ and $c = 4 \geq 2$

$$d\Phi_o(t)/dt = -c\, k_o\, \Phi_o(t) + (k_g + k_{pol}\, p(t))\Phi_1(t) + u_1$$

I-1. $1 \leq k \leq c = 4$

$$d\Phi_k(t)/dt = k_o \sum_{p=0}^{k-1} \Phi_p - (c\,k_o + k_g + k_{pol}\,p(t))\Phi_k(t) + (k_g + k_{pol}\,p(t))\Phi_{k+1}(t)$$

$$d\Phi_1(t)/dt = k_o\,\Phi_o - (c\,k_o + k_g + k_{pol}\,p(t))\Phi_1(t) + (k_g + k_{pol}\,p(t))\Phi_2(t) + u_2$$

$$d\Phi_2(t)/dt = k_o\,(\Phi_o + \Phi_1) - (c\,k_o + k_g + k_{pol}\,p(t))\Phi_2(t) + (k_g + k_{pol}\,p(t))\Phi_3(t) + u_3$$

$$d\Phi_3(t)/dt = k_o\,(\Phi_o + \Phi_1 + \Phi_2) - (c\,k_o + k_g + k_{pol}\,p(t))\Phi_3(t) + (k_g + k_{pol}\,p(t))\Phi_4(t) + u_4$$

$$d\Phi_4(t)/dt = k_o\,(\Phi_o + \Phi_1 + \Phi_2 + \Phi_3) - (c\,k_o + k_g + k_{pol}\,p(t))\Phi_4(t) + (k_g + k_{pol}\,p(t))\Phi_5(t) + u_5$$

I-2. $c(=4) + 1 \leq k \leq L-c\,(=9-4),\quad k=5$

$$d\Phi_k(t)/dt = k_o \sum_{p=1}^{c=4} \Phi_{k-p} - (c\,k_o + k_g + k_{pol}\,p(t))\Phi_k(t) + (k_g + k_{pol}\,p(t))\Phi_{k+1}(t)$$

$$d\Phi_5(t)/dt = k_o\,(\Phi_{5-1} + \Phi_{5-2} + \Phi_{5-3} + \Phi_{5-4}) - (c\,k_o + k_g + k_{pol}\,p(t))\Phi_5(t) + (k_g + k_{pol}\,p(t))\Phi_6(t) + u_6$$

I-3. $L(=9) - c(=4) + 1 \leq k \leq L-1\,(=9-1=8)\quad k=6,7,8$

$$d\Phi_k(t)/dt = k_o \sum_{p=1}^{c=4} \Phi_{k-p} - [(L-k)\,k_o + k_g + k_{pol}\,p(t)]\Phi_k(t) + (k_g + k_{pol}\,p(t))\Phi_{k+1}(t)$$

$$d\Phi_6(t)/dt = k_o\,(\Phi_{6-1} + \Phi_{6-2} + \Phi_{6-3} + \Phi_{6-4}) - [3k_o + k_g + k_{pol}\,p(t)]\Phi_6(t) + (k_g + k_{pol}\,p(t))\Phi_{6+1}(t) + u_7$$

$$d\Phi_7(t)/dt = k_o\,(\Phi_{7-1} + \Phi_{7-2} + \Phi_{7-3} + \Phi_{7-4}) - [2k_o + k_g + k_{pol}\,p(t)]\Phi_7(t) + (k_g + k_{pol}\,p(t))\Phi_{7+1}(t) + u_8$$

$$d\Phi_8(t)/dt = k_o\,(\Phi_{8-1} + \Phi_{8-2} + \Phi_{8-3} + \Phi_{8-4}) - [k_o + k_g + k_{pol}\,p(t)]\Phi_8(t) + (k_g + k_{pol}\,p(t))\Phi_{8+1}(t) + u_9$$

setting
$a_{11} = -c\,k_o,\ a_{12} = k_g + k_{pol}\,p(t),\ a_{22} = -(c\,k_o + k_g + k_{pol}\,p(t))$
$a_{77} = -[3\,k_o + k_g + k_{pol}\,p(t)],\ a_{78} = a_{12}$
$a_{88} = -[2\,k_o + k_g + k_{pol}\,p(t)],\ a_{89} = a_{12}$
$a_{99} = -[k_o + k_g + k_{pol}\,p(t)],\ a_{910} = a_{12}$, we have
$d\Phi_o(t)/dt = -c\,k_o\,\Phi_o(t) + a_{12}\Phi_1(t)$
$d\Phi_1(t)/dt = k_o\,\Phi_O + a_{22}\Phi_1(t) + a_{12}\Phi_2(t)$
$d\Phi_2(t)/dt = k_o\,(\Phi_O + \Phi_1) + a_{33}\Phi_2(t) + a_{12}\Phi_3(t)$
$d\Phi_3(t)/dt = k_o\,(\Phi_O + \Phi_1 + \Phi_2) + a_{44}\Phi_3(t) + a_{12}\Phi_4(t)$
$d\Phi_4(t)/dt = k_o\,(\Phi_O + \Phi_1 + \Phi_2 + \Phi_3) + a_{55}\Phi_4(t) + a_{12}\Phi_5(t)$
$d\Phi_5(t)/dt = k_o\,(\Phi_4 + \Phi_3 + \Phi_2 + \Phi_1) + a_{66}\Phi_5(t) + a_{12}\Phi_6(t)$
$d\Phi_6(t)/dt = k_o\,(\Phi_5 + \Phi_4 + \Phi_3 + \Phi_2) + a_{77}\Phi_6(t) + a_{12}\Phi_7(t)$
$d\Phi_7(t)/dt = k_o\,(\Phi_6 + \Phi_5 + \Phi_4 + \Phi_3) + a_{88}\Phi_7(t) + a_{12}\Phi_8(t)$
$d\Phi_8(t)/dt = k_o\,(\Phi_7 + \Phi_6 + \Phi_5 + \Phi_4) + a_{99}\Phi_8(t) + a_{12}\Phi_9(t)$

Appendix II

Thus we set vector state equation

$$\partial x(t)/\partial t = \mathbf{A}\,x + \mathbf{B1}\,w + \mathbf{B2}\,U \tag{A1}$$

The elements of matrix $\mathbf{A} = [\,a_{ij}\,]$ is given in Appendix I. U is a matrix for the control input, w is a matrix for the noise. **B1** and **B2** are the weighting matrices to characterize the relative amounts of disturbing noises and control inputs that are acting on the different species.

Vector equation for the controlled output z is

$$z = \mathbf{C1}\,x + \mathbf{D12}\,u \tag{A2}$$

$\mathbf{C1} = [\,q_{ij}\,]$ and $\mathbf{D12} = [\,q_{kl}\,]$ are given in Appendix. The equation for input to the observer is

$$y = \mathbf{C2}\,x + \mathbf{D21}\,w \tag{A3}$$

$\mathbf{C2} = [\,s_{ij}\,]$ and **D21** is an unit diagonal vector. The elements of **B1, B2, C1, C2, D12** and **D21** were easily obtained. The elements in **B1** and **D21** describe the changes in the energy of noise (DB). For the standard state, we selected the physiological state and all the elements in **B1, B2** and **D21** were normalized to unity. Since even under the physiological circumstance, there are a lot of biophysical mimetics to disturb the replication processes. Hence we did not set the elements in **B1** and **D21** zero. Elements q_1 to q_9 in **C1** signify relative weights of the state variables on the controlled outputs z. Elements q_{10} to q_{18} in **D12** signify relative weights of the control inputs on the output z. Elements s_1 to s_9 in **C2** signify those of the state variables on the inputs for observer y.

H-infinity norm

The present problem to minimize control of noise on the Na channel gating process is formalized by the ordinal H-infinity control as [4]:

"**Given a finite value γ, synthesize an internally stabilizing proper controller K(s) such that the closed-loop transfer matrix from noise w to output z, T_{zw} satisfies the H infinite norm $\|T_{zw}\|\infty < \gamma$.**"

The norm of a transfer function is defined by

$$\|T_{zw}(j\omega)\|\infty = \sup_{0\,\leq\omega\leq\infty} \sigma\{\,T_{zw}(j\omega)\,\} \tag{A4}$$

σ is the largest singular value of the transfer function $T_{zw}(j\omega)$. The H∞ can also be described by the ratio of output z(t) against the input w(t) signals, namely the induced norm such that

$$\|T_{zw}(j\omega)\|\infty = \sup_{0 \leq \omega \leq \infty} [\{\int \infty_z T(t)z(t)dt\}/\{\int \infty_w T(t)w(t)dt\}] = \sup_{0 \leq \omega \leq \infty} [\|z\|2/\|w\|2] \quad (A5)$$

The H∞ norm is a performance index for the system in response to the worst external disturbances.

Assumptions for H-infinity control and their biological significance
The assumptions are as follows [4]:

[A1]. $(A, B2)$ can be stabilized and $(C2, A)$ is detectable
[A2]. $(A, B1)$ can be stabilized and $(C1, A)$ is detectable.
[A3]. $C1^T D12 = 0$ and $B1 D21^T = 0$
[A4]. $D12$ has full column rank with $D12^T D12 = I$ and $D21$ has full row rank with $D21 D21^T = I$.
    T denotes transposition.

Condition [A1] is necessary for the system to be stabilized by output feedback. This assumption is quite natural and essential for all biological systems, since biological systems always achieve some kind of stable state in order to preserve the most economic biological performance. In such cases, deviation from the target state must be minimized since excessive change of a biological system will be harmful to the system. Any non-converging system oscillation must be suppressed so as to attain a given target state in which all biological operations converge to the physiological state that must be the best state in which the biological system can live.

Condition [A2] is made for computational technical reasons. The stabilization condition of $(A, B1)$ describes that adequate control makes the system stable. Adequate control can be produced through observation y which is expressed by the detectability of $(C1, A)$.

[A1] and [A2] guarantee that the control and filtering Riccati equations associated with a related H-infinity problem admit positive semi-definite stabilizing solutions.

Condition [A3] is the orthogonality assumption. This simplifies the mathematical treatment.

The rank assumption **[A4]** guarantees that the H-infinity problem is non-singular. Since we treat the physiological replication state transition under the influence of noise, the system is assumed to behave non-singularly. This must be quite natural for biological systems. Without this condition, the system will oscillate or be chaotic.

The full column rank condition with $D12^T D12 = I$ indicates that the number of controlled output variables z exceeds the number of control input variables u. When the number of inputs is greater than the number of controlled output variables, the system is over-commanded and cannot achieve a unique state.

The full row rank condition with $D_{21} D_{21}^T = I$ expresses the condition that the number of noise entities exceeds the number of observed outputs y. This describes that the noise extends to the entire species of the system, namely the worst-case disturbance. This might be the case because there are kinds of biochemical noise that would act on several steps involved in DNA repair processes [1]. Unless this condition is satisfied when the number of elements in y exceeds those of w, the system is in an over-information state meaning that there will be some insignificant information that will not participate in effective DNA replication. Since DNA replication is the most effective and economic system, such insignificance may not actually be true. The assumptions for $D_{11} = 0$ and $D_{22} = 0$ simplify the computation. The case of $D_{22} \neq 0$ can be converted to canonical form by a linear fractional transformation of the controller K(s).

Mathematical process of the H-infinity control.
The related Hamltonian matrices are

$$H\infty = [\, A \quad \gamma^{-2} B_1 B_1^T - B_2 B_2^T \quad -C_1^T C_1 \quad -A^T \,] \tag{A6}$$

for the state variables and

$$J\infty = [\, A^T \quad \gamma^{-2} C_1^T C_1 - C_2^T C_2 \quad -B_1 B_1^T \quad -A \,] \tag{A7}$$

for the observers. Satisfying the following three conditions under assumptions [A1] to [A4]:

I. $H\infty \in \text{dom}(\text{Ric})$ and $X\infty = \text{Ric}(H\infty) > 0$ (A8a)

II. $J\infty \in \text{dom}(\text{Ric})$ and $Y\infty = \text{Ric}(\infty) > 0$ (A8b)

III. $\rho(X\infty\, Y\infty) < \gamma^2$ (A8c)

we have an internally stabilizing controller K(s) that renders:

$$\|T_{zw}\|\infty < \gamma \tag{A9}$$

The state-space realization of the central controller K_{sub} takes the form:

$$K_{sub} = \left[\begin{array}{c|c} A\infty & -Z\infty\, L\infty \\ \hline F\infty & 0 \end{array} \right] \tag{A10}$$

where

$$A\infty = A + \gamma^{-2} B_1 B_1^T X\infty + B_2 F\infty + Z\infty\, L\infty\, C_2 \tag{A11a}$$

$$F\infty = -B_2^T X\infty \tag{A11b}$$

$$L\infty = -Y\infty \, C^{2T} \qquad (A11c)$$

$$Z\infty = (I - \gamma^{-2} Y\infty \, X\infty)^{-1} \qquad (A11d)$$

Here, $X\infty$ is the solution to the algebraic Riccati equation [4]:

$$A^T X\infty + X\infty \, A + X\infty \gamma^{-2} B1 \, B1^T X\infty - X\infty \, B2 \, B2^T X\infty + C1^T C1 = 0 \qquad (A12)$$

where an element of $X\infty$ takes the symmetric form $X_{ij} = X_{ji}$, while $Y\infty$ is the solution to the adjoint algebraic Riccati equation [4]:

$$A \, Y\infty + Y\infty \, A^T + Y\infty \gamma^{-2} C1^T C1 \, Y\infty - Y\infty \, C2^T C_2 \, Y\infty + B1 \, B1^T = 0 \qquad (A13)$$

where an element of $Y\infty$ takes the symmetric form $Y_{ij} = Y_{ji}$. The set of all the internally stabilizing controllers rendering $\|T_{zw}\|\infty < \gamma$ can be parameterized as:

$$K(s) = FL(M\infty, Q) \qquad (A14)$$

Where $Q \in RH\infty$ and $\|Q\|\infty < \gamma$ and $M\infty$ has the following state-space realization

$$M\infty = \begin{bmatrix} A\infty & -Z\infty \, L\infty & Z\infty \, B_2 \\ F\infty & 0 & I \\ -C_2 & I & 0 \end{bmatrix} \qquad (A15)$$

This set of controllers equals the set of all transfer matrices from y to u. The variables in a state vector of the observers was set as:

$$x^{\wedge T} = [\, x10, \ x11, \ \text{-----}, \ x18 \,]^T \qquad (A16)$$

The vector equation of the observer $x^\wedge$ is given by

$$\partial x^\wedge / \partial t = A \, x^\wedge + B1 (\gamma^{-2} B1^T X\infty \, x^\wedge) + B2 \, u + Z\infty \, L\infty \, (C2 \, x^\wedge - y) \qquad (A17)$$

$$U = F\infty \ x^\wedge = -B2^T X\infty \ x^\wedge \qquad (A18)$$

To close the feed back loop, y can be related to the state variable x by:

$$y = -C2 \, x \qquad (A19)$$

Here are some explanations about several related terms [4]:

$\gamma^{-2} B1^T X\infty \, x^\wedge$ is the worst noise.
$\gamma^{-2} B1^T X\infty \, x$ is the worst-case disturbance.
$Z\infty \, L\infty$ is the optimal filter gain (OFG) to estimate the optimal control input $U = F\infty \, x$ in the presence of the worst-case disturbance.

High Resolution Detection and Coating Method for the Functionalisation of Nanometre Separated Gold Electrodes with DNA

C. WÄLTI, R. WIRTZ and M. PEPPER

Semiconductor Physics Group, Cavendish Laboratory, University of Cambridge, Cambridge, UK

W. A. GERMISHUIZEN

Department of Chemical Engineering, University of Cambridge, Cambridge, UK

A. P. J. MIDDELBERG

Department of Chemical Engineering, University of Cambridge, Cambridge, UK

A. G. DAVIES

Semiconductor Physics Group, Cavendish Laboratory, University of Cambridge, Cambridge, UK and School of Electronic and Electrical Engineering, University of Leeds, Leeds, UK

ABSTRACT

Here, we report on a new technique for selectively functionalising closely spaced gold electrodes of separation below 50 nm with different thiolated oligonucleotides using a local, selective electrochemical desorption of a molecular protection layer followed by the subsequent adsorption of the oligonucleotides onto the exposed surface. This technique does not rely on the use of a local probe such as an AFM tip to functionalise a surface locally, but is a bottom-up approach to coat different areas of a surface uniquely with different oligonucleotide monolayers. We demonstrate that the surface-bound oligonucleotides retain their unique molecular recognition and self-assembly properties and so functionalise the electrode array.

INTRODUCTION

It is generally accepted that molecular recognition and self-assembly techniques can, in principle, be exploited to assemble complex nanoscale molecular objects from their building blocks, and could also provide a reliable approach for the attachment of molecular structures to metal electrodes [1]. This procedure may ultimately allow the integration of complex self-assembled structures into conventional devices. Many biological molecules are known to self-assemble complex structures, using very efficient and selective molecular recognition processes. In particular, the selective self-assembly and molecular recognition properties inherent to DNA might be exploited to engineer complex molecular networks on a nanometer scale. The possible exploitation of biological processes for self-organization and self-assembly has therefore generated considerable interest. However, the coating of nanoscale electrodes with appropriate anchor molecules to allow for subsequent self-assembly is proving to be extremely challenging.

A number of techniques are available for introducing oligonucleotides or other anchor molecules locally onto a surface, but none of these simultaneously meet the requirements of resolution, speed and the ability to coat different electrodes uniquely. Among these are microdrop dispensing systems [2], micromachining [3] and microcontact printing [4], and nanografting [5]. Recently, a variant of the nanografting technique, dip-pen nanolithography, has been reported [6,7]. This technique uses an atomic force microscope (AFM) tip, coated with the anchor molecules, as a pen to draw onto the surface. The resolution of this coating technique is also on the nanometre scale, but a high level of stability and solubility of the anchor molecules is required. However, both techniques rely on a local probe (AFM) to coat the surface which is an inherently slow and sequential process. Similarly, a number of techniques are available for detecting DNA bound to metal surfaces, e.g. quartz crystal balance, cyclic voltammetry [8], or electrochemical capacitance measurements [9]. However, none of these provide the spatial resolution required for nanotechnological applications. Other methods, such as AFM [10] or surface plasmon resonance [11] provide appropriate spatial resolution, but involve very high equipment costs.

Here, we report on a high resolution detection technique to assess and quantify the monolayer of oligonucleotides on the gold electrodes [12]. The method is based on the optical detection of a colorimetric substrate precipitated onto the gold electrode. Biotinylated oligonucleotides are hybridised to the complementary, surface-bound, thiolated oligonucleotides. The presence of the biotin is detected with an alkaline phosphatase conjugated antibody [13] that binds specifically to the biotin label. Upon overlaying the antibody with a specific substrate solution, the alkaline phosphatase conjugated to the antibody causes a precipitation of dark blue colour pigments out of the solution onto the gold surface. This technique is widely applied on non-metallic substrates in Southern blotting, but the fact that it is applicable on metal surfaces makes this technique very useful in molecular electronics and molecular nanotechnology. Secondly, we report a new technique for selectively functionalising closely spaced gold electrodes of separation below 50 nm with different thiolated oligonucleotides using a local, selective electrochemical desorption of a molecular protection layer followed by subsequent adsorption of the oligonucleotides onto the exposed surface [14]. This technique does not rely on the use of a local probe such as an AFM tip to functionalise a surface locally, but is a bottom-up approach to coat different areas of a surface uniquely with different oligonucleotide monolayers. We demonstrate that the surface-bound oligonucleotides retain their unique molecular recognition and self-assembly properties and so functionalise the electrode array.

HIGH-RESOLUTION DETECTION METHOD

Experimental Details

A series of glass microscope slides was coated with a 40 nm thick gold layer on top of a 10 nm adhesive layer of nickel-chromium (NiCr). The glass slides were cleaned by immersing in 'piranha etch' (30% hydrogen peroxide, 70% sulphuric acid; caution:

'Piranha etch' reacts violently with organic materials) for one hour, followed by rinsing in deionised water, ethanol and again deionised water. Single stranded DNA of sequence Y ($^{5'}$AGG TCG CCG CCC$^{3'}$–thiol; the thiol is linked to the DNA via a carbon C_6 linker) were dissolved in 10 mM tris(hydroxymethyl)aminomethane, 1 mM EDTA (TE solution) and 1 M NaCl solution of pH 8. 20 µl of this solution was pipetted onto the cleaned gold slide, rinsed off with deionised water after 60 min, and immersed into 1 mM 6-mercapto-1-hexanol (MCH) for another 60 min to remove any non-specifically bound DNA, i.e. any DNA bound to the gold surface not by a sulfur-gold bond.

Biotinylated single stranded oligonucleotides with a sequence complementary to the thiolated oligonucleotide bound to the gold electrode (Y_{RC}: $^{5'}$GGG CGG CGA CCT$^{3'}$–biotin) were dissolved in TE-solution and 1 M NaCl, and pipetted onto the gold coated glass slide. The glass slides were incubated for 90 min to allow the complemetary strands to hybridise before the excess oligonucleotides were rinsed off with tris-buffered saline (TBS).

The gold surface was then immersed in a TBS solution containing 1% of bovine serum albinum (BSA) for 120 min to block any patches of the gold electrode not coated by the oligonucleotide monolayer. Subsequentially, the slides were rinsed several times in TBS containing 0.05% of the detergent Tween 20 (TBS/Tween) and immersed in a 1:1000 dilution of monoclonal anti-biotin antibody conjugated with alkaline phosphatase in TBS/Tween for 60 min. Before the sildes were exposed to the colorimetric substrate 5-bromo-4-chloro-3-indolyl phosphate/nitro blue tetrazolium (BCIP/NBT) they were rinsed several times with TBS/Tween and pure TBS, respectively. The BCIP/NBT was applied to the gold surface for 20 min before the slides were thoroughly washed in deionised water.

Dark blue colour pigments precipitate out of solution onto the surface upon cleavage of the BCIP/NBT by alkaline phosphatase. Unlike in previously reported enzymatic colorimetric detection methods [9], here the colour pigments precipitate onto the surface and thus yield spatially resolved information on the presence of the biotinylated oligonucleotides on the surface. Quantitative information on the amount of oligonucleotides detected were obtained by measuring the reflectivity of the slides using a commercial fluorescence spectrophotometer. Scans over the wavelength range between 300 nm and 700 nm were performed and the average reflection intensity calculated.

Results and Discussion

In order to quantify the reflection intensity and thus the amount of detected DNA, a dimensionless binding index $I_B = (I - I_0) / I_0$ is defined, where I denotes the average reflection intensity of the investigated slide, and I_0 the average reflection intensity of a clean gold slide, both averaged over the whole investigated spectral range from 300 to 700 nm. Figure 1 shows the binding index I_B as a function of the concentration of the thiolated oligonucleotide bound to the surface (left) and as a function of the concentration of the biotinylated complementary oligonucleotide hybridised to the surface bound thiolated DNA (right), respectively. With the detection limit defined at $I_B = 0.5$, the minimum concentration of thiolated oligonucleotides bound to the surface that can be detected is about 100 nM. A much lower limit of only about 2 nM is obtained for the detectable concentration of hybridised biotinylated DNA.

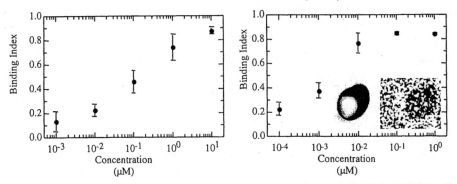

Figure 1. Left: Binding index vs. concentration of the surface-bound thiolated DNA. The complementary biotinylated oligonucleotide concentration was 1 μM. Right: Binding index vs. concentration of the hybridised biotinylated oligonucleotide. The complementary concentration was 10 μM. In both plots, the points represent an average of 3 experiments with the error bars showing the standard deviation. The inset shows an optical image of a 36 μm diameter droplet of 1 nM biotinylated oligonucleotide solution on a gold surface covered with complementary oligonucleotides (10 μM), and an optical image of the same area after enzymatic detection.

In a recent study, the detection limits of fluorescence and colorimetric detection methods were compared [15] and a limit of 100 amol on a spot with 400 μm diameter was reported. Reducing the area of the spot to 400 μm, and assuming the droplet shape to be a semi-sphere, our technique can detect as few as 30 amol when using a concentration of 2 nM in the droplet. However, even smaller spots with a diameter of 36 μm can be detected. The inset in Fig 1(b) shows a droplet of 1 nM biotinylated oligonucleotide solution on a surface covered with complementary thiolated oligonucleotides and the same area after the procedure. The contour of the droplet is clearly visible, demonstrating the high sensitivity of this detection method.

The colorimetric detection method offers a rather high spatial resolution, owing to the fact that the colour pigments precipitate onto the surface rather than being dissolved. This is shown in Figure 2, where two opposing electrodes, separated by a gap of less than 50 nm, are shown. The right electrode was coated with a monolayer of the oligonucleotide Y (10 μM) whereas the left electrode is only coated with a layer of MCH (for details see below). The enzymatic detection method resulted in a substantial darkening of the right electrode, where complementary DNA is present, while the uncoated electrode remained clear. This clearly shows the very high spatial resolution offered by this enzymatic colorimetric detection method.

FUNCTIONALISATION OF GOLD ELECTRODES WITH nM-RESOLUTION

Experimental Details

A series of opposing gold electrodes was fabricated on a Si/SiO$_2$ wafer using a two-step shadow evaporation technique [16]. In the first step, a series of electrodes of separation 35 μm comprising a 35-nm-thick Au layer on top of a 10 nm adhesive layer of Ni/Cr was created by standard UV photolithography, metal evaporation, and lift-off. In the second step, the wafer was tilted appropriately in the evaporator and 5 nm of Ni/Cr followed by 17 nm of Au was deposited in stripes connecting the opposing electrodes. However, because the wafer was tilted, the edges of the existing electrodes closest to the evaporation source shadowed the surface from the evaporation beam leading to the formation of sub-50-nm-sized gaps between opposite electrodes. The electrode arrays were cleaned as described above.

The electrochemical measurements and desorption were all performed in 100 mM phosphate buffer at pH 10 using a standard three electrode setup, where a high purity platinum wire was used as the counter electrode. The electrochemical potentials are reported against a Ag/AgCl reference electrode.

Two thiolated oligonucleotides of sequence X (^{5}'CAG GAT GGC GAA CAA CAA GA$^{3'}$–thiol) and Y were used to selectively functionalise the electrode arrays. The functionalisation was assessed and quantified by hybridising complementary, biotinylated oligonucleotides of sequence X_{RC} (^{5}'TCT TGT TGT TCG CCA TCC TG$^{3'}$–biotin) and Y_{RC} and detecting the presence of the biotin by using the very sensitive colorimetric detection method described above.

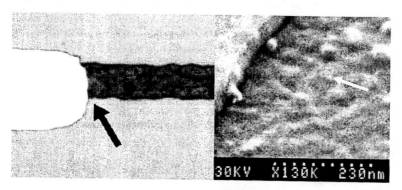

Figure 2. Left: Two opposing gold electrodes separated by a gap of less than 50 nm (indicated by the arrow). The right electrode was coated with the thiolated oligonucleotide, Y, while the left electrode was only coated with a monolayer of MCH. A 1 μM solution of the biotinylated oligonucleotide, Y_{RC}, was applied to the two electrodes. Subsequent enzymatic detection leads to a substantial darkening of the right electrode, demonstrating the very high spatial resolution of the detection as well as the coating technique. Right: Scanning electron micrograph (SEM) of the region between the two electrodes. The image was taken after all experiments were completed, i.e. after coating and colorimetric detection. The gap is indicated by the arrow.

Results and Discussion

A method to functionalise a set of closely spaces electrodes with different thiolated oligonucleotides has been developed. The method comprises 5 steps: formation of a molecular masking layer; selective electrochemical desorption; adsorption of thiolated anchor molecules onto the exposed areas; re-establishing the molecular masking layer in the functionalised areas; repeat selective desorption/adsorption.

Monolayers of thiol-compounds including the small molecule MCH, can form on a gold surface by chemisorption [17, 18]. MCH chemisorption is very efficient and the resulting gold-sulphur bond leads to a stable monolayer acting as a mask against subsequent adsorption of the longer thiolated anchor molecules. A monolayer of MCH forms on the electrode array by immersing the wafer in a 1 mM solution of MCH for 60 min. The masking properties of this layer has been investigated by others [18] and confirmed by us [14].

Electrochemical studies have shown that the gold-sulphur bond can be cleaved reductively at an electrochemical potential of about –1V vs. Ag/AgCl, depending on the pH of the electolyte [19,20]. Complete desorption of the molecular masking layer from a particular electrode was obtained by applying an electrochemical potential of – 1.4 V vs. Ag/AgCl for two minutes, while keeping all other electrodes at open circuit. The desorption process was monitored by cyclic voltammetry (CV). Following the desorption, the electrode array was rinsed thoroughly in deionised water and blow-dried.

The thiolated anchor molecule can be chemisorbed onto the exposed electrode the same as the molecular masking layer. In our case, the electrode is immersed in a solution containing the thiolated oligonucleotide for 60 min and thoroughly rinsed in deionised water. The high spatial resolution of the coating technique is demonstrated in Figure 2. The smaller right electrode has been selectively desorbed before the whole electrode was exposed to a 10 µM solution of the thiolated oligonucleotide Y. The biotinylated oligonucleotide Y_{RC} was applied to the whole array to allow for hybridization to the complementary DNA where present, and the presence of the biotin was subsequently detected by the colorimetric detection technique. Although the gap between the two electrodes is too small to be resolved by optical microscopy, it is very apparent that only the right electrode contained biotin and thus a layer of the anchor molecule Y.

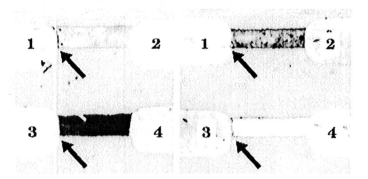

Figure 3. Optical image of an electrode arrays in which electrodes 1 and 4 are coated with oligonucleotides X and electrodes 2 and 3 with oligonucleotides Y. Arrows indicate the nanometer size gaps separating opposing electrodes. Left: When the array is challenged with biotinylated oligonucleotides X_{RC}, followed by the anti-biotin antibody detection process, electrodes 1 and 4 darken, confirming the presence of surface-bound oligonucleotide X. Right: When the array is challenged with biotinylated oligonucleotides Y_{RC}, followed by the anti-biotin antibody process, electrodes 2 and 3 darken, confirming surface-bound oligonucleotide Y.

It has been shown that adsorption of thiolated oligonucleotides onto a gold surface does not lead to a compact and stable monolayer [18, 21], as the DNA does not only bind via the strong sulphur–gold bond but also via amine-bonds [22]. In order to coat further electrodes with different oligonucleotides, the layer on the already functionalised electrodes must be compact and stable to prevent cross-contamination. However, immersing the partly functionalised layer in a 1 mM solution of MCH for 60 min replaces the non-specifically bound oligonucleotides and fills the gaps with MCH molecules [18]. This results in a oligonucleotide–MCH monolayer with similarly good masking characteristics.

To coat further electrodes, the selective desorption and adsorption processes are repeated and the protection capabilities of the coated electrodes are re-established as described above. Figure 3 shows two sets of electrodes in which always two opposing electrodes are separated by very small gaps. The SEM images of the gap areas are very similar to the one shown in Figure 2. Both sets of electrodes were functionalised in the same way: electrode 1 and 4 are coated with the thiolated oligonucleotide X and electrode 2 and 3 with oligonucleotide Y. However, the left set has been exposed to the biotinylated oligonucleotide X_{RC} while the right set has been exposed to oligonucleotide Y_{RC}. The colour change due to the detection method confirms that the thiolated oligonucleotides X and Y bind to the desired electrodes only and demonstrates that the technique can be used to deposit different oligonucleotides selectively onto sub-50 nm-separated electrodes. We note the anti-biotin antibody detection not only shows the required coating was achieved, but also that the bound thiolated oligonucleotides which can act as anchor molecules in nano-assembly, remain intact and can hybridize with their complementary counterparts.

CONCLUSIONS

We have presented a highly sensitive method to detect the hybridisation of single stranded, biotinylated oligonucleotides onto a gold electrode functionalised with complementary, thiolated single stranded DNA. The technique does not involve the detection of fluorescence nor other expensive infrastructure, and we have demonstrated the very high spatial resolution of the method to be better than 50 nm.

Furthermore, we have demonstrated a method based on the electrochemical desorption of a molecular protection layer to selectively functionalise a series of very closely spaced electrodes with separations of less than 50 nm. A set of electrodes has been successfully coated with two different oligonucleotides and, by using the enzymatic colorimetric detection method to detect for the successful hybridisation of complementary biotinylated oligonucleotides, we have demonstrated that the self-assembly properties of the DNA monolayer remained intact. The technique, which is compatible with large scale bottom-up applications, is fast and does not require expensive infrastructure.

Used separately or in combination, the two tools presented are very valuable for molecular nano-electronics, nano-assembly, and other emerging applications in the field of molecular-based nanotechnology as well as for biotechnology applications.

ACKNOWLEDGEMENT

We thank D. A. Williams and J. E. Cunningham for help in taking the SEM pictures and we gratefully acknowledge fruitful discussions with D.M.D. Bailey. This research was in part funded by the EPSRC. CW acknowledges financial support of the Schweizerische Nationalfonds zur Förderung der wissenschaftlichen Forschung. RW thanks the Gottlieb Daimler- and Karl Benz-Stiftung and the George and Lillian Schiff Foundation. WAG and AGD acknowledge the Cambridge Commonwealth Trust and the Royal Society, respectively.

REFERENCES

[1] J.M. Lehn, 1995. Supramolecular Chemistry: Concepts and Perspectives (VCH, Weinheim).
[2] G. Yershov et al., 1996. DNA analysis and diagnostics on oligonucleotide microchips. Proc. Natl. Acad. Sci. USA, 93, 4913–4918.
[3] N.L. Abbott et al., 1994. Using micromachining, molecular self-assembly, and wet etching. Chem. Mater., 6, 596–602.
[4] A. Kumar, G.M. Whitesides, 1994. Patterned condensation figures as optical diffraction gratings. Science, 263, 60–62.
[5] S. Xu, G. Y. Liu, 1997. Nanometer-scale fabrication by simultaneous nanoshaving and molecular self-assembly. Langmuir, 13, 127–129.
[6] R.D. Piner et al., 1999. "Dip-pen" nanolithography. Science, 283, 661–663.
[7] L.M. Demers et al., 2002. Direct patterning of modified oligonucleotides on metals and insulators by dip-pen nanolithography. Science, 296, 1836–1838.
[8] M. Yang et al., 1998. Adsorption kinetics and ligand-binding protperties of thiol-modified double-stranded DNA on a gold surface. Langmuir, 14, 6121–6129.
[9] H. Berney et al., 2000. A DNA diagnostic biosensor: development, characterisation and performance. Sensor. Actuat. B-Chem., 68, 100–108.
[10] S.M. Lindsay et al., 1992. Potentiostatic deposition of DNA for scanning probe microscopy. Biophys. J., 61, 1570–1584.
[11] H. Schaumann et al., 2000. Mechanical stability of single DNA molecules. Biophys. J., 78, 1997–2007.
[12] R. Wirtz et al., 2003. High sensitivity colorimetric detection of DNA hybridisation on a gold surface with high spatial resolution. Nanotechnol., 14, 7–10.
[13] M.S. Blake et al., 1984. A rapid, sensitive method for detection of alkaline phosphatase conjugated anti-antibody on western blots. Anal. Biochem., 136, 175–179.
[14] C. Walti et al., 2003. Direct selective functionalisation of nanometre separated gold electrodes with DNA oligonucleotides. Langmuir, 19, 981–984.
[15] I. Alexandre et al., 2001. Colorimetric silver detection of DNA microarrays. Anal. Biochem., 295, 1–8.
[16] Philipp et al., 1999. Shadow evaporation method for fabrication of sub 10 nm gaps between metal electrodes Microelectron. Eng. 46, 157–160.

[17] A. Ulman, 1996. Formation and structure of self-assembled monolayers. Chem. Rev., 96, 1533–1554.
[18] T.M. Herne, M.J. Tarlov, 1997. Characterization of DNA probes immobilized on gold surfaces. J. Am. Chem. Soc., 119, 8916.
[19] C.A. Widrig et al., 1991. The electrochemical desorption of n-alkanethiol monolayers from polycrystalline Au and Ag electrodes. J. Electroanal. Chem., 310, 335–359.
[20] D.E. Weisshaar et al., 1992. Thermodynamically controlled electrochemical formation of thiolate monolayers at gold - characterization and comparison to self-assembled analogs. J. Am. Chem. Soc., 114, 5860–5862.
[21] R. Levicky et al., 1998. Using self-assembly to control the structure of DNA monolayers on gold: A neutron reflectivity study. J. Am. Chem. Soc., 120, 9787.
[22] D.V. Leff et al., 1996. Synthesis and characterization of hydrophobic, organically- soluble gold nanocrystals functionalized with primary amines. Langmuir, 12, 4723–4730.

Chapter 13: Nanotubes

Magnetic Properties of Doped Silicon Nanotube

A. K. SINGH, T. M. BRIERE, V. KUMAR and Y. KAWAZOE
Institute for Materials Research, Tohoku University, Sendai, Japan

ABSTRACT

In the present work we have explored the magnetic properties of transition metal doped Si nanotubes formed by stacking of hexagonal rings of Si by first-principles calculations. Earlier study of the doping of non-magnetic metal atoms has been shown to stabilize finite and infinite nanotubes of Si [Nano Letters 2, 1243 (2002)]. Doping with Fe and Mn has been found to lead to magnetism in finite and infinite Si nanotubes whereas Ni- and Co-doped nanotubes are mostly non-magnetic. Infinite nanotubes doped with Fe have magnetic moment comparable to bulk Fe. Further results are given on the effects of doping with more than one transition metal. Nanotubes doped with both Fe and Mn have higher binding energies compared to nanotubes doped with only Fe or Mn. High magnetic moments on the metal atoms are preserved in this case as well. Such high moment nanotubes could be used as nanoscale magnets.

INTRODUCTION

Current interest in finding suitable components for miniature electronic devices has led to a focus on the bottom-up approach in which nanoparticles and nanowires can be used as building blocks. The recently developed metal-encapsulated clusters of Si and Ge are attractive for such applications [1–5]. Recently it has been shown that such clusters can be assembled to grow nanotubes of silicon stabilized by metal atoms [6,7]. In this work, we show from first principles total energy calculations that mixing two different transition metal (TM) atom dopants can lead to enhancement of binding energy and magnetic moments.

In previous work [6] the stability of finite undoped silicon nanotubes was studied by stacking 6-membered hexagonal rings. The optimized structures of these nanotubes show a tendency for agglomeration to 3-D structures with tetrahedral-like coordination, showing a clear preference of silicon for sp^3 bonding. It was found that a finite nanotube of silicon could be stabilized when doped with Be [6] atoms or TM (Mn, Fe, Co, Ni) [2]. These nanotubes can be considered to arise from an assembly of metal encapsulated $Si_{12}M$ clusters. Doping of Si_{12} with Be results in a chair-shaped structure whereas TM-doped structures are mostly hexagonal prisms. Packing two or more $Si_{12}Be$ units leads to a transformation of the chair-shaped structures to hexagonal shape. Considering two, three, and four units of $Si_{12}M$ (M = metal atom), the rings are generally stable and hexagonal in shape. Increasing the number of M dopant atoms generally leads to an increase in the binding energy. However, the magnetism of the nanotubes depends on both the type of dopant and the total number of atoms. The finite nanotubes have small to moderate HOMO-LUMO gaps.

The stability of doped infinite nanotubes has been studied using a unit of 12 Si and one M and another unit of 24 Si and four M atoms. The structures were optimized with respect to the cell size along the nanotube axis. In the case of the infinite nanotube with stoichiometry $Si_{24}M_4$ we have found that the presence of a TM atom increases the binding energy of the system relative to the pure nanotube or the tube doped with Be. Interestingly, the positions of the Mn- and Fe-doped nanotubes are quite similar to those of Be, but for the Co- and Ni-doped systems, the TM atom is located directly at the center of the Si ring. As in the case of finite nanotubes, the magnetism of the infinite nanotubes depends on the dopant, with the Fe-doped systems having higher net moments, and the Mn-, Co-, and Ni-doped systems having lower or zero net moments. Density of states analysis and band structure calculations shows these tubes are generally metallic.

METHOD

Our efforts have been to see the effect of doping with two different kinds of TM atoms to explore the possibility of forming a magnetic superlattice structure. As shown previously, metal encapsulation leads to the sp^2 bonding of Si allowing formation of an hexagonal structure. We performed first principles calculations using a plane-wave method [8] incorporating density functional theory within the generalized gradient approximation for the exchange and correlation energy [9]. The cutoff energy depends on the TM atom, with a minimum of 227 eV for Mn and a maximum of 242 eV for Ni. Sampling of 15 k-points along the nanotube axis was used to optimize the infinite nanotubes. Forces were converged to 0.001 eV/Å.

RESULTS AND DISCUSSION

Following our previous work, we doped the nanotubes with more than one kind of TM atom. Since the nanotubes doped with Fe, Mn, Co and Ni have already been studied [7], we chose these TM atoms as the dopants for the present study as well. We used the stoichiometry $Si_{24}A_2B_2$ with an ABAB arrangement of atoms. The mixing energy of the tubes was calculated by subtracting half the sum of the energies of tubes $Si_{24}A_4$ and $Si_{24}B_4$ from the total energy of $Si_{24}A_2B_2$. If this energy is negative then the mixed system has a higher probability of formation. Table 1 shows the energies of nanotubes doped with two different TMs.

Table 1. Energies and magnetic moments of different nanotubes. Mixing energy = Energy of $Si_{24}A_2B_2$ − 1/2 (Energy of $Si_{24}A_4$ + Energy of $Si_{24}B_4$). The magnetic moment represents the net magnetic moment of the unit cell. The magnetic moments of $Si_{24}Mn_4$ and $Si_{24}Fe_4$ are 9.12 μ_B and 9.40 μ_B. $Si_{24}Ni_4$ and $Si_{24}Co_4$ are nonmagnetic.

System ($Si_{24}A_2B_2$)	Energy of $Si_{24}A_2B_2$ (eV)	Energy of $Si_{24}A_4$ (eV)	Energy of $Si_{24}B_4$ (eV)	Mixing Energy (eV)	Magnetic Moment (μ_B)
$Si_{24}Fe_2Co_2$	-147.16	-149.69	-145.60	0.48	5.73
$Si_{24}Fe_2Ni_2$	-144.25	-149.69	-140.15	0.67	3.56
$Si_{24}Mn_2Co_2$	-148.79	-152.29	-145.60	0.15	9.86
$Si_{24}Mn_2Fe_2$	-151.16	-152.29	-149.69	-0.13	8.40
$Si_{24}Mn_2Ni_2$	-145.60	-152.29	-140.15	0.62	8.40
$Si_{24}Ni_2Co_2$	-142.69	-140.15	-145.60	0.18	0.36

Except for nanotubes doped with Fe and Mn, the mixing energy is positive, indicating segregation behavior and low probability to form compared to singly-doped nanotubes. Nanotubes doped with Mn and Fe have a negative mixing energy indicating this system may well form experimentally. The structure of this tube is symmetric (Figure 1), wherein the TMs are located on the tube axis and not directly between two silicon rings but closer to one ring than the other.

The magnetic behavior of a nanotube doped with Fe and Mn is quite different than its pure counterparts, as the Mn-doped nanotube possesses almost degenerate ferro- and anti-ferro-magnetic coupled states. It is ferromagnetically-coupled. Like the singly-doped nanotubes, this nanotube also possesses high local magnetic moments, though the total magnetic moment is lower than that of either the Mn- or Fe-doped nanotubes. The local mean moment of Mn ($2.5\mu_B$) is higher than the local mean moment of Fe ($1.7\mu_B$). The local moment of Mn is slightly enhanced relative to the singly-doped system; however, the local moment of Fe is reduced, leading to a lowering of total magnetic moment of system.

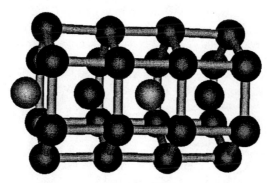

Figure 1. Optimized structure of $Si_{24}Mn_2Fe_2$.

As in the earlier study of TM-doped nanotubes [7], charge transfer is again from the TM atom to the Si chain but unlike the singly-doped case the charge transfer is not uniform, but higher for the middle atoms than the corresponding terminal atoms. The charge transfer from the one pair of TM atoms is higher than the singly-doped cases; however, the charge transfer from the other pair of TM atoms is almost the same as the single atom cases. This aspect is very important from the point of view of conduction through this nanotube, as conduction could occur through the silicon nanotube or the TM atom chain. The band structure shows crossing of the bands at the Fermi level for the both spin-up and spin-down components, suggesting metallic behavior for this tube.

CONCLUSIONS

We have studied superlattice structures of the Si nanotube with stoichiometry $Si_{24}A_2B_2$ (A,B = Mn, Fe, Co, and Ni), and have found it possible to control the magnetic behavior of Si nanotubes through appropriate doping. Only $Si_{24}Mn_2Fe_2$ has a negative mixing energy. This nanotube has high local magnetic moments preferring ferromagnetic coupling. The local moment of Fe in this nanotube is reduced in comparison with the Fe-doped nanotube, leading to a lower magnetic moment compared to singly-doped systems. The nanotube is also metallic. The high magnetic moment and metallic character may be important for spintronics devices and nanomagnets.

ACKNOWLEDGEMENTS

The authors thankfully acknowledge the support of the staff of the Center for Computational Materials Science, IMR, Tohoku University for the use of the SR8000/H64 supercomputer facilities. AKS is also thankful for the support of Monbusho. VK gratefully acknowledges the hospitality of IMR, Tohoku University.

REFERENCES

[1] V. Kumar and Y. Kawazoe, 2001. Phys. Rev. Lett. 87, 045503.
[2] V. Kumar and Y. Kawazoe, 2002. Phys. Rev. B 65, 073404.
[3] H. Hiura, T. Miyazaki and T. Kanayama, 2001. Phys. Rev. Lett. 86, 1733.
[4] S. M. Beck, 1989. J. Chem. Phys. 90, 6306.
[5] V. Kumar and Y. Kawazoe, 2003. Phys. Rev. Lett. 90, 055502.
[6] A.K. Singh, V. Kumar, T.M. Briere, and Y. Kawazoe, 2002. Nano Letters 2, 1243.
[7] A.K. Singh, T.M. Briere, V. Kumar, and Y. Kawazoe, 2003. submitted for publication.
[8] A.K. Singh, T.M. Briere, V. Kumar, and Y. Kawazoe, 2003. submitted for publication; G. Kresse, J. Furthmüller, 1996. Phys. Rev. B 54, 11169; G. Kresse, J. Furthmüller, 1996. Comput. Mat. Sci. 6, 15.
[9] J.P. Perdew, 1991. in Electronic Structure of Solids '91, Ed. P. Ziesche and H. Eschrig, Akademie Verlag, Berlin.

Computer Simulation of Single Wall Carbon Nanotube Crystals

V. KUMAR
Institute for Materials Research, Tohoku University, Sendai, Japan and Dr. Vijay Kumar Foundation, 45 Bazaar Street, Chennai 600 078, India

M. H. F. SLUITER
Laboratory for Advanced Materials, Institute for Materials Research, Tohoku University, Sendai, Japan

Y. KAWAZOE
Institute for Materials Research, Tohoku University, Sendai, Japan

ABSTRACT

Ever since their discovery, carbon nanotubes have amazed researchers with their unusual and often unsuspected properties. Single wall carbon nanotubes are often found in the form of bundles. In an ideal description, these bundles can be called crystals in which tubes align to a triangular lattice. Such materials promise unprecedented features at the nanometer scale, such as pores of constant diameter between and within individual nanotubes with relatively weak chemical reactivity of walls. This is expected to give high selectivity for adsorbed molecules and physical properties that might be tuneable with the diameter of the nanotubes and the adsorbents. Physical properties of individual single wall carbon nanotubes have been predicted with great success in numerous works using semi-empirical tight binding descriptions or ab initio (local density) formulations. However, prediction of the properties of nanotube crystals is challenging for several reasons. In nanotube crystals there are simultaneously both strong intratubular bonds and weak intertubular interactions that defy simple semi-empirical parameterisations, yet the scale of the problem is such that ab initio approaches become very slow. In this paper we illustrate how the subtle interaction between nanotubes gives rise to computationally very demanding structural relaxation processes and the surprisingly difficult task of determining the ground state of single wall carbon nanotube crystals. We also show that symmetry of nanotubes plays an important role in deformations that occur upon crystallisation.

INTRODUCTION

Single wall carbon nanotubes are currently of great interest because of their potential in nano-technologies such as electronic devices, catalysis, electron emitters, etc. As the nanotubes often occur in the form of bundles, it is important to understand the changes in the structure as well as properties of nanotubes upon the bundle formation. The mechanical properties of single wall carbon nanotube bundles (SWCNTBs) are particularly important since following deformation, important changes may occur in their properties. Recent experiments show [1-4] a structural transformation occurs in SWCNTBs under hydrostatic pressure, but the nature of this transition is unclear. Venkateswaran et al. [1] reported the disappearance of radial breathing modes between 150 and 200 cm^{-1} in Raman spectra at around 1.5 GPa. They interpreted this to be due to a hexagonal distortion of initially cylindrical nanotubes. Tang et al. [2] have reported a similar result from synchrotron X-ray diffraction and an irreversible transformation around 5 GPa. Peters et al. [3] reported a reversible structural transformation at ~1.7 GPa from Raman spectroscopy that has been interpreted from classical molecular dynamics simulation to be due to a structural phase transition from near hexagonal to monoclinic. These different results could arise from different chiralities and diameters of nanotubes in different samples of the bundles.

We consider crystals of nanotubes for the purpose of calculations as periodic boundary conditions can be used. When nanotubes are brought together to form a bundle or a crystal, the symmetry of the nanotube is generally broken. For example, the widely studied (10,10) SWCNTs cannot keep its 10-fold rotational symmetry upon crystal formation. Therefore, structural changes are bound to occur in the nanotubes when they form bundles or crystals. If the rotational symmetry of the nanotubes is commensurate with that of the underlying lattice, then the transformation is such that it is commensurate with the lattice such as hexagonal deformation of the nanotubes from a cylindrical form. However, in other cases, the nanotubes deform to an oval shape as shown by us earlier [5-7]. The relative orientations of the nanotubes are more difficult to obtain since proper calculation may require many nanotubes per cell which is computationally very demanding. A large number of studies have been performed using empirical approaches [3,8]. However, the interaction energy between the nanotubes is quite small (about 10 meV/atom) and requires high accuracy. As we shall show, the convergence of the structural optimization is a very slow process particularly for nanotubes having rotational symmetry incompatible with the lattice. Further, the coupling between the nanotubes is affected by the application of pressure due to changes in the electronic structure. Therefore, a proper understanding of the pressure effects needs *ab initio* calculations. We report here results of such an *ab initio* study of the pressure effects on SWCNTBs of (10,10) and (12,12) type that represent nanotubes incommensurate and commensurate with the hexagonal lattice, respectively.

METHOD OF CALCULATION

We use an ultrasoft pseudopotential plane wave method [9] and the local density approximation for the exchange-correlation energy. The latter gives a good description of the interaction between the graphene layers. In a crystal, nanotubes arrange in a triangular lattice with weak inter-tube interactions similar to graphite. In experiments, bundles generally have nanotubes of different chiralities and with slightly different diameters. We consider here an ideal case where all the nanotubes are identical. Our earlier calculations [5-7] considered one nanotube per cell only. However, we have also carried out calculations with two nanotubes per cell for the (10,10) nanotubes and obtained lower energies as well as better relative orientations of the nanotubes. Our calculations show a very slow dynamics of the nanotubes and a very demanding effort for structural optimization. As there are 40 and 48 atoms per unit cell per (10,10) and (12,12) nanotube, respectively, the cell dimensions perpendicular to the nanotube axis are large. Therefore, we perform k-space integrations using 15 k-points along the nanotube axis. Structural optimization is considered converged when the force on each ion becomes less than 1 meV/A. The optimizations are performed without any constraints on the shape and size of the cell using the conjugate gradient method.

RESULTS AND DISCUSSION

Earlier calculations [5] on (10,10) nanotube crystals showed slight deviation from hexagonal lattice even at zero pressure as the (10,10) nanotubes are not commensurate with the 3-fold symmetry of the lattice. The nanotubes remain nearly cylindrical as shown in Figure 1a. As the pressure is increased, deviations from hexagonal lattice increase further. There is a competition between the cost of deformation, the attractive interaction between the nanotubes and the volume change. At 1.4 GPa, a sudden transition occurs to a monoclinic phase in which the nanotubes have transformed into an oval shape (Figure 1b).

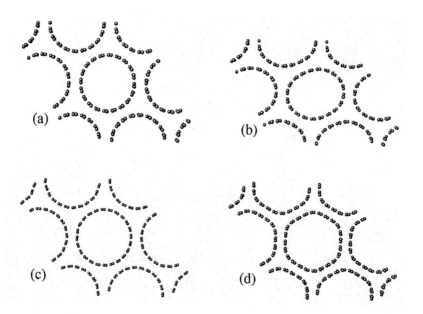

Figure 1. (10, 10) SWCNTB (a) at zero and (b) at 1 GPa hydrostatic pressure; (12, 12) SWCNTB (c) at zero and (d) at 6 GPa hydrostatic pressure. In these calculations only one nanotube per cell is considered.

This is due to asymmetric interactions between the nanotubes. Further increase in pressure leads to additional flattening of the nanotubes enhancing interaction between the nanotubes. The lattice vector along the nanotube axis remains nearly the same over the whole pressure range. In order to locate the transition pressure, further calculations at lower pressures were done. It was found that the monoclinic structure was retained even down to zero pressure. In addition, the enthalpy was found lower in this phase than in the nearly hexagonal phase. Therefore, within LDA, the monoclinic phase is more favorable. Figure 2 shows the variation of the lattice parameters perpendicular to the nanotube axis as a function of pressure. The zero pressure lattice parameter is in good agreement with experimental results for 13.55 Å diameter nanotubes.

Figure 3 shows the difference of the enthalpy in the near hexagonal and monoclinic structures. These results indicate a transition to cylindrical shape at a negative pressure.

Further calculations have been performed on crystals with two (10,10) nanotubes in a cell. Going from one nanotube to two nanotubes per cell gives additional freedom to optimize the relative orientations as well as the relative placements along the nanotube axis. The optimizations were performed without any symmetry constraint and allowed nanotube shape change as well as the change of the lattice. The optimized structure is shown in Figure 4. The optimization process is very slow.

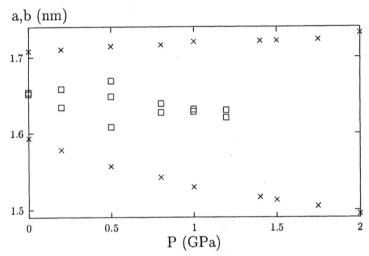

Figure 2. Variation of the lattice parameters perpendicular to the nanotube axis of the (10,10) bundles as a function of pressure. The squares correspond to increasing pressure. After the transition, crosses correspond to the monoclinic phase. The results are obtained by considering one nanotube per cell.

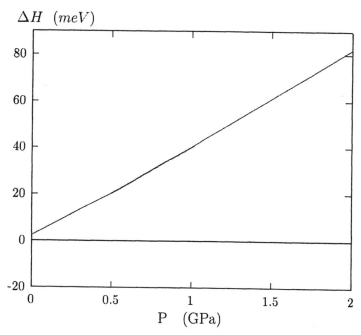

Figure 3. The difference in the enthalpy of (10, 10) bundles in the hexagonal and the monoclinic phases as a function of pressure.

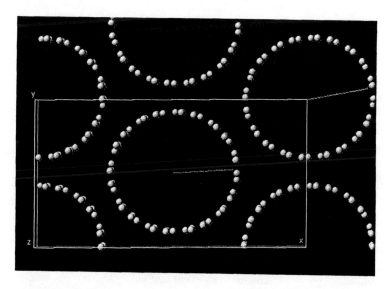

Figure 4. Optimized structure for two (10, 10) nanotubes per unit cell at zero pressure. Note the slight rotation (6.77°) of the center and corner nanotubes with respect to each other as indicated by the red lines. The largest and smallest radii are almost the same for the center and corner nanotubes at about 0.706 and 0.642 nm.

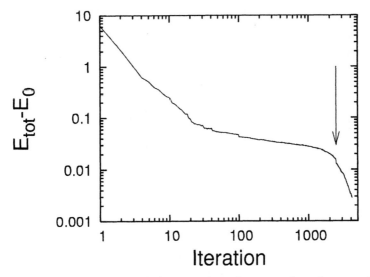

Figure 5. Total energy (eV) relative to estimated converged total energy for 2 (10, 10) nanotubes per cell as a function of structural iteration count. Note the rapid decrease during the first 10 iterations, followed by a very slow decline during the next several thousand iterations. Discontinuities exist because of restarting, most noticeably at 100 and 2460 iterations.

Figure 5 shows the change in energy as a function of the logarithm of the number of iterations. Initially the energy decreases rapidly which is related to the quick optimization of the appropriate lattice constant. After that, the crucial optimization of the relative orientation starts which slows down the optimization. The gain in energy in going from one nanotube to two nanotubes per cell is 58 meV. However, this is a significant fraction of the 338 meV binding energy per nanotube of the (10,10) nanotube crystal with one nanotube per cell. This is so because most of the relative orientations of hexagons between neighbouring nanotubes become graphite-like. The nature of the deformation remains the same as with one nanotube per cell.

In order to understand the dependence of the structural transitions on the diameter of the nanotubes as well as the symmetry of the nanotubes, we performed calculations on (12,12) crystals up to 6 GPa. These are found to undergo only a polygonization such that the nanotubes transform into a hexagonal shape with increasing pressure, with the lattice remaining hexagonal (Figure 1c and d). The optimization of the structure is much faster. Further studies on zig-zag nanotubes show similar oval shape transformation for (n, 0) bundles with n ≠ 3m (m, integer) while n = 3m bundles undergo polygonization into a hexagonal shape. A detailed report on these results will be published separately. An important point that emerged from our calculations is that even nanotubes with smaller diameters such as (7,0) are as deformed as (11,0) nanotubes. This is in contrast to earlier predictions [10] that large diameter nanotubes (~25 Å diameter) only become polygonalized upon crystal formation. Therefore, the symmetry of the nanotubes with respect to the lattice as well as the interactions between nanotubes play important roles in determining the structure of the bundles of nanotubes. Experimentally it was found [8] that even nanotubes with diameter of about 15 Å are deformed. Our results show that even much smaller nanotubes are deformed upon bundle formation.

The electronic structure of bundles changes with pressure and during the transformation. It is well known that the conduction and valence bands in SWCNTBs are closest at $k_z \sim 1/3$. In (10,10) SWCNTBs at zero pressure, the conduction and valence bands touch near k = -0.2 \ 0.2 \ 1/3 (in dimensionless units) while elsewhere in the $k = k_x \setminus k_y \setminus 1/3$ plane a gap exists. At 2 GPa in the monoclinic structure the valence band moves up most near k = -0.4 \ 0.2 \ 1/3, while the conduction band moves up least at $k = 0 \setminus 0 \setminus 1/3$. The movement of the valence band is greater than that of the conduction band so that the material becomes metallic with a finite density of states at the Fermi level. For the (12,12) SWCNTBs at zero pressure there is a band crossing along the K-H direction. Under hydrostatic pressure the conduction band moves up more than the valence band and the indirect gap in other directions increases. These results have been discussed in detail in Ref. [7].

CONCLUSIONS

In summary, we have studied the structures of the (10,10) and (12,12) crystals as zero as well as non-zero hydrostatic pressures from *ab initio* calculations. Our results using no constraints on the cell shape and size suggest a transition from near hexagonal to a monoclinic structure at around 1.3 GPa for (10,10) nanotube bundles with increasing pressure such that the nanotubes transform into an oval shape. However, when the pressure is decreased, the monoclinic phase is found to be lower in energy even at zero pressure. This is due to the asymmetric interactions between the nanotubes. Even at intermediate pressures, deviations from hexagonal lattice are significant. When two nanotubes per cell are considered, the energy of the system is lowered and the relative orientations of the hexagons of the neighbouring nanotubes become similar to the one in graphite. The relaxation process of these (10,10) nanotube bundles is very slow as it involves rotation of the nanotubes. The (12,12) bundles, however, undergo polygonization into a hexagonal shape and the lattice remains hexagonal. Similar behavior of symmetry dependent phase transitions has been obtained for the zig-zag nanotube crystals. These results would clarify the disparate results obtained in experiments.

ACKNOWLEDGEMENT

VK thankfully acknowledges the kind hospitality at the Institute for Materials Research, Tohoku University and the support from JSPS. Part of this work was performed under the inter-university cooperative research program of the Laboratory for Advanced Materials, Institute for Materials Research, Tohoku University. The authors gratefully acknowledge the Center for Computational Materials Science at the Institute for Materials Research for allocations on the Hitachi SR8000 supercomputer system.

REFERENCES

[1] U.D. Venkateswaran, A.M. Rao, E. Richter, M. Menon, A. Rinzler, R.E. Smalley, and P.C. Eklund, 1999. Probing the single-wall carbon nanotube bundle: Raman scattering under high pressure, Phys. Rev. B59, 10928–10934.

[2] J. Tang, L.-C. Qin, T. Sasaki, M. Yudasaka, A. Matsushita, and S. Iijima, 2000. Compressibility and Polygonization of Single-Walled Carbon Nanotubes under Hydrostatic Pressure, Phys. Rev. Lett. 85, 1887–1889.

[3] M.J. Peters, L.E. McNeil, J.P. Lu, and D. Kahn, 2000. Structural phase transition in carbon nanotube bundles under pressure, Phys. Rev. B61, 5939–5944.

[4] U.D. Venkateswaran, E.A. Brandsen, U. Schlecht, A.M. Rao, E. Richter, I. Loa, K. Syassen, and P.C. Eklund, 2001. High Pressure Studies of the Raman-Active Phonons in Carbon Nanotubes, Phys. Stat. Sol. (b) 223, 225–236.

[5] M.H.F. Sluiter, V. Kumar, and Y. Kawazoe, 2002. Symmetry-driven phase transformations in single-wall carbon-nanotube bundles under hydrostatic pressure, Phys. Rev. B65, 161402-1–161402-4.

[6] V. Kumar, M.H.F. Sluiter and Y. Kawazoe, 2002. Structural and electronic transitions in single wall carbon nanotube bundles under pressure, Physica B323, 199–202.

[7] M.H.F. Sluiter, V. Kumar, and Y. Kawazoe, 2002. Electronic structure of single wall carbon nanotube bundles under compression as compared to graphite and hexagonal graphene stacking, Physica B323, 203–205.

[8] M. J. López, A. Rubio, J. A. Alonso, L.-C. Qin, and S. Iijima 2001. Novel Polygonized Single-Wall Carbon Nanotube Bundles, Phys. Rev. Lett. 86, 3056–3059.

[9] G. Kresse, J. Furthmüller, 1996. Efficiency of ab-initio total energy calculations for metals and semiconductors using a plane-wave basis set, J., Comp. Mat. Sci., 6, 15–50.

[10] J. Tersoff and R. S. Ruoff, 1994. Structural Properties of a Carbon-Nanotube Crystal, Phys. Rev. Lett. 73, 676–679.

A Study of the Pressure Dependence of the Raman Spectrum of C_{60} and Single-Walled Nanotubes in Methanol-Water Mixtures

M. EL-ASHRY and M. AMER
Department of Mechanical and Materials Engineering, Wright State University, Dayton, OH

J. F. MAGUIRE
Air Force Research Laboratory, WPAFB, Materials & Manufacturing Directorate, Polymer Branch, Dayton, OH, USA

ABSTRACT

We report the results of a study of adsorption of small molecules on the surface of buckminsterfullerene, C_{60} and on single walled nanotubes (SWNT). The pressure dependence of the Raman spectrum was investigated over the range 0-100 k bar in methanol-water mixtures that were used as the pressure-transmitting-fluid (PTF) in a diamond anvil cell. It is found that the spectral shift and its pressure derivative are sensitive to both the applied pressure and to the composition of the PTF. These observations are consistent with an explanation that involves preferential adsorption onto the surface of the C_{60}. In particular, the notion of C_{60} collapse need not be invoked to explain the observations.

INTRODUCTION

Since their discovery in the mid 1980s, buckminsterfullerenes have been a subject of intense research interest. They have the potential to provide new classes of structural materials, sensors, electronic devices, and many other applications [1]. It has not been so widely recognized that these molecules, including the closely related single walled nanotubes, provide unique new experimental models with which to study the statistical mechanics and condensed matter physics of thermodynamically small systems in low dimensionality. A thermodynamically "small" system is defined as a system in which the correlation length is of the same order as the system dimension. Small systems are ubiquitous in nature and include processes that occur at the cell wall and the interactions that govern the formation of long-range structures (sometimes called "self-assembly") in the mesoscopic regime [2]. They also occur in three common physical situations. First, near a critical point the correlation length diverges giving rise to marked changes in the properties like the isothermal compressibility and diffusivity [3]. Second, *within* an interface the properties of matter are very different from bulk matter. For example, if we take a mixture of oil and water then there will be a little water in the oil-rich phase and a little oil in the water-rich phase. However, inside the thin interface (~5 nm), between the two bulk phases, there is a huge concentration gradient in which there is a region containing a 50% homogeneous mixture of oil and water, a substance that has very unusual properties but does not exist in any other situation. Here the range of the density correlation function is of order ten molecular diameters, which is about the same as the thickness of the interface [4]. Finally, when particles (e.g. colloids) are very large on the scale of molecules but still much smaller than the size at which the laws of continuum macroscopic mechanics might hold, they are said to be mesoscopic and may display unique and unusual behavior.

It is the ratio of the range of the correlation length to the characteristic size of the system that determines whether the interaction can be integrated in the thermodynamic limit, or whether the system is small and might display unusual or interesting behavior. The key point is that it is the *ratio* that is important and not the absolute length or size of the system. It turns out that small interfacial and colloidal systems often have correlation lengths in the range of nanometers and this has given rise to the unfortunate term "nanosystem" or "nanomaterial", but it is quite possible for entropic transitions to govern the structure of matter over distance scales of many microns; hence, it should not be thought that the structure or properties of "nanomaterials" are in any fundamental or absolute way related to a particular subdivision of the meter.

In the present work, a C_{60} molecule presents a small number of graphite-like adsorption sites (20 hexagons and 12 pentagons) to solvent molecules. This solvent interaction gives rise to a large change in the spectral response of the C_{60}, hence the latter can act as a probe of the surface interaction offering a unique opportunity for direct spectroscopic investigation of the thermodynamics and statistical mechanics of thermodynamically small interfacial systems. This is important because there are relatively few experimental systems that can serve as models for such systems. For example, the adsorption of argon on graphite has provided much insight into gas phase adsorption and the existence of commensurate and incommensurate phases at low temperature [5,6]. However, when a liquid phase is in contact with a solid surface the bulk response from both phases dominates the signal and little of the interfacial interaction can be discerned from direct spectroscopic observation through the liquid. The nature of the adsorbed state has had to be inferred from thermochemical and related studies. On the other hand, a C_{60} or a single walled carbon nanotube (SWNT) has high symmetry, a well-defined structure and a huge polarizability. Experimentally, the spectral response of C_{60} or SWNTs may be readily observed against the bulk solvent background so that studies of liquids in contact with this class of surface may be conducted in a straightforward fashion using bulk systems.

EXPERIMENTAL PROCEDURE

C_{60} samples (99.95%) were obtained from Aldrich Chemical Company and were used without further purification. Samples were prepared by dissolving a small amount (~1 mg) of sample into solution (~20 cm^{-3}) of methanol and water having different mole fractions of methanol of 1, 0.8, and 0.66. Typically samples were sonicated for about 5 hours to assure adequate dispersion and dissolution. The sample was then transferred to a diamond anvil cell. Spectra were recorded over the range 0~100 k bar using a Renishaw Model 2000 spectrometer equipped with an objective of focal length 50 mm that enabled probing of the scattering volume through the 0.6 mm diamond window of the cell.

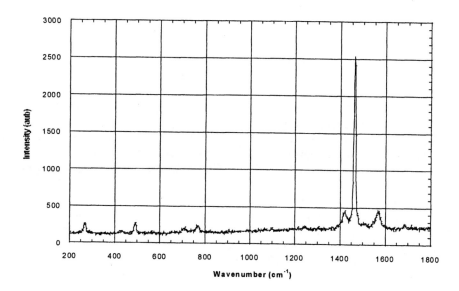

Figure 1. Raman spectrum of C_{60} Molecules.

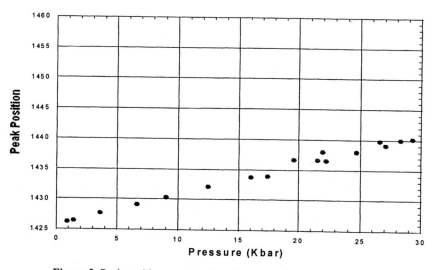

Figure 2. Peak position as a function of pressure for pure methanol as a PTF.

RESULTS AND DISCUSSION

Figure 1 shows a typical Raman spectrum for C_{60}. Here the peaks at 1420 cm^{-1} and 1469 cm^{-1} have been assigned previously [7] to the H_g "pentagon shear" and A_g "pentagon pinch" mode respectively. Figure 2 shows a plot of the 1420 cm^{-1} peak position in pure methanol over the range 1~30 k bar. Note the marked plateau in the peak position observed around 25 k bar. Figure 3 shows the shift of the same peak over the a pressure range 0 k bar to 100 k bar measured for pure methanol and 0.8 methanol mole fraction as pressure transmission fluid (PTF). Now a well-defined plateau develops over the pressure range 18 k bar to 40 k bar. A second plateau is also observed for the pure methanol sample in the range 40 k bar to 90 k bar depending on the composition of the PTF.

The behavior remained unchanged as the composition of the pressure transmission fluid was changed to 0.66 methanol mole fraction. Figure 3 shows a plot of the low plateau onset pressure as a function of methanol mole fraction in the pressure transmission fluid. It is clear that the peak maximum is a strong function of the solvent chemical composition. Moreover, a plateau region of the sort observed in Figure 3 is strongly suggestive of an adsorption process in which methanol and water molecules compete for hexagonal and/or pentagonal sites on the surface of the C_{60}. A plausible rationale for the observed behavior, which is currently under investigation, is as follows.

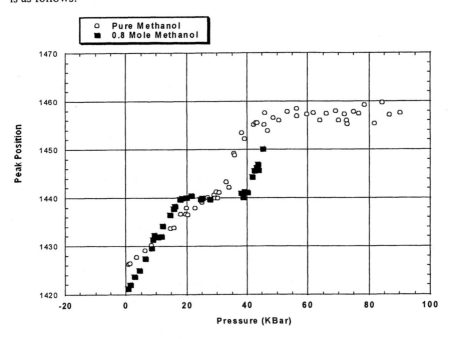

Figure 3. Peak position as a function of pressure for pure methanol and 0.8 mole methanol/water mixture as pressure transmission fluid.

Upon increasing the pressure, water and methanol molecules are adsorbed forming a mobile phase on to the surface of the buckyball until it is saturated at monolayer coverage. Then the onset of the first plateau in the peak position is observed. Further increasing the pressure does not increase adsorption (and hence peak position) but causes a phase transition, possibly to a pinned or solid-like phase. If the pressure is further increased, adsorption of a second adsorbed layer is initiated as indicated by the further change in the peak position followed by formation of a second layer (onset of the second plateau) that also goes through a phase transition as indicated by the second plateau. It is important to note that the onset pressure and length of the plateau observed at 1440 cm^{-1} peak position is strongly dependent on the PTF chemical composition as shown in Figure 4. This observation further supports a preferential surface adsorption explanation of the observed behavior.

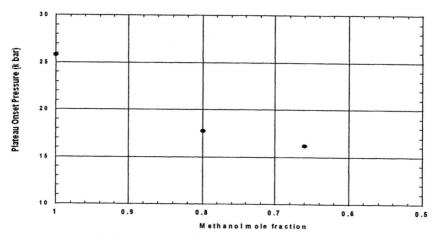

Figure 4. Plateau starting pressure as a function of PTF composition.

It is important to note that in previous studies on single walled carbon nanotubes under pressure [8-10] a discontinuity of slope as well as plateau region were observed. There has been considerable discussion in the literature attributing this discontinuity and plateau to a rather implausible reversible collapse of the nanotube [8,11]. The observed transition, in both buckyballs and nanotubes should be associated with a first order transition in the surface adsorbed phase.

ACKNOWLEDGEMENT

The authors thank the Air Force Research Laboratory for support of this work.

REFERENCES

[1] K.M. Kadish, R.S. Ruoff, 2000. Fullerenes: chemistry, physics, and technology, Wiley Interscience, N.Y.
[2] A.K. Boal, F. Ilhan, J.E. Derouchey, T.T. Albrecht, T.P. Russell, Y.M. Rotello, 2000. Nature 404, 746–748.
[3] J.S. Rowlinson, 1983. Faraday Lecture, The Molecular Theory of Small Systems, Chemical Society Reviews, Imperial College, London.

[4] I.M. Svishchev, M. Zassetsky, A. Yu, 2000. J. Chem. Phys., 113(17), 7432–7436.
[5] N.Y. Matos, G.E. López, 1998. J. Chem. Phys. 109(3), 1141–1146.
[6] E. Kierlik, M.L. Rosinberg, 1991. Phys. Rev. A, 44 (8) 5025–5037.
[7] G.A. Kourouklis, S. Ves, and K.P. Meletov, 1999. Physica B, 265, 214–222.
[8] U.D. Venkateswaran, A.M. Rao, E. Richter, M. Menon, A. Rinzler, R.E. Smalley, P.C. Eklund, 1999. Phys. Rev. B, 59(16) 10928–10934.
[9] P. V. Teredesia, A. K. Sood, S. M. Sharma, S. Karmakar, S. K. Sikka, A. Govindaraj, C. N. R. Rao, 2001. Phys. Stat. Sol., 223, 479–487.
[10] J.R. Wood, H.D. Wagner, 2000. Appl. Phys. Lett. 76(20) 2883–2885.
[11] J.R. Wood, M.D. Frogley, E.R. Meurs, A.D. Prins, T. Peijs, D.J. Dunstan, D.H. Wagner, 1999. J. Phys. Chem. B 103 10388–10392.

Chapter 14: Nano-Scale Surface Phenomena

Fundamental Studies of Large Organic Molecules on Metallic Substrates by High Resolution STM

F. ROSEI

Institut National de la Recherche Scientifique–Énergie, Matériaux et Télécommunications, Université du Québec, Varennes, Québec, Canada

ABSTRACT

Large organic molecules are of great current interest because of their fundamental physical and chemical properties. At the same time, they may lead to interesting applications in the emerging field of nanotechnology, since they are considered the basic building blocks for developing novel devices that operate on the nanometer scale, including for example nanomechanical and nanoelectronic systems. Complex molecule-surface and molecule-molecule interactions are responsible for various phenomena, such as surface diffusion and molecular self-assembly into ordered structures. In addition, upon adsorption the surface of choice may rearrange dramatically to accommodate different molecular conformations. In turn, the strong interaction with the substrate often leads to conformational changes within the adsorbed molecules, often also changing their properties. Here I will illustrate a few selected examples of the interaction between large organic molecules with metal surfaces, also outlining some perspectives for future developments of this exciting field of research.

INTRODUCTION

In the past decade the field of nanoscience has grown explosively. While we are just beginning to understand the functionalities that can be accessed through the use of nanostructured materials, the tremendous potential of "nano" approaches[1] to revolutionize the ways in which materials are processed is already apparent. Presently atoms, molecules, clusters and nano-particles can be used as functional building blocks for fabricating advanced and totally new materials. The optimal size of these

---

[1] By *nano* here we mean that at least one of the three dimensions is smaller than 100 nm, with a consequent, strong size effect resulting from the reduction in dimension.

unit components depends on the particular property to be engineered: by altering the dimensions of these building blocks, controlling their surface geometry, chemistry and assembly, it will be possible to engineer functionalities in unprecedented ways.

At the start of the new millennium we are therefore confronted with the need and desire to learn more about the atomic scale structure of matter. Besides our intrinsic interest in fundamental science in fact, the ultimate goal in this context is to develop functional devices that can operate on the nanoscale. This obsessive trend towards miniaturization is driven partly by an analogous trend in the semiconductor industry, with the aim of developing ever smaller and faster computers [1], and partly by the desire to develop novel devices of other types, for example micro-electro mechanical systems (MEMS), gas and chemical sensors or even biomedical equipment. Examples include mechanical devices that operate below the micron length scale that may perform surgical or similar operations in blood vessels or other vital organs of the human body. For all these reasons, it is expected that Nanotechnology will have a much greater impact on our modern society than the Si integrated circuit (which led to the "electronic revolution" of the 20th century), since it will be applicable to many different fields of human activity, including medicine, security, mechanics and telecommunications.

Clearly, proper tools must be used to study properties of materials and surfaces on the nanometer length scale. Whenever we push the limits of instrumentation and develop new probes, novel and unexpected phenomena may suddenly appear.

In the following I will present a short description of some recent results, which reveal several features of the complex interaction between largish molecules and metal substrates. The focus of this brief overview is on experiments performed using Scanning Tunnelling Microscopy (STM) [2]. Finally, I will outline a few possible perspectives for the future, which will strongly rely on the development of new instrumentation.

INTERACTION BETWEEN COMPLEX ORGANIC MOLECULES AND METAL SUBSTRATES

Upon adsorption on a surface, molecular ordering is in general controlled by a delicate balance between intermolecular forces and molecule-substrate interactions. Under certain conditions, these interactions can be controlled to some extent, and sometimes even tuned by the appropriate choice of substrate material and symmetry. Several studies have indicated that, upon molecular adsorption, surfaces do not always behave as static templates, but may rearrange dramatically to accommodate different molecular species [3]. In this context, the scanning tunnelling microscope (STM) [2] has proved to be a very powerful tool for studying the atomic-scale structure of surfaces, and for investigating adsorbate–surface interactions.

In Figs. 1A and C we show the chemical structure of two related complex molecules, decacyclene (DC, $C_{36}H_{18}$) and hexa-*tert*-butyl decacyclene (HtBDC, $C_{60}H_{66}$), respectively. These molecules consist of the same aromatic π system, which adsorbs parallel to the Cu substrate. Compared to DC, besides the central π system HtBDC has six *tert*-butyl "spacer" groups surrounding its aromatic core [4,5].

At temperatures below 150 K, molecular mobility is frozen and individual molecules can be imaged by STM. HtBDC molecules are found in two symmetry equivalent adsorption conformations where each molecule appears as six lobes (Figure 1D) [4,5] arranged in a distorted hexagon with threefold rotational symmetry. From molecular dimensions and by interplay with elastic scattering quantum chemistry (ESQC) calculations [6] it is inferred that the lobes correspond to tunneling through single *tert*-butyl groups. This confirms, as expected, that HtBDC adsorbs with the aromatic board parallel to the substrate.

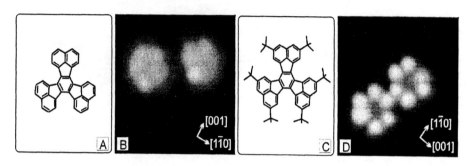

Figure 1. (A) Chemical structure of DC. (B) STM image of DC (5×5 nm^2, V_t=-1215 mV, I_t= -0.11 nA). (C) Chemical structure of HtBDC. (D) STM image of HtBDC (5×5 nm^2, V_t=1250 mV, I_t=0.34 nA).

The DC molecules are also found in two symmetry equivalent adsorption conformations (Figure 1B). However, they do not show a pronounced internal structure as in the case of HtBDC, since the *tert*-butyl groups are missing in this case. A close inspection of STM images leads to the conclusion that DC molecules also prefer the adsorption geometry with the aromatic board parallel to the surface. This is believed to be caused by the strong interaction between the surface and the aromatic π system.

Similarly, the Lander molecule ($C_{90}H_{98}$) consists of an aromatic π-system (I will refer to it as "board" in the following) with four spacer legs that elevate the board from the substrate (Figs. 2A and B) [6,7]. In Figs. 2C and D, a low temperature (100 K), high-resolution STM image is shown of a single Lander molecule deposited on a Cu(110) surface with corresponding calculated images (Figs. 2E and F) [6,7]. Lander molecules are imaged by STM as four lobes arranged in either a rhomboidal or rectangular geometry.

From ESQC calculations [6] it is inferred that the two different molecular shapes correspond to two possible geometrical conformations of the molecule on the surface, one with the four legs arranged parallel and the other with its legs arranged antiparallel to one other (Figs. 2A and B). The four lobes in the images correspond to tunnelling through spacer groups of the molecule.

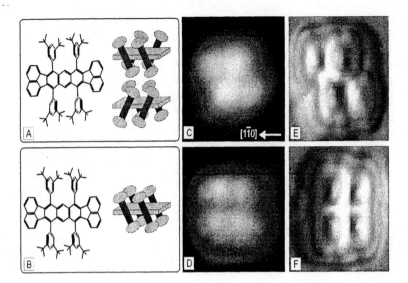

Figure 2. Chemical structure and space-filling model of the Lander molecule in (A) rhomboidal and (B) rectangular shape. STM image of a Lander molecule on Cu(110) at 100 K (2×2 nm^2, V_t=-1000 mV, I_t=-0.21 nA) with (C) rhomboidal shape and (D) rectangular shape. (E and F) ESQC-calculated images of rhomboidal (chiral) and rectangular (achiral) Lander molecules respectively, using the same tunnelling parameters as in experiments. The aromatic board of the Lander molecule is parallel to the close-packed Cu direction indicated in the image. Adapted from Ref. [7], with permission.

These are two good examples of how high-resolution STM has recently permitted unprecedented new insights into a number of fundamental processes related to largish molecular interactions with surfaces. Other examples include molecular diffusion, adsorbate bonding on surfaces and molecular self-assembly. In addition to the normal imaging mode, as I will describe in the following pages, the STM tip can also be employed to manipulate single atoms and molecules in a bottom-up fashion. In this way nanostructures can be engineered with atomic precision to study surface quantum phenomena of fundamental interest.

SURFACE DIFFUSION PROCESSES

In the last decade, development of fast-scanning STM prototypes has led to major advances in studying surface diffusion processes on an atomic/molecular scale. Diffusion of adatoms has been observed directly by field ion microscopy for metals on metals [8] and by STM for various adsorbates [9,10] on a wide range of surfaces. However, detailed studies on surface mobility of complex molecules are still lacking [11].

Diffusion processes allow molecules to meet and interact on a substrate and eventually form molecular nanostructures and self-assembled monolayers. Therefore, in order to employ molecules as elementary building-blocks in self-organized nanodevices it is *essential* to gain quantitative information on their mobility. For example, this may lead to first-principles design of molecules that do not diffuse on the surface at room temperature (RT). Studying the behaviour of large molecules at lower temperatures is in fact appealing from a fundamental viewpoint, however it is apparent that in technological applications the temperature of operation must be RT.

In the following, the results of a recent study on comparative diffusion of two related large molecules on a Cu(110) surface are discussed. This work shows that molecular surface diffusion depends critically on molecule-surface interactions and that molecular diffusivity can be tuned to some extent by appropriately designing the chemical structure of the molecule [5]. Surface diffusion of large organic molecules was studied by acquiring time-resolved STM movies, i.e., series of sequential STM images from the same area of the sample, using a fast-scanning, home-built STM [5, 9]. Both molecules used in this work, Decacyclene (DC) and hexa-di-tert-butyl Decacyclene (HtBDC), were observed to diffuse along the close-packed $[1\bar{1}0]$ direction of the Cu(110) substrate as indicated in Figure 3 [5]. They are therefore constrained to move in only one dimension, because of the anisotropy of Cu(110).

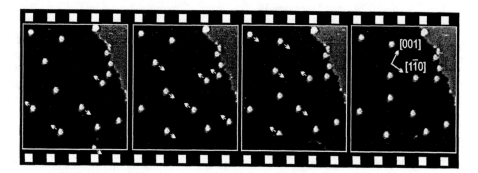

Figure 3. Still photographs extracted from an STM movie of HtBDC molecules imaged as bright spots on Cu(110) at $T = 194$ K. Image size is 50×50 nm^2. Movies consist of 400-2500 image-to-image observations of individual, diffusing molecules (a complete movie can be seen by clicking on the first frame in the above figure or by visiting http://www.phys.au.dk/camp/movies/ht79.mpg). Molecular displacement can be clearly discerned; arrows indicate the direction in which the molecules have moved in the successive image. From Ref. [5], with permission.

A surface diffusion process can be simply described as a random-walk performed by an adsorbate hopping with a rate h between adjacent adsorption sites on a given two-dimensional substrate. Even while being bound to the surface, in fact the adsorbate may jump between different adsorption sites provided that it has enough energy. This excess energy is normally supplied by surface phonons provided the temperature is high enough. Diffusion processes are generally described in terms of three parameters: (i) To hop between nearest neighbour sites, an adsorbate has to overcome a barrier on the potential energy surface—the activation energy for diffusion, E_d; (ii) The frequency with which an adsorbate tries to overcome the energy barrier is called the attempt frequency, h_0; (iii) Finally, the root mean-squared jump length λ contains information about whether the adsorbate performs jumps between nearest neighbour adsorption sites only or can also perform long jumps spanning multiple lattice spacings.

Moreover, the diffusion constant $D = \langle(\Delta x)^2\rangle/2t$ is defined as the molecular mean-squared displacement, $(\Delta x)^2$, during the image acquisition time t. Both the hopping rate h and the diffusion constant D have a temperature dependence that is generally described by the Arrhenius law [12]:

$$h = h_0 \exp(-E_d / kT) \quad \text{and} \quad D = D_0 \exp(-E_d / kT) \qquad (1)$$

Here h_0 and D_0 are the prefactors for hopping rate and diffusion constant respectively, k is Boltzmann's constant, and T is the temperature. The activation energy of diffusion E_d as well as the prefactors are be extracted from Eq. 1 by means of the so-called Arrhenius analysis, in which $\ln h$ or $\ln D$ is plotted versus $1/kT$. To carry out a quantitative analysis in terms of D, it is necessary to evaluate molecular displacements Δx between consecutive images. The molecular mean-squared displacement, $(\Delta x)^2$, can be calculated directly from displacement distributions [11].

From the Arrhenius plots of h and D for both DC and HtBDC, the activation energies and prefactors were determined to be 0.60 eV (HtBDC) and 0.72 eV (DC). The lower activation barrier for diffusion of HtBDC compared to DC is attributed to the presence of six *tert*-butyl spacer groups on HtBDC. The effect of these groups, as mentioned previously, is to effectively increase the distance between the aromatic π-system and the surface, which are generally believed to have a strong attractive interaction.

Another important parameter for diffusion is the average jump length of molecules between different adsorption sites. In the simplest picture, adsorbate migration occurs by random jumps between nearest neighbour sites [8]. Long jumps (spanning several lattice sites) are also believed to contribute to diffusion processes in the case of weak adsorbate-substrate interaction [12, 13], but experimental evidence for such events is still quite limited and restricted to the mobility of metal adatoms [14].

A new, simple approach to determine root mean-square jump lengths λ in surface diffusion has recently been proposed. It is based on the fundamental relation [5]:

$$\langle(\Delta x)^2\rangle = \lambda^2 \, ht \qquad (2)$$

This equation can be used to determine the root mean-squared jump length, if $\langle(\Delta x)^2\rangle$ and h are measured independently from sequences of consecutive STM images. Surprisingly, it was found that long jumps play a dominant role in the diffusion of DC and HtBDC on Cu(110) [5]. Remarkably, the root mean-squared jump lengths are determined (see Figure 4 B) to be as large as $\lambda = 3.9\pm0.2$ and 6.8 ± 0.3 Cu nearest neighbour distances for DC and HtBDC, respectively [5].

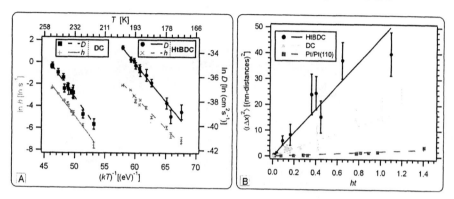

Figure 4. (A) Arrhenius plots of hopping rates h (grey) and tracer diffusion constants D (black) for DC (left) and HtBDC (right). Lines are best fits to the Arrhenius expression for h and D (see Eq. 1). (B) Plot of mean-squared displacement $\langle(\Delta x)^2\rangle$ versus ht for HtBDC and DC. Also shown are data for the diffusion of Pt on Pt(110) [28]. Straight lines are best fits to Eq. 2. yielding root mean-squared jump lengths. From Ref. [5], with permission.

This finding is in strong contrast to results from metal-on-metal diffusion, where root mean-squared jump lengths close to unity are found. In Figure 4B, data for diffusion of Pt on Pt(110) are shown with $\lambda = 1.1$ Pt nearest neighbour distances. Note that Pt adatoms preferentially jump between nearest neighbour sites, and the contribution of long jumps is only of the order of 10%, in excellent agreement with earlier results [9].

This elaborate quantitative analysis of surface diffusion shows it is possible to tailor molecular diffusion properties by designing molecules with the appropriate chemical structure. By raising the aromatic board common to DC/HtBDC away from the surface with spacer groups in the case of HtBDC, this molecule has a diffusion constant that is approximately four orders of magnitude higher compared to that of its related DC molecule on the same surface [5]. Although microscopic insight into the origin of long jumps is still missing, clearly the higher diffusivity for HtBDC is both due to larger root mean-squared jump lengths and a lower activation barrier for diffusion.

SURFACE RESTRUCTURING PROCESSES: MOLECULAR MOLDING ACTION AT STEP EDGES

When depositing large organic molecules onto surfaces at the technologically relevant temperature range (about RT), they are often observed to diffuse very rapidly, and it is not possible to resolve individual molecules. In some cases it has however been observed that when molecules are adsorbed at RT, they may anchor to the surface by inducing a restructuring of the topmost metal layer. Together with other effects, this significantly reduces their diffusivity.

In a previous paragraph we described the conformational changes of Lander molecules upon deposition on a surface at low temperatures. Upon deposition of the Lander at RT, molecules adsorb on the surface diffusing readily towards step edges, as shown in Figure 5a.

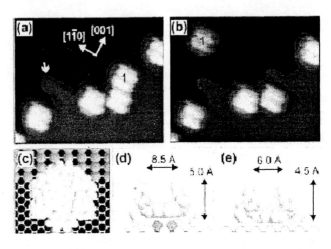

Figure 5. (a) Before manipulating the Lander molecule labeled "1" from a Cu(110) step edge, a nano-structure present from a previous manipulation experiment is indicated by an arrow, (b) After manipulation of the molecule along the 110 direction, the nanostructure underneath becomes visible (tunneling parameters for manipulation; $V = -55$ mV, $I = -1.05$ nA), (c) Model of the Lander on the nanostructure, showing that the board is parallel to the nanostructure (Note that the length of the nanostructure extends beyond the molecule), Cross-sectional side view of the conformation of a Lander molecule on a nanostructure, and Cross-sectional side view on a flat terrace. Numbers refer to the distance between and height of the spacer groups calculated from STM images.

Surprisingly, it has been found that when molecules move away from the anchoring positions at step edges, a restructuring of monoatomic Cu steps induced by individual docked molecules is revealed [6, 7]. To unravel this phenomenon, the STM tip was used as a tool to manipulate individual Lander molecules. The manipulation is performed by lowering the tip towards the surface in a controlled manner either by decreasing or increasing the tunnelling voltage or current, respectively. This exerts a force on surface adsorbates, and by tuning the magnitude and direction of this force, the tip can push or pull molecules across the surface [15–18]. Finally, withdrawing the tip by resetting the tunnelling voltage/current to imaging values terminates interactions between molecules and tip. By controlling the tip position, it is possible to manipulate individual molecules one at a time along a predefined path, leaving the rest of the scan area unperturbed.

A manipulation sequence of the Lander molecule is shown in Figure 5A, in which one Lander molecule is removed from the step edge. A characteristic metal nanostructure appears at the site where molecules were previously attached (attachment site in Figure 5B).

The overall process for anchoring a Lander molecule to the Cu nanostructure can be described as follows. When the molecule diffuses towards step edges at RT, it reshapes fluctuating Cu step adatoms into a nanostructure. It is favorable for the Lander to anchor to the nanostructure at the step edge because the gain in energy by adsorbing the molecule on the nanostructure relative to a flat terrace site is higher than the energy required to create the structure itself. The dimension of the board and leg fits such that two atomic rows can be accommodated between the legs under the board. This leads to a favorable interaction between the π-system and the Cu atoms underneath. The dimensions and shape of the molecule form a perfect template for the double row of Cu atoms to be stabilized when the temperature is lowered.

From ESQC calculations, further insight was gained into the conformational changes of the molecules on the terrace and the nanostructure (Figures 5D and 5E). The molecule's central board is strongly attracted to the surface because of the large π-system facing the metal substrate [6,7]. This introduces a severe constraint on the legs on a flat terrace leading to an out-of-plane distortion of each leg-board σ bond. This σ bond almost restores its planarity relative to the board because when the Lander is anchored to the structure, its central board is lifted up by more than 0.1 nm relative to the surface (Figures 5D and 5E). This reduces the steric constraint on the leg-board σ bond leading to an increased width (0.83 vs. 0.63 nm) and height (0.50 vs. 0.45 nm) of the Lander in the STM image in good agreement with experimental findings [6,19].

The adsorption of Lander molecules at low temperatures (150K) does not lead to a restructuring of Cu step edges. At low temperatures in fact, the molecules simply anchor to a step edge or onto the terrace. In this case, the mobility of Cu kink atoms at the step edge is not high enough for the template to be effective. It is therefore concluded that this process is thermally activated.

PERSPECTIVES: THE ADVANCED LASER LIGHT SOURCE (ALLS)

The Advanced Laser Light Source (ALLS) is an international infrastructure project[2] that will enable the combination of any or all of the most advanced laser technologies to exploit the interaction between electromagnetic radiation and gaseous or condensed matter [20].

The ALLS effort will be based on a new state-of-the-art multi-beam femtosecond laser system that will be built at INRS–EMT in Varennes (QC). The basic goal is to provide the Canadian and international research communities with new tools that have the capability to image changing molecular structures. The approach is based on recent advances in ultrafast laser technology to provide a new generation light source for molecular imaging and other applications.

The principal concept of ALLS is to use a plurality of laser interactions with the shortest light pulse in the world, spanning the x-ray to IR spectrum with sufficient

---

[2] The present estimate of the overall budget is 21 Million Canadian dollars.

peak power in this range to manipulate matter at will and probe its dynamics. Production of X-Rays for biomedical applications is also envisaged [21].

The ALLS project will bring under one roof, a full suite of world-class technologies in laser-molecule stimulation, manipulation and diagnostics. These are technologies in which many of the researchers involved in developing ALLS have already played leading roles. The ALLS project will be a unique infrastructure worldwide.

In this context, I intend to employ the extraordinary capabilities of ALLS in combination with a state-of-the-art, variable temperature STM. Femtosecond laser pulses (e.g. mid IR-wavelength) will be used to excite vibrational modes of complex molecules deposited on a suitable substrate, or to "pump" high-barrier pathways of surface reactions [22]. Both systems will be characterized by STM. Photo-assisted manipulation will allow assembly and characterization of nanostructured samples on a mesoscopic scale. After tuning coherent light pulses at the proper quantized vibrational energies, the STM will be used to selectively break or form chemical bonds. This project should complement existing research efforts in which the STM is used to induce simple chemical reactions on metal surfaces on an atomic scale [23–25].

In particular, I plan to explore the deposition and characterization of complex organic molecules on semiconductor surfaces, an effort which is also being thoroughly investigated by other groups [26–30].

SUMMARY

In this contribution, the ability to image surfaces with atomic resolution and to displace individual atoms and molecules in a controlled manner has been demonstrated making the STM a unique tool to study interactions between complex molecules and metal surfaces. Moreover, by using fast-scanning STM prototypes, it is possible to analyze quantitatively the surface diffusion of such large molecules. Notably these molecules perform very long jumps during the diffusive motion even at low temperatures [31].

I have described how the bonding and ordering of molecules on surfaces are indeed governed by molecule-substrate interactions [3–7]. In this context, I have discussed STM results that demonstrate how the anchoring of complex molecules and the subsequent self-assembly of molecular nanostructures on a metal surface may be associated with a local disruption of the uppermost surface layer directly underneath the molecules. Furthermore I have illustrated how a surface can undergo a restructuring process in order to accommodate a specific molecular geometry [4,6], and conformational changes may occur within individual molecules [6,7]. Restructuring of surface terraces or step edges provides preferential adsorption sites to which the molecules can anchor [4,6,7]. This body of results effectively demonstrates that molecule-surface interactions may be the driving force for self-assembly of molecules on surfaces.

Finally, I have outlined some possible future developments that rely in part on advances in instrumentation.

REFERENCES

[1] M.A. Reed, J.M. Tour, 2000. Scientific American, 6, June.
[2] G. Binnig, H. Röhrer, 1999. Rev. Mod. Phys., 71, S324.
[3] F. Rosei, R. Rosei, 2002. Surface Science, 500, 395.
[4] M. Schunack, L. Petersen, A. Kuhnle, E. Laegsgaard, I. Stensgaard, I. Johannsen, F. Besenbacher, 2001. Phys. Rev. Lett., 86, 456.
[5] M. Schunack, T.R. Linderoth, F. Rosei, E. Laegsgaard, I. Stensgaard, F. Besenbacher, 2002. Phys. Rev. Lett., 88, 156102.
[6] F. Rosei, M. Schunack, P. Jiang, A. Gourdon, E. Laegsgaard, I. Stensgaard, C. Joachim, F. Besenbacher, 2002. Science 296, 328.
[7] M. Schunack, F. Rosei, Y. Naitoh, P. Jiang, A. Gourdon, E. Laegsgaard, I. Stensgaard, C. Joachim, F. Besenbacher, 2002. J. Chem. Phys., 117, 6259.
[8] G.L. Kellogg, 1994. Surf. Sci. Rep., 21, 1.
[9] T.R. Linderoth et al., 1997. Phys. Rev. Lett., 78, 4978.
[10] B.S. Swartzentruber, 1996. Phys. Rev. Lett., 76, 459.
[11] J. Weckesser et al., 2001. J. Chem. Phys., 115, 9001.
[12] K.D. Dobbs, D. J. Doren, 1992. J. Chem. Phys., 97, 3722.
[13] S.-M. Oh et al., 2002. Phys. Rev. Lett., 88, 236102.
[14] J.L. Brand et al., 1990. J. Chem. Phys., 92, 5136.
[15] D.M. Eigler, E.K. Schweizer, 1990. Nature, 344, 524.
[16] L. Bartels, G. Meyer, K.-H. Rieder, 1997. Phys. Rev. Lett., 79, 697.
[17] Ph. Avouris, 1995. Acc. Chem. Res., 28, 95.
[18] F. Moresco et al., 2001. Appl. Phys. Lett., 78, 306.
[19] T. Zambelli et al., 2001. Chem. Phys. Lett., 348, 1.
[20] A. Cavalleri, C. Toth, C.W. Siders, J.A. Squier, F. Raksi, P. Forget, J.C. Kieffer, 2001. Phys. Rev. Lett., 87, 237401.
[21] J.C. Kieffer, A. Krol, Z. Jiang, C.C. Chamberlain, E. Scalzetti, Z. Ichalalene, 2002. Appl. Phys., B 74, S75.
[22] M. Dürr, A. Biedermann, Z. Hu, U. Höfer, T.F. Heinz, 2002. Science, 296, 1838.
[23] H.J. Lee, W. Ho,1999. Science, 286, 1719.
[24] L.J. Lauhon, W. Ho, 2000. Phys. Rev. Lett., 84, 1527.
[25] S.W. Hla, L. Bartels, G. Meyer, K.H. Rieder, 2000. Phys. Rev. Lett., 85, 2777.
[26] P.H. Lu, J.C. Polanyi, D. Rogers, 1999. J. Chem. Phys., 111, 9905.
[27] P.H. Lu, J.C. Polanyi, D. Rogers, 2000. J. Chem. Phys., 112, 11005.
[28] G. Lopinski, D. Moffatt, D. Wayner, R. Wolkow, 2000. J.Am. Chem. Soc., 122, 3548.
[29] P. Kruse, E.R. Johnson, G.A. DiLabio, R.A. Wolkow, 2002. Nano Lett., 2, 807.
[30] G.P. Lopinski, D.D.M. Wayner, R.A. Wolkow, 2000. Nature, 406, 48.

Dielectrophoretic Manipulation of Surface-Bound DNA

W. A. GERMISHUIZEN
Department of Chemical Engineering, University of Cambridge, UK

C. WÄLTI, P. TOSCH, A. E. COHEN, R. WIRTZ and M. PEPPER
Semiconductor Physics Group, Cavendish Laboratory, University of Cambridge, UK

A. P. J. MIDDELBERG
Department of Chemical Engineering, University of Cambridge, UK

A. G. DAVIES
Semiconductor Physics Group, Cavendish Laboratory, University of Cambridge, UK
and School of Electronic and Electrical Engineering, University of Leeds, UK

ABSTRACT

Here, we present a study on the influence of frequency and amplitude of the electric field on the length of the DNA molecules which are stretched out under the influence of the dielectrophoretic force and torque. •-DNA fragments (48 and 25 kilobases) were attached to an array of gold electrodes via a terminal thiol bond and the orientation and stretching characterised as a function of frequency (0.03–1.2 MHz) and electric field (0.2 -1.5 MV/m). A distinct change in behaviour was observed for both fragment types at about 100 kHz. At frequencies below 100 kHz, a change in polarisation of the DNA causes the fragments to collapse onto the electrodes. For frequencies above 100 kHz, the polarisability of the DNA decreases with increasing field which leads to a decrease in the dielectrophoretic force and torque and therefore to a decrease in stretching length with increasing frequency. A maximum length of 20 μm for the 48kb fragment and 10 μm for the 25kb fragment was observed around 300 kHz for DNA molecules attached to the electrodes. These findings are in good agreement with the calculated length of DNA with a high intercalator density.

INTRODUCTION

DNA is increasingly being investigated for application in molecular electronic devices [1] and advanced bio-analytical systems [2–5], mainly owing to the pronounced self-assembly properties of the DNA. In most applications, positioning and orientation of the DNA is essential while manipulation of the DNA on a molecular scale is required. Dielectrophoresis is a non-contact manipulation technique enabling positioning and orientation of DNA molecules in nanotechnological devices with a high spatial resolution. The method is becoming increasingly popular because of the high degree of controllability and the low equipment cost. However, a detailed understanding of the dielectrophoresis force and DNA orientation as a function of frequency and amplitude of the electric field is required to tap the full potential of the technique.

Dielectrophoresis is the response of a dielectric particle, such as DNA, to the force that is exerted on the particle by a non-uniform ac electric field [6]. Double-stranded DNA is a 2 nm thick molecule that can be several micrometers long, and consists of two strands of nucleotide bases connected to a sugar-phosphate backbone. In solution, the phosphate groups on the DNA backbone dissociate to form a negatively charged molecule surrounded radially by a cloud of positively charged counter-ions [3,5,7]. This makes the DNA highly polarizable, and a dipole is induced in the molecule when an electric field is applied. In a non-uniform field there is an imbalance between the forces on different charges of the particle, which results in a net movement. The DNA is attracted towards higher electric field gradients, i.e. towards the electrode edges, and is also orientated parallel to the electric field lines as a result of a torque that is exerted on the dipole by the electric field [6–9] and is therefore stretched out.

The response of the DNA to the dielectrophoretic force and torque needs to be well understood and characterised in order to facilitate reproducible manipulation using this technique. The dielectrophoretic force and torque, and thus the orientation and stretching of the DNA, depend on the dielectric properties of the particle and the medium, as well as the frequency and magnitude of the ac electric field. The orientational response of the DNA dipole to the applied ac electric field is increasingly phase shifted with increasing frequency, therefore reducing its size in the frequency range from 100 kHz to 1 MHz so the magnitude of the dielectrophoretic force is expected to decrease over this region [3,7]. Below about 100 kHz, a change in the DNA polarization occurs and the band collapses onto the electrode, as observed in [7]. Thermal randomisation effects dominate the small dielectrophoretic forces obtained using low electric fields and/or high frequencies. High electric fields increase the dielectrophoretic force, but will result in local Joule heating of the liquid which can disrupt the stretching.

The orientation of the DNA as a function of frequency, electric field, pH and cation concentration has been studied in detail using fluorescence intensity [5] and anisotropy [7] measurements to quantify the accumulation of DNA around the electrodes and the orientation of the DNA relative to the electric field lines. However, a detailed understanding of dielectrophoresis has not yet been established, and a detailed investigation of the length of stretched DNA as a function of applied ac electric field is required. Most dielectrophoresis experiments reported in the literature focused on concentrating DNA from solution around electrode edges, rather than the stretching of immobilised DNA using dielectrophoresis. When applying an ac electric filed, the DNA is orientated parallel to the electric field lines, which leads to a bright fluorescent band around the electrodes as shown in Figure 1. The width of the band of stretched DNA depends on whether the DNA was immobilised onto the electrode prior to dielectrophoresis or whether it was concentrated around the electrode edges from solution [8].

In this work, we measured the length of two different long fragments of DNA (48 and 25 kilobases (kb)) as a function of electric field magnitude and frequency. Furthermore, we determined the electric field and frequency limits for effective manipulation of DNA and compared the stretching of the two DNA molecules.

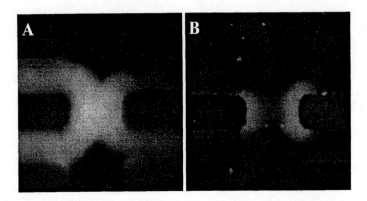

Figure 1. Dielectrophoretic stretching of λ-DNA with a 100 kHz, 0.8 MV/m ac electric field. (A) Dielectrophoresis of unbound λ-DNA in a 5 ng/μl solution. The electric field was applied to the right electrode, while the left electrode was earthed, and the others left floating. (B) Dielectrophoresis of λ-DNA immobilized via a thiol bond onto the gold electrodes.

MATERIALS AND METHODS

A gold microelectrode array was used to generate the strong electric fields required for the dielectrophoresis experiments. The electrodes were 20 μm in width, 15 μm apart, and 30 μm from the opposite electrodes. Prior to use the wafer was cleaned by washing it in "piranha" solution (30% H_2O_2, 70% H_2SO_4) for 1 hour, followed by rinsing in deionised water, ethanol and again in deionised water, and finally air-drying. All chemical reagents were purchased from Sigma-Aldrich, unless otherwise noted.

λ-DNA is a double-stranded circular DNA that contains two single-stranded nicks, twelve bases apart. Upon heating to 65°C, the base pairs between the two nicks dissociate and the DNA molecule linearises to form a molecule with two single-stranded complementary ends.

The 25 kb fragments were obtained by digesting λ-DNA with NheI (New England Biolabs) and XbaI (Roche) simultaneously. The 25 kb DNA fragments were isolated by separating individual restriction fragments by electrophoresis in a 0.5% agarose gel with a 1x Tris-Acetate-EDTA (TAE) buffer. A well was cut in front of the 25 kb fragment, and filled with a 0.5% low melting point TAE agarose gel. Electrophoresis continued until the 25 kb DNA was within this low melting point gel. The piece containing the 25 kb fragments was cut out and digested with GELase (Epicentre), followed by ethanol precipitation.

The 48 and 25 kb DNA, (each at 50 ng/μl), were labeled with a fluorescent intercalator (YOYO-1, Molecular Probes, Eugene, USA) at an intercalator to basepair ratio of 1:8 and was then immobilized onto the gold electrodes. Both the 48 and 25 kb DNA was immobilized with one end on the gold electrodes, using a multistep procedure [8].

An AC voltage was applied across two opposing electrodes of the array to generate the electric field used for dielectrophoresis of the DNA. The potentials were generated by a 20 kHz – 1.1 MHz sine-wave generator and amplified with a custom-built amplifier. The electric field referred to in this work is the applied voltage divided by the distance between the electrodes (30 μm).

The orientation and stretching of the DNA in response to the electric field was observed with a fluorescence microscope (BX60, Olympus), using a 50x objective. Images of the stretched DNA were captured using a cooled CCD camera (150CL, Pixera) and analysed with Scion Image. Length measurements were carried out by calibrating the software using a calibrated image.

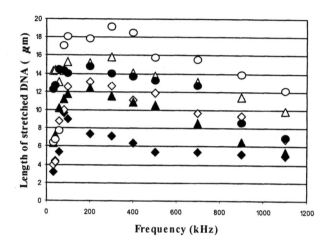

Figure 2. Length of elongated λ-DNA upon dielectrophoretic stretching as a function of frequency at different electric fields: (diamonds) 0.5 MV/m, (triangles) 0.8 MV/m, (circles) 1.2 MV/m. The solid symbols refer to dielectrophoresis of surface-bound DNA and open symbols refer to the dielectrophoresis of unbound DNA in solution.

RESULTS

The length of the stretched surface-bound 48 kb DNA (λ-DNA) was investigated as a function of electric field magnitude and frequency. The length of the DNA molecules was measured from the edge of the electrode to the edge of the fluorescent band (see Figure 1), directly between the opposing electrodes. Figure 2 shows the length of dielectrophoretically stretched λ-DNA as a function of frequency for a set of different electric fields. Images were captured about 10 seconds after each frequency adjustment to allow the stretching to reach steady-state. For comparison, dielectrophoretic stretching of unbound λ-DNA in solution (5 ng/μl) using clean electrodes was performed.

Figure 3 shows the length of dielectrophoretically stretched λ-DNA as a function of electric field for a set of different frequencies. The frequency was held constant while the electric field was varied and images were taken ~10 seconds after each electric field adjustment to allow the stretching to reach a steady-state value.

The effect of the dielectrophoretic stretching length of DNA as a function of molecule length is shown in Figure 4. The length of the stretched surface-bound 25 kb fragments is compared to that of surface-bound λ-DNA (48 kb) for different frequencies at fixed electric fields. The electric field was kept constant and the frequency varied. In a separate experiment, the dielectrophoretic stretching of unbound 25 kb DNA in solution (5 ng/μl) using clean electrodes was investigated.

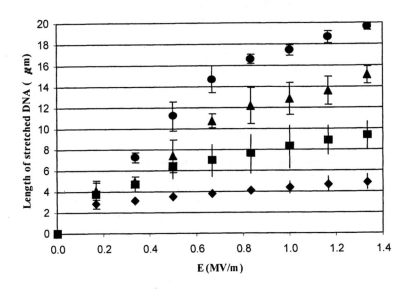

Figure 3. Length of elongated surface-bound λ-DNA upon dielectrophoretic stretching as a function of electric field at different frequencies: (squares) 100 kHz, (circles) 300 kHz, (triangles) 500 kHz, (diamond) 700 kHz.

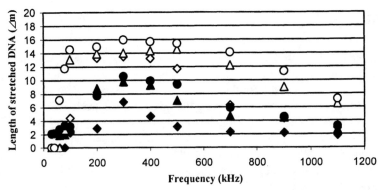

Figure 4. Length of elongated and immobilised 25 kb DNA upon dielectrophoretic stretching as a function of frequency at different electric fields: (diamonds) 0.5 MV/m, (triangles) 0.8 MV/m, (circles) 1.2 MV/m. The solid symbols refer to dielectrophoresis of surface-bound 25 kb DNA and open symbols refer to the dielectrophoresis of unbound 25 kb DNA in solution.

DISCUSSION

In most dielectrophoresis experiments reported in the literature, the DNA is free in solution and upon applying an ac electric field the DNA stretches out, aligns parallel to the electric field because the dominant dipole moment occurs axially along the DNA, and moves towards the electrode edge, where the electric field gradient is the highest [3-9]. In our experiments however, the DNA was tethered to the gold surface and the dielectrophoretic force and torque only stretched out and aligned the DNA molecules parallel to the electric field lines, forming a band-like structure around the electrode (see Figure 1).

The width of the band decreases with increasing frequency above 100 kHz owing to the decreasing dielectrophoretic force and torque which is due to the decrease of the induced dipole. The reorientation of the dipole on the DNA lags behind the change in polarity of the electrodes in the AC field, therefore reducing its size in the frequency range from 100 kHz to 1 MHz. So, the magnitude of the dielectrophoretic force is expected to decrease over this region. Below about 100 kHz, there is a change in the DNA polarization and the band collapses onto the electrode as observed in [7]. The natural contour length of λ-DNA is 16.5 μm. but the high concentration of the intercalating fluorophores causes an expansion of the DNA molecule to about 20μm [10] which agrees with our observations for the maximum length of tethered DNA at a frequency of around 200–400 kHz at 1.2 MV/m (Figures 2 and 3). In this range the force on the DNA reaches a maximum as the dipole moment on the DNA decreases with an increase in frequency.

The stretching of λ-DNA increases with electric field, but is limited. Below 0.2 MV/m no stretching or accumulation of DNA at the electrode edges was observed (data not shown). In this range the dielectrophoretic force was dominated by thermal randomisation. Above 1.5 MV/m electrothermal forces disrupt the band of stretched DNA over the whole frequency range investigated. Joule heating results in local density, viscosity, permittivity and conductivity changes, which is temperature-dependent. A dielectric force exists across a permittivity gradient and a Coulomb force across a conductivity gradient, and these frequency-dependent electro-thermal forces result in fluid flow.

The stretching length of the 25 kb DNA fragments shows a very similar behaviour as a function of frequency as the 48 kb λ-DNA. A change in behaviour occurs at about 100 kHz, below which a change in polarisation causes the DNA molecules to collapse onto the electrodes. For frequencies above 100 kHz, the stretching of the DNA decreases with increasing frequency as a result of the increasing phase shift between the changing polarity of the electrodes and the induced dipole. A maximum stretching length of about 10 μm was observed, which is again in good agreement with the calculated length of a 25 kb DNA fragment with very high intercalator density [10].

CONCLUSIONS

We have investigated the stretching of surface-bound λ-DNA (48 kb) and 25 kb DNA fragments as a function of electric field magnitude and frequency, and compared the results with those for the dielectrophoresis of DNA in solution. We used the width of the band of stretched λ-DNA around the electrodes to quantify stretching due to dielectrophoresis and found a maximum length of 20 μm at electric fields between 0.2 and 1.5 MV/m at frequencies of about 300 kHz, which is in good agreement with the calculated length of the fluorescent-labeled λ-DNA used in the experiments. Very similar results were obtained for the 25 kb DNA fragment. A maximum length of about 10 μm was observed at electric fields between 0.2 and 1.5 MV/m at frequencies of about 300 kHz. Finally, below 0.2 MV/m the stretching is overshadowed by thermal randomisation, and above 1.5 MV/m it is limited by electrothermal effects. Dielectrophoresis of the surface-bound DNA lead to reproducible results while DNA free in solution led to substantially different results in the same experiment, suggesting that the DNA was not bound onto the electrode edge, but rather located close to the electrodes in solution.

ACKNOWLEDGEMENT

This research was in part funded by the EPSRC. For financial support W.A.G. acknowledges the Cambridge Commonwealth Trust, C.W. acknowledges the Schweizerische Nationalfonds zur Förderung der wissenschaftlichen Forschung and R.W. thanks the Gottlieb Daimler and Karl Benz-Stiftung. A.G.D. acknowledges the Royal Society.

REFERENCES

[1] C.M. Niemeyer, 2002, Nanotechnology: Tools for the biomolecular engineer, Science 297 62–63.
[2] M.J. Heller, A. H. Forster, and E. Tu, 2000, Active microelectronic chip devices which utilize controlled electrophoretic fields for multiplex DNA hybridization and other genomic applications. Electrophoresis, 21, 157–164.
[3] M. Washizu et al., 1995, Applications of electrostatic stretch-and-positioning of DNA. IEEE Trans. Ind. Appl., 31, 447–456.
[4] M.P. Hughes, 2000, AC electrokinetics: applications for nanotechnology. Nanotechnol., 11, 124.
[5] C.L. Asbury and G. Van den Engh, 1998, Trapping of DNA in nonuniform oscillating electric fields, Biophys. J. 74 1024–1030.
[6] H.A. Pohl, 1978, Dielectrophoresis (Cambridge: Cambridge University Press).
[7] S. Suzuki et al., 1998, Quantitative analysis of DNA orientation in stationary AC electric fields using fluorescence anisotropy. IEEE Trans. Ind. Appl., 34, 75.

[8] W.A. Germishuizen et al., 2002, Dielectrophoretic addressing of surface-bound DNA, submitted to Nanotechnology.
[9] F. Dewarrat, M. Calame and C. Schönenberger, 2002. Orientation and Positioning of DNA Molecules with an Electric Field Technique, Single Mol. 3 189–193.
[10] M. Daune, 1999. Molecular Biophysics, Oxford University Press, New York.

Comparative Study on Interactions of Polypeptides of Bacteriorhodopsin in the Langmuir-Blodgett Film Based on Hydrogenated Amorphous Silicon Thin Film

Y. TSUJIUCHI, J. SUTO, K. GOTO and S. SHIBATA
Department of Materials Science and Engineering, Akita University Tegata-Gakuen, Akita, Japan

M. IHARA
Institute of Multidisciplinary Research for Advanced Materials, Tohoku University, Aoba, Sendai, Japan

H. MASUMOTO and T. GOTO
Institute of Materials Research, Tohoku University, Aoba, Sendai, Japan

ABSTRACT

In order to study the effect of hydrogen bond of biopolymer in Langmuir-Blodgett (LB) films, two synthesized polypeptides coding the region of an alpha helix A or B of bacteriorhodopsin with different size of loop region were prepared and mixed with retinoic acid. The mixtures were deposited as the LB films based on a quartz substrate and a hydrogenated amorphous silicon (a-Si:H) film and. The UV-VIS spectrums and FTIR reflection absorption spectrums of hybridized films were analyzed. We found that the polypeptide coding the region of an alpha helix B of bacteriorhodopsin with long size of loop region exhibited the oriented interaction with the retinoic acid on a-Si:H.

INTRODUCTION

The alpha helix in proteins plays various roles in the stability [1, 2] and the flexibility of the protein. The alpha helix itself is constructed from hydrogen bonds [2], ionic bonds and so on. Because of its stability, the alpha helix of a protein has the possibility to become a 3D fabrication process not only as a bio-memory or bio-sensor, but also as a bio-computer. Furthermore, the hydrogen bond plays another role with respect to proton transfer. So it is important to consider the possibility and effects of hydrogen on the artificial fabrication of organic/inorganic hybridized membranes or films.

We have demonstrated [3] that the structural properties of hydrogenated amorphous silicon (α-Si:H) films prepared by ECR plasma-sputtering reveal the possibility for a highly-oriented surface structure of an LB film of retinoic acid on a hydrogenated amorphous silicon film. We have also reported on a novel spectroscopic property of the LB film of retinoic acid and polypeptides[3] including the alpha helix region of bacteriorhodopsin.

Bacteriorhodopsin (bR), which is an energy-conversion membrane protein, exhibits a light-induced proton current in only one molecule [4,5]. bR consists of seven alpha helices (A~G) and linker loop regions which can be refolded from three partial polypeptides including helix A, helix B, helices C~G respectively. It also exhibits photo-reactive properties [6]. Photoisomerization of all-trans-retinal to 13-cis-retinal occur in femtoseconds order [7] in bR molecules supplying the energy that is necessary for structural change and sequentially generating proton transfer in 11 milliseconds between several amino acids. It is possible that events like this may be applicable to optical data storage and transfer or to biosensors [8]. For advanced use of this biopolymer to diversify three-dimensional fabrication, an improvement is required in the precision to control ion transfers such as protons in organic/inorganic hybridized systems.

We have examined how amino acid sequences or hydrogen in organic/inorganic hybridized systems can lead to a functional effect using UV-VIS spectroscopy. In this paper, we compare the polypeptide which codes the alpha helix region A and B of bR, and is influenced by light absorption of retinoic acid on α-Si:H films.

EXPERIMENTAL DETAILS

Preparation of α-Si:H Film

Silicon crystal wafers with no impurity (purchased from Nilaco Co. Japan) were processed to create a cylindrical target (diameter 100 mm, height 55mm) using ECR plasma sputtering[9]. Silicon and quartz were used as substrates. Process gas containing 0%, 10%, 20%, 30%, 49% concentration of H_2 respectively were made up by mixing 100% Ar gas and 49% H_2 plus 51% Ar gas. Before sputtering, the reaction chamber was evacuated to 1.0×10^{-7} torr. Microwave power was controlled to 500-800W and the target RF power was controlled to 300-500W. Current in the ECR magnetic coil was controlled at 21 amps. Current in the mirror magnetic coil was held at 12 amps. The distance between target and substrate was held at 160 mm. Sputtering was set to one hour. After cooling the reaction chamber, a second sputtering was performed for one hour and so on. Total sputtering times were set for one hour, two hours, or three hours. This was so the apparatus could be stabilized thermally. The substrate was not heated. Substrate temperatures reached a maximum of 70 °C. After sputtering, the deposited film was allowed to cool down to room temperature within 30-60 minutes. The amorphous structure of the deposited film was confirmed by X-ray diffraction using a JOEL JDX 3530 diffractometer. The existence of Si-H bond in the films was confirmed by IR transmission measured with a SHIMADZU IR8200 IR spectrometer. The surface of the film was scanned using a SII 300HV atomic force microscope.

LB Film Preparation

Retinoic acid (purchased from Sigma Co. Japan) was diluted in chloroform to a concentration of 2 mM. Two synthesized polypeptides containing the coding region of alpha helix A or B and additional amino acid sequence shown in Figure 1 were prepared using a SHIMADZU PSSM-8 peptide synthesizer and purified using the ISCO 2350 high performance liquid chromatography system. Then these polypeptides were diluted in chloroform to the concentration of 0.4 mM and mixed with retinoic acid. The ratio of polypetide and retinoic acid was 1 to 2.

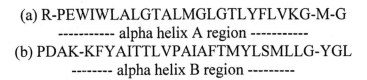

Figure 1. (a) amino acid sequence of polypeptide coding the helix A region of bacteriorhodopsin (bR), (b) amino acid sequence of polypeptide coding the helix B region.

Five layer LB films of these mixtures were made using a moving wall system (a hand-made system using a linear motor LUS2F250ASMA-3 (Oriental Motor Co. Ltd Japan) and a computer controlled system). The surface pressure of the Langmuir film was monitored by polystyrene float displacement with a 0-700mg weight change. Film speed was set to 0.01mm/sec.

MEASUREMENT OF UV-VIS ABSORPTION AND FTIR REFLECTION ABSORPTION OF LB FILEMS

The FTIR reflection absorption spectrum of the LB film was measured using a Shimadzu IR Prestage21 spectrometer. The UV-VIS absorption spectrum of the LB film was measured using a Shimadzu UV 1200 spectrometer. For the sample for UV-VIS measurement, an LB film of retinoic acid and polypeptides was used. This was prepared by moving the substrate up and down five times based on the α-Si:H film prepared by using a process gas containing 20% hydrogen. Deposition duration was set to 30 minutes. These condition were previously determined to optimize the measurement [3].

RESULTS AND DISCUSSION

FTIR Reflection Absorption of LB Film on a-Si:H Film

The FTIR reflection absorption spectrums of five layers of LB film that contained polypeptides A or B and retinoic acid based on a quartz substrate are shown in Figure 2 and 3 respectively. Absorptions around 670 cm-1 and 3700 cm-1 are attributed to –OH groups from water molecules. Comparing the attribution of peptide A and peptide B to absorption from 1225 to 1335 cm^{-1} (alkyl groups in the polypeptide), the sharp peaks observed in Figure 2 indicates a more oriented interaction of retinoic acid and polypeptide B occurred.

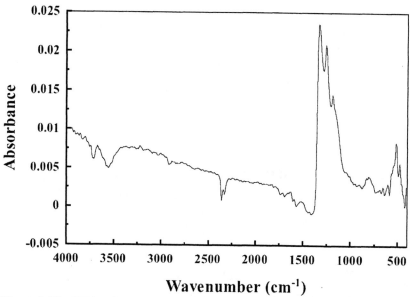

Figure 2. The FTIR reflection absorption spectrums of five layers of LB film which contains polypeptides A and retinoic acid based on a quartz substrate.

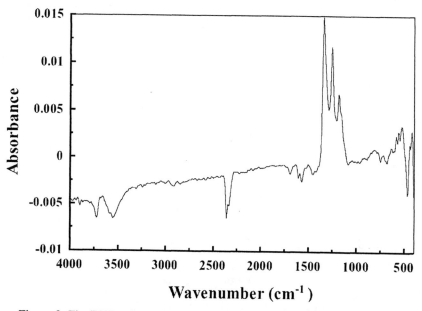

Figure 3. The FTIR reflection absorption spectrums of five layers of LB film which contains polypeptides B and retinoic acid based on a quartz substrate.

UV-VIS Absorption Spectrum of the Hybridized Film

The UV-VIS absorption spectrums of LB films which contain polypeptides A or B and retinoic acid based on a quartz substrate are shown in Figure 4 and Figure 5 respectively.

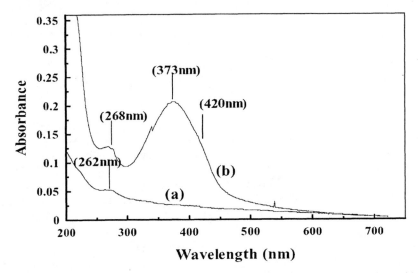

Figure 4. The UV-VIS absorption spectrums of LB films which contains polypeptides A and retinoic acid based on a quartz substrate, (a) the reference spectrums which contain only the polypeptide, (b) the spectrums which contain the polypeptide A and retinoic acid.

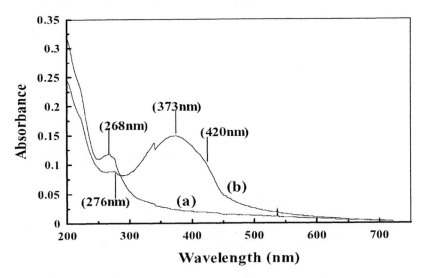

Figure 5. The UV-VIS absorption spectrums of LB films which contains polypeptides B and retinoic acid based on a quartz substrate., (a) the reference spectrums which contain only the polypeptide, (b) the spectrums which contain the polypeptide B and retinoic acid.

Figure 4(a) and 5(a) are the reference spectrums that contain only the polypeptide. Figure 4(b) and 5(b) are spectrums that contain the polypeptide and retinoic acid. In both cases, absorption peaks around 373 nm were observed. These are attributed to the intrinsic absorption of retinoic acid. Furthermore, another absorption peak around 420 nm was observed. These are attributed to the interaction between retinoic acid and polypeptide because similar absorption peaks have not been observed when only the retinoic acid used. Polypeptide B is seen to have a broader spectrum than that of polypeptide A.

Comparing Figure 5(b) with 4(b), the background level of the absorption spectrum of the sample with polypeptide B is lower than that of polypeptide A indicating a more-oriented interaction of retinoic acid and polypeptide B which coincides with observations described above.

The UV-VIS difference absorption spectrums of a LB film containing only retinoic acid on an α-Si:H film was calculated from absorption data using an LB film on the α-Si:H film and data for only α-Si:H as shown in Figure 6(a). The difference spectrum of a LB film containing retinoic acid and polypeptide B based on an α-Si:H film is shown in Figure 6(b). Figure 7 is the difference between the difference spectrums in Figure 6(b) and 6(a).

As we described in a previous paper [3], the absorption peak around 400nm is due to retinoic acid. It was 40nm wider than the intrinsic value of the maximum absorption wavelength of retinoic acid i.e., 360nm. The absorption peak around 500nm was entirely different than the intrinsic absorption of retinoic acid. This may be due to a greater change in the electrostatic environment around retinoic acid.

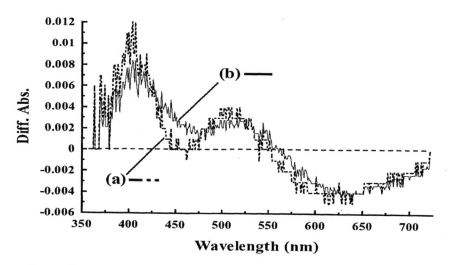

Figure 6. (a) The UV-VIS difference absorption spectrums of a LB film which only contains retinoic acid based on an α-Si:H film calculated from the absorption data using LB film on the α-Si:H film and the data using only α-Si:H, (b) The UV-VIS difference absorption spectrums of a LB film which contain retinoic acid and polypeptide B based on an α-Si:H film calculated from absorption data using LB film on the α-Si:H film and data using only α-Si:H.

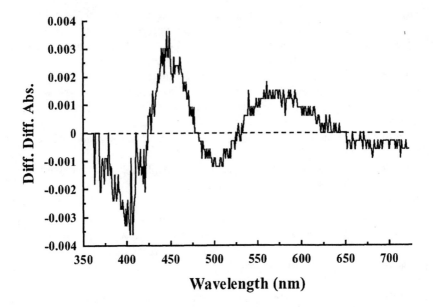

Figure 7. The difference between the difference spectrums of Figure 6(b) and Figure 6(a).

The maximum peak absorption around 560nm shown in Figure 7 indicates greater red shift of light absorption attributed to the interaction between retinoic acid and polypeptide B.

CONCLUSIONS

In the Langmuir-Blodgett films of retinoic acid and polypeptides A or B including in the alpha helix region of bR, we observed from UV-VIS absorption spectra and FTIR reflection absorption spectra that polypeptide B interacts with retinoic acid in a more-oriented manner than polypeptide A. The UV absorption spectrum of the Langmuir-Blodgett film of retinoic acid and polypeptide B based on an hydrogenated amorphous silicon film show a greater red-shift in the absorption spectrum than that we have reported previously. This phenomenon may be attributed to interaction between polypeptides of the alpha helix and retinoic acid and a-Si:H film structure in a combined manner.

ACKNOWLEDGEMENT

This work was partly supported by a Grant-in-Aid from the Ministry of Education and Science and Culture of Japan. This study was carried out as a part of "Ground Research Announcement for Space Utilization" promoted by Japan Space Forum.

REFERENCES

[1] K.R. Shoemaker, P.S. Kim, E.J. York, J.M. Stewart, R.L. Baldwin, 1987. Nature, 326, 653.
[2] D. Sali, M. Bycroft, A.R. Fersht, 1988. Nature, 335, 740.
[3] Y. Tsujiuchi, J. Suto, K. Goto, S. Shibata, M. Ihara, H. Masumoto, T. Goto, 2003. Thin Solid Films, in press.
[4] J.C. Arents, H.V. Dekken, K.J. Hellingwerf, H.V. Westerhoff, 1981. Biochem., 20, 5144.
[5] H. Merz, G. Zundel, 1986. Biochem. & Biophys. Res. Comm., 138, 819.
[6] M. Kataoka, T.W. Kahn, Y. Tsujiuchi, D.M. Engelman, F. Tokunaga, 1992. Photochem. & Photobio. 56, 895.
[7] M. Ben-Nun, F. Molnar, H. Lu, J.C. Phillips, T.J. Martinez, K. Schulten, 1998. Farady Discuss, 110, 447.
[8] J. Li, J. Wang, L. Jiang, 1994. Biosensor & Bioelectronics, 9,147.
[9] L. Jastrabik, L. Soukup, L.R. Shaginyan, A.A. Onoprienko, 2000. Surface and Coatings Technology, 123, 2–3, 261.

Nano-Granular Co-Zr-O Magnetic Films Studied by Lorentz Microscopy and Electron Holography

D. SHINDO, Z. LIU and G. YOUHUI
Institute of Multidisciplinary Research for Advanced Materials, Tohoku University, Sendai, Japan

S. OHNUMA and H. FUJIMORI
The Research Institute for Electric and Magnetic Materials, Sendai, Japan

ABSTRACT

Magnetic domain structures of $Co_{69}Zr_8O_{23}$, $Co_{71}Zr_{12}O_{17}$ and $Co_{71}Zr_{13}O_{16}$ as-sputtered films have been extensively studied by Lorentz microscopy and electron holography. Although all these films consist of an amorphous phase and hcp Co nano-particles, their magnetic properties, especially the direction and the magnitude of the magnetic anisotropy field, strongly depend on the sputtering reactive atmosphere. Lorentz microscopy clarifies these characteristic domain structures. A regular wide domain in $Co_{69}Zr_8O_{23}$ indicates an in-plane magnetization distribution while maze domains in $Co_{71}Zr_{12}O_{17}$ and $Co_{71}Zr_{13}O_{16}$ show a perpendicular magnetization feature. The difference in magnetic anisotropy fields between $Co_{71}Zr_{12}O_{17}$ and $Co_{71}Zr_{13}O_{16}$ is qualitatively described by the contrasts of Lorentz microscope images and the magnetization distribution observed by electron holography. Thus, it is pointed out that electron holography and Lorentz microscopy are useful ways to gain understanding about the magnetic domain structure and their magnetic properties such as direction and magnitude of magnetic anisotropy of advanced magnetic films.

Introduction

Development in electrical devices has led to an urgent need for magnetic devices that operate at higher frequencies. Thus larger values of magnetic anisotropy field (H_k), saturation magnetization (B_s) and electrical resistivity (ρ) are desired for special high frequency soft magnetic materials [1], as the permeability (μ) is remarkably influenced by eddy current losses and magnetic resonance at high frequencies. Recently, excellent magnetic properties have been obtained in soft magnetic Co-O based films with nanogranular structure. By selecting a specific composition in the Co-Zr-O system, various thin films with different high frequency properties can be obtained [2-4]. In order to well-understand the soft magnetic properties, the relationship between microstructure and magnetic structure needs to be clarified. In this paper, in addition to the microstructure analyzed by high-resolution transmission electron microscopy (HRTEM), the distributions of the magnetic domain walls and the lines of magnetic flux in the Co-Zr-O films with different compositions were investigated by Lorenz microscopy and electron holography respectively.

Experimental Details

Three kinds of Co-Zr-O films were prepared on water-cooled glass substrates (Corning 7059, 0.5mm in thickness) by a RF magnetron sputtering in different reactive atmospheres (O_2+Ar), using Co-Zr alloy targets [3,5]. The compositions of specimens analyzed by SEM-EDS with a windowless detector were $Co_{69}Zr_8O_{23}$, $Co_{71}Zr_{12}O_{17}$ and $Co_{71}Zr_{13}O_{16}$. HRTEM and Lorentz microscopy were carried out with a 400kV electron microscope (JEM-4000EX) and a high-voltage electron microscope (JEM-1250) respectively, while electron holograms were observed with a 300kV electron microscope (JEM-3000F) installed with a field-emission gun and a bi-prism. The magnetic properties were measured by vibrating sample magnetometer (VSM).

Results and Discussion

Magnetization loops of three specimens are shown in Figure 1 and some magnetic properties are presented in Table 1. By adjusting the direction of the applied magnetic field in the film plane, a square loop can be observed for $Co_{69}Zr_8O_{23}$ (Figure 1(a)), but loop shapes do not change very much for $Co_{71}Zr_{12}O_{17}$ and $Co_{71}Zr_{13}O_{16}$ (Figures 1(b) and 1(c)). This explicitly indicates that $Co_{69}Zr_8O_{23}$ has an in-plane magnetic anisotropy while the demagnetization processes for $Co_{71}Zr_{12}O_{17}$ and $Co_{71}Zr_{13}O_{16}$ are dominated by a perpendicular magnetic anisotropy. However, the magnetic hysteresis and remenance infer that the easy axes are not exactly along the films normal direction. The effective anisotropy constants estimated by the loops are 95, 1250 and 960 Oe for $Co_{69}Zr_8O_{23}$, $Co_{71}Zr_{12}O_{17}$ and $Co_{71}Zr_{13}O_{16}$ respectively.

Table 1. Magnetic properties of three specimens.

Sample	Coercivity (Oe)	Saturation induction (kG)	Anisotropy Field (Oe)	Easy axis
$Co_{69}Zr_8O_{23}$	0.15	11.9	95	In-plane
$Co_{71}Zr_{12}O_{17}$	50	11.4	1250	Perpendicular
$Co_{71}Zr_{13}O_{16}$	67	11.7	960	Perpendicular

Figures 2(a), 2(b) and 2(c) are electron diffraction patterns and HRTEM images of $Co_{69}Zr_8O_{23}$, $Co_{71}Zr_{12}O_{17}$ and $Co_{71}Zr_{13}O_{16}$ thin films respectively. In each ED pattern, there exist a halo ring and Debye-Scherrer rings corresponding to an amorphous phase and a crystalline phase respectively. All the Debye-Scherrer rings in diffraction patterns can be indexed by the hexagonal structure of Co whose space group is $P6_3/mmc$. Thus it is considered that the crystals in these films are basically hcp Co. From HRTEM images, it is clear that all these films have a nanogranular structure, and Co nanocrystals are indicated by arrows are dispersed in the amorphous matrix. From Figure 2, it can be seen that there is no definite crystal orientation relationship among the Co grains. This infers that the magnetic properties of these nano-granular thin films are governed by an induced magnetic anisotropy.

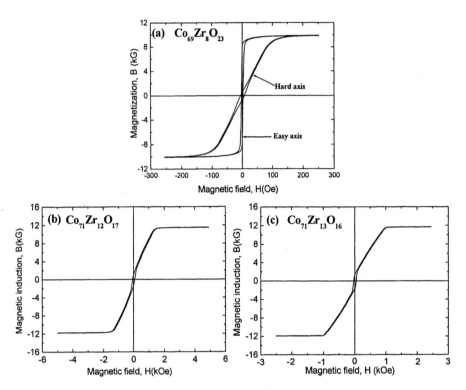

Figure 1. Magnetization loops of three specimens.

Figures 3(a), 3(b), and 3(c) show Lorentz micrographs of three specimens taken with the Fresnel method. The image contrast of the three Lorentz micrographs are sharply different. In the image of $Co_{69}Zr_8O_{23}$ (Figure 3a), the bright and dark lines indicated by arrowheads correspond to the magnetic domain walls. From the monotonous distribution of the magnetic domain walls, it is seen that the magnetization is basically in the film plane in agreement with the magnetic measurements. On the other hand the micrographs of $Co_{71}Zr_{12}O_{17}$ (Figure 3b) and $Co_{71}Zr_{13}O_{16}$ (Figure 3c) show an image contrast of the so-called maze domain wherein magnetization is perpendicular to the film plane. Theoretically when the magnetization vector is parallel to the electron beam, no contrast is expected in a Lorentz microscope image. In the case of $Co_{71}Zr_{12}O_{17}$ and $Co_{71}Zr_{13}O_{16}$, the contrasts are considered to be induced by a clockwise and an anti-clockwise domain wall, as shown in the sketch (Figure 3(d)). The in-plane component of magnetization enhances the image contrasts. It is interesting to note that the white and black contrasts in the image of $Co_{71}Zr_{13}O_{16}$ (Figure 3(c)) are much wider that those in the image of $Co_{71}Zr_{12}O_{17}$ (Figure 3(b)), i.e., the domain wall of $Co_{71}Zr_{13}O_{16}$ is wider than that of $Co_{71}Zr_{12}O_{17}$. According to equation $\delta = 2\pi\sqrt{A/K}$, where A is the exchange constant and K is the anisotropy constant, the wall width is inversely proportional to $K^{1/2}$. Thus, the thin image contrast of $Co_{71}Zr_{12}O_{17}$ reflects a high magnetic anisotropy being consistent with the measurement in Table 1 as explained in Figure 3(d).

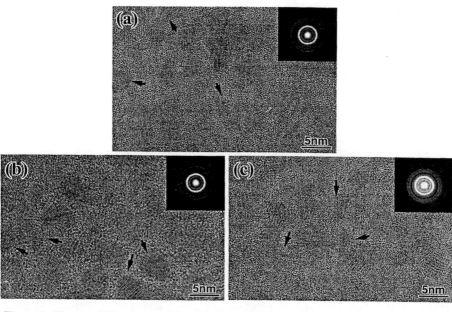

Figure 2. Electron diffraction patterns and HRTEM images of (a) $Co_{69}Zr_8O_{23}$, (b) $Co_{71}Zr_{12}O_{17}$ and (c) $Co_{71}Zr_{13}O_{16}$ thin films.

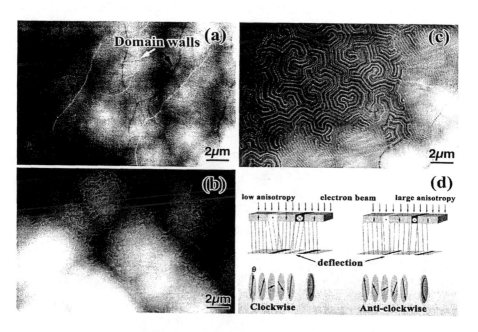

Figure 3. Lorentz micrographs of $Co_{69}Zr_8O_{23}$ (a), $Co_{71}Zr_{12}O_{17}$ (b) and $Co_{71}Zr_{13}O_{16}$ (c) thin films. The image contrasts induced by the walls and in-plane components are sketched in (d).

The characteristics of magnetic anisotropy in the three specimens are also clarified in the magnetic flux distribution by electron holography. Figures 4(a), 4(b) and 4(c) are the Lorentz microscope images (top) and the reconstructed phase images (bottom) obtained with the Fourier transform from the holograms of $Co_{69}Zr_8O_{23}$, $Co_{71}Zr_{12}O_{17}$ and $Co_{71}Zr_{13}O_{16}$ respectively. Since the holograms are observed near specimen edges, the specimens are thinner than those of Lorentz microscope images in Figure 3. The contour lines correspond to magnetic flux while the arrows in Figure 4(a) indicate the direction of magnetic flux. Between two adjacent contour lines, there is a constant flux of h/e (=4×10^{-15}Wb) [6]. Note that the lines of magnetic flux in Figure 4(a) are straight and the density is larger than those of the other specimens. Also, strong stray fields were seen outside the specimens. These magnetic flux features demonstrate the in-plane magnetization of $Co_{69}Zr_8O_{23}$ consistent with the observation of sharp magnetic domain walls in Figure 3(a).

On the other hand, the magnetization orientation changes gradually to form closure domains in the reconstructed phase images of $Co_{71}Zr_{12}O_{17}$ (Figure 4(b)) and $Co_{71}Zr_{13}O_{16}$ (Figure 4(c)). Note that at the center of the closure domains, magnetization is perpendicular to the film plane, sharply different from $Co_{69}Zr_8O_{23}$ (Figure 4(a)). As well in the image of $Co_{71}Zr_{12}O_{17}$, the area where the density of the lines of magnetic flux is much lower is indicated by the arrowhead. In this region, the magnetization vector is considered to be tilted off the film plane consistent with the large magnitude of magnetic anisotropy field perpendicular to the film plane as shown in Table 1. These findings demonstrate that electron holography and Lorentz microscopy are very useful to clarify the direction and magnitude of the magnetic anisotropy field of advanced magnetic materials such as nano-granular Co-Zr-O films.

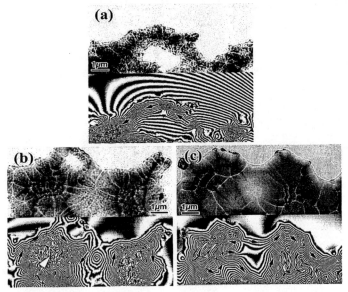

Figure 4. Reconstructed phase images of (a) $Co_{69}Zr_8O_{23}$, (b) $Co_{71}Zr_{12}O_{17}$ and (c) $Co_{71}Zr_{13}O_{16}$ obtained from the holograms with the Fourier transform.

Conclusion

Results of TEM analysis on $Co_{69}Zr_8O_{23}$, $Co_{71}Zr_{12}O_{17}$ and $Co_{71}Zr_{13}O_{16}$ thin films can be summarized as follows:
1) All the films consist of an amorphous phase and hcp Co particles several nanometres in diameter but the magnetic properties, especially the direction and magnitude of the magnetic anisotropy field have been found to be different from one other.
2) Lorentz microscopy reveals regular wide domains in $Co_{69}Zr_8O_{23}$, while maze domains are observed in $Co_{71}Zr_{12}O_{17}$ and $Co_{71}Zr_{13}O_{16}$ with different contrast being well correlated to the different magnitude of magnetic anisotropy fields perpendicular to the film plane.
3) Electron holography directly visualizes the magnetization distribution in these films which correlates to the direction and magnitude of the magnetic anisotropy field of the three films.

Acknowledgment

The authors thank Dr. T. Ohsuna and Prof. K. Hiraga, Institute for Materials Research, Tohoku University for their support on phase reconstruction. They also thank Professor T. Masumoto, Research Institute for Electric and Magnetic Materials, for his warm encouragement. The work was supported by Special Coordination Funds for Promoting Science and Technology on "Nano-hetero Metallic Materials" from the Science and Technology Agency and by a Grant-in-Aid for Scientific Research (B) from the Ministry of Education, Culture, Sports, Sci. and Tech. of Japan.

References

[1] A. Hosono, Y. Shimada, 1988. J. Magn. Soc. Japan, 12, 295.
[2] S. Ohnuma, H. Lée, N. Kobayashi, H. Fujimori, T. Masumoto, 2001. IEEE Trans. on Mag., 37, 2251.
[3] S. Ohnuma, H. Fujimori, S. Mitani, T. Masumoto, 1996. J. Appl. Phys., 79, 5130.
[4] M. Ohnuma, K. Hono, E. Abe, H. Onodera, 1997. J. Appl. Phys., 82, 5646.
[5] Z. Liu D. Shindo, S. Ohnuma and H. Fujimori, 2003. J. Magn. Magn. Mater. (in press)
[6] D. Shindo, Y. G. Park, Y. Yoshizawa, 2002. J. Magn. Magn. Mater., 238, 101.

Electrochemomechanical Deformation of Polypyrrole Film in Complex Buffer Media

S. S. PANDEY, W. TAKASIMA and K. KANETO
Graduate School of Life Science and Systems Eng., Kyushu Institute of Technology, Wakamatsu, Japan

ABSTRACT

Direct measurement of electrochemomechanical deformation of electrodeposited polypyrrole has been investigated using cyclic voltammetry in a special electrochemical cell and laser displacement meter. Quantitative evaluation of the magnitude of deformation in different electrolytic media such as simple salt, simple buffer, and complex buffer has been estimated. Dependence of the magnitude of deformation upon NaCl concentration has revealed that polypyrrole shows actuation performance in the dynamic concentration range of 1 mM to 1M having maximum performance at 0.1M. Our results indicate that polypyrrole film acting as soft electrochemical actuator operates successfully in simple as well as complex electrolytic media without appreciable loss in deformation magnitude as well as electrochemical activity.

INTRODUCTION

The soft actuators exhibiting inherent exotic motions like bending, twisting, expanding etc. in the thin films have a great potentiality in the area of Biomimetic science and technology such as artificial muscles, human robotics, prosthetics, micro-valves, microtweezers and anti-vibration systems [1]. Reversible dimensional changes in the soft materials like ionic polymers; hydrogels and electrically conducting polymers have attracted much attention in the recent past due to their potential application in the area of Biomimetic devices [2]. All inherently electro active conductive polymers (ECPs) undergo reversible oxidation/reduction reaction, brought about by the application of small electrical potential (1-3V) resulting into appreciable degree of electro-mechanical actuation, which is about 100 times smaller, compared to the potential required for actuators based on piezoelectricity and static electricity [3]. The change in volume of the conducting polymers resulting from the reversible redox reaction is a key phenomenon towards the fabrication of soft actuators. In the organic conjugated polymers the driving force for actuation is being generated by electrochemomechanical deformation (ECMD).

Electrochemical actuators based conducting polymers show expansion larger than several % with large contraction force, which is more than 10 times, compared that of skeletal muscles, quick response compared to that of gel based actuators and low voltage operation [4,5]. A good deal of efforts have been made towards optimisation of performance of soft actuators based on conducting polymers by suitable method of film preparation [6,7], judicious selection of subtle actuator design [8-10] and development of efficient measurement system [11] pertaining to the measurement of magnitude of deformation (strain) and work capacity of the actuator. In most of the cases soft actuators based on conducting polymers has been demonstrated in simple ionic electrolytes such as sodium chloride (NaCl), sodium nitrate etc. without having much focus on magnitude of deformation in complex media [12,13]. We had recently reported that Soft actuators based on Polypyrrole (PPy) works successfully in a wide concentration range of NaCl having maximum efficiency at concentration of about 100 mM [14].

Application of soft actuators in the area of biomimetic devices is based on their success in complex electrolytic media like mixtures of salts, buffers, blood and serum etc. Lupu et al [15] recently reported on using conducting polymer electrodes to determine ascorbic acid in 0.1M sodium phosphate buffer (PB), a common constituent of foods as well as a famous antioxidant in the human body. Phosphate buffer has also been used for the construction of conducting polymer based biosensor as well as immobilization of antibodies in conducting polymers [16, 17]. Tris-HCl buffer has been used for the fabrication of conducting polymer-modified electrodes to determine the well-known redox protein cytochrome c by Lu et al [18]. Potassium phosphate buffered saline (KPBS) has been used for the construction and electrochemical evaluation of DNA biosensor by Evtugyn el al [19]. Biological buffers like Ringer buffer solution (RBS), balance salt solution (BSS) and Krebs Ringer Hepes buffer (KRHB) etc., are useful for the in-vitro cell culture, enzyme essays, and some elecrophoretic applications at physiological pH ranges [20-22]. Use of conducting polymers under such complex electrolytic media and their electrochemical activities has not received much study which seems to open a fruitful direction pertaining to enhancing potential applications of organic conducting polymers. In this paper, our recent results on ECMD measurements on PPy soft actuators in simple and complex buffer media are presented.

EXPERIMENTAL DETAILS
Sample Preparation
PPy freestanding films were prepared by electrodeposition from an aqueous solution containing 0.15M of freshly distilled pyrrole and 0.25M of dodecylbenzene sulphonic acid (DBSA). Electrodeposition was performed on a stainless steel working electrode at a current density of 1 mA/cm^2 in a 3-electrode cell assembly with Pt foil and Ag/AgCl used as counter and reference electrodes, respectively. Aqueous electrolyte solution containing pyrrole monomer was purged with dry nitrogen gas for 15 min prior to initiation of electrodeposition. Galvanostatic electropolymerization was carried out for 15 min leading to the formation of DBSA doped polypyrrole (PPy) film having a thickness of 15-20 μm. Electrodeposited PPy film thus obtained on SS Steel working electrode was peeled off and used for ECMD analysis.

Electrolyte and Buffer Preparation

Sodium chloride (NaCl), Potassium chloride (KCl) and Magnesium chloride ($MgCl_2$) aqueous solutions were prepared by dissolving calculated amount of respective salts in deionised water. 0.1M sodium phosphate buffer (PB) was prepared from 0.2M solution of sodium dihydrogen phosphate (dibasic) and 0.2M disodium hydrogen phosphate (monobasic). 0.1M PB of pH 7.2 was prepared by taking 28 ml of monobasic and 72 ml of dibasic solution followed by addition of 100 ml of deionised water. A 0.1M potassium phosphate buffered saline (KPBS) was prepared by taking 0.9 ml of 0.2M potassium dihydrogen phosphate, 4.1 ml of 0.2M dipotassium hydrogen phosphate and 2.0 gm of NaCl making the total volume of 250 ml with deionised water. Ringer buffer solution (RBS) was prepared by dissolving 0.66 gm of NaCl, 15 mg of KCl, 15 mg of calcium chloride ($CaCl_2$) and 10 mg of sodium hydrogen carbonate ($NaHCO_3$) in 100 ml of deionised water under stirring. Finally pH of the solution was adjusted to 7.4 with $NaHCO_3$. A balanced salt solution (BSS) was prepared by dissolving 35 mg $CaCl_2$, 0.10 gm KCl, 15 mg potassium dihydrogen phosphate, 25 mg $MgCl_2$, 25 mg magnesium sulphate ($MgSO_4$), 2.0 gm NaCl, 87.5 mg $NaHCO_3$ and 12.5 mg of disodium hydrogen phosphate in 250 ml deionised water. First ingredients were dissolved one at a time in 200 ml water on magnetic stirrer and volume was finally made up to 250 ml. Krebs-Ringer-Hepes buffer (KRHB) with bovine serum albumin (BSA) was prepared by first dissolving 1.78 gm NaCl, 91 mg KCl, 75 mg $MgSO_4$, 25 mg $CaCl_2$ and 0.715 gm of 2-[4-(2-hydroxyethyl)-1-piperazinyl] ethanesulphonic acid (HEPES) in 250 ml deionised water and adjusting the pH to 7.4. 0.25 gm of BSA was then added and finally pH was again adjusted to 7.4.

Measurement Set up

The experimental set up for the ECMD and cyclic voltammetry (CV) has been reported previously [23]. The schematic representation of our unique electrochemical cell used for ECMD and CV evaluation is shown in Figure 1. The electrochemical cell consists of two platinum (Pt) wires, one Ag wire and two polytetrafluoroethylene (PTFE) thin tubes fixed via a PTFE plug. One of the Pt wire inserted in the centre of PTFE plug was used to hold the sample which behaves as the working electrode (WE). Another Pt wire connected with rolled Pt foil (to enhance the surface area) on the bottom was used as counter electrode (CE). An Ag/AgCl wire was used as reference electrode (RE). A long PTFE tube reaching up to the bottom of the cell was used as a guide to inject and eject the electrolyte solution while another PTFE tube was designed to act as an air leak. A thin Pt wire connected with the bottom of the sample with polyimide tape was passed through the bottom of the capillary of the glass cell whose terminal was ended like a hook to hold a tray containing a fixed weight to straighten the wire. A strip of PPy film with an effective dimension of 10 x 2 mm^2 was mounted in the electrochemical cell. Finally dimensional changes occurred in the film as result of reversible redox reaction was measured with the help of laser displacement meter (Keyence LB-040/LB-1000). The cyclic voltammetry (CV) was done with the aid of a potentiostat (Hakuto Denko HA-501) and a function generator (Hakuto Denko HP-105). All the data were supplied as analogue voltage, which were then converted into digital data using an A/D converter and were acquired by a personnel computer with an interval of 10 ms. For all the experiments CV was measured at the scan rate of 5 mV/s. The magnitude of deformation was defined by $\Delta l/l_o$, where l_o and Δl are original length and change in the film length (stroke), respectively.

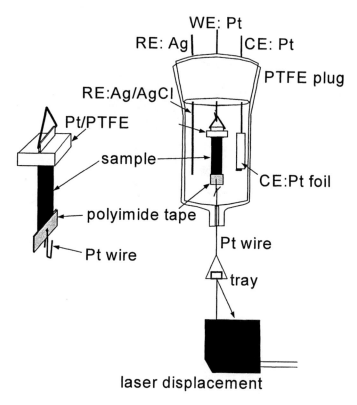

Figure 1. Schematic drawing of the electrochemical cell for direct CV and ECMD measurement.

RESULTS AND DISCUSSION

Figure 2 shows the CV characteristics of a PPy-DBSA film in 1M NaCl solution. The film was subjected to 20 potential cycles in a potential window of −850 mV to + 500 mV at a sweep rate of 5 mV/s to make the film active for facile ion transport across the film. This pre-treated active film was used for further CV and ECMD measurements in different electrolytes. It can be seen from the diagram that there is a gradual stabilization of CV response and after about 10 potential cycles a stable redox reaction occurs having peaks at −0.29 and −0.16 V vs. Ag/AgCl respectively. Occurrence of redox peaks in the negative potential region suggests this behaviour is associated with (de) insertion of cations (Na^+) leading to cathodic expansion. This behaviour is reasonable considering the bulky nature of the DBSA anion which would have difficulty taking part in insertion and deinsertion during redox cycling. Takashima et al [24] have also reported that PPy-DBSA films show only cathodic

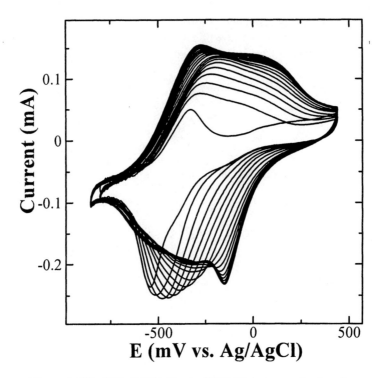

Figure 2. CV of PPy-DBSA film in 1M NaCl during pre-treatment.

expansion by judicious selection of electrolytes during ECMD and CV measurements.

ECMD and CV characteristics of PPy-DBSA film in NaCl solution after 20 cycles of pre-treatment are shown in Figure 3. Note that upon dilution, the peaks shift to a lower potential along with an enhancement of magnitude of ECMD as Scaarup et al [25] have also observed in the shift of redox peaks for DBSA-doped PPy in an NaCl solution. This proves that the dominant mobile ionic carriers are cations (Na^+) which is in accord with the results obtained by Takashima et al [24] indicating that PPy-DBSA films shows only cation-driven cathodic expansive behaviour.

Studies pertaining to the concentration dependence of CV and ECMD find application in both fundamental and technological areas. On the one hand, the concentration dependence of CV is useful in determining the major ionic carriers during redox reactions while on the other hand, the concentration dependence of ECMD provides a versatile application potential as a soft actuator. Recently, Scaarup et al [25] and Takashima et al [14] discussed the concentration dependence of CV of PPy-DBSA film in NaCl solution and proved that the majority of ionic carriers are cations (Na^+) with the trend in redox potential shift in accord with the Nernst Equation. Figure 4 shows the dependence of magnitude of ECMD on concentration of NaCl as the electrolyte.

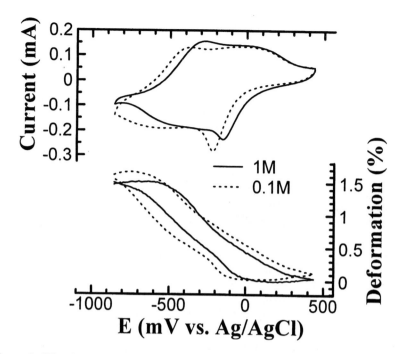

Figure 3. CV and ECMD characteristics of PPy-DBSA freestanding film NaCl solution.

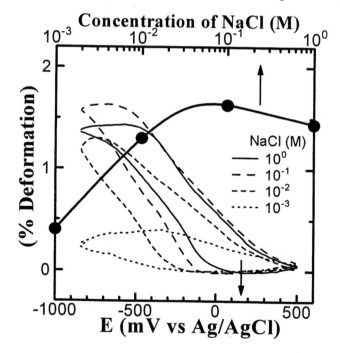

Figure 4. NaCl concentration dependence of ECMD in PPy-DBSA freestanding film.

The dependence of the ECMD magnitude on NaCl concentration reveals that PPy-DBSA actuates in a concentration range of 1 mm to 1M with a maximum around 0.1M. The concentration of Na^+ ions in seawater and blood is 1M and 0.1M respectively indicating the technological potential for PPy-DBSA soft actuators in the area of marine science as well as bio-medical science.

Application of soft actuators in marine science depends on successful operation of such actuators in a number of aqueous solutions of salts as well as salt mixtures. The six most abundant ions in seawater are chloride (Cl^-), sodium (Na^+), sulphate (SO_4^{2-}), magnesium (Mg^{2+}), calcium (Ca^{2+}) and potassium (K^+). These ions make up about 99% of the total salt content of sea water. ECMD and CV behaviours of PPy-DBSA film observed in various aqueous electrolytes consisting of simple salts are shown in Fig. 5. A perusal of CV characteristics of PPy-DBSA film in 0.1M aqueous solution of simple salt indicate that it maintains the electrochemical activities with well defined redox peaks in the presence of monovalent cations (Na^+, K^+), divalent cation (Mg^{2+}) and equimolar mixture of monovalent alkali metal cations. Solutions of 0.1M NaCl show a maximum magnitude of deformation (1.7%) for PPy-DBSA film while in the case of 0.1M KCl, $MgCl_2$ as well as 0.1M equimolar mixtures of NaCl and KCl, this is reduced and is estimated to be ~1.2%, which is still appreciably good. This indicates that the actuator devices based on PPy-DBSA films are expected to operate in seawater. It is interesting to mention here that seawater having 3.5% salinity has concentrations of Na^+, Mg^{2+} and K^+ ions of 1M, 0.1M and 0.02M respectively, where the magnitude of % deformation in PPy-DBSA film is approximately the same.

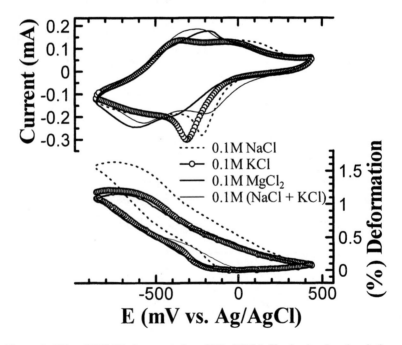

Figure 5. CV and ECMD characteristics of PPy-DBSA film in simple salt solution.

Buffers are basically aqueous solution having partially neutralized weak acids and bases exhibiting very small changes in the pH upon small addition of strong acids or bases. Buffers are being commonly used in applications like chromatography, electrophoresis and cell culture, which allows the scientists to investigate the behaviour of biomolecules in a specific pH range. Conducting polymers have been widely used for immobilization of biomolecules like enzyme, whole cells, and antibodies etc., while phosphate buffer has been used as a medium to dissolve biomolecules in the physiological pH ranges [16, 17] while, there are very little report about electroactivity of conducting polymers in buffer media [15]. Garner et al [26] have also reported the use of polypyrrole composite material for endothelial cell growth. Fig. 6 shows the ECMD and CV characteristics of PPy-DBSA film performed in the 0.1M simple buffers such as PB and KPBS along with the 0.1M mixtures of NaCl and KCl for comparison.

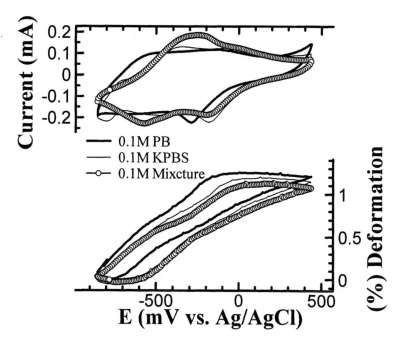

Figure 6. ECMD and CV characteristics of PPy-DBSA film in simple buffer solutions.

The study of electroactivity of PPy-DBSA in buffer media and ECMD is an important step towards the active use of conducting polymers in general and PPy in particular in the area of biomedical science. A perusal of Figure 6 corroborates that the PPy-DBSA film shows well-defined redox activity in simple buffer systems such as PS and KPBS similar to that observed for simple salts as shown in Figure 5. At the same time the magnitude of deformation in PS was found to be higher (~1.3%) than for simple salts suggesting that PPy-DBSA is a good material to construct new-generation sensors and actuators in the future.

Studies pertaining to the ECMD and CV characteristics of PPy-DBSA freestanding films in 0.1M simple buffers where such films show good characteristics similar to that observed in the case of aqueous solution of simple salts provided us with the impetus to check the performance of electroactivity and magnitude of deformation of such soft actuators in more complex buffer systems such as RBS, BSS, and KRHB, etc. Such complex buffer systems have a number of cations and anions that are commonly used for cell culture, in-vitro studies, pathogenesis such as phacoemulsification [27], incubation of proteins [28], etc. The performance of PPy-DBSA soft actuators in terms of CV characteristics and magnitude of ECMD is shown in Figure 7. It is interesting to see that all three complex buffers such as RBS, BSS and KRHB all show well-defined redox peaks at around –500 mV and –300 mV vs. Ag/AgCl reference electrodes respectively, suggesting the maintenance of electroactivity of PPy-DBSA film in such complex buffer media. Apart from the appearance of well-defined redox peaks, PPy-DBSA freestanding films show good actuation behaviour having magnitude of deformation of about 1.3%. Thus the observation of appreciable electroactivity in complex buffer along with good ECMD characteristics in PPy freestanding films indicate that PPy-DBSA can be used to construct new generations of bio-compatible smart sensors and actuators.

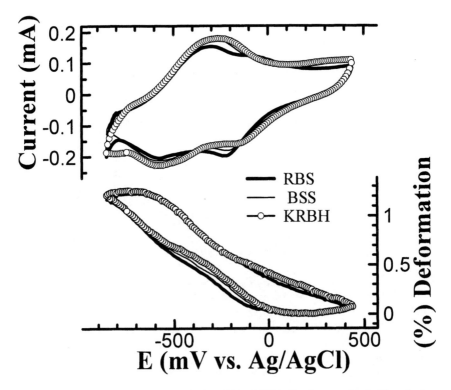

Figure 7. CV and ECMD characteristics of PPy-DBSA film in complex buffer solutions.

CONCLUSIONS

A direct ECMD and CV characteristic of PPy-DBSA freestanding films has been evaluated in aqueous electrolytes containing simple salts, simple buffers and complex buffers. It has been found that the pre-treatment of DBSA-doped PPy films in a 1M NaCl solution, gives stable ECMD and CV responses after 10 potential cycles. The NaCl concentration-dependence of the ECMD characteristics shows the best response around 0.1M NaCl having magnitude deformation of about 1.7%. PPy-DBSA soft actuators work well in 0.1M PB and 0.1M KPBS having similar electrochemical response compared to that observed using simple salts with about 1.3% deformation magnitude. In the case of using complex buffers such as RBS, BSS and KRHB, PPy-DBSA shows a well-defined redox peak having good ECMD characteristics with a magnitude of deformation of ~1.3%. This suggests that PPy-DBSA can be used to fabricate biocompatible sensors and actuators.

ACKNOWLEDGEMENT

One of the authors (SSP) would like to thank Japan Society for Promotion of Science (JSPS) for providing financial support and opportunity to work in Japan.

REFERENCES

[1] R.H Baughman, 1996. Conducting polymer artificial muscles, Synth. Met, 78, 339–353.
[2] K. Kaneto, Y. Sonoda, W. Takashima 2000. Direct measurement and mechanism of electrochemo-mechanical expansion and contraction in polypyrrole films, Jpn. J. App. Phys., 39(10), 5918–5922.
[3] K. Kaneto, M. Kaneko and W. Takashima 1995. Response of chemomechanical deformation in polyaniline film on variety of anions, Jpn. J. Appl. Phys., 34(7A), L837–L840.
[4] A. Della Santa, D. De Rossi and A. Mazzoldi 1997. Performance and work capacity of a polypyrrole conducting polymer linear actuator, Synth. Met., 90, 93–100.
[5] John D. Madden, R.A. Cush, T.S. Kanigan and I.W. Hunter, 2000. Fast contracting polypyrrole actuators, Synth. Met., 113, 185–192.
[6] T. Okamoto, K. Tada and M. Onoda, 2000. Bending machine using anisotropic polypyrrole films, Jpn. J. Appl. Phys., 39(5A), 2854–2858.
[7] S.S. Pandey, W. Takashima, M. Fuchiwaki and K. Kaneto, 2003. Effect of film morphology on the actuation behaviour in polypyrrole films, Synth. Met. (accepted for publication)
[8] A.S. Hutchinson, T.W. Lewis, S.E. Moulton, G.M. Spinks and G.G. Wallace, 2000. Development of polypyrrole based electromechanical actuators, Synth. Met., 113, 121–127.
[9] E.W.H. Jager, O. Ingans and I. Lundstrom, 2000. Microrobots for micrometer size objects in aqueous media: potential tools for single cell manipulation, Science, 238, 2335–2338.

[10] W. Takashima, S.S. Pandey, M. Fuchiwaki and K. Kaneto, 2003. Bi-ionic actuators by polypyrrole films, Synth. Met. (accepted for publication).
[11] W. Takashima, M. Fukui, M. Kaneko and K. Kaneto, 1995. Electrochemomechanical deformation of polyaniline films, Jpn. J. Appl. Phys., 34(7B), 3786–3789.
[12] T.W. Lewis, S.E. Moulton, G.M. Spinks and G.G. Wallace, 1997. Optimization of polypyrrole based actuator, Synth. Met., 85, 1419–1420.
[13] M. Fuchiwaki, W. Takashima, S.S. Pandey, K. Kaneto, 2003. Proposal of novel actuators using conducting polymer laminates, Synth. Met., (in press).
[14] W. Takashima, S. Pandey, K. Kaneto, 2003. Cyclicvoltammetric and electrochemomechanical characteristics of freestanding polypyrrole films in diluted media, Thin Solid Films, (in press).
[15] S. Lupu, A. Mucci, L. Pigani, R. Seeber and C. Zanardi, 2002. Polythiophene derivative conducting polymer modified electrodes and microelectrodes for determination of ascorbic acid. Effects of possible inteferents, Electroanalysis, 14(7-8), 519–525.
[16] J. Vidal, E. Gracia, J. Castillo, 1999. In situ preparation of over-oxidized PPy/o-PPD bilayer biosensor to determine glucose and cholesterol in serum, Sens. & Act. B, 57, 219–226.
[17] A.I. Minett, J.N. Barisci and G.G. Wallace, 2002. Immobilization of anti-Listeria in a polypyrrole film, Reactive and Functional Polymers, 53, 217–227.
[18] W. Lu, G.G. Wallace and A.A. Karayakin, 1998. Use of Prussian blue/Conducting polymer modified electrodes for the detection of Cytochrome c, Electroanalysis, 10(7), 472–476.
[19] G. Evtugyn, A. Mingaleva, H. Budnikov, E. Stoikova, V. Vinter and S. Eremin, 2003. Affinity biosensors based on disposable screen printed electrodes modified with DNA, Analyrica Chimica Acta, 479, 125–134.
[20] B. Janic, T.M. Umstead, D.S. Phelps and J. Floros, 2003. An in vitro cell model system for the study of the effects of the ozone and other gaseous agents on phagocytic cells, J. Immunological Methods, 272, 125–134.
[21] R. Araya, H. Gomez-Mora, R. Vera and J.M. Bastidas, 2003. Human spermatozoa motility analysis in a Ringer's solution containing cupric ions, Contraception, 67, 161–163.
[22] Y. Masuda, T. Oguma and A. Kimura, 2002. Biphasic effects of oxethazaine, a topical anesthetic on intracellular ca2+ concentration of PB12 cells, Biochem. Pharm., 64, 677–687.
[23] W. Takashima, S. Pandey, M. Fuchiwaki, K. Kaneto, 2002. Cyclic stepvoltammetric analysis of cation- and anion–driven actuation in polypyrrole films, Jpn. J. Appl. Phys., 41(12), 1–5.
[24] W. Takashima, S. Pandey, K. Kaneto, 2003. Investigation of bi-ionic contribution to enhance bending actuation in polypyrrole films, Sensors & Actuators. B:Chemical, 89, 48–52.
[25] S. Skaarup, K. West, L.M.W.K. Gunaratne, K.P. Vidanapathirana and M.A. Careem, 2000. Determination of Ionic Carriers in Polypyrrole, Solid State Ionics, 136-137, 577–582.

[26] B. Garner, A. Georgevich, A.J. Hodgson, L. Liu and G.G. Wallace, 1999. Polypyrrole-heparin composites as stimulus-responsive substrates for endothelial cell growth, J. Biomed. Mater. Res., 44, 121–129.
[27] M. Lautenschlager, M. Holtje, B. Von Jagov, R.W. Veh, C. Harms, A. Bergk, U. Dirnagl, G. Ahnert-Hilger and H. Hortnagl, 2000. Serotonin uptake and release in developing cultures of rat embryonic raphe neurons: age and region specific differences, Neurosci., 99(3), 519–527.
[28] J A. Davison, T. Chylack Jr. 2003. Clinical application of the Lens Opacities Classification System III in the performance of phacoemulsification, J. Cataract Refract. Surg, 29, 138–145.

Nano-Metallization on Surface Nano-Sized Granules of Polytetrafluoroethylene Matrix

M. S. KOROBOV, G. Y. YURKOV and S. P. GUBIN
N.S. Kurnakov Institute of General and Inorganic Chemistry, Russian Academy of Sciences, Moscow, Russia

ABSTRACT

Nanometer-sized structures are an intermediate form of matter that fills the gap between atoms/molecules and bulk materials. Often, these types of structures exhibit exotic physical and chemical properties different from those observed in bulk three-dimensional materials. The interdisciplinary fields of mesoscopic and nano-scale systems are important for the field of fundamental physics, as well as for some new technologies. It is well known that some metallic nanoparticles are thermodynamically unstable and chemically very reactive. For many technical applications, materials with nanoparticles embedded in a matrix are important. In this paper in order to stabilize nanoparticles, matrices consisting of organic polymers such as polytetrafluorinethylene (teflon), polyethylene, and polypropylene are used.

A universal method to introduce metal-containing nanoparticles in a teflon matrix has been developed to allow fabrication of large amounts (kilogram-scale) of nanoparticle polymer composites. Encapsulation was done by thermal decomposition of the metal-containing compounds (MRn; M = Fe, Cu, Ni; R = CO, HCOO, CH3COO) in a dispersion system of polytetrafluorinethylene in mineral oil or in a solution-melt of polyethylene or polypropylene in mineral oil. The optimum conditions are developed for decomposition of MCC in order to introduce highly reactive nanoparticles into the polymeric matrix with a concentration of 2 to 10 % by weight. Samples containing 4 %Fe, 8 %Fe, 5 %Cu, and 5 %Ni were investigated in detail. For subsequent characterization, we used TEM, EXAFS, X–ray emission spectroscopy and Mossbauer spectroscopy. The morphology and particle size distributions were investigated using a high-resolution electron microscope (TEM). TEM studies showed that the particles were dispersed on the surface of nanosized granules of teflon matrix and in the bulk volume of polyethylene or polypropylene matrixes with very narrow log-normal sized bimodal distributions centered at 6±1.6

nm. The granules of polytetrafluorinethylene had sizes in the interval 50–200 nm. Electronic paramagnetic-resonance for Fe-containing samples showed the nanoparticles to be superparamagnetic at room temperature. This data shows that Fe-containing particles have very small size. As well, this EPR method showed the availability in samples with a concentration of 4 %Fe contained Feoxide and α-Fe, but in samples with a concentration of 8 %Fe, only Fe oxides are present. Mossbauer spectroscopy for Fe-containing samples made it possible to determine and control the "phase" constitution of the composites and also to evaluate the particle size. According to all available data in samples prepared from FeRn (R = CO, HCOO, CH3COO), the composites contained α-Fe, Fe2O3, Fe3O4. Our experiments demonstrate that it is possible to stabilize and characterize nanoparticles of various compositions using a polytetrafluorinethylene, polyethylene, or polypropylene matrix. The teflon matrix, on surfaces where nanoparticles synthesize, plays an active role in determining the composition and physical properties in addition to providing a means for particle dispersion. The research demonstrates the significant potential of MCC thermodestruction processes for synthesis of appreciable quantity of nanoparticles containing polymer materials.

INTRODUCTION

Nanometer size structures are an intermediate form of matter, which fills the gap between atoms/molecules and bulk materials. Often, these types of structures exhibit exotic physical and chemical properties different from those observed in bulk three-dimensional materials. This interdisciplinary field of mesoscopic and nano-scale systems is important to fundamental physics, as well as for some new technologies. As is well known, some of the metallic nano–particles are thermodynamically unstable and chemically very reactive. For many technological applications, materials with nanoparticles embedded in a matrix are impotant. In this paper for the stabilization of nano-particles the matrices of the organic polymers as polytetrafluorinethylene (teflon), polyethylene, polypropylene it is used.

The high reactivity of nanoparticles and their tendency toward spontaneous compaction accompanied by deterioration of basic physical properties make stabilization be a major challenge in fabrication of materials based on metal nanoparticles. The best-developed method of stabilization is embedding nanoparticles in polymer matrices [1]. The known "friability" of the structures of most of partially crystalline carbon-chain polymers forms the basis for the method of introducing metal-containing nanoparticles into solution melts of polymers in hydrocarbon oils [2]. However, this method is inapplicable to "hard" polymer matrices, such as polytetrafluoroethylene. At the same time, creation of materials that combine opposite properties of its components at the nanolevel is one of the rapidly developing strategies of modern materials science. For example, it is expedient to augment the high thermal and chemical stability of polytetrafluoroethylene with the electrical conductivity on a semiconductor scale, magnetization by means of introducing metal nanoparticles, or optoelectronic properties by introducing so-called quantum dots, nanoparticles of some metal sulfides and selenides.

At first glance, it would seem that poor solubility and swelling ability of polytetrafluoroethylene in solvents hinder the introduction of different nanoparticles in this matrix and their uniform distribution over its bulk. In searching for the solution of this problem, we have focused our attention of ultradispersed polytetrafluoroethylene (UPTFE). UPTFE was fabricated by a thermal gas dynamic method, which is suitable for commercial production of UPTFE [3]. The essence of the method is in formation of a finely dispersed UPTFE powder upon the thermodestruction of the block polymer.

EXPERIMENTAL DETAILS

Fe-containing nanoparticles on the surface nanosizing polytetrafluoroethylene were prepeared by new methods on based a standart "cluspol" technique [4]. This is universal method, which allows the fabrication of large amounts (kilogram-scale) of polymer nanoparticle composites. The encapsulation was done by thermal decomposition of metallcontaining compounds (MRn; M = Fe, Cu, Ni; R = CO, HCOO, CH3COO) in dispersion system of polytetrafluorinethylene in mineral oil; solution-melt polyethylene or polypropylene in mineral oil. We are discovered that granules ultradispersion polytetrafluoroethylen make fluidized bed on the surfase mineral oil. This effect we are used for nanometallization of nanogranules. An appropriate amount of metallcontaining compounds was added to the high-temperature dispersion system of polytetrafluoroethylene in mineral oil with vigorous stirring. The residual oil was removed by washing with benzene in a Sohxlet apparatus. The resultant powder was dried in wacuum and stored in air. The optimum conditions are developed for the decomposition of MCC in order to introduce highly reactive nanoparticles into the polymeric matrix with concentration of 2-10 wt. %.

The particle size of iron oxide was determined by transmission electron microscopy (TEM) on a JEOL JEM-100B high-resolution microscope. The powder was dispersed ultrasonically in ethanol and then spread over a carbon substrate.

Extended Fe K x-ray absorption fine structure (EXAFS) measurements were made with a bench-scale spectrometer [5] equipped with a focusing quartz analyzer. The x-ray source used was a Mo-BSV21 finefocus x-ray tube (20 kV, 30 mA). The sample was thoroughly ground with Apiezon, and the mixture was sandwiched between mylar films. The absorber thickness was such that the ratio of incident to transmitted intensity was about 3. The absorption spectra $\mu_i(E)$ were processed by a standard procedure [6]. Fourier analysis of the EXAFS curves, $k2(\chi)k$, of the samples and standards (α-Fe and y-Fe2O3) was carried out in the range k = 2.7-12.5 Å-1.

Mossbauer spectra were recorded on an MC1101E high-speed spectrometer at room temperature (without thermostatic control) and liquid-nitrogen temperature. The source used was rhodium-shielded "Co with an activity of 10 mCi (ZAO Tsiklotron). The isomer shift was measured relative to a-Fe. The spectra were analyzed by a least squares fitting routine using the UNTVEM v.4.50 program.

X-ray difraction (XRD) measurements were made on powder and pressed samples with a DRON-3 difractometer *(CuKa* radiation, scan speed of 2°/min). Peak positions were determined with an accuracy of ±0.05°.

Electron paramagnetic resonance (EPR) spectra were measured with a Varian E-4 X-band spectrometer equipped with a Varian-E257 nitrogen-sweep variable-temperature facility. From the measured spectra, we determined the effective peak-to-peak width ΔHpp, peak-to-peak height App, and signal intensity $I \cong App(\Delta Hpp)2$.

RESULTS AND DISCUSSION

Scanning tunneling and transmission electron microscopy [7-8] shows that powder particles several hundreds of nanometers in size have a regular spherical shape and an intricate inner structure composed of tinier particles; they are prone to conglomeration. Inasmuch as UPTFE has a highly developed surface (~1000 m2/g) and the above-mentioned surface functional groups, it is very active and can be stabilized as stable suspensions in some solvents.

In this work, metal (Fe, Co, Ni, Cu) formates, acetates, oxalates, ammines, and carbonyls were used as MCCs; thermodestruction of these compounds has been comprehensively studied [2]. Metal content of the resulting composite samples was 5–20 wt %.

The size of granules of UPTFE in the range from 100-300 nm are shown in Figure 1. Metal-containing nanoparticles are arranged in bunches at the surface of nanoglobules (Figure 2). The mean size of these nanoparticles is below 6 nm, and the size distribution is rather narrow. The particle shape often differs from the spheroidal shape typical of metal-containing nanoparticles in polyethylene and other polymeric matrices. In some cases, a clearly pronounced polyhedral form of particles is observed. However, most of the powders, except those containing copper nanoparticles, are amorphous as probed by X-ray diffraction.

The XRD pattern of the Fe-containing nanoparticles on the surface of UPTFE shows a strong low-angle peak from polytetrafluoroethylene and weak, broad peaks from an Fe-containing phase, indicative of small particle size (Table 1).

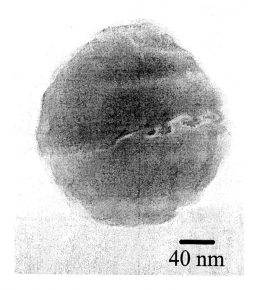

Figure 1. TEM microphotograph of granule of ultradispersed polytetrafluoroethylene.

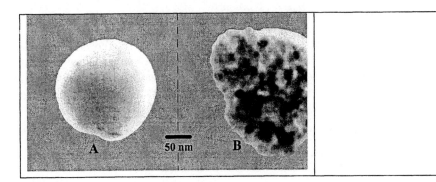

Figure 2. TEM microphotograph of Fe-containing nanoparticles (obtained from Fe(III) formate) at the surface of nanograins (d=250 nm) of ultradispersed polytetrafluoroethylene. The particle size was determined to be 6.4 ± 1.4 nm.

The particle composition is rather complex. Previously [9], we showed that, in the course of formation of Fe nanoparticles in the matrix of the copolymer of ethylene and tetrafluoroethylene (fluoroplastic-40), the matrix is partially defluorinated to produce the FeF2 phase at the surface of the particles. It is not improbable that an analogous process takes place in the matrix under study: as shown by Mössbauer spectra (Figure 3), the resulting nanoparticles contain, in addition to the metal and metal oxide and carbide phases, iron(II) ions, which are presumably incorporated in iron fluoride. Magnetic measurements give the strongest evidence of the formation of nanoparticles in the material. The fact is that a decrease in the size of a magnetic particle results, under definite conditions, in its single-domain state. In the single-domain state, domain walls are lacking and the external magnetic flux is present, because in a particle whose size is smaller that than some critical value, creation of the magnetostatic energy is more favorable.

As judged from the EPR spectral shape, we may assume that the particles in the sample are mainly super-paramagnetic at room temperature; this points to their relatively small dimensions and single-domain magnetic structure. However, the internal magnetic energy of each of these particles is sufficient for the entire material to be strongly attracted by a magnet. Therefore, we developed the method of nanometalization of ultradispersed polytetrafluoroethylene; the resulting powdery material contains nanoparticles of complex composition with physical characteristics typical of such particles. The composite powders can be converted by known methods into compact block polytetrafluoroethylene with uniformly distributed metal-containing nanoparticles.

ACKNOWLEDGEMENT

This work was supported by the Russian Foundation for Basic Research (Project Nos. 01–03–32955, 02–03–32435, 02–03–06153), ISTC (project no. 1991), and by the complex program of the Russian Academy of Sciences "Nanomaterials and Supramolecular Systems."

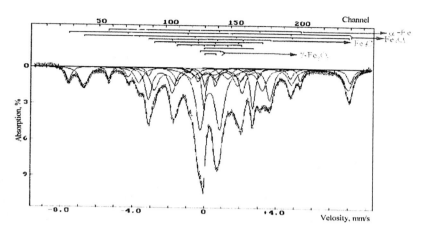

Figure 3. Mathematical model to describe the Mössbauer spectrum of Fe-containing nanoparticles deposited on ultradispersed polytetrafluoroethylene for the sample containing 14.03 wt % Fe.

REFERENCES

[1] A.D. Pomogailo, A.S. Rozenberg, and I.E. Uflyand, 2000. Nanochastitsy metallov v polimerakh (Metal Nanoparticles in Polymers), Moscow: Khimiya, 2000.
[2] S.P. Gubin and I.D. Kosobudskii, 1983. Usp. Khim., 52, 1350–1364.
[3] V.M. Buznik, A.K. Tsvetnikov, L.A. Matveenko, 1996. Khimiya v Interesakh Ustoich. Razvitiya, 4, 489–496.
[4] I.D. Kosobudskii and S.P. Gubin. 1985. Metallic Clusters in Polymer Matrices: A New Type of Metal-Pilled Polymers, Vysokomol. Soedin., 27(3), 689–695.
[5] A.T. Shuvaev, B.Yu. Khel'mer and T.A. Lyubeznova, 1988. Spectrometric Facility for DRON-3 Xray Diffractometers for Investigation of the Short-Range Order in Crystalline and Disordered Systems, Prib. Tekh. Eksp., 3, 234–237.
[6] D.L. Kochubei, Yu.A. Babanov, K.I. Zamaraev et al., 1988. Rentgenospektral'nyi metod issledovaniya struktury amorfnykh tel: EXAFS-Spektroskopiya (EXAFS Spectroscopy: A Tool for Probing the Structure of Amorphous Materials), Novosibirsk: Nauka.
[7] V.M. Buznik, A.K. Tsvetnikov, B.Yu. Shikunov and V.V. Pol'kin, Perspekt. Mater., 2, 69–72.
[8] V.G.Kuryavyi, A.K. Tsvetnikov and V.M. Buznik, 2001. Perspekt. Mater., 3, 57–62.
[9] I.D. Kosobudskii, S.P. Gubin, V.P. Piskorskii et al., 1985. Vysokomol. Soedin., 4, 689–695.

Table 1. XRD data for Fe-containing compounds (14,03 wt. % Fe) on the surface UPTFE.

№	D, Å	I/I$_0$, %	d, Å	I/I$_0$, %	d, Å	I/I$_0$, %	d, Å	I/I$_0$, %	d, Å	I/I$_0$, %
	This work		JCPDS PDF data							
			γ-Fe$_2$O$_3$ (№ 39-1346*)		α-Fe$_2$O$_3$ (№ 33-664*)		β-FeOH (№ 34-1266*)		Fe$_5$O$_7$(OH)·4H$_2$O (№ 29-712*)	
1	4,900	100								
			3,740	5	3,684	30	3,728	5		
			3,411	5			3,333	100		
			2,953	35						
2	2,667	15	2,6435	2	2,700	100	2,6344	25		
3	2,521	20	2,5177	100	2,519	70	2,5502	55	2,5	100
							2,3559	9		
			2,2320	1	2,207	20	2,2952	35	2,21	80
							2,1038	7		
							2,0666	7		
4	2,022	20	2,0886	16	2,0779	3	1,9540	20	1,96	80
					1,8406	40				
			1,7045	10	1,6941	45	1,7557	15	1,72	50
5	1,605	15			1,6033	5	1,6434	35		
			1,6073	24	1,5992	10				
			1,5248	2	1,4859	30	1,5155	9	1,51	70
							1,5034	5		
6	1,475	15	1,4758	34	1,4538	30	1,4456	15	1,48	80
					1,3115	10				
					1,3064	6				
			1,2730	5	1,2592	8				

Chapter 15: Nano-Clusters

Cluster Expansion Method for Chemisorption Geometries

M. H. F. SLUITER and Y. KAWAZOE
Institute for Materials Research, Tohoku University, Sendai, Japan

ABSTRACT

While determining ground state configurations is well established in the field of alloy theory with particular application to substitutional metallic alloys in bulk, in the case of chemisorption such searches are difficult because of the large number of possible configurations as well as complications such as reconstruction. In bulk alloys there is a systematic method for finding the most stable states, this is the so-called cluster expansion method. In that approach the energy is written as an expansion in terms of cluster probabilities and effective cluster interactions. Here, we examine if the cluster expansion method might be applicable also to ground state searches with lower dimensionality. If such a cluster expansion can be truncated at sufficiently small cluster sizes, ground states can be determined then based on a small number of first-principles total energy calculations only. We shall examine and illustrate the method with a particularly challenging case, determination of the most stable configurations for chemisorption of hydrogen on isolated graphene sheets.

INTRODUCTION

Chemisorption and physisorption at surfaces are an important area of research in view of the many practical applications such as in catalysis. When an isolated atom or molecule is absorbed on to a surface there are usually only a few high symmetry configurations that are likely to be stable. While relaxations and reconstructions make simulating the absorption of a single species onto a surface non-trivial, the number of distinct topologies remains rather limited. However, when many species absorb this is no longer the case. The interactions between absorbed species in addition to the surface—species interactions make that very large periodic cells can become ground states even when the interactions themselves are rather short-ranged. This is a well-known phenomenon in the study of the ground states of bulk substitutional alloys [1]. Fortunately, for bulk alloys there is an algorithm that has been highly successful for finding the most likely ground states [2]. The algorithm is based on a cluster expansion of the energy of an alloy [3] and it is described in detail below. Studies on bulk alloys have shown that the cluster expansion converges most rapidly when strain effects are small [4]. The question in applying the cluster expansion algorithm for the determination of ground state configurations in absorption problems then is whether the large strains can be handled. To test this a rather severe example was selected: the case of hydrogen chemisorption on to a single graphene sheet. In the graphene sheet the carbon valence orbitals are in the sp^2 hybridised state associated with in-plane bonds, whereas after chemisorption sp^3 hybridisation occurs which demands tetrahedral coordination. Clearly, chemisorption must cause very large strains in the initially flat graphene sheet. Moreover, assuming a horizontally positioned graphene sheet there are 3 possibilities for every carbon atom, no hydrogen, hydrogen above, and hydrogen below, making for a very large number of configurations even when a modest number of carbon sites is considered only. Chemisorption of hydrogen is also of interest in its own right and several studies have been devoted to it [5]. It should be remarked however, that the large hydrogen absorption capacity reported for carbon nanotube materials [6] is based on physisorption of H_2 molecules, rather than on chemisorption of single H atoms. However, as physisorption of H_2 on to graphene involves only minor strains, such a calculation is not nearly as stringent a test as chemisorption of H.

CLUSTER EXPANSION ALGORITHM

The cluster expansion assumes that the energy can be written as a rapidly converging sum over cluster contributions where contributions from larger clusters become negligible. A practical example of this idea is common in organic chemistry where the formation enthalpy of a molecule can be expressed to a good approximation as a sum of contributions from nearest neighbor pairs (read bonds) only. In the case of ethane one would count one C-C pair and six C-H pairs and estimate that the formation energy with respect to the isolated atoms is given as the sum of the single C-C bond energy and the six C-H bond energies. Contributions from larger clusters such as triplets usually provide only minor corrections and are neglected. In the case of bulk alloys an expansion in terms of nearest neighbor pairs only can explain some of the most commonly occurring ground states, but generally more distant pairs and clusters involving more than just two atoms have to be considered. For a bulk alloy with a fixed underlying crystal structure each configuration can be specified purely in terms of the occupancy of each site. In actuality one might represent the occupancy of a site **i** as a vector on dimension **m**, where **m** is the number of components, e.g. in case there are atom types A, B, and C the occupancy is

σ_i = (100), (010), (001) if there is an A, B or C atom, respectively, at site **i**.

Concentrations **C** of the atomic species then are given simply the average of σ over all sites **i**, the expectation value of σ

$$C = <\sigma_i>$$

The correlations are expressed as products of occupancies [3] e.g. a nearest neighbor pair correlation between sites i and j is given by

$$\xi_{ij}^{PQ} = <\sigma_i^P \sigma_j^Q>$$

where **P** and **Q** indicate the atomic species (A,B, or C) and σ_i^P is the element of the vector σ_i corresponding to species **P**. In the thermodynamic limit a description in terms of σ is normally useless because the specification of any arbitrary configuration requires an infinite number of occupancies. Fortunately, we are interested in ground states, which tend to have periodic structures with unit cells with a limited number of sites only so that such a configuration can be represented in terms of just three translation vectors and only a limited number of occupancies. In such a unit cell it is also feasible to actually count the number of pairs and other clusters of a given type and determine exactly the correlations ξ. The cluster expansion of the energy [3] is given simply by

$$E^s = \sum_\alpha \xi_\alpha^s J_\alpha$$

where E^s is the energy per site for configuration (structure) s, the sum runs over clusters α and J_α is the effective cluster interaction (ECI) associated with cluster α. It should be noted that while the ECIs are generally unknowns, the energy of configurations can be computed with theoretical models such as electronic density functional theory. The correlations for a configuration can be found simply by inspection. Therefore, provided that the summation over clusters alpha can be truncated, the ECI can be obtained by inverting the cluster expansion. To be of use, the cluster expansion must converge with cluster size, so that retaining in the sum terms corresponding to compact clusters with a few sites only gives a sufficiently accurate representation of the energy. As the chemical example for computing the formation enthalpy illustrated, this can be the case. However, in bulk alloys strain and relaxation are known to require longer ranged interactions and for superlattice structures in the long wave limit a reciprocal space treatment is needed [4]. Fortunately, we are interested in chemisorbed configurations with relatively small periodicities so that a set of ECIs pertaining to compact clusters can be expected to suffice. The issue then is to select a small but adequately accurate set of clusters to represent the energies of configurations. However, it is not really necessary to represent accurately all configurations, only those that stable, or almost stable. For very unstable configurations we can allow large errors because in reality such configurations will not occur. Restricting ourselves to accurately representing the most stable configurations only allows us to use a cluster expansion with yet a smaller set of clusters. This is evident in the chemical model for formation energy of molecules also. The chemical model does not describe nearly as well radicals, and other atomic assemblies in which the atomic valences are not respected, as stable molecules.

Once a choice is made of which clusters to use in the cluster expansion and additionally a set of configurations is selected and the corresponding energies have been computed the ECI are obtained by inversion. Straightforward matrix inversion can be applied when the matrix ξ is square, that is, when the number of unknown ECI to be solved is equal to the number of configurations for which energies have been computed. This method is not very good because the cluster expansion is truncated and the neglected ECIs 'pollute' the ECIs retained in the cluster expansion. Therefore there should be more configurations than ECIs so that the inversion process can average out over the neglected ECIs. Such an overdetermined system is most reliably inverted with singular value decomposition (SVD). This method incidentally also allows us to simultaneously examine how significant the contribution is from a particular ECI. This means that we could start out with trying to obtain more ECIs than there are configurations. Naturally, SVD can determine out of a set of n^s configurations at utmost n^s ECI, but generally not all of these ECI will much contribute to an accurate representation of the energy. Usually a nearest neighbor pair ECI contributes much, but e.g. a cluster with more than 4 sites could be expected to contribute very little. Some of the n^s ECIs, which do not much improve the fit between actual E^s and the value obtained from the cluster expansion, can then simply be omitted and a new SVD inversion is done using the most important ECIs only. Again there may be ECIs that contribute little to improving the fit and one can select an even smaller set of ECIs.

Thus, SVD allows one to find the most significant ECIs from a set that is larger than the set of configurations and in this way it helps with selection of ECIs. Now we have approximate values for the ECIs, they are not exact because of the limited set of energies and the limited set of clusters employed. It is important to note here that now that approximate ECI are known, the energy of any configuration can be predicted with the cluster expansion requiring as only extra information a knowledge of the correlations of this configuration. As the latter is purely geometrical information it can be tabulated easily for a very large set of configurations including all those possible up to some given unit cell size. As was just discussed, the reverse is possible too, from the knowledge of the energies of some configurations one can extract what the ECIs are.

It is this feature, of easily computing energies from ECIs and ECIs from energies that is at the heart of the cluster expansion algorithm for ground state searching. In the algorithm one starts by computing from *ab initio* the energies of some configurations, usually configurations with small unit cells so that the *ab initio* calculations can be done speedily. One then extracts ECIs and uses these ECIs to compute the energies of all the tabulated configurations. Usually, one then discovers that the cluster expansion predicts that there are configurations with lower energies than the ones included in the prior *ab initio* calculation.

The newly predicted stable configurations are now computed by the *ab initio* method. As the cluster expansion is only approximate, certainly at the initial stages, the *ab initio* energies will not coincide with the cluster expansion values. Now a new larger set of *ab initio* energies is available. This larger set is used to extract ECIs. These ECIs will be more accurate than the previously extracted ECIs because the new ones are based on a larger data set. One now repeats the previous steps, the new ECIs are again used to compute energies of all tabulated configurations and new stable configurations are predicted by the cluster expansion.

As the steps are repeated, an increasing number of configurations are computed by *ab initio*, but gradually fewer and fewer new stable configurations are found because the ECIs become increasingly accurate. Then the energies as given by the *ab initio* method and by the cluster expansion method approach each other. When no new stable configurations are found by cluster expansion the latter may be converged. This can be verified by repeating the above algorithm using a different starting set of configurations, or by checking how well the cluster expansion is able to predict the energy of a configuration that was not used to extract the ECIs. The latter, a predictive error test [2], can be carried out conveniently as follows.

Now that one has a set of n^s configurations for which the energy was computed *ab initio*, one removes one configuration from this set and extracts the ECIs from the set of n^s-1 configurations. The ECIs are then used to predict the energy of the removed configuration and this value is compared with the known *ab initio* value. Naturally, the two results should be close, if they are not; the cluster expansion is not likely to be converged. Then, one must consider using a larger set of clusters in the cluster expansion and as this larger set of clusters has more degrees of freedom, usually more *ab initio* calculations are going to be needed as well. When one has computed the predictive error by removing each configuration in turn, and one is satisfied with the predictive error for every configuration, and moreover one finds the same ground states even when one starts with a completely different set of configurations, then the search for the ground state configuration is assumed to have converged.

Experience shows that convergence is reached when the number of actual *ab initio* calculated configurations is several orders of magnitude smaller than the number of tabulated configurations considered in the cluster expansion. As the *ab initio* calculations tends to be computationally demanding and as this algorithm is rather easily implemented in a computer program, this method of searching for the ground state can be orders of magnitude faster. Conversely it can consider many more configurations than a straightforward trial and error search by *ab initio* calculation. Actual calculations for bulk binary alloys show that a few tens of configurations [2,7] suffice for the highly sensitive test of a composition-temperature phase diagram.

COMPUTATIONAL DETAILS AND METHOD

Ab initio calculations of the hydrogen chemisorbed graphene configurations were performed within the local density approximation using the Vienna *ab initio* simulation program (VASP) [8]. An orthorhombic unit cell was used in which the volume was held constant at 203 Å^3. This resulted in a separation distance of the graphene layers of about 10 Å, which was verified to be sufficient to not pollute the configuration energies with spurious inter-graphene layer contributions. The exchange-correlation functional with generalized-gradient corrections [9] was used. The calculations were performed using fully non-local optimized ultrasoft pseudopotentials [8] with energy cut-offs of 287 (200) eV and augmentation charge cutoff energies of 550 (400) eV for C (H). The cutoff energy for the wavefunctions has been set at 360 eV. Integrations in reciprocal space used a 6x6x3 Monkhorst-Pack [10] grid in the 1st Brillouin zone, giving rise to 27 k points in the irreducible section.

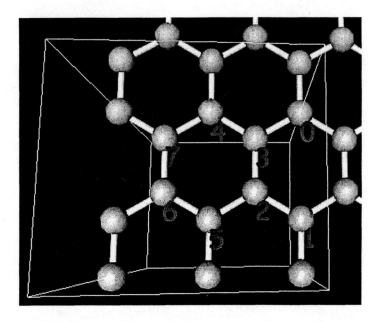

Figure 1. Top view of the orthorhombic cell with 8 carbon atoms. For pure graphene, the lattice parameters are 4.26, 4.92, and 9.68 Å.

The Hermite-Gauss smearing method of Methfessel and Paxton [11] of order 1 has been used to accelerate k point convergence. The **a** and **b** lattice parameters of the orthorhombic cell and the internal coordinates were optimized for all configurations using the conjugate gradient method. Structural optimizations were considered converged when the greatest magnitude of the force on any atom was less than 1 meV/Å.

The orthorhombic cell with 8 carbon atoms, see figure 1, allows for $3^8 = 6561$ configurations. The configurations were given an identification number **Q** computed with

$$Q = \Sigma_i \sigma_i 3^i \qquad i = 0,1,...,7$$

where the position of the sites is as in figure 1 and the occupancy σ takes value 0,1,2 when H is above, H is below, or when there is no H associated with the site i. E.g. configuration 0000 corresponds to the case where is a hydrogen atom above each carbon atom, while in configuration 3280 there is a hydrogen atom below each carbon, and configuration 6560 indicates a graphene sheet without hydrogen.

Application of symmetry reduces the number of configurations to just 528. However, even 528 configurations are a daunting number for direct *ab initio* calculations especially considering that each configuration needs to be carefully structurally optimized. It is important to note that although the system consists of hydrogen and carbon atoms, each configuration can be represented by the occupancies σ_i of the carbon atoms only since all hydrogen atoms form only a single bond with one particular carbon atom. The correlations for the 528 configurations could be easily tabulated using a computer program. The maximal clusters considered for the cluster expansion are the hexagon and the centered triangle. These maximal clusters were generated on the principle of the complete set of most compact clusters that are found within a sphere of a given radius.

Vul and de Fontaine [12] gave a theoretical justification for this principle. We use a radius of 2.5 times the nearest neighbor carbon-carbon distance (which is about 1.4 Å). In Figure 1 the hexagon can be recognized in the sites labeled 2 through 7 and the sites 1, 2, 3, and 5 represent the centered triangle. Considering all possible subclusters including the so-called empty cluster, this gives a total of 15. However, as there are three species, the number of distinctly independent ECIs is 104.

RESULTS AND DISCUSSION

SVD was used to determine which of the 104 ECIs are most significant for representing the energy. After the cluster expansion had converged, meaning no new ground states could be found and the predictive error reached a value of 12 % of the range of formation enthalpies, about 46 ECIs were found to contribute significantly. At that point total energies for about 56 structures had been computed from *ab initio* which is almost one order of magnitude less than a full calculation of all the possible ground states with the periodicity considered here. The large strains are clearly reflected in the large number of ECIs that are required to describe the energetics, the relatively large predictive error and also in the rather large values of ECIs associated with multi-site clusters.

The most stable structures are shown in Figure 2. It is important to note that all these structures have 1:1 hydrogen carbon ratios. The most stable structure, 0820, is completely in accordance with intuition because each carbon atom can adopt the ideal sp^3 bonding angles without any strain. However, the next most stable structures are not so easily guessed. Structures 0984 and 0328 are 56 and 103 meV/carbon atom above the 0820 structure. Both 0984 and 0328 are buckled. In the case of the former, the buckling puts one half of a carbon hexagon above the original graphene sheet and the other half below it, giving rise to a corrugated pattern where the grooves accommodate some of the strain involved in the sp^2 to sp^3 conversion. Structure 0328 also relieves some of the strain, but in this case by dimerization along one direction.

The most interesting result is that none of the structures with partial hydrogen coverage can be stable with respect to pure graphene, structure 6560, and the fully covered structure 0820 as is shown in Figure 3. This means that when hydrogen is chemisorbed onto graphene, there is no gradual homogeneous covering of the surface, but rather islands must form of pure 0820 in a sea of unhydrogenated graphene. Such islands have an excess energy associated with their perimeter so that chemisorption must be accompanied by classical nucleation and growth phenomena.

The energy released by chemisorption at zero temperature can be roughly estimated by calculating the energy of the H_2 molecules in a large cell where inter H_2 interactions vanish. The zero temperature chemisorption energy which neglects all zero point librations and vibrations is about 0.2 eV per hydrogen atom relative to H_2 gas in near-vacuum. This value agrees well with the reported value [5].

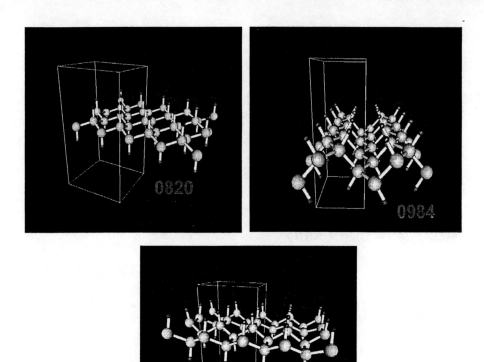

Figure 2. Most stable hydrogenated graphene layers as determined with the cluster expansion.

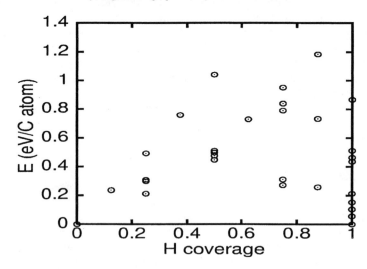

Figure 3. Chemisorption energy of hydrogen onto graphene. Pure graphene and structure 0820 have been taken as reference points.

CONCLUSIONS

It has been demonstrated that an efficient cluster expansion technique can be applied to chemisorption and physisorption at surfaces and that important physical insights can be gleaned from such an analysis. Even for chemisorption at graphene, where strain effects are extremely large, one can obtain a cluster expansion that identifies the most stable surface structures. In the case of hydrogen onto graphene it was shown that a fully covered structure with alternating above and below plane hydrogen attachment is the most stable. This can be readily rationalized on the basis of strains associated with the sp^2 to sp^3 conversion for the carbon hybridization. Moreover such an analysis shows that partial coverage is not favored. Therefore, chemisorbed regions must form islands, which as a consequence must exhibit nucleation and growth phenomena also.

REFERENCES

[1] D. de Fontaine, 1979. Configurational Thermodynamics of Solid Solutions in Solid State Physics, Vol. 34, ed. H. Ehrenreich, F. Seitz and D. Turnbull, Academic Press, N.Y., 73–274.

[2] M.H.F. Sluiter, Y. Watanabe, D. de Fontaine, and Y. Kawazoe, 1996. A First-Principles Calculation of the Pressure Dependence of Phase Equilibria in the Al-Li System, Phys. Rev. B 53, 6137–6149.

[3] D. de Fontaine, 1994. Cluster Approach to Order-Disorder Transformations in Alloys in Solid State Physics, vol. 47, ed. by H. Ehrenreich and D. Turnbull, Academic Press, New York, 33–176.

[4] D.B. Laks, L.G. Ferreira, S. Froyen, and A. Zunger, 1992. Efficient cluster expansion for substitutional systems, Phys. Rev. B 46, 12587–12605.

[5] J.S. Arellano, L.M. Molina, A. Rubio, M.J. López, J.A. Alonso, 2002. Interaction of molecular and atomic hydrogen with (5,5) and (6,6) single-wall carbon nanotubes, J. Chem. Phys. 117, 2281–2288

[6] Y. Ye, C. C. Ahn, C. Witham, B. Fultz, J. Liu, A. G. Rinzler, D. Colbert, K. A. Smith, and R. E. Smalley, 1999. Hydrogen adsorption and cohesive energy of single-walled carbon nanotubes, Appl. Phys. Lett 74, 2307–2309.

[7] M. Sluiter and Y. Kawazoe, 2001. Bondlengths and Phase Stability of Silicon-Germanium Alloys under Pressure, Mat. Trans. JIM 42, 2201–2205.

[8] G. Kresse and J. Furthmüller, 1996. Efficiency of ab-initio total energy calculations for metals and semiconductors using a plane-wave basis set, Comp. Mat. Sci. 6, 15–50.

[9] J.P. Perdew and Y. Wang, 1992. Accurate and simple analytic representation of the electron-gas correlation energy, Phys. Rev. B 45, 13244–13249.

[10] H.J. Monkhorst and J.D. Pack, 1976. Special points for Brillouin-zone integrations, Phys. Rev. B 13, 5188–5192.

[11] M. Methfessel and A.T. Paxton, 1989. High-precision sampling for Brillouin-zone integration in metals, Phys. Rev. B 40, 3616–3621.

[12] D.A. Vul and D. de Fontaine, 1994. On the Choice of a Maximal Cluster in the Cluster Variational Method, Mat. Res. Soc. Symp. Proc. Vol. 291, eds. J. Broughton, P.D. Bristowe, J.M. Newsam, Materials Research Society, Materials Park, 401–406.

All-Electron *GW* Calculations for Small Clusters

S. ISHII
Institute for Materials Research, Tohoku University, Sendai, Japan

K. OHNO
Department of Physics, Yokohama National University, Yokohama, Japan

Y. KAWAZOE
Institute for Materials Research, Tohoku University, Sendai, Japan

ABSTRACT

Density functional theory (DFT) is known to be excellent in reproducing ground state properties, but it does not correctly reproduce excited state properties. One needs a theory beyond DFT to investigate the excited states. One way of elucidating the correct one-particle excitation spectra is to use the GW approximation. Most ab-initio GW codes have been applied successfully to crystals, but, for clusters, only a few papers have been published.

In the present study, we have developed a new ab-initio GW code that employs an all-electron mixed-basis approach where the wave function is represented by plane waves and atomic orbitals as a basis set. We have applied this code to small clusters such as alkali-metal (Li_n, Na_n, K_n, n = 2, 4, 6, 8) and silicon (Si_n, n = 4, 5, 6) clusters.

The resulting GW quasi-particle energies, such as ionization potentials and electron affinities, agree well with available experimental data. For silicon clusters, we fond it very important to take into account the difference in the atomic geometry between neutral- and negatively-charged clusters to reproduce correctly the experimentally measured electron affinities.

INTRODUCTION

The local density approximation (LDA) [1] and the generalized-gradient approximation (GGA) based on density functional theory (DFT)[2] are very good approximation to study ground state properties such as atomic structures and cohesive energies. However, these calculation methods fail in reproducing excited state properties such as the HOMO (highest occupied molecular orbital)-LUMO (lowest unoccupied molecular orbital) gap of molecules and clusters, the band gap of semiconductors and insulators, and the absorption cross-section. For example, the band gap of semiconductors and insulators such as silicon crystals estimated by these methods are smaller than those of the experimental values by about 30-50%. In order to evaluate correctly the excitation spectra by *ab-initio* methods, one has to go beyond DFT. One of the theories to do it is the *GW* approximation (GWA) introduced first by Hedin[3], who considered, from the viewpoint of the many-body quantum theory, the lowest order self-energy diagram given by the product of the one-particle Green's function G and the dynamically screened Coulomb interaction W:

$$\Sigma(\mathbf{r},\mathbf{r}',\omega) = \frac{i}{2\pi}\int d\omega' G(\mathbf{r},\mathbf{r}';\omega+\omega')W(\mathbf{r},\mathbf{r}';\omega')e^{i\eta\omega'}, \qquad (1)$$

where $\mathbf{r}$ ($\mathbf{r}'$) and ω (ω') denote the positions and frequencies, respectively. η is positive infinitesimal. W is usually evaluated within the random phase approximation (RPA). Σ is non-local in space and it is frequency dependent.

The state-of-the-art GW calculations are performed mainly for crystals such as silicon and germanium and succeeding in reproducing the experimental band gaps, which is not reproduced by the LDA[4]. However, they used the pseudopotential approach. In addition, only a few papers are published for clusters. Onida et al. performed a GW calculation for sodium tetramer employing the pseudopotential approach.[5]. Rohlfing and Louie performed GW calculations for Si_nH_m [6]. Saito et al. performed the GW calculations for sodium and potassium clusters employing the jellium-background model [7]. (Their method [7] is not ab-initio but applicable for only the clusters with closed shell structures, i.e., magic number clusters). We have developed a new GW code using the all-electron mixed-basis approach and applied it to several clusters. In the present paper, we overview and discuss the GW quasiparticle energies of alkali-metal clusters (sodium, lithium, and potassium) [8,9], and silicon clusters [10].

METHODOLOGY

The all-electron mixed-basis approach is a natural extension of the pseudopotential mixed-basis approach to take into account the core electrons fully. The present approach uses both plane waves and atomic orbitals as a basis set. The all-electron mixed-basis approach is successfully applied to crystals, molecules, and clusters [11]. We use Herman-Skillman atomic code to obtain atomic orbitals. In atomic core region the wave function is divided into logarithmic mesh [12].

In Figure 1 we show the atomic configurations of (a) lithium, (b) sodium, (c) potassium, and (d) silicon clusters, respectively, studied in the present study. These are referred to the Ref.13 Except for silicon clusters of which structures are obtained by using the Gaussian98 program.

In the calculations, we employ an fcc supercell with a cubic edge of 50, 50, 70, and 20 a.u. for sodium, lithium, potassium, and silicon clusters, respectively because the wave functions of potassium (silicon) clusters are delocalized (localized) than other clusters studied in the present calculations. These values are chosen carefully to obtain convergence of absolute LDA energy levels. In the LDA calculation we employ 3 and 5 Ry to achieve a good convergency within the error of 0.01 eV for alkali-metal clusters and silicon clusters, respectively. We also introduce a spherically truncated Coulomb potential [5] to avoid interaction between the cells.

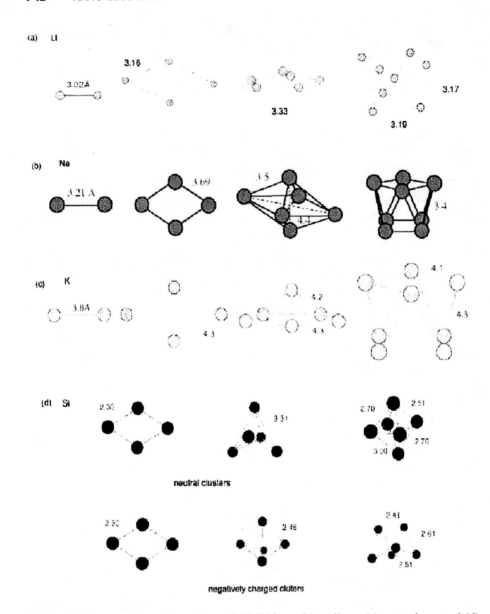

Figure 1. The atomic configurations of (a) lithium, (b) sodium, (c) potassium, and (d) silicon clusters studied in the present calculations. Notice that we employ both neutral and negatively charged geometries to perform the GW calculations for silicon clusters (see text).

In the GW evaluations of electron self-energy, we employ both the generalized-plasmon pole (GPP) model explained elsewhere [4] and numerical integration with frequency to evaluate the correlation part of the self-energy for alkali-metal clusters and employ the former for silicon clusters. The contour of the frequency integral is chosen on the positive real axis (0<ω'<14 eV). The frequency interval of 0.25 eV are necessary and sufficient with the condition of δ=0.01 a.u.

All calculations are performed for only the Γ point, corresponding to q = 0. This is sufficient when the supercell is chosen sufficiently large. In our calculation, we confirmed that our result does not change even if we introduce q-point sampling. The size of the dielectric matrix given by Eq. 4 is chosen to be 645x645 (corresponding to about 1 Ry). We found it necessary and sufficient for all cases.

In the GWA, the electron self-energy is defined by eq.(1). As mentioned above W is evaluated within the RPA, defined by using the dielectric response function $\varepsilon_{G,G'}(q,\omega)$.

$$W_{G,G'}(q,\omega) = \varepsilon_{G,G'}^{-1}(q,\omega)v(q+G'), \tag{2}$$

$$\varepsilon_{G,G'}(q,\omega) = \delta_{G,G'} - v(q+G)P_{G,G'}(q,\omega), \tag{3}$$

$$P_{G,G'}(q,\omega) = \sum_{k}\sum_{n}^{occ}\sum_{n'}^{emp} \frac{\langle n,k|e^{-i(q+G)\cdot r}|n',k+q\rangle\langle n',k+q|e^{i(q+G')\cdot r'}|n,k\rangle}{E_{k,n} - E_{k+q,n'} - \omega - i\delta_{k-q,n'}} \cdot [f(E_{k,n}) - f(E_{k+q,n'})] \tag{4}$$

where $f(E)$ is a Fermi-Dirac distribution function. G (G') is a reciprocal lattice vector and $V(q)$ is a coulomb potential in Fourier space. The self-energy operator is divided into two parts. One is the exchange part:

$$\Sigma_x(r,r') = \frac{i}{2\pi}v(r-r')\int G(r,r';\omega')e^{i\omega'\eta}d\omega' \tag{5}$$

and the diagonal matrix element of this operator becomes

$$\Sigma_{x,n} = \langle \psi_n(r)|\Sigma_x(r,r')|\psi_n(r')\rangle$$
$$= -\int dr \int dr' \sum_m \frac{\psi_n^*(r)\psi_m(r)\psi_m^*(r')\psi_n(r')}{|r-r'|} \tag{6}$$

This is the Fock exchange energy evaluated by using LDA wave functions while the other term represents a correlation term given by

$$\Sigma_c(r,r') = \frac{i}{2\pi}\int d\omega' e^{i\eta\omega'}G(r,r';\omega+\omega')[W(r,r';\omega') - v(r-r')], \tag{7}$$

and its diagonal matrix element becomes

$$\Sigma_{c,n} = \rangle n,\mathbf{k} | \Sigma_c(\mathbf{r},\mathbf{r}') | n,\mathbf{k} \langle$$

$$= \frac{i}{2\pi} \sum_{n'} \sum_{\mathbf{q}} \sum_{\mathbf{G},\mathbf{G}'} \rangle n,\mathbf{k} | e^{i(\mathbf{q}+\mathbf{G})\cdot\mathbf{r}} | n',\mathbf{k}-\mathbf{q} \rangle \langle n',\mathbf{k}-\mathbf{q} | e^{-i(\mathbf{q}+\mathbf{G}')\cdot\mathbf{r}'} | n,\mathbf{k} \langle \int_0^\infty d\omega' [W_{\mathbf{G},\mathbf{G}'}(\mathbf{q},\omega') - \delta_{\mathbf{G},\mathbf{G}'}] v(\mathbf{q}+\mathbf{G})] \quad (8)$$

$$\times \left[\frac{1}{\omega + \omega' - E_{\mathbf{k}-\mathbf{q},n'} - i\delta_{\mathbf{k}-\mathbf{q},n'}} + \frac{1}{\omega - \omega' - E_{\mathbf{k}-\mathbf{q},n'} - i\delta_{\mathbf{k}-\mathbf{q},n'}} \right]$$

with the help of $W(\omega) = W(-\omega)$, where $\mathbf{G}(\mathbf{G}')$ represents reciprocal lattice vector.

On the other hand, when one employs the GPP model, one can skip the integration with frequency and the diagonal matrix element of the correlation part becomes

$$\Sigma_{c,n} = -\sum_{n'}^{occ} \sum_{\mathbf{q}} \sum_{\mathbf{G},\mathbf{G}'} \rangle n,\mathbf{k} | e^{i(\mathbf{q}+\mathbf{G})\cdot\mathbf{r}} | n',\mathbf{k}-\mathbf{q} \rangle \langle n',\mathbf{k}-\mathbf{q} | e^{-i(\mathbf{q}+\mathbf{G}')\cdot\mathbf{r}'} | n,\mathbf{k} \langle \frac{\Omega^2_{\mathbf{G},\mathbf{G}'}(\mathbf{q})}{(\omega - E_{\mathbf{k}-\mathbf{q},n'})^2 - \tilde{\omega}^2_{\mathbf{G},\mathbf{G}'}(\mathbf{q})} v(\mathbf{q}+\mathbf{G}')$$

$$+ \frac{1}{2} \sum_{n'} \sum_{\mathbf{q}} \sum_{\mathbf{G},\mathbf{G}'} \rangle n,\mathbf{k} | e^{i(\mathbf{q}+\mathbf{G})\cdot\mathbf{r}} | n',\mathbf{k}-\mathbf{q} \rangle \langle n',\mathbf{k}-\mathbf{q} | e^{-i(\mathbf{q}+\mathbf{G}')\cdot\mathbf{r}'} | n,\mathbf{k} \langle \frac{\Omega^2_{\mathbf{G},\mathbf{G}'}(\mathbf{q})}{\tilde{\omega}_{\mathbf{G},\mathbf{G}'}(\mathbf{q})[\omega - E_{\mathbf{k}-\mathbf{q},n'} - \tilde{\omega}_{\mathbf{G},\mathbf{G}'}(\mathbf{q})]} v(\mathbf{q}+\mathbf{G}'). \quad (9)$$

Here $\Omega_{\mathbf{G},\mathbf{G}'}(\mathbf{q})$ and $\tilde{\omega}_{\mathbf{G},\mathbf{G}'}(\mathbf{q})$ are determined by the generalized f-sum rule and given by using the plasma frequency ω_p and the charge density ρ:

$$\Omega^2_{\mathbf{G},\mathbf{G}'}(\mathbf{q}) = \omega_p^2 \frac{(\mathbf{q}+\mathbf{G})\cdot(\mathbf{q}+\mathbf{G}')}{|\mathbf{q}+\mathbf{G}|^2} \cdot \frac{\rho(\mathbf{G}-\mathbf{G}')}{\rho(0)} \quad (10)$$

$$\tilde{\omega}^2_{\mathbf{G},\mathbf{G}'}(\mathbf{q}) = \frac{\Omega^2_{\mathbf{G},\mathbf{G}'}(\mathbf{q})}{\delta_{\mathbf{G},\mathbf{G}'} - \varepsilon^{-1}_{\mathbf{G},\mathbf{G}'}(\mathbf{q},\omega=0)} \quad (11)$$

Therefore, under this model, one can skip the frequency integration in the evaluation of the self-energy.

Finally, the GW quasiparticle energies are given by

$$E^{GWA}_{\mathbf{k},n} = E^{LDA}_{\mathbf{k},n} + \frac{1}{1 - [\partial \Sigma(E)/\partial E]_{E=E^{LDA}_{\mathbf{k},n}}} \rangle n,\mathbf{k} | \Sigma(E^{LDA}_{\mathbf{k},n}) - V^{LDA}_{xc} | n,\mathbf{k} \langle, \quad (12)$$

where V_{xc}^{LDA} represent the exchange-correlation potential of the LDA.

Although the Dyson equation should be solved self-consistently in principle, we perform GW calculations as a first-order perturbation of the LDA because of the following reasons. Self-consistent GW calculations do not always guarantee the f-sum rule, which is related to charge neutrality [14]. In addition, in such a calculation one needs add the vertex correction [14].

RESULTS AND DISCUSSION
Alkali-metal clusters

In the present subsection, we discuss alkali-metal clusters (Li_n, Na_n, K_n, n=2,4,6,8). Alkali-metal clusters are most simple clusters, which are very good example to test the program. For potassium and sodium clusters, the *GW* calculation employing the jellium model was performed by Saito et al.[7]. As a test calculation of exchange part given by Eq. (6) (Σ_x), we performed *GW* calculation for an isolated sodium atom. The value of the HOMO level is about –6.99 eV, which is very good agreement with the results obtained using the Herman-Skillman code (-7.01 eV). In the calculation, the core contribution is about 0.8 eV, which cannot be ignored at all. On the other hand, the core contribution to the correlation part is within 0.1 eV and therefore we omitted this calculation. Local-field corrections play a significant role in the evaluation of the correlation part of the self-energy, $\Sigma_{c,n}$. In fact, when we employ 65x65 matrix size of **G** vectors (corresponding to almost 0 Ry) in eq.(8), the value of $\Sigma_{c,n}$ for the HOMO (LUMO) level is about –0.05eV (-0.37eV), while the correct value is –0.73eV (-0.66eV).

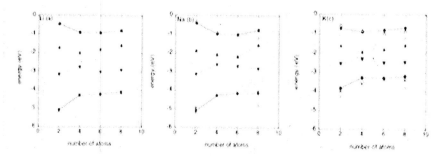

Figure 2. The HOMO and LUMO energy obtained by the GWA (closed square for HOMO and closed circle for LUMO) and LDA calculations (closed triangle down for HOMO and closed triangle up for LUMO) for (a) lithium clusters, (b) sodium clusters, and (c) potassium clusters, compared with negative sign of the available experimental ionization potentials (cross) [15], electron affinities (open triangle) [15], and *GW* quasi-particle energies using jellium-background model [7].

The absolute value of the HOMO (LUMO) energy is an ionization potential, IP (electron affinity, EA). In Figure 2, we show the cluster-size-dependence of the HOMO and LUMO energies obtained by the present LDA and *GW* calculations for (a) lithium, (b) sodium, and (c) potassium clusters, compared with experimental IPs, EAs [15], and the value of the ref. [7]. For the HOMO state, although the LDA eigenvalues underestimate the experimental IPs by about 30–50%, the present *GW* quasiparticle energies are in good agreement with experimental values and better agreement than that of the *GW* jellium-model. Similarly, for the LUMO state, although LDA eigenvalues overestimate the experimental EAs by about 150-200%, the present *GW* quasiparticle energies are in good agreement with experimental EAs very much. The HOMO-LUMO gap obtained by the LDA is smaller than that of the *GW* and experimental values by about 30–50% for all clusters studies here. The

HOMO-LUMO gap of Na_2 and Na_8 is larger than that of other clusters because those clusters are magic number clusters.

Let us discuss the validity of the GPP model. When one compares the GPP model and numerical integration with frequency, one can compare the correlation part of the self-energy because the exchange part is independent of frequency (see Eq. 8 and Eq. 9). In Table 1, we show the results obtained by the GPP model and numerical integration with frequency for sodium clusters. Note that the GPP model reproduces the results of the numerical integration within an error of about 0.15 eV.

Table 1. The value of correlation part of self-energy ($\Sigma_{c,n}$ in units of eV) for the HOMO and LUMO levels of sodium clusters employing either the numerical integration with frequency or GPP model.

Cluster size	Level	Numerical integration	GPP model
Na_2	HOMO	-0.73	-0.71
Na_2	LUMO	-0.66	-0.66
Na_4	HOMO	-1.01	-0.90
Na_4	LUMO	-1.35	-1.23
Na_6	HOMO	-0.86	-0.69
Na_6	LUMO	-1.53	-1.49
Na_8	HOMO	-0.85	-0.70
Na_8	LUMO	-1.40	-1.25

The GPP model assumes a δ-function type excitation such that excitations occur at the energy of $\tilde{\omega}_{G,G'}(q)$. In fact, the contribution from the detail local structures of W is not important as pointed out in ref. 4. However, Eq. 9 does not always have a real solution for all combinations of G, G', and q. Sometimes, it leads to imaginary numbers of $\tilde{\omega}_{G,G'}(q)$, which means breakdown of the GPP model. We simply ignore such combinations of G, G', and q [4]. If there are so many such combinations, the GPP model may become worse. In the present calculation, we found that such cases are about 1/3 to 1/2 of the total combinations. It is surprising, however, that the GPP model nevertheless reproduces the results obtained by numerical integration fairly well.

Silicon Clusters

Si-clusters are of interest from the viewpoint of not only physics, but also applications for wide band gap engineering. The bonding nature of Si-clusters differs from that of alkali-metal clusters. We use the Gaussian98 program with B3PW91 exchange-correlation potential and 6-311g* basis set to get ground state geometry. The configurations agree well with previous work [16].

Table 2 shows the results of the present LDA eigenvalues and GW quasiparticle energies, using the neutral structures, as compared to experimental IPs and EAs.[17]. For the tetramer, IP and EA obtained by the GW calculations agree well with experimental values. For the LUMO state of pentamer and hexamer, agreement is poor. As reported previously [16], the structure of Si_5 and Si_6 changes dramatically if one electron is added to the neutral cluster. We have confirmed this fact. The structure of anions is shown in Figure 1(d). The bond length of the pentamer changes by about 10%. Because of this, we performed the GW calculations employing a negatively charged configuration but a *neutral condition*. As a result, the GW quasiparticle energies of the LUMO level change dramatically by about 1eV in good agreement with experimental EAs.

Table 2. Cluster-size dependence of LDA eingenvalues, GW quasiparticle (QP) energies for the HOMO and LUMO levels obtained by present calculations, compared with negative sign of experimental ionization potentials and electron affinities [17].

Cluster	Level	LDA eigenvalue	GW QP energy	Experimental value
Si_4	HOMO	-5.56 eV	-7.42	-7.5
Si_4	LUMO	-4.50	-1.92	-2.0
Si_5	HOMO	-5.86	-7.57	-7.8
Si_5	LUMO	-3.82	-1.17	------
$Si_5^{(-)}$	HOMO	-5.48	-7.83	-7.8
$Si_5^{(-)}$	LUMO	-5.12	-2.90	-2.5 to -3.5
Si_6	HOMO	-5.59	-7.57	-7.7
Si_6	LUMO	-3.39	-1.25	------
$Si_6^{(-)}$	HOMO	-5.59	-7.77	-7.7
$Si_6^{(-)}$	LUMO	-4.57	-2.50	-2.2 to -2.6

CONCLUSIONS

We have developed a new GW code employing the all-electron mixed-basis approach and applied it to the small clusters (Li_n, Na_n, K_n, Si_n clusters). In the present approach, wave function is expanded using both plane waves and atomic orbitals as a basis set.

For alkali-metal clusters, GW quasiparticle energies of the HOMO and LUMO levels agree well with available experimental data. In calculating the correlation part of the self-energy, the GPP models reproduce the results of the numerical integration with frequency. The HOMO-LUMO gap obtained by the GW calculations is larger than that of the LDA calculations and greatly improved.

For all silicon clusters studied in the presesnt paper, IPs obtained the present study are in good agreement with experimental data. For EAs, it is very important to take into account the atomic structure relaxation between neutral and negatively charged clusters to reproduce the experimental IPs and EAs except Si_4, because the structurural difference between Si_4 and Si_4^- is very small. Obtained results for IPs and EAs are in good agreement with experimental data.

ACKNOWLEDGMENT

The authors thank the Center for Computational Materials Science of the Institute for Materials Research, Tohoku University for the support of the SR8000 supercomputing facilities. V.K. gratefully acknowledges the hospitality at the Institute for Materials Research. The authors also thank Prof. Steven G. Louie for valuable discussions of alkali-metal clusters and Prof. Vijay Kumar for valuable discussions of silicon clusters. The study was performed through Special Coordination Funds for Promoting Science and Technology from the Ministry of Education, Culture, Sports, Science and Technology of the Japanese Government.

REFERENCES
[1] W. Kohn and L. J. Sham, 1965. Phys. Rev. 140, 1133.
[2] P. Hohenberg and W. Kohn, 1964. Phys. Rev. 136, 864.
[3] L. Hedin, 1965. Phys. Rev. 139, A796.
[4] M. Hybertsen and S. G. Louie, 1986. Phys. Rev. B 34, 5390.
[5] G. Onida, L. Reining, R. Godby, R. Sole, and W. Andreoni, 1995. Phys. Rev. Lett. 75, 818.
[6] M. Rohlfing and S. G. Louie, 1998. Phys. Rev. Lett. 80, 3320.
[7] S. Saito, S. B. Zhang, S. G. Louie, and M. L. Cohen, 1990. J. Phys.:Condens. Matter 2, 9041.
[8] S. Ishii, K.Ohno, Y. Kawazoe, and S. G. Louie, 2001. Phys. Rev. B63, 155104.
[9] S. Ishii, K.Ohno, Y. Kawazoe, and S. G. Louie, 2002. Phys. Rev. B65, 245019.
[10] S. Ishii, K.Ohno, V. Kumar, and Y. Kawazoe, 2003. submitted to Phys. Rev. B.
[11] See, for example, K. Ohno, K. Esfarjani, and Y. Kawazoe, 1999. Computational Materials Science, Solid-State Sciences, Vol. 129, Springer, Berlin, and references therein.
[12] F. Herman, S. Skillman, 1963. Atomic Structure Calculations, Prentice-Hall, Englewood Cliffs, NJ.
[13] U. Rothlisberger and W. Andreoni, 1991. J. Chem. Phys. 94, 8129; A.K. Ray, 1989. Solid State Commun. 71, 311; I. Boustani et al., 1987. Phys. Rev. B 35, 9437; M.W. Sung, R. Kawai, and J. H. Weare, 1994. Phys. Rev. Lett. 73, 3552.
[14] D. Tamme, R. Schepe, and K. Henneberger, 1999. Phys. Rev. Lett. 83, 241.
[15] A. Hermann, E. Shumacher, and L. Woste, 1978. J. Chem. Phys. 68, 2327; K.M. McHugh, J. G. Eaton, L.H. Kidder, and J. T. Snodgrass, 1989. J. Chem. Phys. 91, 3792.
[16] N. Benggeli and J. R. Chelikowsky, 1995. Phys. Rev. Lett. 75, 493.
[17] K. Fuke et al., 1993. Sup. Z. Phys. D 26, S204; O. Cheshnovsky et al., 1987. Chem. Phys. Lett. 138, 119; M. Maus et al., 2000. Apply. Phys. A 70 535.

Theoretical Study of Unimolecular Rectifying Function of a Donor-Spacer-Acceptor Structure Molecule

H. MIZUSEKI, Y. KIKUCHI and K. NIIMURA
Institute for Materials Research, Tohoku University, Sendai, Japan

C. MAJUMDER
Novel Materials and Structural Chemistry Division, Bhabha Atomic Research Center, Mumbai, India

R. V. BELOSLUDOV, A. A. FARAJIAN and Y. KAWAZOE
Institute for Materials Research, Tohoku University, Sendai, Japan

ABSTRACT

Recently, molecular electronics has attracted much attention as a "post-silicon technology" for future nanoscale electronic devices. In molecular electronic devices, one of the most important requirements is the realization of a unimolecular rectifier. In the present study, the geometric and electronic structure of TTF(SMe)4-TCNQ and DDOP-C-BHTCNQ, leading candidates for a unimolecular rectifying device, has been investigated theoretically using ab initio quantum mechanical calculations. The results suggest that in such donor-acceptor molecular complexes while the lowest unoccupied orbital is concentrated on the acceptor subunit, the highest occupied molecular orbital is localized on the donor subunit. The approximate potential differences for the optimized TTF derivative have been estimated at the B3LYP/6-311g++(d,p) level of theory achieving good agreement with experimentally reported results.

INTRODUCTION

Molecular electronic devices have attracted attention as a "post-silicon technology" for future applications in advanced computer electronics [1]. Realization of a unimolecular rectifying function is one of the most fundamental requirements in a molecular device. A quarter of a century ago, Aviram and Ratner [2] first proposed rectification using a single molecule. This work was followed by a number of experimental results [3–4] and several theoretical studies have been published [5–7]. In order to realize an efficient unimolecular rectifier a D (donor sub-unit) - Spacer - A (acceptor sub-unit) structure has been proposed, with which it is necessary to induce an effective charge separation and transfer. A molecule should have roughly the properties of a bulk-effect solid-state p-n junction diode. For the spacer, σ and π bonds have been introduced. σ bonds have the potential for strong charge separation. On the other hand, π bonds have delocalized orbitals and good conductivity. One of

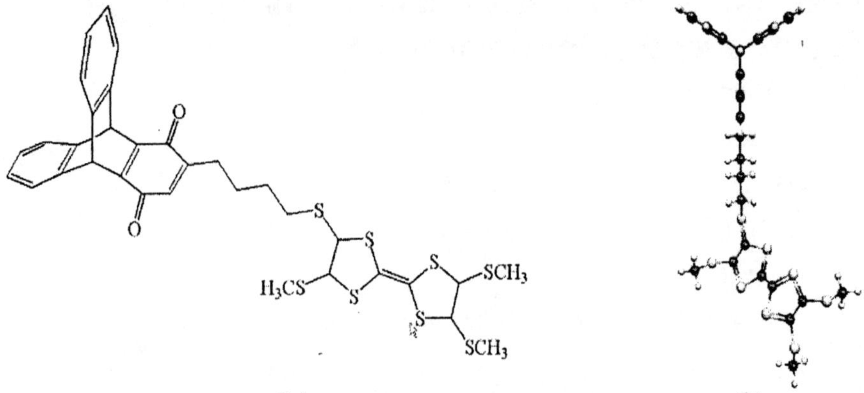

Figure 1. (a) Chemical and (b) Stable structures of the TTF(SMe)4-TCNQ molecule.

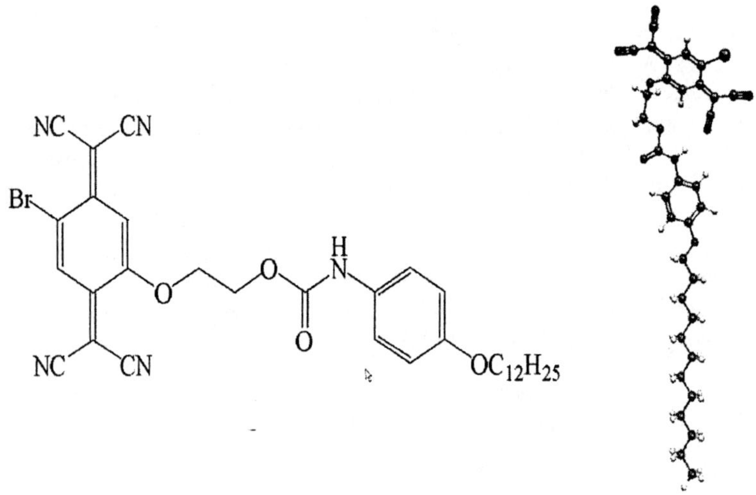

Figure 2. (a) Chemical and (b) Stable structures of p-dodecyloxyphenyl carbamate of 2-bromo, 5(2'hydroxyethoxy)tetracyanoquinodimethan (DDOP-C-BHTCNQ).

the purposes of this work is to explore a little further the role of the spacer in the operation of the rectifier.

From the viewpoint of practical applications, many research group proposed and measured the I-V characteristics for the TTF(SMe)4-TCNQ (See Figure 1) [8–10] and p-dodecyloxyphenyl carbamate of 2-bromo, 5(2'hydroxyethoxy)tetracyanoquinodimethan (DDOP-C-BHTCNQ) (See Figure 2) [11] as a molecular rectifier. These molecules have a Donor-sigma-Acceptor structure. TTF(SMe)4-TCNQ is composed of a TCNQ moiety for the acceptor with a TTF moiety providing the donor function. DDOP-C-BHTCNQ is composed of a TCNQ moiety for the acceptor with a DDOP moiety providing a weak donor function. In this present work, the geometric and electronic structure of this candidate for a unimolecular rectifier is studied in order to examine its rectifying function. In the next section, we explain in detail the numerical method used and the results for the donor—spacer—acceptor structure molecule.

MODEL AND NUMERICAL METHOD

The total energy calculations were performed using density functional theory [12] and the exchange and correlation energies were calculated using a hybrid functional, which we used in preference to the Hartree-Fock (HF) method. This is because accurate descriptions of the LUMO states are very important, since the incoming electrons are assumed to pass through the molecule. Therefore, the use of a hybrid function in the DFT is fully justified. Several successful applications of molecular devices using hybrid functions have been reported [12,13]. All the calculations were performed using the Gaussian98 program [14] at the B3LYP theory level. The B3LYP/6-31(d) was used to obtain the stable structure of the TTF(SMe)4-TCNQ and DDOP-C-BHTCNQ molecules. After optimization of the structure, the 6-311++g(d,p) basis set was used, augmented by appropriate polarization functions.

RESULTS AND DISCUSSION

To estimate the electron transport through this molecule, we analyzed the spatial distribution of the frontier orbitals (HOMO and LUMO), providing a strategy by which the rectifying properties of the donor—spacer—acceptor molecule could be understood. The results suggest that in donor-acceptor molecular complexes such as this, the lowest unoccupied orbital is concentrated around the acceptor sub-unit, while the highest occupied molecular orbital is localized on the donor sub-unit. From Figure 3, it is clear that the LUMO+1 is delocalized on the donor side. This can be attributed to the localization of the HOMO and LUMO energy levels on the donor and acceptor sides of the Donor—Sigma—Acceptor molecular complex, respectively [5-7]. Figure 3 suggests that the potential drop ΔE_{LUMO} across the TTF(SMe)4-TCNQ molecule is determined by the difference between E_{LUMO} and the ΔE_{LUMO+K} for an unoccupied orbital localized on the opposite (donor) side of the molecule from the LUMO. The potential drop in a vacuum can be explained as the difference in the LUMO energies between the donor and acceptor molecules when they are widely separated ($\Delta E_{LUMO}(\infty) = E_{LUMO}(donor) - E_{LUMO}(acceptor)$) [5].

During the measurement for the conductivity of DDOP-C-BHTCN LB film, a Schottky barrier could easily form between Mg as an electrode and TCNQ; therefore, the molecular origin of asymmetries seen in the measured currents of DDOP-C-BHTCNQ was put into doubt [15]. DDOP has a weak donor function, nevertheless, Figure 4 suggests that DDOP-C-BHTCNQ posses the same strategy for a rectifier function. The feature and the difference of the energy levels have been affected by the functional group of the donor and we can see the same strategy in rectifier molecules with the same reported localization [7].

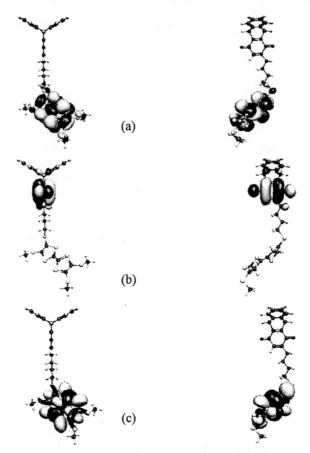

Figure 3. Orbital spatial orientation of (a) HOMO, (b) LUMO, and (c) LUMO+1 for the TTF(SMe)4-TCNQ molecule. Left: Front view. Right: Side view.

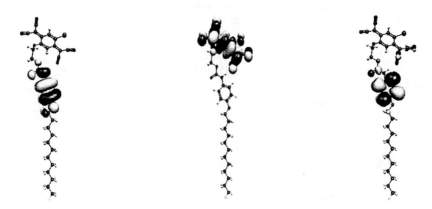

Figure 4. Orbital spatial orientation of (a) HOMO, (b) LUMO, and (c) LUMO+7 for the DDOP-C-BHTCNQ molecule. From LUMO+1 to LUMO+6 are localized on the acceptor subunit.

CONCLUSION

The geometry and electronic structure of neutral TTF(SMe)4-TCNQ and DDOP-C-BHTCNQ molecules have been calculated using density functional theory. All calculations were done using a Gaussian software package. The effects of substituents in these molecules were analyzed based on spatial distribution of frontier orbitals. While the occupied orbitals are localized on donor subunits, the unoccupied orbitals are localized on acceptor subunits. As previously reported the tendency to localize frontier orbitals of other donor-spacer-acceptor supramolecules is the same [5].

ACKNOWLEDGEMENT

The authors express their sincere thanks to the Center for Computational Materials Science at the Institute for Materials Research, Tohoku University for continued support of the HITAC SR8000-G1/64 super-computing facilities. This study was funded by the Ministry of Education, Culture, Sports, Science and Technology of Japan.

REFERENCES

[1] Y. Wada, M. Tsukada, M. Fujihira, K. Matsushige, T. Ogawa, M. Haga and S. Tanaka, 2000. Jpn. J. Appl. Phys. 39, 3835 and references therein.
[2] A. Aviram, M.A. Ratner, 1974. Chem. Phys. Lett. 29, 277.
[3] A.S. Martin, J.R. Sambles, G.J. Ashwell, 1993. Phys. Rev. Lett. 70, 218.
[4] F.R.F. Fan, J.P. Yang, L.T. Cai, D.W. Price, S.M. Dirk, D.V. Kosynkin, Y.X. Yao, A.M. Rawlett, J.M. Tour, A.J. Bard, 2002. J. Am. Chem. Soc. 124, 5550.
[5] C. Majumder, H. Mizuseki, Y. Kawazoe, 2001. J. Phys. Chem. A 105, 9454.
[6] H. Mizuseki, K. Niimura, C. Majumder, Y. Kawazoe, 2003. Comput. Mater. Sci., in press.
[7] H. Mizuseki, N. Igarashi, C. Majumder, R. Belosludov, A. Farajian,, Y. Kawazoe, 2003. Thin Solid Films, in press.
[8] S. Scheib, M.P. Cava, J.W. Baldwin, R.M. Metzger, 1998. Thin Solid Film, 327–329, 100.
[9] S. Scheib, M.P. Cava, J.W. Baldwin, R.M. Metzger, 1998. J. Org. Chem. 63, 1198.
[10] R. Metzger, 1999. J. Mater. Chem. 9, 2027.
[11] N. Geddes, J. Sambles, D. Jarvis, W. Parker, D. Sandman, 1990. Appl. Phys. Lett. 56, 1916.
[12] R. Parr, W. Yang, 1989. Density-Functional Theory of Atoms and Molecules, Oxford Press: NY.
[13] J.M. Seminario, A.G. Zacarias, J.M. Tour, 2000. J. Am. Chem. Soc. 122, 3015.
[14] Gaussian 98, 2001. Revision A.11.1, Gaussian, Inc., Pittsburgh PA.
[15] N. Geddes, J. Sambles, D. Jarvis, W. Parker, D. Sandman, 1992. J. Appl. Phys. 71, 756.

Fully-Coordinated Silica Nanoclusters: Building Blocks for Novel Materials

S. T. BROMLEY, M. A. ZWIJNENBURG, E. FLIKKEMA
and TH. MASCHMEYER
Laboratory of Applied Organic Chemistry and Catalysis, DelftChem Tech,
Technical University of Delft, The Netherlands

ABSTRACT

All-silica materials span a remarkably large spectrum of different polymorphs and densities unknown to any other single class of solids. Furthermore, largely due to this diversity of form, silica has a large range of important applications including microelectronic devices, optical fibres, catalysts, absorbants and composite plastics. The structural basis for the vast majority of these silicas is a three-dimensional network of corner-sharing SiO_4 tetrahedra. In contrast silica tetrahedra can also participate in edge-sharing, giving rise to closed rings containing two silicon atoms and two oxygen atoms. Such two-rings provide one way in which our new concept of fully-coordinated silica nanoclusters [1] could be realised. Based on a genetic algorithm approach to generate energetically low lying silica clusters, combined with our a newly developed silica interatomic potential, specifically designed for nanoscale silica, and finally on high level density functional theory (DFT) calculations, we further argue that two-ring-containing nanoclusters could allow for the formation of a whole new class of silica materials with novel topologies and properties.

INTRODUCTION

Silica-based materials are typically continuous three-dimensional networks of Si-O-Si links, forming both crystalline and amorphous structures (e.g. quartz, zeolites, glass). The use of discrete silica-based clusters to improve and functionalise existing materials has to some small extent been exploited via silsesquioxane ($Si_NO_{3N/2}X_N$) clusters, used, for instance, in the strengthening of polymers. Furthermore, molecular crystals of only silica-containing clusters also exist, such as in the crystalline packings of silsesquioxane clusters. In all these materials the silicon atoms sit at the centres of tetrahedral sites having a single linking atom (usually an oxygen atom) between any two silicons. In contrast, it is also known that bridges between silicon atoms in silica may also contain two oxygen atoms forming a small so-called two-ring. Such two rings can be formed chemically by the oxidation of clusters containing Si=Si bonds [2]. It is also known that two-rings can be formed on the surface both of amorphous and crystalline silicas at elevated temperatures (>800K) by either (i) the condensation of pairs of vicinal surface hydroxyl groups [3], and/or (ii) by the thermodynamic rearrangement of the pure silica surface [4]. Furthermore, it is also known, by a discovery in 1954 of a synthetic form of silica, known as silica-w, that materials may be based solely on two-rings [5], see Figure 1.

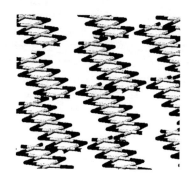

Figure 1. Schematic view of the synthetic fibrous silica material silica-w looking along the two-ring chains.

Red indicates oxygen atom positions and yellow silicon atom positions. Silica-w, is a polymer-like material, formed from aligned two-ring chains of silica interacting through non-bonding dispersive interactions forming long fibres. The density of this fibrous material, is found to be lower than most other ordered silicas based on three coordinated corner-sharing networks, and furthermore has a particularly low refractive index for a pure silica crystalline polymorph (1.41). The main problem of this material for applications (e.g. in optical fibres/coatings) is its instability towards "water corrosion". It is known quite generally that two-rings are quite strained and readily broken under dissociative chemisorption of water and some other reagents, although this process can be simply reversed by heating. Furthermore, two-rings are stable under oxygen, carbon dioxide, nitrogen and hydrogen [3] containing atmospheres. Thus, although two-ring are often energetically/structurally favoured, and of potential interest in applications, their reactive susceptibility to water, together with their usual surface accessibility often causes them to be fairly short lived states.

In this work we suggest that fully-coordinated silica nanolusters, and other two-ring containing silica clusters, may be a way forward to forming useful stable two-ring-containing silica materials.

METHODOLOGY AND RESULTS

In the case of silica-w the building blocks are the edge-sharing chains. From density functional (DF) calculations it is known that for $(SiO_2)_N$ N=1-6 nanocluster chains of two-rings are the most thermodynamically stable form of silica, lower in energy than any other stoichiometric nanocluster [6]. For N>6 we have shown through high-level DFT calculations that other two-ring-containing structures are lower in energy than the chains [1,7]. In particular we have showed that novel fully-coordinated silica nanoclusters, i.e. with no reactive non-bridging oxygens, can be lower in energy than the corresponding chains and also that they are more reactively stable. For example, we demonstrate that the deformation of $(SiO_2)_N$ chains into fully-coordinated molecular rings is energetically preferred for N>11 [1], see figures 2 and 3. In this work we further show that other types of fully-coordinated silica clusters are lower in energy and more reactively stable than the molecular rings.

Via the use of a genetic algorithm approach to generate energetically low lying silica clusters, combined with our newly developed silica interatomic potential, specifically designed for nanoscale silica, and finally by using high-level DFT calculations, we describe the thermodynamically favoured structures of nanocluster silica—$(SiO_2)_N$ N=6-24—and compare them with the energies and properties of the corresponding fully-coordinated nanoclusters. On the basis of these calculations we further argue that certain combinations of two-ring-containing silica nanoclusters could allow for the formation of a whole new class of silica materials with novel topologies and properties. We speculate on possible structures of new materials based upon the union of such silica nanoclusters and, on their reactive/thermal stability and potential properties and applications.

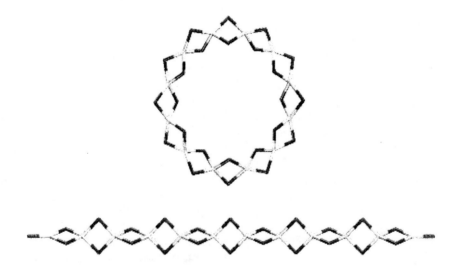

Figure 2. Schematic diagram of the structures of the N=12 $(SiO_2)_N$ molecular ring and chain. Red indicates oxygen atom positions and yellow silicon atom positions.

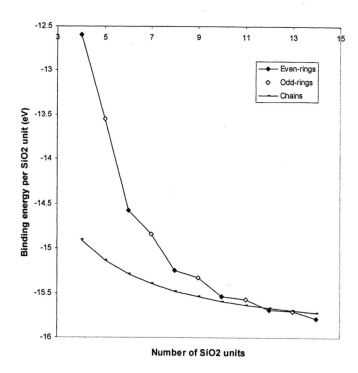

Figure 3. Binding energy change of 2-ring chains and rings with increasing number of SiO_2 units. For $(SiO_2)_N$ N>11, fully-coordinated rings are more energetically stable than Si=O terminated chains.

REFERENCES

[1] S.T. Bromley, M.A. Zwijnenburg, Th. Maschmeyer, 2003. Phys. Rev. Lett., 035502, 90.
[2] S. Willms, A. Grybat, W. Saak, M. Weidenbruch, H. Marsmann, 2000. Z. Anorg. Allg. Chem., 626, 1148.
[3] A.M. Ferrari, E. Garrone, G. Spoto, P. Ugliengo, and A. Zecchina, 1995. Surf. Sci. 323, 151.
[4] C.M. Chaing, B. R. Zegarksi, L. H. Dubois, 1993. J. Phys. Chem. 97, 6948.
[5] A. Weiss, 1994. Z. Anorg. Allg. Chem., 276, 95.
[6] T.S. Chu, R. Q. Zhang, H.F. Cheung, 2001. J. Phys. Chem. B, 105, 1705.
[7] E. Flikkema and S.T. Bromley, 2003. Chem. Phys. Lett. (submitted).

Reductive Selective Deposition of Ni-Zn Nanoparticles onto TiO$_2$ Fine Particles in the Liquid Phase

H. TAKAHASHI, Y. SUNAGAWA, S. MYAGMARJAV, K. YAMAMOTO, N. SATO and A. MURAMATSU
Institute of Multidisciplinary Research for Advanced Materials, Tohoku University, Sendai, Japan

ABSTRACT

Among various methods to synthesize nanometer-sizes particles [1], liquid phase reduction is one of the easiest procedures since nanoparticles can be obtained from any material that is soluble in a specific solvent [2]. However, as-prepared particles are rather unstable since tremendous aggregation sometimes occurs using this method [3]. To solve this problem, selective deposition of Zn-Ni nanoparticles onto TiO$_2$ particles in a liquid phase and their catalytic activities were studied.

INTRODUCTION

Among various methods to synthesize nanometer-size particles [1], the liquid phase reduction method is one of the easiest procedures, since nanoparticles can be directly obtained from any precursor compounds soluble in a specific solvent [2]. However, as-prepared particles are rather unstable since tremendous aggregation sometimes occurs in the solution [3]. To solve this problem, selective deposition of Ni-Zn nanoparticles onto TiO$_2$ fine particles in the liquid phase and their catalytic activities were studied.

EXPERIMENTAL DETAILS

Figure 1 shows a flowchart and a device for synthesis of bimetallic nanoparticles by liquid phase reduction. Nickel acetylacetonate (Ni(AA)$_2$) and zinc acetylacetonate (Zn(AA)$_2$) were co-dissolved in 2-propanol (50ml) with a ratio of Zn/Ni ranging from 0 to 1.0 where [Ni] was constant at 5x10^{-3} mol/dm^3. TiO$_2$ particles (0.125g, Ishihara Ind., STO1) were dispersed in the Ni-Zn solutions in 4-neck flasks and heated at 355K under a continuous flow of N$_2$ for 30 minute. Ni and

Zn were promptly reduced by addition of 1.5×10^{-1} mol/dm³ NaBH$_4$ in 2-propanol (10ml). The hydrogenation activity of 1-octene (5ml, H$_2$ gas flow rate: 45 ml/min) on as-prepared particles was tested and the product analyzed by gas chromatography (GC, Shimadzu Co., Ltd. GC-14B system). Fresh and used nanoparticles were characterized by high-resolution transmission electron microscopy (HR-TEM, JEOL Co., Ltd. JEM-200CX), X-ray diffraction (XRD, Rigaku Co., Ltd. CuKα 40KV, 30mA) and Electron Spectroscopy for Chemical Analysis (ESCA, ULVAC-PHI ESCA 5600□Al Kα, 187.85eV□45°). Filtered solutions were measured using an Inductively Coupled Plasma Atomic Emission Spectrometer (ICP, Shimadzu Co., Ltd. ICPS-1000III) in order to analyze the yield of Ni and Zn, independently.

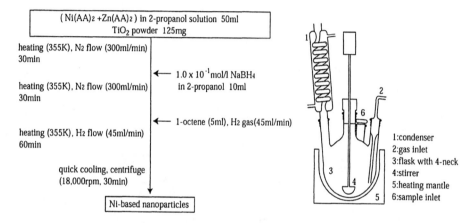

Figure 1. A flowchart and a device for the nanoparticles synthesis in this work.

RESULTS AND DISCUSSION

Figure 2 shows the ICP results of filtered solutions. The residual amounts of Ni in solutions are 0 to 1.0%, and these of Zn are 0.5 to 2.5%. This result suggested that almost all metal species was deposited onto TiO$_2$. The color of the particles synthesized was black in the case of Zn/Ni<0.5, while it was dark gray in the case of Zn/Ni=1.0. These results represent that more large amount of reducing agent is need to the reduction of Zn in the case of Zn/Ni=1.0, nevertheless the amount of reducing agent is fully larger than the total amount of metals in this experimental condition. However, the color of the Zn/Ni=1.0 particles was not changed when the amount of reducing agent increased to 2 and 3 times larger. From the results mentioned above, it can be considered that metallic Ni is grown from the adsorbed Ni species on TiO$_2$ by their reduction and that nascent state hydrogen atom formed at the surface of Ni or reducing agent (BH$_4^-$) adsorbed on the Ni catalytically reduced the Zn species, and ZnO is easily synthesized by the hydrolysis of Zn metal synthesized because of its less-reductive nature (ΔHf,$_{ZnO}$= -348.28KJ/mol). The detailed research is now in progress.

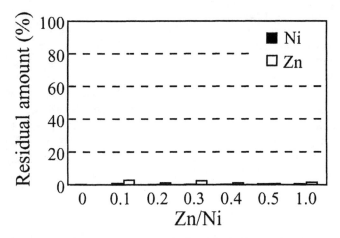

Figure 2. Residual amount of Ni and Zn in the solutions.

Figure 3 shows the TEM micrographs of (a) TiO_2 and as-prepared Ni-Zn particles on the TiO_2 with a ratio Zn/Ni of (b) 0.2, and (c) 1.0. As the size of Zn/Ni particles was estimated to be (b) 5-6 nm and (c) 1–2 nm, the particle size appears to decrease with increasing amounts of Zn added.

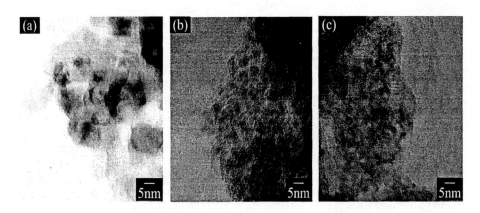

Figure 3. HR-TEM micrographs of (a) TiO_2 support and Zn/Ni nanoparticles with Zn/Ni: (b) 0.2 and (c) 1.0.

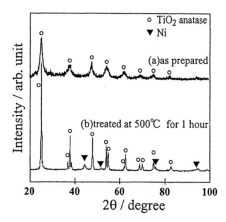

Figure 4. XRD spectra of the Zn/Ni nanoparticles before and after heat treatment.

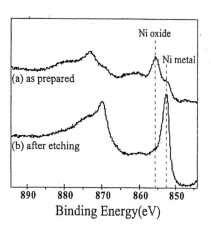

Figure 5. ESCA spectra for the Ni $2P_{3/2}$ electron of the Zn/Ni nanoparticles before and after etching.

Figure 4 shows the XRD spectra of the Zn/Ni nanoparticles on the TiO_2 (a) as-prepared and (b) heat treated at 500°C for 1 hour in vacuum. In the case of as-prepared sample only the broad peaks corresponding to TiO_2 (anatase structure) are clearly observed. From the results of XRD and TEM micrographs, TiO_2 particles used in this experiment was anatase-type crystalline fine particles. After the heat treatment, the broad peaks corresponding to anatase become sharp and new peaks appeared centered at $2\theta = 44°$, $52°$, $76°$ and $93°$ corresponding to Ni metal (fcc). This result indicates that the Ni metal is crystallized by heat treatment from the amorphous phase in the initial state. Thus it can be concluded that the synthesized bimetallic nanoparticles have an amorphous-like structure. Identification of the Zn state is difficult in this experiment since the peak positions of Zn and ZnO in ESCA spectra are near to that of Ni and TiO_2.

Figure 5 shows the ESCA spectra for the Ni $2P_{3/2}$ electron of Zn/Ni=1.0 particles on the TiO_2. The peaks corresponding to Ni oxide and Ni metal are clearly observed [4-6], while Ni metal predominates after etching with Ar^+. This result implies that the Ni-Zn nanoparticles were metallic, but that their surface was oxidized under atmospheric conditions.

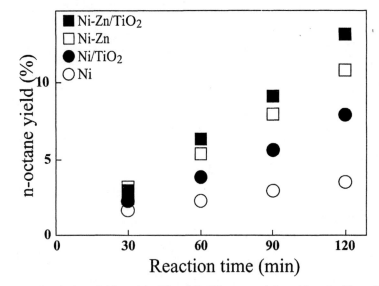

Figure 6. Catalytic activities of the Ni and Zn/Ni nanoparticles with and without TiO_2.

The hydrogenation activity of Zn-added TiO_2-supported Ni nanoparticles was evaluated through 1-octene hydrogenation. Figure 6 shows the change in the yield of n-octane converted from 1-octene with time on-stream. The good linearity between the yield and time indicates no change in surface characteristics of the catalysts while 1-octene hydrogenates to n-octane. The hydrogenation activity increases in the sequence: Ni < Ni/TiO_2 < Ni-Zn < Ni-Zn/TiO_2. The catalytic activity of Ni-Zn/TiO_2 was 3 times higher than that of the unsupported Ni nanoparticles. In consideration of our earlier work [3], the electron transfer between Ni and Zn, such as $Ni^{\delta-}$- $Zn^{\delta+}$, also occurs in this system. Thus, Zn plays the role of an electron donor to Ni nanoparticles so the catalytic activity of Ni is increased when Zn is used. Moreover, Zn addition works to decrease the particle size of the prepared particles thus increasing the total catalytic surface area. The rate of aggradation became low when TiO_2 was used, so the catalytic activity increased over the tests without TiO_2. Namely, we found an outstanding effect of Zn addition and the stabilization of Ni nanoparticles by deposition onto TiO_2.

CONCLUSION

Ni-Zn nanoparticles synthesized in this experiment were bimetallic, but their surface was oxidized in atmospheric conditions. The particle size decreased with an increase in the amount of Zn added. The catalytic activity of Ni-Zn/TiO_2 was 3 times higher than unsupported Ni nanoparticles. It can be conclude that Zn addition plays an outstanding effect in increasing catalytic activity with stabilization of Ni nanoparticles when they are deposited onto TiO_2.

REFERENCES

[1] For example,
   a. P.N. Barnes, P.T. Murray, T. Haugan, R. Rogow, G.P. Perram, 2002. Physical C-Superconductivity and its Applications, 377, 578–584.
   b. K. Wegner, B. Walker, S. Tsantilis, S.E. Pratsinis, 2002. Chem. Eng. Sci. 57, 1753–1762.
   c. B. Xia, K. Okuyama, I.W. Lenggoro, 2001. Advanced Materials, 13, 1744–1744.
[2] A. Muramatsu, S. Shitara, H. Sasaki, S. Usui, 1990. Shigen-to-sozai, 106, 805–810.
[3] H. Takahashi, A. Muramatsu, E. Matsubara, Y. Waseda, 2002. Shigen-to-sozai, 118, 211–216.
[4] A. Lebugle, U. Axelsson, R. Nyholm, N. Martensson, 1981. Phys. Scr., 23, 825–827.
[5] T. Dickinson, A. Povey, P. Sherwood, 1977. J. Chem. Soc. Faraday Trans., 173, 332–343.

Author Index

Abdallah, E., 685
Abe, J. M., 247
Aizawa, N., 730
Amer, M., 869
Anderson, C., 336

Baratoff, A., 701
Barker, D. G., 359
Barth, T., 277
Belosludov, R. V., 423, 704, 781, 786, 949
Belosludov, V. R., 423, 704
Benedict, M., 175
Berner, S., 701
Bick, A., 412
Bonifazi, G., 301, 316
Briere, T. M., 857
Bromley, S. T., 954
Bystrzejewski, M., 721

Cagin, T., 502
Campbell, J. J., 359
Chandrasekaran, M., 646
Chehun, V. P., 508
Chen, M., 130
Cherukuri, H., 459
Chonan, S., 261
Cohen, A. E., 888
Cudzilo, S., 721

d'Amore, M., 100
Davey, K. J., 359
Davies, A. G., 845, 888
de Wild, M., 701
Di Stasio, S., 316

Dong, K.-J., 537

El-Ashry, M., 869
Eldeeb, H., 183, 685
Elsadek, H., 685

Falkus, J., 149
Farajian, A. A., 781, 786, 949
Fichera, M., 807
Flikkema, E., 954
Frayman, Y., 160, 199
Fujimori, H., 904
Fujisaki, K., 217, 448, 606
Furukawa, T., 606
Furuya, Y., 495, 820

Gachet, S., 721
Galal, M., 183
Gao, W., 459
Gaskov, A. M., 508
Gerdes, M., 277
Germishuizen, W. A., 845, 888
Ghomshei, M. M., 401
Giancontieri, V., 301
Goddard, W. A., III, 502
Goldberg, A., 412
Goto, K., 896
Goto, T., 896
Grauer, M., 277
Greene, J., 12
Grimes, R. W., 523
Grinchenko, V. T., 508
Gubin, S. P., 801, 922
Guentherodt, H.-J., 701

Harib, K. H., 470
Hasegawa, H., 622
Hayase, T., 261
Heifets, E., 502
Hirayama, H., 230, 827
Hodgson, P. D., 160
Holmes, R. J., 359
Hong, S.-I., 511
Huczko, A., 721
Hwang, G. S., 502
Hyötyniemi, H., 28

Ichinoseki, K., 786
Ihara, M., 896
Ikemoto, I., 730
Il'chenko, V. V., 508
Inerbaev, T. M., 423, 704
Inoue, A., 585
Ishii, S., 742, 749, 940
Ishii, T., 730

Jin, J.-H., 511
Jung, T. A., 701

Kachitvichyanukul, V., 470
Kadowaki, A., 487
Kajiwara, S., 495
Kanehama, R., 730
Kaneto, K., 910
Kawazoe, Y., 3, 423, 704, 742, 749, 781, 786, 857, 861, 931, 940, 949
Khanin, V. V., 801
Khomutov, G. B., 801
Kikuchi, K., 730
Kikuchi, R., 545
Kikuchi, T., 495
Kikuchi, Y., 949
Kisanuki, S., 661
Kishimoto, S., 599, 637
Kislov, V. V., 801
Kodama, T., 730
Korobov, M. S., 922
Kravchenko, A. I., 508
Kroto, H. W., 721
Kubota, T., 495
Kudoh, J., 423, 704
Kumar, V., 857, 861
Kusiak, J., 149

La Mantia, A., 807
Lang, B., 343
Lange, H., 721
LeClair, S., 439
Lee, G. C., 559
Lee, J. H., 559
Li, J.-Y., 537
Libertino, S., 807
Linkens, D. A., 130
Liu, F.-X., 537
Liu, R.-S., 537
Liu, Z., 904
Luo, Y., 793

Maguire, J. F., 141, 175, 523, 869
Majumder, C., 949
Maschmeyer, Th., 954
Masuda-Jindo, K., 545
Masumoto, H., 896
Maximov, I. A., 801
Meech, J. A., 3, 326, 336, 343, 376, 401
Middelberg, A. P. J., 845, 888
Min, N.-K., 511
Miyasaka, H., 730
Miyata, M., 502
Mizoguchi, S., 622
Mizuseki, H., 781, 786, 949
Montelius, L., 801
Montemagno, C. D., 761
Monthioux, M., 721
Mori, M., 412
Morisato, T., 742, 749
Mun, C. C., 646
Muramatsu, A., 958
Myagmarjav, S., 958

Nagasaka, Y., 661
Nakadate, H., 502
Nakajima, H., 495
Nakamatsu, K., 247
Naruse, T., 261
Nassar, S., 183
Neuser, P., 277
Niimura, K., 949
Nishi, Y., 487, 576

Oguri, K., 487
Ohno, K., 742, 749, 940
Ohnuma, S., 904

Ohtsuki, T., 742, 749
Oka, S., 439
Okazaki, T., 495, 820

Pakalnis, R., 343
Palma, V., 110
Pandey, S. S., 910
Pappalardo, M., 100
Park, Y. C., 559
Paszek, J., 721
Patten, J., 459
Pellegrino, A., 100, 110
Pepper, M., 845, 888
Pharr, G., 459
Phillips, P. L., 359
Phillpot, S. R., 523
Pietrzkiewicz, P., 149
Pietrzyk, W., 149
Pirzada, M., 523

Rahman, M. R., 470
Reichert, O., 277
Robustelli, S., 110
Rolfe, B. F., 160, 199
Rosei, F., 877
Russo, P., 100

Saito, F., 570
Samuelson, L., 801
Sato, H., 786
Sato, N., 958
Sato, T., 570
Sato, Y., 502
Sato-Ilic, M., 65
Satou, S., 606
Scattergood, R., 459
Schelling, P. K., 523
Schmidt, J. J., 761
Seno, T., 247
Sergeyev-Cherenkov, A. N., 801
Serranti, S., 301
Sharif Ullah, A. M. M., 470
Sharp, V., 359
Shibata, S., 896
Shiga, K., 742, 749
Shima, S., 614
Shindo, D., 904
Shinya, N., 599, 637

Shinya, Y., 820
Shorokhov, V. V., 768
Shyan, J. Y. M., 646
Sickafus, K. E., 523
Singh, A. K., 857
Sluiter, M. F., 423, 742, 749, 861, 931
Smetanin, M. V., 801
Smith, G., 336
Snigirev, O. V., 801
Soldatov, E. S., 768, 801
Song, Z., 599, 637
Spencer, S. J., 359
Stewart, B., 336
Stewart, L. S., 86
Stuff, G., 277
Sugiura, K.-I., 730
Sugiyama, S., 286
Sunagawa, Y., 958
Suto, J., 896
Suyatin, D. B., 801
Suzuki, A., 247
Suzuki, D., 217
Suzuki, H., 701
Suzuki, S., 622

Takagi, T., 793
Takahashi, H., 958
Takasima, W., 910
Takefuji, Y., 679
Tamoto, S., 820
Tanahashi, Y., 820
Tanaka, M., 261
Tonegawa, A., 487
Tosch, P., 888
Tromans, D., 376
Tsujiuchi, Y., 896
Tsumori, F., 614

Uehara, M., 502
Ulansky, R., 326

Van Hung, V., 545
Van Le, T., 77
Villecco, F., 100, 110
Volpe, F., 301

Walaszek-Babiszewska, A., 119
Wälti, C., 845, 888
Walton, D. R. M., 721

Wang, J., 570
Waseda, Y., 622
Webb, G. I., 160
Wirtz, R., 845, 888
Wreesmann, C., 336
Wutting, M., 495

Yabe, H., 576
Yakuwa, K., 793
Yamaguchi, N., 487
Yamahira, T., 820
Yamamoto, K., 958
Yamashita, M., 730

Yan, J., 459
Yasuda, N., 614
Yin, S., 570
Yokoyama, M., 820
Youhui, G., 904
Yuki, H., 749
Yun, D.-H., 511
Yurkov, G. Y., 922

Zhang, Q., 570
Zheng, C.-X., 537
Zhu, Y. Q., 721
Zwijnenburg, M. A., 954

About the Editors

John A. Meech is Professor of Mineral Processing in the Department of Mining Engineering at the University of British Columbia in Vancouver, BC, Canada. He is Director of CERM3—the Centre for Environmental Research in Minerals, Metals, and Materials. CERM3 is a multidisciplinary research group set up to conduct sustainable research for the mining industry in all aspects of the environment. Dr. Meech was one of the original founders of IPMM in 1997, together with Dr. Tara Chandra from the University of Woolongong, Dr. Michael Smith from the University of Calgary, and Dr. Steve LeClair from the Materials Directorate of AFRL/ML. A Fellow of the Canadian Institute of Mining and Metallurgy, he was CIM Distinguished Lecturer in 2000. He is a Registered Professional Engineer in British Columbia and serves on the IPMM Executive as President.

Yoshiyuki Kawazoe is Professor of Materials Design by Computer Simulation and Director of the Center for Computational Materials Science at the Institute for Materials Research at Tohoku University in Sendai, Japan. Dr. Kawazoe has published over 450 papers and articles on first principles modeling of materials at the atomic and electronic level. His laboratory houses the 16th most powerful supercomputer in the world (as of summer 2001)—an Hitachi SR8000/H64 Model G1 TeraFLOPs machine with 64 nodes, each with eight 1.8 GFLOP CPUs. He is a founding member of the Asian Consortium for Computational Materials Science (ACCMS). Together with Dr. Meech, he was General Co-Chair for IPMM'03 scheduled for Sendai and Matsushima from May 18 to May 23, 2003. He serves on the IPMM Executive as Secretary/Treasurer.

Vijay Kumar is Visiting Professor in the Center for Computational Materials Science of the Institute for Materials Research at Tohoku University in Sendai, Japan since 1999. He conducts research and directs a research team in the fields of electronic structures, clusters and nanomaterials, surfaces and interfaces, segregation in alloys, order-disorder transitions, chemisorption, quasicrystals, fullerenes and nanotubes, ab-initio molecular dynamics and computational

condensed matter in general. In 2000, he founded the Dr. Vijay Kumar Foundation in Chennai, India. In 2004, he received the ACCMS Award for his dedication to scientific research, his continued help in organizing and contributing to the establishment of the Asian Consortia for Computational Material Science.

John F. Maguire is Chief of the Polymers Group in the Materials and Manufacturing Directorate at the US Airforce Research Laboratories at Wright-Paterson Airforce Base in Dayton, Ohio. He has been a research leader in the Manufacturing Technology Division Materials Process Design Branch since 1998. In 2001, Dr. Maguire received the IPMM J. Keith Brimacombe Award in recognition of his contribution to research in soft and interfacial matter and to the development of materials processing and new techniques in computer simulation and molecular dynamics. He is a member of several professional societies, including the American Chemical Society, American Association for the Advancement of Science, New York Academy of Sciences, UCLA Chemists and Biochemists Club, and Faraday Society. Maguire also received the Air Force Office of Scientific Research "STAR Team" Award in 2000. He is Vice-President of IPMM.